The Heart Cell In Culture

Volume I

Editor

Arié Pinson, D.Sc.
Senior Investigator
Laboratory for Myocardial Research
Institute of Biochemistry
Hebrew University - Hadassah Medical School
Jerusalem, Israel

CRC Press, Inc.
Boca Raton, Florida

Library of Congress Cataloging-in-Publication Data

The Heart cell in culture.
  Includes index.
  1. Heart cells. 2. Cell culture. I. Pinson,
Arié, 1931-   [DNLM: 1. Cells, Cultured.
2. Myocardium--cytology. WG 280 H4365]
QP114.C44H43   1987   612′.17′0724   87-21864
ISBN 0-8493-4696-7 (set)

International Standard Book Number 0-8483-4696-7 (set)
International Standard Book Number 0-8493-4697-5 (Volume I)
International Standard Book Number 0-8493-4698-3 (Volume II)
International Standard Book Number 0-8493-4699-1 (Volume III)

Library of Congress Card Number 87-21864
Printed in the United States

PREFACE

At the turn of the century, cell culture and other in vitro methods were developed and have since proven invaluable in research on differentiation, specific function, and metabolism of various tissues, which are often difficult to study in vivo due to the complexity of the interactions between different tissues and between tissues and body fluids.

In 1912, Burrows made the pioneering discovery that cardiac explants produced muscle cells that were capable of dividing and beating with a regular rhythm, thus demonstrating the myogenic nature of heart contraction. The tryptic method of dissociating cells was first described by Rous and Jones in 1916, and revived almost 40 years later by Moscona. In 1955, Cavanaugh isolated and grew heart cells from the chick embryo. Twenty-five years ago, Harary and Farley prepared the first cultures of neonatal rat heart cells. This method of culturing mammalian neonatal heart cells has proved to be a crucial step in the application of heart cell culture techniques to various research areas dealing with myocardial biology at the cellular level. As a direct consequence of this, since the early 1970s, there has been about a sixfold increase in the number of scientific papers per year dealing with heart cells in culture. Clearly, a summary of the progress in this field is long overdue.

My primary aim as the editor of *The Heart Cell in Culture* was not to provide a laboratory manual, although techniques are discussed when relevant. Rather, each chapter presents a critical review, demonstrating the progress made during the last 25 years as well as perspectives for the future.

The Heart Cell in Culture is presented in three multidisciplinary volumes, spanning a wide spectrum of topics and presenting state-of-the-art reviews. It is difficult to divide such a book into distinct parts, but it has been presented in three sections, one in each volume: the first on the heart cell culture system — historical background, the development of culturing techniques, and the function of such cells; the second on the basic biochemistry of the myocardial cell in culture; and the third on applied research in pharmacology and pathology of cardiac cells, including a chapter on cultured adult heart cells, which is becoming a popular tool in cardiac research. The minireview chapters in the series were selected with two main objectives in mind: to complement full-length comprehensive reviews in the same field by the addition of new dimensions and scope, in which case they immediately follow the main chapter, or to consider topics which are relatively new or show promise for the future development of research in cardiac biology. Indeed, I expect that these chapters should be extremely stimulating for this reason.

I hope that people working in the field of heart research will find useful information in these volumes, and that they will foster better understanding of cardiac biology and stimulate further progress in this important field of research.

I would like to thank Dr. R. Goldberg for her assistance with the thankless task of language editing, corrections to the text, and proofreading. I am grateful to all the contributors who made this book, *The Heart Cell in Culture,* possible.

Arié Pinson
Jerusalem, Israel

THE EDITOR

Dr. Arié Pinson, D.Sc., is a Senior Investigator at the Hebrew University-Hadassah Medical School and permanent visiting professor at the Faculty of Life and Environmental Sciences of the University of Dijon in France.

He studied Clinical Chemistry and Biology at the Univesity of Geneva in Switzerland. He earned his M.Sc. degree from the Hebrew University of Jerusalem in 1967 and his D.Sc. for a thesis on the Metabolism of Palmitic and Erucic Acid in Rat Heart Cell Cultures from the Univesity of Dijon in France in 1975.

Since 1975, the research carried out at Dr. Pinson's Laboratory for Myocardial Research has contributed toward establishing cultures of cardiac myocytes as a standard technique for studying the biochemistry of the heart in both health and disease. In a broader sense, these studies, often conducted in collaboration with leading scientists overseas, have advanced knowledge on cardiac cells in culture as a tool in the fields of development, differentiation, biochemistry, physiology, and medical research.

Dr. Pinson has been awarded two prizes for distinguished research by the Hebrew University. He has also held long-term fellowships from INSERM — Institut National de la Santé et de la Recherche Médicale (France), CNRS — Centre National de la Recherche Scientifique (France), DGRSRT — Délégation Générale à la Recherche Scientifique et Technique (France), ZWO — The Netherlands Society for the Advancement of Pure Research, and annual summer fellowships from INSERM.

A recognized authority in his field, he has frequently been invited to give plenary lectures at international conferences: at the Biochemical Society (U.K.) and at several ISHR (International Society for Heart Research) meetings and at other meetings focusing on cardiomyocytes.

Dr. Pinson has been invited to many seminars at universities and research institutes in Holland, Germany, France, and Israel. In 1984, he founded the IGHR (the Israeli Group for Heart Research), which is affiliated to the ISHR.

Dr. Pinson's current research interests include the biochemistry of cultured cardiomyocytes with particular emphasis on lipid and glucose metabolism, function and structure of sarcolemma, metabolism of high energy phosphates, and the use of this system as a model for elucidation of anoxic injury and iron overload in the heart.

CONTRIBUTORS

VOLUME I

Shigetoyo Amano, Ph.D.
Department of Immunology and Biology
Kanazawa University
Ishikawa, Japan

Pierre Athias
Chef de Travaux
Laboratory of Physiology
Faculty of Medicine
Dijon, France

Janis M. Burt, Ph.D.
Assistant Professor
Department of Physiology
University of Arizona College of
  Medicine
Tucson, Arizona

David E. Clapham, M.D., Ph.D.
Assistant Professor
Department of Medicine and Physiology
Harvard Medical School
Boston, Massachusetts

Kiyota Goshima, Ph.D.
Associate Professor
Institute of Molecular Biology
Nagoya University
Nagoya, Japan

Alain Grynberg
Chargé de Recherches
I.N.R.A.
Dijon, France

Isaac Harary, Ph.D.
Professor
Department of Biological Chemistry
UCLA
Los Angeles, California

Ian A. Hendry, Ph.D.
Department of Pharmacology
John Curtin School of Medical Research
Canberra, ACT, Australia

C. Russell Horres, Ph.D.
Adjunct Assistant Professor
Department of Physiology
Duke University Medical Center
Durham, North Carolina

Tomoichi Ishikawa, D.Sc.
Department of Anatomy
Kochi Medical School
Kochi, Japan

Hiroyuki Kaneko
Department of Biology
Osaka City University
Osaka, Japan

Gania Kessler-Icekson, Ph.D.
Senior Investigator
Rogoff-Wellcome Medical Research
  Institute
Beilinson Medical Center
Petah-Tiqva, Israel

Syoyu Kobayashi
Institute of Molecular Biology
Nagoya University
Nagoya, Japan

Eiji Kurimoto
Institute of Molecular Biology
Nagoya University
Nagoya, Japan

Melvyn Lieberman, Ph.D.
Professor
Department of Physiology
Duke University Medical Center
Durham, North Carolina

Prudent Padieu
Professor
Department Biochimie Médicale
Faculté de Médecine
C.H.R.U.
Dijon, France

Arié Pinson, D.Sc.
Senior Investigator
Department of Biochemistry
Hebrew University-
  Hadassah Medical School
Jerusalem, Israel

Kazuko Uno, Ph.D.
Senior Investigator
Institut Pasteur de Kyoto
Kyoto, Japan

David Piwnica-Worms, Ph.D.
Clinical Fellow
Department of Radiology
Harvard University
Boston, Massachusetts

Shigeo Wakabayashi
Department of Etiology and
  Pathophysiology
National Cardiovascular Center
Osaka, Japan

David M. Wheeler, Ph.D.
Resident
Department of Anesthesiology
University of Alabama
Birmingham, Alabama

In memory of my parents,
Zeev and Pessia Pinson

To my wife and children,
Yona, Gavriel, Halléli, and Shira

To Professor Prudent Padieu

TABLE OF CONTENTS

VOLUME I

TABLE OF CONTENTS

VOLUME II

Chapter 1

HISTORICAL INTRODUCTION

Isaac Harary

The articles in this book contain much historical background as to the development of the heart cell in culture, providing logical historical frameworks for our present day concepts. I thought, therefore, that in this historical introduction it would be of added interest to examine the past events in this field from the viewpoint of how the past events were shaped by the time in which they occurred. New discoveries are either ignored or well received, according to the scientific and nonscientific contemporary environment. In the case of the heart cell in culture, both played a major role in progress of the field.

M. T. Burrows in 1912[1] was examining isolated tissues of embryonic chick heart, and noticed that migrating from the periphery were single beating heart cells. From our vantage point this was a marvelous and exciting observation. What was the period like? What was Burrows' reaction and what sort of impact did his observation have? The time was an interesting one, as it was at the beginning of modern experimental biology. We were still at a time when science had very limited meaning to everyday life and thought, except perhaps for the impact of Darwinism. Life scientists were struggling with how to approach their discipline in more quantitative terms. Their attempts were best represented by the controversy between two philosophical approaches: vitalism and mechanism. Vitalism was a view which regarded living systems as purposive and directed by a vital force. Consequently, for vitalists any disruption of the life form, for the purpose of study, reduced it to a meaningless collection of artifacts. Thus meaningful data could only be gotten from a study of the intact organism. On the other hand, the mechanists, who were in the minority, perceived the living form as the sum of its components. For them, form only existed after the fact and had no life separate from the parts. Thus a study of the components was, of course, meaningful. They suggested that new qualitities were perhaps created by the summation, but only as a theoretically predictable (mostly in retrospect) resultant of the attributes of the parts. These positions were the extremes at that time. Most scientists worked somewhere in between but leaned more toward the inviolability of the organism. Most believed that the only meaningful observations were derived from a study of the whole organism, and consequently viewed with distrust data obtained from study of the separated components.

All this was not really new in 1912. This attitude predominated in the biological sciences for as long as we have records. Aristotle was the father of vitalism. He believed that there was an actuality, which he termed the "entelechy", which organized the body. Aristotle even speculated about cell culture. Wondering about form and function, he noted that one could not produce little hands by reducing a whole hand into component parts. This reflected a profound belief in the inviolability of form. What sort of effect did this all have?

The work of Burrows was largely ignored. Little advantage was taken of his observations. He suggested that the beating of single heart cells supported the myogenic theory of the heart beat, but he did not continue studies in this area. He had the means to utilize the system for many meaningful observations, but he did not. We can only speculate that he didn't think they would be really significant. His was work that came into an era not ready for it. The period seems important. For example, how little attention was paid to the following, and I quote from a book written by C. J. J. Galois[2] in 1812. "If one could substitute for the heart a kind of injection of arterial blood, either natural or artificial, one would succeed easily in maintaining alive indefinitely any part of the body whatsoever." Truly a thought which was a stranger in its time. Both Galois' speculation and Burrows' later observation had to wait until we were ready.

Interest in isolated systems was sporadically maintained over the early decades of the twentieth century. But nothing much came of these few studies. It is instructive to quote from some comments made at a meeting by Dr. Raymond C. Parker in 1964.[3]

> Tissue culture was introduced by . . . Dr. Ross Harrison . . . in 1907. In 1912 . . . Alexis Carrel . . . isolated his famous strain of cells from a chick embryo heart . . . which soon consisted solely of connective tissue. In 1916 Dr. Peyton Rous of the Rockefeller Institute introduced trypsin as a means of dissociating discrete tissue cells prior to cultivation. During my time there, we saw much of Dr. Rous, but we did not unfortunately take advantage of his useful discovery. In 1924 another old chief of mine . . . Dr. Albert Fischer . . . noted that separated fragments, taken from the same embryo heart, pulsated with independent rhythm, but when the two fragments became joined by growth they pulsated synchronously. This was a charming observation and may well have inspired a sonnet or two . . . but we still had to wait many years . . . for meaningful observations. . . . Some of these should have been made 40 years ago. The techniques were available, all that was lacking was insight . . . and I share the guilt.

It took not only insight but the proper frame of mind in the proper time and the application of new and old technology to advance the field. What was needed was the means for reducing the tissue to single viable cells in large amounts, and techniques for maintaining them and growing them for long periods, but more importantly what was needed was the perception that this would be an important and fruitful technique which could produce cells that had meaningful biological function. That came about in 1952 when the technique for separation of chick embryonic tissues was achieved by digestion of the tissue with proteolytic enzymes such as trypsin, and the cultivation of the cells in a medium which consisted of metabolites, substrates, essential cofactors, vitamins, and serum.[4] This technique allowed successful primary culturing of cells from tissues. Until that point, the observation of Burrows remained a kind of curiosity. The culturing of viable cells was a pioneering breakthrough. From these primary cultures, cell lines were established. However, the situation was still not really ripe for exploration. Controversy over whether cells which survived in culture were representative of the cells in the intact tissue lingered for some time, and for good reason, for fundamental changes in the cells and overgrowth by nontissue specific cells could and did change the cell population in culture. During that decade most of the research done on cultured cells was not in the area of specific function. The cells were used as a kind of mammalian *E. coli,* in that large amounts of them could be obtained for the study of general problems such as amino acid[5] and carbohydrate metabolism.[6]

The trypsinization technique applied to the embryonic chick heart in 1955[7] and to the postnatal rat heart in 1960[8] was truly important, for it produced cells in culture which continued to function. Beating heart cells were thus isolated from the heart and maintained in culture. The most fascinating aspect of these preparations was of course the spontaneous contraction of the single cells indicating that they functioned, and it was clear that they did not need the intervention and control of the nervous system for beating. Equally interesting was the observation that the single cells beat with different rhythms which only became synchronous as the cells grew into contact. This was true for both embryonic chick and neonatal rat heart cells.[7,9] That postnatal single mammalian cells could be cultivated was confirmation that the heart was composed of mononucleated cells, an observation first made in the intact heart by electron microscopic examination.[10]

The first observation in the chick heart was made in 1955. However, nothing much was done with it for at least 5 years. It was almost as if the system again appeared to contemporary scientists as a curiosity. Why such a delay? It was 43 years after Burrows' observation and long after the sway of vitalism in the field of embryology. What was the atmosphere like? Different of course, but perhaps not so different. Many of us remember those times as containing a difference of opinion between classical physiology and the newer enzymology. Many biologists, while respecting the information gotten from the study of isolated enzymes, did not believe that much of importance could be learned about the whole living system from studies of the properties of isolated enzymes. One may remember one popular metaphor likened such studies to an analysis of a book, such as the Bible, for example, by putting it into a Waring blender and anlayzing the composition of the pages and the ink. In the main, most cardiac physiologists also did not believe that much could be learned about heart function from a cell so rudely torn away from the parent tissue. Even if it turned out that the cells were similar to those in the heart, they somehow did not feel that its study would yield information pertinent to the properties of the whole heart. On the other hand, in the fifties, biochemists paid little attention to cell or tissue cultures.

At that time, in 1960, it therefore seemed that the first priority was to establish that the cultured cell could function as a specialized entity which could yield useful biochemical information. There was no doubt about the functioning of the heart cell. It did beat. Yet that single property did not establish that cultured cells were similar enough to the in vivo parent to permit them to be a significant model. It was, therefore, essential to determine if other known heart properties were shown by the cells. Since the cells eventually lost their specific function, studies were also motivated by the need to learn how to maintain their properties. Much of the work at that time centered around these concerns. The questions asked concerned the reaction to drugs and hormones, their metabolism, structure, and how to maintain specific function. The attempt during this period was not to get new information about heart function, but more to assess the essential properties of the cells. For this reason their reactions to drugs were tested and they were shown to react to hormone and drug additions in the predicted way. Acetycholine, eserine, ouabain[8] and catecholamines[11,12] affected the beating rate. This set the scene for many later studies in this area.

The heart cells proved also useful for the study of the regulation of beating rhythm. Arrhythmia was early observed in the isolated cells[13] and later studied[14] to great advantage. Many arrhythmic patterns such as tachycardia, flutter, and fibrillation were found to be similar to those seen in the whole heart. Ouabain affected the rhythm in a manner similar to the whole heart. Moreover, it became clear that new information on the mechanism of drug action could be obtained from cell studies that could not be gotten from the whole heart. From a study of cultured quail ouabain sensitive cells in contact with mouse ouabain insensitive cells, it was shown that the ouabain increase of arrhythmia depended on the inhibition of the Na, K, ATPase pump.[15]

Pacemaker cells, cells which beat spontaneously and stimulated nonbeating cells to beat, were observed in both the chick embryo[7] and rat heart culture.[8] A varying percent of the single cells beat, ranging from 10[16] to 70%[17] depending on conditions such as the K level.[18] Earlier studies gave an indication that the rate of beating of two or more cells in contact was determined by the fastest beating cell.[19,20,21] This conclusion was derived from separation of connected cells[19] or by the observation of two cells growing together. This seemed reasonable, since it was consistent with conclusions drawn from the intact heart. But more detailed studies with cultured cells indicated that the rate was a result of both the fastest pacemaker cells and other cells driven by them, and did not necessarily only reflect the fastest beating cell.[22] The regularlity and beating rate of a monolayer of heart cells could be determined by the interaction of several pacemaker centers which were less regular and beat more slowly.[22,23]

The beating of the chick cells depended upon the presence of myofibrils, which in chick embryo cells could only be visualized by use of fluorescein labeled antimyosin.[24] Electron microscopic examination of cultured rat heart cells showed myofibrils in all stages of assembly and development, with helical polysomes running parallel to the myofibrils.[25] Gap junctions or desmosomes were seen in cultures beating synchronously, but were not present in the single cell cultures,[26] indicating in vitro development. The first studies of muscle specific proteins in these cultured cells demonstrated a 4- to 5-fold increase in myosin ATPase in one week and also that active cell division was necessary for this increase.[27]

There was an active synthesis of myosin[28] in these cells which was dependent on a heat stable factor in the serum.[29]

In the area of the relation of mitosis to development, the cultured heart cells made an important contribution. Beating rat heart cells were observed to divide in culture,[30] indicating that unlike skeletal muscle cells, mitosis could occur in heart cells which contained adult specific proteins. This was also later seen by electron microscopic examination of the chick embryo heart, in which myofibrils were observed in cells with mitotic figures.[31]

It emerged that the biochemistry of the cells could be studied, and the first results showed that the metabolism of the cultured cells was similar to that of the intact heart. The neonatal rat heart cells preferentially metabolized fatty acids, with a respiratory quotient of 0.78.[32] Inhibitor studies of mitochondrial oxidation and glycolysis using such inhibitors as dinitrophenol, oligomycin, iodoacetate, 2-deoxyglucose, and fluoroacetate clearly demonstrated that both fatty acid oxidation and glycolysis contributed to the maintenance of the ATP level.[33,34] Furthermore, the muscle specific role of creatine phosphate as a backup for ATP was also demonstrated in these cells.[35] The role of lipids for function was shown by the observation that cells incubated in a lipid-free medium stopped beating and were restored by the addition of long chain fatty acids.[36] The ATP level in the cells was not affected by the lack of medium lipid. The possible role of the lipids in membrane function was indicated by a later observation[37] that the beating of myocardial cells could be stopped by treatment with phospholipase C and restored with lysophosphatides. These and later investigations in several laboratories[38,39,40,41] clearly established the validity of cultured heart cells as models for the study of cardiac metobolism.

Perhaps the observation that generated most faith in these cells as bona fide heart cell models was the finding that they could generate action potentials and that electrophysiological studies could realistically be done with them.[42-44]

These topics will be more fully developed in the chapters that follow. I have briefly mentioned them in the context of a historical review of the early period of the heart cell in culture. At the outset, as I have described, most interest was focused on proving that the cells could be useful models. But the real value of the heart cell culture as an investigative tool was demonstrated many times in permitting experiments that were not possible in the intact heart. It provided a unique opportunity by providing a system which allowed direct experimental manipulation of the cell population and direct control of the environment without intervention and interference by other systems present in the whole body. As a further more recent example, the effect of the several heart cell types, i.e., endothelial, fibroblastic, and muscle cells, on each other could be examined more directly by separating and recombining them in different proportions.[45,46] Another example of cellular manipulation is the formation from the cells of multicellular strands[42] and aggregates[48] which served as models for electrophysiolical studies.

The heart cell system is relatively young and still not completely accepted as a valid model for the study of cardiac function. It is hoped that the following articles and future research will further establish its validity and increase its usefulness.

REFERENCES

1. **Burrows, M. T.,** Rhythmische Kontractionen der isolieren Herzmuskelzellen ausserhalb des Organismsus, *Munchen Med. Wschr.,* 59, 1473, 1912.
2. **Le Gallois, C. J. J.,** *Experience sur les Principe de la Vie,* D'Hautel, Paris, 1812.
3. **Parker, R. C.,** Introductory comments to studies on individual heart cells, *Circulation Res.,* 14 and 15 (Suppl 2), 195, 1964.

4. **Moscona, A.,** Cell suspension from organ rudiments of chick embryos, *Exp. Cell Res.,* 3, 535, 1952.
5. **Eagle, H.,** Amino acid metabolism in mammalian cell cultures, *Science,* 130, 432, 1959.
6. **Eagle, H., Barber, S., Levy, M., and Schultze, H. O.,** The utilization of carbohydrates by human cell cultures, *J. Biol. Chem.,* 233, 551, 1958.
7. **Cavanaugh, M. W.,** Pulsation, migration, and division in chick embryo heart cells in vitro, *J. Exp. Zool.,* 28, 212, 1955.
8. **Harary, I. and Farley, B.,** In vitro studies of isolated beating heart cells, *Science,* 131, 1674, 1960.
9. **Harary, I. and Farley, B.,** In vitro organization of single beating rat heart cells into beating fibers, *Science,* 132, 1839, 1960.
10. **Sjostrand, F. S. and Anderson-Cedergren, E.,** Intercalated discs of heart muscle, in *Structure and Function of Muscle,* Vol. 1, Bourne, G. H., Ed., Academic Press, New York, 1960, 421.
11. **Wollenberger, A. and Halle, W.,** Einfluss von Reserpin und Dichloroisoproterenol auf die durch Adrenalin und Digitoxin hervorgerufenen Wirkungen an Kulturen spontan schlangenden isolierten Zellen des embryonalen Huhnerherzens, *Monatsber Deut. Akad. Wiss. Berlin,* 5, 38, 1963.
12. **Harary, I., Renaud, J-F., Sato, E., and Wallace, G.,** Ca ions regulate cyclic AMP and beating in cultured heart cells, *Nature,* 261, 60, 1976.
13. **Wollenberger, A.,** Rhythmic and arrhythmic contractile activity of single myocardial cells cultured in vitro, *Circulation Res.,* 14 and 15 (Suppl. 2), 184, 1964.
14. **Goshima, K.,** Arrhythmic movements of myocardial cells in culture and their improvement with antiarrhythmic drugs, *J. Mol. Cell. Cardiol.,* 8, 217, 1976.
15. **Goshima, K. and Wakabayoshi, S.,** Inhibition of ouabain induced arrhythmias of ouabain-sensitive myocardial cells (quail) by contact with ouabain-resistant cells (mouse) and its mechanism, *J. Mol. Cell. Cardiol.,* 13, 75, 1981.
16. **Harary, I. and Farley, B.,** In vitro studies on single beating rat heart cells. I. Growth and organization, *Exp. Cell Res.,* 29, 451, 1963.
17. **Lehmkuhl, D. and Sperelakis, N.,** Transmembrane potentials of trypsin dispersed chick heart cells cultured in vitro, *Am.J. Physiol.,* 205, 213, 1963.
18. **De Haan, R. L.,** Regulation of spontaneous activity and growth of embryonic chick heart cells in tissue culture, *Dev. Biol.,* 16, 216, 1967.
19. **Garofolini, L.,** Rhythmical contractions of single heart muscle cells in tissue culture in vitro, *J. Physiol. Proc.,* 63, v, 1927.
20. **Harary, I. and Farley, B.,** *In vitro* studies on single beating rat heart cells. II. Intercellular communication, *Exptl. Cell Res.,* 29, 466, 1963.
21. **Goshima, K.,** Synchronized beating of and electrotonic transmission between myocardial cells mediated by heterotrophic strain cells in monolayer culture, *Exptl. Cell Res.,* 58, 420, 1969.
22. **De Haan, R. L., and Hirackow, R.,** Synchronization of pulsation rates in isolated cardiac myocytes, *Exp. Cell Res.,* 70, 214, 1972.
23. **Jongsma, H. J., Tsjernina, L., and de Brujne, J.,** The establishment of regular beating in populations of pacemaker cells. A study with tissue culture rat heart cells, *J. Mol. Cell. Cardiol.,* 15, 123, 1983.
24. **Holtzer, H., Abbott, J., and Cavanaugh, M. W.,** Some properties of embryonic cardiac myoblasts, *Exp. Cell Res.,* 16, 595, 1959.
25. **Cedergren, B. and Harary, I.,** In vitro studies on single beating rat heart cells. VI. Electron microscopic studies of single cells., *J. Ultrastruct. Res.,* 11, 428, 1964.
26. **Cedergren, B. and Harary, I.,** In vitro studies on single beating rat heart cells. VII. Ultrastructure of the beating cell layer., *J. Ultrastruct. Res.,* 11, 443, 1964.
27. **Lewis, H. and Harary, I.,** In vitro studies on single beating rat heart cells in culture. XIV. Reversible changes in the level of myosin adenosine triphosphatase, *Arch. Biochem. Biophys.,* 142, 501, 1971.
28. **Desmond, W. and Harary, I.,** In vitro studies of beating heart cells in culture. XV. Myosin turnover and the effect of serum, *Arch. Biochem. Biophys.,* 151, 285, 1972.
29. **Casanello, D., Bieleski, R. L., Desmond, W., and Harary, I.,** In vitro studies of beating heart cells in culture. XVII. The stimulation of myosin synthesis by a heat stable serum fraction., *Arch. Biochem. Biophys.,* 156, 570, 1973.
30. **Mark, G. E. and Strasser, F. F.,** Pacemaker activity and mitosis in cultures of newborn rat heart ventricle cells, *Exp. Cell Res.,* 44, 217, 1966.
31. **Manasek, F. J.,** Mitosis in developing cardiac muscle, *J. Cell Biol.,* 37, 191, 1968.
32. **Fujimoto, A. and Harary, I.,** Studies in vitro on single beating rat heart cells. IV. The shift from fat to carbohydrate metabolism in culture, *Biochim. Biophys. Acta,* 86, 74, 1964.
33. **Harary, I. and Slater, E. C.,** Studies in vitro on single beating heart cells. VIII. The Effect of oligomycin, dinitrophenol, and ouabain on the beating rate, *Biochim. Biophys. Acta,* 99, 227, 1964.
34. **Seraydarian, M. W., Harary, I., and Sato, E.,** In vitro studies of beating heart cells in culture. XI. The ATPase level and contraction of the heart cells, *Biochim. Biophys. Acta,* 162, 414, 1968.

35. **Seraydarian, M. W., Sato, E., and Harary, I.,** In vitro studies of beating heart cells in culture. XIII. The effect of 1-fluoro-2,4-dinitrobenzene, *Arch. Biochem. Biophys.,* 138, 233, 1970.
36. **Harary, I., McCarl, R., and Farley, B.,** Studies in vitro on single beating rat heart cells. IX. The restoration of beating by serum lipids and fatty acids, *Biochim. Biophys. Acta,* 115, 15, 1966.
37. **Goshima, K.,** Beating of myocardial cells in monolayer culture. Inhibition by phospholipase C and restoration by lysophosphatides, *Exp. Cell Res.,* 67, 352, 1971.
38. **Warshaw, J. B. and Rosenthal, M. D.,** Change in glucose oxidation during growth of embryonic heart cells in culture, *J. Cell Biol.,* 52, 283, 1972.
39. **Pinson, A. and Padieu, P.,** Erucic acid oxidation by beating heart cells in culture, *FEBS Lett.,* 39, 88, 1974.
40. **Pinson, A., Frelin, C., and Padieu, P.,** Palimitate oxidation by beating heart cells in culture, *Recent Adv. Stud. Cardiac Struct. Metab.,* 12, 667, 1976.
41. **Ross, P. D. and McCarl, R. L.,** Oxidation of carbohydrates and palmitate by intact cultured neonatal rat heart cells, *Am. J. Physiol.,* 246, (Heart Circ. Physiol. 15), H389, 1984.
42. **Fange, R., Persson, H., and Hesleff, T.,** Electrophysiologic and pharmacologic observations on trypsin-disintegrated embryonic chick hearts cultured in vitro, *Acta Physiol. Scand.,* 38, 173, 1956.
43. **Grill, W. E., Rumery, R. E., and Woodbury, J. W.,** Effects of membrane current on transmembrane potentials of cultured chick embryo heart cells, *Am. J. Physiol.,* 197, 733, 1959.
44. **Lehmkuhl, D. and Sperelakis, N.,** Transmembrane potentials of trypsin dispersed chick heart cells cultured in vitro, *Am. J. Physiol.,* 205, 1213, 1963.
45. **Gross, W. O.,** Fibroblast-myocyte interactions in in vitro cardiomyogenesis, *Exp. Cell Res.,* 142, 341, 1982.
46. **Schroedl, N. A. and Hartzell, C. R.,** Myocytes and fibroblasts exhibit functional synergism in mixed cultures of neonatal rat heart cells, *J. Cell. Physiol.,* 117, 326, 1983.
47. **Purdy, J. E., Lieberman, M., Roggeveen, A. E., and Kirk, R. G.,** Synthetic strands of cardiac muscle: formation and ultrastructure, *J. Cell Biol.,* 55, 563, 1967.
48. **Seperlakis, N.,** Cultured heart cell reaggregate model for studying problems in cardiac toxicology, in *Cardiovascular Toxicology,* Van Stee, E. W., Ed., Raven Press, New York, 1982, 57.

Chapter 2

TECHNIQUES FOR CULTURING HEART CELLS

Arié Pinson, Prudent Padieu, and Isaac Harary

TABLE OF CONTENTS

I. INTRODUCTION

Tremendous progress has been made in the field of heart cell cultures since 1912 when Burrows made his pioneering observation that beating heart cells developed in embryonic chick heart explants grown on plasma.[1] In 1955, Cavanaugh noted synchronized beating upon cell contact in dissociated chick embryo heart cells.[2] Since 1960, Harary and Farley[3,4] have been working on cardiomyocytes from postnatal mammals. Subsequent improvements in techniques and equipment have made heart cell culture a major research tool in cardiac biology.

At the present time, the variety of techniques available is overwhelming and undoubtedly accounts for the great variation in results in the literature. This has led to caution in the interpretation of data. In this chapter, we will discuss the methods for culturing cardiac cells, pointing out the considerable gaps in our knowledge, since optimal requirements for tissue disaggregation and nutrition have yet to be established.

II. CULTURING HEART CELLS

A. Culture Systems

In vitro models have been derived from organ cultures,[5] tissue explants,[1,6] reaggregated single adult or embryonic cells (primarily in electrophysiological research[7-9]), and from single cell cardiomyocyte cultures. The latter are the topic of this chapter. Such cultures have been derived from chick embryo (CE), mouse embryos, new born rats (NBR), postnatal hamsters and human fetuses (see Table 1).

B. Cell Dissociation

Moscona reestablished Rous and Jones's long forgotten technique of tissue dissociation using trypsin.[10-11] The exact conditions employed in this step are crucial in achieving successful cultures. Various proteolytic enzymes have been used with highly conflicting results (see Table 1). Cell damage by the proteolytic enzymes was minimized by carrying out the procedure in several steps and collecting the free cells after each of these in a Ca^{2+}, Mg^{2+}, and serum containing medium, which effectively inhibits further proteolysis.

The optimal enzyme requirements for cell dissociation vary with the type of tissue and the age of the animals used. In addition, it is well established that crude enzyme preparations are more efficient than purified ones. Indeed, as early as 1958, it was recognized that contaminating enzymes in crude trypsin contributed to tissue dissociation.[12] Highly purified collagenase is a poor dispersing agent; so, clearly, impurities in cruder preparations render this process more efficient.[13] Other enzymes, e.g., pronase and elastase, have also been employed for making rat or chick embryo heart cell suspensions (see Table 1).[14-18] A commercial preparation of proteolytic enzymes called Viokase® (Viobin Corp., Montecello, Ill.)[19] has been used very successfully and is more consistent in its effect than crude trypsin, which tends to vary in its composition. Most crude enzyme preparations contain other proteolytic enzymes in varying quantities. Crude trypsin, for example, also contains chymotrypsin and elastase.[20] There are also significant variations between different batches of commercially available crude enzymes in their ability to dissociate heart cells — some of them causing severe cell damage.[21] Some workers even claim that trypsin preferentially damages myocardial cell.[22-23] Thus Masson-Pévet et al.[24] compared cardiomycytes prepared using crude trypsin and crude collagenase in the dispersion step by ultrastructural methods, just after cell isolation and again after a monolayer had formed. They showed that whereas there were no signs of cell injury in collagenase dissociated cells, trypsin treated cells were severely damaged. This is evidently reflected in cell recovery time — 2 days after collagenase and 1 week after trypsin isolation. Gross et al.[25] found that trypsinized preparations of

Table 1

SUMMARY OF TECHNIQUES FOR PREPARATION OF HEART CELLS IN CULTURE AS REPORTED IN THE LITERATURE

Animal	Disaggregation	Media	Sera and growth-promoting substances	Culture support	Seeding density	Ref.
NBR	Trypsin	Puck	10% HuS 10% FBS	Glass	—	3, 4
CE	0.1% trypsin in saline + glucose	Eagle's MEM or m-199	10% FBS; 2% CE — extracts	Synthetic strands	10^5—10^6 cells/60 mm P.d. (coated with collagen)	7—9
NBR	Trypsin in saline "A"	Balanced solution + amino acids + vitamins	20% bovine serum	P.d. 60 mm	4—5 × 10^6	20
NBR	Trypsin 0.05% in Puck's solution or collagenase Worthington 450 U/mℓ	HAM F10	10% H.S. 10% FBS	Falcon P.d. (50 mm)	5 × 10^5 cells/mℓ	24
CE NBR, NBM	0.25% trypsin in Rinaldini's Ca^{2+} free buffer solution	Hank's	0.5% lactalbumin hydrolysate + 0.1% YE + 20% inactivated cals serum (in Ref. 25 + 10% FBS)	Riedel chamber	2.5—8 × 10^5 cells/mℓ	25, 60
NBR	Crude trypsin 0.1% or purified trypsin 0.01% in buffered saline	Balanced solution + amino acids + vitamins	10% FBS	60 mm Falcon	4.5 × 10^6 cells/ P.d.	28
NBR	0.1% Trypsin in HAM F10	Ham F10	10% HuS; 10% FBS	Falcon P.d., 35 or 60 mm	2 × 10^6 cells/60 mm P.d.	29, 30, 56
CE	0.5% Trypsin in tyrode solution	DM 40% + CM 60%		30 mm sealable glass dishes	ND	34, 35

Table 1
SUMMARY OF TECHNIQUES FOR PREPARATION OF HEART CELLS IN CULTURE AS REPORTED IN THE LITERATURE

NBR	Trypsin + elastase + chymotrypsin in balanced saline	Eagle's MEM with Earle's salts	10% BS		2 or 5 $\times$ 10^6 60 mm P.d.	37
NBR	0.125% Trypsin in Tyrode solution	Grey's balanced salt solution + 1% MEM + vitamins and amino acids	10% HuS 10% FBS	Rose chamber	2-4 $\times$ 10^5 cells/ chamber	39
NBR	0.125% Trypsin in Hank's BSS w/o Mg^2 + Ca^{2+}	MEM + Grey's BSS + vitamins	10% HuS 10% FBS	Rose chamber (glass)	1—2 $\times$ 10^5/mℓ	40, 47
CE	0.05% Trypsin in CMF	M-199 20% BSS 69%	4% HS	Falcon P.d.	10^5/plate	42, 62
NBR	0.25% Trypsin in PBS	MEM	0.06% Y.E. + 0.25% lactalbumin hydrolysate + 10% FBS	Falcon 10 cm	5.5 $\times$ 10^5 — 1 $\times$ 10^6 cells/ml	43
NBR	Viokase	CMRL	10% FBS 10%HuS	Falcon P.d.	Variable	55
CE	Trypsin 0.05%	M-199 20% BSS 69%	4% HS 6% Ce extract	Falcon P.d.	2—4 $\times$ 10^5/plate	61
NBR	—	SM 20-I	10% FBS	16.5 cm^2 Muller flask (glass)	2.4 $\times$ 10^6/flask	63

Table 1 (continued)
SUMMARY OF TECHNIQUES FOR PREPARATION OF HEART CELLS IN CULTURE AS REPORTED IN THE LITERATURE

Animal	Disaggregation	Media	Sera and growth-promoting substances	Culture support	Seeding density	Ref.
NBR	37 mg% Trypsin in buffered salt solution	MEM	10% HS	100 mm Falcon P.d.	5×10^5 cells/mℓ 5×10^6 cells/ P.d.	64
NBR	Collagenase 0.5 mg/mℓ	BSS + HAM F10	10% FBS	Scintillation vials (glass)	ND	65
CE	Trypsin 0.05% or DM8	80% Eagle's BSS + M199 20%	2% HS + 4% FBS	Aggregates transported to Falcon P.d.	Aggregates	66
Different species	0.25% Trypsin in Tyrode's solution	88% Eagle's basal medium	10% FBS	Glass	1×10^6/mℓ (aggregates)	67
Mouse embryo	Trypsin	Eagle's MEM	8% FBS	35 mm P.d.	2×10^5/P.d.	68
NBR	0.1% Trypsin in Ham F10	HAM F10	10% FBS 10% Hus (A or O)	Falcon P.d.	1.5×10^6 60 mm P.d.	69
NBR	0.2% Viocase or 0.25% pancreatin 4×	90% CMRL 1415 ATM	5% FBS 5% HS		$5-10 \times 10^4$ 35 mm P.d.	70
HuF	Trypsin: 0.025% or 0.04%					71

Notes: CE, chick embryo. BS, beef serum. BSS, balanced salt solution. CMF, calcium-magnesium-free solution. CMRL, Connaught Medical Research Laboratories. DM, Eagle's MEM + Earle's salts + nonessential amino acids. HuF, human fetuses. FBS, fetal bovine serum. HuS, human serum. HS, horse serum. MEM, minimal essential medium. NBM, newborn mice. NBR, newborn rats. P.d., petri dishes. YE, yeast extracts. ND, not determined.

dispersed cells showed a loss of myofibril alignment in the mitochondria, only traces of the Z-lines remained, while intercalated discs could no longer be detected. They concluded that isolated cells in culture had to begin redifferentiation from a very early stage and that this process required at least three days.[25] Other workers share the view that trypsin penetrates the myocardial cells and digests myofibrils.[23-26]

Despite the problems mentioned above, crude trypsin remains the most widely used enzyme for cell dissociation. The concentration in standard use varies from 0.05 to 0.1%. Speicher and McCarl have reported a procedure for purifying crude trypsin preparations,[27] leading to a mixture enriched in chymotrypsin, elastase, and trypsin, which affected cell dissociation at much lower concentrations, (0.01%). This purified mixture of enzymes, free of toxic materials, caused less damage to the isolated cells.[28] The possibility of routinely using this technique should clearly be investigated.

C. The Enzyme Solution

Table 1 presents a summary of the enzyme mixtures used by various workers, and these differences clearly contribute to confusion in the field, as the state of cell preservation varies greatly. As mentioned by many workers, the purity of the water also influences the success of cytocultures, and in somes cases, it may be necessary to specially purify water for this purpose.[29]

One may successfully use 0.1% trypsin (Sigma grade III) in a solution of Ham's F10 (Ca^{2+} and Mg^{2+} free) medium at pH 7.2 to 7.4 at 32 to 35°C, or 0.1% Viokase in phosphate buffered saline (PBS) also Ca^{2+} and Mg^{2+} free. Our data show that glucose during tissue degradation and the shortest possible trypsinization cycles (10 to 20 min) give optimal cell preservation.

D. Media and Sera

As Table 1 indicates, a large variety of media are in use in various laboratories. In addition, the frequency with which the medium is changed varies from group to group. Cell division, growth, and metabolism depend on the exact composition of the medium (the growth factors, etc.) and on the length of the lag periods between media changes when less substrate may be available to the cells.

It is no simple matter to attempt to reproduce in vitro conditions resembling the finely balanced, complex neurohumoral mechanism that exists in vivo. In cultured cells, isolated from neurohumoral influences, cell division, differentiation, and metabolism are controlled by the composition of the extracellular medium, contact between cells, and the relative populations of different types of cells. The frequency of medium changes is critical in this respect.[29,30] Certain substrates, such as glucose and fatty acids (FAs), are completely taken up by the cells within 12 hr of a medium change.[31,32] At this stage, cells adapt metabolically, either by storing excess glucose in the form of glycogen or by exporting it from the cells as lactate.[29,30,33] It has also been shown that in cultures grown in Ham's F10, isoleucine is rapidly exhausted from the medium, so that the rate of protein synthesis becomes dependent on the amount of isoleucine available in the amino acid pool or formed as a product of degradation.[30]

Sera from different sources, sometimes supplemented with growth factors, are commonly added to media (see Table 1). From 5 to 20% serum, from such widely different sources as the rat fetus, bovine fetus or newborn horse, and human, which may be supplemented with embryo extracts, is added to media. Human serum tends to be unreliable, since the source cannot be controlled, and batches with toxic factors can occur. The growth adaptation and cell population are strongly influenced by the composition of the medium. The optimal conditions for growth and maintenance of heart cell cultures have yet to be determined.

E. Conditioned Media

Some workers supplemented culture media with conditioned media, obtained by incubating either chick heart cell cultures or fragments for 4 days with fresh medium.[34,35] On the one hand, conditioned media retarded cell growth (the population doubling time was tripled — 72 hr vs. 23 hr in fresh medium) probably because the conditioned media added are exhausted of substrates; whereas, on the other hand, cell attachment and spreading was enhanced, and spontaneous beating was maintained for longer time periods.[34] The contractile and spreading promoting factors were shown by Gordon and Brice[35] to be a proteoglycan and a protein containing material, respectively. Recent work has shown that cultured heart cells secrete a nerve growth promoting factor (see Chapter 6). Are the effects of conditioned medium on heart cell growth mediated via factors similar to those promoting the growth of neurones? It would certainly be worthwhile studying how the components of conditioned media influence cell growth.

F. Defined Media

Media of this nature were devised in order to gain better control over the growth factors present in the system. Thus, sera and embryo were not used as sources of growth factors since they contain many unknown factors. Rather, known quantities of materials, such as insulin and butyryl AMP, were added to promote culture growth. This offers a promising approach for gaining further understanding of the undefined process of cell growth under clearly defined conditions (see Chapter 3).

G. Plating, Plating Density, and Cell Populations in Cultures

At the present time sterile, plastic tissue culture quality Petri dishes are used for plating, which are usually treated specifically to allow attachment and growth of the cells.

In the 1960s glass plates were more commonly employed for this purpose. There seems to be no particular advantage in either of these types of plates, except for the fact that the plastic ones are disposable. Plastic dishes may contain metals, e.g., Pb, which may have a significant effect on certain parameters.[36]

Seeding density has recently been shown to affect cell growth, division, and population by Speicher et al.,[37] and biochemical differentiation criteria in terms of enzymatic activities by Yagev et al.[38] Plating density is, therefore, a major factor affecting relative populations of different cells, which was not always taken into account by researchers, although it varied by up to an order of magnitude in different studies (see Table 1). Indeed, many erroneous conclusions were reached in early research work because neither seeding density nor the overgrowth of nonmuscle cells were taken into account. Cells isolated from newborn rats consist of two populations: (1) About 80% of the initial cell suspension, with a long cell cycle, are cardiomyoblasts, which rapidly lose their ability to divide once they become fully differentiated cardiomyocytes;[37-41] and (2) About 20% of the initial cell suspension are nonmuscle cells (NMCS), with a much shorter cell cycle, maintaining their ability to divide throughout, usually referred to as "fibroblasts", but also containing other types of cells, particularly endothelial cells.[37-41] Seeding at high density (10^6 cells/mℓ) assures that the culture reaches confluency and contact inhibition within 24 hr, thus minimizing cell division in NMCS and preventing them from overgrowing the cardiomyocytes, as [^{3}H]thymidine incorporation studies show.[37,38] A low seeding density, on the other hand, allows rapid fibroblast proliferation. The relative populations of different types of cells changes within a few days of seeding.[37-41] Thus high density seeding is an easy way of maintaining a constant cell population in cultures for a sufficiently extended period of time. An alternative method, first proposed in the early 1970s, is the "selective adhesion" technique, which depends on the different rates of attachment of various types of cells to the substratum.[42,43]

H. Cell Division

Myocardial cells may be distinguished from NMCs in culture by their "dense" cytoplasm, well developed mitochondria and Golgi system, and the presence of myofibrils and intercalated discs and their spontaneous contraction activiity. The average mitotic cycle is 2.5-fold longer in myocardial cells than in NMCs.[40,41]

The general rule, that differentiated cells no longer divide, has limited applicability to myocardial cells, where the cardiomyoblast which can synthesize myosin and also divide is an intermediate state between the presumptive myoblast and the adult postmitotic cardiomyocyte. It is also difficult to define the last cell division in such a system.[44] It may well be that relatively undifferentiated myocardial cells, premyoblasts, predominate in these cultures (as reported by Masse and Harary[45]), since differentiated myocytes and fully developed myoblasts would be expected to be less resistant to cellular dissociation. The fact that under optimal culture conditions the most characteristic expression of differentiation, automatic synchronous beating, occurs within 24 hr, a much shorter time period than reported by Gross et al.,[25] supports this premise. The postnatal rat heart contains cells in different stages of development. In one day postnatal rats, roughly 55% of the cells are dividing cardiomyoblasts containing myosin; however, this percentage decreases to 40% by day 4.[46] This mirrors the situation in the intact animal, in which all the cardiac muscle cells are postmitotic by day 21.

I. Cell Lines

As shown by Kasten and Yip[47] myocardial cells may be "banked" by multiple trypsinization-freezing-thawing cycles. Kimes and Brandt[48] claimed to have established a clonal muscle cell line from embryonic rat heart tissue. However, since then there have been no subsequent reports on this system. Another cell line, the Girardi cell line derived from human atrial appendage, has long lost any resemblance to the parent cells.[49] Since it has been shown that the percentage of postmitotic cells increases with time in culture, it is unlikely that a line of cardiomyocytes can be developed unless a means of preventing the loss of cell division, thus maintaining the intermediate cardiomyoblast, can be found. It is not possible to establish a cell line with postmitotic cells.

J. Selecting Cell Populations

The interpretation of biochemical and pharmacological data is complicated by the presence of two types of cells in the medium, myocardial and nonmuscle cells, mainly because NMCs proliferate more quickly and may eventually overgrow the culture. Several methods have been developed in order to overcome this problem.

The selective adhesion technique takes advantage of the observation that cells of the fibroblast type become more readily attached to the substratum and then spread over the surface while the myoblasts retain their round shape for longer, and therefore remain in suspension.[38,42,43] Replating of cells at the critical point allows preparation of a population that consists essentially of myocytes.[38] The main difficulty with this procedure lies in deciding on the length of replating time - if this phase is too brief NMCs will preponderate, while if the preplating time is too long, myoblast attachment to the substrate surface will be considerable. Alternatively, cultures consisting primarily of cardiac endothelial cells may be produced easily by adding factors derived from tumor cells, which selectively stimulate endothelial cell growth.[50]

Other techniques are based on selective elimination of the NMCs by growing the cells in a serum free medium (see Chapter 3), by treatment with 7-β-OH cholesterol (see Volume III, Chapter 30), or with the Ca^{2+} ionophore, A23187, which lyses cardiac NMCs.[51] (see Chapter 30).

Another technique is by using DNA synthesis inhibitors.[52,53] This has been applied to the cell culture system with BUdR (5-Bromodeoxyuridine) followed by UV irradiation which kills all the dividing cells.[53] However, all these chemical methods suffer from the disadvantage that they may damage myocardial cells, e.g., Ca ionophores in disrupting Ca^{2+} flux across cell membranes[54] may also have other adverse effects. In addition, the elimination of one type of cell population from the Petri dish surface while the myocytes are already in a nondividing state may lead to a nonconfluent culture, which would give rise to significant errors in studies using the Petri dish as a means of expressing data.[38]

III. CULTURING TECHNIQUES*

A. Equipment

The basic equipment for a tissue culture unit includes: (1) an isolated UV-sterilized area, (2) a sterile hood, (3) an inverted phase microscope, (4) an incubator at 37°C with an air/ CO_2 (95%:5%) flow, (5) sterile plastic disposable laboratory ware (e.g., Petri dishes, flasks, pipettes and tubes), (6) an autoclave for sterilizing glassware and filters, (7) "Trypsinators" (a 25 mℓ in vitro flask containing a teflon-covered plastic bar attached to the stopper) and (8) filter holders 25 and 142 mm in diameter suitable for small and large volumes, respectively.

B. Reagents and Materials

The growth medium routinely used is Ham's F10 culture medium[57] supplemented with 10% horse serum and 10% fetal bovine serum, $CaCl_2.2H_2O$ at 135 g/ℓ, penicillin at 200,000 units/ℓ and 0.2 g/ℓ streptomycin. Another medium used is CMRL 1415ATM (Connaught Medical Research Laboratories, Willowdale, Ontario, Canada). Ham's F10 requires incubation in a 5% CO_2 atmosphere.[58] CMRL is a nonbicarbonate buffer and therefore can be used without CO_2 in the atmosphere.[55] F10 can be used with Hepes buffer and thus avoid the need for CO_2. "Solution H" used for mincing and washing the hearts and for the preparation of the trypsin solution (Sigma, grade III, 0.1% w/v) that should be dissolved in Ham's F10 culture medium in the absence of Ca^{2+} and Mg^{2+}.

C. Isolation of Cells and Preparation of Cultures

Heart cell cultures are prepared by the dissociation of newborn rat hearts into single cells (see Scheme 1). Sterile conditions are employed throughout. One-day-old rats are killed by decapitation and allowed to bleed. The animals are immersed in 70% ethanol and then held from the back in order to stretch the chest skin, thus revealing the position of the heart. The chest is opened by a transverse cut with sterile scissors and the hearts are then aseptically removed with forceps, attempting to take the ventricles only. Excised hearts are placed in a Petri dish containing "Solution H" and then minced into the smallest fragments possible, with one or two washings. Solution H is then replaced by a trypsin-containing solution and the preparation is then transferred to a trypsinator. 10 to 15 mℓ of trypsin are required for 30 to 50 hearts. Trypsinizations are performed at 32 to 35°C with a stirring rate of 150 to 200 rpm for 15 to 20 min. Fragments are then allowed to settle and the supernatant is removed after each trypsinization. The first 2 to 3 trypsinizations, essentially containing cell debris, red blood cells, pericardial and endothelial cells, are rejected. Cells from subsequent trypsinizations are collected in 30 mℓ sterile tubes (Sterilin, England). A few mℓ of growth medium are added and the cells then centrifuged at 1,000 rpm. The pellet is then resuspended in a small volume of growth medium. Trypsinizations are repeated until all the fragments are dissociated.

Cells in each tube are suspended by repeated aspirations in a sterile pipette and all the fractions combined in a sterile 250 mℓ flask (Nunc, Denmark), passing them through a

---

* These techniques have been described in a number of publications.[4,38,53,55,56]

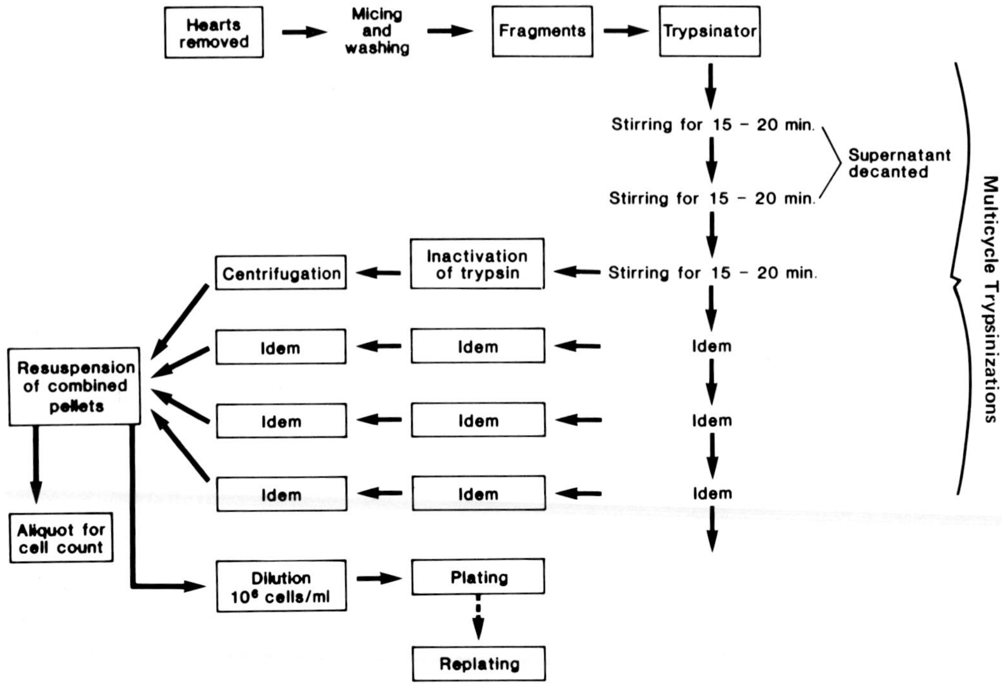

SCHEME 1. Schematic representation of culture preparation methods.

sterile mesh in order to exclude explants. Cell suspensions are diluted with growth medium to a final density of 1×10^6 cells/mℓ and seeded onto Petri dishes — either 2 mℓ onto a 35 mm diameter Petri dish (Falcon 3001) or 5 mℓ onto a 60 mm diameter dish (Falcon 3002,® Falcon Plastics, Los Angeles, Calif. or other types according to the type of surface required). After seeding, Petri dishes are shaken gently and linearly in order to achieve a more uniform distribution, and if the cultures are in Ham's F10 they are incubated at 37°C in an atmosphere of 95% air per 5% CO_2. After 24 to 36 hr, an almost confluent layer of beating cells is formed. The medium is changed every 2 days.

D. Separation of Cells

Enrichment of cells with respect to either muscle or nonmuscle cells may be achieved by plating the post-trypsinized cell suspension for one hour at 37°C. Petri dishes are then shaken and the unattached cells transferred to a new Petri dish. Under these conditions, since NMCs display more rapid attachment, the first Petri dish contains an almost pure population of NMCs, whereas the second one contains a highly enriched myoblast population (see Scheme 1).

IV. SOME REPRESENTATIVE DATA AND COMMENTS

The importance of seeding density may be demonstrated by microscopic observation of a heavily seeded culture (10^6 cells/mℓ) at day 6 (see Figure 1A), which primarily contains beating muscle cells as compared to a lightly seeded culture (Figure 1B), which contains a large number of mitotic cells. This was confirmed by [³H]thymidine uptake studies — a heavily seeded culture was compared to a culture seeded with only half the number of cells (Figure 2). Only NMCs continued to divide, although at a lower rate, probably due to space limitation. The activity of various marker enzymes may also serve as an index of cell populations in culture. Thus LDH activity increases linearly in both myocyte and NMC

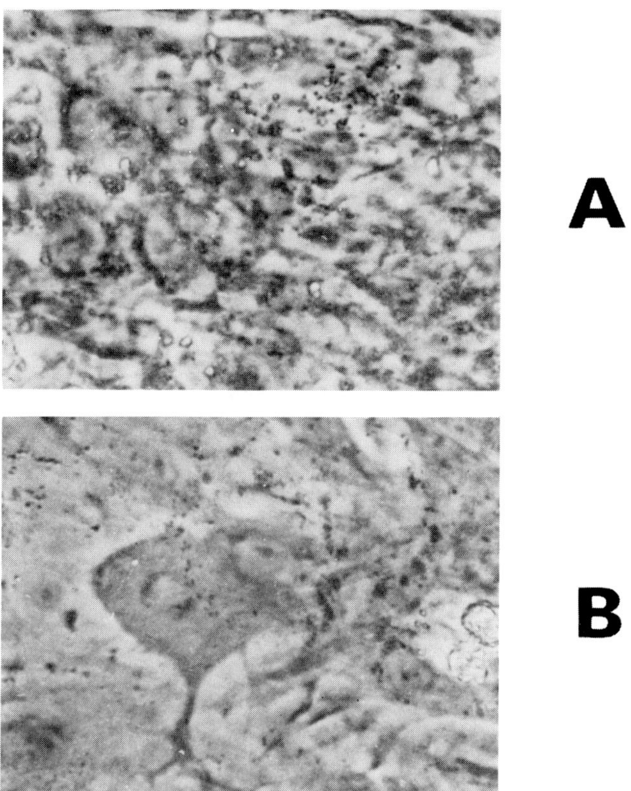

FIGURE 1.   Six-day-old heart cell cultures as observed under a microscope. (A) Heavily seeded culture, 9 × 10⁶ cells/mℓ; and (B) lightly seeded culture, 4.5 × 10⁶ cells/mℓ. (Reproduced from Yagev, S., Heller, M., and Pinson, A., Changes in cytoplasmic and lysosomal enzyme activities in cultured rat heart cells: the relationship to cell differentiation and cell population in culture, *In Vitro*, 20, 893, 1984. With permission.)

enriched cultures (Figure 3); however, LDH activity is threefold higher in muscle cells as compared with NMCs. The evolution of CPK activity follows a similar pattern, but is only twofold higher in muscular cells. In cultures of mixed population, however, a different pattern of enzymatic activities was observed — up to the sixth day in culture they followed the same pattern as cultures enriched with myocytes; however, the values subsequently level off at an activity higher than the one observed in NMCs. By the twelfth day, enzymatic activities closely resembled those in cultures of NMCs. This leveling off was due to overgrowth of NMCs with reduced enzyme activities, leading to an apparent overall decrease in activity as shown in Figure 3. Lysosomal enzymes, such as β-galactosidase and acid phosphatase, followed a similar pattern; however, higher activities were found in NMCs, in this case. It is evident from these data that the nature of the cell population, as expressed by the various enzyme activities, largely depends upon the initial seeding density — high initial densities leading to more or less homogeneous populations of myocytes up to the sixth day of culture. Enzymatic activities of a culture depend on its age, which in turn governs the nature of the cell population. Consequently, analysis of different enzyme activities yields a reasonable estimate of the relative populations of the different types of cells in the culture. The replating technique enables effective separation between muscle and nonmuscle cells. In addition, highly dense initial plating of cells also leads to a rather homogeneous population of myocytes up to the sixth day in culture.

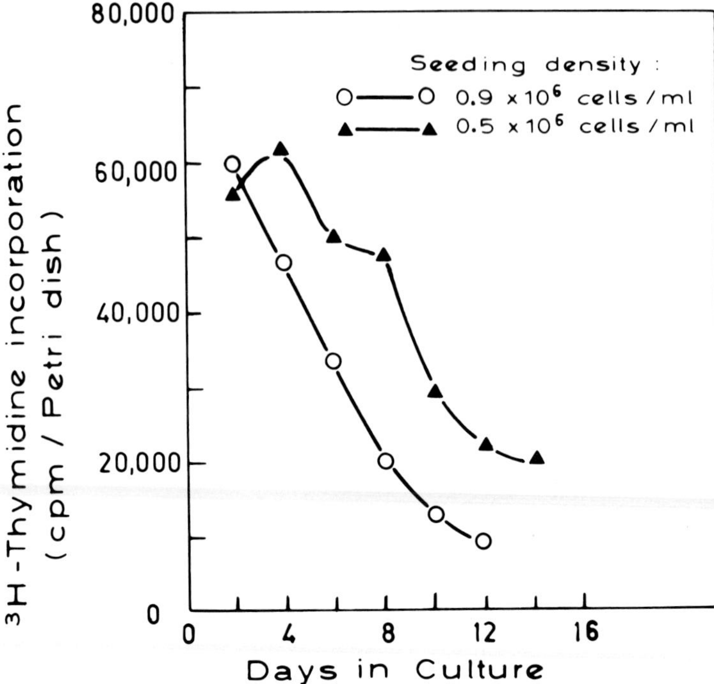

FIGURE 2.   [³H] Thymidine incorporation into TCA insoluble cellular components. Cells were pulse labeled for 24 hr with [³H]thymidine. Proteins were precipitated after homogenization with TCA. (Reproduced from Yagev, S., Heller, M., and Pinson, A., Changes in cytoplasmic and lysomal enzyme activities in cultured rat heart cells: the relationship to cell differentiation and cell population in culture, *In Vitro*, 20, 893, 1984. With permission.)

The importance of plating density had previously been shown by Speicher et al.[37] They also found that cell division, as measured by [³H]thymidine incorporation, takes longer in cultures seeded at low (4 × 10⁵ cells/mℓ) rather than at high (1 × 10⁶ cells/mℓ) densities. However, the rate of protein synthesis was similar in both types of culture. They concluded, as we did too, that comparing data from two cultures seeded at different densities amounts to a comparison of two systems which hardly have anything in common. Furthermore, the importance of standarization not only applies to plating density but also to all other aspects of handling the culture. Since the cells are in a closed system, the Petri dish, a steady state is not maintained in the same sense as it is in vivo, as many of the substrates become exhausted within the first 12 to 20 hr,[31,32] whereas the cells continue to export metabolites into the medium. This serves to emphasize the importance of the frequency of medium changes and of the time lag between the last medium replacement and the use of the cells for experimental purposes, which may influence the nutritional and physiological state of the cells. It is certainly worthwhile, even from the economic point of view, concentrating research efforts on finding the true optimum media conditions, since some substrates are almost certainly exhausted too rapidly and sera levels of 20% are undoubtedly excessive. Finally, at the present time, the best method for selecting a homogeneous myocyte population in cultures is the replating method combined with a high seeding density and taking the cells from the area closest to the apex. Very little is known about the effects of chemical inhibition of NMC growth and it should therefore be treated with caution.

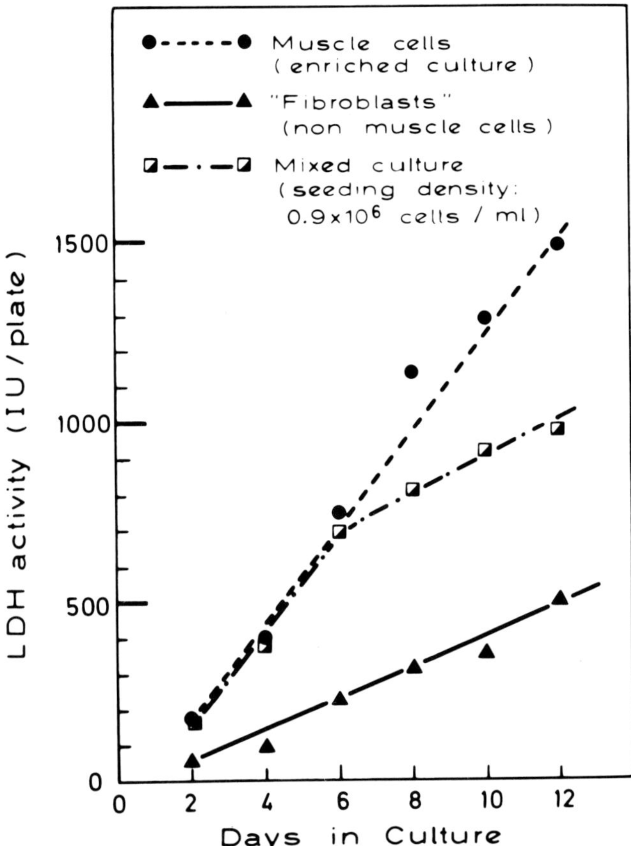

FIGURE 3.   LDH activity in cardiac cells as a function of culture age. The total activity, in units per plate, was assayed according to Wroblewski and La Due[59] in cultures of mixed type and in those enriched with either muscle or nonmuscle cells (fibroblasts). (Reproduced from Yagev, S., Heller, M., and Pinson, A., Changes in cytoplasmic and lysomal enzyme activities in cultured rat heart cells: the relationship to cell differentiation and cell population in culture, *In Vitro*, 20, 893, 1984. With permission.)

REFERENCES

1. **Burrows, M. T.,** Rhythmical activity of isolated heart muscle cell in vitro, *Science*, 36, 90, 1912.
2. **Cavanaugh, M. W.,** Pulsation migration and division in dissociated chick heart cells in vitro, *J. Exp. Zool.*, 128, 573, 1955.
3. **Harary, I. and Farley, B.,** In vitro studies of single isolated beating heart cells, *Science*, 131, 1674, 1960.
4. **Harary, I. and Farley, B.,** In vitro studies on single beating rat heart cells. I. Growth and Organization, *Exp. Cell Res.*, 29, 451, 1963.
5. **Wildenthal, K.,** Factors promoting the survival of beating of intact foetal mouse hearts in organ culture, *J. Mol. Cell. Cardiol.*, 1, 101, 1970.
6. **Goshima,. K.,** Synchronized beating of myocardial cells mediated by FL cells in monolayer culture and its inhibition by trypsin-treated FL cells, *Exp. Cell Res.*, 65, 161, 1971.
7. **Lieberman, M., Roggeveen, A., Purdy, I. E., and Johnson, E. A.,** Synthetic strands of cardiac muscle: Growth and physiological implications, *Science*, 175, 909, 1972.
8. **Purdy, J. E., Lieberman, M., Roggeveen, A. E., and Kirk, R. G.,** Synthetic strands of cardiac muscle: formation and ultrastructure, *J. Cell Biol.*, 55, 563, 1972.

9. **Horres, C. R., Lieberman, M., and Purdy, J. E.,** Growth orientation of heart cells on nylon monofilaments: determination of the volume-to-surface area ratio and intracellular potassium concentration, *J. Membr. Biol.,* 34, 313, 1977.
10. **Moscona, A.,** Cell suspensions from organ rudiments and aggregation of cells from organ rudiments of the early chick embryo, *J. Anat.,* 86, 287, 1952.
11. **Rous, P. and Jones, F. S.,** A method for obtaining suspension of living cells from the fixed tissues and for the plating of individual cells, *J. Exp. Med.,* 23, 549, 1919.
12. **Rinaldini, L. M.,** An improved method for the isolation and quantitative cultivation of embryonic cells, *Exp. Cell. Res.,* 16, 477, 1958.
13. **Kono, I.,** Roles of collagenases and other proteolytic enzymes in the dispersal of animal tissues, *Biochim. Biophys. Acta.,* 178, 397, 1969.
14. **Weinstein, D.,** Comparison of pronase and trypsin for detachment of human cells during serial cultivation, *Exp. Cell. Res.,* 43, 234, 1966.
15. **Houba, V.,** The use of pronase for dispersing cells, *Experientia,* 23, 572, 1967.
16. **Levinson, C. and Green, J. W.,** Cellular injury resulting from tissue disaggregation, *Exp. Cell Res.,* 39, 309, 1965.
17. **Cavanaugh, D. J., Berndt, W. O., and Smith, T. E.,** Dissociation of heart cells by collagenase, *Nature,* 200, 261, 1963.
18. **Smith, T. E. and Berndt, W. O.,** The establishment of beating myocardial cells in long-term culture in fluid medium, *Exp. Cell Res.,* 36, 179, 1964.
19. **Yasumura, Y., Tashjian, A. H., Jr., and Sata, G.,** Establishment of four functional clonal strains of animal cells in culture, *Science,* 154, 1186, 1966.
20. **Speicher, D. W. and McCarl, R. L.,** Pancreatic enzyme requirements for the dissociation of rat hearts for culture, *In Vitro,* 10, 30, 1974.
21. **Pine, L., Taylor, G. C., Miller, D. M., Bradley, G., and Wetmore, H. R.,** Comparison of good and bad lots of trypsin used in the production of primary monkey kidney cells: a definition of the problem and comparison of certain enzymatic characteristics, *Cytobios,* 2, 197, 1969.
22. **Wollenberger, A.,** Rhythmic and arrhythmic contractile activity of single myocardial cells cultured in vitro, *Circ. Res.,* 15 (Suppl. 2), 1984, 1964.
23. **Kasten, F. H.,** Electron microscope studies of the combined effects of trypsinization and centrifugation on rat heart cells, *J. Cell Biol.,* 31, 131A, 1966.
24. **Masson-Pévet, M., Jongsma, H. J., and De Bruijne, J.,** Collagenase and trypsin-dissociated heart cells: a comparative ultrastructural study, *J. Mol. Cell. Cardiol.,* 8, 747, 1976.
25. **Gross, W. O., Müler, C. A. M., and Schlotman, E. H. M.,** Loss of differentiation features in trypsin separated heart muscle cells, *Anat. Embryol.,* 151, 341, 1977.
26. **Fischmann, D. A. and Moscona, A. S.,** Reconstitution of heart tissue from suspensions of embryonic myocardial cells: ultrastructural studies on dispersed and reaggregated cells, in *Cardiac Hypertrophy,* Alpert, N. R., Ed., Academic Press, New York, 1971, 125, 139.
27. **Speicher, D. W. and McCarl, R. L.,** Isolation and characterization of the proteolytic enzyme component from commercially available crude typsin, *Anal. Biochem.,* 84, 205, 1978.
28. **Speicher, D. W. and McCarl, R. L.,** Evaluation of a proteolytic enzyme mixture isolated from crude typrsins in tissue diaggregation, *In Vitro,* 14, 849, 1978.
29. **Padieu, P., Frelin, C., Pinson, A., Charbonné, F., and Athias, P.,** Effect of environmental factors and tissue culture methodology in producing and studying cultured cardiac cells, *Recent Adv. Stud. Card. Struct. Metab.,* 12, 609, 1978.
30. **Frelin, C. and Padieu, P.,** Pleiotypic response of rat heart cells in culture to serum stimulation, *Biochimie,* 58, 953, 1976.
31. **Pinson, A., Degrès, J., and Heller, M.,** Partial and incomplete oxidation of palmitate by cultured beating cardiac cells from neonatal rats, *J. Biol. Chem.,* 254, 8331, 1979.
32. **Frelin, C.,** The growth of heart cells in culture. Evidence for a multiple activation of the pleitypic program, *Biochimie,* 60, 627, 1978.
33. **Anastasia, J. V. and McCarl, R. L.,** Effects of cortisol on cultured rat heart cells. Lipase activity, fatty acid oxidation, glycogen metabolism and ATP levels as related to the beating phenomenon, *J. Cell Biol.,* 57, 109, 1973.
34. **Gordon, H. P. and Brice, M. C.,** Intrinsic factors influencing the maintenance of contractile embryonic heart cells in vitro. I. The heart muscle conditioned medium effect, *Exp. Cell Res.,* 85, 303, 1974.
35. **Gordon, H. P. and Brice, M. C.,** Intrinsic factors influencing the maintenance of contractile embryonic heart cells in vitro. II. Biochemical analysis of heart muscle conditioned medium, *Exp. Cell Res.,* 85, 311, 1974.
36. **Pinson, A.,** et al., unpublished data, 1985.
37. **Speicher, D. W., Peace, J. N., and McCarl, R. L.,** Effects of plating density and in culture on growth and cell division of neonatal rat heart primary cultures, *In Vitro,* 17, 863, 1981.

38. **Yagev, S., Heller, M., and Pinson, A.,** Changes in cytoplasmic and lysosomal enzyme activities in cultured rat heart cells: the relationship to cell differentiation and cell population in culture, *In Vitro,* 20, 893, 1984.
39. **Mark, G. and Strasser, F. F.,** Pacemaker activity and mitosis in cultures of newborn rat heart ventricle cells, *Exp. Cell Res.,* 44, 217, 1966.
40. **Kasten, F. H.,** Rat myocardial cells in vitro: mitosis and differentiated properties, *In Vitro,* 8, 128, 1972.
41. **Clark, W. A. R. and Fischmann, D. A.,** Analysis of population cytokinetics of chick myocardial cells in tissue culture, *Dev. Biol.,* 97, 1, 1983.
42. **Polinger, I. S.,** Separation of cell types in embryonic heart cell culture, *Exp. Cell Res.,* 63, 78, 1970.
43. **Blondel, B., Roijen, I., and Cheneval, J. P.,** Heart cells in culture: a simple method for increasing the proportion of myoblasts, *Experientia,* 27, 356, 1971.
44. **Holtzer, H.,** Myogenesis, in *Cell Differentiation,* Schjeide Q. A. and de Vellis, I., Eds., Van Nostrand Reinhold, New York, 1970, 476.
45. **Masse, M. J. and Harary, I.,** Role of cell division in the cytodifferentiation of rat heart cells in culture, *Biochimie,* 56, 1581, 1974.
46. **Masse, M. J. O. and Harary, I.,** The use of fluorescent antimyosin and DNA labeling in the estimation of the myoblast and myocyte population of primary rat heart cell cultures, *J. Cell Physiol.,* 106, 165, 1981.
47. **Kasten, F. H. and Yip, D. K.,** Reamination of cultured mammalian myocardial cells during multiple cycles of trypsinization freezing-thawing, *In Vitro,* 9, 246, 1974.
48. **Kimes, B. W. and Brandt, B. L.,** Properties of a clonal muscle cell line from rat heart, *Exp. Cell Res.,* 98, 367, 1976.
49. **Girardi, A. J., Warren, J., Goldman, C., and Jeffries, B.,** *Proc. Soc. Exp. Biol. Med.,* 98, 18, 1958.
50. **Fenselau, A., and Mello, R. J.,** Growth stimulation of cultured endothelial cells by tumor cell homogenates, *Cancer Res.,* 36, 3269, 1976.
51. **Kaneko, H. and Goshima, K.,** Selective killing of fibroblast-like cells in cultures of mouse heart cells by treatment with CA ionophore A 23187, *Exp. Cell Res.,* 142, 407, 1982.
52. **Chacko, S.,** The effect of BUdR on the emergence of cardiac muscle cells in the developing embryo, *J. Cell Biol.,* 55, 36A, 1972.
53. **Masse, M. J. O. and Harary, I.,** The use of 5-Bromodeoxyuridine and irradiation for the estimation of the myoblast and myocyte content of primary rat heart cell cultures, *J. Cell Physiol.,* 105, 147, 1980.
54. **Pfeiffer, D. R., Taylor, R. W., and Lardy, H. A.,** Ionophore A 23187: cation binding and transport properties, *Ann. N. Y. Acad. Sci.,* 307, 402, 1978.
55. **Harary, I., Hoover, F. and Farley, B.,** The isolation and cultivation of rat heart cells, *Methods in Enzymology XXXII,* Flusha, S. and Packer, L., Eds., 1974, 704.
56. **Pinson, A. and Padieu, P.,** Les cultures de cellules isoleés: techniques, croissance, et application, *Cah. Nutr. Diet.,* 9, 237, 1975.
57. **Ham, R. G.,** An improved nutrient solution for diploid Chinese hamsters and human cell lines, *Exp. Cell Res.,* 29, 515, 1963.
58. **Healy, G. M. and Parker, R. C.,** An improved chemically defined basal medium (CMRL-1415) for newly implanted mouse embryo cells, *J. Cell Biol.,* 30, 531, 1966.
59. **Wroblewski, F. and La Due, J. S.,** Lactic dehydrogenase activity in blood, *Proc. Soc. Exp. Biol. Med.,* 90, 210, 1955.
60. **Gross, W. O., Schöf-Ebner, E., and Bucher, O. M.,** Technique for the preparation of homogeneous cultures of isolated heart muscle cells, *Exp. Cell Res.,* 53, 1, 1968.
61. **De Haan, R. L. and Gottlieb, S. H.,** The electrical activity of embryonic chick heart cells isolated in tissue culture singly or in interconnected cell sheets, *J. Gen. Phyisol.,* 52, 643, 1968.
62. **Polinger, I. S.,** Identification of cardiac myocytes in vivo and in vitro by the presence of glycogen and myofibrils, *Exp. Cell Res.,* 76, 243, 1973.
63. **Wollenberger, A., Karsten, U., and Kössler, A.,** Cultivation of nonproliferating rat heart cells in a Roller system at oxygen tension down to 0.66 mm Hg, *Cardiology,* 56, 224, 1971/2.
64. **Lau, Y. H., Robinson, R. B., Rosen, M. R., and Bilezikian, J. P.,** Subclassification of β-Adrenergic receptors in cultured rat cardiac myoblasts and fibroblasts, *Circ. Res.,* 47, 41, 1980.
65. **Higgins, T. J. C., Allsopp, D., and Bailey, P. J.,** The effect of extracellular calcium concentration in Ca-Antagonist drugs on enzyme releases and lactate production by anoxic heart cell cultures, *J. Mol. Cell. Cardiol.,* 12, 909, 1980.
66. **Clapham, D. E., Shrier, A., and De Haan, R. L.,** Junctional resistance and action potential delay between embryonic heart cell aggregates, *J. Gen. Physiol.,* 75, 633, 1980.
67. **Nag, A. C. and Cheng, M.,** Intercellular adhesion: coconstruction of contractile tissue by cells of different species, *Science,* 208, 1150, 1980.
68. **Goshima, K., Wakabayashi, S., and Masuda, A.,** Ionic mechanism of morphological changes of cultured myocardial cells on successive incubation in media without and with Ca^{2+}, *J. Mol. Cell Cardiol.,* 12, 1135, 1980.

69. **Perissel, B., Charbonné, F., Moalic J. M., and Malet, P.,** Initial stages of trypsinized cell culture of cardiac myoblasts, ultrastructural data, *J. Mol. Cell. Cardiol.,* 12, 63, 1980.

70. **Norwood, C. R., Casteneda, A. R., and Norwood, W. I.,** Heterogeneity of rat cardiac cells of defined origin in single cell culture, *J. Mol. Cell. Cardiol.,* 12, 201, 1980.

71. **Halbert, S. F., Bruderer, R., and Thompson, A.,** Growth of dissociated beating human heart cells in tissue culture, *Life Science,* 13, 969, 1973.

71. **Halbert, S. F., Bruderer, R., and Thompson, A.,** Growth of dissociated beating human heart cells in tissue culture, . *Life Science,* 13, 969, 1973.

Chapter 3

CARDIOMYOCYTES GROWN IN SERUM-FREE MEDIUM

Gania Kessler-Icekson

TABLE OF CONTENTS

I. BACKGROUND

Cultures of heart muscle cells are meant to provide a simple model for studies of factors regulating myocardial functions at the cellular level, free of the complex in vivo environment. It has been generally accepted that cultured heart muscle cells, like most other cells, require the presence of serum in the growth medium.[1,2] Serum serves as a universal substitute for interstitial fluids, contributing to the medium numerous biologically active substances such as nutrients, hormones, growth factors, carrier molecules, and many other ill-defined components.[1-4] Variations between serum preparations due to differences in species, age, and physiological state of the donor animal impair the uniformity of culture conditions. Moreover, in the case of heart cell cultures, serum leads to a decrease in the proportion of myocytes due to enhancement of nonmyocyte proliferation.[5] It is apparent, therefore, that the goal of a simple and fully defined culture environment can be achieved only when serum is replaced by the minimal number of well-defined substances.

Early attempts to develop serum-free chemically defined media were partly successful and further progress was only made following advances in the field of hormones and growth factors.[6-8] These advances along with an improved understanding of the nutritional demands of the cell, have enabled the successful replacement of serum in the culture media of most cells, including cardiomyocytes.[9-12]

The composition of serum-free hormonally supplemented media proposed for myocardial cells in the last few years will be summarized below and further evaluated with regard to cell attachment, growth, survival, and function. In addition, the contribution of serum-free media to in vitro studies of cardiomyocyte biology will be discussed.

II. COMPOSITION OF CARDIOMYOCYTE SERUM-FREE MEDIA AND ROLES OF THE MAJOR COMPONENTS

The first serum-free hormonally supplemented medium for cultured cardiomyocytes was proposed by Claycomb in 1980.[12] Since then several other formulations have been published for embryonic, neonatal, and adult heart cells, as summarized in Table 1.[13-18]

Coating of culture dishes with components of the extracellular matrix, such as collagen or fibronectin, is a prerequisite for efficient cell attachment and spreading in the absence of serum.[12-14] Coating of culture dishes with collagen and subsequently with fibronectin improves the survival and long-term performance of the cells. On the other hand, dishes coated with biomatrix, synthesized in vitro by endothelial cells, prove inadequate. Cardiomyocytes sink into the matrix, intercellular contacts are not established, and beating capacity is poor.[30] Some investigators circumvent dish coating by preincubation of dishes in serum containing medium or even by plating the cells in such a medium for the first few hours of culture.[15-17] The latter procedure exposes the cells to a short-term influence of serum which is not always desirable.

Rich nutritional mixtures are generally chosen as the basal media, since the medium serves as the sole source of low molecular weight nutrients to which hormones and a few essential supplements are added. None of the reviewed reports include dose response curves for optimal supplement concentration and it is assumed that each group of investigators has found its formulation satisfactory for cardiomyocytes.

Cardiac myocytes, like skeletal myoblasts, require the presence of the serum glycoprotein, fetuin, in the culture medium.[12-15,18,19] The exact function of fetuin is obscure but most people associate it with cell attachment and spreading. The possibility that the observed activity resides in a contaminant rather than in fetuin itself has still not been proven.[20] If so, it must be a very potent impurity to remain effective in highly purified fetuin preparations and over a wide range of fetuin concentrations (Table 1).[19]

Table 1
COMPOSITIONS OF CARDIOMYOCYTE SERUM-FREE MEDIA

Dish coating	Basal medium	Fetuin (mg/mℓ)	Insulin (µg/mℓ)	Glucocorticoids (µM)	Transferrin (µg/mℓ)	BSA (%)	Selenium (ng/mℓ)	Ascorbic acid (mg/mℓ)	Ref.
Fn	MEM	2.5	25	0.1	—	—	—	—	12
Fn	DMEM/F12	0.25	5	13(5 µg/mℓ)	7.5	1	5	0.02 mg/mℓ	13
C	DMEM/F12	1	25	0.1	25	—	—	—	14
—	M199	—	10	—	10	—	—	—	16
S	M199	—	0.06(10^{-8})	—	—	0.2	—	—	17
—	F12	0.25	5	—	5	1	5	0.02 mg/mℓ	18

Notes: A summary of formulations published up to summer 1984. BSA — bovine serum albumin; Fn — fibronectin; C — collagen; S — serum; MEM — minimum essential medium, Eagle; DMEM — Dulbecco modified Eagle's medium; F12 — Ham's F12 nutritional mixture; M199 — medium 199.

Insulin is present in all serum-free combinations.[4,10,11] In addition to its regulatory effects on fatty acid and glycogen synthesis, it probably acts as a weak somatomedin analog.[19] Glucocorticoids help sustain the beating capacity of the cells but they restrict cell growth.[13,21] Transferrin is included as an iron carrier and as a detoxifier for trace amounts of toxic metals.[4,10] Albumin serves as a carrier for fatty acids which are sometimes conjugated to it before use.[13,17,18] The trace element, selenium, protects cell cultures against oxidative damage.[9] Ascorbic acid may be beneficial, although its susceptibility to oxidation sheds doubt on its effectiveness.[9]

The addition of growth factors, such as epidermal growth factor (EGF), stimulates cell proliferation, apparently of nonmycotes.[13] It is therefore advisable to consider carefully the application of growth factors to cardiomyocyte cultures. Glutamine, for example, is omitted from the medium in order to limit nonmyocyte proliferation.[12,13]

The quality of all medium components including water is of major importance in serum-free media as they lack the detoxifying ability of serum.

III. EVALUATION OF CARDIOMYOCYTES GROWN IN SERUM-FREE MEDIUM

A. Cell Attachment and Spreading

Plating efficiency of cardiomyocytes on precoated dishes ranges between 40 to 70%, satisfactory values for any primary cell culture.[12-14] Examination of cultures by phase contrast microscopy and scanning electron microscopy reveals well spread cells (Figure 1, a and b).[12-15] In contrast to their behavior in the presence of serum, nonmyocytes do not overgrow the cultures under serum-free conditions.[12-15]

B. Cell Survival and Growth

Culture survival under serum-free conditions ranged from 2 to 14 weeks, an adequate life span for most kinds of experiments.[12-15] Cell growth occurs mainly in the first week of culture, manifested by increased protein content and cell number (Figure 2).[13,14] Accordingly, maximal rates of amino acid and thymidine incorporation are observed during the first 5 days of culture.[14]

Many of the formulations for serum-free media do not support growth rates as high as those observed in serum supplemented media.[13,14] This may imply that the newly designed media require further optimization, or alternatively, that serum imposes abnormal doses of growth promoting factors on the cells. As already mentioned, the addition of EGF enhances protein accumulation and prolongs the growth period, responses which are attributed primarily to the nonmyocyte population.[13]

C. Retention of Cardiac Specific Traits

The essence of myocyte function is contraction. Many cellular events are involved in contraction, some of which have been examined in serum-free cultures. Rate of spontaneous cell beating has been found to be maximal during the first week of culture, decreasing thereafter.[13] The administration of catecholamines triggers positive chronotropic response, typical to β-adrenergic sensitization,[14] and hydrocortisone induces higher beating rates.[13] Contraction of cultured adult myocytes is dependent on electrical stimulation.[17] Values of transmembrane potentials (ca., -80 mV) are close to values measured in vivo.[15] In addition, glycogen deposits and cross-striated myofibrils are detected in the cells (Figure 1, c and d).

The phenotypic expression of cardiac muscle also includes specific isozymes of creatine kinase (CK) and of myosin. The proportions of the three CK isozymes in serum-free cultures

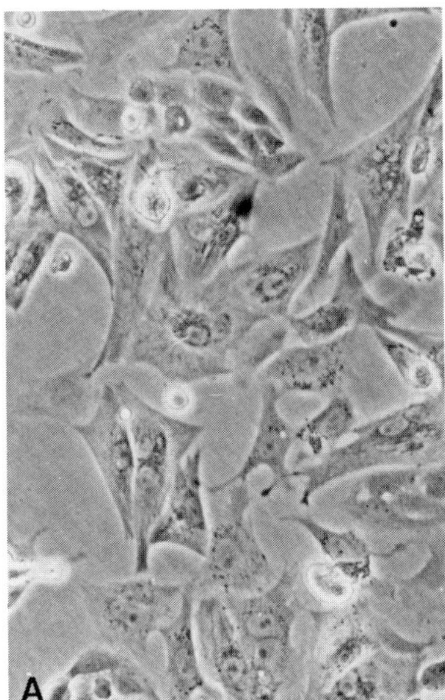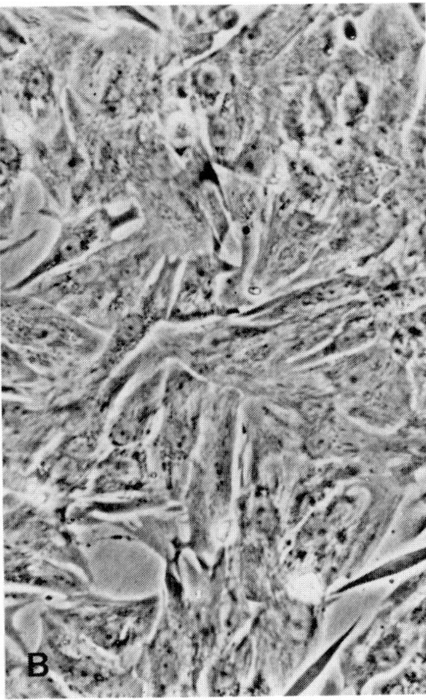

FIGURE 1. Morphology and cytochemical staining of cultured cardiomyocytes. All cultures were grown as described by Kessler-Icekson et al.[14] (A,B) — Phase contrast micrographs of beating cultures. (A) 1-day-old culture ($\times 200$); (B) 3-day-old culture ($\times 200$). (C,D) — Cytochemical staining for glycogen and cross-striations.[31] (C) periodic acid Schiff's reagent (PAS) 3-day-old culture ($\times 250$), arrows point out glycogen accumulations; (D) phosphotungstic acid hematoxylin (PTAH), 3-day-old culture ($\times 450$).

are similar to those of the neonatal rat heart.[14] Most of the enzymatic activity is contributed by CK-MM and CK-MB, CK-MM being the predominant fraction. CK-BB always remains the minor isozyme, never increasing with age of culture, thus providing a quantitative support to earlier qualitative observations of the absence of nonmyocyte overgrowth in serum-free media, mentioned above.[14]

Myosin of neonatal rat myocytes grown in serum-free medium is predominantly of V3 type, the form prevailing in embryonic and newborn rat hearts (Figure 3a).[18] It is actively synthesized in culture, but no significant transition towards the V1 form has been observed. An increase in the proportion of V1 isomyosin is obtained in the presence of horse serum (Figure 3, b and c), suggesting that the serum contains maturation factor(s). T4 may be one such factor since it stimulates the replacement of V3 by V1 isomyosin in serum-free cultures of embryonic rat heart myocytes, as it does in the heart of the developing rat.[18]

IV. CONCLUDING REMARKS AND FUTURE PERSPECTIVES

Serum-free media which support the growth and maintenance of cardiomyocyte cultures are now available for cells derived from embryonic, neonatal, and adult heart. The cells retain their differentiated state in culture, reflecting in vitro many properties of myocardial cells in vivo. The media proposed so far all have a great deal in common and yield highly reproducible cultures in well-defined nutritional conditions. The main drawbacks of the media described are a low threshold for toxic effects and the use of fetuin which may introduce unknown impurities.

Serum-free cultures have particular advantages for studies on the interaction of hormones and drugs with the heart muscle cell. The effects of triiodothyronine and hydrocortisone on the activity of enzymes involved in energy metabolism and the interrelations of the two hormones and of insulin in exerting their effects have thus been studied.[22,23] Another example

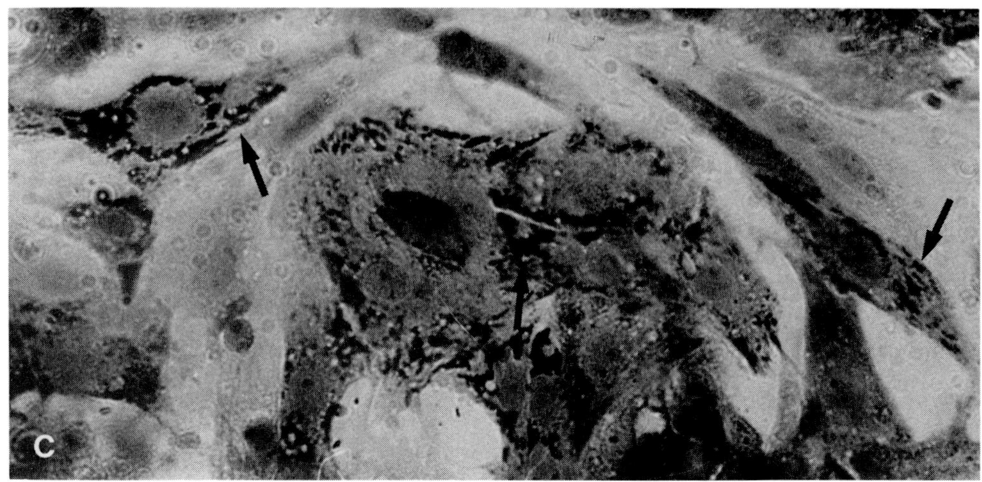

FIGURE 1C.

FIGURE 1D.

is the recent demonstration of the transition from V3 to V1 isomyosin in response to thyroxine treatment in serum-free cultures, a simple system for the study of thyroid hormone regulation of heart muscle myosins.[18] Likewise, the stimulation of growth by norepinephrine provides an in vitro model for the investigation of catecholamine-induced myocardial hypertrophy at the cellular level.[16] The exploration of serum components and medium conditioning factors which affect cardiomyocyte metabolism as well as the elucidation of the nutritional demands of the cell should be facilitated in these systems.[23,24] The characterization and eventual purification of substances released by cardiomyocytes which mediate other cells activities should now be feasible.[25] In addition, the availability of cultures containing a more homogenous myocyte population should be generally useful for all biochemical studies.[26,27]

To conclude, serum-free hormonally defined media may be used in almost all studies performed on cultured cardiomyocytes and the advantages of such media may open new horizons for investigators in the field.

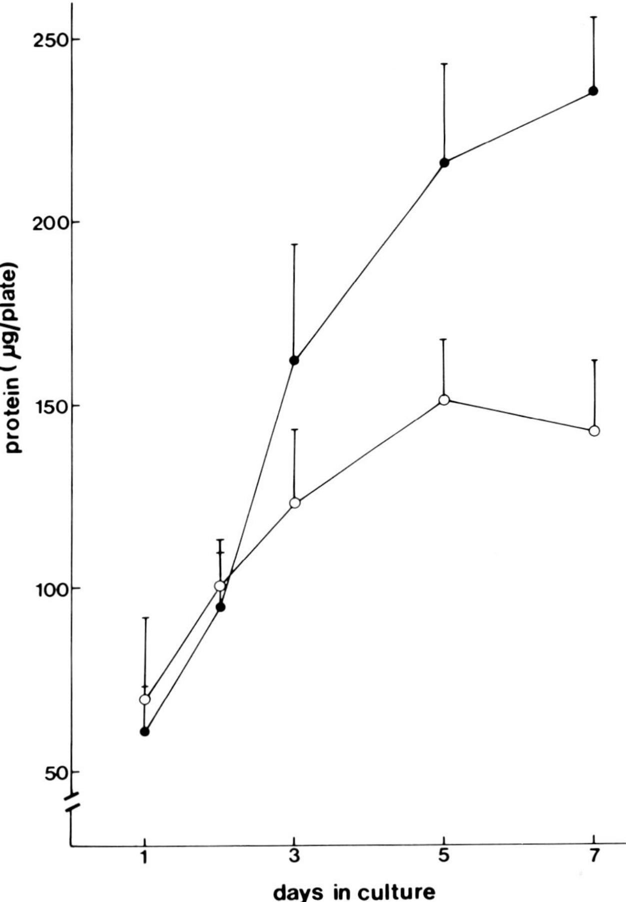

FIGURE 2.   Protein accumulation in cardiomyocyte cultures. A comparison between serum-free (O-O) and serum-supplemented (●-●) cultures. (From Kessler-Icekson, G., Sperling, O., Rotem, C., and Wasserman, L., Cardiomyocytes cultured in serum-free medium: growth and creatine kinase activity, *Exp. Cell Res.*, 155, 113, 1984. With permission of Academic Press.)

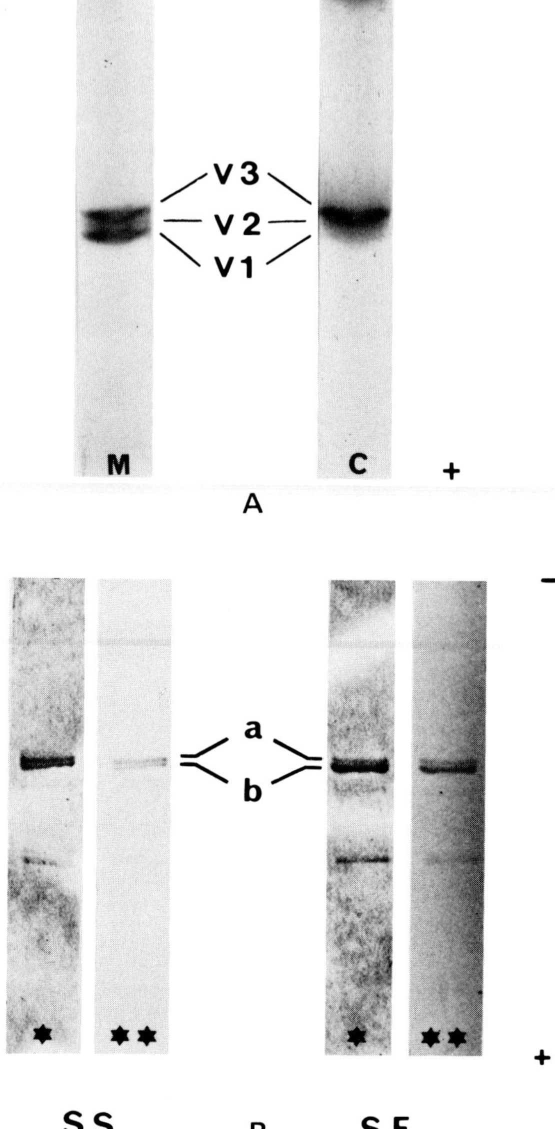

FIGURE 3.   Isoforms of myosin heavy chains. (A) Electrophoresis under nondenaturing conditions of myosin obtained from 3-day-old serum-free cultures. V1, V2, and V3 native myosin isoforms.[29] M — markers, C — cultures. (B) Electrophoresis of myosin heavy chains from 4-day-old cultures, labeled with ^{35}S-methionine for the previous 48 hr, performed on 5% gels under denaturing conditions.[28] (C) Densitometric scanning of the autoradiograms in (B). SF — Serum-Free, SS — Serum Supplemented; a — heavy chain α of myosin V1; b — heavy chain β of myosin V3; ★ — Coomassie staining; ★★— Autoradiogram. Note that in SF cultures the β chain is dominant, whereas in SS cultures α chain predominates. The scanning profiles demonstrate the relative synthesis of α and β chains in both cultures.[32]

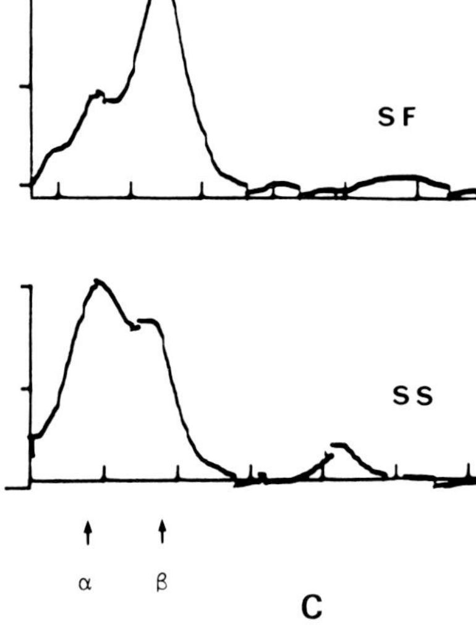

FIGURE 3C.

REFERENCES

1. **Harary, I., McCarl, R., and Farley, B.,** Studies in vitro on single beating rat heart cells. XI. The restoration of beating by serum lipids and fatty acids, *Biochim. Biophys. Acta,* 115, 15, 1966.
2. **Frelin, C. and Padieu, P.,** Pleiotypic response of rat heart cells in culture to serum stimulation, *Biochimie,* 58, 953, 1976.
3. **Frelin, C.,** The regulation of protein turnover in newborn rat heart cell cultures, *J. Biol. Chem.,* 255, 11149, 1980.
4. **Barnes, D. and Sato, G.,** Serum-free cell culture: a unifying approach, *Cell,* 22, 649, 1980.
5. **Van der Laarse, A., Hollar, L., Kokshoorn, L. J. M., and Witteveen, S. A. G. J.,** The activity of cardiospecific isoenzymes of creatine phosphokinase and lactate dehydrogenase in monolayer cultures of neonatal rat heart cells, *J. Mol. Cell. Cardiol.,* 11, 501, 1979.
6. **Lieberman, I. and Ove, J.,** Growth factors for mammalian cells in culture, *J. Biol. Chem.,* 234, 2754, 1959.
7. **Hale, W. and Wollenberger, A.,** Differentiation and behavior of isolated embryonic and neonatal heart cells in a chemically defined medium, *Am. J. Cardiol.,* 25, 292, 1970.
8. **Gospodarowicz, D.,** Fibroblast and epidermal growth factors: their uses in vivo and in vitro in studies on cell functions and cell transplantation, *Mol. Cell. Biochem.,* 25, 79, 1979.
9. **Ham, R. G. and McKeehan, W. L.,** Media and growth requirements, in *Methods in Enzymology,* Vol. 58, Jakoby, W. B. and Pastan, I. H., Eds., Academic Press, New York, 1979, 44.
10. **Barnes, D. and Sato, G.,** Methods for growth of cultured cells in serum-free medium, *Anal. Biochem.,* 102, 255, 1980.
11. **Bottenstein, J., Hayashi, I., Hutchings, S., Masui, H., Mather, J., McClure, O. B., Ohasa, S., Rizzino, A., Sato, G., Serrero, G., Wolfe, R., and Wu, R.,** The growth of cells in serum-free hormone supplemented media, in *Methods in Enzymology,* Vol. 58, Jakoby, W. B. and Pastan, I. H., Eds., Academic Press, New York, 1979, 94.
12. **Claycomb, W. C.,** Culture of cardiac muscle cells in serum-free media, *Exp. Cell. Res.,* 131, 231, 1980.
13. **Mohamed, S. N. W., Holmes, R., and Hartzell, C. R.,** A serum-free chemically defined medium for function and growth of primary neonatal rat heart cell cultures, *In Vitro,* 19, 471, 1983.
14. **Kessler-Icekson, G., Sperling, O., Rotem, C., and Wasserman, L.,** Cardiomyocytes cultured in serum-free medium: growth and creatine kinase activity, *Exp. Cell. Res.,* 155, 113, 1984.
15. **Kessler-Icekson, G., Wasserman, L., Yoles, E., and Sampson, S. R.,** Characterization of cardiomyocytes cultured in serum-free medium, in *Homonally Defined Media, A Tool in Cell Biology,* Fischer, G. and Weiser, R. J., Eds., Springer-Verlag, Berlin, 1983, 383.

16. **Simpson, P., McGrath, A., and Savion, S.,** Myocyte hypertrophy in neonatal rat heart cultures and its regulation by serum and by catecholamines, *Circ. Res.,* 51, 787, 1982.
17. **Piper, H. M., Probst, I., Schwartz, P., Hutter, F. J., and Spiekermann, P. G.,** Culturing of calcium stable adult cardiac myocytes, *J. Mol. Cell. Cardiol.,* 14, 397, 1982.
18. **Nag, A. C. and Cheng, M.,** Expression of myosin isozymes in cardiac muscle cells in culture, *Biochem. J.,* 221, 21, 1984.
19. **Florini, J. R. and Ewton, D. Z.,** Insulin acts as a somatomedin analog in stimulating myoblast growth in serum-free medium, *In Vitro,* 17, 763, 1981.
20. **Rizzino, A. and Sato, G.,** Growth of embryonal carcinoma cells in serum-free medium, *Proc. Natl. Acad. Sci., USA,* 75, 1844, 1978.
21. **McCarl, R. L. and Margossian, S. S.,** Restoration of beating and enzymatic response of cultured rat heart cells to cortisol acetate, *Arch. Biochem. Biophys.,* 130, 321, 1969.
22. **Freerksen, D. L., Schroedl, N. A., and Hartzell, C. R.,** Triiodothyronine depresses the NAD-linked glycerol-3-phosphate dehydrogenase activity of cultured neonatal rat heart cells, *Arch. Biochem. Biophys.,* 228, 474, 1984.
23. **Freerksen, D. L., Schroedl, N. A., and Hartzell, C. R.,** Control of enzyme activity levels by serum and hydrocortisone in neonatal rat heart cells cultured in serum-free medium, *J. Cell. Physiol.,* 120, 126, 1984.
24. **Bauriedel, G., Werdan, K., Krawietz, W., and Erdmann, E.,** Vanadate antagonizes detrimental effects of serum deprivation in cultured heart muscle cells, in *Hormonally Defined Media, A Tool in Cell Biology,* Fischer, G. and Wieser, R. J., Eds., Springer-Verlag, Berlin, 1983, 380.
25. **Norrgren, G., Ebendal, T., and Wikstrom, H.,** Production of nerve growth stimulating factor(s) from chick embryo heart cells. Use of cytodex 3 microcarriers and serum-free media, *Exp. Cell. Res.,* 152, 427, 1984.
26. **Zoref-Shani, E., Kessler-Icekson, G., Wasserman, L., and Sperling, O.,** Characterization of purine nucleotide metabolism in primary rat cardiomyocyte cultures, *Biochim. Biophys. Acta,* 804, 161, 1984.
27. **Piper, H. M., Hutter, J. F., and Spiekermann, G.,** Relation between enzyme release and metabolic changes in reversible anoxic injury of myocardial cells, *Life Sci.,* 35, 127, 1984.
28. **Hoh, J. F. Y., McGrath, P. A., and Hale, P. T.,** Electrophoretic analysis of multiple forms of rat cardiac myosin: effects of hypophysectomy and thyroxine replacement, *J. Mol. Cell. Cardiol.,* 10, 1053, 1977.
29. **Carraro, U. and Catani, C.,** A sensitive SDS-PAGE method separating myosin heavy chain isoforms of rat skeletal muscles reveals the heterogeneous nature of the embryonic myosin, *Biochim. Biophys. Res. Comm.,* 116, 793, 1983.
30. **Kessler-Icekson, G.,** unpublished.
31. **Wasserman, L. and Kessler-Icekson, G.,** unpublished.
32. **Schlesinger, H. and Kessler-Icekson, G.,** unpublished.

Chapter 4

GROWTH AND DEVELOPMENT

Isaac Harary

TABLE OF CONTENTS

I. INTRODUCTION

Growth and development of the heart encompasses all levels of cardiac function including anatomy, physiology, pharmacology, and biochemistry. To make this more manageable, this chapter will be limited to a review of the role of cell division and growth, the initiation of the synthesis of adult specific macromolecules as exemplified by myosin and the metabolic changes occurring during late fetal and early postnatal life. Emphasis will be placed on the shift from the relative anaerobiosis of the embryo to the aerobic environment of the newborn, with a view towards describing the main changes and exploring the effect that the shift might have on the observed developmental events.

All efforts will be made to focus on the heart cell in culture; however, most of the information we have to date, is derived from in vivo studies. Furthermore, data derived from the in vitro system only make sense when related to the whole animal. Thus, no clear separation should be made between these two systems and consequently, studies from both will be used in this discussion.

II. CELL DIVISION AND MYOSIN SYNTHESIS

In many developing systems ranging from slime mold to mammalian tissues, a major mechanism of development seems to involve an incompatibility between mitosis and terminal development. Rapidly dividing cells in these developing systems do not produce adult specific proteins; nor does cell division play a role in the adult state.

A. Cell Division
1. Cessation of Mitosis
Adult mammalian muscle cells do not divide. The regeneration of skeletal muscle is a response to injury and results from the initiation of cell division in presumptive muscle cells known as satellite cells. In heart muscle, injury does not lead to regeneration of the muscle tissue, but rather to the repair of the damaged area with connective tissue.

Early observations in the embryonic skeletal muscle and the heart muscle systems have in general substantiated the functional separation of mitosis from synthesis of adult specific proteins. In the chick embryo, early microscopic studies of the brachial myotomes demonstrated that multinucleated myotubes do not synthesize DNA and that no mononucleated cells which synthesized DNA contained myosin.[1] In addition, an overall decline of mitotic activity preceded the appearance of myosin by about 4 days.[2] In cultures of chick embryo muscle cells, no muscle specific proteins existed in cells that had pronounced mitotic activity.[3] Thus, both in vivo and in vitro observations indicated that cell division ceases before the initiation of myosin synthesis. The actual cellular process was most clearly delineated in cell culture where it was shown that single presumptive myoblasts ceased cell division, fused into multinucleated myotubes and subsequently initiated synthesis of adult specific muscle proteins.[4] The problem was, however, whether cell division cessation is an obligatory event. With the temporal proximity of fusion and loss of cell division it was difficult to determine which of these events was causally related to the initiation of myosis synthesis. In fact, one of the early studies of the mitosis problem in the heart was motivated by this question.[5] It was thought, since cardiac muscle cells do not fuse, but evince decreased mitotic activity during the increase in myosin content, that a closer study of mitosis and myosin synthesis could shed more light on their relationship. However, as will be noted later, the relation of mitosis to myosin synthesis in the heart is not similar to that in skeletal muscle and thus did not help discriminate between the role of fusion and cessation of cell division in the skeletal muscle system.

It was the use of culture techniques which helped solve this problem, for it was only in skeletal muscle cultures that these two processes could be experimentally separated. Fusion was inhibited by decrease of the Ca concentration.[6] The loss of fusion did not prevent the initiation of myosin synthesis, leading to cultures with mononuculeated cells containing myosin.[7-9] One of these reports[8] indicated that these myoblasts were postmitotic. In the mononucleated cells, prevented from fusing, the uptake of (^{3}H) deoxythymidine was severely depressed during the period of myosin synthesis. Thus, though these cells did not fuse, DNA synthesis did cease, and if fusion was not necessary for the initiation of myosin synthesis it appeared that cessation of DNA synthesis might be obligatory.

a. *In Vivo Studies*

The presence of active cell division in the developing, actively contracting, heart made it difficult to determine whether mitosis had to cease before myosin synthesis began in the single cell, or whether in fact mitosis could occur in myosin containing cells. The mitotic index of embryonic rat heart showed a steady decline from 35% in the 14-day-old embryo to 10% at birth.[10] However, the developing heart consists of many cell types such as fibroblasts, endothelial, smooth muscle, nerve, and blood cells. Thus it was difficult, without specific identification, to determine which of the heart cells contributed to the mitotic activity and to observe the relation of DNA synthesis to the initiation of synthesis of muscle specific molecules. Electron microscopic examination of the developing chick heart helped clarify this problem in demonstrating that chick cardiac muscle cells containing myofibrils also could contain mitotic figures.[11] Rat hearts were found to be similar. Cardiac muscle cells with myofibrils and mitotic figures were observed in colchicine treated animals.[5] They appeared during the early postnatal period, but disappeared after 4 weeks.[12] In both chick and rat, growth of the heart during the developmental period appears to result from an increase in the total number of cells and is accompanied by an increase in total DNA.[13,5] Cell division of cardiac muscle cells seems to be confined to the developmental periods, and though dividing cells in this period contain myosin, the fully differentiated cardiac muscle cell does not normally divide.[14] Thus the ratio of protein to DNA remains constant during gestation. After birth, growth is mainly due to increase in cell size.[14,15] In the chick, DNA accumulates throughout embryonic life, but reaches a plateau just before the chick hatches. After hatching, the DNA level remains constant, but protein continues to accumulate, so that by 3 months, the protein-to-DNA ratio is four times higher than the embryonic value. Similarly, the rat heart cells increase in size by virtue of increased cellular components so that the adult cell is eight times larger than the embryonic cell.

The decrease in the rate of DNA synthesis in the heart correlates with both the decreased rate of DNA accumulation and the decreased rate of cell division in the postnatal rat. Cell division ceases by the 17th day after birth.[16] In the chick, the rate of DNA synthesis also decreased, but, in contrast to the rat, it decreases to almost zero upon hatching.[15] The decrease in mitotic activity and DNA synthesis corresponds to a decrease of the soluble (cytoplasmic) DNA polymerase α.[14] It has been reported that this enzyme is related to the rate of DNA synthesis in skeletal muscle cells[17] and other proliferating cells. The soluble DNA polymerase present in the 100,000 g supernatant of chick heart homogenates falls rapidly during embryonic and postnatal development. Although it is more active with a denatured DNA than with a native DNA primer, its cellular activity mirrors the total cellular DNA synthesis and the mitotic index and is thus believed to be related to cellular DNA replication.[18]

Thymidine kinase, a key enzyme in DNA replication, also shows a high degree of correlation with the cellular proliferation rate and DNA synthesis activity in many developing and regenerating tissues.[19,20,21] It catalyzes the phosphorylation of thymidine, and is the first step in the so-called salvage pathway of deoxythymidine synthesis. The activity of this enzyme also correlates with cellular proliferation in the heart.[22] In the rat heart it decreases with postnatal development, reaching zero by 15 to 17 days after birth. The 70-fold decrease observed in the period, from 3 days before birth to 17 days after birth, correlates well with the decline and cessation of cardiac DNA synthesis and the loss of cyoplasmic DNA polymerase.

b. In Vitro Studies

There are few studies which deal with cardiac cell differentiation in culture. Although the cell culture system provides many advantages, such as ease of control of the environment, manipulation of cell population, possible separation of cellular events, and theoretically unlimited amounts of tissue, it is obvious that it also presents certain limitations. Many organ specific characteristics such as shape and overall structure cannot be directly studied in culture. In addition, the cell culture system differs from the in vivo system. However, this difference can provide advantages in that the actual changes can give clues as to the mechanism of control of the heart cell characteristics. Thus, despite the problems presented by the reduction of the tissue into its component cells and the culturing of the cells in an artificial environment, many studies have proved that much useful information can be gained and some studies can only be performed in this kind of system.

The developing cell population of cell cultures mirrors the composition of the intact heart, although the proportion of the different cells may vary and this variation may have an effect on the muscle cells. Both postnatal rat heart cells and embryonic chick heart cell cultures containing developing cells. The role of cell division has been studied in both of these systems. Beating cells were seen to divide in culture,[23] though it was reported at that time to be a rare event[24] and it was not clear whether such a division was directly dependent on the pathway of development. In addition, the evidence from the whole heart was largely electron microscopic, which dealt with individual cells and did not yield a quantitiatve estimation of the importance of mitosis in myosin-containing cells. A quantitative estimation of the number of dividing myosin cells in the intact heart and in cultures was necessary. Experiments in which cell division was inhibited have shown that cell division is important for the development of cells capable of myosin synthesis in culture. Cultures of neonatal rat heart cells have a low level of myosin and increase their total myosin content up to five times in 4 to 6 days.[25] This increase is dependent on cell division and the presence of serum in the medium. Colchicine, in the presence or absence of serum, prevents the normal rise in myosin; this indicates that rapidly dividing precursor cells or cells containing myosin are necessary for the normal increase in total myosin. Evidence that myosin-containing rat heart cells do indeed divide was derived from BrdU-treated cultures.[26,27] Only dividing cells incorporate BrdU into their DNA. Thus, visible-light irradiation affects primarily the dividing cells and causes the release of destructive molecules from damaged cells. BrdU-treated rat heart cell cultures lose a substantial amount of myosin, and it was concluded that a large number of myosin-containing cells had incorporated BrdU. Studies in the intact rat embryo heart did indeed demonstrate that cells with myofilaments contain mitotic figures,[28] and chick heart muscle cells contain myofibrils and exhibit prominent chromosomes.[11] In chick heart cultures, a high percentage of cells contain myosin and DNA labeled with (^{3}H)deoxythymidine. Recent studies substantiate these findings. Cells were cultivated from 14-day-old rat embryo hearts and 80% of the cells contained both myosin (visualized with fluorescent antibody to myosin) and exhibited DNA synthesis.[24] DNA synthesis was also shown to occur in cultured chick heart cells.[30] Further studies[31] indicated that contracting cells, containing cross-striated myofibrils, can undergo one to five doublings.

A more quantitative estimation was obtained by studying rat heart cell cultures derived from different developmental stages.[32] Using staining with fluorescent anti-myosin and labeling with tritiated thymidine, an estimation of the number of myosin containing cells with no DNA label, myocytes, and of cells with myosin and DNA label, myoblasts, was made in cultures derived from different stages of the developing animal. In this way it was possible to determine the actual percent of myoblasts and myocytes in the various stages. The cultures were derived from 14- and 17-day postcoital (dPC) rat embryos and from 1- to 4-day postnatal (dPN) rates. The percentage of nondividing cells increased with development and the age of the postnatal rat. The percentage of nondividing cells in 14 dPC equaled 21%, 17 dPC equaled 25%, 1 dPN equaled 44% and 4 dPN equalled 60%.

More recently a study of the growth rate was made in cultures of chick heart cells in order to compare them with the intact heart.[33] The mitotic cycle rate, the generative fraction, and the intermitotic period and doubling time of generating cells were calculated. A loss similar to that found in the whole heart was observed. For example, 90% of the myocytes derived from the 8-day-old embryonic chick heart was capable of being labeled. But after 17 days, the value dropped to 5%. In addition, the myocyte growth rate declined rapidly and doubling time increased. The intermitotic period showed only a slight increase. The indication is that, in culture, the majority of muscle cells withdraw from a diminishing proliferation pool of cells which continue to divide at a diminished rate. The values obtained were similar to those reported in the intact chick heart.[34] These studies showed that the rate of withdrawal of myocytes from the cell cycle was influenced by the stage of embryonic development.

The factors and conditions which influence hyperplasia and hypertrophy remain to be elucidated. The cell culture model provides a valuable means for study of this problem. Manipulation of the medium allows experimental examination of the processes. Recent studies in neonatal rat heart cells have provided evidence that fetal calf serum has a marked effect on myocardiac cell hypertrophy, giving a fourfold increase in cell size when serum concentration increases from zero to 10%.[35] Earlier studies in the same system showed that a heat stable factor is present in fetal calf serum which stimulates myosin synthesis.[36,37] These results indicate that humoral factors such as hormone and serum growth factors may control protein synthesis and hypertrophy.

2. Reversibility of Loss of Mitosis

It seems clear that an abrupt or gradual cessation of mitosis precedes full expression of adult specific molecules. Moreover, most evidence indicates that the loss of DNA synthesis is irreversible, especially in the adult tissue. The clearest evidence in the skeletal muscle system is derived from experiment in the L_6 muscle cell line.[38] These L_6 cells in culture lose their capacity to be cloned as they proceed to the fusion state.[39] Since cloning conditions are the optimum for growth and multiplication of the L_6 cells, their inability to undergo mitosis and increase in number indicates that the loss of proliferative capacity, seen in these cultures, is irreversible. However, a recent report[40] indicates that such myocytes may be able to reenter the cell cycle. Quail myocytes, synchronized in mitosis and allowed to enter a terminal G_1 during which virtually all the cells initiated myosin synthesis, were challenged with a high growth promoting medium. A marked increase in cells synthesizing DNA occurred in those cells which already contained myosin. The difference between these experiments[40] and the cloning experiments[38] may be due to differences in time of testing and also differences in the composition of the medium.

The heart is classified as a muscle tissue without regeneration powers since defects are repaired by scar formation. However, increase in ploidy in postnatal rat hearts has been widely reported.[41] The recent development of methodology for culture of adult rat myocardial cells[42,43] has presented an opportunity to examine postnatal DNA synthesis in culture. Cells isolated and cultivated from the adult heart go through a process of disorganization and reorganization which may or may not be similar to embryonic neonatal development. Recently it has been observed that these terminally differentiated cells may contain many nuclei. They were seen to incorporate (^{3}H)thymidine into their replicating DNA[44] and also to reacquire activity of enzymes needed to synthesize DNA.[45] Thus, some DNA synthesis occurs in both the intact heart and in cultured adult cells. However, under normal circumstances, the adult heart is postreplicative and these observations do not change the unfortunate fact that the heart does not regenerate. Nevertheless these infrequent events in the intact heart and in adult cell in vitro indicate that the mechanism for DNA synthesis is potentially present in these cells and it may be possible to discover the conditions necessary to actually reverse the loss and promote active hyperplasia in adult heart.

3. Possible Mechanisms for Loss of Mitosis

Developing heart cells decrease cell division when thrust into an aerobic environment. However, it is not obligatory that all cell division cease before adult-specific proteins begin their synthesis, although the two events seem related. Several questions arise from the observations already discussed.

1.   Are the two events, cessation of DNA synthesis and initiation of myosin synthesis, e.g., casually related or are they both controlled (in opposite directions) by a specific or general condition?
2.   What are the mechanisms of cessation of cell division and the initiation and maintenance of specific cell protein synthesis? Does the shift to an aerobic environment play a role and by what means does this occur?
3.   How much of the cellular structure can be established while cell division still occurs?

a. Myofibrillar Structure

It has been suggested that the contractile elements themselves may act as a physical barrier to cell division. On this basis one might expect that the myofibrillar structure might be dispersed before or during mitosis to allow the physical manipulation necessary for cell division to occur. A study of the state of the organization of the myofibrils before, during, and after mitosis in cultures has been helpful in exploring the problem.

It has been reported, on the one hand, that Z-lines and myofibrils are dispersed in mitotic cells[46,47] and, on the other hand, that they may not change at all during cell division.[48] Electron micrographs of chick cells in metaphase had clear, well-organized myofilaments with intact Z-bands. It would appear, however, that most muscle cells do not undergo mitosis after filaments became organized into well-defined myofibrils.[48] In chick heart culture, only 2 to 4% of the total mitotic cells contained definitely striated myofibrils, while 20 to 40% of the mitotic cells contained fibrillar material that bound fluorescent antimyosin. In contrast, 51% of all interphase cells contained myofibrils.[49]

Study of the cell cycle and generation time of dividing muscle cells is difficult in a mixed-population culture. One way to accomplish this is to measure (by autoradiography) the rate of appearance of (^{3}H)deoxythymidine into the chromosomes of muscle cells in mitosis. This requires a technique for identifying muscle cells in the mixed culture. Mitotic figures are difficult to see in muscle cells, and most staining techniques for revealing chromosomes are incompatible with the attachment of antimyosin. To solve this problem mitotic cells were identified as the cells that could be collected by agitating the culture, thus removing the rounded cells that are not securely attached to the substratum.[31] Using this technique, rough estimates of G_2, G_1, and length of mitosis have been made. However, it has been reported that, unlike many other cells, cardiac muscle cells do not necessarily become rounded during mitosis.[23,31,49] The cells remain flattened and attached and are difficult to remove by agitation. The flattened condition of these cells during mitosis may be an adaptation to allow cell division without disturbing the cell cytoplasm organization and to permit the myofibrils to remain intact. Thus only heart muscle cells capable of rounding, presumably with little myofibrillar organization, can be collected by agitation and scored as mitotic cells. This selection of underdeveloped cells could lead to an underestimation of the cells in mitosis and could also lead to the erroneous conclusion that only cells with little myofibrillar organization can divide.

b. O_2, NAD+, Poly(ADPR)

The shift from an anaerobic to a postnatal aerobic environment may play a significant role in the loss of mitotic activity. This change must have a profound implication for the heart. It has been suggested that O_2 and the redox potential play possible roles in the regulation of the shifts in enzyme profiles, changes in metabolism, and DNA synthesis. Correlated with the change from an anaerobic glycolytic state to a postnatal one with a high respiratory rate is an increase in mitochondria and mitochondrial enzymes, a decrease in glycogen, and an increase in fatty acid oxidation. The concentration of NAD+ seems to be an important factor. NAD+ itself has been shown to inhibit cell division while NADH has little effect.[50] The action of O_2 on cell proliferation may thus be partly mediated through its effect on the NAD+ to NADH ratio. However, an actual increase in the total NAD+ would play a more important role. This has been observed in the developing rat.[51] NAD+ concentration expressed on a DNA or per cell basis is $5.2 \times$ higher in the adult cardiac muscle than in the 1-day postnatal rat heart. Addition of NAD+ directly to tissue slices of ventricular muscle caused an inhibition of DNA synthesis. Nuclei (or chromatin from nuclei) preincubated with NAD+ serve as a poor template for transcription. One possible mechanism for the action of NAD+ is its action as a substrate for formation of poly(adenosine diphosphate ribose) [Poly(ADPR)]. The enzyme poly(ADPR) polymerase is present in the eukaryotic cell nucleus and catalyzes the transfer of ADPR moiety of NAD+ to chromatin, adding successively to form a homopolymer poly(ADPR), and releasing nicotinamide. The early work in this field has been reviewed.[52,53]

Evidence has accumulated that suggests that poly(ADPR) may regulate cell proliferation. The poly(ADPR) polymerase activity of cardiomyocyte nuclei diminish synchronously with the decrease in capacity to synthesize DNA.[54] It has also been demonstrated that poly(ADPR) chromatin is a poorer template for DNA synthesis than chromatin itself.[55] In the developing heart, the loss of DNA synthesis and almost complete loss of DNA polymerase α has been temporally correlated with the increase in poly(ADPR) which doubles in the neonatal rat from day 1 to day 16, with the adult value somewhat higher.[51,56] The increase in poly(ADPR) polymerase occurs at the same time as the increase in the substrate. Thus NAD+ concentration may play a direct role in controlling DNA synthesis. As indicated, nuclei preincubated with NAD+ serve as a poor template for transcription.[50]

A cell that has lost its ability to divide should show this by a change in mRNA synthesis. This suggests a possible connection between poly(ADPR) addition to nuclear proteins and nucleic acid synthesis. Administration of T_3, which induces ventricular enlargement, has been shown to augment mRNA synthesis at the same time as the poly(ADPR) polymerase activity decreases.[57] The increase in mRNA synthesis is also correlated with a decrease in poly(ADPR) nuclear protein adduct[58] in cardiomyocytes, but not in heart nonmuscle cells. It appears, therefore, that poly(ADPR), its polymerase, and substrate may be involved in the regulation of DNA and RNA synthesis thus controlling cell division.

c. Cyclic AMP, Polyamines

cAMP has been widely observed to affect cell proliferation. In the rat, cardiac cAMP concentration increases during late fetal and early neonatal development.[59,60] The adenyl cylase activity is also higher in the adult.[61,62] Both dibutyryl cAMP and isoproterenol inhibit DNA synthesis in differentiating cardiac muscle. Injection of dibutyryl cAMP into newborn rats not only decreases DNA synthesis in the heart but increases poly(ADPR) and decreases DNA polymerase.[63,64] The level of cAMP rises and that of cGMP declines in cells which approach confluency and stop dividing. The opposite is true in actively dividing cells.[65]

Addition of fetal calf serum or growth promoting factors to confluent or quiescent cell cultures causes a rapid decrease in cycAMP and an increase in cGMP,[66,67] which suggests that a fall in the cAMP/cGMP ratio might trigger cell proliferation. It has been reported that the polyamines, spermine, spermidine, and putrescine correlate positively with cell growth[68] and may modulate cyclic nucleotide concentration. The polyamines reduce cAMP content,[69] possibly by increasing the cAMP dependent phosphodiesterase activity. Addition of polyamines to confluent and serum restricted heart cell cultures lowers the cAMP content two-fold, but increases the cGMP content up to threefold[70] possibly by reducing the rate of cGMP content degradation. The action of polyamines may be due mainly to their effect on cGMP phosphodiesterase, since they have been reported to decrease the activity of both soluble and particulate cGMP phosphodiesterase.[70] This would lead to an increase in cGMP and indirectly to an increase in cAMP phosphodiesterase activity by virtue of the action of cGMP which has been shown to increase the activity of cAMP phosphodiesterase. Such action would lead to a decrease in the cAMP level.

Other agents may also exert their mitogenic effort through their effect on cAMP. For example, it has been reported that Ca^{2+} increases cell proliferation in HeLa cells,[71] possibly by affecting the intracellular concentration of cAMP. Since the degradation of cAMP is catalyzed by a Ca^{2+}-stimulated phosphodiesterase and activator protein, an increase in Ca^{2+} should lead to a decrease in cAMP. It has been demonstrated in cultured rat heart cells that a decrease in Ca^{2+} in the medium caused a doubling of intracellular cAMP.[72]

The polyamines may also affect growth by mechanisms other than their effect on cyclic nucleotides. They can modify histones by enhancing acetylation and may in this way affect DNA synthesis. Spermine enhances acetylation of arginine rich histone fractions[73] and also increases RNA polymerase activity associated with nuclear chromatin.[74] In erythropoiesis, histone acetylation has been correlated with increased cell division.[75] In addition, an increase in spermidine is associated with increased cell replication in rapid growth neoplasms.[76] In chick embryo heart cells in culture, an increase in O_2 caused a considerable decrease in the uptake of (^{3}H)deoxythymidine into DNA and a decrease in the acetylation of histones. It would be of interest to trace the spermine and spermidine content and the degree of acetylation in the developing heart as cell proliferation decreases and also to investigate the effect of cAMP on these parameters.

The changes in cyclic nucleotides have been correlated with the maturation of the adrenergic neurons.[17,78] There is also an increase in norepinephrine observed in the heart which reflects the density of these neurons.[79,80] It has been suggested[44] that the temporal relationship of adrenergic innervation, cAMP increases, and loss of DNA synthesis may be involved in cellular proliferation.

B. Initiation of Myosin Synthesis

The first stage of heart myocyte development culminates with a cessation of mitosis. The second stage begins during the period of decreased cell division and is characterized by the initiation of the synthesis of adult muscle specific macromolecules. One of these, myosin, is a crucial protein for muscle contraction and despite its presence in nonmuscle cells[81] it is undoubtedly a major specific participant in specialized muscle structure and function. As such it has proved to be a useful marker and many studies on its nature and the initiation and control of its synthesis in the developing heart have been carried out.

1. In Vitro Studies

Most of the attention has been focused on the intact heart. Few studies have been carried out in the heart cell cultures. The culture system is handicapped by the mixed nature of the cell population and the lack of clearly separated cellular stages of development.

To date, all useful heart cell cultures are primary explant and as such are a mixture of several cell types, including muscle cells at various stages of development and nonmuscle cells such as fibroblasts and endothelial and epithelial cells which may overgrow the culture. A pure line of developing myocytes is desirable. Most attempts to clone the heart cells and to establish a stable heart cell line have not been successful, because in order to clone, one needs a cell that can continue to divide. As we have seen the presumptive heart muscle cell divides for a limited period of time and then develops into the functional myocyte that has terminated mitosis. Needed for successful cloning is the ability to recognize the presumptive muscle cell and to prevent its development and thus maintain its capacity to divide. Failing this, primary cultures can still be useful. One may take advantage of the differential ability of muscle cells to adhere to culture dishes to enrich the muscle cell population.[82] One may also estimate the effect of overgrowth of the cultures by other cell types by inhibiting cell division and studying the function in a nondividing culture. This can be done by eliminating the serum that contains essential cell division factors or by using cell division inhibitors, such as colchicine.[25] Bromodeoxyuridine (BrdU) followed by irradiation can be used to kill dividing cells and produce a rather pure culture of mature nondividing myocytes.[26,27] The system can, therefore, be useful if one follows muscle-specific functions in these enriched cultures, but all results must be interpreted in light of the nature of cell population. As previously mentioned, myosin synthesis has been demonstrated in rat heart cultures[36,37] that is responsive to serum factors. Myosin increases rapidly in cultured rat heart cells, and the increase is inhibited by cell division inhibitors[25] or by the selective destruction of dividing cells.[26] Under these conditions, the rate of increase in myosin is much lower, reflecting the rate of synthesis in the nondividing cells. In contrast, the rapid increase of myosin from neonatal cultures must therefore be a result either of the proliferation of presumptive myoblasts or of dividing muscle cells, both of which could lead to an increased number of myosin-producing cells in culture.

2. In Vivo Studies

a. Early Appearance

Cardiac muscle cells differentiate very early. At about the 10-somite stage, the chick heart beats spontaneously and rhythmically. At this stage the myocardium consists of a pure population of muscle cells.[83] The evidence that myosin-containing cells can divide and that adult cells cease cell division indicates at least three stages in heart development: (1) the dividing myoblast without myosin, (2) the dividing muscle cell with myosin, and (3) the adult myocyte, containing myosin, incapable of cell division. Cell division of myosin-containing cells is limited, but this limited capacity may be sufficient to add all the new muscle cells acquired during growth.[50] Alternatively, the division of presumptive myoblasts may be the source of many of the new muscle cells. To study the role of dividing cells in the development of cardiac muscle cells in the chick embryo,[84] BrdU was injected directly into the chick embryos. BrdU added before stage 7 stopped the development of the heart; when added to stage 7 and 8, it led to smaller and more weakly beating hearts. In mesodermal cells, prepared from stage 5, BrdU stopped the development of beating cells. BrdU had no effect on the beating cells or the formation of myofibrils in cells after stage 8. BrdU incorporated in cells prior to stage 7 appears to suppress the expression of the cardiac phenotype. BrdU affects cells into which it has been incorporated, and the results may indicate that beyond stage 8, BrdU in the dividing muscle cell does not affect the capacity to synthesize myosin. Myosin, therefore, increases in cultures or in the whole animal as a result both of the production of more myosin-producing cells and of the increase of myosin in these cells.

b. Isozyme Shifts

Early observations indicated that the molecular forms of myosin changed during development of skeletal muscle[85] and cardiac muscle.[86] It appeared that cardiac myosin existed in an embryonic form which corresponded to the myosin of slow muscles and in an adult form similar to the fast muscle myosin. It was also observed that the Ca ATPase activity of chick myosin changed during development.[87] As the chick developed from the 8-day embryo to 1-day posthatching, the Ca ATPase was stimulated at progressively lower Ca concentrations.

Immunological properties of myosin also changed with time during development. Antibodies to skeletal myosin reacted with both cardiac and skeletal muscle myosin at the early stage of the embryo. Cardiac antimyosin reacts in the same way; however, cross-reaction decreased with development and disappeared after hatching.[88] The myosin synthesized by polysomes prepared from embryonic chicken breast muscle also cross-reacted. As the embryo grew, the synthesis of embryonic cardiac myosin (or slow myosin) decreased and the fast myosin increased. Myosin has been prepared from polysomes obtained at various stages of development of the embryo heart, and the change in the immunological properties corresponds to the changes in the intact heart.

Further advances in this area have been made possible by the discovery that atrial and ventricular cardiac myosin differed in primary structure. The atrial and ventricular myosins isolated from calf heart possessed heavy chains with different primary structure.[89] Earlier reports indicated that purified atrial myosin had a greater Ca ATPase activity and also differed in LC_1 with respect to amino acid composition and electrophoretic mobility when compared to ventricular isomyosin.[90] However, the adult atrial light chain is identical to that of the fetal ventricle and also to that of the skeletal muscle.[91] Atrial myosin electrophoretically isolated from the heart and muscle of the chick had a similar Ca ATPase activity to that of the fast skeletal pectoralis myosin, and ventricular myosin was similar in this regard to that from slow anterior latissimus dorsi muscle.[9] More recently the evidence is that these isomyosins are not completely separated from each other. Heterogeneity was demonstrated immunochemically is muscle fibers of both the atrial and ventricular rabbit myocardium.[93] Fluorescent antibodies to atrial myosin stained all atrial muscle fibers and a minor proportion of right ventricular muscle fibers of the bovine heart. In contrast, all the fibers of the left ventricle were unreactive. An intermediate response was seen in the rabbit where many fibers were stained, and some were completely negative in both the right and left ventricles. Developmental changes in distribution were also observed. Atrial antimyosin reacted strongly with fibers of the rabbit ventricle up to the fourth postnatal week. The overall proportion of completely unreactive fibers became progressively greater with time.

Two molecular forms of myosin heavy chains have been isolated from rabbit ventricle by affinity chromatography and characterized by use of monoclonal antibodies.[94] They were shown to coexist in different proportions in the developing rabbit. One form, HC_α, constitutes 20% of the ventricular myosin in the 20-day embryonic rabbit, rising to 50% during the first 4 weeks after birth, falling to 12% at 12 weeks and to virtually zero at 1 year. At this time virtually all the heavy chain is HC_β, the second isomyosin. On the basis of increasing electrophoretic mobility three isomyosins were early identified and named V_1, V_2, and V_3.[95] V_1 with the highest Ca ATPase activity is probably equivalent to HC_α and the atrial heavy chain while V_3 is similar to HC_β or the ventricular heavy chain. Thus Ca ATPase activity begins at a lower level and passes through a peak during the early weeks after birth. Correspondingly, immunofluorescent antibody to atrial myosin heavy-chain reacted strongly with fibers of the rabbit ventricle up to the fourth postnatal week. The overall proportion of completely unreactive fibers became progressively greater during subsequent stages. Since atrial myosin has a higher Ca ATPase, these results also correlate with the loss of HC_α or V_1.

A survey of the V_1 and V_3 myosins has revealed species differences showing that there is not a strict classification of the heavy chain of myosin into an embryonic and adult form.

In all species examined, V_3 was essentially the fetal form with V_1 appearing around the time of birth. In adult mice V_1 remained, while in rabbits and pigs returned to V_3 after 3 or 4 weeks. Adult dogs, beef, and human ventricular myosin were all composed of the V_3 form only.[96]

The cellular distribution of V_1 and V_3 also varied with age. All isolated myocytes from 3-week-old rats stained with fluorescent antibodies to V_1. From adults, 50% of the myocytes stained with anti-V_1, 10% with anti-V_3 and 40% with both.[97] The V_1 and V_3 forms were homogeneously distributed throughout the cell.

Of interest to this developmental change is the effect of the thyroid hormone on the myosin content of the heart. After thyroxine treatment the rabbit ventricle, fibers became more reactive with fluorescent anti-atrial myosin.[93] In addition, using monoclonal antibodies to HC_α and HC_β it was found that the myosin isolated from throxine treated rabbits reacted with the HC_α antibody. This compares with the normal ventricular myosin which is mostly HC_β.[98] Thus thyroxine treatment increases the Ca ATPase activity of the heart by increasing the proportion of HC_α. Other physiological properties are associated with this shift in isomyosins. Thyroxine treatment of the rabbit heart leads to an increase in the rate of force, development, tension development and heat production.[99]

Thyroxine appeared to affect the synthesis rate of the myosin heavy chain. Administration of tetraiodothyronine to the rabbit causes a three-fold increase in the synthesis rate of HC_α and a comparable decrease in that of the HC_β.[100] Utilizing cDNA clones, that contain cardiac myosin heavy chain mRNA sequences, it has also been shown that two different cardiac myosin heavy chain genes are expressed in the adult rat.[101]

Studies of the appearance and the turnover of myosin do not give information about whether the controlling step is transcription or translation. In skeletal muscle, for example, myosin is not present in monomucleated myoblasts although myosin mRNA is present.[102] The control of mRNA synthesis and its translation remains to be explored. In the heart, even less is known about these steps. The turnover of myosin has been studied in neonatal rat heart cultures,[36] in chick embryos,[103] and in various tissues of the same and other adult animals,[104] but this give very little information on the initiation of myosin synthesis and its control. A recent study of the mRNAs in the myocardium of the 12, 14, and 17-day fetal, 5-day neonatal, and 3-day adult mice indicated a developmental pattern.[105] Several mRNAs appear to be abundant in the late fetal heart while others, including two that were coded for myosin light chain 1 and 2, were most abundant in the adult heart. Although late fetal polypeptides were abundant in cell-free translation products, they were not visible in the extracts of the 17-day-old fetal or adult heart. It is possible that these mRNA directed polypeptides may be metabolized too rapidly or alternatively, the immediate product may undergo modification in vivo which does not occur in the reticulocyte lysate used for in vivo translation.

III. CHANGES IN METABOLISM

A. Glycolysis

The growth and development of the heart is accompanied by profound metabolic changes reflecting the shift from an anaerobic to aerobic state. The adult heart has a preference for fatty acids as an energy source.[106] In contrast, the early developing heart is mainly glycolytic. Glucose is the most effective substrate for prolonging the beating of isolated mouse hearts.[107] The decreased use of glucose is accompanied by a decline in glucose uptake.[108] The shift from glycolysis to fatty acid oxidation is indicated by increased oxygen utilization. The contractile rate of the 11-day-old embryo rat heart is independent of O_2 but by the 12th day, a greater need for O_2 develops.[109]

A similar result was obtained with late fetal mouse heart in organ culture.[110] Deprivation of oxygen and glucose to mimic ischemic conditions resulted in a loss of 50% of the ATP and a cessation of beating. If oxygen and glucose are reintroduced after an hour, the beating returned. This is consistent with the results in the 13-day-old embryonic rat heart, where no substrate can maintain the beating under anaerobic conditions.[109] The isolated beating rat heat is resistant to O_2 deprivation on the 10th day of gestation, but becomes increasingly dependent thereafter. The rate of glycolysis falls progressively as O_2 dependence increases.[111]

A direct measurement of the ability of the heart to utilize glucose as a substrate has been made on chick heart cell cultures derived from the 7-day and 13-day embryo. Using uniformly labeled glucose, the specific activity of the CO_2 formed from the cells isolated from the 13-day heart was 50% lower than that in 7-day cells.[112]

Consistent with these changes is the decrease in hexokinase and aldolase in the newly hatched chick as compared to the embryo.[113] In addition, the P/O ratio increases. All of these changes indicate that the newly hatched chick function is similar to the adult. The chick, in this regard, differs from the newborn rat. The neonatal rat heart starts with a strong glycolysis, but cell division and development continue after birth. The glucose depletion in 2- to 3-day and 5- to 6-day-old cultured rat heart cells was measured and it was shown that in this 3-day period, the glucose depletion rate and production of lactate in the older cells falls to one third of the value of the 2- to 3-day-old cell.[114]

Neonatal rat heart cells in culture utilize glycolysis for the maintenance of beating. Inhibition of oxidative phosphorylation or anaerobiosis does not affect their beating or the ATP level for several hours[115] despite the indication that fatty acid oxidation is the main energy source.[116] The inhibition of glycolysis by deoxyglucose or iodoacetate also does not depress the beating rate for at least 1 hr, during which period the ATP level is maintained. However, inhibition of both glycolysis and respiration leads to an immediate cessation of beating with a drop of only 15 to 20% in ATP level. The observation that inhibition of each metabolic source of ATP alone is not enough to decrease the ATP level indicates that the heart cell has a compensatory mechanism for maintaining the ATP level. Also, although it appears that each inhibitor alone has no effect on the metabolism, the ability of each inhibitor to potentiate the effect of the other indicates that this is not so.

Recent observation made in chick heart cultures show some differences. Deoxyglucose or iodoacetate quickly decrease contractility and the high energy phosphate level, but not in the presence of Krebs cycle intermediates such as acetate.[117] As observed in the rat cultures, energy derived from glycolysis, although not essential for beating when oxidative phosphorylation is operative, contributes importantly to the maintenance of high energy phosphate and contraction during inhibition of oxidative phosphorylation. It has been estimated that during normal myocardial function, glycolysis supplies 4% of the total ATP,[118] but during severe cellular hypoxia, 80% of the ATP utilized is derived from glycolysis.[119] This is further evidence that both glycolysis and oxidative phosphorylation supply energy for the heart, and that these two processes are interdependent in order to maintain the high energy phosphate level.

B. Metabolic Differences in Cultured Cells

An important question can be raised at this point as to how the metabolism of cultured cells compares to that of the in vivo cell. One problem could be that of O_2 availability, which has been proposed to be insufficient under culture conditions. However, direct measurement of the partial pressure of O_2 in the heart cell culture medium has been shown to be comparable to the intact heart tissue.[120] O_2 tension has been suggested to play a role in the induction of oxidative enzymes. As an example, hypoxia and reduced O_2 tension lead to reduced cytochrome concentration in cultured fibroblasts[121] and in hearts of newborn rats.[122]

The differences between the cultured cells and those in the intact heart may be a result of differences in work performed in these two systems. Differences in cell population and cellular overgrowth by one cell type may also be responsible. It is possible that the heart muscle cells and fibroblasts interact and have important effects on each other both in metabolism and development. Cinematographs of chick heart cultures have demonstrated the interaction of these cell types.[123] Fibroblasts and myocytes affect each others movement and shape, and these changes contribute to the development of synchronously pulsating muscle cell units. Separation of the myocytes and fibroblasts from the neonatal rat heart by differential plating has allowed a closer analysis of their metabolic differences.[124] The glucose and palmitate utilization and the glycogen content of myocytes is greater than that of the fibroblasts, but surprisingly, glucose utilization by mixed cells, derived as such from the heart, was greater than that by the myocytes alone. This suggests that a synergism between the cell types can exist. In contrast, the fibroblasts may possibly interfere with myocyte development. Cultured rat myocytes purified with the use of BrdU had a greater number of cells with cross-striations than the control mixed cell population. In addition, the maximum response of the beating rate of the mixed culture to isoproterenol was 20 to 30% of that of the purified culture.[125] Undoubtedly, the cells in culture can differ in metabolic and physiologic activity from those in the intact heart depending on conditions.[126] This is only a detriment if one is rigidly attempting to study intact heart behavior using the culture system. The differences that exist can be most instructive, since they provide information on the mechanism of control of normal function. The ease of manipulation of cells in culture have demonstrated the possible role of O_2 in the development of metabolic pathways and the possible synergism between the cell types present in the intact heart.

C. Isozymes of Lactic Dehydrogenase

We have emphasized the possibility that the shift to aerobiosis may play a role in the differentiation of the heart. Increased vascularization during gestation and the increased availability of oxygen at birth may have a developmental impact. We have already seen that · this may have an effect on decreased DNA synthesis and possibly lead to changes in the enzyme profile of the heart. It is perhaps not a coincidence that the direction of change fostered by a high pO_2 is consistent with the need by the heart for a highly aerobic metabolism.

The aerobic environment may also affect the transcription of specific proteins. Evidence for this exists in the control of lactic dehydrogenase (LDH). Lactic dehydrogenase is an ubiquitous enzyme that serves a general role in carbohydrate metabolism. It exists in isozymic forms whose kinetic properties are tailored to fit different physiological roles.[127] In many cases one may correlate the presence of a so-called heart or H form, also called LDH_1, with highly aerobic organisms or tissues and the muscle or M form, LDH^5 with more anaerobic tissues. The H form favors the funneling of pruvate metabolism toward its complete oxidation to CO_2 and H_2O. It is inhibited by high levels of pruvate, which results in directing pyruvate away from reduction to lactate. By contrast, the M form is not as inhibited by pyruvate and allows high levels of pruvate to be reduced to lactate. In addition, the H form has a greater affinity for lactate, thus leading to a more efficient reverse reaction. A consequence of regulation by pyruvate is that under normal in vivo conditions, the enzyme is prevented from reducing pyruvate to lactate while maintaining its full capacity to oxidize lactate. The H form is adapted for more substantial sustained conversion of energy to ATP.

The adult rat heart has a predominance of the H form. The embroynic rat heart LDH is in the M form, and this shifts to the H form during development. In contrast, the human embryo heart contains the H form, which persists in the adult.[128] In the chick, the embryonic type in both breast muscle and heart is the H form, and maturation leads to a shift to the M form in breast muscle only.[129] The two forms are under independent gene control.

When embryonic chick heart cells are cultured they begin with 86% in the H form. But after 6 days in culture the ratio shifts, and 78% is in the M form. This is in sharp contrast to the in vivo situation. The shift in culture could be ascribed to the relative anaerobiosis of the culture conditions. Increasing partial pressure of oxygen (pO_2) and adding citric acid cycle intermediates partially prevents the increase in M form,[130] which indicates that the redox potential in the cell plays a role in the kind of LDH formed. A similar observation was made in chick eggs aerated with different pO_2 values. Increase in pO_2 lowered the M form in the intact developing heart.[131] These changes are not confined to the muscle. Oxygen also influences the LDH pattern of lymphocytes[132] and fibroblasts.[133] An anaerobic environment favors the M form, increased cell division, and collagen synthesis in the fibroblasts.

Criticism of the direct effect of O_2 on LDH isozymes has been leveled recently on the basis that the changes with LDH pattern in neonatal rat heat cells in culture were consistent with the changes in the fibroblast and myocyte population.[134] However, the percentage of muscle cells was estimated by using the values of the specific activity of heart specific isoenzymes. A myocyte measure independent of the isoenzyme profile would have been more convincing.

D. Mitochondrial and Fatty Acid Oxidation

Studies in many different cell types appear to indicate that in some way O_2 and the redox potential may affect the transcription of the LDH gene in addition to DNA synthesis. We can follow this increase in oxidative metabolism by examining the shift toward increases in mitochondrial oxidative function and fatty acid oxidation. The total area of the mitochondrial[135] and respiratory enzymes increases during fetal development. In the rat from the 10th day to the 14th day of gestation, mitochondrial ATPase, succinic dehydrogenase (SDH), NADH oxidase, and cytochrome *c* oxidase increase in a fashion parallel to the increase in the number of mitochondrial cristae.[136] The respiration rate of the rabbit heart is low in midgestation, but increases at birth.[137] Alterations in the spectral properties of the mitochondrial cytochromes during development may reflect increasing ability to conduct electron-transport-associated reactions. Cytochrome *a* and *b* absorption bands appear to become more prominent in the late fetal mitochondria of the rabbit heart. SDH, cytochrome *c* oxidase, and cytochrome *c* reductase have also been reported to increase[138] in the rat embryo.

The most clear-cut studies have been performed in the neonatal rat, which is known to continue its heart development for as many as 30 days after birth. NADH-cytochrome *c* reductase, succinic cytochrome *c* reductase, and monoamine oxidase increase with age of the rat. The increase was 3- to 5-fold from 10 to 60 days of age.[139]

Though changes occur with development, it appears that the fetal heart mitochondria have efficient oxidation and phosphorylation capabilities.[137,140] An increase in the P/O ratio in the rat has been reported.[121] Since this data has been derived from tissue slices, it is questionable. Thus oxidation phosphorylation may not be limiting.

Fatty acid oxidation via the mitochondrial enzymes is the principal metabolic source of high energy phosphate in the adult heart. A respiratory quotient of 0.78 demonstrates this.[114] The lower capacity of the fetal heart to oxidize fatty acids may be due to lowered activity of fatty acid enzymes.

It has been suggested that the decreased capacity of newborn rats to utilize palmitate may result from low tissue levels of carnitine and decreased carnitine palmitoyltransferase.[141] Fetal bovine hearts are less able to oxidize palmitoyl CoA than the calf heart, although both utilize palmitoyl carnitine equally well.[142] These observations suggest that the fetal bovine heart has a decreased biosynthesis of palmitoyl carnitine. Several enzyme activities involved in this sequence have been investigated.[143]

Palmitoyl-CoA synthetase activity in the rat heart is low in the fetus and increases 6- to 10-fold during the first postnatal week. In the chick heart, however, this enzyme peaks between 13 to 16 days of development and then drops after hatching, to values of approximately 50 to 70% of the peak. Acetyl-CoA synthetase and carnitine palmitoyltransferase behave in the same way. In the rat they begin at a low value and increase in the postnatal period. In the chick they reach the peak value at 13 to 15 days of development. Coincident with the postnatal rise in fatty acid activation and carnitine palmitoyltransferase activity in the developing rat, the oxidation of palmitoyl-CoA plus carnitine and palmitoyl carnitine increases from barely measurable levels at birth to adult levels by 30 days of age. The deficiency in the fetus appears to be at the level of transport of the fatty acids across the mitochrondial membrane.

In this chapter several aspects of heart growth and development have been reviewed, to which the heart cell in culture has made important contributions. As we have learned more about the heart cell system, it has become more useful. For example, we have recently reached the point where the mechanism of the control of mitosis, the cell cycle, and of myosin gene expression, may best be studied in the cell culture system. There is no reason not to expect that the same may become true of many other problems relating to cardiac function.

REFERENCES

1. **Holtzer, H., Marshall, J., and Finck, H.,** An analysis of myogenesis by use of fluorescent antimyosin, *J. Biophys. Cytol.,* 3, 705, 1957.
2. **Herman, H., Marchok, A., and Baril, E. F.,** Growth rate and differentiated function of cells, *Natl. Cancer Monograph,* 26, 303, 1967.
3. **Stockdale, F. E. and Holtzer, H.,** DNA systhesis and myogenesis, *Exp. Cell Res.,* 24, 508, 1961.
4. **Konigsberg, I. R.,** Clonal analyses of myogenesis, *Science,* 140, 1273, 1963.
5. **Weinstein, R. B. and Hay, E. D.,** Deoxyribonucleic acid synthesis and mitosis in differentiated cardiac muscle cells of chick embryos, *J. Cell Biol.,* 47, 310, 1970.
6. **Shainberg, A., Yagil, G., and Yaffe, D.,** Control of myogenesis in vitro by Ca^{++} concentration in nutritional medium, *Exp. Cell Res.,* 58, 163, 1969.
7. **Merlie, J. P. and Gros, F.,** In vitro myogenesis. Expression of muscle specific function in the absence of cell fusion, *Exp. Cell Res.,* 97, 406, 1976.
8. **Moss, P. S. and Strohman, R. C.,** Myosin synthesis by fusion-arrested chick embryo myoblasts in cell culture., *Develop. Biol.,* 48, 431, 1976.
9. **Vertel, B. A. and Fischman, D. A.,** Myosin accumulation in mononucleated cells of chick muscle cultures., *Develop. Biol.,* 48, 438, 1976.
10. **Rumiantsev, P. P.,** DNA synthesis and nuclear division in embryonal and postnatal histogenesis of myocardium (autoradiographic study), *Arkh. Anat. Gistol. Embriol.,* 47, 59, 1964.
11. **Manasek, F. J.,** Mitosis in developing cardiac muscle, *J. Cell Biol.,* 37, 191, 1968.
12. **Sasaki, R., Morishita, T., and Yamagata, S.,** Mitosis of heart muscle cells in normal rats, *Tohoku, J. Exp. Med.,* 96, 405, 1968.
13. **Fischman, D. A., Doyle, C. M., and Zak, R.,** DNA synthesis in chick cardiac muscle. Comparative observation in vivo and in vitro growth, in *Developmental and Physiological Correlates of Cardiac Muscle,* Vol. 1, Lieberman, M. and Sano, T., Eds., Raven Press, New York, 1975, 67.
14. **Zak, R.,** Development and proliferative capacity of cardiac muscle cells, *Circulation Res.,* 35, (Suppl. 2), 17, 1974.
15. **Doyle, C. M., Zak R., and Fischman, D. A.,** The correlation of DNA synthesis and DNA polymerase activity in the developing chick heart, *Develop. Biol.,* 37, 133, 1974.
16. **Claycomb, W. C.,** DNA synthesis and DNA polymerase activity in differentiating cardiac muscle, *Biochem. Biophys. Res. Commun.,* 54, 715, 1973.
17. **O'Neil, M. and Strohman R. C.,** Studies of the decline of deoxyribonucleic acid polymerase activity during embryonic muscle cell fusion in vitro., *Biochemistry,* 9, 2832, 1970.
18. **Stockdale F. E.,** Changing levels of DNA polymerase activity during the development of skeletal muscle tissue in vivo, *Develop. Biol.,* 21, 462, 1970.
19. **Machovich, B. and Greengard, O.,** Thymidine kinase in rat tissues during growth and differentiation, *Biochem. Biophys. Acta,* 286, 375, 1972.
20. **Pegoraro, L. and Baserga, R.,** Time of appearance of deoxythymidylate kinase and deoxythymidylate synthetase and of their templates in isoproterenol-stimulated deoxyribonucleic acid synthesis, *Lab. Invest.,* 22, 266, 1970.

21. **Taylor, A. T., Stafford, M. A., and Jones, O. W.,** Properties of thymidine kinase partially purified from human fetal and adult tissue, *J. Biol. Chem.,* 247, 1930, 1972.
22. **Gilette, P. C. and Claycomb, W. C.,** Thymidine kinase activity in cardiac muscle during embryonic and postnatal development, *Biochem. J.,* 142, 685, 1974.
23. **Mark, G. E. and Strasser, F. F.,** Pacemaker activity and mitosis in cultures of newborn rat heart ventricle cells, *Exp. Cell Res.,* 44, 217, 1966.
24. **Kasten, F. H.,** Rat myocardial cells in vitro: mitosis and differentiated properties, *In Vitro,* 8, 128, 1972.
25. **Lewis, H. and Harary, I.,** In vitro studies of beating heart cells in culture. XIV. Reversible changes in the level of myosin adenosinetriphosphatase, *Arch. Biochem. Biophys.,* 142, 501, 1971.
26. **Masse, M. J. and Harary, I.,** Role of cell division in the cytodifferentiation of rat heart cells in culture, *Biochimie,* 56, 1581, 1974.
27. **Masse, M. J. and Harary, I.,** The use of 5-bromodeoxyuridine and irradiation for the estimation of myoblast and myocyte content of primary rat heart cell cultures, *J. Cell Physiol.,* 105, 197, 1980.
28. **Chacko, S.,** Ultrastructural observations on mitosis in myocardial cells of the rat embryo, *Am. J. Anat.,* 135, 305, 1972.
29. **Masse, M. J. and Harary, I.,** Primary cultures of embryonic rat heart cells, *In Vitro,* 17, 388, 1981.
30. **Polinger, I. S.,** Growth and DNA synthesis in embryonic chick heart cells in vivo and in vitro, *Exp. Cell Res.,* 76, 253, 1973.
31. **Goode, D.,** Mitosis of embryonic heart muscle cell in vitro. Immunofluorescence and ultrastructral study, *Cytobiologie,* 11, 203, 1975.
32. **Masse, M. J. and Harary I.,** The use of fluorescent antimyosin and DNA labeling for the estimation of myoblast and myocyte population of primary rat heart cell cultures, *J. Cell Physiol.,* 106, 165, 1981.
33. **Clark, Jr., W. A. and Fishman, D. A.,** Analyses of population cytokinetics of chick myocardial cells in tissue culture, *Develop. Biol.,* 97, 1, 1983.
34. **Jeter, J. R. and Cameron, I. L.,** Cell proliferation patterns during cytodifferentiation in embryonic chick tissues, liver, heart, and erythrocytes, *J. Embryol. Exp. Morphol.,* 25, 405, 1971.
35. **Simpson, P., McGrath, A., and Savion, S.,** Myocyte hypertrophy in neonatal rat heart cultures and regulation by serum and catecholamines, *Circ. Res.,* 51, 787, 1982.
36. **Desmond, W., Jr., and Harary, I.,** In vitro studies of beating heart cells in culture. XV. Myosin turnover and the effect of serum, *Arch. Biochem. Biophys.,* 151, 285, 1972.
37. **Casanello, D., Bieleski, R. L., Desmond, W., and Harary, I.,** In vitro studies of beating heart cells in culture. XVII. The stimulation of myosin synthesis by a heat stable serum fraction, *Arch. Biochem. Biophys.,* 156, 570, 1973.
38. **Yaffe, D.,** Retention of differentiation potentialities during prolonged cultivation of myogenic cells, *Proc. Nat. Acad. Sci. USA.,* 61, 477, 1968.
39. **Delain, D. Wahrmann, J. P., and Gros, F.,** Influence of external diffusable factors on myogenesis of the cells of the line L$_6$, *Exp. Cell Res.,* 131, 217, 1981.
40. **Devlin, B. H. and Konigsberg, I. R.,** Reentry into the cell cycle of differentiated skeletal myocytes, *Develop. Biol.,* 95, 175, 1983.
41. **Rumyantsev, P. P.,** Reproduction differenzierung and regeneration non kardiomyocytes (Russian), *Nauka Akademie der Wissenschaften der UdSSR* (Leningrad), 1982.
42. **Jacobson, S. L.,** Culture of spontaneously contracting myocardial cells from adult rats, *Cell Structure and Function,* 2, 1, 1977.
43. **Nag, A. C. and Cheng, M.,** Adult mammalian cardiac muscle cells in culture, *Tissue Cell,* 13, 515, 1981.
44. **Claycomb, W. C.,** The adult rat cardiac muscle cell in culture reinitiates DNA replication and acquires multiple nuclei, *J. Cell Biol.,* 91, 342a, 1981.
45. **Claycomb, W. C.,** Cardiac muscle cell proliferation and cell differentiation in vivo and in vitro, in *Advances in Experimental Medicine and Biology,* Vol. 161, Spitzer, J. J., Ed., Plenum Press, New York, 1983, 249.
46. **Hay, D. A. and Low, F. N.,** The fine structure of progressive stages of myocardial mitosis in chick embryos, *Am. J. Anat.,* 134, 175, 1972.
47. **Rumiantsev, P. P.,** Electron microscope study of the myofibril partial disintegration and recovery in the mitotically dividing cardiac muscle cells, *Z. Zellforsch. Mikroskop. Anat.,* 129, 141, 1972.
48. **Kelly, A. M. and Chacko, S.,** Myofibril organization and mitosis in cultured cardiac muscle cells, *Develop. Biol.,* 48, 421, 1976.
49. **Chacko, S.,** DNA synthesis, mitosis and differentiation in cardiac myogenesis, *Develop. Biol.,* 35, 1, 1973.
50. **Claycomb, W. C.,** Inhibition of DNA synthesis in differentiating cardiac muscle by NAD, *FEBS Letters,* 61, 231, 1976.
51. **Claycomb, W. C.,** Poly(adenosine diphosphate ribose) polymerase activity and nicotinamide adenine dinucleotide in differentiating cardiac muscle, *Biochem. J.,* 154, 387, 1976.

52. Seminar on Poly (ADP-ribose) and ADP-ribosylation of protein, *Proc. J. Biochem. Tokyo*, 77, 1P, 1975.
53. **Sugimura, T.**, fPoly(adenosine diphosphate ribose), *Prog. Nucleic Acid Res. Mol. Biol.*, 13, 127, 1973.
54. **Jackowski, G. and Kun, E.**, Age dependent variation of rates of polyadenosine-diphosphoribose synthesis by cardiomyocyte nuclei and the lack of correlation of enzymatic activity with macromolecular size distribution of DNA, *J. Biol. Chem.*, 256, 3667, 1981.
55. **Burzio, L. and Koide, S. S.**, A functional role of polyADPR in DNA synthesis, *Biochem. Biophys. Res. Commun.*, 40, 1013, 1970.
56. **Claycomb, W. C.**, Possible involvement of cyclic AMP, NAD, and poly (ADP ribose) polymerase in the control of DNA synthesis in differentiating cardiac muscle, *J. Cell Biol.*, 67, 71a, 1975.
57. **Jackowski, G. and Kun, E.**, The influence of triiodothyronine on polyadenosine diphosphoribose polymerase and RNA synthesis in cardiomyocyte nuclei, *J. Mol. Cell. Cardiol.* 14 (Suppl. 3), 65, 1982.
58. **Jackowski, G. and Kun, E.**, The effect of in vivo treatment with triiodothyronine on the in vitro synthesis of poly(ADP) ribose adducts by isolated cardiocyte nuclei and the separation of poly(ADP)-ribosylated proteins by phenol extraction and electrophoresis, *J. Biol. Chem.*, 258, 12,587, 1983.
59. **Novak, E., Drummond, G. I., Skala, J., and Hahn, P.**, Developmental changes in cyclic AMP, protein kinase, phosphorylase kinase, and phosphorylase in liver, heart and skeletal muscle of the rat, *Arch. Biochem. Biophys.*, 150, 511, 1972.
60. **Kim, G. and Silverstein, G.**, Developmental changes in guanyl cyclase and cyclic 3', 5'-guanosine monophosphate in rat lung and heart, *Fed. Proc.*, 34, 231, 1975.
61. **Brus, R.**, Adenyl cyclase activity in developing rat heart, *Polish J. Pharmacol. Pharm.*, 26, 337, 1974.
62. **Yaunt, E. A. and Clark, Jr., C. M.**, Rat heart adenyl cyclase activity during postnatal development: effects of glucagon and 5'-guanyl imidodiphosphate, *Fed. Proc.*, 35, 423, 1976.
63. **Claycomb, W. C.**, Biochemical aspects of cardiac muscle differentiation. DNA synthesis and nuclear and cytoplasmic DNA polymerase activity, *J. Biol. Chem.*, 250, 3229, 1975.
64. **Claycomb, W. C.**, Biochemical aspects of cardiac muscle differentiation. Possible control of DNA synthesis and cell differentiation by adrenergic innervation and cAMP., *J. Biol. Chem.*, 251, 6082, 1976.
65. **Schonhofe, P. S. and Peters, D. H.**, Role of cycle nucleotides in cultured cells, in *Cyclic 3'5' Nucleotide: Mechanism of Action*, Cramer, H. and Schultz, J., Eds., John Wiley & Sons, London, 1977, 107.
66. **Kram, R.**, Cyclic AMP and Cyclic GMP concentration in serum and density restricted fibroblast cultures, *Proc. Natl. Acad. Sci. USA*, 72, 1063, 1975.
67. **Rutland, P. S., Seeley, M., and Seifert, W. E.**, Cyclic GMP and cyclic AMP levels in normal and transformed fibroblasts, *Nature*, 251, 417, 1974.
68. **Jonne, J., Poso, H., and Raiva, A.**, Polyamines in rapid growth and cancer, *Biochem. Biophys. Acta*, 473, 241, 1978.
69. **Clô, C. Tantini, B., Coccolini, M. N., and Caldarera, C. M.**, Involvement of polyamines in cyclic AMP metabolism in heart cell cultures, *Adv. Polyam. Res.*, 3, 333, 1981.
70. **Clô, C., Tantini, B., Pignatti, C., and Caldarera, C. M.**, Increased cyclic GMP content in confluent and serum restricted heart cell cultures exposed to polyamines., *J. Mol. Cell. Cardiol.*, 15, 139, 1983.
71. **Dulbecco, R. and Elkington, T.**, Induction of growth in resting fibroblastic cell cultures by Ca^{++}, *Proc. Natl. Acad. Sci. USA*, 72, 1584, 1975.
72. **Harary, I., Renaud, J. F., Sato, E., and Wallace, G.**, Calcium ions regulate cyclic AMP and beating in cultured heart cells, *Nature*, 261, 60, 1976.
73. **Caldarera, C. M., Casti, A., Guarnieri, C., and Moruzzi, G.**, Regulation of ribonucleic acid synthesis by polyamines. Reversal of spermine by inhibition of methylglyoxal bis(guanylhydrazone) of ribonucleic acid synthesis and histone acetylation in the rabbit heart, *Biochem. J.*, 152, 91, 1975.
74. **Moruzzi, G., Barbiroli, B., Moruzzi, M. S., and Tadolini, B.**, The effect of spermine on transcription of mammalian chromatin by mammalian deoxyribonucleic acid-dependent ribonucleic acid polymerase, *Biochem. J.*, 146, 697, 1975.
75. **Ruiz-Carillo, A., Wangh, L. J., Littan, V. C., and Allfrey, V. G.**, Changes in histone acetyl content and in nuclear non-histone protein composition of avian erythroid cells at different stages of maturation, *J. Biol. Chem.*, 249, 7358, 1974.
76. **Heby, O., Marton, L. J., Wilson, C. B., and Martinez, H. M.**, Polyamines: a high correlation with cell replication, *FEBS Lett.*, 50, 1, 1975.
77. **Mirkin, B. L.**, Ontogenesis of the adrenergic nervous system: functional and pharmacologic implication, *Fed. Proc.*, 31, 65, 1972.
78. **Burnstock, C. and Cesta, M.**, *Adrenergic Neurons. Their Organization Function and Development in Peripheral Nervous System*, John Wiley & Sons, New York, 1975.
79. **Glawinski, J., Axelrod, J., Kopin, I., and Wurtman, R. J.**, Physiological disposition of H^3-norepinephrine in the developing rat, *J. Pharmacol. Exp. Ther.*, 146, 48, 1964.
80. **Iversen, L. L., De Champlain, U., Glawinski, J., and Axelrod, J.**, Uptake, storage, and metabolism of norepinephrine in tissues of the developing rat, *J. Pharmacol. Exp. Ther.*, 157, 509, 1967.

81. **Weber, K. and Groeschel-Stewart, U.,** Antibody to myosin; the specific visualization of myosin-containing filaments in nonmuscle cells, *Proc. Natl. Acad. Sci., USA,* 71, 4561, 1974.

82. **Harary, I., Hoover, F., and Farley, B.,** The isolation and cultivation of rat heart cells, in *Biomembranes, Part B, Methods in Enzymology,* Vol. 32, Fleisher, S. and Packer, L., Eds., Academic Press, New York, 1974, 740.

83. **Manasek, F. J.,** Embryonic development of the heart. I. A. light and electron microscopic study of myocardial development in the early chick embryo., *J. Morphol.,* 125, 329, 1968.

84. **Chacko, S. and Joseph, X,** The effect of 5-bromodeoxyuridine (BrdU) on cardiac muscle differentiation, *Develop. Biol.,* 40, 340, 1974.

85. **Perry, S. V.,** Biochemical adaptation during development and growth in skeletal muscle, in *Physiology and Biochemistry of Muscle as a Food, 2. Proceedings of an International Symposium 1969,* Briskey, E. J., Cassens, R. J., and March, E. B., Eds., University of Wisconsin Press, Madison, Wis., 1970, 537.

86. **Sreter, F. A., Balin, M., and Gergely, J.,** Structural and functional changes of myosin during development. Comparison with adult fast, slow, and cardiac muscle, *Develop. Biol.,* 46, 317, 1975.

87. **Radha, E.,** Embryonic differentiation in the nucleoside triphosphatase activities of myosin from the fast, slow, and cardiac muscles of chick, *Enzyme,* 18, 327, 1974.

88. **Masaki, T.** Differentiation of muscle proteins, *Advan. Neurol. Sci.,* 19, 801, 1975.

89. **Flink, I. L., Rader, J. H., Banerjee, S. K., and Morkin, E.,** Atrial and ventricular cardiac myosins contain different heavy chain species, *FEBS Lett.,* 94, 125, 1978.

90. **Wickman-Coffelt J. and Srivastava, S.,** Differences in atrial and ventricular myosin light chain LC_1, *FEBS Lett.,* 106, 207, 1979.

91. **Whalen, R. G., Sell, S. M., Eriksson, A., and Thornell, L. E.,** Myosin subunit types in skeletal and cardiac tissues and their developmental distribution, *Develop. Biol.,* 91, 478, 1982.

92. **Dalla Libera, L., Sartore, S., and Schiaffino, S.,** Comparative analysis of chicken atrial and ventricular myosins, *Biochem. Biophys. Acta,* 581, 283, 1979.

93. **Sartore, S., Gorza, L., Pierobon, Bormioli, S., Dalla Libera, L., and Schiaffino, S.,** Myosin types and fiber types in cardiac muscle 1. Ventricular myocardium, *J. Cell Biol.,* 88, 226, 1981.

94. **Chizzonite, R. A., Everett, A. W., Clark, W. A., Jakovcic, S., Rabinowitz, M., and Zak, R.,** Isolation and characterization of two molecular variants of myosin heavy chain from rabbit ventricle, *J. Biol. Chem.,* 257, 2056, 1982.

95. **Hoh, J. Y., McGrath, P. A., and Hale, P. T.,** Electrophoretic analysis of multiple forms of rat cardiac myosin: effects of hypophysectomy and thyronine replacement, *J. Mol. Cell. Cardiol.,* 10, 1053, 1977.

96. **Lompre, A. M., Mercadier, J. J., Wisnewsky, C., Bouveret, P., Pantaloni, C., D'Abis, A., and Schwartz, K.,** Species and age dependent changes in the relative amounts of cardiac myosin isoenzyzmes in mammals, *Develop. Biol.,* 84, 286, 1981.

97. **Samuel, J-L., Rappaport, L., Mercadier, J-J., Lompre, A-M., Sartore, S., Triban, C., Schiaffino, S., and Schwartz, K.,** Distribution of muscle isozymes within single cardiac cells — an immunohisto-chemical study, *Circ. Res.,* J2, 200, 1983.

98. **Zak, R., Chizzonite, R. A., Everett, A. W., and Clark, W. A.,** Study of ventricular isomyosins during normal and thyroid hormone induced cardiac growth, *J. Mol. Cell. Cardiol.,* 14(Suppl. 3), 111, 1982.

99. **Luten, R. Z., Martin, B. J., and Alpert, N. R.,** Altered myosin peptides from thyrotoxic and pressure overloaded hypertrophied rabbit hearts, *Fed. Proc.,* 40, 618A, 1981.

100. **Everett, A. W., Clark, W. A., Chizzonite, R. A., and Zak, R.,** Change in synthesis rates of α and β myosin heavy chains in rabbit heart after treatment with thyroid hormone, *J. Biol. Chem.,* 258, 2421, 1983.

101. **Mahdovi, V., Periasany, M., and Nadel-Grinard, B.,** Molecular characterizations of two myosin heavy chain genes expressed in the adult heart, *Nature,* 297, 659, 1982.

102. **Buckingham, M. E., Caput, D., Cohen, A., Whalen, R. G., and Gros, F.,** The synthesis and stability of cytoplasmic messenger RNA during myoblast differentiation in culture, *Proc. Natl. Acad. Sci. USA,* 71, 1466, 1974.

103. **Coleman, J. R. and Coleman, A. W.,** Muscle differentiation and macromolecular synthesis, *J. Cell. Physiol.,* 72(Suppl. 1), 19, 1969.

104. **Wikman-Coffelt, J., Zelis, R., Fenner, C., and Mason, D. T.,** Studies on the synthesis and degradation of light and heavy chains of cardiac myosin, *J. Biol. Chem.,* 248, 5206, 1973.

105. **Ouellete, A. J., Croall, D. E., Van Ness, J., and Ingwall, J. S.,** Developmental regulation of mRNA in mouse heart, *Proc. Natl. Acad. Sci. USA,* 80, 223, 1983.

106. **Scott, J. C., Finkelstein, L. J., and Spitzer, J. J.,** Myocardial removal of free fatty acids under normal and pathological conditions, *Am. J. Physiol.,* 203, 482, 1962.

107. **Wildenthal, K.,** Studies of foetal mouse hearts in organ culture: metabolic requirements for prolonged function in vitro and the influence of cardiac maturation on substrate utilization, *J. Mol. Cell. Cardiol.,* 5, 87, 1973.

108. **Clark, C. M., Jr.,** Carbohydrate metabolism in the isolated fetal rat heart, *Am. J. Physiol.,* 220, 583, 1971.

109. **Cox, S. S. and Greenberg, D. L.,** Metabolite utilization by isolated embryonic rat hearts in vitro, *J. Embryol. Exp. Morphol.,* 28, 235, 1972.
110. **DeLuca, M. A., Ingwall, J. S., and Bittle, J. A.,** Biochemical response of myocardial cells in culture to oxygen and glucose deprivation, *Biochem. Biophys. Res. Commun.,* 59, 749, 1974.
111. **Tanimura, T. and Shepard, T. H.,** Glucose metabolism by rat embryos in vitro, *Proc. Soc. Exp. Biol. Med.,* 135, 51, 1970.
112. **Warshaw, J. B. and Rosenthal, M. D.,** Change in glucose oxidation during growth of embryonic heart cells in culture, *J. Cell Biol.,* 52, 283, 1972.
113. **Radha, E. and Kirshnamoorthy, R. V.,** Metabolic differentiation of fast, slow, and cardiac muscles of chick during development, *Comp. Biochem. Physiol. Part A,* 50, 423, 1975.
114. **Schroedel, N. A., Hartzell, C. R., Ross, P. D., and McCarl, R. L.,** Glucose metabolism, insulin effects, and developmental age of cultured neonatal rat heart cells, *J. Cell. Physiol.,* 113, 231, 1982.
115. **Seraydarian, M., Sato, E., Savageau, M., and Harary, I.,** In vitro studies of beating heart cells in culture. XII. The utilization of ATP and phosphocreatine in oligomycin and 2-deoxyglucose inhibited cells, *Biochim. Biophys. Acta,* 180, 264, 1969.
116. **Fujimoto, A. and Harary, I.,** Studies in vitro on single beating rat-heart cells. IV. The shift from fat to carbohydrate metabolism in culture, *Biochim. Biophys. Acta,* 86, 74, 1964.
117. **Doorey, A. J. and Barry, W. H.,** The effects of inhibition of oxidative phosphorylation and glycolysis on contractility and high energy phosphate content in cultured chick heart cells, *Circ. Res.,* 53, 192, 1983.
118. **Neeley, J. R., Whitmer, J., and Rovetto, M. J.,** Effect of coronary blood flow on glycolytic flux and intracellular pH in isolated rat hearts, *Circ. Res.,* 37, 733, 1975.
119. **Hill, M. L. and Mayer, S. E.,** Total ischemia in dog hearts in vitro. 1. Comparison of high energy phosphate production, utilization, and depletion in vitro vs. severe ischemia in vivo, *Circ. Res.,* 49, 892, 1981.
120. **Frelin, C.,** Amino acid metabolism by newborn rat heart cells in monolayer cultures, *J. Mol. Cell. Cardiol.,* 12, 479, 1980.
121. **Pious, D. A.,** Induction of cytochromes in human cells by oxygen, *Proc. Natl. Acad. Sci. USA,* 65, 1001, 1970.
122. **Hallman, M.,** Changes in mitochondrial respiratory chain proteins during perinatal development. Evidence of the importance of environmental oxygen tension, *Biochim. Biophys. Acta,* 253, 360, 1971.
123. **Gross, W. O.,** Fibroblast-myocyte interactions in in vitro cardiomyogenesis, *Exp. Cell Res.,* 142, 341, 1982.
124. **Schroedl, N. A. and Hartzell, C. R.,** Myocytes and fibroblasts exhibit functional synergism in mixed cultures of neonatal rat heart cells, *J. Cell. Physiol.,* 117, 326, 1983.
125. **Simpson, P. and Savion, S.,** Differentiation of rat myocytes in single cell cultures with and without proliferation of nonmyocardial cells: Cross striations, ultrastructure and chronotropic response to isoproterenol, *Circ. Res.,* 50, 101, 1982.
126. **Ross, P. D. and McCarl, R. L.,** Oxidation of carbohydrates and palmitate by intact and cultured neonatal rat heart cells, *Am. J. Physiol.,* 246, H389, 1984.
127. **Everse, J. and Kaplan, N. O.,** Lactate dehydrogenases: structure and function, *Adv. Enzymol.* 37, 61, 1973.
128. **Fine, I. H., Kaplan, N. O., and Kuftinec, D.,** Developmental changes of mammalian lactic dehydrogenases, *Biochemistry,* 2, 116, 1963.
129. **Cahn, R. D., Kaplan, N. O., Levine, L., and Zwilling, L. E.,** Nature and development of lactic dehydrogenases, *Science,* 136, 962, 1962.
130. **Cahn, R. D.,** Developmental changes in embryonic enzyme patterns: the effect of oxidative substrates on lactic dehydrogenase in beating chick embryonic heart cell cultures, *Develop. Biol.,* 9, 327, 1964.
131. **Lindy, S., and Rajasalmi, M.,** Lactate dehydrogenase isozymes of chick embryo: response to variations of ambient oxygen tension, *Science,* 153, 1401, 1966.
132. **Hellung-Larsen, P. and Andersen, V.,** Lactate dehydrogenase isoenzymes of human lymphocytes cultured with phytohaemagglutinin at different oxygen tensions, *Exp. Cell Res.,* 50, 286, 1968.
133. **Shroeder, J., Upson, R., and Chvapil, M.,** LDH isoenzyme pattern, lactate/pyruvate, and ultrastructure of fibrinogenic and epithelial cell lines exposed to 3% oxygen, *Exp. Cell Res.,* 94, 227, 1975.
134. **Van Der Laarse, A., Hollaar, L., Kokshoorn, L. J. M., and Witteveen, S. A. G. J.,** The activity of cardio-specific isoenzymes of creatine phosphokinase and lactic dehydrogenase in monolayers of cultures of neonatal rat heart cells, *J. Mol. Cell. Cardiol.,* 11, 501, 1979.
135. **Hirakow, R. and Gotob, T.,** A quantitative ultrastructural study on the developing heart, in *Developmental and Physiological Correlates of Cardiac Muscle,* Vol. 1, Lieberman, M. and Sano, T., Eds., Raven Press, New York, 1975, 37.
136. **Mackler, B., Grace, R., and Duncan, H. M.,** Studies of mitochondrial development during embryogenesis in the rat, *Arch. Biochem. Biophys.,* 144, 603, 1971.

137. **Sordahl., L. A., Crow, C. A., Kraft, G. H., and Schwartz, A.,** Some ultrastructural and biochemical aspects of heart mitochondria associated with development: fetal and cardiomyopathic tissue, *J. Mol. Cell. Cardiol.,* 4, 1, 1972.

138. **Skala, J., Drahota, Z., and Mahn, P.,** Succinate dehydrogenase activity of rat heart muscle mitochondria during development, *Physiol. Bohemoslov.,* 19, 15, 1970.

139. **Abo Zeit-Har, S. A. and Drahota, Z.,** The development of mitochondrial oxidative enzymes in rat heart muscle, *Physiol. Bohemoslov.,* 24, 289, 1975.

140. **Warshaw, J. B.,** Cellular energy metabolism during fetal development. I. Oxidative phosphorylation in the fetal heart, *J. Cell Biol.,* 41, 651, 1969.

141. **Wittels, B. and Bressler, R.,** Lipid metabolism in the newborn heart, *J. Clin. Invest.,* 44, 1639, 1965.

142. **Warshaw, J. B.,** Cellular energy metabolism during fetal development. III. Deficient acetyl-CoA synthetase, acetylcarnitine transferase and oxidation of acetate in the fetal bovine heart, *Biochim. Biophys. Acta,* 223, 409, 1970.

143. **Warshaw, J. B.,** Cellular energy metabolism during fetal development. IV. Fatty acid activation, acetyl transfer, and fatty acid oxidation during development of the chick and rat, *Develop. Biol.,* 28, 537, 1972.

Chapter 5

GENESIS OF MYOCARDIAL CELLS FROM MOUSE TERATOCARCINOMA CELL LINES

Kazuko Uno, Tomoichi Ishikawa, and Shigetoyo Amano

TABLE OF CONTENTS

I. INTRODUCTION

A teratocarcinoma is a tumor composed of a variety of differentiated cell types derived from three germ layers and from stem cells which are known as embryonal carcinoma. Teratocarcinomas occur spontaneously or can be experimentally induced in mice,[1-3] in the latter case by transplanting either genital ridge or embryos at an early stage of development into extrauterine sites by the technique of Stevens.[4-6] Thanks to the pioneering work of Stevens, many in vivo teratocarcinoma lines have been established. These are retransplantable as subcutaneous solid tumors. When a solid teratocarcinoma is dissociated and injected into the peritoneal cavity of an adult mouse, it is converted into ascites and forms cell aggregates termed "embryoid bodies";[7-8] these are also retransplantable.

Recently, pluripotent stem cell lines directly derived from early mouse embryos were established by culturing them in vitro.[9] They differentiated into a variety of tissues in the same manner as teratocarcinomas in vitro and in vivo. This should greatly expedite the realization of genetically marked pluripotent stem cell lines in which it has previously proven difficult to induce teratocarcinomas.

Since these stem cells adopt the embryonal nature of their origin, they are capable of differentiating into a variety of cell types. Their pluripotency has been proven — a single embryonal carcinoma cell can give rise to a teratocarcinoma with multiple cell types.[10,11] Surprisingly, malignant teratocarcinoma stem cells could form virtually all the tissues of a normal tumor-free animal, even functional germ cells, when placed in the appropriate environment.[12,13] The idea that embryonal carcinoma cells are closely analogous to cells in the normal embryo is based on these observations.

Established teratocarcinoma stem cell lines thus provide an alternative to pluripotent mammalian embryonic cells in cases where it is difficult to obtain adequate quantities of normal embyronic material. With the aid of teratocarcinomas, it may even be possible to study embryonic development in mammals without using embryos. However, to our knowledge, there are very few reliable experimental systems in which the pathways of cell differentiation from a very primitive and multipotent stage can be manipulated. The purpose of this chapter is to describe some of the teratocarcinoma stem cell lines that have been isolated and the ways in which they are being used to study early mammalian development, especially cardiomyogenesis.

II. ESTABLISHMENT OF PLURIPOTENT CELL LINES

A number of clonal cell lines of pluripotent embryonal carcinomas have been isolated from mouse teratocarcinomas.[14-19] Most of them are feeder-dependent, i.e., maintenance of their embryonic nature and pluripotency is hard to achieve in the absence of non dividing feeder cells and frequent subculture. In vitro embryoid body lines adapted to the conditions of culture have also been established.[20,21] Although these cells have a similar structure to the in vivo ones, they grow well floating in the culture medium without a feeder layer (Figure 1). Most of the cell lines of this type isolated so far are malignant and form tumors when injected subcutaneously. If a line is pluripotent, a variety of somatic tissues may be found in the tumor.

In general, the embryonal carcinoma cells of all these lines share common characteristics; cells in culture are epitheloid in appearance and relatively small, with a diameter of about 13 μm. Their cytoplasm is relatively scant compared to the nucleus which usually contains one or two prominent nucleoli. In the cytoplasm, organelles are rare except for a large number of free ribosomes and a few mitochondria. Although most mouse tumor cell lines generally have an abnormal chromosome number, embryonal carcinoma cell lines usually retain a diploid or near diploid chromosome number, even after long periods of culture.[22]

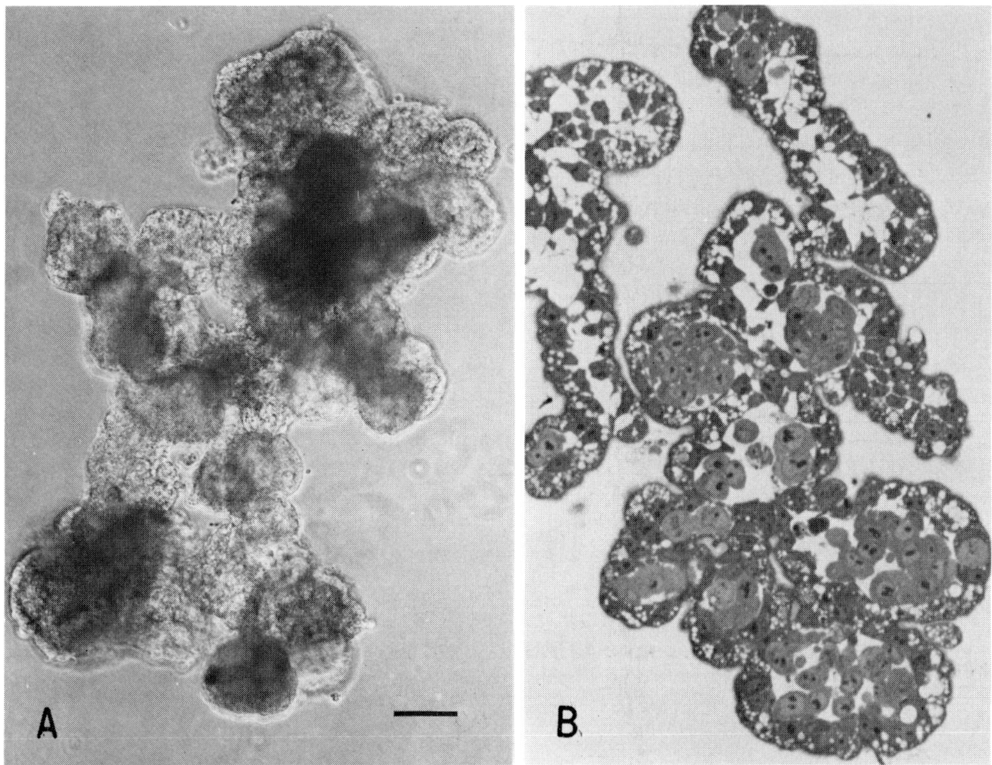

FIGURE 1.   Morphology of an in vitro embryoid body line. (A) A phase contrast micrograph of a large in vitro embryoid body. The bar indicates 100 μm. (B) A micrograph of a 1 μm section. Note that each large in vitro embryoid body consists of a number of small embryoid bodies which are budding but still connected with one another. Cores of embryonal carcinoma cells and larger surrounding endodermal cells with many vacuoles are shown. (From Amano, S., Uno, K., and Hagiwora, A., Cardiac muscle cell differentiation in vitro from a characteristic cell line isolated from mouse teratocarcinoma, *Dev. Growth Differ.*, 20, 43, 1978. With permission.)

Besides pluripotency in differentiation, embryonal carcinoma cells have a number of properties in common with early embryo cells. High levels of alkaline phosphatase activity were detected in both types of cells.[23,24] Electron microscopy also showed that embryonal carcinoma cells differentiate in a manner closely resembling the early mouse embryo.[25] The embryonal carcinoma cells have several antigens in common with early mouse embryo cells, but they either become progressively restricted to certain tissues or they disappear during the course of development.[26-28] As exemplified in these studies, an embryonal carcinoma cell line may provide in vitro material for studying the mechanism of cell differentiation.

III. FORMATION OF EMBRYOID BODIES IN VITRO

Pluripotent embryonal carcinoma cells may be useful for studying the determination of cell differentiation in vitro as they are able to differentiate into any somatic cell type in culture. Despite considerable efforts, early attempts to obtain differentiation into multiple cell types in vitro met with only limited success. Over the past few years, however, several laboratories have succeeded in differentiating multiple types of cells in vitro from embryonal carcinoma lines.[29,30] In these studies it has been stressed that the formation of embryoid body structure is an obligatory intermediate phase for the occurrence of determination and differentiation in vitro.

Martin and Evans[30-33] reported embryoid body formation in a culture from embryonal carcinoma cells plated in a bacterial grade dish. When pluripotent clonal embryonal carcinoma cells are plated in the absence of feeder cells, they initially form small colonies firmly attached to the substrate. As the cells multiply, these homogeneous colonies become less firmly attached to the substrate and separate to form clumps of cells. The outer layer of these clumps differentiates to form endoderm. This is the first differentiative event that takes place in embryonal carcinoma cells. Some embryonal carcinoma cell lines form embryoid bodies which remain simple when kept in suspension. They can differentiate into a variety of cell types only when the embryoid bodies are allowed to attach to the substratum. On the other hand, certain embryonal carcinoma cell lines (e.g., PSA-1) will differentiate in an embryo-like manner when cultured in suspension. They continue to differentiate into complex "cystic" structures.[30-34] A cystic embryoid body develops one or more cavities. During the development of the cavity, we observed a great deal of cellular debris, presumably as a result of cell death. This process is analogous to the formation of the proamniotic cavity and embryonic ectoderm in the 6-day-old embryo. Cavity formation is followed by mesoderm formation. A mesodermal layer forms between the outer endoderm and inner ectoderm layer. Further differentiation was not observed in suspension culture. If allowed to reattach, the culture becomes complex and a variety of differentiated cell types appear. Cystic embryoid bodies show a wider spectrum of differentiation than simple embryoid bodies, their development paralleling the early postimplantation phase of embryonic development.

Clonal "nullipotent" embryonal carcinomas have also been isolated.[33] They are morphologically similar to the pluripotent ones, but are unable to differentiate either in vivo or in vitro. These nullipotent lines, cultured in the same way, formed aggregates but no endodermal layer appeared. From these results it was concluded that the first event of differentiation in pluripotent embryonal carcinoma cells in vitro is endodermal cell layer formation, and the formation of embryoid bodies is an obligatory intermediate phase for the occurrence of determination and differentiation in vitro. This event is closely comparable to the development of the inner cell mass in early mouse embryos.[33,34]

IV. CARDIOMYOGENESIS IN VITRO

As a rule, a pluripotent embryonal carcinoma line gives rise to a variety of cell types when it differentiates in vivo or in vitro. It is, however, only under inductive conditions that embryonal carcinoma cells predominantly differentiate into a specific cell type in culture. Cardiomyogenesis in vitro will be discussed in the section that follows.

We isolated pluripotent in vitro embryoid body lines (SEBIII and derivatives) which grow suspended in the culture medium, are not differentiated, and are composed of only endodermal cells.[20,21] When plated in a collagen coated plastic dish they adhere to the substrate but no new cell types appear in the culture. We found, however, that an intraperitoneal "sojourn" induces cardiomyogenesis (Table 1).[20,35] These in vitro embryoid bodies were injected intraperitoneally into a strain 129 mouse, collected from the peritoneal cavity after a week, and plated in collagen coated plastic dishes, most of them adhering to the substrate. Several days later, a number of rhythmically contracting colonies were observed in the culture (Figure 2). Many cell types appeared in the same dish, e.g., neural tissues, keratinized epithelium, and blood islands.

Although the external appearance of these contracting colonies varied, ultrastructural studies invariably demonstrated cardiac muscle cells in them, thus proving in vitro cardiomyogenesis. In these myocardial cells, myofilaments ran parallel and Z-bands were clearly observable (Figure 3). The muscle cells did not fuse with each other but a well-developed intercalated disc was clearly demonstrated at the area of contact between two muscle cells.

Bustan and Herz[36] performed ultrastructural studies on the transition phases from em-

Table 1
CORRELATION BETWEEN
CARDIOMYOGENESIS AND CYSTIC
EMBRYOID BODY FORMATION
FOLLOWING A SEVEN-DAY
INTRAPERITONEAL "SOJOURN"

In vitro embryoid body lines	Frequency of cardiac muscle differentiation[a]	Number of cystic embryoid bodies[b]
SKEB	0.62 ±0.46	34.7
SSEB	0.61 ±0.19	ND
SEB III	0.27 ±0.13	ND
ASEB	0.003 ±0.004	3.4
ODEB	0	0.4

Note: ND; Not Determined. Values are expresssed in percentages.

[a] The frequency of cardiac muscle differentiation was defined as the number of contracting colonies observed during 14 days of in vitro culture divided by the number of embryoid bodies in the culture dish.

[b] The number of cystic embryoid bodies was determined using three photographs of randomly selected 1 μm sections.

bryonal carcinoma cells to myocardial cells. The first sign of differentiation is that primitive myofibers dispersed into a rather random orientation within cells which are otherwise indistinguishable from embryonal carcinoma cells. These cells show ultrastructural similarities to splanchnic mesodermal cells, the precursors of myocardial cells.[37,38]

Figure 4 shows the time course of the appearance of spontaneously contracting colonies in culture after a previous intraperitoneal "sojourn" of the in vitro embryoid bodies for 3, 7, or 14 days.[35] Two conclusions may be drawn from these results. First, irrespective of the duration of an intraperitoneal "sojourn", the embryoid bodies never initiate spontaneous contracting as long as they remain in the peritoneal cavity. Second, most contracting colonies do not appear immediately after transfer to an in vitro culture, but appear several days later. They are greatest in number after about 1 week in culture irrespective of the duration of the intraperitoneal "sojourn" (3, 7, or 14 days). Therefore, it is clear that the embryoid bodies differentiate into precardiac muscle in the peritoneal cavity, but are unable to begin spontaneous contraction there. This only starts after they become attached to a substrate and are in culture for several days.

McBurney et al.[38,40] reported differentiation induced by drugs such as dimethyl sulfoxide (DMSO) and retinoic acid, in a P19 pluripotent embryonal carcinoma line. When a suspension of dispersed embryonal carcinoma cells was plated onto bacterial-grade dishes, the cells did not become attached to the substrate, but adhered to one another to form aggregates. These aggregates were cultured in suspension for 4 to 5 days in the presence of DMSO or low concentrations of retinoic acid ($10^{-9}\mu M$ and then plated onto tissue culture-grade plastic dishes where they adhered to the substrate. After a few days, most plated aggregates mainly consisted of rhythmically contracting cardiac muscle cells. These cultures contained actin and myosin which are characteristically only present in skeletal and cardiac muscle cells. The aggregates treated with a high concentration ($10^{-7}\mu M$) of retinoic acid differentiated in the same way into many types of neural tissues.[40] When monolayers, rather than aggregates,

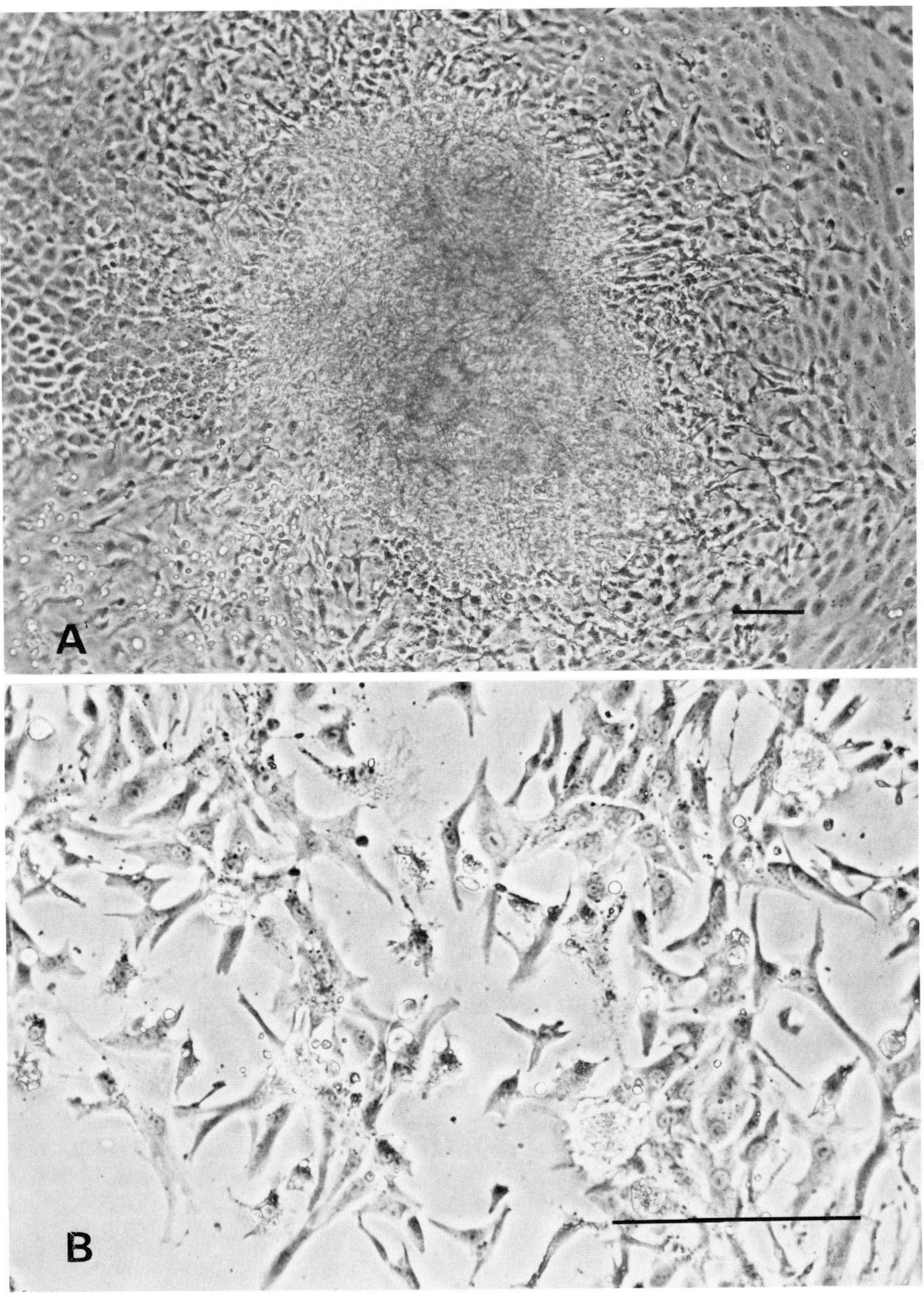

FIGURE 2.   (A and B) Phase contrast micrographs of rhythmically contracting colonies in cultures following an intraperitoneal "sojourn". The bar indicates 100 μm.

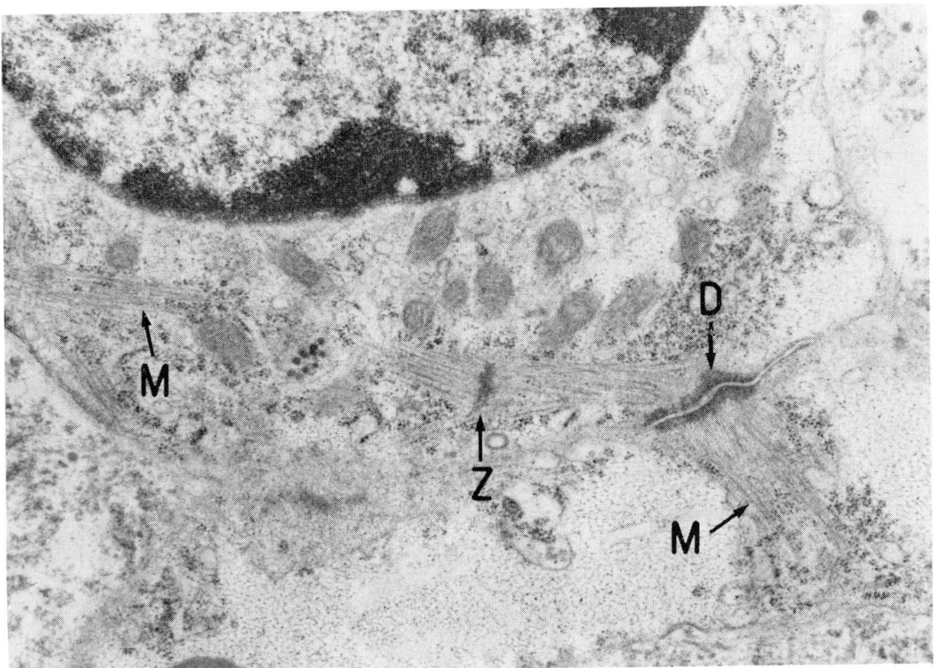

FIGURE 3. An electron micrograph of myocardial cells in a rhythmically contracting colony in culture. Myofilaments (M) with Z bands (Z) are shown in the cytoplasm. An intercalated disc (D) is seen in the area in which the two cells make contact. (From Amano, S., Uno, K., and Hagiwora, A., Cardiac muscle cell differentiation in vitro from a characteristic cell line isolated from mouse teratocarcinoma, *Dev. Growth Differ.*, 20, 46, 1978. With permission.)

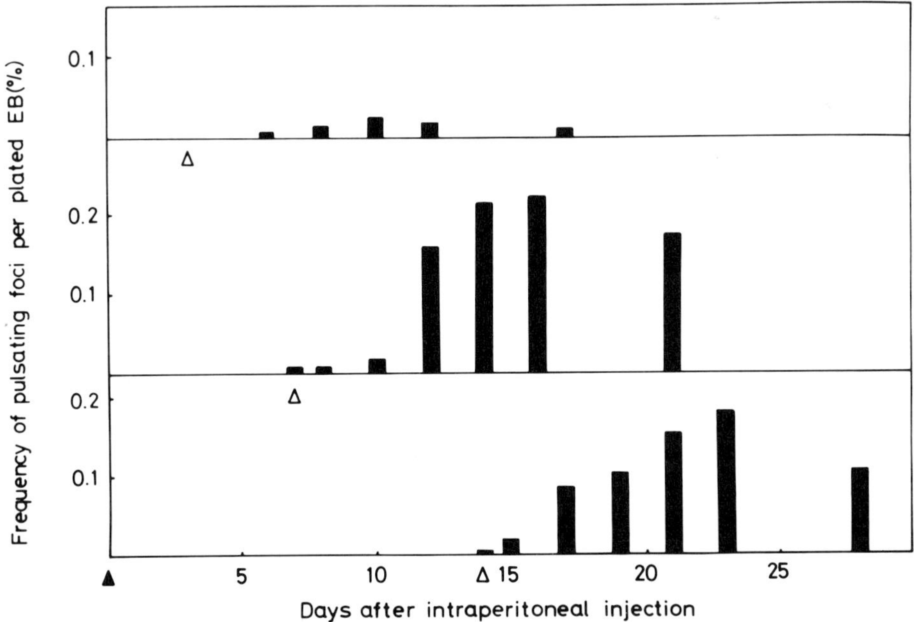

FIGURE 4. Time course of appearance of rhythmically contracting colonies in cultures of in vitro embryoid bodies following intraperitoneal "sojourns" of 3, 7, and 14 days. ▲; day on which in vitro embryoid bodies were injected to the peritoneal cavity. △; day on which the embryoid bodies were collected following intraperitoneal "sojourns" and plated on to collagen coated culture dishes. (From Uno, K. and Amano S., Induction of myocardiogenesis from mouse teratocarcinoma, *Dev. Growth Differ.*, 20, 270, 1978. With permission.)

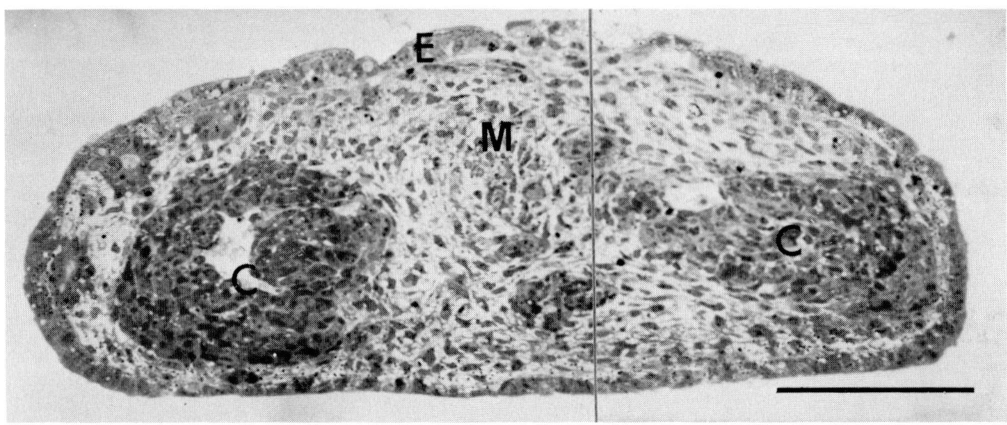

FIGURE 5.   Micrograph of a histological section of a rhythmically contracting colony. Two islets of myocardial cells (C) and the mesenchymal cells (M) surrounding them may be observed. Larger endodermal cells (E) envelop this colony. The bar indicates 100 μm. (From Uno, K. and Amano, S., Induction of myocardiogenesis from mouse teratocarcinoma, *Dev. Growth Differ.*, 20, 272, 1978. With permission.)

of embryonal carcinoma cells were treated with drugs, no new cell type appeared in the culture. From these results it was concluded that not only drug treatment but also aggregate formation is necessary for differentiation to occur. Although these workers did not mention the fact, the endodermal cells which appeared in the aggregates and embryoid body structure had probably been present in the suspension culture for 4 to 5 days, since as soon as the aggregates attached to the substrate endodermal cells appeared in the culture.

McBurney and his colleagues[40] also demonstrated that adrenalin enhances contraction of the myocardial cells differentiated from a pluripotent embryonal carcinoma line induced with DMSO. When 10 μM adrenalin was added to the contracting colonies, the contraction frequently increased by 2 to 2.5-fold, confirming that β-adrenergic receptors are now present in these cardiac cells.

In the contracting colonies, mesenchymal cells were also frequently observed (Figure 5). The extent of myocardial differentiation differed in the various in vitro embryoid body lines, although they were all treated in a similar manner (Table 1). Those which preferentially formed myocardial cells often derived from cells which had been in the form of cystic embryoid bodies during the intraperitoneal "sojourn" (Figure 6).[21] Cystic embryoid bodies were also observed in cultures of a clonal embryonal carcinoma line.[30] Mesoderm formation was frequently observed in these cystic embryoid bodies (Figure 6B), so that apparently cardiomyogenesis is closely related to mesoderm formation per se in mouse teratocarcinomas.

V. CONCLUSIONS AND PERSPECTIVES

Differentiation in pluripotent embryonal carcinoma cells in vitro consists of three main steps: (1) embryoid body formation, (2) induction of differentiation in these embryoid bodies, and (3) attachment to the substratum. Although differentiation into many cell types generally occurs in these cultures, the proportion of particular cell types varies largely depends upon processes occurring in the second step. An intraperitoneal "sojourn", addition of DMSO, or treatment with a low concentration of retinoic acid greatly increased the frequency of myocardial differentiation, and addition of a high concentration of retinoic acid increased the proportion of neural tissues relative to other cell types.[20,35,39,40]

Embryoid body formation is apparently a crucial step in differentiation, since monolayer cultures of embryonal carcinoma cells cannot be induced to undergo cardiomyogenesis when

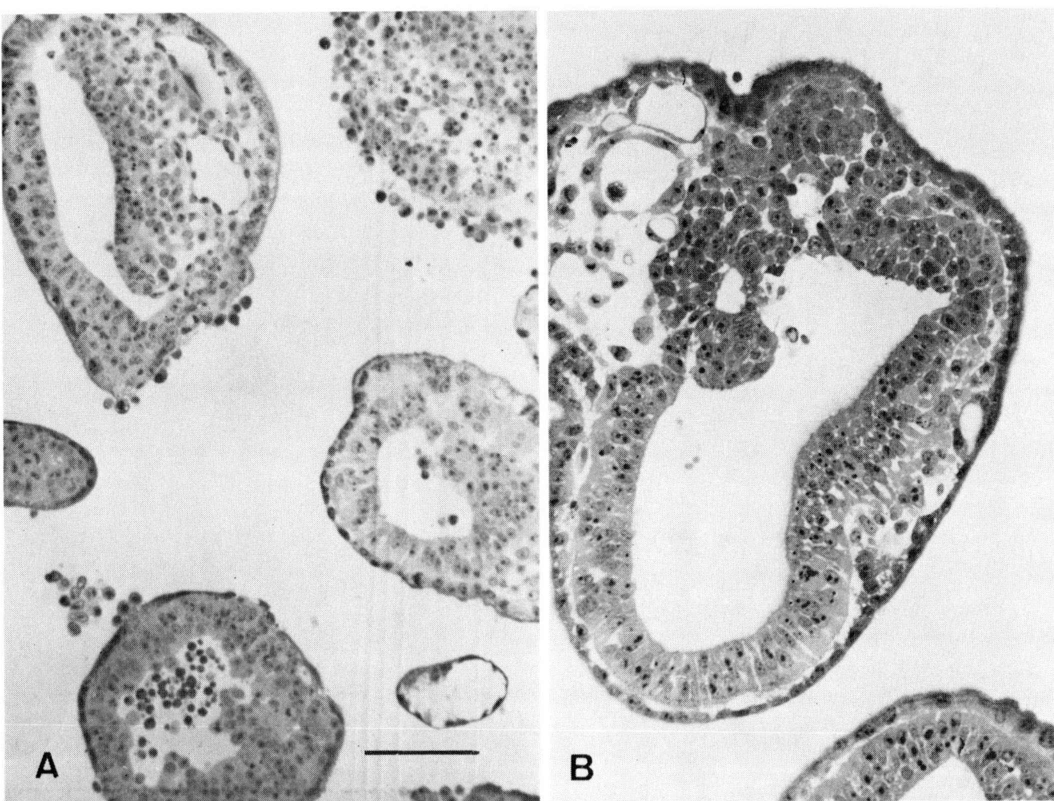

FIGURE 6. (A and B) Micrographs of a 1 μm section of a cystic embryoid body formed during an intraperitoneal "sojourn". (B) Mesodermal cells are intercalated between ectodermal (outer layer) and endodermal cells (inner layer) from the primitive streak-like area. The bar indicates 100 μm.

treated in a similar manner.[39] The embryonal carcinoma cells within an embryoid body are firmly enveloped by a layer of endodermal cells in contrast to those in monolayer cultures. Following embryoid body formation, cavity formation may be observed, and it has been pointed out that this structure closely resembles the 5- to 6-day-old mouse embryo. Furthermore, within the cystic embryoid bodies, mesoderm formation was often observed to originate at the primitive streak-like area between the endodermal layer and the ectodermal layer.[33,34] The mechanism of mesoderm formation in this case is very similar to embryogenesis. It has been demonstrated repeatedly, that attachment to a substratum greatly enhanced differentiation of embryonal carcinoma cells; however, its exact role is not yet clearly understood.

In contrast to early embryo cells, the embryonal carcinoma cell lines from mouse teratocarcinoma are much more easily manipulated in culture, forming homogeneous cells in large quantities. Several attempts have also been made to introduce specific genes into embryonal carcinoma.[41] The expression of the transferred marker genes elucidates the role of these genes in the differentiation of embryogenesis. The mechanism of cardiomyogenesis in teratocarcinomas is similar to that occurring in normal embryogenesis. Thus, teratocarcinomas can be used for studying the process of myocardial determination and differentiation in vitro.

REFERENCES

1. **Stevens, L. C. and Little, C. C.,** Spontaneous testicular tumors in an inbred strain of mice, *Proc. Natl. Acad. Sci. USA,* 40, 1080, 1954.
2. **Stevens, L. C.,** Studies on transplantable testicular teratomas of strain 129 mice, *J. Nat. Cancer Inst.,* 20, 1257, 1958.
3. **Stevens, L. C.,** Embryology of testicular teratomas in strain 129 mice, *J. Nat. Cancer Inst.,* 23, 1249, 1959.
4. **Stevens, L. C.,** The biology of teratomas, *Adv. Morph.,* 6, 1, 1967.
5. **Stevens, L. C.,** The development of teratomas from intratesticular grafts of tubal mouse eggs, *J. Embryo. Exp. Morph.,* 20, 329, 1968.
6. **Stevens, L. C.,** The development of transplantable teratocarcinomas from intratesticular grafts of pre- and postimplantation mouse embryos, *Develop. Biol.,* 21, 364, 1970.
7. **Pierce, G. B. and Dixon, F. J.,** Testicular teratomas. II. Teratocarcinoma as an ascitic tumor, *Cancer,* 12, 584, 1959.
8. **Pierce, G. B., Dixon, F. J., and Verney, E. L.,** Teratocarcinogenic and tissue-forming potentialities of the cell types comprising neoplastic embryoid bodies, *Lab. Invest.,* 9, 583, 1960.
9. **Martin, G. R.,** Isolation of a pluripotent cell line from early mouse embryos cultured in medium conditioned by teratocarcinoma stem cells, *Proc. Natl. Acad. Sci. USA,* 78, 7634, 1981.
10. **Kleinsmith, L. J. and Pierce, G. B., Jr.,** Multipotentiality of single embryonal carcinoma cells, *Cancer Res.,* 24, 1544, 1964.
11. **Stevens, L. C.,** Embryonic potency of embryoid bodies derived from a transplantable testicular teratoma of the mouse, *Develop. Biol.,* 2, 285, 1960.
12. **Mintz, B. and Illmensee, K.,** Normal genetically mosaic mice produced from malignant teratocarcinoma cells, *Proc. Natl. Acad. Sci. USA,* 72, 3585, 1975.
13. **Stewart, T. A. and Mintz, B.,** Successive generations of mice produced from an established culture line of euploid teratocarcinoma cells, *Proc. Natl. Acad. Sci. USA,* 78, 6314, 1981.
14. **Kahan, B. W. and Ephrussi, B.,** Developmental potentialities of clonal in vitro cultures of mouse testicular teratomas, *J. Nat. Cancer Inst.,* 44, 1015, 1970.
15. **Rosenthal, M. D., Wishnow, R. M., and Sato, G. H.,** In vitro growth and differentiation of clonal populations of multipotential mouse cells derived from a transplantable testicular teratocarcinoma, *J. Nat. Cancer Inst.,* 44, 1001, 1970.
16. **Evans, M. J.,** Isolation and properties of a clonal tissue culture strain of pluripotent mouse teratoma cells, *J. Embryo. Exp. Morph.,* 28, 163, 1972.
17. **Jacob, H., Boon, T., Gailland, D., Nicolas, J. F., and Jacob, F.,** Teratocarcinoma de la souris: isolement, culture at properietes de cellules a potentialites multiples, *Ann. Microbial. (Institut Pasteur),* 124B, 269, 1973.
18. **Jami, J. and Ritz, E.,** Multipotentiality of single cells of transplantable teratocarcinomas derived from mouse embryo graft, *J. Nat. Cancer Inst.,* 52, 1547, 1974.
19. **McBurney, M. W. and Rogers, B. J.,** Isolation of male embryonal carcinoma cells and their chromosome replication patterns, *Develop. Biol.,* 89, 503, 1982.
20. **Amano, S., Uno, K., and Hagiwara, A.,** Cardiac muscle cell differentiation in vitro from a characteristic cell line isolated from mouse teratocarcinoma, *Develop. Growth Differ.,* 20, 41, 1978.
21. **Uno, K.,** A correlation between the capacity of cavity formation and the subsequent differentiation of teratocarcinoma embryoid body lines, *J. Embryol. Exp. Morph.,* 69, 127, 1982.
22. **Cronmiller, C. and Mintz, B.,** Karyotypic normalcy and quasi-normalcy of developmentally totipotent mouse teratocarcinoma cells, *Develop. Biol.,* 67, 465, 1978.
23. **Damjanov, I., Solter, D., and Skreb, N.,** Teratocarcinogenesis as related to the age of embryos grafted under the kidney capsule, *Wilhelm Roux' Arch.,* 167, 288, 1971.
24. **Bernstine, E. G., Hooper, M. L., Grandchamp, S., and Ephrussi, B.,** Alkaline phosphatase activity in mouse teratoma, *Proc. Natl. Acad. Sci. USA,* 70, 3899, 1973.
25. **Takeuchi, I. K., Watanabe, T., and Uno, K.,** Nucleolus-like bodies in embryonal carcinoma cells of the embryoid bodies isolated from mouse teratocarcinoma, *Exp. Cell Res.,* 143, 467, 1982.
26. **Gachelin, G., Delarbre, C., Coulon-Morelec, M-J., Keil-Dlouha, V., and Muramatsu, T.,** F9 antigens: a reevaluation, in *Teratocarcinoma and Embryonic Cell Interactions,* Muramatsu, T., Gachelin, G., Moscona, A. A., and Ikawa, Y., Eds., Academic Press, New York, 1982, 121.
27. **Andrews, P. W., Knowles, B. B., Cossu, G., and Solter, D.,** Teratocarcinoma and mouse embryo cell surface antigens: Characterization of the molecule(s) carrying the SSEA-1 antigenic determinant, in *Teratocarcinoma and Embryonic Cell Interactions,* Muramatsu, T., Gachelin, G., Moscona, A. A., and Ikawa, Y., Eds., Academic Press, New York, 1982, 103.

28. **Hatta, K., Okada, T. S., and Takeichi, M.,** A monoclonal antibody disrupting calcium-dependent cell-cell adhesion of brain tissues: possible role of its target antigen in animal pattern formation, *Proc. Natl. Acad. Sci. USA,* 82, 2789, 1985.

29. **Martin, G. R. and Evans, M. J.,** Differentiation of clonal lines of teratocarcinoma cells: formation of embryoid bodies in vitro, *Proc. Natl. Acad. Sci. USA,* 72, 1441, 1975.

30. **Martin, G. R. and Evans, M. J.,** Multiple differentiation of clonal teratocarcinoma stem cells following embryoid body formation in vitro, *Cell,* 6, 467, 1975.

31. **Martin, G. R.,** Teratocarcinomas as a model system for the study of embryoenesis and neoplasia, *Cell,* 5, 229, 1975.

32. **Martin, G. R.,** Teratocarcinomas and mammalian embryogenesis, *Science,* 209, 768, 1980.

33. **Martin, G. R.,** Teratocarcinoma stem cells provide a model system for the study of early mammalian development in vitro, in *Teratocarcinoma and Embryonic Cell Interactions,* Muramatsu, T., Gachelin, G., Moscona, A., A., and Ikawa, Y., Eds., Academic Press, New York, 1982, 3.

34. **Martin, G. R., Wiley, L. M., and Damjanov, I.,** The development of cystic embryoid bodies in vitro from cloned teratocarcinoma stem cells, *Develop. Biol.,* 61, 69, 1977.

35. **Uno, K. and Amano, S.,** Induction of myocardiogenesis from mouse teratocarcinoma, *Develop. Growth Differ.,* 20, 269, 1978.

36. **Bustan, H. and Hertz, A.,** *In vitro* determination and differentiation of mouse teratocarcinoma toward myocardiogenesis, *Develop. Biol.,* 59, 1, 1977.

37. **Manasek, F. J.,** Embryonic development of the heart. I. A light and electron microscopic study of myocardial development in the early chick embryo, *J. Morphol.,* 125, 329, 1968.

38. **Viragh, S. and Challice, C. E.,** Origin and differentiation of cardiac muscle cells in the mouse, *J. Ultrastruct. Res.,* 42, 1, 1973.

39. **McBurney, M. W., Jones-Villeneauve, E. M. V., Edward, M. K. S., and Anderson, P. J.,** Control of muscle and neuronal differentiation in a cultured embryonal carcinoma cell line, *Nature,* 299, 165, 1982.

40. **Edwards, M. K. S. and McBurney, M. W.,** The concentration of retinoic acid determines the differentiated cell types formed by a teratocarcinoma cell line, *Develop. Biol.,* 98, 187, 1983.

41. **Kondoh, H., Takahashi, Y., and Okada, T. S.,** Differentiation-dependent expression of the chicken-crystallin gene introduced into mouse teratocarcinoma stem cells, *EMBO J.,* 3, 2009, 1984.

Chapter 6

TROPHIC FACTORS OF HEART CELLS IN CULTURE RESPONSIBLE FOR GROWTH AND DIFFERENTIATION OF PERIPHERAL NEURONS

I. A. Hendry

TABLE OF CONTENTS

I. INTRODUCTION

The heart is innervated by sympathetic, parasympathetic, and sensory neurons. The normal growth and development of these neurons into the heart, the control of their ramification within the heart and the formation of correct synaptic connections is largely controlled by factors produced by the heart cells. Tissue culture provides a convenient way in which to study the production and effects of such factors controlling neuronal development.

Explants or dissociated neuronal cultures can be grown, in co-culture with heart as either explant[1] or dissociated cells,[2] or in the presence of medium conditioned by growth over heart cells in culture,[3] or in the presence of extracts of heart tissue.[4] These cultures allow the identification of any factor present in the heart that can influence the growth of the neuronal cultures.

Cardiac tissue can affect the growth and development of the neurons that innervate it in several ways. Factors can stimulate the growth of neurites in a directed or nondirected manner. The survival of neurons can be enhanced, providing a mechanism for the selection of appropriate connecting neurons. Finally, the phenotype of neurons can be altered so that the type of neurotransmitter utilized may change.

II. NEURON SURVIVAL FACTORS AND NEURITE OUTGROWTH PROMOTING FACTORS

Cultured heart cells have been shown to produce many molecules that influence neurite outgrowth and survival from the autonomic and sensory ganglia.

Neurite promoting factors may not be studied in the absence of neuronal survival factors, and so these two phenomena will be taken together, although it is possible to demonstrate the presence of pure neuronal survival factors, pure neurite promoting factors, and factors with both properties.

Early studies, where explant cultures of peripheral ganglia were grown together with explants of cardiac and other tissues showed that there could be directed outgrowth of nerve fibers to target tissues across the collagen substrate of the culture vessel.[5] When given equal opportunity to grow towards two different tissues, axons from sympathetic ganglion explants appeared to be preferentially directed towards the tissue that normally received the densest innervation. For example, when a small piece of atrium was placed in culture together with an explanted sympathetic ganglion and a small piece of vas deferens, there was a strong fiber outgrowth to both target tissues, but the greatest amount was to the atrium. Not only was there direct growth to the atrium explants, but many fibers appeared to change direction to grow towards the explants.[6] When growing nerve fibers were observed with time lapse photography, it was seen that preferential growth of sympathetic nerve fibers around myocardial cells occurred frequently. A close association between nerve and muscle formed and was maintained for several weeks. These myocardial cells appeared morphologically more differentiated than those in which there was no association, although in all cases the myocardial cells were contracting.[7]

This type of directed growth can be brought about by two different mechanisms. The first requires the production of some kind of gradient that influences neurite growth. The production and release by the heart cells of a soluble neurite growth promoting agent could give rise to a gradient of such a factor in the culture medium (Figure 1;2).[6,8-12] The growing fibers would then tend to grow towards the source of such a gradient, as the elongation rate of neurites is related to the local level of growth factor sensed by the axonal growth cone (Figure 1; 2 to 5).[12,13] It is not likely, however, that this type of gradient could be maintained easily in the normal fluid culture medium, and so such mechanisms could only work over a very short distance and for a short period of time. This type of response was seen when

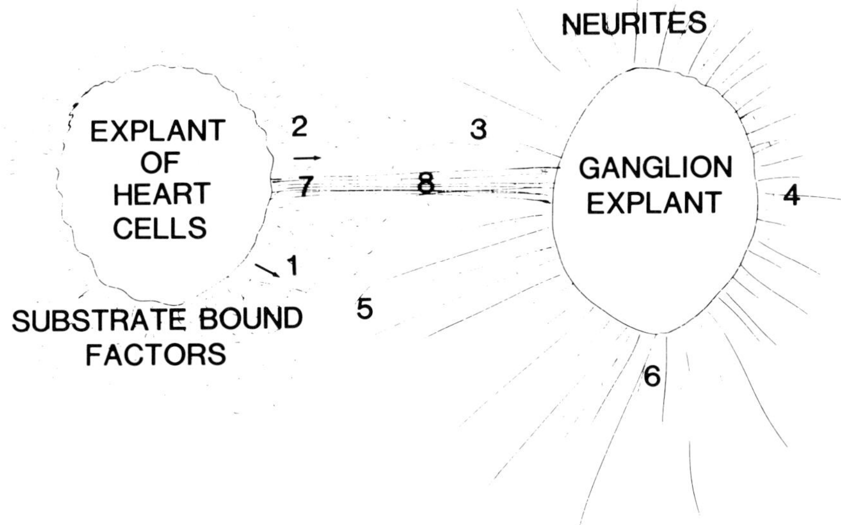

FIGURE 1. Schematic diagram of mechanisms for directed growth between autonomic ganglion and cardiac explants. (1) Neurites grow on a substrate bound material released from cardiac explant as a gradient. (2) Neurites grow up a gradient of growth factor released from cardiac explants. (3) Neurites grow more strongly in presence of higher levels of growth factors. (4) Neurites grow less strongly in shadow of ganglion where concentration of growth factor may be lower. (5) Neurites change direction on meeting substrate bound adhesion molecules or gradient of growth factor. (6) Neurites grow radially in presence of uniform level of growth factor in culture medium. (7) Neurites grow in bundles on bridge of substrate bound molecules between the two explants. (8) Neurites making contact with cardiac explants are reinforced at expense of other branches and bundle formation occurs.

nerve growth factor ejected from the barrel of a glass pipette was placed in the vicinity of a growing axon.[14] More stable gradients can be achieved using semisolid media such as agar or methocel,[15,16] where mixing of the medium cannot destroy the gradient so readily. This situation may mimic more closely the situation in the developing embryo.

A more likely mechanism in culture is the production and release by the heart cells of a factor that can bind to the substrate so that a stable gradient of the factor is produced (Figure 1;1). If this factor can promote the growth of neurites[3,17] or can promote adhesion of neurite growth cones[15,17] to the substratum then neurites will readily grow up such a gradient. The growth between explants of ganglia and cardiac tissue will be strongest over the region of the substrate where such factors are bound and this of course will be between the two explants (Figure 1;7). This could give rise to the bundle formation often observed in such cultures.[18]

The second mechanism results in apparent directed growth because of the selective maintenance or positive reinforcement of neurites making contact with the target tissue. In this case the initial outgrowth of the neurites from the ganglia would be radially in a random direction until some of these neurites by chance make contact with the cultured heart cells (Figure 1;6). At this stage, these neurites are reinforced at the expense of neurite outgrowth in other directions, either by interaction with cell surface molecules such as the cell-adhesion molecule (CAM)[19,20] or by the action of a soluble trophic agent (Figure 1;8).[21,22] This could then give the appearance of directed growth and also lead to the bundle formation seen in the cultures.[18]

There is evidence that to some degree all these possible mechanisms play a role in directed growth in culture. Soluble survival and neurite outgrowth factors are clearly present and are found in conditioned medium. Extracts of heart tissue contain several factors that promote the survival and neurite outgrowth of sympathetic, parasympathetic, and sensory neurons grown in dissociated cell culture[23-27] or explant culture.[28-30] The levels of factors influencing the survival of sympathetic, parasympathetic, and sensory neurons increase if the nerve supply to the heart is interfered with by vagotomy or chemical sympathectomy.[31] When heart cells are placed in culture, they are of course denervated, and thus it would be expected that they would also produce an increased level of survival factors. The best documented of the neuronal survival and neurite outgrowth factors is the nerve growth factor (NGF).[32] This factor has been shown to be in much higher levels in cultured tissues than the same tissues in vivo.[33,34]

The nerve growth factor is a protein that was originally discovered as a diffusable substance able to promote neurite outgrowth from chick sensory and sympathetic neurones.[35] It was first found in sarcomas,[36] then in snake venom,[37] salivary glands of male mice,[38] prostrate glands in guinea pigs[39] and in bull semen.[40-42] All these sources were extremely rich in the protein, but later studies have indicated that many tissues contain low levels of this protein.[43-44] It is most likely that the NGF has two different types of function in its interaction with sensitive neurons.[45] As a neurotrophic molecule, NGF acts on the cell to provide an anabolic stimulus.[46,47] In this role NGF is made and released in the target tissue to be taken up by the ingrowing nerve terminals and transported back to the neuronal perikaryon[48] where it has an anabolic effect, promoting the survival of neurons and making appropriate target contacts.[22,49] This role has been called that of a ''retrophin'' (a retrogradely transported neurotrophic factor).[50,51] The other function appears to be promotion of the terminal sprouting and arborization of axons as they ramify in the tissues.[52] By growing neurons in multiwell chambers, it has been shown that NGF has a direct effect on nerve terminals. Nerve fibers will only enter chambers containing NGF and will retract if the NGF is removed in spite of the continued presence of NGF in the chamber where the cell bodies are located.[13,53] Thus NGF has a local direct effect on growing axons.

NGF is present in heart tissue[44,54] and is synthesized by heart cells in culture.[55] It is present in medium conditioned by growth over heart cells in culture. This is measured by the nerve fiber outgrowth from NGF sensitive neurons of the chick dorsal root or sympathetic ganglia grown in the conditioned medium. The nerve fiber outgrowth could be inhibited by antiserum to NGF.[33,56] The requirements for exogenous NGF for the survival of sympathetic neurons could be replaced by cardiac cells or medium conditioned by cardiac cells, and this effect was also blocked by antiserum to NGF.[57,58] In addition, NGF has been directly measured using sensitive immunoassays.[59] The cultured heart cells released most of the NGF they made into the medium and only contained about 2% of the total NGF produced in 24 hr. No difference was found in the ability to make NGF in hearts from male or female, young or old animals, indicating that heart cells can make NGF throughout their entire life.[55]

The finding that heart cells released nearly all the NGF they make is extremely important, as it may explain why heart tissue contains very low amounts of NGF in vivo.[60] Most of the NGF made by heart cells would be released and this free NGF could be rapidly taken up by the innervating sympathetic and sensory nerve fibers in which it would be removed from the cardiac tissue by the process of retrograde axonal transport.[48] Denervation of the heart would prevent this removal process and the levels of NGF would immediately build up in the tissue.[31,34]

The usual preparation of NGF is from the male mouse submandibular gland[38] in which it occurs as a high molecular weight complex called 7S NGF, which has a molecular weight of 140,000 and three subunits called α, β and γ.[61-63] In this complex, all of the nerve growth promoting activity is contained in the β subunit which has a molecular weight of 26,518 as calculated from its complete sequence.[64] The molecular weight and isoelectric point of the NGF made in cultured cardiac tissue shows that heart NGF is physicochemically similar to the mouse submaxillary gland β-NGF and does not occur as 7S NGF.[55]

Heart cells in culture and heart cell conditioned medium contain factors other than NGF that promote the survival of sympathetic,[65,67] parasympathetic,[10,66,68-70] and sensory[66,67] neurons, and in most cases the conditioned medium also promoted neurite outgrowth from these neurons. Parasympathetic neurons of the chick ciliary ganglion are influenced by at least two factors in heart cell conditioned medium. One factor binds to the culture plate and is absorbed out of the medium when the conditioned medium is incubated over polyornithine coated culture dishes. This factor promotes neurite outgrowth if the neurons are maintained alive by other survival factors, but does not provide survival activity by itself. The other factor remains in the medium after the polyornithine absorption and acts on the neurons to promote survival but has no effect on fiber outgrowth.[3]

The particular substrate is very important for both neuronal survival and neurite outgrowth. Rat sympathetic neurons readily attach and initiate neurite outgrowth when the dissociated cells are plated out on a preexisting layer of cardiac myocytes.[71,72] These cells, however, can be killed by a variety of harsh methods, namely — heat, trichloracetic acid, ethanol, formaldehyde, and glutaraldehyde — and all of these procedures preserve to some extent the ability to promote attachment and neurite outgrowth of the sympathetic neurons. The fiber outgrowth usually followed the orientation of the underlying mat of cells.[16] Using scaning electron microscopy, it can be seen that the heart cells and axon processes grow on the upper surface of the collagen substrate. On aligned collagen, fibroblasts grow mainly in the direction of the alignment axis and the nerve fibers also show orientation preference along this axis.[73] Thus, not only are neurons sensitive to the type of surface over which they travel, but they also can receive directional information from it. Microexudates of heart cells in culture also promote neurite outgrowth and this is abolished by the proteolytic enzyme, trypsin, and the carbohydrate reagent, periodate, showing that the active components are glycoproteins.[16] Thus cultured heart cells contain on their surface and excrete onto the substrate substances that promote neuronal adhesion and neurite outgrowth and which lead to directed growth.

III. SPECIFYING FACTORS

During development, neurons become progressively restricted in their possible phenotypes, largely due to the various environmental factors that act on the differentiating nerve cell.[74,76]

There are two possible mechanisms by which developmental factors may regulate the neurotransmitter phenotype of a group of neurons. The first is by promoting the selective survival of precursor cells which are already destined to make a particular transmitter. In Figure 2A, this is demonstrated schematically for the events that are believed to occur in chick sympathetic ganglia in culture. NGF promotes the survival of neurons destined to make noradrenaline as their neurotransmitter and a factor in heart cell conditioned medium promotes the survival of precursor cells destined to make acetylcholine as their neurotransmitter. The second mechanism is by selectively promoting the development of new neurotransmitter enzymes in a pluripotent precursor cell by a change in the direction of its differentiation. In Figure 2B this is demonstrated schematically, as the events are believed to occur in the neonatal rat superior cervical ganglion grown in culture. The immature neuron expresses adrenergic characteristics, but under the influence of various developmental factors it may become either cholinergic or adrenergic.

This second mechanism is the significant one for the determination of neuronal phenotype and these factors do not appear to alter neuronal survival or growth, but rather play a role as specifying, instructional, or developmental factors.[75,77] Their presence appears to guide the developmental pathway in the culture environment by influencing the neuronal differentiation pattern. These developmental factors presumably also act in vivo to provide an appropriate stimulus for the selection of transmitter and receptor type by the neurons. They

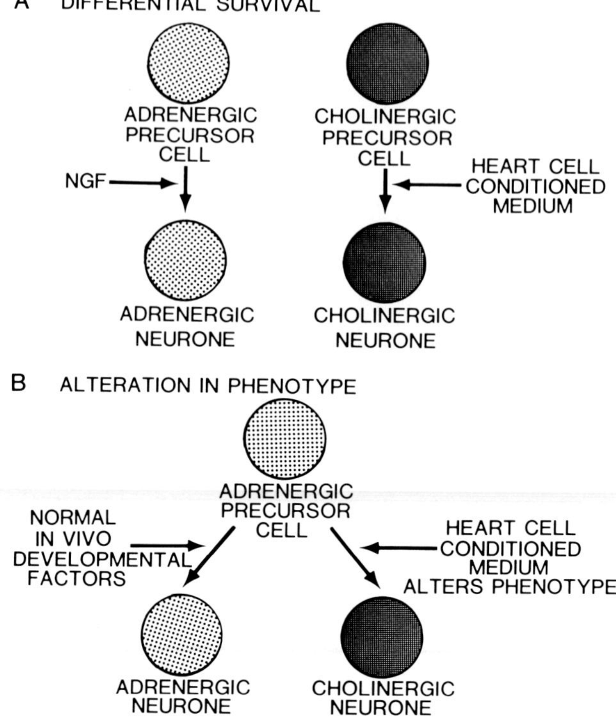

FIGURE 2.   Schematic diagram of mechanisms for alterations of phenotype in cultures of neurons. (A) Two populations of neurons are present initially and different factors promote the survival of subpopulations of neurons destined to make different neurotransmitters. In this case the steps determining the final phenotype have already occurred. (B) One population of neurons is present initially and various environmental factors directly influence the final phenotype expressed. This is the function of true developmental specifying factors.

also may provide a mechanism for regulation by the postsynaptic cell of the transmitter type of the presynaptic neuron.

In early experiments where explants of sympathetic ganglia were grown in cultures containing dissociated heart cells, it was observed that the heart cells could be both excited and inhibited after stimulation of the ganglia.[78] This was due to the liberation of both acetylcholine and noradrenaline by the cultured neurons. Other studies using dissociated sympathetic neurons grown on layers of heart cells showed similar results.[79] At first this was believed to be due to survival of preexisting cholinergic neurons (Figure 2A), but by growing the neurons in microcultures so that only one neuron at a time could be stimulated, it became apparent that the formerly adrenergic sympathetic neuron had become cholinergic.[80] This was a real change in phenotype rather than a selective survival of a cholinergic subpopulation of neurons,[72,81] as in Figure 2B. This plasticity was due to a protein, present in the medium, made by nonneuronal cells including heart cells.[74] The molecule responsible for this has been partially purified from heart conditioned medium by a sequence of ammonium sulfate precipitation, DEAE, CM-cellulose, and Sephadex G-100 chromatography. The partially purified factor was active at a protein concentration in the culture medium of 1 μg protein/ mℓ and elutes from Sephadex with an apparent molecular weight of 40 to 45,000. It is very sensitive to periodate and less so to proteolytic enzymes and not at all to neuraminidase. Therefore, it seems that the molecule consists essentially of a carbohydrate on a protein backbone.[82]

In view of the production of this factor by the same nonneuronal cells that normally occur in vivo, the question may be raised as to why sympathetic neurons are adrenergic and not cholinergic in vivo. One difference between cultured neurons and those in vivo is the effect of nerve activity, since there is no presynaptic input in vitro. It was found that depolarizing agents, such as potassium or veratradine, or electrical stimulation could block the action of heart conditioned medium, turning the neurons cholinergic.[83] Furthermore, the chronic depolarization effect could itself be blocked by preventing the entry of calcium into the neurons.[84] Thus it appeared that chronic depolarization mimicking the preganglionic nerve activity was one of the factors that determined the phenotype of cultured sympathetic neurones, and that this effect may be mediated by calcium fluxes. The relevance of these studies to the physiological role of the preganglionic nerve activity was brought into question, however, by the in vivo finding that isolation of sympathetic neurons from preganglionic activity by transection of the preganglionic nerve trunk did not have any effect on the normal neurotransmitter developmental pattern of the sympathetic neurons in rat. In this case there was no increase in the activity of the cholinergic marker enzyme, choline acetyltransferase, after the operation.[85] Thus factors other than nerve activity must be responsible in vivo for the maintenance of the adrenergic nature of the sympathetic ganglia. It has been proposed that cyclic nucleotides play a role in the induction of the cholinergic phenotype, but detailed experiments indicate a lack of correlation between changes in the levels of these nucleotides and the final phenotype of sympathetic neurons cultured in the presence or absence of heart conditioned medium.[86]

Another difference between cultured neurons and neurons developing in vivo is the presence of circulating hormones. Since it is known that steroid hormones affect neuronal development, the role of steroids was investigated. It was found that both glucocorticosteroids and mineralocorticosteroids could prevent the induction of the cholinergic phenotype.[87] Furthermore, the classical glucocorticoid and mineralocorticoid antagonists did not block this effect, but rather acted as agonists in their own right.[88] It was also shown that this effect of steroids was indirect, acting on the nonneuronal heart cells rather than on the neurons.[89] The steroids prevented the synthesis or release of the factor by the cultured heart cells so that medium conditioned by heart cells grown in the presence of steroids did not contain the factor. On the other hand, neurons grown in the presence of steroids and medium conditioned by heart cells grown in the absence of steroids became cholinergic.[89] Thus, the steroids did not antagonize the presence of the factor but prevented its appearance. Hence, it may be thought that the steroids in vivo would keep sympathetic neurons adrenergic by this mechanism. Recent experiments have shown, however, that adrenalectomy in vivo does not result in the appearance of new cholinergic sympathetic neurones. At present, the mechanism of action of this factor in the formation of cholinergic rat sympathetic neurones is unclear. One possible role is in the control of the innervation of the rat forepaw sweat gland, which commences as an adenergic innervation and becomes cholinergic as the animal develops.[90]

Heart conditioned medium, from cultures of chick heart, influences the appearance of neurotransmitters found in cultures of chick sympathetic neurons. In this case, however, it was suggested that the factor acts by permitting the survival of a specific subpopulation of the ganglion neurones that express cholinergic properties,[65] as in Figure 2A. The heart cell conditioned medium results in high levels of choline acetyltransferase in the neuronal cultures, while NGF results in high levels of tyrosine hydroxylase. When both heart conditioned medium and nerve growth factor are added together then the level of choline acetyltransferase and tyrosine hydroxylase are additive. Furthermore, neurons surviving in the presence of NGF and making predominently tyrosine hydroxylase after 48 hr do not commence to express choline acetyltransferase when changed to heart conditioned medium. Thus, these results appear to be more consistent with the chick heart factor selecting a population of neurons that are always cholinergic rather than actually inducing a similar change in the adrenergic to cholinergic phenotype as seen for the rat heart factor. Further evidence that chick heart conditioned medium acts mainly to promote selective survival comes from the finding that pretreatment of polyornithine culture substrates with the conditioned medium increased by up to tenfold the number of chick embryo sympathetic neurons which survive in response to NGF.[91] The substrate bound material does not have survival value alone, but permits the survival of all sympathetic neurons in response to NGF, including those which would normally die on polyornithine in the presence of NGF.[65] It is suggested that such a permissive role could be effected either by the induction of NGF receptors by the substrate-attached material in parallel with the induction of neurite outgrowth,[3,69] or other changes in neuronal properties,[16] or alternatively, by providing conditions necessary for the survival of a subpopulation of sympathetic neurons which cannot be rescued by NGF alone although they possess NGF receptors.[65]

IV. SUMMARY AND CONCLUSIONS

1. Heart cells in culture synthesize and secrete many factors that affect the survival, neurite outgrowth, and developmental patterns of the neurons that innervate the heart.
2. Nerve growth factor is essential for the survival of sympathetic and some sensory neurons and is made and released in large amounts by cultured heart cells.
3. Other survival factors for sympathetic, sensory, and parasympathetic neurons are also made by the cultured cells.
4. Neurite promoting factors that are either soluble or bind to the substrate are released by the cells and are also contained on the cultured heart cell surface.
5. Developmental factors that control the phenotypic fate of neurons are released into the medium and their production is under steroid hormonal regulation.
6. Other developmental factors that permit the survival of specific subpopulations of neurons have also been demonstrated.
7. The cultured heart cell, therefore, provides a useful tool in order to study the intractions of a target tissue with the neurons destined to innervate it.

REFERENCES

1. **Ebendal, T., Jondell-Kylberg, A., and Soderström, S.,** Stimulation by tissue explants on nerve fiber outgrowth in culture, *Zoon,* 6, 235, 1978.
2. **Chamley, J. H., Campbell, G. R., and Burnstock, G.,** An analysis of the interactions between sympathetic nerve fibers and smooth muscle cells in tissue culture, *Dev. Biol.,* 33, 344, 1973.
3. **Adler, R. and Varon, S.,** Cholinergic neurotrophic factors: V. Segregation of survival and neurite promoting activities in heart conditioned medium, *Brain Res.,* 188, 437, 1980.
4. **Ebendal, T., Belew, M., Jacobson, C-O., and Porath, J.,** Neurite outgrowth elicited by embryonic chick heart: partial purification of the active factor, *Neurosci. Lett.,* 14, 91, 1979.
5. **Chamley, J. H. and Dowell, J. J.,** Specificity of nerve fiber "attraction" to autonomic effector organs in tissue culture. *Exp. Cell Res.,* 90, 1, 1975.
6. **Chamley, J. H., Goller, I., and Burnstock, G.,** Selective growth of sympathetic nerve fibers to explants of normally densely innervated autonomic effector organs, *Dev. Biol.,* 31, 362, 1973.
7. **Mark, G. E., Chamley, J. H., and Burnstock, G.,** Interactions between autonomic nerves and smooth and cardiac muscle cells in tissue culture, *Dev. Biol.,* 32, 194, 1973.
8. **Charlwood, K. A., Lamont, D. M., and Banks, B. E. C.,** Apparent orientating effects produced by nerve growth factor, *Nerve Growth Factor and its Antiserum,* Zaimis, E. and Knight, J., Eds., Athlone Press, London, 1972, 102.
9. **Coughlin, M. D.,** Target organ stimulation of parasympathetic nerve growth in the developing mouse submandibular gland, *Dev. Biol.,* 43, 140, 1975.

10. **Ebendal, T. and Jacobson, C.-O.**, Tests of possible role of NGF in neurite outgrowth stimulation exerted by glial cells in heart explants in culture, *Brain Res.*, 131, 373, 1977.
11. **Pollack, E. D. and Liebig, V.**, Differentiating limb tissue affects neurite growth in spinal cord cultures, *Science*, 197, 899, 1977.
12. **Ebendal, T.**, Stage dependent stimulation of neurite outgrowth exerted by nerve growth factor and chick heart in cultured embryonic ganglia, *Dev. Biol.*, 72, 276, 1979.
13. **Campenot, R. B.**, Local control of neurite development by nerve growth factor, *Proc. Natl. Acad. Sci. USA*, 74, 4516, 1977.
14. **Gundersen, R. W. and Barrett, J. N.**, Neuronal chemotaxis: chick dorsal-root axons turn toward high concentrations of nerve growth factor, *Science*, 206, 1079, 1979.
15. **Letourneau, P. C.**, Cell-substratum adhesion of neurite growth cones, and its role in neurite elongation *Exp. Cell. Res.*, 124, 127, 1979.
16. **Hawrot, E.**, Cultured sympathetic neurones: effects of cell derived and synthetic substrata on survival and development, *Dev. Biol.*, 74, 136, 1980.
17. **Letourneau, P. C.**, Cell-to-substratum adhesion and guidance of axonal elongation, *Dev. Biol.*, 44, 92, 1975.
18. **Chamley, J. H., Campbell, G. R., and Burnstock, G.**, Dedifferentiation, redifferentiation and bundle formation of smooth muscle cells in tissue culture — the influence of cell number and nerve fibers, *J. Embryol. Exp. Morph.*, 32, 297, 1974.
19. **Edelman, G. M.**, Cell adhesion molecules, *Science*, 219, 450, 1983.
20. **Rutishauser, U., Gall, W. E., and Edelman, G. M.**, Adhesion among neural cells of the chick embryo. IV. Role of the cell surface molecule CAM in the formation of neurite bundles in cultures of spinal ganglia, *J. Cell Biol.*, 79, 382, 1978.
21. **Harper, G. P. and Thoenen, H.**, Target cells, biological effects, and mechanism of action of nerve growth factor and its antibodies, *Ann. Rev. Pharmacol. Toxicol.*, 21, 205, 1981.
22. **Hendry, I. A.**, Control in the development of the vertebrate sympathetic nervous system, in *Reviews of Neuroscience*, Vol. 2, Ehrenpreis, S. and Kopin, I. J., Eds., Raven Press, New York, 149, 1976.
23. **Lindsay, R. M. and Tarbit, J.**, Developmentally regulated induction of neurite outgrowth from immature chick sensory neurones (DGR) by homogenates of avian or mammalian heart, liver and brain, *Neurosci. Lett.*, 12, 195, 1979.
24. **Bonyhady, R. E., Hendry, I. A., Hill, C. E., and Watters, D. J.**, An analysis of peripheral neuronal survival factors present in muscle, *Neurosci. Res.*, 13, 357, 1985.
25. **Hill, C. E., Hendry, I. A., and Boneyhady, R. E.**, Avian parasympathetic neurotrophic factors age related increases and lack of regional specificity, *Dev. Biol.*, 85, 258, 1981.
26. **Bonyhady, R. E., Hendry, I. A., Hill, C. E., and Mclennan, I. S.**, Characterization of a cardiac muscle factor required for the survival of cultured parasympathetic neurones, *Neurosci. Lett.* 18, 197, 1980.
27. **Bonyhady, R. E., Hendry, I. A., Hill, C. E., and McLennan, I. S.**, Characterization of a cardiac muscle factor required for the survival of cultured parasympathetic neurones, *Neurosci. Lett.*, 18, 197, 1980.
28. **McLennan, I. S. and Hendry, I. A.**, Parasympathetic neuronal survival induced by factors from muscle, *Neurosci. Lett.*, 10, 269, 1978.
29. **Ebendal, T. and Jacobson, C. O.**, Tissue explants affecting extension and orientation of axons in cultured chick embryo ganglia, *Exp. Cell. Res.*, 195, 379, 1977.
30. **Ebendal, T. and Belew, M.**, Chick heart factor controlling neurite extension, *Eur. J. Cell Biol.*, 22, 409, 1980.
31. **Boon, Y. H., Hendry, I. A., Hill, C. E., and Watters, D. J.**, Effects of cardiac denervation on levels of neuronal survival factors for cultured autonomic neurones, *Dev. Brain Res.*, 12, 154, 1984.
32. **Levi-Montalcini, R. and Angeletti, P. U.**, Nerve growth factor, *Physiol. Rev.*, 48, 534, 1968.
33. **Harper, G. P., Al-Saffar, A. M., Pearce, F. L., and Vernon, C. A.**, Production of nerve growth factor in vitro by tissues of the mouse, rat and embryonic chick, *Dev. Biol.*, 77, 379, 1980.
34. **Ebendal, T., Olson, L., Seiger, A., and Hedlund, K. O.**, Nerve growth factors in the rat iris, *Nature (London)*, 286, 25, 1980.
35. **Levi-Montalcini, R.**, The nerve growth factor, *Ann. N.Y. Acad. Sci.*, 118, 149, 1964.
36. **Cohen, S. and Levi-Montalcini, R.**, Purification and properties of a nerve growth promoting factor isolated from mouse sarcoma 180, *Cancer Res.*, 17, 15, 1957.
37. **Cohen, S. and Levi-Montalcini, R.**, A nerve growth-stimulating factor isolated from snake venom, *Proc. Natl. Acad. Sci. USA*, 42, 571, 1956.
38. **Cohen, S.**, Purification of a nerve growth promoting protein from the mouse salivary gland and its neurocytotoxic antiserum, *Proc. Natl. Acad. Sci. USA*, 46, 302, 1960.
39. **Harper, G., Bonde, Y. A., Burnstock, G., Carstains, J. R., Dennison, M. E., Suda, K., and Vernon, C. A.**, Guinea pig prostate is a rich source of nerve growth factor, *Nature*, 279, 160, 1979.

40. **Harper, G. P., Glanville, R. W., and Thoenen, H.,** The purification of nerve growth factor from bovine seminal plasma. Biochemical characterization and partial amino acid sequence, *J. Biol. Chem.,* 257, 8541, 1982.

41. **Harper, G. P. and Thoenen, H.,** The distribution of nerve growth factor in the male sex organs of mammals, *Neurochemistry,* 34, 893, 1980.

42. **Harper, G. P. and Thoenen, H.,** Nerve growth factor: biological significance, measurement and distribution, *Neurochemistry,* 34, 5, 1980.

43. **Hendry, I. A.,** Developmental changes in tissue and plasma concentrations of the biologically active species of nerve growth factor in the mouse, by using a two-site radioimmuno assay, *Biochemistry,* 128, 1265, 1972.

44. **Korsching, S. and Thoenen, H.,** Nerve growth factor in sympathetic ganglia and corresponding target organs of the rat: correlation with density of sympathetic innervation, *Proc. Natl. Acad. Sci. USA,* 80, 3513, 1983.

45. **Hill, C. E. and Hendry, I. A.,** Differences in sensitivity to nerve growth factor of axon formation and tyrosine hydroxylase induction in cultured sympathetic neurones, *Neuroscience,* 1, 489, 1976.

46. **Bradshaw, R. A.,** Nerve growth factor, *Ann. Rev. Biochem.,* 47, 191, 1978.

47. **Thoenen, H. and Barde, Y. A.,** Physiology of nerve growth factor, *Physiol. Rev.,* 60, 1284, 1980.

48. **Hendry, I. A., Stöckel, K., Thoenen, H., and Iversen, L. L.,** The retrograde axonal transport of nerve growth factor, *Brain Res.,* 68, 103, 1974.

49. **Thoenen, H., Barde, Y. A., and Edgar, D.,** The role of nerve growth factor (NGF) and related factors for the survival of peripheral neurones, *Adv. Biochem. Psychopharmacol.,* 28, 263, 1981.

50. **Hendry, I. A., Bonyhady, R. E., and Hill, C. E.,** The role of target tissues in development and regeneration — retrophins, *Birth Defects,* 19, 263, 1983.

51. **Hendry, I. A. and Hill, C. E.,** Retrograde axonal transport of target tissue-derived macromolecules, *Nature (London),* 287, 647, 1980.

52. **Bjerre, B., Björklund, A., and Mobley, W.,** A stimulating effect by nerve growth factor on the regrowth of adrenergic nerve fibers in the mouse peripheral tissues after chemical sympathectomy with 6-hydroxy-dopamine, *Z. Zellforsch. Mikrosk. Anat.,* 46, 15, 1973.

53. **Campenot, R. B.,** Regeneration of neurites on long term cultures of sympathetic neurones derived of nerve growth factor, *Science,* 214, 579, 1981.

54. **Kanakis, S. J., Hill, C. E., Hendry, I. A., and Watters, D. J.,** Sympathetic neuronal survival factors change after denervation, *Dev. Brain Res.,* 352, 197, 1985.

55. **Furukawa, Y., Furukawa, S., Satoyoshi, E., and Hayashi, K.,** Nerve growth factor secreted by mouse heart cells in culture, *J. Biol. Chem.,* 259, 1259, 1984.

56. **Ebendal, T., Jordell-Kylberg, A., and Soderström, S.,** Stimulation by tissue explants on nerve fiber outgrowth in culture, in *Formshaping Movements in Neurogenesis,* Johnson, C. O. and Ebendal, T., Eds., Almqvist and Wiksell, Stockhölm, 1978, 235.

57. **Chun, L. L. Y. and Patterson, P. H.,** Role of nerve growth factor in the development of rat sympathetic neurones in vitro. III. Effect on acetylcholine production, *Cell Biol.,* 75, 712, 1977.

58. **Chun, L. L. Y., and Patterson, P. H.,** Role of nerve growth factor in the development of rat sympathetic neurones in vitro. I. Survival, growth and differentiation of catecholamine production, *Cell. Biol.,* 75, 694, 1977.

59. **Furukawa, S., Kamo, I., Furukawa, Y., Adazawa, S., Satoyoshi, E., Itoh, K., and Hayashi, K.,** A highly sensitive enzyme immunoassay for mouse β nerve growth factor, *Neurochemistry,* 40, 734, 1983.

60. **Harper, G. P., Pearce, F. L., and Vernon, C. A.,** The production and storage of nerve growth factor in vivo by tissues of the mouse, rat, guinea pig, hamster, and gerbil, *Dev. Biol.,* 77, 391, 1980.

61. **Varon, S., Nomura, J., and Shooter, E. M.,** Isolation of mouse nerve growth factor in high molecular weight form, *Biochemistry,* 6, 2202, 1967.

62. **Varon, S., Nomura, J., and Shooter, E. M.,** Reversible dissociation of the nerve growth factor protein into different mouse subunits, *Biochemistry,* 7, 1296, 1968.

63. **Varon, S., Nomura, J., and Shooter, E. M.,** Subunit structure of a high molecular weight form of the nerve growth factor from mouse submaxillary gland, *Proc. Natl. Acad. Sci. USA,* 57, 1782, 1967.

64. **Angeletti, R. H. and Bradshaw, R. A.,** Nerve growth factor from mouse submaxillary gland: amino acid sequence, *Proc. Natl. Acad. Sci. USA,* 68, 2417, 1971.

65. **Edgar, D., Barde, Y-A., Thoenen, H.,** Subpopulations of cultured chick sympathetic neurones differ in their requirement for survival factors, *Nature,* 289, 294, 1981.

66. **Helfand, S. L., Riopelle, R. J., and Wessells, N.,** Non-equivalence of conditioned medium and nerve growth factor for sympathetic parasympathetic and sensory neurones, *Exp. Cell Res.,* 113, 39, 1978.

67. **Varon, S., Skaper, S. D., and Manthorpe, M.,** Trophic activities for dorsal root and sympathetic ganglionic neurones in media conditioned by Schwann and other peripheral cells, *Brain Res.,* 277, 3, 1981.

68. **Helfand, S., Smith, G. A., and Wessels, N. K.,** Survival and development in culture of dissociated parasympathetic neurons from ciliary ganglia, *Dev. Biol.,* 50, 541, 1976.

69. **Collins, F.,** Induction of neurite outgrowth by a conditioned medium factor bound to the culture substratum, *Proc. Natl. Acad. Sci. USA,* 75, 5210, 1978.
70. **Coughlin, M. D., Bloom, E. M., and Black, I. B.,** Characterization of a neuronal growth factor from mouse heart-cell-conditioned medium, *Devel. Biol.,* 82, 56, 1981.
71. **O'Lague, P. H., MacLeish, P. R., Nurse, C. A., Claude, P., Furshpan, E. J., and Potter, D. D.,** Physiological and morphological studies on developing sympathetic neurones in dissociated cell culture, *Cold Spring Harbor Symp. Quant. Biol.,* 40, 399, 1976.
72. **Furshpan, E. J., MacLeish, P. R., O'Lague, P. H., and Potter, D. D.,** Chemical transmission between rat sympathetic neurones and cardiac myocytes developing in microcultures: evidence for cholinergic, adrenergic and dual-function neurones, *Proc. Natl. Acad. Sci. USA,* 73, 4225, 1976.
73. **Ebendal, T.,** Scanning electron microscopy of chick embryo nerve fibers and heart fibroplasts on collagen sustrata in vitro, *Zoon,* 2, 99, 1974.
74. **Patterson, P. H.,** Environmental determination of autonomic neurotransmitter functions, *Ann. Rev. Neurosci.,* 1, 1, 1978.
75. **Varon, S. S. and Bunge, R. P.,** Trophic mechanisms in the peripheral nervous system, *Ann. Rev. Neurosci.,* 1, 327, 1978.
76. **Varon, S.,** Neural growth and regeneration: a cellular perspective, *Exp. Neurol.,* 54, 1, 1977.
77. **Berg, D. K.,** New neuronal growth factors, *Ann. Rev. Neurosci.,* 7, 149, 1984.
78. **Purves, R. D., Hill, C. E., Chamley, J. H., Monk, G. E., Fry, D. M., and Burnstockm, G.,** Functional autonomic neuromuscular junctions in culture, *Plfügers Archiv.,* 350, 1, 1974.
79. **O'Lague, P. H., Obata, K., Claude, P., Furshpan, E. J., and Potter, D. D.,** Evidence for cholinergic synapses between dissociated rat sympathetic neurones in cell culture, *Proc. Natl. Acad. Sci. USA,* 71, 3602, 1974.
80. **Landis, S. C.,** Rat sympathetic neurones and cardiac myocytes developing in microcultures: correlation of the fine structure of endings with neurotransmitter function in single neurones, *Proc. Natl. Acad. Sci. USA,* 73, 4220, 1976.
81. **Reichardt, L. F. and Patterson, P. H.,** Neurotransmitter synthesis and uptake by isolated sympathetic neurones in micro-cultures, *Nature (London),* 270, 147, 1977.
82. **Weber, M. J.,** A diffusible factor responsible for the determination of cholinergic functions in cultured sympathetic neurones, *J. Biol. Chem.,* 256, 3447, 1981.
83. **Walicke, P. A., Campenot, R. B., and Patterson, P. H.,** Determination of transmitter function by neuronal activity, *Proc. Natl. Acad. Sci. USA,* 74, 5767, 1977.
84. **Walicke, P. A. and Patterson, P. H.,** On the role of Ca^{2+} in the transmitter choice made by cultured sympathetic neurones, *Neuroscience,* 1, 343, 1981.
85. **Hill, C. E. and Hendry, I. A.,** The influence of preganglionic nerves on the superior cervical ganglion of the rat, *Neurosci. Lett.,* 13, 133, 1979.
86. **Walicke, P. A. and Patterson, P. H.,** On the role of cyclic nucleotides in the transmitter choice made by cultured sympathetic neurones, *Neuroscience,* 1, 333, 1981.
87. **McLennan, I. S., Hill, C. E., and Hendry, I. A.,** Glucocorticosteroids modulate transmitter choice in developing superior cervical ganglion, *Nature (London),* 283, 206, 1980.
88. **McLennan, I. S., Hill, C. E., and Hendry, I. A.,** Pharmacology of the steroid regulation of transmitter choice in cultured rat sympathetic ganglia, *Aust. J. Exp. Biol. Med. Sci.,* 62, 627, 1984.
89. **Fukada, K.,** Hormonal control of neurotransmitter choice in sympathetic neurone cultures, *Nature,* 287, 553, 1980.
90. **Landis, S. C.,** Development of cholinergic sympathetic neurons evidence for transmitter plasticity in vivo, *Fed. Proc.,* 42, 1633, 1983.
91. **Edgar, D. and Thoenen, H.,** Modulation of NGF — individual survival of chick sympathetic neurons, *Dev. Brain Res.,* 5, 89, 1982.

Chapter 7

ION TRANSPORT IN CULTURED HEART CELLS

C. R. Horres, D. M. Wheeler, D. Piwnica-Worms, and M. Lieberman

TABLE OF CONTENTS

I. INTRODUCTION

Ion transport across the cardiac sarcolemma forms the basis for the electrophysiological and contractile activity of the cells. Measurements of ion fluxes, particularly with radioisotopic tracer techniques, have played an important role in the development of our knowledge of specific ion movements across excitable cell membranes. However, results of such studies in naturally occurring cardiac preparations are complex and difficult to interpret, and thus have failed to provide an adequate picture of events at the cell membrane level. Tissue-cultured preparations have provided an extremely useful model system for the study of cardiac cell membrane transport, avoiding some of the complexities and uncertainties inherent to naturally occurring tissues. The purposes of this chapter are, therefore, as follows: (1) to review the development of tissue culture preparations pertinent to ion transport measurements, (2) to describe briefly the various experimental methods which have been utilized to elucidate ion transport, with emphasis on radiotracer techniques, (3) to consider certain results (again largely of the tracer variety) which define the transport of potassium, sodium and chloride in tissue-cultured heart cells, and (4) to discuss the implications of these findings to the understanding of cardiac cellular function.

A. Naturally Occurring Cardiac Muscle vs. Cultured Heart Cells

Under physiologic conditions, identification of the true transmembrane flux of an ionic species has been a formidable task in cardiac muscle. Among the complicating factors are (1) small cells with large surface area to volume ratio, (2) diffusional delays due to a complex extracellular space, (3) heterogeneous cell population, (4) dissection trauma, and (5) absence of uniform steady-state conditions.[1] The collective effect of these limitations has placed severe constraints on most forms of ion transport measurement and specifically on the determination of radiotracer kinetics. In fact, the apparent simplicity of collecting tracer data has been overshadowed by difficulties in the resolution and identification of the cellular components of such data in cardiac muscle.[2,3] Thus, the interpretation of such experiments must be approached with caution. The use of tissue culture preparations of heart cells for radioactive tracer and other transport studies has proven to be a suitable, and perhaps necessary, substitute for measurement of the movement of ions (for review, see Reference 4).

In 1962, Burrows and Lamb[5] reported the first application of tracer flux measurements to cultured heart cells. The advantage cited for using monolayer cultures over naturally occurring preparations was the fact that cells grown as monolayers essentially are in direct contact with the extracellular culture medium. In contrast, naturally occurring or dissected cardiac preparations contain a restricted extracellular space and an overall thickness or diameter of at least several hundred microns. As discussed subsequently, the diffusional effect across such dimensions can profoundly affect the tracer kinetics of rapidly exchanging ions.[6,7] Another advantage of cultured heart cells is their ability to maintain true steady-state conditions in contrast to either isolated cells or freshly dissected preparations, neither of which can be maintained in a functional state for equivalent periods of time.[8] Although extremely thin natural preparations of cardiac muscle can be prepared and exposed to environmental manipulations,[9] the combination of extensive cell injury, limited tissue mass, and inherent diffusional limitations constrain the transport capabilities of these preparations. With respect to populations of single cardiac cells obtained by the availability of improved cell isolation techniques, radiotracer methods are being successfully applied under conditions that theoretically approach the ideal for measuring the transport of ions.[10]

Despite the advantages afforded by confluent cultured cells or monolayers for ionic flux measurements, problems are present due to the possible loss of cells from the plastic or glass substrate during wash procedures[5] as well as the difficulty of uniformly stimulating a sheet or monolayer of cultured heart cells. These problems prompted the development of a novel preparation of cultured cells, namely, the growth-oriented polystrand.[11] The advantages of this preparation include small size (<100 μm diameter), relatively simple extracellular space, mechanical stability, uniform electrical excitability, and the ability to increase the tissue mass in a given volume without a concomitant increase in tissue diameter, simply by increasing the number of windings of the nylon substrate (for detailed description, see section II.A).

The heterogeneity of cell types in naturally occurring heart muscle also carries over to preparations of cultured heart cells. The coexistence of excitable (muscle) and nonexcitable (fibroblast) cells in culture presents considerable difficulty in interpreting data from isotopic flux measurements, because these measurements are nondiscriminatory with respect to cell type. To resolve this problem, tissue culture techniques have been used to grow fibroblast-like preparations (nonmuscle cells) of cardiac origin from which radiotracer and other measurements could be obtained without the complications of heterogeneous cell types. With such information, it was then possible to resolve the respective contributions of the cell types in the heterogeneous contractile preparations.[11-13] Thus, by engineering the growth of cardiac cells, tissue culture can address or minimize several problems confronting ion transport studies in cardiac muscle, e.g., diffusional delays, cellular heterogeneity, and loss of steady-state conditions.

B. Limitations of Tissue Culture Preparations

Although preparations of heart cells in culture have served successfully as a model system for studying ion fluxes in cardiac muscle,[4] questions have been raised over both the technique of tissue culture as well as the use of embryonic and neonatal heart cells in such studies. The following have been listed as cause for concern by several investigators:[14] (1) differences in cell metabolism compared to adult cells, i.e., dependence upon glycolysis as the main energy source, (2) structural and functional dedifferentiation, (3) differences in electrophysiological properties and pharmacological responsiveness to cardioactive agents, toxins, and autonomic neurotransmitters, (4) nonmuscle cell overgrowth leading to an apparent state of dedifferentiation, (5) inability to produce synchronized cell populations or clones of cardiac muscle cells, and (6) absence of functional innervation. Depending on the question under study, these can be legitimate issues which must be considered when planning to apply isotopic methodologies to cultured heart cell preparations.

C. Ion Transport Studies as Applied to Cultured Heart Cells

The potential limitations of working with cultured heart cells are convincingly diminished by the unique advantages afforded by this model system over intact myocardial preparations for measurements of ion transport. Several methods that have been applied successfully to evaluate the transport of ions across cultured heart cell membranes will be considered according to the following general classification: electrophysiological, chemical analysis, physiological responses, and radioisotopic flux.

1. Electrophysiology

The application of voltage-clamp techniques using microelectrodes to various preparations of cultured heart cells has provided a direct and quantitative description of time and voltage dependent ionic currents associated with the cardiac action potential.[1,14-18] However, even under conditions that appear to approach the ideal, interpretation of the rapid changes in ion conductance from the voltage-clamp data has not been straightfoward. Identification of the inward and outward ionic currents[19] and their separation from either the electrogenic transport mechanisms[20-22] or other novel mechanisms that regulate ionic currents, e.g., calcium activated potassium conductance,[23] are unresolved problems. Nevertheless, microelectrode techniques have been proven successful in demonstrating, unequivocally, that the Na-K pump and Na,Ca exchange transport mechanisms in cultured heart cells are electrogenic.[4,22]

Whereas Werrlein[24] demonstrated the feasibility of using ion-sensitive potassium microelectrodes in a noncardiac cultured cell line, our laboratory was the first to measure intracellular ionic activities (Na,K,Cl) in conjunction with studies of the transport properties of cultured embryonic chick heart cells.[25,26] Despite reservations about the relative size of the electrodes to the small size of the cells[27,28] and the reliance on the effectiveness of cell-to-cell coupling to correct for membrane potentials,[29] the ion-selective electrode technique can be suitable for accurate measurements of net ionic fluxes in the absence of significant intracellular buffering, and, under special conditions, for measurement of unidirectional fluxes.[30] Furthermore, the ability to monitor intracellular ionic activity both in quiescent and actively contracting cultured heart cells under steady-state and non-steady-state conditions is a significant achievement that undoubtedly will provide important new information about the electrochemical properties of cardiac muscle. As technological improvements result in the availability of electrodes with faster response times and increased selectivity, ion-selective electrodes will certainly extend our research capabilities to measure rapid changes of intracellular ion activities in cardiac cells.

2. Chemical Analysis

The accuracy of intracellular ion concentrations determined by analytic chemical methods, e.g., using atomic absorption spectrophotometry (Na, K) or coulometric titration (Cl), depends upon a precise measurement of the extracellular compartment. This problem is particularly important when determining the intracellular concentration of ions that are highly concentrated in extracellular space. The simple geometry and short extracellular diffusion distances of cultured heart cell preparations have enabled us to report comparable values for intracellular ions determined by isotopic equilibrium methods and either flame photometry (Na, K) or coulometric titration (Cl).[7,12,13] Unfortunately, chemical methods require destruction of the cells and thus prevent the monitoring of continuous or transient changes in intracellular ions from the same cell culture. Another limitation of the analytic method is the inability to distinguish between the distribution of bound and free ions within a cell, a problem of greater significance for ions such as calcium — nearly all of which is bound or complexed or stored in intracellular compartments.[28]

Direct measurements of intracellular calcium in cultured heart cells using metallochromic and fluorescent indicators have been successfully achieved in recent years.[31,32] Although these methods confirmed that only a small fraction of the total intracellular calcium is freely ionized, several technical problems have been identified that place constraints on their quantitative accuracy and applicability to transient measurements.[33,34] Methods involving the null-point procedure, though feasible for static measurements, require exposure of cells to very low extracellular concentrations of the ion in question, e.g., calcium, and ultimately to cell disruption by permeabilizing the cell membrane so that data points can be obtained and extrapolated to obtain the level of free intracellular calcium.[35] Nevertheless, as refinements emerge in the development of optical indicators, we may expect their widespread adaptation to research with cultured cells, particularly because of the rapid response time, applicability to cell populations of small diameter and the potential to localize signals within a cell under direct observation.

Two analytic methods that appear to offer considerable promise for measuring intracellular ion contents are nuclear magnetic resonance (NMR) and electron probe microanalysis (EPMA). The noninvasive NMR technique can be applied to measurements of free Mg, Na, K, Ca, and H in living cells under various physiological conditions.[36] However, the need to have a relatively large number of cells at high densities under suitable conditions for long periods of time may limit the success of this technique for studying the regulatory processes of ion transport in cultured heart cells. EPMA is a method, based on X-ray spectroscopy, which when used in conjunction with the preparatory technique of cryo-ultramicrotomy can analyze the chemical composition (elements) of cells and cell organelles *in situ*.[37,38] Results obtained with cultured heart cells in response to metabolic inhibition suggest the relative importance of the mitochondria in regulating intracellular calcium content.[31,39] However, the EPMA method cannot distinguish between free and bound calcium and does not appear to have the sensitivity required to detect the very low level of cytoplasmic calcium. Regardless of future technical developments, our ability to measure the ion content of subcellular structures already is contributing to our understanding of the regulatory processes associated with ion transport in cultured heart cells, as well as other cell types in culture.

3. Contractility

The relationship between ion transport and the contractile response of heart cells, grown as monolayers or aggregates, can be evaluated using a noninvasive optical technique that monitors both the rate and amplitude of movement.[40,41] The effects of low temperature, rapid pacing, Na-K pump inhibition and superfusion with solutions of varying ionic composition have provided the basis for postulating that sodium-calcium exchange regulates contractions of embryonic chick heart cells in a manner similar to that of the adult myocardium.

4. Radioisotopic Flux

Geometrically thin and simple tissue cultured preparations, unaffected by the trauma of dissection, have radiotracer kinetics which are not as complex as those in naturally occurring cardiac preparations and may be more reasonably interpreted. With documentation of the cell types present, cardiac cells in culture can serve as a model system for studying tracer fluxes, specifically K, Na, and Cl, and offer the prospect of assigning physiologic significance to the multiple kinetic components frequently found in such studies. Ions such as calcium, whose major fraction is bound or compartmentalized within cells may reveal complex radiotracer kinetics even in geometrically simple preparations.[27,42,43] Thus, the accuracy in interpretation of the measurement of such ions is limited.[29,33] The following paragraphs present a thorough discussion of both the theoretical basis of radiotracer measurements and the application of tracer techniques to cultured heart cells.

The investigation of membrane processes involved in regulating electrochemical gradients, ionic conductances, and metabolic exchange requires methodologies for evaluating ionic movements with the cell in the steady state, as well as in non-steady state conditions. As there is in steady-state conditions with no net ion content change (i.e., ionic influx is exactly balanced by efflux), a method is needed for measuring unidirectional ion movements without altering the ionic gradients. Radioactive tracer techniques have been widely applied for this purpose, as these measurements reveal the transport of a specific ion over a period of time during which the cells are physiologically stable. Tracer measurements also provide information about the transmembrane flux of an ion whether governed by electrogenic, electrodiffusive, or electroneutral mechanisms.

The successful application of radiotracers to flux measurements requires that the assumptions of the mathematical process used to convert the measured parameter (tracer kinetics) to the physiologic parameter (transmembrane flux) are in fact achieved. Primary to these assumptions is that the cell membrane is the major source of limitation to tracer movement. This principle is relatively easy to visualize, but often difficult to achieve experimentally. If the cell membrane is relatively impermeable to an ionic species (i.e., presents a large resistance to tracer movement) and other extra- and intracellular barriers do not affect the movement of tracer to an appreciable degree, then the kinetics of tracer exchange from the preparation will accurately reflect the cell membrane characteristics. On the other hand, should the membrane be highly permeable or possess rapid exchange processes, and should extracellular barriers represent significant limitations to tracer exchange, the experimental results will not reflect events at the cell membrane.

a. Computation of Transmembrane Fluxes

The classic formalism for calculating a transmembrane flux from tracer kinetics was developed by Harris and Burn[44] and Keynes and Lewis.[45] It requires the determination of three cellular characteristics: the rate constant of tracer exchange (from either an efflux or influx curve), the cell volume-to-surface area ratio, and the intracellular concentration of the species of interest. The equation to which we refer is

$$M = k \frac{V}{A} C_i \tag{1}$$

where k = the rate constant for tracer exchange, V = cell volume, A = cell surface area, C_i = intracellular concentration of the ion of interest, and M = transmembrane flux of the ion. It was developed on the basis of the number of assumptions, which include: (1) a single cellular compartment which is effectively homogeneous, (2) a cell membrane (compartment border) which is rate-limiting to the movement of the species, (3) a medium (extracellular space) of tightly-controlled and constant composition, and (4) steady-state conditions except for the tracer itself. An alternative method for computing a transmembrane flux using tracers requires the measurement of tracer uptake during a time interval in which no significant amount of tracer is transported back out of the cells. In effect, this is a determination of the initial slope of tracer uptake and can be formally stated as follows:

$$M_i = \frac{dY_i}{dt} \frac{V}{A} \frac{Ce}{Ye}$$

$$t \to 0 \tag{2}$$

where M_i is the transmembrane influx, Y_i is the intracellular tracer content, t is time, V/A is the cell volume to surface ratio, and Ce/Ye is the inverse of tracer specific activity in the medium. This method does not require a determination of the intracellular concentration of the species of interest. However, it is essential that the uptake measurement be made in a time that is short relative to the half time of tracer exchange, since tracer efflux is assumed to be nil. To maintain an error of less than 10%, the measurement time interval must be limited to approximately one sixth of the half-time for tracer exchange. For species with rapid transmembrane movements, such a requirement may prove experimentally unfeasible. The assumptions of a rate-limiting membrane and a tightly controlled extracellular space (constant specific activity) also pertain to the initial tracer uptake method.

b. Effects of Restricted Extracellular Space

Given the assumptions required for the analysis of radiotracer transport data, practical experimental factors pertinent to cardiac tissue render tracer techniques extremely difficult. The small volume of the cardiac cell dictates rapid exchange kinetics for a given flux level (see Equation 1). In addition, the tissue morphology, a syncytium of tightly packed cells with a well-organized extracellular space, creates a region of slow diffusion between the cell membrane and the bulk external medium. The net result of these two factors is that in many experimental preparations the cell membrane is not rate limiting to tracer movement, and thus there has been considerable difficulty in relating tracer kinetics to membrane fluxes. This problem of a diffusion-limited preparation with a restricted extracellular space has also been described in terms of tracer re-uptake or "backflux" during an efflux experiment. Thus the tracer concentration in a restricted extracellular space during an efflux is finite, and thus some tracer may undergo reuptake. The standard analysis assumes zero extracellular tracer and no re-uptake. The experimentally determined tracer kinetics will thus be slower than the true kinetics. A reverse but analogous problem may occur during an influx experiment.

Using a mathematical model which accounts for diffusion in the extracellular space of a cylindrical muscle bundle,[46] an estimate can be made of the errors seen in experimentally determined tracer kinetics. As shown in Figure 1A, as the preparation diameter increases, the error in the measured flux increases. Also, as the true transmembrane flux increases, so does the error (Figure 1B). These considerations illustrate the need for extremely thin preparations when dealing with tracer fluxes in cardiac cells. Tissue culture preparations, as noted previously, have been designed so as to achieve minimal diffusion distances while maintaining sufficient tissue mass to permit accurate resolution of tracer exchange characteristics.

c. Multicompartment Analysis

Rarely, if ever, does a biological system adhere to the simplistic model of a single compartment with exchange into a well-stirred bathing medium. In the discussion above on diffusion limitations, we have seen that pseudocompartments can exist in series with the cells and complicate the analysis of results. Other, nonartifactual compartments can also complicate interpretation of results. This is especially relevant to naturally occurring cardiac preparations in which a multiplicity of functionally and anatomically distinct cells comprise the tissue. Even if the cells possessed the same membrane characteristics, the varying V/A would add different kinetic components to the tracer exchange. Although one cell type may predominate, all the cells contribute to the experimental result.

A further complication of cellular heterogeneity is induced by experimental manipulation of the preparation. In naturally occurring preparations, cut edges present an undefined zone of damaged cells which will contribute to the kinetics of exchange. Metabolically compromised regions may exist in natural preparations and in some large tissue aggregates due to diffusional limitations. Analysis of tracer exchange in freshly disaggregated single cell suspensions must quantitatively account for the contribution of the cells damaged by the isolation procedure.

Tissue culture preparations should be viewed with no less concern regarding the existence of heterogeneous cell types. Although one may argue that the effects of disaggregation damage have lessened after a period in culture, variant cell types, especially fibroblast-like

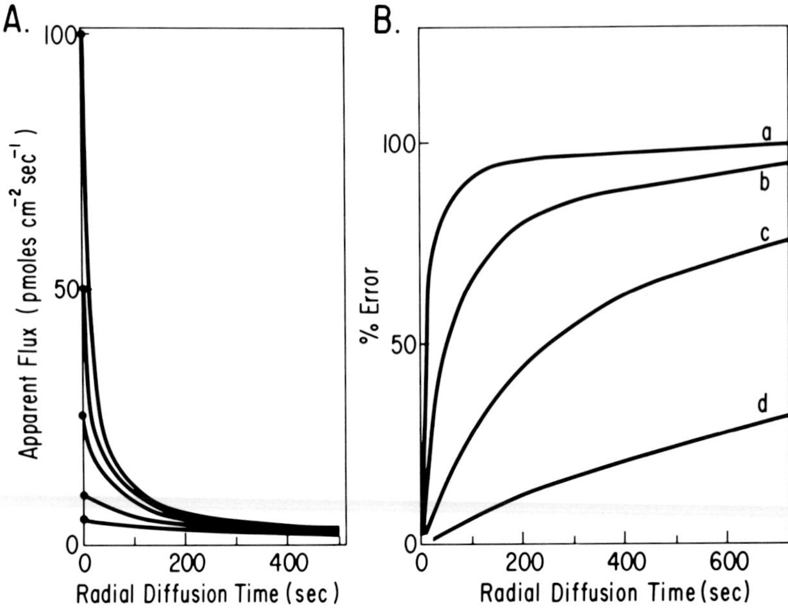

FIGURE 1. Graphical representation of solutions of radial diffusion equation. (A) Apparent potassium flux as a function of radial diffusion time. The latter is given by the square of the preparation radius divided by extracellular diffusion coefficient (r^2/D). True membrane flux values for each curve indicated by dots at zero radial diffusion time. (B) Percentage error of measured flux values of (a)100, (b)25, (c)5, and (d)1 pmol $\cdot$ cm^{-2} $\cdot$ sec^{-1}. Parameters used in both computations: V/A $= 1 \times 10^{-4}$ cm, packing fraction $= 0.75$, $K_i = 150$ mM, and $K_o = 5.4$ mM. (Modified after Horres, C. R. and Lieberman, M., Compartmental analysis of potassium efflux from growth oriented heart cells, *J. Mem. Biol.*, 34, 331, 1977.)

cells, can proliferate in culture more rapidly than the cells of interest, thereby creating a time-dependent population to evaluate.

An approach to the analysis of complex tracer exchange data due to multiple cell types is through multicompartmental analysis. The technique assumes that the tracer exchange is composed of a number of compartments exchanging in parallel. Compartment sizes and exchange constants are assumed to be different, but fixed, in each experiment. Mathematically, the expression for tracer efflux as a function of time, when all compartments are assumed loaded to tracer equilibrium, might appear as:

$$Y_i(t) = \frac{Ye}{Ce}(C_1e^{-k_1t} + C_2e^{-k_2t} + C_3e^{-k_3t}) \tag{3}$$

where $Y_i(t)$ = tracer content as a function of time (t), C_n = ion content of compartment n, k_n = rate constant of tracer exchange of compartment n, and Ye/Ce = specific activity of the tracer loading solution. If the compartment sizes are similar and the rate constants at least several-fold different, then the contributions of the various compartments can be readily visualized in a semi-log plot of an efflux curve. An example is shown in Figure 2. However, if the rate constants are similar, their determination by curve-fitting, regardless of method, is prone to significant errors. Assuming that the intracellular concentrations and rate constants of the V/A can all be experimentally resolved, then the transmembrane fluxes for each cell type can be computed using Equation 1.

True compartmentation in a biologic system can exist in series as well as in parallel. An

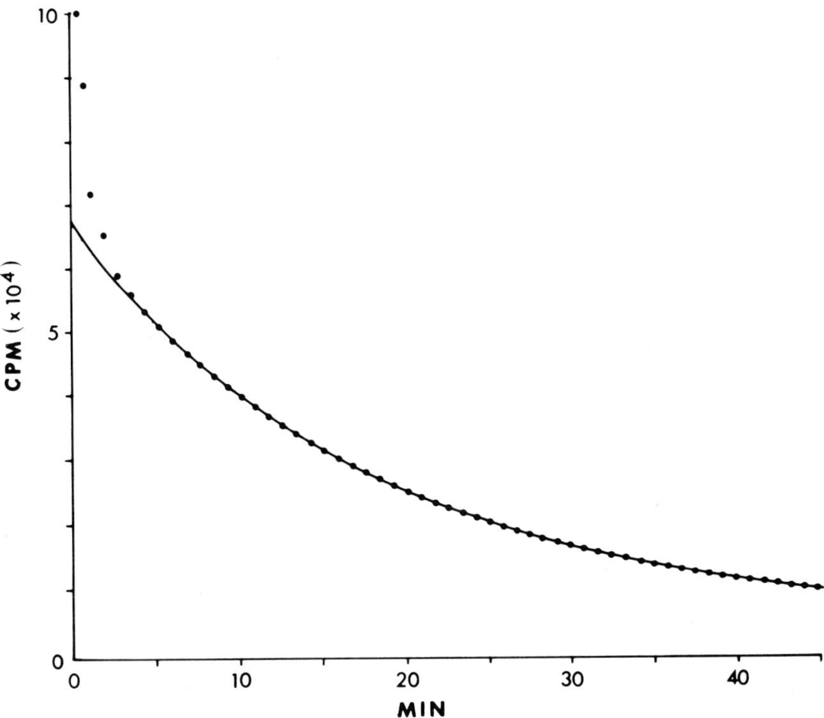

FIGURE 2.   Analog computer curve fit for ^{42}K efflux kinetics of a contractile preparation of heart cells. Conditions: K_o = 5.4 mM, 37.5°C, 150 beats/min. Parameters: Initial conditions of fast and slow compartments are 48,400 cpm and 19,000 cpm, respectively. Rate constants: fast compartment (.062 min^{-1}), slow compartment (.015 min^{-1}). (Modified after Horres, C. R. and Lieberman, M., Compartmental analysis of potassium efflux from growth oriented heart cells, *J. Mem. Biol.*, 34, 331, 1977.)

intracellular organelle wholly contained within the cytoplasm may behave as a series compartmental system. However, only those intracellular compartments which exchange the ion of interest with the cytoplasm more slowly than the exchange of that ion between cytoplasm and extracellular space can be detected by tracer kinetic analysis in intact cells or tissue. A more rapidly exchanging intracellular compartment is effectively hidden. Application of series compartmental analysis to an experimental situation requires assumptions regarding conditions and compartmental definition (rate limiting compartment boundaries, effectively uniform composition within compartments, constant external medium composition, and steady-state conditions except for the tracer) just as in the single compartment case. The tracer kinetics for an efflux experiment in a series compartment system is of the same multiexponential form as in the parallel case. However, unlike the parallel situation, the parameters (rate constants and compartment sizes) determined from the experimental curve do not directly reflect the true parameters of the compartment. Rather, the parameters of a series system interact in a complex fashion to produce the tracer kinetic curve. Huxley[47] has derived a series of equations in which the experimentally derived quantities (generated by fitting a tracer efflux curve to a multiexponential function of time) are used to compute the true compartment parameters. As the rate constants of the efflux components become numerically closer, the differences between the constants stripped from the experimental curve and the true constants become larger.* For example, given an ideal efflux experiment

---

*   This statement slightly oversimplifies the situation. The percentage difference between experimental and true constants in a series system with adjacent compartments of equal size actually maximizes when the ratio of the true faster and slower rate constants is approximately 1.7. Of course with rate constants such as this, accurate separation of the two exponential components from an experimental curve is unrealistic.

in a series system with two compartments and an external medium, if the true efflux rate constants differ by a factor of 10 and compartment sizes are equal, then the faster rate constant derived experimentally will be 8% greater than the true constant and the slower experimental rate constant will be 7% less than true. If the true efflux rate constants differ by a factor of three, then the percentage differences between experimental and true rate constants will be 17% for the faster and 15% for the slower rate constant. If the compartment sizes are unequal, the differences may be much larger. When a series compartment system can reasonably be presumed to exist, such corrections can legitimately be applied to tracer kinetic data. However, the greater the correction required, the greater the uncertainty produced.

A tracer kinetic curve which exhibits a multiexponential character provides no direct evidence as to the parallel or series nature of the compartments. Other observations must be brought to bear to determine the probable compartment arrangement. Multiple components, do, however, imply independent pools of the ion in question. There may be many different transport mechanisms at a compartment boundary, but as long as all operate on the same pool, each compartment (or pool) will produce only one exponential component in the flux curve. Furthermore, the method to calculate the transsarcolemmal ion flux is dependent on whether a parallel or a series compartment model is chosen. If multiple cell types are present (parallel case), then the flux for each is calculated independently using the intracellular ion concentration and surface-to-volume ratio of each cell type. If a separate intracellular compartment is presumed (series case), then the sarcolemmal flux is calculated using the ion content of a rapidly exchanging compartment only, since the slowly exchanging internal compartment does not directly participate in sarcolemmal exchange.

II. METHODOLOGY

A. Preparation of Cultured Heart Cells

In our laboratory, suspensions of cardiac cells are obtained by enzymatic disaggregation of either 10- to 11-day-old chick embryo hearts or primary mass cultures of embryonic heart cells. Intact embryonic hearts are the source of cells for contractile preparations, whereas noncontractile preparations are produced by seeding culture dishes or growth chambers with cells obtained from quiescent primary cultures of preplated cells. The methods for producing primary cell cultures is as follows:[11-13]

1.   Whole hearts are dissected from chick embryos, mechanically minced and disaggregated with 0.025 to 0.1% trypsin in Ca, Mg free Hank's salt solution.
2.   Cells are separated by agitation and gentle pipetting.
3.   At periodic intervals of 6 to 10 min, the entire disaggregating solution is aspirated and immediately added to an equal volume of ice-cold medium containing 5 to 10% fetal bovine serum to quench the enzyme activity.
4.   The resultant suspensions obtained from four disaggregation cycles are combined and filtered through Nitex cloth (60 μm mesh) to remove residual clumps.
5.   The suspension is centrifuged at 140 g for 6 min, and following aspiration of the supernatant, the cells are resuspended to obtain the desired cell concentration.
6.   The percent of muscle cells in the suspension is increased by preplating for 1 hr.[48]
7.   A new cell pellet is prepared from the muscle-enriched supernate and appropriate dilutions are made.
8.   Cell viability is assessed by the trypan-blue exclusion method.
9.   Cells are cultured either conventionally to confluency (mass culture) or as a synthetic strand or cultured cluster or polystrand, depending on the experimental requirements.

1. Mass Cultures

Mass cultures of contractile heart cells are obtained by seeding 35 mm culture dishes with 1.5×10^6 cells in a medium that contains 5% fetal bovine serum, 2% chick embryo extract and a modified Medium 199 with Earle's salt solution (bicarbonate buffer). The cells are incubated for only 3 days at 37°C in 96% air 4% CO_2 to reduce the risk of residual nonmuscle cell overgrowth. Preparations of nonmuscle cells are obtained by subculturing the cells that preferentially adhere to the culture dish surface during the preplating procedure. Cell cultures grown to confluency whether in plastic culture dishes or on cover slips have been found suitable for studying ion fluxes and contents as well as the effects of cardioactive agents on contractility of cardiac cells.[5,27,31,40,42,43,49] The advantages and disadvantages of these preparations have already been stated. In brief, although the effects of diffusion are minimized, the cells are not easily stimulated with electrodes and they tend to detach from the substrate during continuous perfusion.

2. Synthetic Strands

Synthetic strands of cultured heart cells can be grown to provide short (ca. 100 μm), narrow (50 to 100 μm) preparations to which the analytic technique of voltage clamp can be applied.[1] Dissociated embryonic chick heart cells are growth oriented within channels of varying length that are cut in an agar-coated substrate. Analysis of the time sequence of the current-voltage relationships throughout the clamp steps suggested that the membrane current consisted of two components: a transient current composed largely of sodium ions and a steady-state current dependent on potassium ions. However, these studies were limited in scope because the preparations are not ideally suited for ion substitution studies, as they become dislodged during continuous perfusion.

3. Cultured Clusters

By altering the geometry of the cylindrical strand to that of a spherical aggregate (60 to 80 μm diameter), cultured clusters come closest to the ideal (single spherical cell) in minimizing problems associated with using a point source of current with voltage-clamp methods. Clusters of cardiac cells are obtained by modifying the methods used to prepare synthetic strands.[15] Using a 27 gauge needle, a 20 μm opening is exposed beneath the agar-coated culture dish and preformed small spherical aggregates (50 to 200 μm diameter) are able to attach and become functionally competent. Although these preparations are well-suited for certain physiologic and pharmacologic interventions, they too are mechanically unstable under constant rapid perfusion.

4. Polystrands

Growth orientation in tissue culture has been utilized successfully to obtain polystrand preparations with suitable dimensions and stability for ionic flux studies.[11] Muscle enriched cells are seeded in specially designed chambers in which the cells reaggregate and become aligned as an annulus of 30 to 40 μm around a nylon core (20 μm). Tissue mass is increased by allowing the cells to grow along a parallel array of 70 segments (5 mm long) formed by a continuous wrapping of nylon monofilament around a silver wire (Figure 3). This maneuver enhances the resolution of tracer and chemical measurements because the tissue mass is increased without increasing strand diameter. These preparations are electrically stimulatable and are ideally suited for rapid perfusion experiments. Whether they can reproducibly be grown to the dimensions required for voltage-clamp stability remains to be determined.

In summary, growth orientation in tissue culture provides the means to develop preparations of minimal complexity. The method also allows the investigator to vary the seeding density of a cell suspension so that tracer kinetics of preparations of varying dimensions can be studied to evaluate the effects of tracer reflux. Recently, isolated feline cardiac

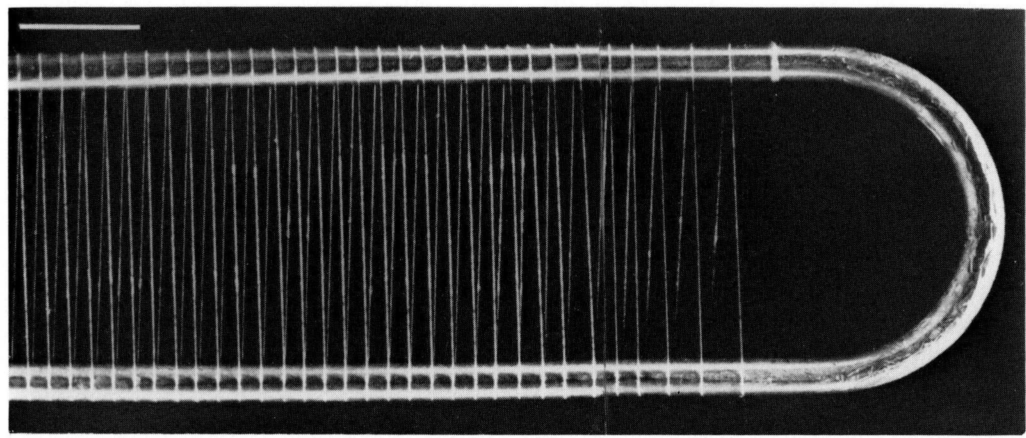

FIGURE 3.   Photomicrograph of a segment of the polystrand preparation. Central portion of the nylon substrate that wraps around the silver wire is the region of cell growth as indicated by the thickened portion of the strand. Scale: 3.8 mm. (Modified after Lieberman, M., et al., Cardiac muscle with controlled geometry. Application to electrophysiological and ion transport studies, in *Excitable Cells in Tissue Culture*, Nelson, P. G. and Lieberman, M., Eds., Plenum Press, New York, 1981, 379.)

myocytes were used to measure unidirectional transmembrane potassium fluxes.[10] Improvements in methods to isolate myocytes[50,51] offers encouragement for future studies requiring stable cell preparations free of diffusional limitations.

B. Special Techniques
1. Volume to Surface Area

Cell volume and surface area play a significant role in tracer exchange kinetics since, as previously noted, given the same transmembrane flux, cells with a low V/A will have relatively rapid exchange, while those with higher V/A (such as spherical cells) will have slower kinetics. In organized tissue, measurement of cell V/A is best accomplished by stereological techniques that involve analysis of high resolution electron micrographs with a grid matrix. The theory of the technique and its application to cardiac muscle has been described by Page et al.[52,53] Correction should be applied for tissue orientation (anisotropy) as discussed by Eisenberg et al.[54,55] by making measurements at several grid angles. The success of the technique is largely dependent on the ability to obtain micrographs with cell morphology in the native state.

An alternative approach to V/A determination consists of independently measuring cell volume and cell surface area. With disaggregated cells, packed cell volume corrected for trapped fluid with extracellular markers can be correlated with cell count prior to centrifugation. Cell surface area presents a more difficult parameter to estimate due to the numerous membrane foldings and projections of cardiac cells. However, disaggregated embryonic cardiac cells when exposed to hypotonic solutions, become spherical,[56] and thus enable the use of light microscopic techniques to measure diameter from which calculations of surface area can be obtained from the geometrical relation: $A = 4\pi r^2$. Based on the calculated values for cell volume and surface area of freshly disaggregated 19-day embryonic chick hearts, the data resulted in a value for V/A of $0.91 \pm 0.09 \times 10^{-4}$ cm, which compared favorably with a value of $1.24 \pm 0.07 \times 10^{-4}$ cm found for the same age hearts by electron micrographic stereological techniques.[56] With embryonic heart cells in culture, two distinctly different cell types are seen in electron micrographs: spindle-shaped muscle cells and flattened fibroblast-like cells. The V/A values of the two cell types differ markedly, with a V/A of

1.06 $\pm$ 0.25 $\times$ 10^{-4} cm for muscle cells and 0.48 $\pm$ 0.08 $\times$ 10^{-4} cm for fibroblast-like cells as determined stereologically.[11] These values reflect the fact that the flattened fibroblast cells project more surface area for the volume contained in the cell than do the more cylindrical muscle cells.

2. Cell Water

An essential parameter in the calculation of intracellular concentration of a species of interest is the measurement of cell water. With cells in culture, water measurements are particularly difficult to determine on the traditional basis of wet/dry weight differences. Because the usual blotting techniques are not possible with cultured cells, adhering culture media can be a significant proportion of the wet weight. In addition, culture dishes are unsuitable for tare containers as their mass greatly exceeds that of the cell culture. The polystrand preparation offers considerable advantage over cultures grown on dishes for cell water determinations. Orientation of the strands permits the use of a new blotting technique to remove excess fluid and thereby enhance the resolution of cell wet weight.[11] This method involves briefly submerging the preparation into a test tube filled with fluorocarbon FC-80 (3M Company, St. Paul, Minn.) and gently tapping the tube which allows the less dense aqueous media to rise to the surface where it can be removed with cotton-tipped applicators. The nylon strand on which the preparation is grown is an ideal tare container for the cells and can be rapidly slipped off the growth support for weighing after blotting. Typical nylon strand weights are 0.3 mg, whereas preparation wet weights are typically 2 to 3 mg. Moisture uptake of the nylon during culture is minimal. Because water loss from the tissue is rapid after removal from the blotting solutions, we employ an extrapolation technique from wet weight data points gathered over a 10 min period in a high humidity environment to determine weight on removal from the blotting medium.

Since tissue blotting removes only the adherent water, wet weight must be corrected for residual extracellular water by using a suitable impermeant marker such as ^{125}I-iothalamic acid.[11] Using the blotting technique described above, we were able to reduce extracellular water to about 50% of total preparation wet weight in polystrand preparations. After obtaining wet weight, preparation dry weight was measured after storage overnight at 110°C. Cell dry weight was determined by subtracting the tare weight of the nylon substrate. Cell water, deduced from the difference between cell wet weight and cell dry weight and expressed as a percentage of cell wet weight, was found to be 81.1 $\pm$ 1.4% for 12 preparations of mixed contractile/noncontractile cells and 78.2 $\pm$ 5.0% for 8 preparations of noncontractile cells. As the two values are not significantly different, the muscle cells in the mixed population can be assumed to be equivalent in cell water to the nonmuscle cells. These values were confirmed by determining the osmotically active component (73%) based on changes in strand volume.[57] Comparable values were obtained for intact 6- to 8-day embryonic chick hearts (80.5%)[56] and cultured aggregates of chick heart (80.3%).[58] Using the percentage of cell water, it is possible to calculate concentration from ion content per cell dry weight.

Ion activity in chick embryo ventricular muscle has also been determined with ion-specific electrodes,[59] but considerable care must be taken to eliminate artifacts contributed by interfering ions, membrane potential measurement, and electrode damage.[28,30] The correlation of activity with concentration depends on the extent of ionic binding occurring in the cytoplasm. From the standpoint of the computation of tracer fluxes, the concentration is more relevant than chemical activity, provided that the ions exchange from any intracellular binding sites more rapidly than they exchange across the cell membrane.

3. Tracer Flux Measurements with Rapidly Exchanging Compartments

As is apparent from the foregoing discussions, measurement of the transmembrane tracer flux of a rapidly exchanging ion requires a tissue with minimal extracellular diffusion

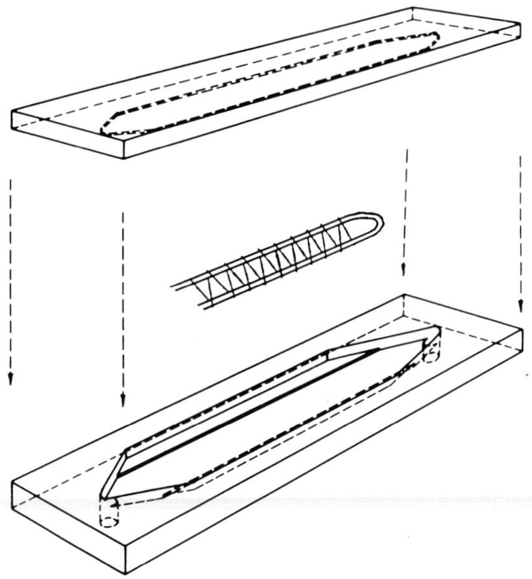

FIGURE 4.   Flux chamber with polystrand. Silicone rubber potting compound (Sylgard 184, Dow Corning, Midland, Mich.) is used to fabricate flux chamber. For details, see Wheeler et al.[12]

distances. Even with such a preparation, the handling of solutions and tissue must also be sufficiently expedient so as not to obscure the true transmembrane flux. In the case of an efflux experiment, the clearance of tracer ions from the solution surrounding the preparation must be at least several-fold more rapid than the true transmembrane exchange. For an influx experiment, at the completion of the uptake period, the extracellular space and surrounding solution must be rapidly cleared of tracer while the intracellular contents are retained.

We have designed a tracer efflux apparatus which can accurately measure a tracer kinetic component with a rate constant of at least 3 min^{-1} in the polystrand preparation.[12,60] The flux chamber which accepts the polystrand was designed to have the minimum volume required to accommodate the preparation so that solution turnover would be as rapid as possible. The chamber volume is 1 mℓ, and typical fluid flow through the chamber is 24 m$\ell \cdot min^{-1}$. Two silicone rubber halves which conveniently clamp and seal together form the flux chamber (Figure 4). A platinum plate electrode is embedded in each half to provide for field stimulation of the polystrand at a controlled rate. The chamber is mounted on a sliding frame which has three working positions (Figure 5). One is an unobstructed position where the chamber can be opened and preparations inserted or removed. The second places the chamber under a microscope and over a light source so that the tissue can be viewed without opening the chamber. The third position locates the chamber between two lead-shielded NaI scintillation crystals which directly monitor the radioactivity in the preparation during the washout.

Polystrands are isotopically loaded under controlled conditions outside the flux chamber. Transfer of preparations to the flux chamber is accomplished by immersing the polystrand in a tube of fluorocarbon liquid (FC-80) and blotting the excess radioactive solution from the top of the denser fluorocarbon. After two very brief rinses in cold solution (to dilute any remaining adherent loading solution which could contaminate the area near the flux chamber), the polystrand is placed in the chamber, the chamber closed and moved to the counting position, and perfusion of the chamber begun. Just upstream from the flux chamber

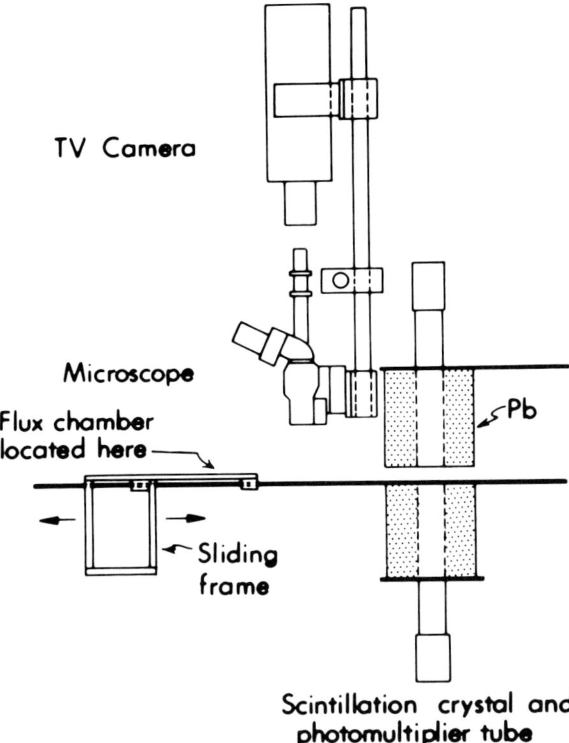

FIGURE 5.   Side view of experimental apparatus illustrating relationship between flux chamber, sliding frame, microscope, and scintillation crystals. Shaded area indicates lead shield. (Modified after Wheeler, D. M., Sodium Tracer Kinetics and Transmembrane Flux in Tissue-Cultured Chick Heart Cells, Ph.D. thesis, Duke University, Durham, N. C., 1981.)

is a four-way valve to provide for solution changes without disruption of flow or counting. This feature proved most useful in separating extracellular from intracellular ^{24}Na efflux by allowing a rapid switch of superfusion solution from one at 6°C (for extracellular washout) to one at 37.5°C (transmembrane efflux).

The above apparatus and methods offer several advantages in tracer efflux determination in addition to rendering the measurement of rapid fluxes feasible. A single polystrand preparation can generate an entire efflux curve, unlike most techniques using monolayers of cells.* Also, given the appropriate data collection and display equipment, the efflux curve is generated in real time, allowing for immediate feedback and for rapid and efficient data processing. This feature is not possible with methods based on effluent collection and counting.

The tracer influx of a rapidly exchanging ion has been measured using cells grown on small glass or plastic coverslips.[5,27,40] (see also Figure 8). These facilitate the rapid passage of tissue through isotopic loading and wash solutions so that early time points on the uptake curve can be obtained. Similar experiments can be carried out with cells grown in culture dishes or flasks:[62-64] however, the wash procedures in these situations are somewhat more difficult.

---

*   Langer and co-workers[61] have devised a flow cell for superfusion of monolayers of cells grown on scintillator discs. Radioactivity is directly monitored during tracer washout. Two limitations pertain: (1) chamber temperature cannot be maintained at 37°C, so experiments are performed at 24°C; (2) solution turnover within the flow cell is slower than in our apparatus and may not be sufficient to measure very rapid exchanges.

An alternative to tissue-cultured preparations for ion transport work is the use of freshly disaggregated cardiac cells. Such preparations of isolated cardiac cells in suspension have proven useful for morphologic, electrophysiologic, metabolic, and receptor studies. However, their development for ion transport determinations remains in its infancy. The techniques used for procuring such cells have recently been reviewed in several forums,[65,66] so only a few points relevant to ion transport studies will be noted here.

Early methods of disaggregation produced cells which maintained their morphology and function only in the presence of micromolar calcium concentrations. Upon exposure to physiologic calcium concentrations, these "Ca-intolerant" cells would go into irreversible contracture, losing their intracellular organization and eventually their functional integrity. Most recent investigations have utilized methods which produce "Ca-tolerant" myocyte suspensions (for review, see Reference 8). The majority of these cells maintain their structure and function in physiologic concentrations of calcium. However, a portion of these suspensions do consist of rounded cells in contracture. Although these rounded cells deteriorate relatively rapidly, in some studies they are found to have a grossly intact sarcolemma.[67] Methods to separate the functional, rod-shaped myocardial cells in suspension from the rounded cells using Ficoll and albumin gradients have been described,[51,68] but the efficacy of these techniques in producing preparations suitable for ion transport studies remains to be established. Clearly, the presence of a population of damaged cells (in contracture) with altered ion contents and transport mechanisms compromise the unique interpretation of ion transport measurements. Specifically, the ion contents will reflect the sum of the contents of both cell types (normal and damaged) whereas the flux studies may reveal two kinetically definable compartments. Other concerns regarding myocardial cell suspensions include alterations in ion transport mechanisms (even in rod-shaped cells) secondary to the enzyme treatment and low Ca required for cell isolation, cell fragility, and the non-steady-state nature of some preparations.

In spite of the above difficulties, there are potential advantages of isolated cardiac cell preparations. In a well-stirred suspension there is very little extracellular diffusion limitation to transport, and the cells can be rapidly and uniformly exposed to tracers and drugs. Transport can be stopped by centrifugation through an oil phase, although markers are required for determination of total water and intracellular water. Unlike tissue culture preparations, which generally originate from embryonal or neonatal tissues, adult hearts can be used to provide isolated cell suspensions. Furthermore, recent results show that some methods of isolation produce cells with Na and K contents comparable to both intact tissue and cultured preparation.[10,69,70] Potassium fluxes in such cells have been found to be in the range of 20 to 25 $pmol \cdot cm^{-2} \cdot sec^{-1}$,[10] a value only slightly larger than that found in most tissue-cultured preparations.

III. RESULTS AND DISCUSSION

A. Potassium Transport

The potassium ion plays an important role in the electrophysiology of cardiac muscle. Resting membrane potential in intact cardiac tissue, isolated cells, and tissue-cultured preparations respond in a predictable manner to changes in external potassium concentration. This characteristic has generated significant interest in investigating the exchange of potassium across the cell membrane. Although pioneering work was conducted in 1962 by Burrows and Lamb,[5] it has only been within the last decade that an understanding of the advantages of tissue culture in studying potassium transport has been appreciated.

1. Potassium Efflux

We have examined in detail potassium tracer influx and efflux kinetics in the polystrand

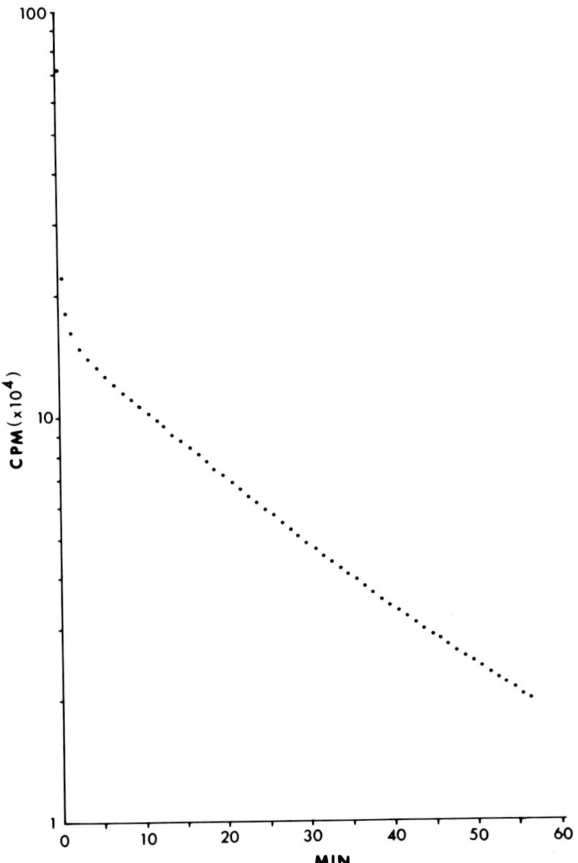

FIGURE 6.   Time course of ^{42}K efflux from a contractile polystrand paced at 150/min. After correcting for isotopic decay and background, tissue radioactivity is expressed logarithmically in counts per minute (ordinate). Time (in minutes) from the onset of perfusion with nonradioactive solution (K_o = 5.4 mM) at 37.5°C (abscissa). (Modified after Horres, C. R. and Lieberman, M. Compartmental analysis of potassium efflux from growth oriented heart cells, *J. Mem. Biol.*, 34, 331, 1977.)

preparation. Efflux kinetics are the most readily obtained because of the need during an influx experiment to remove the tracer from the bathing solution to quantitate the amount of tracer which has entered the cells. We have utilized two experimental approaches in the evaluation of potassium-42 efflux kinetics. The first is based on collecting effluent in test tubes from a superfused preparation over timed intervals and then counting the tracer in each tube in a well-type scintillation counter. Residual counts in the preparation at the end of the perfusion are also obtained, and the total tracer in the preparation as a function of time is calculated by back addition from the end to the beginning. The second technique involves direct monitoring of tracer content by placing the preparation in a superfusion apparatus mounted between gamma detection crystals as described above in section IIB3.

The potassium exchange kinetics of the polystrand were independent of experimental technique. A representative result for potassium efflux is shown in its raw form in Figure 6. The first phase is the most rapid and represents a composite of the removal of tracer from the adhering film of liquid and from the interstices between cells. If superfusion is rapid

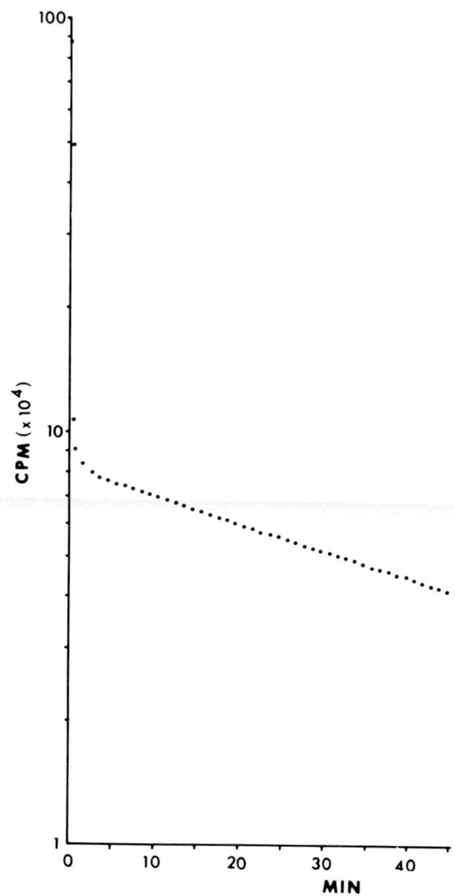

FIGURE 7. ^{42}K efflux kinetics of a noncontractile (nonmuscle) preparation. K_o = 5.4 mM; 37.5°C. (Modified after Horres, C. M. and Lieberman, M., Compartmental analysis of potassium efflux from growth oriented heart cells, *J. Mem. Biol.*, 34, 331, 1977.)

and diffusion unrestricted, these spaces exchange with a half-time on the order of seconds. These extracellular exchanges are little affected by reduction in temperature, because of the relative insensitivity of the diffusion coefficient of ions in water from physiologic temperatures down to the freezing point when compared to the marked temperature sensitivity of biologic transport processes. The next phase of exchange occurs predominately in cardiac muscle cells in the culture and is followed by a phase reflecting the slowly exchanging nonmuscle cells.

Recognition that the exchange from cardiac muscle cells represented the more rapid of the two cellular compartments was made by manipulating the population ratios of the polystrand. Although it is difficult to obtain highly purified cardiac muscle cell populations in tissue culture, preparations of nonmuscle cells derived from secondary cultures of embryonic cardiac tissue are readily obtained. Figure 7 shows that tracer effluxes from these preparations match the result of the slow compartment of the mixed preparations. Our experience has shown that the population ratio can vary considerably depending on factors including media type, enzyme sources, skill of the tissue culture technician, enrichment techniques, and more importantly, age of the culture. The potassium exchange from the muscle cells appears to remain fairly constant in culture, but the proportion of nonmuscle cells increases with time.

Table 1
POTASSIUM EFFLUX KINETICS OF CULTURED HEART CELLS

Preparation	Steady-state efflux rate constant (min^{-1})	Ref.
Chick embryo heart muscle-polystrand	0.067	7
Chick embryo heart fibroblast-polystrand	0.015	7
Chick embryo heart fibroblast-monolayer	0.026	5
Neonatal rat heart fibroblast-monolayer	0.018	71
Neonatal rat heart muscle-monolayer	0.058	63
Neonatal rat heart muscle-scintillation disc	0.016	72
Neonatal rat heart muscle-scintillation disc	0.064	61

Table 2
POTASSIUM UPTAKE IN CULTURED HEART CELLS

Preparation	Rate of Uptake (nmol · mg protein^{-1} · min^{-1})[a]	Ref.
Chick embryo heart muscle	32	75
Chick embryo heart muscle	22	76
Chick embryo fibroblast	10	76
Neonatal rat heart muscle	15.8	64
Neonatal rat fibroblast	6.5	64
Guinea pig embryo fibroblast	14.3	64
Chick embryo fibroblast	14.9	64

[a]   The experimental data did not allow for the quantitation of K influx, contrary to the studies cited in References 73 and 74.

A summary that illustrates the differences of potassium efflux rate constants for primary and secondary culture of heart muscle cells (muscle and fibroblast respectively) is shown in Table 1. When data is also available for cell volume to surface area and potassium concentration, the exchange constants can be converted to flux values. In the polystrand preparation, a flux of 15.7 pmol · cm^{-2} · sec^{-1} is obtained for the muscle cell component and 1.9 pmol · cm^{-2} · sec^{-1} for the fibroblast-like cells.[7] These data, in combination with an assessment of the electrochemical gradient for potassium, have enabled us to determine the potassium permeability of the cardiac muscle cell membrane (for details, see Reference 53).

2. Potassium Influx

Although potassium influx experiments are somewhat more difficult to perform and evaluate than efflux measurements, experiments with the polystrand preparations have yielded values for potassium influx of 14.6 ± 4.4 pmol · cm^{-2} · sec^{-1} that approximate the rate of potassium efflux from these cells.[73] Recent studies with tissue cultured neonatal rat heart cells reported a potassium influx value of 13 pmol · cm^{-2} sec^{-1}.[74] Exposure to ouabain (10^{-4} M) caused a reduction of potassium influx by 72% in the polystrand preparation[73] and 75% in tissue cultured neonatal rat heart cells.[63]

Several investigators have used tracer uptake as an index of potassium influx in cultured heart cells (Table 2). However, in the absence of a complete analysis of cell volume to surface area and multicompartment involvement, the results do not provide a basis for quantitatively evaluating potassium influx. Nonetheless, differences between heart muscle and fibroblast cultures are apparent when data are compared from the same species.

B. Sodium Transport
1. Determination of Sodium Content
a. Methodologic Issues

The major concern in measurement of the intracellular sodium concentration in tissue-cultured heart cells is identical to that in intact tissue: adequate clearance of extracellular sodium prior to tissue digestion or counting, or alternatively, proper correction for extracellular sodium via extracellular space measurements. The difficulty with the former approach is the possible loss of intracellular Na during washing, while the latter introduces the errors inherent to extracellular space determinations. In tissue culture preparations, most investigators have utilized the technique of clearance of extracellular Na, taking advantage of the minimal extracellular diffusion distances and the relatively open extracellular space of these preparations. Specific extracellular clearance protocols vary between investigators, but the washing solution is generally chilled and contains some calcium to minimize the loss of intracellular Na. Regarding the efficiency of washing techniques, Barry and Smith[43] have demonstrated that four washes of 4 sec each at 4°C effectively clear the extracellular space of ^{51}Cr-EDTA in monolayers of chick embryo heart cells grown on glass coverslips. The extracellular space of the polystrand preparation is cleared of ^{24}Na with a rate constant of ≥ 10 min^{-1} at 6°C when the individual strands average 100 μ diameter or less.[60] Thus, the polystrand preparation can be cleared of >99% of its extracellular Na within 30 sec. The use of a cold washing solution in order to retain intracellular Na appears justified from the data of Burrows and Lamb,[5] which reveals a ^{24}Na efflux rate constant of 0.08 min^{-1} at 2 to 3°C in confluent cultures of noncontractile cells derived from chick embryo heart. We have also found that in contractile polystrand preparations,[12] the intracellular ^{24}Na efflux at 6°C has a rate constant of approximately 0.1 min^{-1}. This rate constant corresponds to a 5% or less loss of intracellular Na in 30 sec.

b. Intracellular Na Concentration in the Steady State

Values reported for sodium content and concentration in tissue-cultured heart cells are shown in Table 3. A wide variety of methods and tissues are included in this tabulation, but nearly all values fall between 11 and 20 mM. The highest value (reported by McDonald and De Haan[58]) was determined by a method dependent on measurement of extracellular space in a cell pellet and subsequent correction for extracellular Na. All other investigators utilized a rinsing procedure for clearance of extracellular Na. The lowest values are those of Werdan *et al.*[64] in rat heart cells. If one assumes a cell water content of 4 to 7 $\mu\ell \cdot$ (mg prot)$^{-1}$ (see[77,78,80]), the resultant [Na]$_i$ is 4 to 7 mM given the measured value of 27 nmol $\cdot$ (mg prot)$^{-1}$. No explanation is apparent for this divergence.

There appears to be no systematic difference between Na contents determined by isotopic equilibration and those by tissue digestion and subsequent flame photometry. Thus, all intracellular Na is exchangeable with extracellular Na. Also, the Na concentration of noncontractile cells appears similar to that of heart muscle cells.

c. Na Content Changes in Response to Na-K Pump Inhibition

The maintenance of a large, inwardly directed sodium gradient is widely accepted to be the result of a membrane-associated Na-K pump. Inhibition of the pump results in a dissipation of the usual sodium gradient. The net uptake of sodium immediately following pump inhibition may be taken as an approximation of the steady-state Na efflux due to the pump* if the following assumptions are granted: (1) pump inhibition is complete and essentially instantaneous; (2) measurements are sufficiently rapid so that the *initial* net sodium uptake can be determined; (3) the method of pump inhibition directly affects no other sodium

---

* The contribution of Na-for-Na exchange mediated by the Na-K pump would not be detected by this method.

Table 3
INTRACELLULAR SODIUM CONCENTRATION IN TISSUE-CULTURED HEART CELLS

Preparation	Na Assay[a]	$[Na]_i$	Ref.
Noncontractile chick embryo heart cells	F	16 mM	5
Noncontractile chick embryo heart cells	I	17.6 mM	5
Noncontractile chick embryo heart cells (polystrand)	F	20 mM	12
Noncontractile chick embryo heart cells (polystrand)	I	16 mM	12
Girardi heart cells	?	19 mM	62
Girardi heart cells	I	54 nmol·(mg prot)$^{-1}$	64
Noncontractile rat heart cells	I	12.5 mM	63
Aggregates of chick embryo heart cells	F	33.5 mM	58
Contractile chick embryo heart cells	I	15 mM	77
Contractile chick embryo heart cells	I	19 mM	78
Contractile chick embryo heart cells (polystrand)	F	16 mM	12
Contractile chick embryo heart cells (polystrand)	I	18 mM	12
Contractile chick embryo heart cells	F	83 nmol·(mg prot)$^{-1}$	31
Contractile quail embryo heart cells	F	20 mM	79
Contractile quail embryo heart cells	I	19 mM	79
Contractile neonatal rat heart cells	I	11.7 mM	63
Contractile neonatal rat heart cells	I	27 nmol·(mg prot)$^{-1}$	64

[a]   Assay method: I — isotope equilibration: F — flame photometry.

transport mechanisms; and (4) the contribution of the Na-K pump to the membrane potential is sufficiently small so that no significant changes in Na transport occur when that contribution is removed. Although these conditions can never be strictly met, in tissue culture preparations it is possible to apply inhibitors and determine ion content changes over short time intervals, and thus an approximation of Na pump flux may be obtained.

Results in our laboratory using confluent cultures of contractile chick embryo heart cells[81] show that 3 min after exposure to 10^{-4} M ouabain, $[Na]_i$, as determined by flame photometry, increases from 137 ± 23 to 220 ± 23 nmol · (mg prot)$^{-1}$ (n = 6; Mean ± SEM). The average net Na uptake over this time period is then 28 nmol · (mg prot)$^{-1}$ · min^{-1} or 7 pmol · cm^{-2} · sec^{-1} (using 7($\mu\ell$ cell H$_2$O) · (mg prot)$^{-1}$ and V/A = 1.0 × 10^{-4} cm). In quail myocardial cells,[79] the maximum rate of increase of $[Na]_i$ after exposure to 5 × 10^{-6} M ouabain is 34 nmol · (mg prot)$^{-1}$ · min^{-1} or 8 pmol · cm^{-2} · sec^{-1}. These approximations will likely underestimate the true steady-state Na pump flux because existing data does not reflect the actual initial rate of net Na uptake for the following reasons: neither pump inhibition nor uptake measurement is instantaneous and other transport mechanisms may attenuate the rise in $[Na]_i$.

After a period of Na-K pump inhibition and subsequent rise in $[Na]_i$, if the pump inhibitor is removed, then Na content falls towards its physiologic value. The *initial* rate of sodium content decrease after an *instantaneous* release of pump inhibition will approximate Na efflux due to the pump, assuming that the rate of change in $[Na]_i$ was relatively slow immediately prior to release. This latter condition appears to be achieved in some tissue-cultured heart cell preparations after about 10 min of pump inhibition (see Murphy et al.,[31] Table 2). Based on the presumption that the Na-K pump is not maximally activated at physiologic $[Na]_i$,[2,82] one would predict that the sodium efflux due to the pump after release of pump inhibition should be greater than in the physiologic state. Experiments in which ouabain has been used as the pump inhibitor have generally failed to show this predicted result.[78,79] The finite time required for ouabain release from the cells is the likely explanation for this divergence, even though widely different rates of ouabain release have been obtained in tissue-cultured cells

($T_{1/2}$ = 12 hr in neonatal rat heart cells;[63] $T_{1/2}$ = 2 min in chick embryo heart cells[83]). Using potassium-free solution to inhibit the Na-K pump, Horres et al.[22] have shown that upon restoring the normal extracellular K to confluent cultures of chick embryo heart cells there is rapid restoration of $[Na]_i$ toward control values. The net change in sodium concentration during the first 30 sec following release of pump inhibition was approximately 20 mM. This figure converts to a net Na efflux of 67 pmol · cm^{-2} · sec^{-1} at the initial $[Na]_i$ of 72 mM. Although this result is only an approximation of the Na pump flux, when compared to the values previously noted for steady-state pump flux, it clearly suggests that the Na-K pump in cardiac muscle can increase its rate in response to increased Na_i.

2. Sodium Tracer Flux Measurement
a. Methodologic and Theoretical Considerations

In nearly all tissues, Na tracer flux measurement is plagued by what might be called a low signal-to-noise ratio. That is, extracellular tracer tends to interfere with the detection of the intracellular ion, analogous to problems associated with Na content measurement. This characteristic of sodium distribution creates the need for the use of high specific activity loading solutions and thorough extracellular clearance protocols. Also, Na tracer fluxes will tend to be more rapid than those of potassium.* Thus, points early in the flux curves are required, and precautions must be taken to minimize intracellular tracer loss during the clearance of extracellular space. Tracer reuptake (or backflux) during a Na efflux experiment is reduced (when compared to potassium) by dilution of the effluxing tracer in the relatively high $[Na]_o$, but is accentuated by the more rapid nature of Na fluxes. A model of diffusion in a cylindrical muscle bundle[46] predicts that in a 100 μ diameter strand in physiologic solution, a 3% error in the Na efflux rate constant does not occur until the true transmembrane flux is presumed to be 100 pmole · cm^{-2} · sec^{-1} or greater. Thus, in most tissue-cultured preparations with physiologic $[Na]_o$, tracer re-uptake should not produce significant errors.

b. Sodium Tracer Fluxes in Nonmuscle Cells Derived from Heart Tissue

Steady state, unidirectional sodium fluxes in nonmuscle cells derived from heart tissue, determined in various preparations in different laboratories, fall into a remarkably narrow range. Burrows and Lamb,[5] utilizing secondary cultures of chick embryo heart cells (which were fibroblastic in morphology and noncontractile), found a Na efflux rate constant of 0.59 min^{-1} at 37°C. Given their measurements of $[Na]_i$ and V/A, a flux of 14 pmol · cm^{-2} · sec^{-1} was calculated. In Girardi heart cells (a noncontractile human cell line), Lamb and McCall[62] found a steady state Na efflux of 27 pmol · cm^{-2} · sec^{-1} and Na influx of 25 pmol · cm^{-2} · sec^{-1}. Using their $[Na]_i$ and V/A, the corresponding Na efflux rate constant would be 0.61 min^{-1}. In fibroblastic cells derived from neonatal rat heart, McCall[63] found a steady state Na efflux rate constant of 0.32 min^{-1}. Wheeler et al.[12] measured a rate constant of 0.42 min^{-1} in polystrand preparations consisting of fibroblastic cells derived from chick embryo heart. The Na flux calculated from this value was 6 pmol · cm^{-2} · sec^{-1}. Clearly apparent is the overall consistency of the rate constants with the differences in the calculated fluxes largely the result of variations in the $[Na]_i$ and V/A values, most particularly the latter. The Na efflux in these preparations is markedly sensitive to ouabain, as illustrated in Table 4.

c. Sodium Tracer Fluxes in Cardiac Muscle Cells

Unlike the nonmuscle cell results, there are significant areas of inconsistency when comparing sodium fluxes in different contractile preparations. However, there is general agree-

---

* This assertion results from contrasting distributions of those ions coupled with Equation 1.[45] If the Na and K fluxes are *a priori* presumed to be of the same order of magnitude, then the ratio of the rate of the constants will reflect the inverse of the ratio of the intracellular concentrations.

Table 4
NA EFFLUX RATE CONSTANTS IN NONCONTRACTILE
PREPARATIONS DERIVED FROM HEART TISSUE

Preparation	Steady state Na efflux rate constant	% Change of rate constant in ouabain	Ref.
Chick embryo heart cells	0.59 min^{-1}	—	5
Girardi heart cells	0.61 min^{-1}	-74%	62
Neonatal rat heart cells	0.32 min^{-1}	-50%	63
Chick embryo heart cells (polystrand)	0.42 min^{-1}	-60%	12

ment that steady state, unidirectional sodium fluxes in heart muscle cells are far more rapid than in nonmuscle cells and also more rapid than those predicted on the basis of Na-K pump fluxes alone.

In monolayer cultures of beating neonatal rat heart cells, McCall[63] found a single component Na influx with rate constant of 1.2 min^{-1}. The steady state Na efflux in these cells, however, had two exponential components with rate constants of 1.9 min^{-1} and 0.23 min^{-1}. The rapidly exchanging compartment accounted for approximately 80% of the total intracellular Na. McCall further reports that in potassium-free solution (0K) or in 10^{-2} M ouabain only the more rapid of the two components is affected, declining to 0.84 min^{-1} in 0K and 0.78 min^{-1} in ouabain. McCall concludes that both components are related to cardiac muscle and specifically that the slow component does not reflect fibroblasts because of the ouabain insensitivity of the slow component compared to the marked ouabain sensitivity of Na efflux in pure fibroblast preparations. Also, the rate constant for the fibroblasts (0.32 min^{-1}) is somewhat different from the slow component of contractile preparations (0.23 min^{-1}). McCall's calculation of a steady state sodium efflux in cardiac muscle cells of 28 pmol $\cdot$ cm^{-2} $\cdot$ sec^{-1} is apparently based on a series compartment model.

Another laboratory using contractile preparations of neonatal rat heart cells also obtains a very rapid steady-state sodium influx. Werdan et al.[64] report a half-time for ^{22}Na uptake of 15 to 30 sec. Analyzing their control curve for Na uptake (Figure 6A[64]) and using the 2 min value as an approximate saturation level, a rate constant of 3.1 min^{-1} is found from the first three points on the curve. These authors also show a ^{22}Na influx curve after a 15 min incubation in 10^{-3} M ouabain. As expected, the amount of ^{22}Na taken up at isotopic equilibrium is increased (approx 3$\times$ control), but the *rate* of Na tracer uptake is little different from the control case. McCall[63] also found little effect of ouabain on the rate of Na tracer influx. We have evaluated the steady state Na influx of contractile chick embryo heart cells grown to confluency on plastic rectangles (Figure 8). The rate constant found in these cells is 3.0 min^{-1}, in general agreement with the investigators previously cited.

Extensive investigation of ^{24}Na efflux in the laboratory using chick embryo heart cells in the polystrand preparation also reveals the rapid nature of sodium exchange in the steady state. Like McCall,[63] we find that once extracellular space is cleared of isotope, the ^{24}Na efflux in contractile preparations consists of two distinct exponential components. The rate constants were 3.1 min^{-1} and 0.35 min^{-1}, while the rate constant of the single exponential ^{24}Na efflux in noncontractile preparations[12] was 0.42 min^{-1}. Each of these values is reasonably similar to those obtained by McCall.[63] However, we find that ouabain (10^{-4} M) or potassium-free solution has relatively little effect on the rapid component while significantly inhibiting the slow component. Because the behavior of the slow component matches results seen with pure nonmuscle preparations, we conclude that the slow component represents ^{24}Na exchange from the fibroblastic cells present in contractile preparations. Further evidence for this conclusion resulted from one culture in which the preplating procedure was omitted,

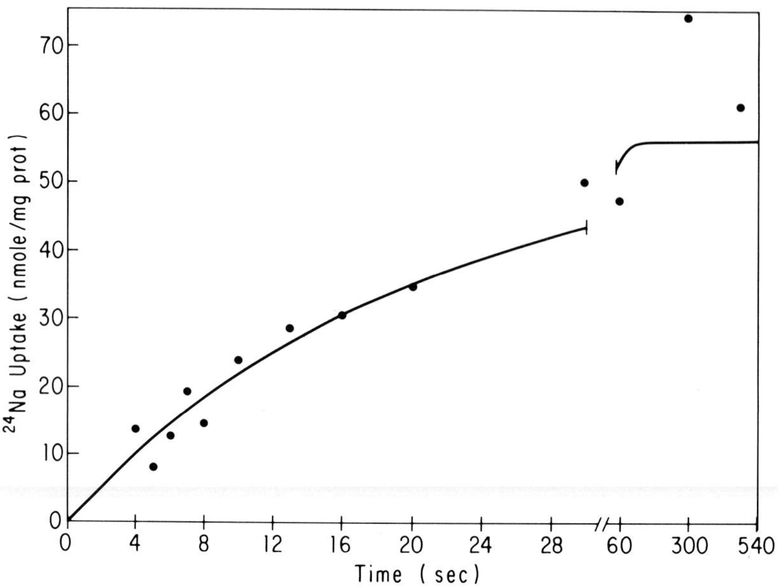

FIGURE 8. ^{24}Na influx curve in contractile chick embryo heart cells grown to confluency on plastic rectangles (approx. 25 mm × 6 mm). Experiment was performed on cells from a single culture, 4 days after culturing. Uptake solution contained: Na^+ 142 mM, K^+ 5.4 mM, Ca^{++} 2.7 mM, Mg^{++} 0.8 mM, Cl^- 126 mM, HCO_3^- 26.2 mM, SO_4^- 0.8 mM, $H_2PO_4^-$ 0.8 mM, Dextrose 5.6 mM, Bovine Serum Albumin 1.4 g·l^{-1}. Nominal specific activity 1.8 Ci·(mole Na)$^{-1}$. Influx carried out at 37.5°C in humidified 4% CO_2/ air environment. Each data point represents one rectangle of cells. After the appropriate isotopic loading time, each rectangle was thoroughly rinsed by passing through four beakers containing chilled, nonradioactive medium identical in composition to that above. The radioactivity of each rectangle was then counted and protein was later determined by the method of Lowry et al.[84] The fitted curve is given by y = 56 (1 = e$^{-3.0\,t}$). Note that several of the later points lie well above the curve, suggesting that a more slowly exchanging component may be present also.

thereby increasing the proportion of nonmuscle cells in the preparations. The magnitude of slow component relative to the fast was greater in preparations from this culture than from all other cultures.

Biedert et al.[78] have grown contractile chick embryo heart cells on glass coverslips for ion transport measurement. They find that >90% of the isotopic sodium uptake is complete within 5 min, consistent with the rapid Na uptake found in other studies. Their Na efflux curves do not attain monoexponential character until 2 min into the washout. However, the efflux chamber which they used has a relatively slow clearance of isotope and does not allow them to resolve the rapid component. Biedert et al.[40] report that the slow component which remains after 2 min of efflux, has a rate constant of 0.59 min^{-1}, and is 34% sensitive to 10^{-6} M ouabain. This result is not unlike the behavior of the slow component in contractile polystrand preparations.

To summarize, the Na flux results in contractile preparations: there is agreement on the rapid nature of Na exchange; the main area of disagreement concerns the identification of the efflux curve components and largely results from different findings regarding the ouabain sensitivity of the components. Different species have been used in these studies (neonatal rat vs. chick embryo) and culture morphology is different (monolayer vs. polystrand). Such factors would be expected to produce quantitative differences only. The qualitative differences are surprising and without explanation at present.

One reported ^{22}Na efflux result in tissue-cultured heart cells does not at first glance fit into the scheme of rapid sodium exchange. Fosset et al.[85] report a single component ^{22}Na efflux with rate constant of 0.26 min^{-1} in confluent, contractile cultures of chick embryo heart cells exposed to 0.5 m*M* ouabain. However, several methodologic points may alter the interpretation of their data. First, the plates of cells were washed twice (in <10 sec) at 37°C prior to the efflux. Second, the culture method apparently did not include a preplating step but the medium was changed at day 2 in culture.[85,86] It is thus likely that these preparations contain a significant percentage of fibroblastic cells. If the Na exchange in cardiac muscle cells is on the order of that found by other investigators, then even a brief rinse at 37°C may result in significant loss of intracellular isotope from these cells. The more slowly exchanging fibroblastic compartment may then dominate the Na efflux curve. A close examination of the ^{22}Na efflux curve presented by Fosset et al.[85] actually suggests multiexponential behavior. A component with rate constant of 0.2 min^{-1}, a value consistent with that of others for noncontractile cells in ouabain, can be formed from the last three points of the curve. There remains, then, a small rapid component which cannot be quantified due to the limited number of early points. We believe, therefore, that methodologic considerations account for this lower value of isotopic sodium exchange.

d. The Use of Isotopic Sodium Uptake as an Index of Flux via the Voltage-Sensitive Na Channel

Several groups have utilized ^{22}Na uptake in tissue-cultured heart cells as a monitor of Na channel activity. The purpose of these experiments is the pharmacologic and biochemical characterization of the Na channel and not the quantitation of Na fluxes under physiologically relevant conditions. Thus, since this area of work involves the use of nonphysiologic incubation conditions and cells which are not in steady state, comparison of the flux results with those obtained in the physiologic steady state is generally not meaningful. Rather, through the use of toxins and other pharmacologic agents, a biochemical and pharmacologic profile of the Na channel in different species and tissues has been established. The interested reader is referred to the work of Lazdunski et al.[87,88] and Catterall et al.[89]

e. Implication of Sodium Tracer Fluxes

The rapid nature of sodium exchange in cardiac muscle reveals the importance of tissue culture preparations in ion transport studies. The rate of clearance of the extracellular space of naturally occurring tissues, both perfused and superfused, is less than the rates of cellular Na exchange found in tissue culture preparations (see Reference 12 for further discussion). Thus, the cellular Na exchange cannot be detected in most intact tissues.

The magnitude of the isotopic sodium uptake and efflux also has significant implications regarding the mechanisms responsible for Na transport in cardiac muscle. Our calculations in the polystrand preparation show a steady state Na efflux approximately 6× the potassium efflux. McCall[63] finds the Na flux (calculated with the component interpretation which produces the lowest flux) to be twice the steady-state potassium flux. Comparison of the K and Na uptake curves of Werdan et al.[64] reveals that Na influx is 3× that of potassium. If one presumes a 3 to 2 Na-K pump stoichiometry in the steady state, then Na efflux due to the pump should approximate the total K flux (since 50 to 75% of the K influx in cardiac muscle is ouabain sensitive). Thus, the total Na efflux at steady state is significantly in excess of the predicted and measured Na efflux due to the Na-K pump. Calculations based on classic assumptions of membrane and passive ion channel properties predict that Na efflux through passive transmembrane diffusion is quite small relative to the measured total fluxes. Thus, for a portion of Na extrusion and likely also for Na influx, mechanisms of Na transport other than the Na-K pump and passive channels or leaks must be entertained.

In the steady state, the net result of all Na transport must be a simple one-for-one Na exchange, but many separate, complex, and potentially interacting components may be present. The nature and relative magnitudes of the mechanisms responsible for the large Na fluxes in cardiac muscle have not been determined. Sodium-calcium exchange has long been hypothesized to be active in the cardiac muscle membrane.[90] One physiologic role is the extrusion of Ca using the Na gradient as its energy source. Thus, to attribute a significant portion of Na efflux to Na-Ca exchange is open to question. However, data from our laboratory does reveal that the ouabain-insensitive Na efflux is reduced in nominally Ca-free solution,[91] raising the possibility that some Na-Ca exchange may be proceeding in the reverse of the direction usually considered physiologic. Sodium-hydrogen exchange has also been demonstrated in cultured cardiac muscle.[92-94] For example, following an intracellular acid load induced by an NH_4Cl prepulse in the polystrand, rapid net Na^+ uptake is accompanied by transmembrane H^+ extrusion in an Na:H ratio of 1:1.[94] Amiloride, an inhibitor of Na/H exchange in several cell types,[95] inhibits net Na^+ uptake and H^+ extrusion as well as the accompanying pH_i recovery following the acid load. Under acid loaded conditions, Na/H exchange in polystrands is capable of mediating up to 54 pmol $\cdot$ cm^{-2} $\cdot$ sec^{-1} of net transmembrane Na influx. In addition, ethylisopropylamiloride (EIPA, a potent amiloride analogue) inhibits 58% of total ^{22}Na uptake in monolayers of cultured chick heart cells.[93] Note that as with Na-Ca exchange, Na-H exchange can primarily be used to account for Na influx. Mechanisms for Na efflux may be the above exchangers operating in a reverse or in a Na-exchange mode or may be entirely different co- or counter-transport processes.

C. Chloride Transport

1. Tracer Flux Measurement

^{36}Cl efflux rate constants from naturally-occurring cardiac muscle preparations[96-98] were reported to be approximately 0.03 min^{-1} and appeared to correlate with the known low Cl conductance of the cardiac cell membrane.[97,99] Applying fast perfusion protocols to rat ventricles, however, Polimeni and Page[100] reported a ^{36}Cl efflux rate constant of 0.29 min^{-1}. This suggested that reduction of extracellular diffusion delays could reveal a more dynamic perspective of Cl transport in cardiac muscle than previously suspected. Indeed, Lamb and McCall[61] showed that a monolayer preparation of dedifferentiated, noncontractile Girardi heart cells had a chloride rate constant of 0.34 min^{-1}.

To evaluate ^{36}Cl efflux in a contractile cultured heart cell preparation with minimal extracellular diffusion delays, polystrand preparations were again chosen. The apparatus described previously for use in ^{24}Na efflux experiments was modified for use with a fraction collector.[101] Preparations were preequilibrated with ^{36}Cl, transferred to the apparatus, and superfused with control solution. Effluent was collected at 26 sec intervals and counted by liquid scintillation techniques. The data were summed to generate efflux curves. Rapid superfusion protocols and judicious use of chilled superfusate successfully separated extracellular from intracellular ^{36}Cl compartments[13] in a manner similar to that described for ^{24}Na efflux.[12] Inhibition of transmembrane efflux in chilled solution concurrent with clearance of extracellular space created a ^{36}Cl isotopic "plateau" on the efflux curve which allowed estimation of intracellular Cl content. The value of 108 nmol·(mg dry weight)$^{-1}$ (25.1 mM, assuming cell water of 4.3 $\mu\ell$ (mg dry weight)$^{-1}$; Horres et al.[11]) obtained with this technique correlated well with the value of 106.7 nmol $\cdot$ (mg dry weight)$^{-1}$ (24.9 mM) obtained by coulometric titration of a tissue extract of polystrands.[13] These compared favorably with values obtained from naturally occurring cardiac preparation by both analytical techniques[102-104] as well as by Cl-selective microelectrodes.[98,99,105,106]

Warm (37°C) superfusion revealed two cellular ^{36}Cl compartments in contractile polystrands: one exchanged with a rate constant of 0.67 min^{-1}, while the other exchanged with a rate constant of 0.18 min^{-1}. Noncontractile polystrands (fibroblasts) have a single ^{36}Cl compartment that exchanged with a rate constant of 0.23 min^{-1}. Combining the rate constant from the muscle compartment with Cl content and V/A ratio indicated a steady-state transmembrane Cl efflux of 30 pmol $\cdot$ cm^{-2} $\cdot$ sec^{-1}, a value ten times that reported in mammalian preparations.[97]

2. Electrophysiological Measurements

To assure that the rapid ^{36}Cl efflux did not reflect unique electrodiffusive properties of the polystrand cell membrane, electrophysiological experiments were undertaken. Rendering the contractile preparation quiescent in K_o-free, 30 mM K_o or 130 mM K_o solution and subsequently switching between 130 mM Cl$_o$ and 15 mM Cl$_o$ showed changes of membrane potential of <2mV.[13] In addition, the polystrand action potential configuration was little affected by changes in Cl$_o$ (methanesulfonate substitution). These experiments confirm that membrane current carried by Cl ions in the polystrand was as low as that reported in mammalian cardiac preparations (e.g., P_{Cl}/P_K = 0.11 in rabbit ventricular muscle[106]). Comparison of the ^{36}Cl efflux rates and electrophysiological results indicated that ~95% of the transmembrane Cl efflux was mediated by electroneutral pathways.[13]

3. Chloride-Dependent Transport Mechanisms

Further investigations to characterize the membrane transport mechanisms responsible for the neutral fluxes of Cl in cardiac muscle have thus far revealed anion exchange and cotransport mechanisms in cultured heart cells. For example, preliminary results in the polystrand preparation indicate that addition to the control solution of 4,4-diisothiocyano-2,2′-disulfonic acid stilbene (DIDS 10^{-4} M), an inhibitor of anion exchange, reduces the ^{36}Cl efflux rate constant by 25%. Such results in cultured heart cells are in accord with those from mammalian cardiac preparations that demonstrate a Cl/HCO$_3$ exchanger.[99,107]

The presence of co-transport in cardiac muscle was first suggested by Aiton et al.,[108] who reported that furosemide, an inhibitor of Cl-dependent cotransport mechanisms in several cell types, blocked 85% of the ouabain-insensitive K influx in monolayers of cultured chick heart cells. This may indicate K/K exchange mediated by a co-transport mechanism.[109] In polystrands, furosemide (10^{-3} M) decreases the ^{36}Cl efflux rate constant from 0.67 min^{-1} to 0.33 min^{-1}. This likely represents an overestimation of co-transport-mediated Cl efflux since high doses of furosemide partially inhibit anion exchange.[110]

Incubation of polystrands in 130 mM K_o solution, which equalizes the transmembrane K gradient, causes net uptake of Cl and K in a 1:1 ratio.[57] High K_o solution also induces a furosemide-sensitive cell volume increase. Removing external Cl prevents K uptake and causes a furosemide-sensitive volume loss. Na, Li, and choline cannot replicate the effect of K under these conditions. In addition, K_o-free solution promotes the loss of Cl against the Cl electrochemical gradient; Cl loss is furosemide sensitive in a dose-dependent manner.[57] In sum, these results in cultured heart cell preparations suggest the presence of a furosemide-sensitive (K + Cl) co-transport mechanism in cardiac muscle. The effect of Na on this cotransport mechanism requires further study.

Thus, heart cells in culture, by reducing extracellular diffusion delays inherent in naturally occurring preparations, have revealed rapid electroneutral Cl-dependent transport mechanisms. Since the electrochemical, electrophysiological, and biochemical data from cultured preparations such as the polystrand are comparable to data from naturally occurring mammalian preparations, it is likely that these rapid sarcolemmal Cl transport mechanisms also exist in natural preparations, but are obscured from detection by complex morphology.

IV. SUMMARY AND IMPLICATIONS

Tissue culture techniques can be useful in developing preparations of heart cells to overcome some of the complexities associated with naturally occurring cardiac muscle. Of particular interest are the advantages gained when using the polystrand preparation, namely mechanical stability, morphologically simple extracellular space and small diameter, with a tissue mass that is adequate for ion content and radiotracer measurements. With this novel preparation of heart cells, such experiments reveal that tracer exchange of K, Na, and Cl across the cardiac cell membrane is rapid compared with previously published values for naturally occurring cardiac muscle. Furthermore, ion content determinations by isotopic methods agree well with values obtained by flame photometry or coulometric titration. Thus, the freely exchangeable nature of these physiologically important ions is confirmed in the cultured heart cell.

By minimizing extracellular diffusion delays, cultured heart cells have revealed a more dynamic perspective of the cardiac cell membrane. As illustrated in this chapter, passive membrane conductive properties contribute but a small fraction of the total membrane fluxes for ions such as Na and Cl. In addition, rates of steady-state electrogenic transport of Na ions are also overshadowed by electrically silent Na transport mechanisms. Modern comprehensive models of membrane transport and regulation of ion content in heart tissue must incorporate conductive, electrogenic, and electroneutral ion gradient-coupled membrane transport mechanisms to fully account for the complex interactions between ions such as Na, K, H, Ca, Cl, and HCO_3 in cardiac muscle.

In conclusion, heart cells in culture enable the rapid exchange of extracellular space and provide the opportunity to obtain realistic transport measurements. These characteristics are essential to study the complex ion transport mechanisms of cardiac cells.

ACKNOWLEDGMENTS

Our special thanks to Ms. Ellen Eatmon for her secretarial assistance. Supported in part by grants HL27105, HL17670, and HL07101 from the National Institutes of Health.

REFERENCES

1. **Lieberman, M., Horres, C. R., Shigeto, N., Ebihara, L., Aiton, J. F., and Johnson, E. A.,** Cardiac muscle with controlled geometry. Application to electrophysiological and ion transport studies, in *Excitable Cells in Tissue Culture,* Nelson, P. G. and Lieberman, M., Eds., Plenum Press, New York, 1981, 379.
2. **Glitsch, H. G., Pusch, H., and Venetz, K.,** Effects of Na and K ions on the active Na transport in guinea auricles, *Pflugers Arch.,* 365, 29, 1976.
3. **Page, E.,** Ion movement in heart muscle: tissue compartments and the experimental definition of driving forces, *Ann. N.Y. Acad. Sci.,* 127, 34, 1965.
4. **Lieberman, M., Horres, C. R., Jacob, R., Murphy, E., Piwnica-Worms, D., and Wheeler, D. M.,** Physiological criteria for electrogenic transport in tissue-cultured heart cells, in *Electrogenic Transport: Fundamental Principles and Physiological Implications,* Blaustein, M. P. and Lieberman, M., Eds., Raven Press, New York, 1984, 181.
5. **Burrows, R. and Lamb, J. F.,** Sodium and potassium fluxes in cells cultured from chick embryo heart muscle, *J. Physiol. (London),* 162, 510, 1962.
6. **Page, E.,** Cat heart muscle in vitro. VII. The temperature dependence of steady state K exchange in presence and absence of NaCl, *J. Gen. Physiol.,* 48, 949, 1965.
7. **Horres, C. R. and M. Lieberman.,** Compartmental analysis of potassium efflux from growth oriented heart cells, *J. Mem. Biol.,* 34, 331, 1977.
8. **Bkaily, G., Sperelakis, N., and Doane, J.,** A new method for preparation of isolated single adult myocytes, *Am. J. Physiol.,* 247, H1018, 1984.
9. **Grochowski, E., Ganote, C. E., Hill, M. L., and Jennings, R. B.,** Experimental myocardial ischemic injury. I. A comparison of Stadie-Riggs and free-hand slicing techniques on tissue ultrastructure, water and electrolytes during in vitro incubation, *J. Mol. Cell. Cardiol.,* 8, 173, 1976.
10. **Silver, L. H. and Houser, S. R.,** Transmembrane potassium fluxes in isolated feline, ventricular myocytes, *Am. J. Physiol.,* 248, H614, 1985.

11. **Horres, C. R., Lieberman, M., and Purdy, J. E.,** Growth orientation of heart cells on nylon monofilament. Determination of the volume-to-surface area ratio and intracellular potassium concentration, *J. Mem. Biol.,* 34, 313, 1977.
12. **Wheeler, D. M., Horres, C. R., and Lieberman, M.,** Sodium tracer kinetics and transmembrane flux in tissue-cultured heart cells, *Am. J. Physiol.,* 243, C169, 1982.
13. **Piwnica-Worms, D., Jacob, R., Horres, C. R., and Lieberman, M.,** Transmembrane chloride flux in tissue-cultured chick heart cells, *J. Gen. Physiol.,* 81, 731, 1983.
14. **Lieberman, M., LeFurgey, A., Murphy, E., Liu, S.,** Cultured heart cells as a model for studying myocardial ischemia, in *Pathobiology of Cardiovascular Injury,* Stone, H. L. and Weglicki, W. B., Eds., Martinus Nijhoff Publishing, The Hague, 1985, 141.
15. **Ebihara, L., Shigeto, N., Lieberman, M., and Johnson, E. A.,** The initial inward current in spherical clusters of chick embryonic heart cells, *J. Gen. Physiol.,* 75, 437, 1980.
16. **Nathan, R. D. and DeHaan, R. L.,** Voltage-clamp analysis of embryonic heart cell aggregates, *J. Gen. Physiol.,* 73, 175, 1979.
17. **Josephson, I. and Sperelakis, N.,** On the ionic mechanism underlying adrenergic-cholinergic antagonism in ventricular muscle, *J. Gen. Physiol.,* 79, 69, 1982.
18. **Clusin, W. T.,** Normal and abnormal automaticity in cultured embryonic cardiac cells, in *Cardiac Arrhythmias — A Decade of Progress,* Harrison, D. C., Ed., G. K. Hall, Boston, 1981, 69.
19. **Carmeliet, E. and Vereecke, J.,** Electrogenesis of the action potential and automaticity, in *Handbook of Physiology — The Cardiovascular System I,* Berne, R. M. and Sperelakis, N., Eds., Williams and Wilkins, Baltimore, 1980, 269.
20. **Glitsch, H. G.,** Electrogenic Na pumping in the heart, *Ann. Rev. Physiol.,* 44, 389, 1982.
21. **Gadsby, D. C.,** The Na/K pump of cardiac cells, *Ann. Rev. Biophys. Bioeng.,* 13, 373, 1984.
22. **Horres, C. R., Aiton, J. F., Lieberman, M., and Johnson, E. A.,** Electrogenic transport in tissue cultured heart cells, *J. Mol. Cell. Cardiol.,* 11, 1201, 1979.
23. **Coraboeuf, E.,** Ionic basis of electrical activity in cardiac tissues, *Am. J. Physiol.,* 234, H101, 1978.
24. **Werrlein, R. J.,** Cells and tissue culture systems, in *The Application of Ion-Sensitive Microelectrodes,* Zeuthen, T., Ed., Elsevier, Amsterdam, 1981, 257.
25. **Liu, S., Jacob, R., Piwnica-Worms, D., and Lieberman, M.,** KCl cotransport in cultured heart cells measured with ion-sensitive microelectrodes, *Biophys. J.,* 47, 462a, 1985.
26. **Liu, S., Jacob, R., and Lieberman, M.,** Electrogenic Na/Ca exchange in cultured chick heart cells demonstrated using ion selective microelectrodes, *Fed. Proc.,* 44, 1580, 1985.
27. **Barry, W. H., Hasin, Y., and Smith, T. W.,** Sodium pump inhibition, enhanced calcium influx via sodium-calcium exchange, and positive inotropic response in cultured heart cells, *Circ. Res.,* 56, 231, 1985.
28. **Thomas, R. C.,** *Ion-Sensitive Intracellular Microelectrodes,* Academic Press, New York, 1978, chap. 1.
29. **Mullins, L. J.,** *Ion Transport in Heart,* Raven Press, New York, 1981, chap. 2.
30. **Lee, C. O.,** Ionic activities in cardiac muscle cells and application of ion-selective microelectrodes, *Am. J. Physiol.,* 241, H459, 1981.
31. **Murphy, E., Aiton, J. F., Horres, C. R., and Lieberman, M.,** Calcium elevation in cultured heart cells: its role in cell injury, *Am. J. Physiol.,* 245, C316, 1983.
32. **Murphy, E., Jacob, R., and Lieberman, M.,** Cytosolic free calcium in chick heart cells: its role in cell injury, *J. Mol. Cell. Cardiol.,* 17, 221, 1985.
33. **Thomas, M. V.,** *Techniques in Calcium Research,* Academic Press, New York, 1982, chap. 1.
34. **Tsien, R. Y.,** Intracellular measurements of ion activities, *Ann. Rev. Biophys. Bioeng.,* 12, 91, 1983.
35. **Murphy, E., Coll, K. E., Rich, T. L., and Williamson, J. R.,** Hormonal effects of calcium homeostasis in isolated hepatocytes, *J. Biol. Chem.,* 255, 6600, 1980.
36. **Gupta, R. K. and Gupta, P.,** NMR studies of intracellular metal ions in intact cells and tissues, *Ann. Rev. Biophys. Bioeng.,* 13, 221, 1984.
37. **Shuman, H., Somlyo, A. V., and Somlyo, A. P.,** Quantitative electron probe microanalysis of biological thin sections. Methods and validity, *Ultramicroscopy,* 1, 317, 1976.
38. **Lechene, C.,** Electron probe microanalysis of biological soft tissues: principle and technique, *Fed. Proc.,* 39, 2871, 1980.
39. **Murphy, E., Lobaugh, L. A., LeFurgey, A., Jacob, R., Wheeler, D. M., and Lieberman, M.,** Mechanisms of Ca transport in embryonic chick heart cells, *Fed. Proc.,* 44, 1595, 1985.
40. **Biedert, S., Barry, W. H., and Smith, T. W.,** Inotropic effects and changes in sodium and calcium contents associated with inhibition of monovalent cation active transport by ouabain in cultured myocardial cells, *J. Gen. Physiol.,* 74, 479, 1979.
41. **Clusin, W. C.,** The mechanical activity of chick embryonic myocardial cell aggregates, *J. Physiol. (London),* 320, 149, 1981.

42. **Langer, G. A. and Frank, J. S.,** Calcium exchange in cultured cardiac cells, in *Developmental and Physiological Correlates of Cardiac Muscle,* Lieberman, M. and Sano, T., Eds., Raven Press, New York, 1975, 117.

43. **Barry, W. H. and Smith, T. W.,** Mechanisms of transmembrane calcium movements in cultured chick embryo ventricular cells, *J. Physiol. (London),* 325, 243, 1982.

44. **Harris, E. J. and Burn, G. P.,** The transfer of sodium and potassium ions between muscle and the surrounding medium, *Trans. Faraday Soc.,* 45, 508, 1949.

45. **Keynes, R. D. and Lewis, P. R.,** The resting exchange of radioactive potassium in crab nerve, *J. Physiol. (London),* 113, 73, 1951.

46. **MacDonald, R. L., Mann, J. E., Jr., and Sperelakis, N.,** Derivation of general equations describing tracer diffusion in any two-compartment tissue with application to ionic diffusion in cylindrical muscle bundles, *J. Theor. Biol.,* 45, 107, 1974.

47. **Huxley, A. F.,** Appendix 2 to "Compartmental methods of kinetic analysis" by Solomon, A. K., in *Mineral Metabolism,* Vol. 1, part A, Comar, C. L. and Bronner, F., Eds., Academic Press, New York, 1960, 163.

48. **Blondel, B., Roijen, I., and Cheneval, J. P.,** Heart cells in culture, a simple method for increasing the proportion of myoblasts, *Experientia,* 27, 356, 1971.

49. **Horres, C. R., Aiton, J. F., and Lieberman, M.,** Potassium permeability of embryonic avian heart cells in tissue culture, *Am. J. Physiol.,* 236, C163, 1979.

50. **Hume, J. R. and Giles, W.,** Active and passive electrical properties of single bullfrog atrial cells, *J. Gen. Physiol.,* 78, 19, 1981.

51. **Powell, T. and Twist, V. W.,** A rapid technique for the isolation and purification of adult cardiac muscle cells having respiratory control and a tolerance to calcium, *Biochem. Biophys. Res. Comm.,* 72, 327, 1976.

52. **Page, E., McAllister, L. P., and Power, B.,** Stereological measurements of cardiac ultrastructures implicated in excitation-contraction coupling, *Proc. Nat. Acad. Sci.,* 68, 1465, 1971.

53. **Page, E.,** Quantitative ultrastructural analysis in cardiac membrane physiology, *Am. J. Physiol.,* 235, C147, 1978.

54. **Eisenberg, B. R., Kuda, A. M., and Peter, J. B.,** Stereological analysis of mammalian skeletal muscle. I. Soleus muscle of the adult guinea pig, *J. Cell Biol.,* 60, 732, 1974.

55. **Eisenberg, B. R.,** Quantitative ultrastructure of mammalian skeletal muscle, in *Handbook of Physiology-Skeletal Muscle,* Peachey, L. D. and Adrian, R. H., Eds., Williams and Wilkins, Baltimore, 1983, 73.

56. **Carmeliet, E. E., Horres, C. R., Lieberman, M., and Vereecke, J. S.,** Developmental aspects of potassium flux and permeability of the embryonic chick heart, *J. Physiol. (London),* 254, 673, 1976.

57. **Piwnica-Worms, D., Jacob, R., Horres, C. R., and Lieberman, M.,** Potassium-chloride cotransport in cultured chick heart cells, *Am. J. Physiol.,* 249, C337, 1985.

58. **McDonald, T. F. and DeHaan, R. L.,** Ion levels and membrane potential in chick heart tissues and cultured cells, *J. Gen. Physiol.,* 61, 89, 1973.

59. **Fozzard, H. A. and Sheu, S. S.,** Intracellular potassium and sodium activities of chick ventricular muscle during embryonic development, *J. Physiol. (London),* 306, 579, 1980.

60. **Wheeler, D. M.,** Sodium Tracer Kinetics and Transmembrane Flux in Tissue-Cultured Chick Heart Cells, Ph.D. thesis, Duke University, Durham, N.C., 1981.

61. **Frank, J. S., Langer, G. A., Nudd, L. M., and Seraydarian, K.,** The myocardial cell surface, its histochemistry, and the effect of sialic acid and calcium removal on its structure and cellular ionic exchange, *Circ. Res.,* 41, 702, 1977.

62. **Lamb, J. F. and McCall, D.,** Effect of prolonged ouabain treatment on Na, K, Cl, and Ca concentration and fluxes in cultured human cells, *J. Physiol. (London),* 225, 599, 1972.

63. **McCall, D.,** Cation exchange and glycoside binding in cultured rat heart cells, *Am. J. Physiol.,* 236, C87, 1979.

64. **Werdan, K., Bauriedel, G., Fischer, B., Krawietz, W., Erdmann, E., Schmitz, W., and Scholz, H.,** Stimulatory (insulin-mimetic) and inhibitory (ouabain-like) action of vanadate on potassium uptake and cellular sodium and potassium in heart cells in culture, *Biochim. Biophys. Acta,* 687, 79, 1982.

65. **Dow, J. W., Harding, N. G. L., and Powell, T.,** Isolated cardiac myocytes. I. Preparation of adult myocytes and their homology with the intact tissue, *Cardiovasc. Res.,* 15, 483, 1981.

66. **Farmer, B. B., Mancina, M., Williams, E. S., and Watenabe, A. M.,** Isolation of calcium tolerant myocytes from adult rat hearts: review of the literature and description of a method, *Life Sci.,* 33, 1, 1983.

67. **Altschuld, R., Gibb, L., Ansel, A., Hohl, C., Kruger, F. A., and Brierley, G. P.,** Calcium tolerance of isolated rat heart cells, *J. Mol. Cell Cardiol.,* 12, 1383, 1980.

68. **Glick, M. R., Burns, A. H., and Reddy, W. J.,** Dispersion and isolation of beating cells from adult rat heart, *Anal. Biochem.,* 61, 32, 1974.

69. **Powell, T., Terror, D. A., and Twist, V. W.,** Electrical properties of individual cells isolated from adult rat ventricular myocardium, *J. Physiol. (London),* 302, 131, 1980.

70. **Isenberg, G. and Klockner, U.,** Isolated bovine ventricular myocytes: characterization of the action potential, *Pflugers Arch.,* 395, 19, 1982.
71. **Cheneval, J. P., Hyde, A., Blondel, B., and Girardier, L.,** Heart cells in culture: metabolism, action potential and transmembrane ionic movements, *J. Physiol. (Paris),* 64, 413, 1972.
72. **Langer, G. A., Frank, J. S., and Philipson, K. D.,** Correlation of alterations in cation exchange and sarcolemmal ultrastructure produced by neuraminidase and phospholipases in cardiac cell tissue culture, *Circ. Res.,* 49, 1289, 1981.
73. **Horres, C. R.,** Potassium Tracer Kinetics of Growth Oriented Heart Cells in Tissue Culture, Ph.D. thesis, Duke University, Durham, N.C., 1975.
74. **McCall, D., Zimmer, L. J., and Katz, A. M.,** Kinetics of thallium exchange in cultured rat myocardial cells, *Circ. Res.,* 56, 370, 1985.
75. **Kim, D., Barry, W. H., and Smith, T. W.,** Kinetics of ouabain binding and changes in cellular sodium content, $42K^+$ transport and contractile state during ouabain exposure in cultured chick heart cells, *J. Pharmacol. Exp. Therap.,* 231, 326, 1984.
76. **Werdan, K., Wagenknecht, B., Zwissler, B., Brown, L., Krawietz, W., and Erdmann, E.,** Cardiac glycoside receptors in cultured-heart cells. I. Characterization of one single class of high affinity receptors in heart muscle cells from chick embryos, *Biochem. Pharmacol.,* 33, 55, 1984.
77. **Couraud, F., Rochat, H., and Lissitzky, S.,** Stimulation of sodium and calcium uptake by scorpion toxin in chick embryo heart cells, *Biochim. Biophys. Acta,* 433, 90, 1976.
78. **Biedert, S., Barry, W. H., Miura, D. S., and Smith, T. W.,** Changes in cellular Na^+, K^+, and Ca^{2+} contents, monovalent cation transport rate, and contractile state during washout of cardiac glycosides from cultured chick heart cells, *Circ. Res.,* 49, 141, 1981.
79. **Goshima, K. and Wakabayashi, S.,** Involvement of an Na^+-Ca^{2+} exchange system in genesis of ouabain-induced arrhythmias of cultured myocardial cells, *J. Mol. Cell. Cardiol.,* 13, 489, 1981.
80. **Wakabayashi, S. and Goshima, K.,** Kinetic studies on sodium-dependent calcium uptake by myocardial cells and neuroblastoma cells in culture, *Biochim. Biophys. Acta,* 642, 158, 1981.
81. **Murphy, E.,** unpublished data.
82. **Yamamoto, S., Akera, T., and Brody, T. M.,** Sodium influx rate and ouabain-sensitive rubidium uptake in isolated guinea pig atria, *Biochim. Biophys. Acta,* 555, 270, 1979.
83. **Lieberman, M., Horres, C. R., Aiton, J. F., Shigeto, N., and Wheeler, D. M.,** Developmental aspects of cardiac excitation: active transport, in *Normal and Abnormal Conduction in the Heart,* Paes de Carvalho, A., Hoffman, B. F., and Lieberman, M., Eds., Futura Publishing, Mt. Kisco, 1982, 812.
84. **Lowry, O. H., Rosenbrough, N. J., Farr, A. L., and Randall, R. J.,** Protein measurements with the Folin phenol reagent, *J. Biol. Chem.,* 193, 265, 1951.
85. **Fosset, M., DeBarry, J., Lenoir, M.-C., and Lazdunski, M.,** Analysis of molecular aspects of Na^+ and Ca^{2+} uptakes by embryonic cardiac cells in culture, *J. Biol. Chem.,* 252, 6112, 1977.
86. **Fayet, G., Couraud, F., Miranda, F., and Lissitzky, S.,** Electrooptical system for monitoring activity of heart cells in culture: application to the study of several drugs and scorpion toxins, *Eur. J. Pharmacol.,* 27, 165, 1974.
87. **Lazdunski, M. and Renaud, J. F.,** The action of cardiotoxins on cardiac plasma membranes, *Ann. Rev. Physiol.,* 44, 463, 1982.
88. **Lazdunski, M., Balerna, M., Barhanin, J., Chicheportiche, R., Fosset, M., Frelin, C., Jacques, Y., Lombet, A., Pouyssegur, J., Renaud, J. F., Romey, G., Schweitz, H., and Vincent, J. P.,** Molecular aspects of the structure and mechanism of the voltage-dependent sodium channel, *Annals N.Y. Acad. Sci.,* 358, 169, 1980.
89. **Catterall, W. A. and Coppersmith, J.,** Pharmacological properties of sodium channels in cultured rat heart cells, *Mol. Pharmacol.,* 20, 533, 1981.
90. **Reuter, H. and Seitz, N.,** The dependence of calcium efflux from cardiac muscle on temperature and external ion composition, *J. Physiol. (London),* 195, 451, 1968.
91. **Murphy, E., Wheeler, D. M., LeFurgey, A., Jacob, R., Lobaugh, L. A., and Lieberman, M.,** Coupled sodium-calcium transport in cultured chick heart cells, *Am. J. Physiol.,* 250, C442, 1986.
92. **Piwnica-Worms, D. and Lieberman, M.,** Microfluorometric monitoring of pH_i in cultured heart cells: Na^+-H^+ exchange, *Am. J. Physiol.,* 244, C422, 1983.
93. **Frelin, C., Vigne, P., and Lazdunski, M.,** The role of the Na^+/H^+ exchange system in cardiac cells in relation to the control of the internal Na^+ concentration, *J. Biol. Chem.,* 259, 8880, 1984.
94. **Piwnica-Worms, D., Jacob, R., Horres, C. R., Lieberman, M.,** Na/H exchange in cultured chick heart cells: pH_i regulation, *J. Gen. Physiol.,* 85, 43, 1985.
95. **Benos, D. J.,** Amiloride: a molecular probe of sodium transport in tissues and cells, *Am. J. Physiol.,* 242, C131, 1982.
96. **Sekel, A. A. and Holland, W. C.,** Cl^{36} and Ca^{45} exchange in atrial fibrillation, *Am. J. Physiol.,* 197, 752, 1959.

97. **Carmeliet, E. and Verdonck, F.,** Reduction of potassium permeability by chloride substitution in cardiac cells, *J. Physiol. (London),* 265, 193, 1977.

98. **Fong, C. N. and Hinke, J. A. M.,** Intracellular Cl activity, Cl binding, and ^{36}Cl efflux in rabbit papillary muscle, *Can. J. Physiol. Pharmacol.,* 59, 479, 1981.

99. **Vaughan-Jones, R. D.,** Nonpassive chloride distribution in mammalian heart muscle: microelectrode measurement of the intracellular chloride activity, *J. Physiol. (London),* 295, 83, 1979.

100. **Polimeni, P. I. and Page, E.,** Chloride distribution and exchange in rat ventricle, *Am. J. Physiol.,* 238, C169, 1980.

101. **Piwnica-Worms, D.,** Electroneutral Ion Gradient-Coupled Membrane Transport in Cultured Chick Heart Cells, Ph.D. thesis, Duke University, Durham, N.C., 1983.

102. **Lamb, J. F.,** The chloride content of rat auricle, *J. Physiol. (London),* 157, 415, 1961.

103. **Page, E.,** Cat heart muscle in vitro. III. The extracellular space, *J. Gen. Physiol.,* 46, 201, 1962.

104. **Caille, J. P., Ruiz-Ceretti, E., and Schanne, O. F.,** Intracellular chloride activity in rabbit papillary muscle: effect of ouabain, *Am. J. Physiol.,* 240, C183, 1981.

105. **Ladle, R. O. and Walker, J. L.,** Intracellular chloride activity in frog heart, *J. Physiol. (London),* 251, 549, 1975.

106. **Fozzard, H. A. and Lee, C. O.,** Influence of changes of external potassium and chloride on membrane potential and intracellular potassium ion activity in rabbit ventricular muscle, *J. Physiol. (London),* 256, 663, 1976.

107. **VanHeel, B., deHemptinne, A., and Leusen, I.,** Analysis of Cl-HCO$_3$ exchange during recovery from intracellular acidosis in sheep cardiac Purkinje strands, *Am. J. Physiol.,* 246, C391, 1984.

108. **Aiton, J. F., Chipperfield, A. R., Lamb, J. F., Ogden, P., and Simmons, N. L.,** Occurrence of passive furosemide-sensitive transmembrane potassium transport in cultured cells, *Biochim. Biophys. Acta,* 646, 389, 1981.

109. **Aiton, J. F., Brown, C. D. A., Ogden, P., and Simmons, N. L.,** K$^+$ transport in tight epithelial monolayers of MDCK cells, *J. Mem. Biol.,* 65, 99, 1982.

110. **Aull, F.,** Potassium chloride cotransport in steady-state ascites tumor cells, does bumetanide inhibit? *Biochim. Biophys. Acta,* 643, 339, 1981.

Chapter 8

CALCIUM FLUXES IN CULTURED MYOCARDIAL CELLS

Janis M. Burt

TABLE OF CONTENTS

I. INTRODUCTION

Definition of the mechanisms involved in the control of calcium fluxes in cardiac tissue is essential to an understanding of excitation-contraction coupling and its modulation by drug therapies and disease. In the past decade, our knowledge in this area has increased dramatically, due, in part to the use of cultured myocardial cells. In the intact heart, there are five compartments for calcium binding and sequestration, including the vascular and interstitial spaces and at least three cellular compartments: the cell surface, mitochondria, and sarcoplasmic reticulum (SR). Resolution of these compartments is difficult under normal conditions and more difficult during interventions which modify the size of the compartments and thereby the normal calcium flux pattern. Cultured cells have the advantage of fewer compartments for calcium; specifically, the vascular and interstitial spaces are eliminated. In addition, the cultured cells can be manipulated directly without interference from the vascular barrier or impinging nerves. These differences have made it possible to expand the scope of possible experimentation and have simplified data analysis.

It is the goal of this review to summarize the advances in our understanding of calcium compartmentalization and in the control of calcium fluxes in cultured cells, and where appropriate, to indicate confirmation of results in intact heart studies. Since most of the studies in this field have utilized either rat or chick cells, the information from these two species will be presented separately and compared. Finally, a unified picture of excitation-contraction coupling will be presented. With regard to the latter, it is useful at the outset to recall the major pathways for transmembranous calcium flux and major sites for calcium binding and sequestration. The major flux pathways include the slow inward current channel, the Na^+-Ca^{2+} exchanger, and a sarcolemmal located Ca-ATPase. In addition there may be a pathway between the junctional SR and the extracellular space. The major calcium binding and sequestration sites include the cell surface, mitochondria, and SR. The studies discussed below are aimed at resolving the relative contributions of these pathways and compartments, and the regulation thereof, to the overall picture of excitation-contraction coupling.

II. CALCIUM FLUXES IN CULTURED NEONATAL RAT MYOCARDIAL CELLS

The use of cultured neonatal rat myocardial cells for the study of calcium fluxes and compartmentalization has been pioneered by Langer and his colleagues. Since the subject was reviewed by him recently,[1,2] this review will center on studies occurring since that time, with references to the earlier studies made only where necessary to present a complete picture.

A. Methods

The technique employed by Langer and colleagues for the study of calcium exchange, the scintillation-disk flow-cell technique,[1] allows for continuous on-line measurement of cell-associated calcium. Two major advantages of the technique are that each experiment serves as its own control and that rapidly exchanging calcium is detectable. The technique involves growth of the monolayer of cells on scintillant-containing plastic disks, i.e., the cells grow on a solid fluor detector. These disks are then assembled into a flow-cell, such that the disks form the walls of the flow-cell, cells face inward. The flow-cell is inserted into a modified Betamate scintillation counter where the photomultiplier tubes are closely apposed to the disks. The flow-cell is perfused sequentially with control and experimental solutions which contain the identical amount of ^{45}Ca. The cell-associated calcium is monitored continuously throughout the control and experimental periods.

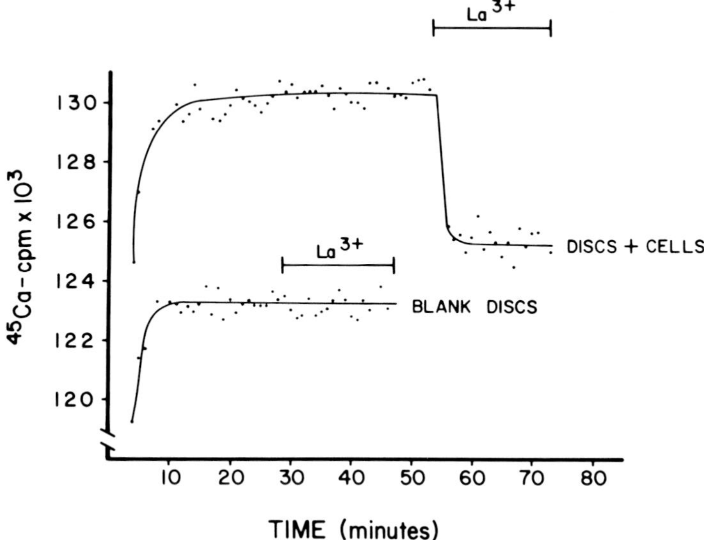

FIGURE 1.   Uptake of calcium by cultured neonatal rat myocardial cells in HEPES-buffered saline: influence of lanthanum. Cells were preincubated in HEPES-buffered saline for 30 min prior to initiation of the uptake period. Uptake was initiated by addition of 1 μCi/mℓ of ^{45}Ca. Flow rate through the flow cell was 10 mℓ/min. The upper curve represents data obtained when cells are present. Note that steady state labeling occurs within 10 to 15 min, the equilibration time of the flow cell. Addition of 1.0 mM lanthanum results in a large decrease in cell associated ^{45}Ca. The lower curve represents data obtained when just the scintillant-containing disks are present. The difference between the two curves represents total cell associated ^{45}Ca. Lanthanum displaces approximately 80% of the total cell associated ^{45}Ca but has no effect on the blank discs.

B. The Surface Compartment

Illustrated in Figure 1 is the pattern of calcium uptake seen when the cells are perfused with balanced saline solution (1 mM Ca^{2+}) buffered by HEPES, with a pH of 7.0 to 7.5 (no HCO$_3^-$ or P$_i$ not present). The striking aspect of the uptake is that a steady state pattern is established as rapidly as the flow-cell exchanges.[3] This suggests that all of the exchangeable calcium is rapidly exchanging. In HEPES buffered saline (1 mM Ca^{2+}), the exchangeable fraction of total cellular calcium is 42%. Addition of La^{3+} results in displacement of 78% of the exchangeable calcium.[4] Since La^{3+} does not penetrate past the membrane[3,5] this displaced ^{45}Ca must be representative of calcium in the surface compartment. The quantity of ^{45}Ca remaining in the cells following La^{3+} displacement accounts for the remaining 22% of the rapidly exchangeable calcium.[4] The location of this non-La^{3+} displaceable calcium within the cell is unknown. Addition of mitochondrial inhibitors,[6] or caffeine[7] midway through an uptake experiment results in no change in the uptake pattern, which suggests that the 22% of exchangeable calcium that is not La^{3+}-displaceable, and therefore presumed intracellular, is not associated with the mitochondria or the sarcoplasmic reticulum. Since contractile activity is dependent on the presence of extracellular calcium, these observations suggest that in HEPES buffered saline (1 mM Ca^{2+}), calcium cycles from the extracellular space to the intracellular space and back without exchanging with the mitochondrial and SR pools. The most likely interpretation of these results is that calcium from the extracellular space directly supports contraction under these circumstances. Similar results are seen for the interventricular septum of the rabbit.[7]

If calcium from the extracellular space does supply the calcium required for activation of the myofilaments, then interventions which result in increased contractility might be expected to augment the calcium content of the surface compartment. Two different observations support this hypothesis. First, the calcium content of the compartment is enhanced when temperature, and thereby the beating rate, is increased from room temperature to 37°C. At 37°C, the rapidly exchangeable component of total cell calcium is increased by 65% and all of the increment is La^{3+}-displaceable.[3] Second, the rapidly exchangeable calcium component of total cell calcium is enhanced by 10 to 15% when the Na^+, K-ATPase is blocked.[4] The increment is distributed between the La^{3+}-displaceable and nondisplaceable compartments just as under control conditions, i.e., 78% La^{3+}-displaceable. No increment in calcium content was seen in the absence of a Na^+ gradient, indicating that the increment is due to a response of the intact cell to an alteration of this gradient. Furthermore, the increment still occurs in the presence of verapamil, indicating that a functional slow inward current channel is not necessary for visualization of the increment. These data suggest that the calcium content of the surface compartment does change under conditions that influence the contractile state of the cell. Furthermore, the data suggest a role for the Na^+-Ca^{2+} exchange system, the only other pathway for influx of calcium, is mediation of the change in contractile state.

With the knowledge that 78% of the exchangeable calcium is in the surface compartment and the conclusion that this calcium supports contraction, three questions arise: (1) to which element(s) of this compartment is the calcium bound, (2) how is the binding regulated during inotropic manipulation, and (3) how is the transmembranous flux of calcium regulated? The surface compartment consists of the glycocalyx and the membrane bilayer. Calcium is most likely to associate with the surface compartment through noncovalent bonds, primarily to negatively charged sites. The glycocalyx contains the negatively charged sugar residue sialic acid. To determine the role of sialic acid in calcium binding and in the regulation of transmembranous flux, Langer and colleagues[3,8,9] treated cells with neuraminidase, which enzymatically removes the sialic acid residues. As a result of this treatment, the permeability of the membrane to calcium increased dramatically, although the permeability to potassium was unchanged. They concluded that the glycocalyx is instrumental to the cardiac cell in the regulation of calcium permeability. This change in permeability as a result of neuraminidase treatment has not yet been confirmed in an intact preparation. Later studies investigating the amount of calcium bound to sialic acid residues of isolated cultured cell membranes revealed that only about 10% of the total La^{3+}-displaceable Ca^{2+} was associated with this element of the membrane.[10]

The lipid bilayer, due primarily to its hydrophobic nature, has traditionally been accorded the role of the permeability barrier of the cell. With regard to calcium binding and exchange, the negatively charged phospholipids of the bilayer are of specific interest. In order to probe the role of the negative phospholipids in calcium binding, a probe which would specifically bind to this component of the membrane, and in so doing displace any bound calcium, was desirable. La^{3+} is a nonspecific displacer of calcium, i.e., it displaces calcium from all elements of the surface compartment, sialic acid and phospholipids included. Polymyxin B is an amphiphile which has five positive charges in the hydrophilic portion of the molecule. It only binds to membranes containing negatively charged phospholipids or phosphatidylethanolamine, which has a negative dipole. It is thought to interact with the membrane by inserting its lipophilic tail into the bilayer and associating its positive head group with negatively charged phospholipids. As such this compound would be predicted to displace any calcium associated with the negatively charged phospholipids. When applied to isolated cultured cell membranes polymyxin B displaces calcium in a dose dependent manner.[11] At 0.1 mM, polymyxin B displaced 42% of the total La^{3+}-displaceable Ca^{2+}. The quantity of calcium displaced by Polymyxin B is enhanced when the negatively charged phospholipid content of the membrane is enhanced, indicating its specificity of action when used at

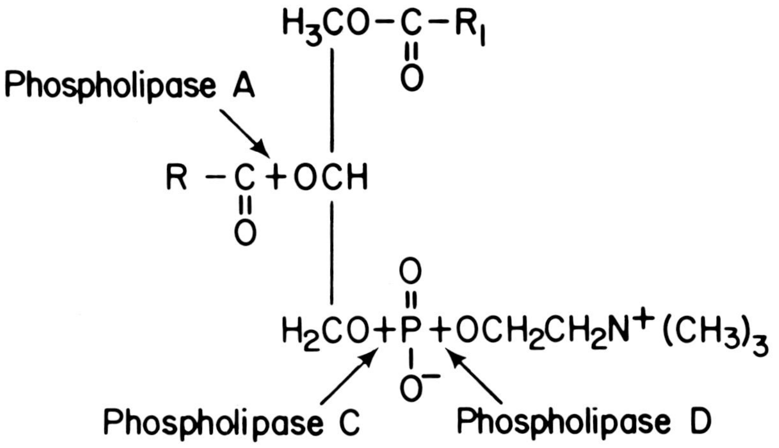

$$H_3CO-\underset{\underset{O}{\|}}{C}-R_1$$

Phospholipase A

$$R-\underset{\underset{O}{\|}}{C}+OCH$$

$$H_2CO+\underset{\underset{O^-}{\underset{\|}{P}}}{\overset{O}{\|}}+OCH_2CH_2N^+(CH_3)_3$$

Phospholipase C   Phospholipase D

FIGURE 2.   Sites of action of various phospholipases on phosphatidyl choline. Phospholipase D removes the choline R-group leaving phosphatidic acid in the membrane. Phospholipase C removes the PO_4 and choline groups leaving a diglyceride in the membrane. Phospholipase A removes the fatty acid in position two leaving lysophosphatidyl choline in the membrane.

appropriate concentrations. When Polymyxin B is applied to intact cultured cells it results in displacement of 18% of the exchangeable calcium.[12] Polymyxin B also causes a reversible slowing of the spontaneous beating activity of the cells and in some cases completely blocks spontaneous electrical activity. These data suggest that the negatively charged phospholipids do, indeed, bind significant amounts of calcium, and that this calcium is important to contractile activation. Further, the data indicate that the negatively charged phospholipids may influence the conductance properties of the membrane.

In order to further probe the role of the lipid bilayer in the binding of calcium and in the regulation of calcium fluxes, the influence of various phospholipases on calcium binding and exchange were examined. Illustrated in Figure 2 are the sites of action of the various phospholipase C, which removes the phosphate and attached R-group from the phosphoglyceride leaving a diglyceride in the membrane, results in an insignificant decrease in the calcium binding of the membrane.[10] The phospholipase C used in these experiments was specific for neutral phosphoglycerides, thus the content of negatively charged sites in the membrane was unchanged. Treatment of intact cells with phospholipase C results in a loss of permeability integrity of the membrane.[9] The phospholipase C treated cell exhibits enhanced permeability to calcium, potassium, and lanthanum, and exhibits aggregated intramembranous particles.

Cells treated with phospholipase A_2, which cleaves the fatty acid in the two position of the phosphoglyceride, exhibit no change in either calcium binding or calcium or potassium permeability.[9] These results suggest that the control of potassium permeability resides with the phosphate head groups of the phosphoglycerides in the membrane, while the control of calcium permeability resides both with the phosphoglycerides and the glycocalyx.

Treatment of isolated cardiac membranes with Phospholipase D (PLD), which cleaves the R-group from the phosphate head group of the phosphoglyceride leaving phosphatidic acid in the membrane, results in an increase in bound calcium, as expected from a treatment which increases the quantity of negatively charged phosphoglyceride in the membrane.[10,11] Similar results are observed when intact cells are treated with the enzyme.[13] The increment in cell-associated calcium following PLD treatment is distributed in the La^{3+}-displaceable and nondisplaceable compartments just as in the untreated control cells, i.e., 78% in the La^{3+}-displaceable compartment. In a parallel series of experiments, neonatal rat papillary muscles when treated with phospholipase D exhibited an increase in contractility. In a separate series of experiments, Philipson and Nishimoto[14] demonstrated enhanced Na^+-Ca^{2+} exchange activity following phospholipase D treatment. These results indicate that the negatively charged membrane phosphoglycerides play a critical role in establishing the calcium content of the surface compartment, in the activity of the Na^+-Ca^{2+} exchanger, and in the overall contractile state of the cell.

The hypothesis that negatively charged species in the sarcolemma act to modulate contractility received further support from recent work by Philipson et al.,[15] in which the influence of charged amphiphiles on contractility and sarcolemma-Ca^{2+} interactions was examined. Treatment of sarcolemmal vesicles with dodecylsulfate, a negatively charged amphiphile, resulted in augmented Ca^{2+} binding and augmented Na^+-Ca^{2+} exchange activity. Rabbit papillary muscles exposed to this compound exhibited enhanced contractility. Treatment with dodecyltrimethylamine, a positively charged amphiphile, resulted in the opposite effects while laurylacetate, a neutral amphiphile, had little or no effect. These results lend further support to the hypothesis that the negatively charged phospholipids play an important role in excitation-contraction coupling in cardiac muscle.

In a recent series of experiments, Fintel[35] examined the influence of calcium in the diffuse double layer on contractility. The calcium content of the diffuse double layer is dependent on the magnitude of the membrane's surface potential, which is in turn dependent on the ionic composition of the bulk extracellular fluid and on the fixed negative surface charge of the cell. When extracellular Na^+ is reduced, the membrane's surface potential should increase and the calcium content of the diffuse double layer should rise concomitantly. The divalent cation, dimethonium, can be used as a specific probe for the diffuse double layer. It serves to screen the surface potential without itself binding to membrane calcium binding sites. Fintel found that dimethonium reduced calcium uptake and tension development only in solutions with reduced sodium content (osmotic replacement with sucrose). Under control conditions and when LiCl was used as the osmotic replacement for Na^+, dimethonium had no effect on calcium uptake or tension development. These results suggest that under normal physiological conditions of extracellular compartment composition, the calcium important to excitation-contraction is bound to the sarcolemma.

In summary, for the surface compartment the results indicate that

1.   A small amount of the total membrane bound calcium is bound to the glycocalyx, approximately 10%. This calcium binding site appears to be important in regulation of membrane permeability to calcium but not potassium.
2.   A large amount of calcium appears to be bound to negatively charged phosphoglycerides in the membrane. When the quantity of negatively charged phosphoglyceride in the membrane is enhanced (or a negatively charged amphiphile added), cell contractility is enhanced as is calcium binding. Alternatively, when the quantity of calcium bound to this component is depressed (or a positively charged amphiphile added), contractility is depressed.
3.   Finally, the phosphate head groups of the phosphoglycerides are important in maintaining the permeability integrity of the membrane to both calcium and potassium.

C. Intracellular Compartments

In all of the studies discussed above, the uptake pattern was that characteristic of cells perfused with HEPES-buffered balanced saline solutions containing 1 mM Ca^{2+} (see Figure 1). Intracellular compartments can be visualized by addition of 10 mM PO_4 to the balanced saline solution[6,7,16] or by reducing the extracellular Ca^{2+} to 0.1 mM. Addition of 10 mM P_i to the balanced saline solution, pH maintained at 7.2, results in a linear increase in cell associated ^{45}Ca, 63.8 μmol/min kg dry wt,[6] see Figure 3. This calcium washes out of the

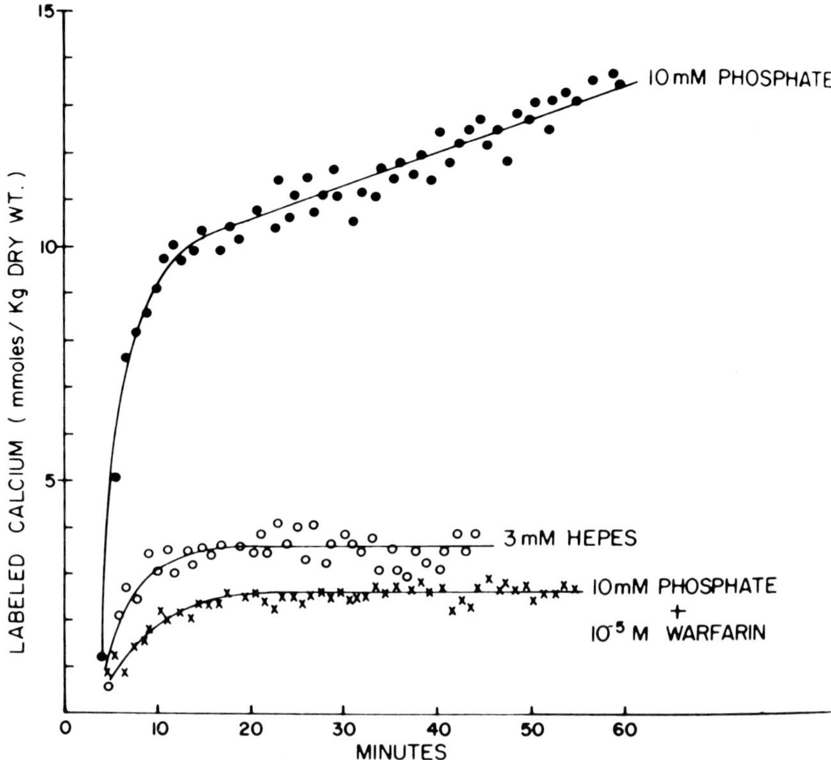

FIGURE 3.   Pattern of Ca^{2+} uptake in the presence of 10 mM P_i, influence of the mitochondrial inhibitor warfarin. Note that after equilibration of the flow cell (approximately 10 min), uptake of ^{45}Ca by the cells (upper curve) is linear. Addition of the mitochondrial inhibitor completely inhibits this linear increase in ^{45}Ca uptake (lower curve). (From Langer, G. A. and Nudd, L. M., Addition and kinetic characterization of mitochondrial calcium in myocardial tissue culture, *Am. J. Physiol.*, 239, H769, 1980. With permission.)

cell with a rate constant much slower than that of the surface compartment. In addition, this calcium is not lanthanum displaceable. These data suggest that this calcium resides in an intracellular compartment. Inhibitors of mitochondrial respiration, warfarin or antimycin A, completely block the linear increment in calcium uptake induced by 10 mM P_i, pH 7.2 (see Figure 3). Thus it appears that this component of calcium uptake is attributable to the mitochondria.

Illustrated in Figure 4 is the pattern of ^{45}Ca uptake seen in 10 mM P_i when the pH is incremented from 7.3 to 7.5. The linear rate of uptake seen when 10 mM P_i is present is enhanced by the increase in pH, on average 185%.[7,16] In this pH range, addition of inhibitors of mitochondrial respiration have no effect on the uptake pattern, and do not prevent the increase in rate of uptake associated with the pH increment. These data suggest that in this pH range the linear uptake of calcium is no longer a simple function of enhanced mitochondrial activity. The magnitude of this pH dependent, P_i supported uptake is dependent on the concentration of available PO_4^{3-}; the increase in pH shifts the equilibrium of H_3PO_4 $\rightleftharpoons H_2PO_4^- \rightleftharpoons HPO_4^{2-} \rightleftharpoons PO_4^{3-}$ towards PO_4^{3-} and thus increases the concentration of this species in solution.

To define the location of the component of calcium added in the presence of 10 mM P_i at pH > 7.2, the caffeine sensitivity of the uptake pattern was tested.[7] Under these conditions, caffeine results in enhanced uptake of calcium, see Figure 5a. In HEPES buffered saline solutions or in 10 mM P_i solutions at pH ≤ 7.2, caffeine has no effect on calcium uptake.

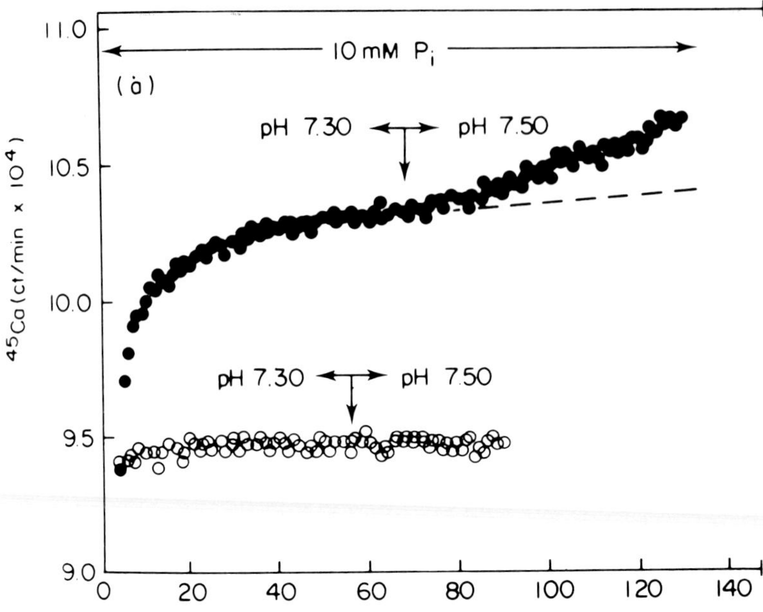

FIGURE 4.   Influence of pH on ^{45}Ca uptake in the presence of 10 mM P$_i$. Upper curve represents cells plus scintillant-containing discs. Lower curve represents the blank (discs without cells). An increment of 0.2 pH units results in a significant increase in the rate of ^{45}Ca uptake while having no effect on the blank. This effect of pH is observed between pH 7.15 and 7.35 as well. (From Ponce-Hornos, J. E., Langer, G. A., and Nudd, L. M., Inorganic phosphate: its effects on Ca exchange and compartmentalization in cultured heart cells, *J. Mol. Cell. Cardiol.*, 14, 41, 1982. With permission.)

Similar results are seen with the adult interventricular septum;[7] at pH $\leq$ 7.2 in 10 mM P$_i$ solutions, or in HEPES buffered solutions (1 mM Ca^{2+}), caffeine has no effect on Ca^{2+} uptake and no sustained effect on rest tension (minor depression of developed tension occurs). In contrast, in 10 mM P$_i$ — pH $\geq$ 7.3, caffeine application results in enhanced calcium uptake and in contracture (see Figure 5b). Whether one presumes that caffeine interferes with calcium accumulation by the SR or enhances release of calcium from the SR to the cytoplasm, the net result is calcium loading of the cytoplasm. Calcium loading of the cytoplasm is confirmed by the presence of contracture and of calcium phosphate crystals in the cytoplasm when viewed by electron microscopy. These results suggest that in the presence of 10 mM P$_i$ at pH values exceeding 7.2, the SR actively accumulates calcium.

To summarize this P$_i$ data, at pH $\leq$ 7.2 the enhanced uptake of calcium appears to be attributable to mitochondrial activity. At pH values exceeding 7.2, enhanced uptake appears to be attributable in part to mitochondrial activity and in part to sequestration by the sarcoplasmic reticulum. The mechanism by which 10 mM P$_i$ induces, in a pH dependent manner, Ca^{2+} sequestering activity by the mitochondria and sarcoplasmic reticulum is not entirely clear. It has been proposed that the phosphate causes a partial inhibition of the Na,K-ATPase, which in turn results in calcium loading of the cell (via stimulation of the Na$^+$-Ca^{2+} exchanger). In the presence of the proton donating anion PO$_4{}^{3-}$, uptake of this excess Ca^{2+} by mitochondria is possible. As pH becomes more alkalotic, more P$_i$ forms, inhibition of the ATPase is strengthened, the calcium load becomes larger, and the sarcoplasmic reticulum is activated.

Intracellular compartments can also be visualized in HEPES-buffered balanced saline solutions by reducing the calcium content to 0.1 mM. Under these conditions uptake and efflux of Ca^{2+} are pH dependent (both enhanced under alkaline conditions). At pH $\leq$ 7.2,

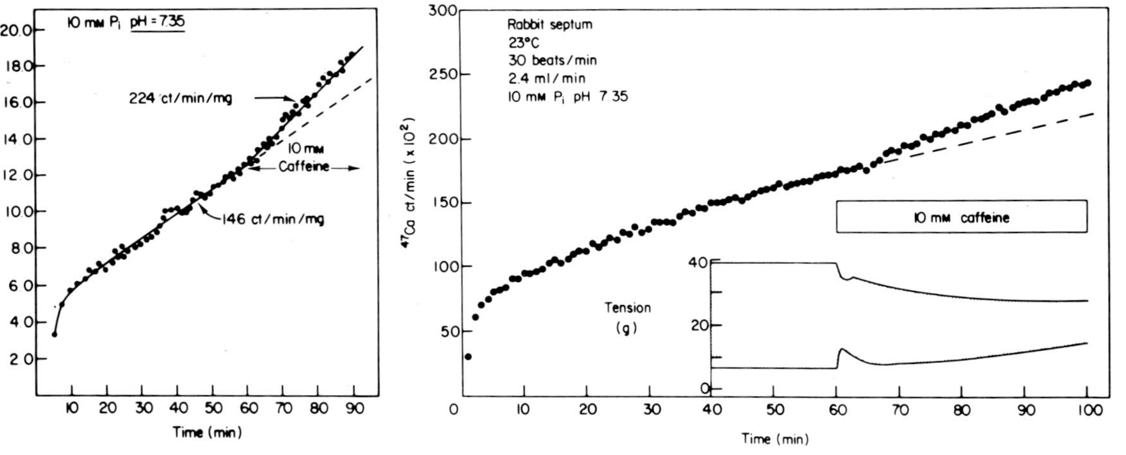

FIGURE 5.   Effect of caffeine on ^{45}Ca uptake by cultured neonatal rat myocardial cells (a) and on ^{47}Ca uptake and contractile activity of the rabbit interventricular septum (b). In (a) it can be seen that the rate of ^{45}Ca uptake is significantly enhanced by the addition of 10 m*M* caffeine. In (b) it can be seen that prior to the addition of caffeine, the septum exhibits stable contractile activity and a linear increase in ^{47}Ca uptake. Exposure to 10 m*M* caffeine results in an immediate transient increase in rest tension and a transient decrease in developed tension. These immediate effects are followed by a more gradual increase in rest tension and decrease in developed tension. The gradual changes are accompanied by enhanced ^{47}Ca accumulation. (From Langer, G. A., Rich, T. L., and Nudd, L. M., Calcium compartmentation in cultured and adult myocardium: activation of a caffeine-sensitive component, *J. Mol. Cell. Cardiol.*, 15, 459, 1983. With permission.)

the uptake of Ca^{2+} is partially inhibited by warfarin, partially by caffeine, and completely inhibited when both agents are present. At pH $\geq$ 7.4 uptake of Ca^{2+} is partially inhibited by warfarin and completely inhibited by caffeine. The mechanism by which reduction of extracellular calcium causes alteration of the sequestering activity of the mitochondria or the sarcoplasmic reticulum is not clear. One possibility is that in the presence of reduced influx there is increased participation of intracellularly stored calcium. As a consequence, turnover of intracellular stores is enhanced and there is a concomitant net uptake of label in these compartments.

D. Mechanisms for Calcium Efflux

There are two identified pathways which may contribute to the efflux of calcium, the sarcolemmal Ca-ATPase and the Na^{+}-Ca^{2+} exchanger, and a possible third pathway, as yet uncharacterized, leading from the SR directly to the extracellular space via the junctional SR. Resolution of these pathways and their contribution to the process of relaxation and calcium efflux is difficult due to the lack of interventions which specifically affect the activity of one pathway without altering the activity of the others. As a consequence, discussion of the available information must be of a much more speculative nature.

The direction in which the Na^{+}-Ca^{2+} exchange system affects net Ca^{2+} movement is dependent on the magnitude of the Na^{+} and Ca^{2+} gradients and the membrane potential.[17] Reduction of the Na^{+} gradient by either reducing the extracellular Na^{+} concentration, or increasing the intracellular Na^{+} concentration results in increased contractility, the magnitude of which is related to the change in the gradient. This increase in contractility has been attributed to the response of the Na^{+}-Ca^{2+} exchanger to pump Ca^{2+} into the cell and Na^{+} out. If the exchanger operates in the opposite direction, i.e., to pump Ca^{2+} out of the cell and Na^{+} in, then reduction of the Na^{+} gradient during a calcium efflux curve should cause a slowing of efflux. When the extracellular Na^{+} concentration is reduced to 33 mM, no slowing of efflux occurs.[18,19] However, when the extracellular Na^{+} concentration is reduced to 12 or 0 mM, Ca^{2+} efflux is significantly slowed.[20] The data indicate that efflux via the Na^{+}-Ca^{2+} exchanger is maximal when extracellular Na^{+} is 33 mM or more. The data also suggest that under normal physiologial conditions, the Na^{+}-Ca^{2+} exchanger plays only a small role or none at all in Ca^{2+} efflux from slowly exchanging compartments. Whether the exchanger plays a role in efflux or rapidly exchanging calcium remains to be determined.

Efflux studies from cells labeled in HEPES buffered solutions (1 mM Ca^{2+}) show one rapidly exchanging and one slowly exchanging compartment, neither of which is influenced by either caffeine or mitochondrial inhibitors. While it is possible that Ca^{2+} cycles through the SR in HEPES buffered solutions (1 mM Ca^{2+}), the absence of any effect of caffeine on the flux pattern strongly argues against this possibility. Similar results are seen in 10 mM P_i buffered solutions when the pH is less than 7.2, namely, that caffeine has no effect on the efflux pattern, again suggesting that the SR pathway is either nonfunctioning or non-existent. One other possibility exists, that the exchange time of the SR is as rapid as the surface compartment and flow cell, and as a consequence is indistinguishable from this compartment.

That a pathway for efflux from the SR does exist is suggested by data in which the presence of the SR compartment can be detected, i.e., 10 mM P_i buffer, pH $\geq$ 7.3. The rate of efflux of Ca^{2+} from cells labeled under these conditions remains at the control rate when caffeine is added midway through an efflux period. When caffeine is present during the labeling period and during the initial portion of the efflux period, Ca^{2+} efflux still occurs at the control rate. However, when caffeine is present during the labeling period and during the initial portion of the efflux period and then removed, Ca^{2+} efflux is significantly enhanced. The best explanation for these results assumes that caffeine inhibits uptake of Ca^{2+} by the SR. With that assumption, the observations[7] suggest that calcium present in the SR at the time of caffeine administration continues to efflux, unaffected by the presence of caffeine, at the control rate. The calcium accumulated by the cell during the period when caffeine is present apparently does not efflux from the cell by alternative mechanisms; rather that calcium apparently must efflux through the SR when the caffeine is removed. Thus it appears that the calcium retained during caffeine inhibition of SR uptake seems to have a very specific route of efflux from the cell via the SR. These explanations do not consider the possible effects of caffeine (10 mM) on the phosphodiesterase system and consequent alterations of cAMP mediated processes.

III. CALCIUM FLUXES IN CULTURED EMBRYONIC CHICK CELLS

A. Methods

The use of chick cells for the study of the regulation of calcium fluxes has been pioneered primarily by Barry and colleagues.[22-24] The technique they developed involves growth of the cells on glass coverslips. Multiple glass coverslips, held in a rack, are then submerged in the uptake solution containing ^{45}Ca. Coverslips are removed at various times and rinsed three times with cold, ^{45}Ca-free solution to remove unbound isotope. The cells are then hydrolysed and aliquots assessed for ^{45}Ca content by liquid scintillation. The data is normalized to protein content which is determined by labeling the cells for 24 hr prior to the uptake experiment with L-(4,5)-^{3}H(N)-leucine and correlating ^{3}H counts per minute to the protein content for that particular culture. The technique has the advantage of being able to resolve intracellular compartments whose exchange time is quite rapid because there is no flow cell to equilibrate. However the technique has the disadvantage that during the wash period, inspite of its brevity, the surface compartment, or lanthanum-displaceable calcium, is washed away. Additionally, the cold washes may cause redistribution of intracellular calcium.

B. Pattern of Uptake

In a CO_2-buffered balanced salt solution, chick cells exhibit two phases of calcium uptake: one fast, with a rate constant of 3.91/min ($t_{1/2}$ = 11 sec) accounting for 1.6 nmol/mg protein, and one slow with a rate constant of 0.069/min accounting for 2.7 nmol/mg protein.[23] Addition of verapamil to the uptake solution, at concentrations which completely inhibit contraction, reduces the content of the rapid phase by 30 to 40%. Calcium uptake can be rapidly blocked by addition of lanthanum, although no evidence for lanthanum-displaceable calcium can be obtained. These data are essentially similar to the observations made in rat cells with the exception of the absence of lanthanum-displaceable calcium. Given the limitations of the technique used for the flux studies, discussed above, the absence of lanthanum-displaceable calcium is not surprising. The rapid phase of calcium uptake observed by Barry and colleagues presumably corresponds to the intracellular component of rapidly exchanging calcium observed by Burt[4] and Langer.[3] Burt[4] and Langer[3] did not specifically look for a change in the size of the intracellular portion of the rapidly exchanging calcium in response to verapamil. However, Langer[19,25] has shown a small, slow decline in calcium uptake in the presence of verapamil in HEPES-buffered solutions (1 mM Ca^{2+}). No change would be expected in the quantity of surface bound calcium, which constitutes 80% of the rapidly exchanging calcium in the rat preparation, in response to verapamil. A 30 to 40% decrease in the intracellular portion of the rapidly exchanging calcium in response to verapamil would represent only a 6 to 8% decrease in the total rapidly exchanging calcium. The small decline in uptake observed by Langer may correspond to this component.

C. Stimulation and Control of Calcium Uptake

Several investigators have found that calcium uptake can be stimulated by interventions which alter the transmembranous Na^+ gradient. Fosset et al.[26] examined the effects of veratridine and ouabain on calcium uptake by chick cells. They found a maximum stimulation of calcium uptake to 25-fold and attributed the increased uptake to activity of the Na^+-Ca^{2+} exchange system. These observations were confirmed and extended to a variety of ages of chick embryos by Pang and Sperelakis.[27] Biedert et al.[21] correlated the positive inotropy caused by ouabain to parallel increases in Na^+, and concomitantly Ca^{2+}, content of the cell. They observed a 12.5% increase in Ca^{2+} content after 5 min. All of these investigations attributed the increase in calcium content to activity of the Na^+-Ca^{2+} exchange system. Similar observations have been discussed above for cultured rat cells.

In later studies the compartmentation of the calcium added by reduction of the Na^+ gradient was examined. Barry et al. found a 25-fold increase in the rapidly exchanging component of cell calcium when extracellular sodium was reduced[23] or ouabain[21] was added to the uptake solution. Murphy et al.[28] observed a 5- to 10-fold increase in cell associated calcium when the Na,K-ATPase was inhibited by removal of extracellular K^+. They did not examine the rapidly exchanging component of cell calcium during this intervention. They did find, however, that 45% of the increase in cell associated calcium was localized to the mitochondria. As observed in the rat cell preparation, these observations suggest that the magnitude of calcium content of the various cellular compartments, rapidly exchanging and mitochondrial compartment, can be altered via inotropic interventions.

Following glycoside inhibition, monovalent ion transport rates return to control levels within 1 min, and calcium content returns to normal levels within 7 min. In contrast, bulk Na^+ content remains elevated and K^+ content depressed for longer periods.[22] These observations suggest a compartment of Na^+, other than that reflected by the bulk Na^+ content, which regulates Ca^{2+} fluxes and thereby contractility. Barry et al.[22] propose that this compartment must be subsarcolemmal, because of its impact on the Na^+-Ca^{2+} exchanger, and that the Na^+ content of this subsarcolemmal space must be regulated by influx rather than efflux.

D. Mechanisms of Calcium Efflux

Barry and Smith[23] have observed that Ca^{2+} efflux is not initially blocked by lanthanum even though uptake is blocked. The time course for blockage of calcium efflux was approximately three minutes where as the time course for inhibition of uptake was five seconds. They argue that if the Na^+-Ca^{2+} exchanger is blocked in the direction of calcium uptake, that it must also be blocked in the direction of calcium efflux and conclude, that as a consequence, the exchanger cannot be responsible for calcium efflux. The experiments documenting this conclusion are not totally satisfactory; however, the conclusion has received further support from this laboratory.[24] They examined the calcium efflux pattern in the presence of zero Na^+ and found that calcium efflux still occurs, in fact the rate was enhanced. This observation certainly argues against the Na^+-Ca^{2+} exchanger being responsible for calcium efflux. The issue is, however, far from resolved. Both Wendt and Langer,[20] for cultured neonatal rat myocardial cells and rabbit septum, and Reuter and Seitz,[29] for guinea pig auricles and ventricular trabeculae of sheep and calf hearts, find a depression of Ca^{2+} efflux when the extracellular Na^+ concentration is reduced. Furthermore, in cultured neonatal rat cells, lanthanum blocks efflux within 10 to 15 sec of application.

Barry et al.[24] further demonstrate that calcium efflux is inhibited when the energy status of the cell is compromised, i.e., by incubating the cells with KCN and 2-deoxyglucose, suggesting that calcium efflux is an energy dependent process. During blockage of energy synthesis, there is an increase in sodium outside dependent calcium efflux, which indicates that the Na^+-Ca^{2+} exchanger can contribute to calcium efflux under abnormal conditions of compromised cellular energy status (and presumed Na^+ and Ca^{2+} load).

A. EXCITATION

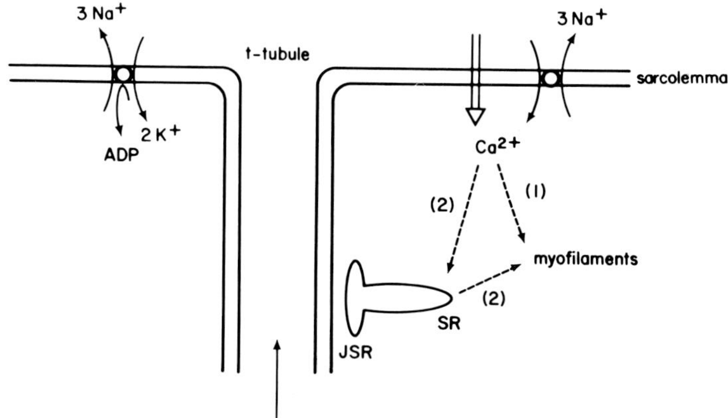

B. RELAXATION

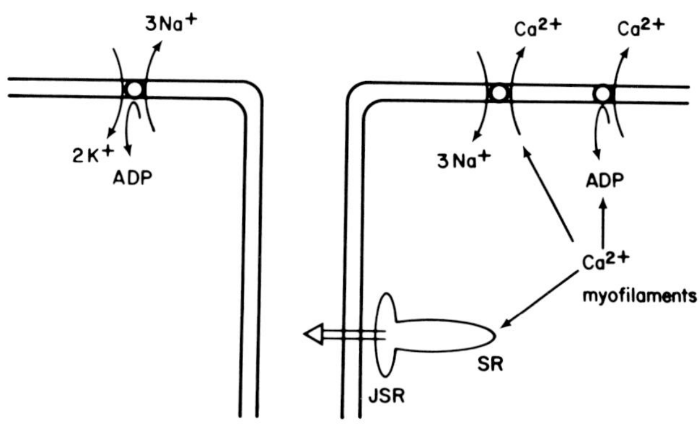

FIGURE 6. Movement of calcium during excitation (A) and relaxation (B) of the heart. (A) Excitation: pathways for calcium influx include the Na^+-Ca^{2+} exchanger and the slow inward current channel. Once inside the cell the calcium may (1) interact directly with the myofilaments, (2) induce release of calcium from the SR (by Ca^{2+} triggered release of Ca^{2+}), or (3) act by a combination of (1) and (2). For cultured neonatal rat myocardial cells and the rabbit interventricular septum, pathway (1) appears to be dominant. For the adult rat pathway (2) appears to be dominant. (B) Relaxation: pathways for calcium efflux include the sarcolemmal Ca-AT-Pase, the Na^+-Ca^{2+} exchanger, and the SR Ca-ATPase in conjunction with a junctional SR (JSR) to extracellular space mechanism. For the cultured neonatal rat myocardial cell and the rabbit interventricular septum the dominant pathway appears to be the sarcolemmal Ca-ATPase. For the adult rat the SR pathway appears to be dominant. The Na^+-Ca^{2+} exchanger appears to play a relatively minor role under normal physiological conditions in Ca^{2+} efflux.

IV. MOVEMENT OF CALCIUM DURING EXCITATION AND RELAXATION

Illustrated in Figure 6 is a scheme for the movement of calcium during excitation-contraction coupling and relaxation. Excitation involves influx of calcium via two pathways, the Na^+-Ca^{2+} exchanger and the slow inward current channel. Calcium influx is modulated by the sialic acid content of the glycocalyx, by the content of negatively charged phospho glycerides in the lipid bilayer, and by the magnitude of the Na^+ gradient, Ca^{2+} gradient and membrane potential. The Na,K-ATPase is included in the drawing because of its importance in maintaining the Na^+ gradient and membrane potential (and Ca^{2+} gradient secondarily). Once calcium has entered the cell its pathway is less certain. In HEPES-buffered solutions (1 mM Ca^{2+}), it appears that the calcium interacts with the myofilaments directly; there does not appear to be intracellular release of Ca^{2+} from the SR, a mechanism thought to operate in many species of mammal.[30] Activity of the SR can be observed when the Na,K-ATPase is blocked (with either 10 mM P_i or high doses of ouabain) and as a consequence, the intracellular sodium and thereby calcium load is elevated. These observations can also be made in the rabbit interventricular septum. It appears that at least for the cultured neonatal rat myocardial cell and the rabbit interventricular septum, normal contractile function can occur with undetectable input by the SR. Recent results suggest that in rabbit papillary muscle,[31] guinea pig atrium,[32] and frog ventricle,[33] enough calcium moves across the membrane to account for activation of the myofilaments without additional intracellular release of calcium. Based on the evidence of Fabiato,[30] one would predict that this would not be the case for adult rat myocardium.

The pathway of calcium movement during relaxation is not well understood. The evidence suggests that calcium efflux is directly energy dependent. Both the sarcolemmal Ca-ATPase and the sarcoplasmic reticulum Ca-ATPase rely on energy directly. It appears that the SR is not active under normal calcium loading conditions (i.e., HEPES-buffered saline containing 1 mM Ca^{2+}) in the cultured rat cells, which would leave the sarcolemmal Ca-ATPase as the primary efflux system. However, the SR pathway and the Na^+-Ca^{2+} exchanger are active under the abnormal condition of elevated intracellular Na^+ and Ca^{2+}.

The pathway for calcium movement during excitation-contraction coupling and relaxation in mammals appears to be defined by a spectrum of possibilities. At one extreme, activating calcium comes predominantly from the extracellular space (cultured neonatal rat, rabbit) while at the other extreme predominantly from the SR (adult rat). Similarly, the major efflux pathway may vary from exclusively the sarcolemmal Ca-ATPase at one extreme, to primarily the SR Ca-ATPase in conjunction with a junctional SR to the extracellular space pathway, at the other extreme. It seems that the species at each extreme retain the mechanisms for all pathways, lending them flexibility as needed under various unusual cellular conditions (Ca^{2+} overload, Na^+ overload, ischemia, etc.).

REFERENCES

1. **Langer, G. A.,** Calcium exchange in myocardial tissue culture, in *Excitable Cells in Tissue Culture*, Nelson, P. G. and Lieberman, M., Eds., Plenum Press, New York, 1981, 359.
2. **Langer, G. A.,** Binding and movement of calcium in the myocardial cell, in *Advances in Myocardiology*, Vol. 3, Chazov, E., Smirnov, V., and Dhalla, N. S., Eds., Plenum Press, New York, 1982, 265.
3. **Langer, G. A., Frank, J. S., and Nudd, L. M.,** Correlation of calcium exchange, structure, and function in myocardial tissue culture, *Am. J. Physiol.*, 237, H339, 1979.
4. **Burt, J. M. and Langer, G. A.,** Ca^{++} distribution after Na^+-pump inhibition in cultured neonatal rat myocardial cells, *Circ. Res.*, 51, 543, 1982.
5. **Langer, G. A. and Frank, J. S.,** Lanthanum in heart cell culture: effect on calcium exchange correlated with its localization, *J. Cell Biol.*, 54, 441, 1972.
6. **Langer, G. A. and Nudd, L. M.,** Addition and kinetic characterization of mitochondrial calcium in myocardial tissue culture, *Am. J. Physiol.*, 239, H769, 1980.
7. **Langer, G. A., Rich, T. L., and Nudd, L. M.,** Calcium compartmentation in cultured and adult myocardium: activation of a caffeine-sensitive component, *J. Mol. Cell. Cardiol.*, 15, 459, 1983.
8. **Langer, G. A., Frank, J. S., Nudd, L. M., and Seraydarian, K.,** Sialic acid: effect of removal on calcium exchangeability of cultured heart cells, *Science*, 193, 1013, 1975.

9. **Langer, G. A., Frank, J. S., and Philipson, K. D.,** Correlation of alterations in cation exchange and sarcolemmal ultrastructure produced by neuraminidase and phospholipases in cardiac cell tissue culture, *Circ. Res.,* 49, 1289, 1981.
10. **Langer, G. A. and Nudd, L. M.,** Effects of cations, phospholipases, and neuraminidase on calcium binding to "gas-dissected" membranes from cultured cardiac cells, *Circ. Res.,* 53, 482, 1983.
11. **Burt, J. M. and Langer, G. A.,** Ca^{2+} displacement by Polymyxin B from sarcolemma isolated by "gas dissection" from cultured neonatal rat myocardial cells, *Biochim. Biophys. Acta,* 729, 44, 1983.
12. **Burt, J. M., Duenas, C. J., and Langer, G. A.,** Influence of Polymyxin B, a probe for anionic phospholipids, on calcium binding and calcium and potassium fluxes of cultured cardiac cells, *Circ. Res.,* 53, 679, 1983.
13. **Burt, J. M., Rich, T. L., and Langer, G. A.,** Phospholipase D increases cell surface Ca^{++}-binding and positive inotropy in rat heart, *Am. J. Physiol.,* 247, H880, 1984.
14. **Philipson, K. D. and Nishimoto, A. Y.,** Stimulation of Na^+-Ca^{2+} exchange in cardiac sarcolemmal vesicles by phospholipase D, *J. Biol. Chem.,* 259, 16, 1984.
15. **Philipson, K. D., Langer, G. A., and Rich, T. L.,** Charged amphiphiles regulate heart contractility and sarcolemma-Ca^{2+} interactions, *Am. J. Physiol.,* 245, H147, 1985.
16. **Ponce-Hornos, J. E., Langer, G. A., and Nudd, L. M.,** Inorganic phosphate: its effects on Ca exchange and compartmentalization in cultured heart cells, *J. Mol. Cell. Cardiol.,* 14, 41, 1982.
17. **Mullins, L. F.,** Generation of electric currents in cardiac fibers by Na/Ca exchange, *Am. J. Physiol.,* 236, C103, 1979.
18. **Langer, G. A., Nudd, L. M., and Ricchiuti, N. V.,** The effect of sodium deficient perfusion on calcium exchange in cardiac tissue culture, *J. Mol. Cell. Cardiol.,* 8, 321, 1976.
19. **Langer, G. A. and Nudd, L. M.,** Calcium compartmentation in cardiac tissue culture: the effects of extracellular sodium depletion, *J. Mol. Cell. Cardiol.,* 16, 1047, 1984.
20. **Wendt, I. R. and Langer, G. A.,** The sodium-calcium relationship in mammalian myocardium: effect of sodium deficient perfusion on calcium fluxes, *J. Mol. Cell. Cardiol.,* 9, 551, 1977.
21. **Biedert, S., Barry, W. H., and Smith, T. W.,** Inotropic effects and changes in sodium and calcium contents associated with inhibition of monovalent cation active transport by ouabain in cultured myocardial cells, *J. Gen. Physiol.,* 74, 479, 1979.
22. **Barry, W. H., Biedert, S., Miura, D. S., and Smith, T. W.,** Changes in cellular Na^+, K^+, and Ca^{2+} contents, monovalent cation transport rate, and contractile state during washout of cardiac glycosides from cultured chick heart cells, *Circ. Res.,* 49, 141, 1981.
23. **Barry, W. H. and Smith, T. W.,** Mechanisms of transmembrane calcium movement in cultured chick embryo ventricular cells, *J. Physiol.,* 325, 243, 1982.
24. **Barry, W. H. and Smith, T. W.,** Movement of Ca^{2+} across the sarcolemma: effects of abrupt exposure to zero external Na concentration, *J. Mol. Cell. Cardiol.,* 16, 155, 1984.
25. **Langer, G. A., Serena, S. D., and Nudd, L. M.,** Localization of contractile-dependent Ca: comparison of Mn and verapamil in cardiac and skeletal muscle, *Am. J. Physiol.,* 229, 1003, 1975.
26. **Fosset, M., deBarry, J., Lenoir, M.-C., Lazdunski, M.,** Analysis of molecular aspects of Na^+ and Ca^{2+} uptakes by embryonic cardiac cells in culture, *J. Biol. Chem.,* 252, 6112, 1977.
27. **Pang, D. C. and Sperelakis, N.,** Veratridine stimulation of calcium uptake by chick embryonic heart cells in culture, *J. Mol. Cell. Cardiol.,* 14, 703, 1982.
28. **Murphy, E., Aiton, J. F., Horres, C. R., and Lieberman, M.,** Calcium elevation in cultured heart cells: its role in cell injury, *Am. J. Physiol.,* 245, C316, 1983.
29. **Reuter, H. and Seitz, N.,** The dependence of calcium efflux from cardiac muscle on temperature and external ion composition, *J. Physiol.,* 195, 451, 1968.
30. **Fabiato, A. and Fabiato, F.,** Calcium-induced release of calcium from the SR of skinned cells from adult human, dog, cat, rabbit, rat, and frog hearts and from fetal and new-born rat ventricles, *Ann. N.Y. Acad. Sci.,* 307, 491, 1978.
31. **Bers, D. M.,** Early transient depletion of extracellular Ca during individual cardiac muscle contractions, *Am. J. Physiol.,* 244, H462, 1983.
32. **Hilgemann, D. W., Delay, M. J., and Langer, G. A.,** Activation-dependent cumulative depletions of extracellular free calcium in guinea pig atrium measured with antipyrylazo III and tetramethylmurexide, *Circ. Res.,* 53, 779, 1983.
33. **Cleemann, L., Pizarro, G., and Morad, M.,** Optical measurements of extracellular calcium depletion during a single heartbeat, *Science,* 226, 174, 1984.
34. **Fintel, M., Langer, G. A., Rohloff, J. C., and Jung, M. E.,** Contribution of myocardial diffuse double-layer calcium to contractile function, *Am. J. Physiol.,* 249, H989, 1985.

Chapter 9

ELECTROPHYSIOLOGICAL STUDIES ON HEART CELLS IN CULTURE

Pierre Athias and Alain Grynberg

TABLE OF CONTENTS

I. INTRODUCTION

The heart cell culture system offers an opportunity of studying cardiac electrogenesis at the cellular level. Since the earliest published electrophysiological investigations in heart cell monolayer,[1] the use of such cell cultures to monitor the electrical activity of cardiac muscle cells has rapidly progressed over the past two decades, in parallel with the increasing variety of the methodologies and culture techniques.

Preparations of heart cells in culture afford the advantages associated with any culture system, such as the absence of neuronal and hemodynamic influences and of diffusion barrier for added drugs and ease in manipulating experimental conditions.[2]

Additionally, other factors make preparations of dissociated myocardial cells attractive for electrophysiological investigations:

1. Cultured cells may be observed continuously under the microscope. The recording microelectrode can thus be accurately positioned at any desired place on the sarcolemmal membrane, favoring successful impalements.
2. Every cultured heart cell is readily accessible to micromanipulation, whereas only the outer cells can be explored in intact heart or heart fragments.
3. In studies on the whole heart, vigorous contractions may dislodge the intracellular microelectrode. In contrast, the amplitude of contraction in individual cultured cardiac myocytes is very small, thus considerably enhancing the stability of the inserted microelectrode.
4. In many cardiac preparations, the presence of narrow intercellular clefts causes the ionic concentrations to fluctuate during activity.[3] In culture conditions, the extracellular space is almost unrestricted and, therefore, the composition of the fluid facing the sarcolemmal membrane is well under control.
5. Free access to heart cells in culture allows the combination of electrophysiological methods and useful complementary techniques, related to the contraction and to the discrete application of drugs.

The purpose of this chapter is to describe the methodology of recording electrical activity in excitable cells in culture, the electrophysiological characteristics of cardiac muscle cells in a variety of dissociated tissue culture systems, and the use of electrophysiological investigation of heart cell cultures in different areas of cardiac research. In this connection, readers may refer to other excellent textbooks and reviews on basic cardiac electrophysiology[4-7] and investigations of cells in culture.[8,9]

II. METHODS AND MATERIALS

A. Cell Culture Mode

From the wide range of culture preparations developed for the maintenance of dissociated cells from heart tissue (see Chapter 2), we present below the methods of cell culture which have been proven to be of interest for electrophysiological studies and summarize the limitation and scope of each system.

1. Sparse Cultures

When dispersed heart cells are plated at low density, i.e., at concentration in the region of or below 3×10^5 cells per milliliter, confluence is delayed. After a few days of growth, cell cultures contain isolated cells and small clusters. The population of muscular cells is composed of cells which contract spontaneously at independent rates, and quiescent cells which contract only when driven by the former cells.[10]

As a consequence, low density cultures are suitable for studies related to cardiac pacemaking. They offer an opportunity to isolate true pacemaker cells,[11] to monitor the establishment of functional contact between cells,[12] and the synchronization of the cellular activity.[13]

However, the most serious limitation of sparse cultures is that they probably represent the most difficult material for electrophysiologic investigation. Isolated cells are much more difficult to impale than cells trapped within a multicellular system. In this connection, it has been suggested that the electrical performance of cells depends upon the density of intercellular contacts.[14]

2. Monolayer Cultures

Disaggregated heart cells seeded at a high density of at least 4×10^5 cells per milliliter form a continuous cellular sheet after 2 to 4 days of growth. The development and organization of these confluent cultures has been described in detail.[10,15]

Electrophysiological recordings are obtained with relative ease from monolayer cultures. Conversely, the number, type, and distribution of the true pacemaker cells as well as the conduction pathway throughout the spontaneously active monolayer are almost inaccessible. Therefore, heart cell monolayers are routinely used as model systems for studying electrogenesis and pharmacology of the cardiac muscle, but the interpretation of the chronotropic actions of drugs in this system suffer from uncertainties concerning the pacemaker site.

Another striking problem is that heart cell cultures contain both myocardial and nonmuscle cells.[15,16] The nonmuscle cells proliferate actively and overgrow the myocardial cells. It has been demonstrated that this decrease in the muscle/nonmuscle cell ratio is correlated to the depressed electrical performances of the myocardial cells.[17] However, the proportion of the myocardial cells can be increased to more than 95% using procedures based on the differential rate of attachment,[18] addition of antimitotic agents,[19] plating at high density,[20] selective killing procedure,[21] or by using serum-free medium.[22]

3. Spherical Aggregates

To obtain aggregation, trypsin-dispersed heart cells are plated at very high density (about 2×10^6 cells per milliliter) so that they rapidly form small clusters. Spherical reaggregation (50 to 150 μm in diameter) is achieved by gyration for 24 to 48 hr or by using a low adherence plating substratum. The core is free of fibroblast-like cells, which are restricted to the surface of the reaggregate.[23]

Spherical reaggregates either contract spontaneously or begin contractions after stimulation. This system has proven useful for the analysis of the ionic membrane properties,[24,25] and for the formation of intercellular junctions.[26]

However, the specific geometry of heart cell reaggregates also displays some of the problems found in the intact heart preparations, i.e., those connected to the restricted extracellular space and the diffusion barriers. In addition, one advantage of the dissociated cell culture system, the ability to directly observe the cell under study, is lost.

4. Synthetic Strands

Some laboratories have developed linearly oriented cultivated cardiac preparations.[27,28] Linear orientation of heart cells disaggregated with trypsin is obtained by plating on agar film-coated dishes, which either contain grooves or palladium deposited in a linear fashion.

The general characteristics of this system are similar to those of the reaggregates. However, a specific feature of the growth-oriented heart cell cultures is the unique opportunity of studying linear electrical properties and therefore conduction processes in the heart.

5. Culture Materials

a. Tissue Source

Since myocardial muscle cells from adult cardiac tissue do not divide in vitro, cultures are prepared using embryonic or neonatal heart tissues. The most frequently used heart cell cultures in electrophysiological studies are prepared from chick embryonic or from neonatal rat hearts.

Chick embryo heart cell cultures have proven useful for the investigation of the physiological changes occurring during the development of the heart.[29]

Neonatal heart cell cultures, especially those from mammalian cardiac tissue, develop differentiated functional properties, and may be considered as a satisfactory model of the mammalian heart.[30,31]

Heart cells can be cultured either from the whole heart or from the different cardiac compartments. It is generally accepted that the behaviour of the cultured heart cells reflects the tissue of origin.[32] Thus, the choice of a particular cardiac tissue as starting material for the cell culture is governed by the problem to be solved. Most frequently, ventricular tissue is used. In this connection, a point of great physiological importance is that cell cultures from the ventricular myocardium may possibly retain cells from certain specialized areas, such as the conducting system,[33] and this must be taken into account for interpretation of data related to pacemaking (see Section III.E. in this chapter).

b. Some Factors Influencing Growth Media

The nutritional needs and incubation conditions of cell cultures have been reviewed (see also Chapter 2 in this volume).[34,35] Nevertheless, certain factors are crucial determinants of well expressed physiological properties of in vitro heart cells.

The essential role of Ca^{2+} ions in the cardiac structure and function has been emphasized.[36-39] According to Coraboeuf in his authoritative review,[4] only the free Ca^{2+} concentration, i.e., the ionic activity, is of functional significance. For this reason, it is important to adjust the Ca^{2+} activity in the growth media of heart cell cultures to the same value as in the plasma of the animal donors.[40] The best possible procedure is to determine the Ca^{2+} activity using an ion-specific electrode. Ca^{2+} activity can also be estimated from calculations which take the amount of protein carried by the serum supplement into account.[41] In all cases, Ca^{2+} activity has to be expressed at a standardized pH (7.40), which also implies careful monitoring of pH in the media.

It is current practice to supplement culture media with antibiotics. Previous experiments have led to the suggestion that the inclusion of antimicrobial agents may have untoward effects on cultured heart cells.[42,43] Thus, it is advisable that the amount of antibiotics in growth media should be as low as possible. In particular, aminoglycoside antibiotics should be avoided since they behave as potent Ca^{2+} blockers at doses routinely used in cell cultures.[43,44]

B. Handling Cultures

The nature and the stability of the cell environment are of paramount importance in physiological experiments. Electrical properties as well as contractility and automaticity of cardiac muscle cells are very sensitive to the external physico-chemical status. As a consequence, electrophysiologists intending to use heart cell cultures should be aware that any variability in the extracellular environment may interfere with the experimental factor being tested.

1. Media

The choice of the bathing fluid for cultured heart cells during the course of electrophysiological manipulations has to meet with the experimental objectives.

When physiological evaluation of heart cells under growth conditions is desired, investigation should be conducted in the growth medium without the need to change the medium, except for routine culture care.

On the other hand, the addition of serum to media would be expected to influence the reproducibility of the results, especially in experiments with hormones and transmitters.[45,46] Moreover, drugs applied may interact with serum components.[47] For these reasons, serum should be omitted from media in heart cell cultures used for physiological and pharmacological studies. Frequently, the heart cell culture is placed in the basal medium used for the growth medium preparation.

In experiments requiring accurate control of ionic concentrations, or where there may be interference from metabolites and vitamins in the basal medium, the bathing fluid should be a simple balanced salt solution, which may or may not be supplemented with glucose.

In this case, the salt mixture included in the initial growth medium can be used, e.g., Puck-F saline alone instead of Ham-F10-containing media.[10]

Due to the aforementioned importance of Ca^{2+} movements in the heart, some Ca^{2+} analogues are currently being used as probes of Ca^{2+}-dependent functions. Unfortunately, many of these form insoluble salts with phosphate, bicarbonate, and sulfate ions. This is the case with Ba^{2+} and La^{3+}. The balanced salt solutions must therefore be modified as follows: (1) Phosphate and bicarbonate buffers should be replaced by their organic alternates, such as HEPES (hydroxyethylpiperazine ethanosulfonic acid); (2) Chloride ions should be used in place of sulfate anions; and (3) The salt mixture must not be supplemented with antibiotics, since some of them are in the sulfate form.

In all circumstances, Ca^{2+} must be carefully adjusted according to the instructions given above (see Section A.5.b). Of course, solutions can easily be modified, but osmolarity must remain constant. Moreover, drastic changes in ionic composition, such as Ca^{2+} withdrawal, may cause cell damage.[36,48]

2. Incubation Chamber

For electrophysiological experimentation, the heart cell cultures must be transferred to a chamber which allows thermal control, insertion of microelectrodes, and visualization of cells under microscope.

This device is usually made of Perspex or metal.[9,14,49] Unfortunately, plastic Petri dishes, because of their shape, never fit in the chamber perfectly.

A good fit can be obtained by making a cast of the dish used as matrix with epoxy resin (e.g., Aremco Products Inc., or Emerson and Cuming Inc.). The best choice is a potting compound featuring high thermal conductivity, such as Stycast 2850-KT® supplied by Emerson and Cuming. A heating element could be included in the wall of the moulded chamber. Holes can easily be drilled after curing. This procedure ensures a tight fit between the inner wall of the chamber and the sides and the bottom of the dish, so that the thermal gradient is minimal.

The temperature should be controlled by heating wires inserted in the wall of the chamber and powered by direct current from a feedback-controlled electronic circuit, the temperature achieved being monitored by thermistors.[49-51] This design allows the temperature to be rapidly changed when required. However, the position of the measuring thermistors must be chosen with care, so as to limit temperature overshoot. Additionally, heating current must be adjusted to a low enough setting in order to avoid overheating.

Thermal control should also be performed by circulating heated water from a thermostatically controlled water bath through tubes embedded in the chamber.[52] This technique ensures a highly stable temperature and a low thermal gradient, but rapid temperature changes are difficult to achieve.

However, temperature stabilization takes a long time to be established with either of these two methods, since the heated chamber has to be thermally equilibrated with the microscope stage on which the chamber is fixed. The microscope stage should also be heated, but the temperature control of the chamber and of the microscope may interfere with each other.

The thermostatically controlled cases supplied by some manufacturers (e.g., Leitz, Zeiss, Nikon) are unsatisfactory for these purposes, because access to culture dish and micromanipulation are too difficult.

3. Perfusion vs. Static Bath

Bathing fluids in which cultured heart cells are maintained during the course of physiological experiments can either be changed continuously or periodically or not changed at all. All these methods have advantages and limitations.

In perfusion studies, solutions are pumped into the chamber using a conventional pump.

Drainage is achieved by applying negative pressure to a capillary tube.[53] This method allows relative quick changes of solutions of known composition. However, constant flow may cause some problems as previously discussed.[9] In addition, a rather large volume of solution is wasted, and this may be a serious limitation when the perfusion fluid contains expensive substances. Furthermore, solutions to be perfused must be preheated. The temperature in the culture dish thus depends upon the accuracy of the preheating, the flow rate as well as the thermostatic control of the chamber, so that thermal variability may be expected. The tolerance of disaggregated cells to a constant flow of medium also has to be examined. Indeed, medium exchanges have been known to produce transitory adverse effects in cultured heart cells.[54]

The alternative method is the static bath. In these conditions cultured cells are well equilibrated in the recording solutions. It is advisable to replace the growth medium by preheated recording solution at least 1 hr before experiments. A static bath is necessary when experiments involving discrete application of substances are to be carried out (see Section D.2.). Whatever the method of drug application (addition to the bath or by microapplication), it is difficult to determine the final concentrations of the drug.

In both static bath and perfusion conditions, fluid in the chamber is in contact with the atmosphere, causing evaporation and gaseous exchanges. This may be prevented by passing a humidified air/CO_2 mixture (95/5%) through the chamber by means of a perforated ring surrounding the chamber. In the static bath, a layer of paraffin oil poured over the medium prevents evaporation, limiting both CO_2 escape and O_2 enrichment.[55] The presence of this oil layer has no influence on the intracellular microelectrode or on the reference electrode. The pH of solutions buffered with HEPES or equivalent buffers are theoretically insensitive to contact with ambient air, but even in this case evaporation has to be controlled. Other methods, e.g., frequent thorough medium changes or periodic addition of water, have proven inadequate.

It is very important to monitor pH throughout the course of experiments. Addition of phenol red can only provide a rough estimate. If possible, microsamples of solution may be collected periodically in capillary tubes and measurements performed in airtight conditions with the aid of a blood-gas analyzer.[56] However, permanent pH monitoring is preferable and this is easily done with oesophageal pH microelectrodes (Microelectrodes Inc., Synectics, Radiometer), using the same reference electrode as for electrophysiological measurements.

4. Visualization and Micromanipulations

The heart cell culture preparation consists of a cellular sheet covered by a few millimeters of fluid. Examination of cells requires the use of an inverted microscope (e.g., Leitz Diavert and Labovert, Nikon Diaphot, Zeiss Invertoscope and ICM-35) with phase contrast optics. It should provide two possible magnifications: about 200 to 300× and 400 to 500×. Due to the geometry of the culture dishes, long-range objectives and condensers are advisable, especially at the higher magnification. Microscopes with built-in photographic tubes are also preferable and must be used for all recordings of cellular contractile activity (see Section D.1.). Equipment should include a low power lamp as well as caloric filters, so as to prevent cell overheating by the light beam.

Microtools, such as intracellular electrodes and micropipettes, are driven by using micromanipulators.[57] As a rule, stereotaxic manipulators may not be used since they are slow to operate. Joystick micromanipulators (e.g., Leitz, Narishige) seem to be a better choice. On the other hand, there are a number of sophisticated, remote-controlled micropositioners.[58,59] These devices, however, are designed to meet with specific experimental requirements, and there is no clear evidence of their suitability for studies in heart cell cultures.

Microscope and micromanipulators must be firmly fixed onto a rigid, vibration-free table stand. Guidelines for preventing mechanical interference have already been detailed.[9,60] A precision balance table or an optical bench system (Oriel Corp.) is suitable. Vibrations can be further attenuated by employing metal base plates (e.g., 0.6 to 0.8 in. duralumin or steel) and synthetic light foam layers, the instruments being fixed onto the uppermost metal slab.

The different parts of the microscope and micromanipulators are often enameled and are not in electrical continuity, and therefore, each of these isolated metal pieces has to be earthed, especially the housing for the microscope lamp.

C. Electrophysiology

The intracellular microelectrode techniques, the recording circuits, and ways of preventing electrical interference are covered in many books and reviews.[8,9,60-63] Therefore, the reader is advised to consult these for basic information on electrophysiological techniques. The book by Purves is the best introductory text,[64] and for this reason is recommended to the new users of microelectrodes.

This section focuses mainly on the principles and techniques of intracellular recording with microelectrodes in contracting muscle cells maintained in culture.

1. Microelectrodes
a. Preparing Micropipettes

Any commercially available two-stage pipette puller, featuring sufficient pulling power, is suitable for making micropipettes (David Kopf, Harvard Apparatus, Narishige, etc.).

Micropipettes are usually drawn from borosilicate glass capillary tubes, although any other glass may be used. Glass fiber-containing capillary tubes have the advantage of allowing very quick filling of the micropipettes at room temperature. When glass fiber should not be included due to the particular experimental objectives, a filling method which involves minimal heating and vibration should be used.[64]

Micropipettes with a tip diameter of 0.5 μm or less are currently recommended. However, since the microelectrode electrical resistance is inversely related to the pipette tip size, the recording electrodes of highest resistance pass currents very poorly. Moreover, the success of impalements is not directly dependent on a clear correlation between the electrode tip diameter and the cell size. It appears to be advisable to select microelectrodes of the lowest resistance which still allows stable recordings from undamaged cells. For intracellular recordings from cultured heart cells, the resistance of microelectrodes should range between 8 and 40 MΩ. Nevertheless, microelectrodes with finer tips may be required for impalements of single isolated cells. Beveling micropipette tips[65] has no definite advantages.

On the other hand, micropipette shape is not very critical. The main requirement is that the shank must be flexible enough to allow the tip to be bent by the contraction twitch of the impaled myocardial cell without damaging it. For this reason, necked or short-tapered micropipettes are unsatisfactory.

b. Solutions Used for Filling Micropipettes

Micropipettes are usually filled with a concentrated (3 M) solution of KCl. Other anions, such as citrate may be used instead of Cl^-, when leakage of choride ions out of the tip is an undesired factor. The diffusional leak of K^+ is of little importance. On the contrary, it must be kept in mind that the use of ions with different diffusion coefficients gives rise to a large, unstable potential at the microelectrode tip.

Electrolytes used for filling solutions should be of the highest purity grade. The commercially available salts often contain many other ions as contaminants. This may be a problem with respect to calcium, whose cytoplasmic activity varies between pCa 8 and pCa 6 during the cardiac cycle.[66] Thus, diffusion of minute quantity of Ca^{2+} from the microelectrode may be expected to interfere with excitation-contraction coupling[38] and with other Ca^{2+}-dependent cellular functions.[67] Contaminant Ca^{2+} may be removed by the complexing agent EGTA (ethylene glycol-bis-α-aminoethyl ether N,N,N',N'-tetracetate).

Since the solution in the microelectrode makes contact with the intracellular phase, good quality water should be used, such as the Millipore-purified type. This solution should be also free of suspended particles and thus filtered through 0.22 μm pores, although sterile conditions are not required.

c. Connection to the Recording System

The electrical continuity between electrolytes and the recording system consists of chloridized wires or pellets.[60] However, the junctions between the bathing solution and the microelectrode and the connecting wires may be of various design.[64]

Heart cell cultures offer the opportunity to obtain stable intracellular potential recordings, from a particular cell over long periods of time exceeding several hours.[30] Therefore, stable junction electrodes are desirable, and, for this reason, electrochemical symmetry is required.

Each junction electrode consists of an Ag/AgCl electrode with an interposed isoosmotic saline-agar bridge. Saline bridges are prepared with 4.5% agar in a 0.9% NaCl solution. The reference electrode may be a bent glass tube (0.12 to 0.16 in. in diameter), one end dipping into the bath and the other end connected to the chloridized Ag wire by isoosmotic saline solution. For connection to microelectrode, the saline-agar bridge may be placed in rigid tube inserted in the microelectrode holder.

2. Cell Impalement

A great variety of techniques have been proposed for achieving good penetration of cells with microelectrodes, generally suited to particular conditions. A gentle tap with a pencil on the angle of the base supporting the instruments remains the easiest way of impaling noncontracting cells.

The impalement of spontaneously beating heart cells in culture presents no particular difficulties, the contraction twitch being even an aid. The microelectrode is driven down with the micromanipulator, until the tip slightly depresses the cell membrane surface. The microelectrode is then lowered further by 1 to 2 μm, puncturing the membrane. One important point is that the microelectrode must be slightly but rapidly raised once the tip has penetrated the cell, this being is the key to satisfactory recordings of membrane potentials.

3. The Electronic Recording System
a. Preamplifier

Intracellular potentials are in the 0.1 V range, which is compatible with the sensitivity of commercially available electronic instruments. On the other hand, living cells are a very low source of current. Therefore, the preamplifier operates as a current amplifier and its voltage gain is generally unity.

For electrophysiological studies of cultured heart cells, the essential specifications of the preamplifier are high input resistance (100 to 1000 times that of the microelectrode), offset control (+/−200 mV minimal range) and neutralization of the input capacity acting as a low-pass filter.

Additional facilities, such as adjustment of leakage current, measurement of the resistance of microelectrodes, and voltage calibration are also advantageous. Bandwidth should range from 0 to 10 kHz or more.

The usefulness of a current injection circuit for stimulating or investigating passive membrane properties needs to be evaluated. Mono- and multilayer cultures of heart cells require such large current, that conventional current injection through the microelectrode may not be used. Large currents may be passed using a so-called breakaway box, but in this case, membrane potential cannot be recorded during the passage of current. On the other hand, current injection can be performed with low-density preparations, such as single isolated cells and small clusters and reaggregates.

Some manufacturers produce satisfactory microelectrode preamplifier (e.g., WPI Instruments, Dagan Corp.). However, designs of simple, easy-to-build preamplifiers are available.[64,68-70]

b. Filtering

A few simple and important rules are applicable for removing noise originating from cultured heart cells electrical signals.

Since resting potential to be recorded is a d.c. signal, only low-pass filters should be used. The upper frequency limit depends on the rise time of the fastest component of the signal to be passed. The upstroke velocity of the mature myocardial cell reaches several hundred V/sec. This depolarization slope allows calculation of the time needed for membrane potential to change from the resting level to the overshoot level. For example, an upstroke velocity of 100 V/sec and an action potential amplitude of 100 mV correspond to a rise time of 1 msec. The upper cutoff frequency is given approximately by the formula, $F = 3\ 500/t$, where t is expressed in ms and F in Hz.

Moreover, the cardiac action potential is characterized by several consecutive phase of different velocity. Therefore, a notch filter which removes certain frequencies from the full bandwidth should not be used, and noise or interference has to be suppressed at the source.

c. Recording Instruments

The cardiac electrical activity can be described as a periodic impulse in which the velocities of the rising and the leading edges are very different. For this reason, several devices for display and recording are required to obtain a full account of the electrophysiological properties of the cultured heart cells.

An oscilloscope is an essential component. It must feature two or three vertical channels and trace capture by electrostatic or preferably by digital storage. Other equipment may be useful, such as integrated calibration facilities, vertical amplifier outputs, X/Y mode and digital readout. Taking into account the frequency range of the biological signals, 1 to 2 MHz oscilloscopes are suitable. Amplifier sensitivity of 5 to 10 mV/unit is sufficient.

An oscillographic direct writing recorder is required for exhaustive storage of the electrical signals. The writing mode should be rectilinear. Conventional thermic and pressure-ink recorders have a narrow bandwidth ranging from d.c. to tens of Hz. Ink-jet recorders (e.g., Siemens) have an upper frequency limit of about 1 kHz. In these both cases, fast upstrokes are not accurately transcribed. The slower potentiometric recorders are only useful for obtaining accurate records of the action potential rates or for other slowly changing events.

The recording of the rising phase of the action potential requires specific devices. This upstroke can simply be photographed on the oscilloscope screen. Digital oscilloscope and transient waveform memories allow direct, high quality copies of the captured traces to be made onto drawing boards using an X-Y plotter. Ultraviolet and electrostatic recorders are also able to record the fastest component of the action potential, but are somewhat expensive.

Due to the composite nature of the cardiac action potential, it is highly useful, if not essential, to store electrophysiological data using a tape recorder. Since the signals to be recorded contain d.c. and low-frequency components, an FM (frequency modulation) tape recorder is required.

An audio monitor signaling electrical activity is a very useful, moderately priced piece of equipment. For use with heart cells, it should produce a bleep for each action potential like a surgical monitor. On the other hand, the audio monitor whose pitch is controlled by voltage can be very disturbing when used with spontaneously active heart cell preparations.

d. Signal Analysis

The parameters characterizing the electrical activity of the cardiac cells are as follows:

1. Amplitude parameters — maximal diastolic potential (MDP) (identifiable with the resting potential, provided that the duration of the diastolic phase is long enough to allow full return of the potential to a stable resting level), overshoot (OS), and action potential amplitude (AP);
2. Time parameters — action potential duration (APD) measured at different degrees of repolarization (e.g., 40-50% or 80-90%) or at certain potential levels (e.g., 0 volt);
3. Slope parameters — upstroke velocity ($\dot{V}$max), slopes of the initial repolarization (plateau phase) and of the final repolarization phase. With respect to the overshoot potential level, an indication should be given as to whether this more positive potential is reached at the end of the fast upstroke (Na^+-dependent) or during the following plateau phase (Ca^{2+}-dependent).

There are three fundamental ways of making these measurements: by hand from original records, or using either analogic circuits or computers. The first method is tedious and time-consuming, but requires no explanation. Therefore, only the latter two will be described.

There are several analogic circuits which provide electrical signals proportional to some of the parameters mentioned above with on-line analysis. The most common one is the well known RC passive circuit giving the first derivative of the potential changes. The values of the resistance and of the capacity should be matched so that the upper frequency limit, defined as F = $1/2 \pi$ RC, does not cause flattening of the fastest component of the signal. A time constant τ = RC of 20 to 40 μsec is suitable for differentiation of rapidly rising action potentials. Differentiating slower action potential or repolarization phases do not require such short time constants.

Differentiation can be performed by a rather simple active circuit.[71] Sample/hold peak detectors are able to record the peak $\dot{V}$max signal within a time period constant with the specifications of slow recorders or measuring devices.[72] Similarly, peak-holding circuits are also designed to provide constant display of maximum positive and negative potentials.[73,74] Simple electronic circuits allow direct measurements of the duration of the action potential to be made.[75,76]

At the present time, the ideal procedure is to use digital computers. This may be expensive and time consuming, since complicated programs have to be developed. However, the materials and algorhythms required for automated analysis of cardiac action potentials are currently available.[77-79] The main practical problem is that this action potential contains both fast and slow components, as mentioned above. Therefore, accurate computer-based analysis should require successive analogic/digital acquisition at different sampling rates. Slightly modified commercially available single-board data-acquisition system,[80] transient waveform memories, or digital oscilloscopes may be used to provide an interface connection to the computer.

3. The Voltage Clamp Method

Voltage clamping is a powerful tool for studying excitability in biological membranes. This technique allows direct measurement of membrane currents caused by membrane ionic fluxes and therefore the calculation of membrane conductance. A number of interesting papers are devoted to the theoretical and practical aspects of the voltage clamp method.[81-84]

The essential requirement for the utilization of the voltage clamp is a homogeneous spatial distribution of current.[82] Large preparations impair the potential uniformity and, therefore, only single isolated cells or small aggregates can be investigated by the voltage clamp technique. Furthermore, since cardiac cells cannot be clamped with a longitudinal internal electrode or with a sucrose gap; microelectrodes must be used.

Generally, two microelectrodes have to be inserted into the same cell, one to pass current and the other to monitor potential.[83] The electrodes must be carefully shielded in order to avoid cross interaction. For voltage clamping, microelectrodes should be selected for fast settling and for their ability to pass large currents. Thus, a compromise must be found between low microelectrode resistance and stable intracellular recording from undamaged cells. Nevertheless, the current-carrying properties of the microelectrode remain the principal limiting factor. A third voltage-recording microelectrode may be useful for testing the degree of spatial nonuniformity in voltage.[85]

When the insertion of two microelectrodes into one cell is impossible, single electrode voltage clamp may be used.[84,85] The same electrode is utilized for both current passing and voltage monitoring. This is possible by alternating voltage sampling and current feedback. Evidently, the response speed of the microelectrode is of major importance. Thus, this technique is well suited for studies of membrane conductance changes which are relatively slow.

For microelectrode voltage clamp, the preamplifier should have good slow rate (>100 V/sec) and the equipment used for displaying and recording should feature extended bandwidth ranging from d.c. up to 10 MHz.

4. The Patch Clamp Method

The patch clamp method is a recently developed electrophysiological technique with sensitivity allowing the recording of currents from a single ionic channel. This technique is based on the ability to form a high resistance seal between the cell membrane and a fire-polished micropipette,[87-89] and is described elsewhere in this Volume (see Chapter 10).

To obtain gigaohm seal, cells must be enzymatically cleaned to remove connective tissue and basement membrane. Interestingly, heart cells are disaggregated by enzymatic treatment before seeding. Thus, they can be studied using the patch clamp technique without additional enzymatic cleaning. Moreover, in the whole cell configuration, patch clamp allows voltage clamp analysis of small cells which cannot be investigated by conventional microelectrode voltage clamp methods.

Apart from certain artifacts associated with the method, the main limitation of the patch clamp relates to the instrumentation required. First, a tape recorder is required to record the large amount of data; second, data analysis, i.e., averaging and estimation of the event duration and amplitude, must be carried out with the aid of computer.

D. Ancillary Techniques

1. Contraction Transducing

Conventional force transducers may not be used for recording contractions in isolated beating cells. In culture systems, contractions are usually picked up by using photoelectrical devices;[90,91] the principles and applications of these methods are described in Chapter 11.

On the other hand, the simultaneous recording of contractions and action potentials in heart cell cultures should follow certain guidelines. Contractions should not be recorded from the cell impaled with the intracellular microelectrodes, since the rigidity of the microelectrode may alter the contraction twitch. For this reason, contractions should be recorded at about 50 to 100 μm from the microelectrode-impaled cell. The resulting error in the measurement of the excitation-contraction delay is negligible (refer to Section III, in this chapter).

In addition, it is recommended that the TV monitor showing the image of the beating cells should be placed well away from the microelectrode circuit to prevent electromagnetic interference.

2. Microiontophoresis and Microperfusion

Microiontophoresis and microperfusion are techniques in which minute quantities of drugs may be applied extracellularly, either in the vicinity of a receptive cell, or even intracellularly.[9,64,92,93]

With microiontophoresis, chemicals are ejected using electrical current, while spontaneous diffusion is counteracted with a retaining current source. Simple current sources are easy to construct.[64,92] Sophisticated programmable multisource current pumps are also commercially available. However, the desirability of such complex systems is questionable, since it is hard to ensure precise and reproducible release of drugs from micropipettes. Microelectrophoresis may only be used with ionized molecules. The amount of drug released is a function of many variables, including the ejection current, time, diffusion, and micropipette size.[64,94]

In the case of the microperfusion technique, drug release is controlled by applying pressure. The regulated pressure source may be controlled manually or via solenoid valves energized by a pulse generator. This method is efficient but rather complex to carry out. Indeed, full control of drug release should require a pressure nulling command and a negative pressure source for reaspiration. An alternative method is the connection of the micropipette to a mineral oil-filled microsyringe, with a plunger activated by a micrometer.[95] The onset of the release is somewhat less steep, although drug release may be carried out in less than 1 sec. The following comments should be noted when using microperfusion:

1. Connections and valves between the pressure source and the micropipette should be noncompliant, high pressure grade without dead volume (e.g., chromatographic accessories).
2. Only capillaries without glass fiber must be used to draw microperfusion pipettes, since the presence of the fiber causes an uncontrollable efflux from the tip.
3. Vehicle for diluting drugs must be of the same composition and pH as the bathing fluid used for recording. The amount of drug released is quantitatively related to the ejection pressure and to the position of the plunger, but the concentration achieved near the recorded cell can be estimated by indirect procedures only.[95]

Generally, the concentration reaches 1 to 10% of that in the micropipette, depending essentially on the tip diameter. We observed that a tip diameter of 20 to 40 μm placed at about 50 μm from the measured cell gives satisfactory results.

Apart from the difficulty in expressing the amount of ejected drug in terms of concentration, the limitation that the microiontophoresis and microperfusion techniques have in common is that they are both not suitable for testing drugs which are expected to have moderate effects. Nevertheless, there is evidence to suggest that electrophysiological investigation coupled to the local application of drugs represents one of the most efficient methods in heart cell culture.

III. RESULTS

A. Passive Electrical Properties

Excitable membranes respond like a circuit composed of passive electronic components. Cardiac sarcolemmal membrane behavior is equivalent to a network containing resistive, capacitive, and inductive elements.[96] The corresponding values of these elements can be determined from the response of the heart cells to a small injected current pulse (Table 1).

The specific membrane resistance (Rm) in sparse cultures of embryonic chick heart cells is measured as about 630 Ohm/cm^2 and the specific membrane capacitance (Cm) as 11 μF/cm^2, and the apparent membrane inductance as about 1.2 H/cm^2.[97] In contrast, Rm and Cm have been measured respectively, as 60 Ohm/cm^2 and 0.3 to 6.5 μF/cm^2 from single isolated heart cells of the neonatal rat,[98] and Rm rises up to several thousand Ohm/cm^2 in monolayer culture.[17]

Membrane responses to current pulses have also been studied in spheroidal aggregates. However, due to reaggregated organization, the specific electrical parameters can be cal-

Table 1

PASSIVE ELECTRICAL PROPERTIES OF HEART CELLS IN CULTURE

Heart cell culture system	Animal source	Specific membrane resistance (Ohm/cm²)	Specific membrane capacitance (µF/cm²)	Membrane inductance (H/cm²)	Ref.
Single isolated	Chick embryo	480—630	11—20	1.2[c]	97
Single isolated	Newborn rat	60	0.3—6.5		98
Monolayer	Newborn rat	5516			17
Aggregate	Chick embryo	13,000[a]/800[b]	2[a]/25[b]		99
Aggregate	Chick embryo	12,000/20,000[a]	1.65—1.24[a]	25—35k[d]	100, 101
Aggregate	Newborn rat	753	0.97		102

[a]   Measured in reference to the total cell surface area.
[b]   Measured in reference to the surface area of the aggregate.
[c]   Apparent.
[d]   Measured.

culated with reference either to the total membrane surface or to the sphere surface. In aggregates of chick embryonic heart cells, Rm and Cm are 12 to 20 KOhm/cm² and 1.2 to 1.5 µF/cm² respectively, relative to total cell membrane surface, and 800 Ohm/cm² and 25 µF/cm², relative to the surface of the aggregate alone.[99-101] Aggregates cultured from neonatal rat heart cells show similar Cm values, but lower Rm values.[102]

Apart from the uncertainties regarding the surface through which current actually flows, this approach can be considered as valid if the aggregate is small enough to be virtually isopotential during the voltage changes.

In addition, the assumption that the myocardial cell membrane behaves like a RLC circuit has further implications. First, the RLC-like impedance acts as a filter towards electrical perturbations. Second, the RLC characteristics of the cardiac sarcolemma are such that small voltage perturbations are able to cause the excitable membrane to oscillate at a resonant frequency of about 1 Hz. Such oscillating behavior may be initiated by the spontaneous voltage fluctuations due to the periodic opening and closure of the membrane channels.[100] This process may provide an explanation for the mechanism of cardiac firing and may also play a role in the generation of abnormal cardiac rhythms.[96,100]

B. Electrical Activity

The electrophysiological characteristics related to action potential observed in heart cell culture are presented in Table 2.

Low-density cultures and, *a fortiori,* single isolated heart cells, feature low resting polarization, reduced action potential amplitude and slow upstroke velocity ($\dot{V}$max).[97,103-106] These electrical properties indicate that the dissociated heart cells in sparse culture revert to an earlier developmental state. Several factors have been proposed to explain this dedifferentiation: irreversible damage by enzymatic attack, cell contact density, and low impalability.[14,107,108] In fact, until now, no fully satisfactory explanation has been suggested. Moreover, recently, nondepressed action potentials have been recorded from single isolated neonatal rat ventricular cell in culture.[109]

Monolayer cultures of heart cells from chick embryo also show the same electrophysiological pattern as embryonic heart.[103,104] Resting potential and spike amplitude are somewhat greater than in sparse heart cell cultures, but $\dot{V}$max remains essentially low. However, it has been demonstrated that this poor electrical differentiation state can be improved by K⁺ elevation, or by the addition of ATP[110] or by using culture media supplemented with lipoprotein-deficient serum.[111]

Table 2
ACTION POTENTIAL PARAMETERS OF HEART CELLS IN CULTURE

Heart cell culture system	MDP (−mV)	OS (mV)	AP (mV)	APD (ms)	Vmax (V/s)	Rate (min⁻¹)	Ref.
Single isolated[a] (7-15)	14	30					103
Monolayer[a]	59		71	150—500	1—20	0—80	97
Monolayer[a] (6)	71.7				31.8		104
Cluster[b] (1-2)	42.4	16.3	58.5		7.9		106
Monolayer[b] (12-18)	65	23		150	14		112
Monolayer[b] (2-3)	67.9	31	98.6		98.6		31
Monolayer[b] (3-5)	74.8		94.3	146.6	90.1	157.8	56
Monolayer[c]	46.8	14.8			1—12		113
Aggregate[a] (7)	78	25		155	96	120	117
Aggregate[a] (11)	62.5	30.1		170.7	82.1		119
Aggregate[a] (14-16)	57—69.5	27.5—31.8	88.4—97	97—132	10.1—29	129—188	120
Aggregate[b] (1-2)	75	21	95	155	80		124
Aggregate[d] (15-16)	65	25.6		98	122		123

Note: Ages of animals expressed in days are given in parentheses. Abbreviations are given in text.

[a] Chick embryo.
[b] Newborn rat.
[c] Adult rat.
[d] Fetal rat.

On the other hand, monolayer cultures from newborn rat heart give rise to conflicting results. They either exhibit electrophysiological characteristics similar to those of fetal heart tissue[112,113] or fast-rising, high amplitude action potentials (Figure 1).[30,31,56] The action potentials recorded from monolayer cultured heart cells of the guinea pig embryo are reported as being similar to those of the newborn rat heart cells under the same culture conditions.[114]

Results from spheroidal aggregates of heart cells from the chick embryo are similarly heterogeneous, depending mainly on the age of the embryos. Aggregates prepared from young embryos (2- to 7-day-old) possess low resting potential and slow-rising action potential.[115,116] Conversely, aggregates from chick embryos at later stage of development show differentiated electrical characteristics,[117-119] with some exceptions.[120] Reaggregated heart cells from young embryos maintain differentiated membrane properties, either when both elevated K⁺ and ATP are used,[121] or when exogenous heart RNA is added.[122]

In contrast, reaggregates from either fetal or newborn rat heart tissue received less attention. In this system, action potentials have rapid rate of rise originating from high resting potentials.[123] Furthermore, lowering the proportion of nonmuscle cells have been shown to favor the development of fully differentiated electrical properties.[124]

To summarize, the electrophysiological data collected from cultured heart cells suggest that:

1. The culture geometry appears to be a factor of major importance for the maintenance of electrical membrane functions.
2. The electrical properties vary greatly between laboratories, even when similar methodology is used.
3. Under certain culture conditions and experimental procedures, cultured heart cells at least retain the same electrophysiological characteristics as in their tissue of origin.

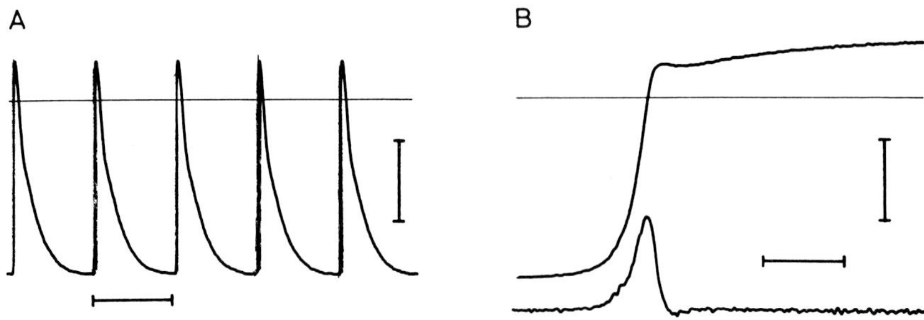

FIGURE 1.   Spontaneous action potentials (A) and the corresponding action potential upstroke (B, upper trace) and dV/dt (B, lower trace) recorded from a single neonatal rat ventricular myocyte monolayer cultured for 4 days. The cell was impaled with a 12 MOhm microelectrode. Note that A shows adult-like shape of the action potentials. Verticals bars: 40 mV (A and B) and 100 V/sec (B). Horizontal bars: 400 msec (A) and 2 msec (B). Horizontal base lines indicate the zero potential level.

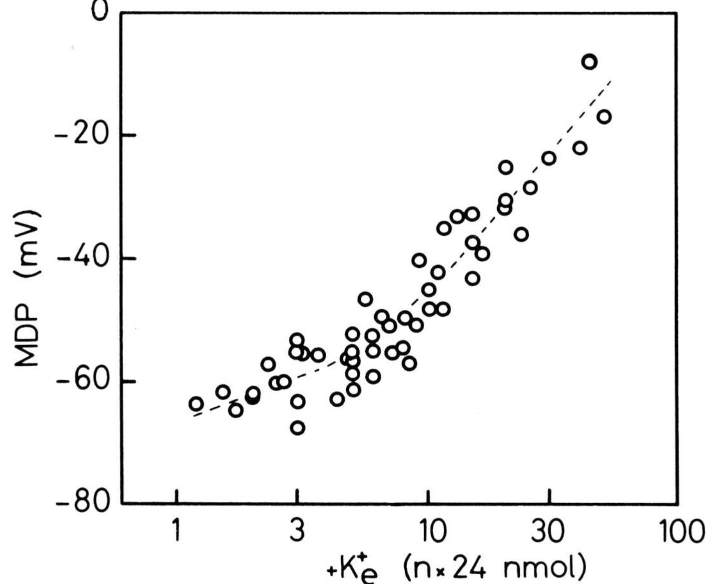

FIGURE 2.   The effect of the microperfusion of a high K^+ solution on the maximal diastolic potential (MDP) of neonatal rat ventricular myocytes in monolayer culture. The elevation of extracellular potassium concentration is expressed as ejected amounts ($+K_e^+$). Data were collected from 16 cells in 3 different culture dishes.

C. Ionic Determinants of Membrane Potential

1. Background Conductances

It is widely accepted that the resting polarization of the myocardial cells in culture is governed by the K^+ gradient via the K^+ membrane permeability in a manner that may be predicted by the Goldman Constant-Field Equation.

Indeed, cultured heart cells depolarize when placed in K^+-rich solutions (Figure 2).[95,115,121] However, the plot of the resting potential versus extracellular K^+ has a linear segment of varying slope. In heart cell cultures with immature electrophysiological characteristics, the slope is in the range of 30 to 53 mV/decade (tenfold change in external concentration),

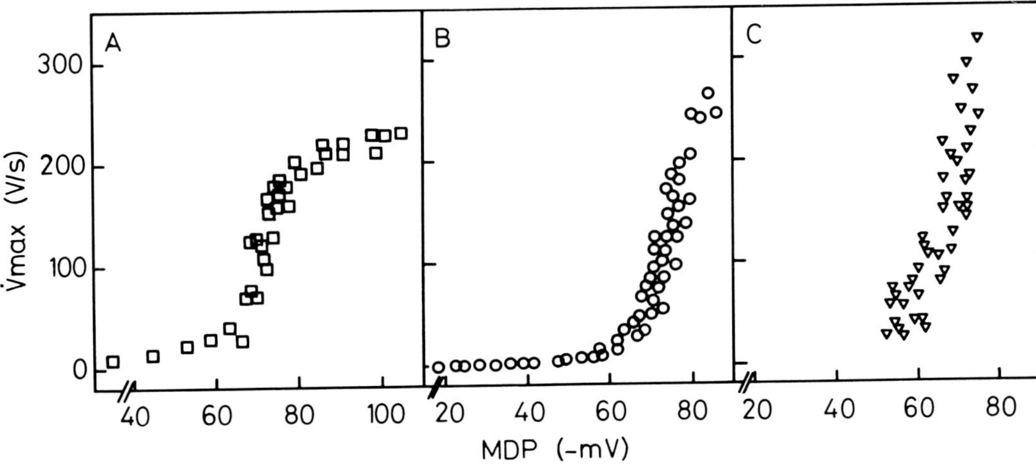

FIGURE 3.   The relationship between the maximal upstroke velocity of the action potential ($\dot{V}_{max}$) and the maximal diastolic potential (MDP) of neonatal rat ventricular myocytes in monolayer culture. The changes in MDP were determined either by current injection through the recording microelectrode (A) or by depolarizing microperfusion of a high K^+ solution (B). In (C), $\dot{V}_{max}$ is plotted versus spontaneous MDP values. Data from 3 cells in (A), from 2 cells in (B), and from other cells in 3 different culture dishes in (C).

suggesting either that the intracellular K^+ content is low or that the Na^+/K^+ permeability ratio is high.[107] In contrast, in heart cells with action potentials typical of mature cells, the measured slope is in the 53 to 60 mV range, as in the late embryonic or adult heart.[95,121] In certain instances, it has been observed that low K^+ paradoxically depolarizes cultured heart cells, probably because of a decrease in K^+ permeability.[97,107]

Moreover, patch clamp experiments conducted on heart cell cultures have allowed direct observation of voltage-sensitive K^+ channels as well as nonspecific Ca^{2+}-activated channels carrying K^+ and Na^+.[125,126]

2. Fast Inward Na^+ Current

Investigations of Na^+ conductance are of key importance, since this membrane property is considered as an index of electrophysiological differentiation.[29] Determinations of the membrane Na^+ permeability are based on voltage dependence, on effects of Na^+ ions and of agents which interact specifically with the fast Na^+ channels and on voltage clamp experiments.

a. Voltage Dependence

The upstroke velocity of the cultured heart cells showing rapidly rising action potentials, hereafter called "fast cells", decreases rapidly upon depolarization, and is completely inhibited at potential less negative than -50 mV.[30,121] This may be shown by depolarization via the microelectrode, by K^+ elevation, or simply by plotting $\dot{V}max$ vs. the spontaneous resting potential (Figure 3). By comparison, the upstroke of the heart cells displaying immature electrical activity, known as "slow cells", decay at much less electronegative resting potentials.[105]

b. Na^+ Manipulation

When the external Na^+ concentration is raised, the action potential overshoot and $\dot{V}max$ are increased in the "fast cells" (Figure 4), supporting the notion that the initial fast depolarization is governed by an inward Na^+ flux.[95,127] In "slow cells", elevation of Na^+ slightly affects $\dot{V}max$ and, in some cases the overshoot.[112,128]

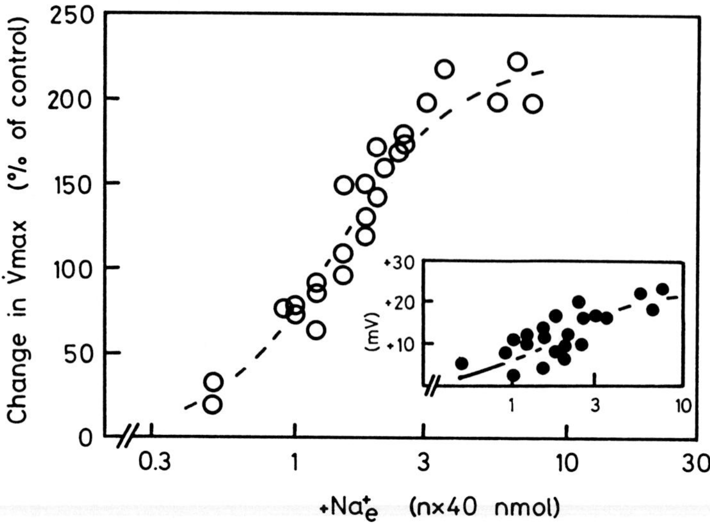

FIGURE 4.   The effects of the microperfusion of a high Na$^+$ solution on the maximal upstroke velocity ($\dot{V}_{max}$) of the action potentials of neonatal rat ventricular myocytes in monolayer culture. Inset: the corresponding changes in the potential reached at the end of the initial depolarization phase of the action potential. The extracellular Na$^+$ elevation is expressed as ejected amounts ($+Na_e^+$). Data were taken from 4 cells.

Furthermore, intracellular Na$^+$ injection mediated by phospholipid liposomes in electrically immature reaggregated heart cells causes only moderate increases in resting potential, and no noticeable change in $\dot{V}$max.[120]

c. Toxins

Tetrodotoxin (TTX) is one of the most widely used toxins for testing fast Na$^+$ channels. Typically, the fast upstroke which characterizes the mature heart cells in culture is totally inhibited by TTX (Figure 5).[95,109,121,123] In contrast, "slow cells" are insensitive to TTX.[97,111,112,117]

Other toxins reputed to be specific for the fast Na$^+$ channels have been used: sea anemone polypeptide toxin ATX$_{II}$ and scorpion toxin, which slow down the inactivation of the Na$^+$ conductance, and veratridine and batrachotoxin, which cause the Na$^+$ channels to be permanently open.[116,119,129] These substances have been shown to cause a sustained TTX-sensitive depolarization of cultured heart cells featuring embryonic-like electrophysiological properties. This supports the idea that functional fast Na$^+$ channels may exist in the inactivated state in these slow cells, although the more recently demonstrated lack of specificity of veratridine and related toxins towards Na$^+$ channels does not encourage this point of view.[130,131] In the same context, "slow cells" can be made TTX-sensitive upon hyperpolarization.[113,132]

d. Voltage Clamp Analysis

The action potential upstroke has been studied in spheroidal aggregates of embryonic chick heart cells using the two-microelectrode voltage clamp method.[24,25,118,133] This approach allows resolution of the voltage and time dependence of the rapid inward Na$^+$ current. It is initiated at potentials more positive than -45 mV and reaches its maximum at about -35 to -20 mV in 1.3 to 2.6 msec. The time constant of the decay of this Na$^+$ current is highly voltage dependent, ranging between 3.5 msec at -55 mV and 0.18 msec at -5 mV, with the intermediate value of 2.14 msec at -40 mV.

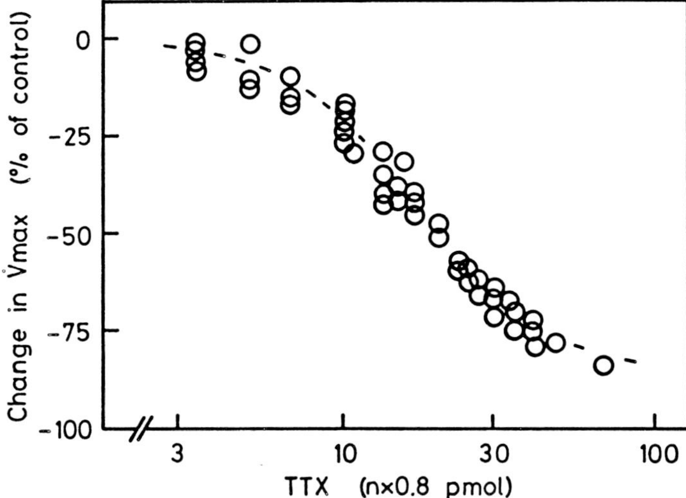

FIGURE 5. The effect of the microperfusion of a tetrodotoxin (TTX)-containing solution on the maximal upstroke velocity ($\dot{V}_{max}$) of action potentials of neonatal ventricular myocytes in monolayer culture. External TTX is expressed as ejected amounts. Data were taken from 4 cells.

The patch clamp method has also been applied for studying whole cell Na^+ current and single channel Na^+ current in cultured cells from the neonatal heart.[134] In the whole cell configuration, TTX-inhibitable Na^+ currents are recorded, displaying time- and voltage-dependence which roughly agree with the above data. Moreover, elementary currents through the fast Na^+ channel of about 1 pA at -10 mV may be measured.

3. Slow Inward Current

The contribution of the slow inward ionic flux, mainly carried by Ca^{2+}, to the myocardial cell depolarization can be studied in naturally occurring "slow cells", in experimentally depolarized cells, in K^+ or TTX-treated cells and in "fast cells".

a. "Slow Cells"

The rate of depolarization of the "slow" myocardial cells in culture was less than 20 V/sec (Table 2) and decays at potentials less negative than -30 mV.[105] Varying external Ca^{2+} affects overshoot, while $\dot{V}max$ may or may not be modified.[112,118] The accompanying changes in action potential duration not being interpretable, due to simultaneous modification in rate. The liposome-mediated increase in internal Ca^{2+} strongly depressed the action potentials of the "slow" reaggregated embryonic heart cells.[120]

One essential characteristic of the slow-rising action potential is that while it is unaffected by TTX, it is inhibited by typical Ca^{2+} antagonistic drugs like verapamil or compound D600.[110,112,113,117,119] These slow spikes can be blocked by addition of cations such as Mn^{2+}, and La^{3+},[106,112,135] although Sr^{2+} and Ba^{2+} support the slow inward current.[95,97] Furthermore, there is evidence that the slow inward current is also carried by Na^+ in cultured heart cells.[97,127,128] Finally, anionic residues on the external surface of the sarcolemma seem to be a crucial determinant of the transmembrane movements of Ca^{2+}.[136]

b. "Fast Cells"

In heart cell cultures from neonatal rat, elevations in external Ca^{2+} increase the plateau height and shortens its duration, while Ca^{2+} reduction has the opposite effects (Figure 6).[55,95]

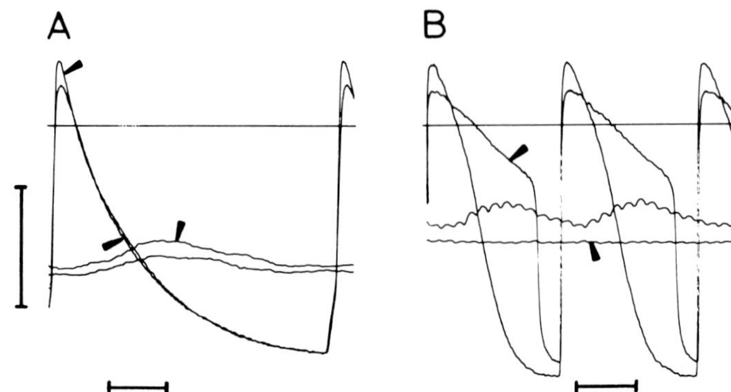

FIGURE 6.   The effects of the microperfusion of a high-Ca^{2+} solution (A) and
an ethylenediaminetetracetic acid (EDTA)-containing solution (B) on the action
potential and the contraction (middle traces) of neonatal rat ventricular myocytes
in monolayer culture. Superimposed records taken before and after (arrows) ejection of 12 nmol Ca^{2+} in (A) and 24 nmol EDTA in (B). Vertical bar: 40 mV in
(A) and (B); the vertical scale for contractions is arbitrary. Horizontal bars: 50
msec (A) and 100 msec (B). Horizontal base lines indicate zero potential level.

Mn^{2+} lowers the plateau height and may also affect $\dot{V}max$. Verapamil and La^{3+} cause a
specific decrease in the plateau height, without significantly affecting the other action potential parameters. D600 reduces the amplitude and duration of the plateau in spheroidal
reaggregates.[117]

c. Voltage Clamp Analysis

Conventional microelectrode voltage clamp of heart cell aggregates allows the demonstration of the fact that the slow component of the inward current reaches its maximum in
the range between -30 and 0 mV in about 7 msec.[25,137] The peak value of the slow inward
current is about 25 $\mu A/cm^2$. This current decays with a time constant which is 10 to 30
times that of the fast Na^+ current, but the dependence on potential of this decay leads to
somewhat conflicting results.

Single Ca^{2+} channel current have been recorded from neonatal cardiac cells in culture.[138]
The peak channel current is about at 2 pA with a mean channel open time of 1 msec. In
agreement with more conventional approaches, Ba^{2+} is able to carry the slow inward ionic
flux.

D. Some Factors Affecting Electrical Activity

Presented here are only some of the methodological conditions that modulate the basal
electrophysiological properties expressed by the isolated myocardial cells in culture.

1. Rate of Activity

It is common knowledge that cardiac action potential parameters, especially the duration,
are linked to the rate.[5] This relationship persists in culture.

An exponential dependence was found between the rate and the duration of the action
potentials in clusters derived from newborn rat ventricles.[105] In monolayer culture, this
frequency dependence can be linearized by plotting the normalized action potential duration
(the ratio of the action potential duration to the cycle length) vs. the cycle length.[139] This
indicates that the action potential duration increases with the cycle length up to a maximum,
and then decreases slightly at the longest cycle length. The other action potential parameters
may also be expected to be frequency dependent, but these particular relationships have not
been experimentally documented.

As a consequence, frequency dependence of electrical properties is assumed to play a significant role, when the experimental parameter is supposed to cause chronotropic effects. However, the rate can be fixed using electrical stimulation. In pharmacological studies, moreover, discrete application procedures (see Section II.D.2 in this chapter) keep the spontaneous rate unchanged under physiological conditions of driving.

2. Age of Culture

Several types of relationships have been described between the electrophysiological characteristics of the cultured heart cells and the age of the culture.

In many instances, TTX-sensitive fast heart cells freshly explanted from cardiac muscle are reported to lose TTX sensitivity rapidly and to develop slow-rising potentials.[104,106,110]

In contrast, other groups have shown that cultured heart cells retain differentiated electrophysiological properties for at least a week.[30,140]

For longer periods in culture, the electrical performance of the myocardial cells are observed to fade[55] because of fibroblast overgrowth,[17] although this latter hypothesis has been challenged.[140]

3. Origin of Tissue

It is well accepted that the spontaneous rate of activity of cultured heart cells reflects the properties of the cardiac tissue from which they are derived.[11] However, this statement has received very little electrophysiological support.

Comparing aggregates of chick embryonic cells taken from the atrium to those of ventricular origin, the only statistically significant difference was in the duration of the action potential.[32] The other action potential parameters are of the same order of magnitude. However, this difference in action potential duration may be caused at least in part by a difference in intrinsic rate of activity.

E. Electrophysiological Assessment of Automaticity and Coupling

1. Identification of "Leading" and "Driven" Cells

The criteria for electrophysiological identification of the pacemaker and nonpacemaker cells have been presented in an earlier review.[97] The "driven" nonpacemaker cells have action potentials showing a sharp inflection in the initial rising phase (see Figure 1), indicating a spread of activation from the adjacent spiking myocardial cells. The inflection point corresponds to the threshold potential level for activation and may be well defined using phase plane representation of the action potential.[55,105] In addition, the rate in nonpacemaker cells is not influenced by voltage.[97]

In comparison, the pacemaker cells show diastolic slow depolarization. The slope of this pacemaker potential is of the order of 10 to 100 mV/sec and increases proportionally to the frequency.[30,97,120] In pacemaker cells, moreover, there is a smooth, progressive transition from the diastolic depolarization to the upstroke. On the contrary, "driven" cells featuring diastolic depolarization show an abrupt rise of the action potential, which indicates driving.

In both cell types, a phase of hyperpolarization may be clearly distinguished after the repolarization phase of the action potential and before the pacemaker depolarization.[97,120]

"Leading" and "driven" cells may also differ in the shape of the action potential. Pacemaker cells are reported to show slow-rising low-amplitude action potentials insensitive to TTX, but which may be inhibited by D600, resembling electrical characteristics of cardiac nodal cells.[30,121] On the other hand, nonpacemaker driven cells have generally more highly differentiated electrophysiological properties. However, these differences in action potentials do not always occur but largely depend upon the culture method.

2. Basis of the Pacemaker Potential

From observations of the effects of changes in external and internal Ca^{2+} and Na^+, it has been postulated that these ions carry currents involved in automaticity of cultured cardiac cells.[112,120,128] Accordingly, Ca^{2+} blockers diminish both rate and amplitude of the action potentials.[141]

The origin of the current causing diastolic depolarization has been studied using voltage and patch clamp analysis. In chick embryonic heart cell aggregates, the pacemaker current is determined by a background current (I_{bg}), a potassium current displaying slow kinetics (I_{K2})and an additional time-dependent component (I_x).[32] In addition, it has been suggested that the spontaneous activity of cultured heart cells may also depend upon a cesium-sensitive Na^+ background current that may be related to the time-dependent pacemaker current (I_f) which develops in various cardiac pacemakers.[142] The diastolic depolarization may also originate from a nonspecific conductance gated by Ca^{2+}.[125]

On the other hand, there is no convincing support for the assumption that the electrogenic Na^+ pump and the Na^+-Ca^{2+} exchange contribute to the pacemaker current.[142]

The phenomenon of overdrive suppression of automaticity has been demonstrated to occur in embryonic heart cell aggregates and seems to be dependent upon the stimulation of an electrogenic Na^+ pump.[143]

3. Intercellular Coupling

The phenomenon of functional coupling has been demonstrated for years in heart cell cultures.[10,12,15] Furthermore, electrical coupling can be mediated by co-cultured nonmuscle cells.[13]

Direct electrophysiological demonstration of electrical junction has been performed in reaggregates and synthetic strands of cultured heart cells. There is a correlation between the development of electronic junctions and the gradual synchronization of the action potentials.[144] The degree of electrotonic interaction depends upon the status of electrophysiological differentiation.[145] The low resistance pathways may correspond to gap junctions observed between adjacent heart cells in culture.[146] However, cell-to-cell low-resistance coupling has also been demonstrated in the absence of specialized gap junctions.[147] The junctional coupling resistance (Rc) is usually measured at a resistance of a few MOhms.[26] Moreover, the action potential delay between adjacent cultured heart cells is quantitatively related to Rc.

Synthetic strands of cardiac muscle cells behave like a single one dimensional cable, characterized by a core resistance of about 180 Ohm/cm.[28] In such preparations, action potentials have been reported to propagate at velocities of 1 to 30 cm/sec.[148]

F. Excitation-Contraction Coupling

The simultaneous transduction of the action potential together with the contraction can easily be performed in heart cells in culture, although this has rarely been done.[55,142,149]

Each action potential is accompanied by a single contraction after a standard delay (Table 3). The reduction of the slow inward current by the addition of compound D600 to the perfusion solution inhibits excitation and contraction in chick embryo heart cell cultures.[141] Use of local microperfusion, on the other hand, has shown that agents chosen in various classes of Ca^{2+} blockers lower both plateau of the action potential and amplitude of the contraction.[55] Since quick drug ejection and withdrawal are possible using pressure microperfusion, the kinetics of the response and the recovery may be accurately determined. Thus, a simple but clear distinction may be made between Ca^{2+} antagonists acting outside the sarcolemma (e.g., La^{3+}, Ca^{2+} chelators) and those which penetrate the myocardial cell (e.g., verapamil, Mn^{2+}).[150]

These data indicate a key role for the slow inward influx of Ca^{2+} in triggering contractions of heart cells in culture. There are also indications that the Na^+-Ca^{2+} exchange system may raise intracellular Ca^{2+} to support contractile activity.[118]

Table 3
ELECTRICAL AND MECHANICAL PARAMETERS OF
NEWBORN RAT HEART CELLS IN MONOLAYER CULTURE

	AP (mV)	APD (ms)	tC (ms)	CD (ms)	Rate (min^{-1})
Mean	89.8	135.9	62.7	199.3	227.2
S.D.	6.3	8.3	7.4	24.4	37.7
Range	81.7—96.9	125.6—147	52—69.1	167.1—235	167.8—258.4

Notes: tC; time to 20% contraction measured from action potential upstroke. CD; contraction duration. Other abbreviations are given in text. The values have been obtained from 72 monolayer cultures of newborn rat heart cells in different dishes. A mean value was calculated for each dish, and then the sample means were averaged together.

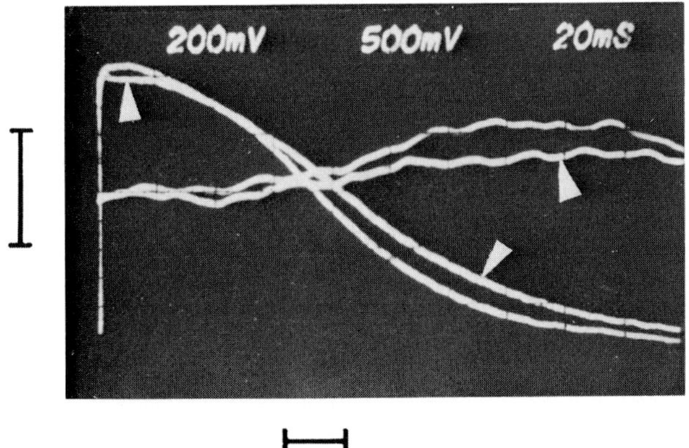

FIGURE 7.   The effects of the microperfusion of a 1-isoproterenol-containing solution on the action potential and contraction (middle traces) of a neonatal rat ventricular myocytes in monolayer culture. Superimposed records taken before (arrows) and 75 sec after the injection of 2.4 nmol isoproterenol. Vertical bar: 40 mV; the voltage scale for contractions is arbitrary. Horizontal bar: 20 msec.

IV. APPLICATIONS

A. Pharmacology

A great number of synthetic and naturally occurring substances have been tested in the heart cell culture system.[151,152] However, there are only few accounts of the electrophysiological effects of the addition of such agents.

1. Adrenergic Processes

The early studies of the action of sympathomimetic drugs have proved to be largely deceptive.[153,154] However, it has been demonstrated that β-adrenergic stimulation by isoproterenol raises the height of the plateau, decreases action potential duration, and increases contraction amplitude (Figure 7).

Furthermore, spheroidal aggregates of chick embryonic heart cells whose fast Na$^+$ channels are blocked by elevation of external K$^+$ or by TTX have been very useful for investigating the Ca^{2+} channel modulation by neurotransmitters and hormones.[29] In these preparations,

Ca^{2+}-dependent slow responses can be induced by β-adrenoceptor agonists, by phosphodiesterase inhibition with methylxanthines, and by adenylate cyclase activation with 5'-guanylimidodiphosphate.[155] In addition, liposome-mediated intracellular injection of protein kinase inhibitors blocks the isoproterenol-induced slow Ca^{2+} current, which suggests that phosphorylation of the slow channels is prerequisite for voltage activation.[156] Yet, experiments using methylxanthines have to be interpreted with care, since their action may in part be unrelated to phosphodiesterase inhibition.[157]

Patch clamp experiments indicate that the cAMP-dependent phosphorylation of the Ca^{2+} channels mainly promotes the open state of the channel, although an increase in the number of Ca^{2+} channels cannot be excluded.[158]

2. Ca^{2+} Antagonists

Agents which depress the slow inward current have been extensively used to differentiate the membrane conductance changes giving rise to the action potential upstroke in cultured heart cells (see above Section III.C.3 in this chapter). In turn, cardiac muscle cell culture is an almost ideal model system for the experimental screening tests of putative Ca^{2+} blockers.

Inhibition of excitability by Ca^{2+} antagonists is classically reported in heart cell cultures which lack fast Na^+ channels.[112,113,119,137,141] Consistently, different types of Ca^{2+} antagonists — bepridil, verapamil, diltiazem, and nifedipine — are efficient in inhibiting the slow action potentials induced in reaggregated chick heart cells in a dose-dependent manner.[159]

In contrast, derivatives of Ca^{2+} blockers of the dihydropyridine class (including nifedipine) have recently been shown to activate the voltage-dependent Ca^{2+} channel in cultured rat cardiac cells.[160] These results are in good agreement with data from patch clamp analysis in comparable preparations.[161]

3. Cardiac Glycosides

Addition of ouabain to heart cell cultures, has no influence on action potentials, whereas inotropic and chronotropic responses are observed. However, the duration of action potential is reduced by supramaximal doses of ouabain.[162]

Ouabain does not induce slow response in TTX-blocked spheroidal aggregates of myocardial cells from the chick embryo. Conversely, elevated concentrations of ouabain inhibit the catecholamine- and methylxanthine-induced slow responses.[163] These findings suggest that the known inotropic effect of cardiac glycosides is unrelated to an increase in the transmembrane Ca^{2+} influx.

4. Local Anesthetics

Local anesthetic agents (lidocaine, procaine, cocaine) have been reported to inbibit the slow responses induced in embryonic chick heart cell aggregates rendered inexcitable by TTX.[164] This indicates that local anesthetics may block the slow channels in the myocardial sarcolemma.

In addition, the extracellular application of a lidocaine derivative (mexiletine) causes a dose-dependent block of the fast inward Na^+ current in cultured heart cells of the newborn rat. There is also evidence that the binding of this substance is voltage-dependent.[165]

B. Models of the Pathological Heart
1. Arrhythmias

Heart cells in culture can be induced to beat asynchronously by the addition of digitalis or aconitine or by low K^+ or high Ca^{2+} concentrations.[166] During arrhythmias, the electrical activity is characterized by irregular discharge of action potentials, multiple oscillations, spontaneous nonperiodic voltage fluctuations, and/or depolarizing afterpotentials.[166,167] These perturbations in membrane potential are possibly related to an increase in intracellular Ca^{2+}.

Furthermore, after addition of antiarrhythmic drugs, the regular action potentials resume with disappearance of oscillatory potentials.

Another approach is based on the study of the perturbing influence of short intracellular current pulses on the spontaneous rhythmic activity of cultured heart cells.[168-170] Depending upon polarity and duration of the current pulse as well as the time of injection of the pulse, spontaneously active myocardial cells react by phase resetting, period doubling, or irregular activity.

Moreover, application of external electrical field has been shown to be correlated to variation in the transmembrane potential and contractile activity of monolayer cultured chick embryonic cells.[171] These arrhythmias resemble those observed clinically after electrical countershock treatment.

Asymmetrical conduction block of propagated action potentials in closed-loop organized synthetic strands of cardiac muscle cells has been shown to induce Wenckebach-type periodicity and reentrant excitation.[172] This mechanism of arrhythmia is known to occur in Purkinje fibers.

Therefore, it is clearly apparent that heart cell cultures are one of the models of choice for studies on the cellular basis of abnormal cardiac rhythms.

2. Metabolic Impairment

Energy deficiency caused by hypoxia, inhibition of energy production, or by factors increasing the rate of energy utilization leads to drastic reduction in resting potential, with a simultaneous decrease in amplitude and duration of the action potentials.[173,174] Under severe energy depletion, spontaneous firing disappears. Therefore, continuous synthesis of ATP is required in order to support cardiac sarcolemmal excitability.[175]

The perturbation of the action potential by metabolic inhibition has been explained in terms of modulation of the Ca^{2+} influx by the submembranous ATP.[173] Furthermore, depolarization produced by impairment of ATP synthesis is not due to a blockade of the Na^+ pump.[149]

3. Miscellaneous

Infection with herpes virus alters electrical properties of embryonic heart cell aggregates.[176] After inoculation, resting potential, overshoot, $\dot{V}max$, and action potential duration are reduced. In addition, herpes virus infection seems to interfere with the formation of coupling junctions.

Dispersed heart cells can be co-cultured with other cell types. The liver epithelial cell line co-cultured with newborn rat heart cells shows periodic voltage oscillations in synchrony with the action potentials of the spontaneously active myocardial cells (Figure 8). This observation suggests that functional nexal junctions form between the transformed cells and the primary cultured heart cells. This may be a basis for the clinically observed electrocardiographic abnormalities in cases of cardiac tumors.

C. Nutritional Studies

In a well-documented survey on nutrition and metabolism in cultured cells, Spector et al.[172] noticed that the functional properties of membranes may be modified by altering the lipid composition of the diet.

Until now, only few studies considered the electrophysiological properties as a functional marker of cultured myocardial cells in nutritional experiments. Since the original studies by Harary et al.,[178] heart cells incubated in a lipid deficient medium have been known to lose their ability to beat. Most of the papers published subsequently referred to beating rate as the unique physiological parameter.

In a collaborative study on erucic acid in heart, it was shown that addition of palmitate

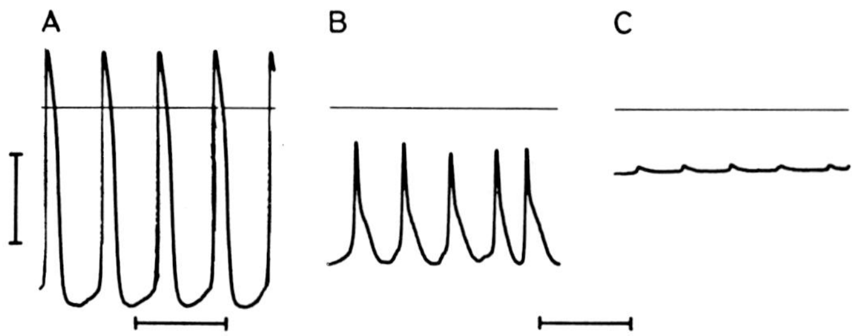

FIGURE 8.   Spontaneous action potentials recorded from neonatal rat ventricular myocytes in control culture (A), co-cultured with a liver epithelial cell line (B), and the membrane potential recorded from an epithelial strain cell in the same co-culture. Vertical bar: 40 mV in (A), (B), and (C). Horizontal bars: 400 msec in (A) and 2 sec in (B) and (C). Horizontal base lines indicate zero potential.

to a complete medium resulted in a slower beating rate and in a decrease in action potential duration in cultured myocardial cells. In contrast erucate supplemented cells exhibited decreased overshoot and beating frequency and increased action potential duration.[52] However, this study was conducted in fatty acid overload conditions and the nutritional interpretation is questionable, since arrhythmogenicity of long chain fatty acids for cultured heart myocytes has been demonstrated.[179,180]

In more physiological fatty acid load conditions, it was shown that linoleate fed cells exhibited an enhanced repolarization slope as compared to controls (i.e., full serum fed) or palmitate treated cells, maximal diastolic potential being unaffected.[181] Unfortunately, there is no biochemical support showing that the modification of membrane fatty acids may be induced by culture media.

Recently, using myocardial cells originating from the same pool and investigated in a standardized basic medium, we observed that cells grown for 1 week before standardization in different sera supplemented media retained very different electrophysiological properties (responsiveness towards standardization, calcium deficiency, and isoproterenol addition). Although the media were very different in their fatty acid profile, the cells showed a rather similar composition of phospholipids.[182]

These preliminary studies show the relevance of studying the electrophysiological parameters in cultured cells in nutritional objectives. Moreover, these methods should soon prove to be of significant interest in determining where the frontier between nutrition and nutritional toxicology lies, in cardiac research.

V. COMMENTS AND FUTURE TRENDS

A. The Importance of the Electrophysiological Criteria

This review presents evidence that heart cell cultures prepared using similar methodologies may exhibit rather different electrophysiological characteristics. As a consequence, interpretation of experiments using heart cell cultures could be entirely different, depending upon the functional pattern of the myocardial sarcolemma. Electrophysiological evaluation of the cultured heart cells under study is thus necessary under each particular set of culture conditions.

B. Interpretations of Cultured Heart Cell Differentiation

Depending on the authors, heart cell cultures are considered to be differentiated either when they retain automaticity or, in contrast, when they do not beat spontaneously. In fact, these two opposite statements are both equally misleading. Strictly speaking, intracellular microelectrode recording of transmembrane potentials only gives an indication of the electrical differentiation of the cells. The fact that a heart cell culture is spontaneously active does not necessarily mean that all the myocardial cells in the culture have acquired abnormal repetitive slow responses. Results reviewed in the preceding sections demonstrate that cultures which retain automaticity may contain a great majority of fully differentiated heart cells. On the other hand, the absence of automatic activity only shows that there are no pacemaker cells in the preparation, but this does not constitute a rigorous criterion *per se* for the maturation of the membrane functions giving rise to the action potential itself. In conclusion, it would be wrong to base statements on the functional development of the cultured cells on observations which are not backed by electrophysiological data.

Moreover, it must be emphasized that the degree of differentiation with respect to the electrophysiological performances may not be linked to the maturation of the other cellular characteristics, e.g., morphology and metabolism.

C. The Relevance of the Model

The electrophysiological characteristics of the normal and pathologically altered cardiac muscle can be approached via heart cell cultures. However, due to the highly simplified organization of this model system, to what extent the information obtained in culture is relevant to the in vivo situation remains an open question. Since electrophysiological investigations refer to single isolated myocardial cells, cultured heart cells can be considered as representing the behavior of the muscular cellular elements of the heart from which they derive. On the contrary, physiological characteristics of the heart in its native state, including tissue components which do not persist in culture (e.g., nerve terminals, and blood vessels), obviously cannot be fully documented. This is particularly true for studies on the origin of the cardiac rhythm and on the response to exogenous physico-chemical factors, such as cardioactive drugs. In summary, the use of dissociated heart cells in culture appears to be valid, at least within the limits of the conditions of the model and further extrapolation to the intact heart can be only speculative.

D. The Basis for Progress
1. Adult Myocardial Cells — A Breakthrough

Development of heart cells in culture implies mitosis and, therefore, hearts from young animals are used. This and other features, such as myofibrillar organization and morphology, have prompted the search for a means of obtaining viable single isolated myocardial cells from the adult heart. In spite of considerable technical difficulties, related to tolerance towards enzymatic attack and Ca^{2+} depletion, some groups have succeeded in this task (see Chapter 23 in Volume III). Under favorable conditions, these cardiac muscle cells retain their adult electrical characteristics after isolation. Such a procedure is perfectly suited to voltage clamp, since in this case there is no further risk of defect in voltage uniformity.

However, the viability of isolated adult heart cells is limited to a few hours in most cases, and many cells are irreversibly damaged by the isolation procedure. In addition, the conventional heart cell culture preparations provide well-differentiated cardiac myocytes with respect to electrophysiological criteria. Therefore, the single isolated adult myocardial cell is, theoretically at least, an ideal model for cardiac muscle cell function, but, in practice, the ultimate choice of a particular preparation is governed by specific experimental objectives.

2. Newly Developing and Well-Established Techniques

It is now possible to monitor the opening and closing of individual membrane ion channels, and it is evident that the highly powerful new technique of patch clamp promises great progress. It is also true that there is a need for direct control of the intracellular ion concentrations modified by the membrane channels, pumps, and exchangers. It is therefore, possible that, in the near future, ion-selective electrodes will attract growing interest. In the same way, cultured heart cells appear to be well suited to the use of the ion-sensitive dyes or indicators (e.g., aequorin for Ca^{2+}), as this preparation naturally allows photometric methods to be applied at the cellular level.

Furthermore, progress may also be expected in the field of signal processing. First, the spectacular improvements in digital computers during the past decade will certainly continue and bring direct benefit to the electrophysiologic community. Second, it is probable that still more could be learned from the electrophysiological recordings in their present form by analysis in the frequency domain, e.g., using the Fourier transform. This method is actually employed for studying several physiological periodic parameters, such as arterial blood pressure and ventricular dimensions. Frequency analysis should also be of interest to cardiocellular electrophysiologists for obtaining rigorous analytical formulation of certain data only indirectly accessible, such as rate variabilities, frequency content of the different phases of the action potential, and also correlations between the electrical and mechanical events, and between action potentials under different conditions.

REFERENCES

1. **Lehmkuhl, D. and Sperelakis, N.**, Transmembrane potentials of trypsin-dispersed chick heart cells cultured in vitro, *Am. J. Physiol.*, 205, 1213, 1963.
2. **Nelson, P. G.**, Nerve and muscle cells in culture, *Physiol. Rev.*, 55, 1, 1975.
3. **Cohen, I. and Kline, R.**, K^+ fluctuations in the extracellular spaces of cardiac muscle. Evidence from the voltage clamp and extracellular K^+ selective microelectrodes, *Circ. Res.*, 50, 1, 1982.
4. **Coraboeuf, E.**, Ionic basis of electrical activity in cardiac tissues, *Am. J. Physiol.*, 234, H101, 1978.
5. **Berne, R. M., Sperelakis, N., and Geiger, S. R.**, The Cardiovascular System, in *Handbook of Physiology*, Section 2, Vol. 1, American Physiological Society, Bethesda, 1979.
6. **Noble, D.**, *The Initiation of the Heartbeat*, 2nd ed., Clarendon Press, Oxford, 1979.
7. **Morad, M. and Tung, L.**, Ionic events responsible for the cardiac resting and action potential, *Am. J. Cardiol.*, 49, 584, 1982.
8. **Nelson, P. G. and Lieberman, M.**, *Excitable Cells in Tissue Culture*, Plenum Press, New York, 1981.
9. **Westbrook, G. L. and Nelson, P. G.**, Electrophysiological techniques in dissociated tissue culture, *Meth. Enzymol.*, 103, 111, 1983.
10. **Harary, I. and Farley, B.**, In vitro studies on single beating rat heart cells. I. Growth and organization, *Exp. Cell. Res.*, 29, 451, 1963.
11. **Marvin, W. J., Chittick, V. L., Rosenthal, J. K., Sandra, A., Atkins, D. L., and Hermsmeyer, K.**, The isolated sinoatrial node cell in primary culture from the newborn rat, *Circ. Res.*, 55, 253, 1984.
12. **De Haan, R. L. and Hirakow, R.**, Synchronization of pulsation rates in isolated cardiac myocytes, *Exp. Cell Res.*, 70, 214, 1972.
13. **Goshima, K.**, Initiation of beating in quiescent myocardial cells by norepinephrine, by contact with beating cells, and by electrical stimulation of adjacent FL cells, *Exp. Cell Res.*, 84, 223, 1974.
14. **Lieberman, M.**, Effects of cell density and low K on action potentials of cultured chick heart cells, *Circ. Res.*, 21, 879, 1967.
15. **Mark, G. E. and Strasser, F. F.**, Pacemaker activity and mitosis in culture of newborn rat heart ventricle cells, *Exp. Cell Res.*, 44, 217, 1966.
16. **Kasten, F. H.**, Rat myocardial cells in vitro: mitosis and differentiated properties, *In Vitro*, 8, 128, 1972.
17. **Hyde, A., Blondel, B., Matter, A., Cheneval, J. P., Filloux, G., and Girardier, L.**, Homo- and heterocellular junctions in cell cultures: an electrophysiological and morphological study, *Prog. Brain Res.*, 31, 283, 1969.
18. **Blondel, B., Roijen, I., and Cheneval, J. P.**, Heart cells in culture: a simple method for increasing the proportion of myoblasts, *Experientia*, 27, 356, 1971.
19. **Masse, M. J. O. and Harary, I.**, The use of 5-bromodeoxyuridine and irradiation for the estimation of the myoblast and myocyte content of primary heart cell cultures, *J. Cell. Physiol.*, 105, 197, 1980.
20. **Speicher, D. W., Peace, J. N., and McCarl, R. L.**, Effects of plating density and age in culture on growth and cell division of neonatal rat heart primary cultures, *In Vitro*, 17, 863, 1981.
21. **Kaneko, H. and Goshima, K.**, Selective killing of fibroblast-like cells in culture of mouse heart cells by treatment with a Ca ionophore A23197, *Exp. Cell Res.*, 142, 407, 1982.

22. **Frelin, C.,** Serum growth factors for rat cardiac non-muscle cells in culture, *J. Mol. Cell. Cardiol.,* 12, 1329, 1980.
23. **Sachs, H. G. and De Haan, R. L.,** Embryonic myocardial cell aggregates: volume and pulsation rate, *Develop. Biol.,* 30, 233, 1973.
24. **Ebihara, L., Shigeto, N., Lieberman, M., and Johnson, E. A.,** The initial inward current in spherical clusters of chick embryonic heart cells, *J. Gen. Physiol.,* 75, 437, 1980.
25. **Nathan, R. D. and De Haan, R. L.,** Voltage clamp analysis of embryonic heart cell aggregates, *J. Gen. Physiol.,* 73, 175, 1979.
26. **Clapham, D. E., Shrier, A., and De Haan, R. L.,** Junctional resistance and action potential delay between embryonic heart cell aggregates, *J. Gen. Physiol.,* 75, 633, 1980.
27. **Purdy, J. E., Lieberman, M., Roggenveen, A. E., and Kirk, R. G.,** Synthetic strands of cardiac muscle. Formation and ultrastructure, *J. Cell Biol.,* 55, 563, 1972.
28. **Lieberman, M., Horres, C. R., Shigeto, N., Ebihara, L., Aiton, J. F., and Johnson, E. A.,** Muscle with controlled geometry. Application to electrophysiological and ion transport studies, in *Excitable Cells in Tissue Culture,* Nelson, P. G. and Lieberman, M., Eds., Plenum Press, New York, 1981, 379.
29. **Sperelakis, N.,** Changes in membrane electrical properties during development of the heart, in *The Slow Inward Current and Cardiac Arrhythmias,* Zipes, D. P., Bayley, J. C., and Elharrar, V., Eds., Martinus Nijhoff Publishers, The Hague, 1980, 221.
30. **Athias, P., Frelin, C., Groz, B., Dumas, J. P., Klepping, J., and Padieu, P.,** Myocardial electrophysiology: intracellular studies on heart cell cultures from newborn rats, *Pathol. Biol.,* 27, 13, 1979.
31. **Robinson, R. B. and Legato, M. J.,** Maintained differentiation in rat cardiac monolayer cultures: tetrodotoxin sensitivity and ultrastructure, *J. Mol. Cell. Cardiol.,* 12, 493, 1980.
32. **Shrier, A. and Clay, J. R.,** Comparison of the pacemaker properties of chick embryonic atrial and ventricular heart cells, *J. Membrane Biol.,* 69, 49, 1982.
33. **Norwood, C. R., Castaneda, A. R., and Norwood, W. I.,** Heterogeneity of rat cardiac cells of defined origin in single cell culture, *J. Mol. Cell. Cardiol.,* 12, 201, 1980.
34. **Paul, J.,** Media for culturing cells and tissues, in *Cell and Tissue Culture,* Paul, J., Ed., Churchill Livingstone, London, 1975, chap. 6.
35. **Waymouth, C.,** Nutritional requirements of cells in culture with special reference to neural cells, in *Cell, Tissue and Organ Cultures in Neurobiology,* Fedoroff, S. and Hertz, L., Eds., Academic Press, New York, 1977.
36. **Singal, P. K., Matsukubo, M. P., and Dhalla, N. S.,** Calcium related changes in the ultrastructure of mammalian myocardium, *Br. J. Exp. Path.,* 60, 96, 1979.
37. **Mc.Donald, T. F.,** The slow inward calcium current in the heart, *Ann. Rev. Physiol.,* 44, 425, 1982.
38. **Wohlfart, B. and Noble, M. I. M.,** The cardiac excitation contraction cycle, *Pharmac. Ther.,* 16, 1, 1982.
39. **Schwarz, W. and Passow, H.,** Ca^{2+}-activated K^+ channels in erythrocytes and excitable cells, *Ann. Rev. Physiol.,* 45, 359, 1983.
40. **Foldes, F. F.,** The significance of physiological (Ca^{2+}) and (Mg^{2+}) for in vitro experiments on synaptic transmission, *Life Sci.,* 28, 1585, 1981.
41. **Marshall, R. W. and Hodgkinson, A.,** Calculation of plasma ionised calcium from total calcium proteins and pH: comparison with measured values, *Clin. Chim. Acta,* 127, 305, 1983.
42. **Payet, M. D., Bkaily, G., Schanne, O. F., and Ruiz-Ceretti, E.,** Influence of streptomycin on spontaneous activity of clusters of cardiac cells from neonatal rats, *Can. J. Physiol. Pharmacol.,* 58, 433, 1980.
43. **Athias, P., De la Chapelle-Groz, B., Variot, N., Ettaiche, M., and Klepping, J.,** Influence of gentamicin on the electrical and contractile activities of cultured cardiac myocytes, *J. Mol. Cell. Cardiol.,* 15, 53, 1983.
44. **Hino, N., Ochi, R., and Yanagisawa, T.,** Inhibition of the slow inward current and the time dependent outward current of mammalian ventricular muscle by gentamicin, *Pflügers Arch.,* 394, 243, 1982.
45. **Frelin, C. and Padieu, P.,** Pleiotropic effect of serum on beating rat heart cell cultures and their modulation by hormones, in *Recent Advances in Studies on Cardiac Structure and Metabolism,* Vol. 12, Kobayashi, T., Ito, Y., and Rona, G., Eds., University Park Press, Baltimore, 1978, 683.
46. **Price, P. J. and Gregory, E. A.,** Relationship between in vitro growth promotion and biophysical and biochemical properties of serum supplement, *In Vitro,* 18, 576, 1982.
47. **Glasson, S., Zini, R., D'Athis, P., Tillement, J. P., and Boissier, J. R.,** The distribution of bound propanolol between the different human serum proteins, *Mol. Pharmacol.,* 17, 187, 1980.
48. **Acosta, D., Ramos, K., and Li-Goldman, C. P.,** Cellular injury of primary cultures of rat myocytes incubated in calcium free medium followed by recovery in calcium, *In Vitro,* 19, 141, 1983.
49. **Bonkowski, L. and Runion, H. I.,** A heated stage and tissue culture chamber for electrophysiology, *Experientia,* 32, 1619, 1976.
50. **Saltz, E. C. and Geller, H. M.,** Simple circuit for laboratory temperature regulation, *Med. Biol. Eng.,* 14, 681, 1976.

51. **Rose, G.,** A temperature controller for in vitro recording chambers, *Br. Res. Bull.,* 10, 713, 1983.
52. **Athias, P., Pinson, A., Frelin, C., Padieu, P., and Klepping, J.,** Comparative study on the effects of exogenous palmitate and erucate on intracellular electric properties of cultured beating heart cells, *J. Mol. Cell. Cardiol.,* 11, 755, 1979.
53. **Ruppersberg, J. P. and Rüdel, R.,** A simple method for controlling the fluid level in a small experimental chamber during slow and rapid fluid exchange, *Pflügers Arch.,* 397, 158, 1983.
54. **Padieu, P., Frelin, C., Pinson, A., Charbonne, F., and Athias, P.,** Effect of environmental factors and tissue culture methodology in producing and studying cultured cardiac cells, in *Recent Advances in Studies on Cardiac Structure and Metabolism,* Vol. 12, Kobayashi, T., Ito, Y., and Rona, G., Eds., University Park Press, Baltimore, 1978, 609.
55. **Athias, P., Groz, B., Surville, J. M., and Klepping, J.,** Simultaneous record of mechanical and electrical activities of rat heart cells in culture, *Biol. Cell.,* 37, 131, 1980.
56. **Ettaiche, M., Athias, P., Variot, N., and Klepping, J.,** Characterization of post junctional adrenoceptor subtypes in isolated rat myocardial cells in culture, *Can. J. Physiol. Pharmacol.,* 63, 1221, 1985.
57. **Kopac, M. J.,** Micromanipulators: principles of design, operation, and application, in *Physical Techniques in Biological Research,* Vol. 5, Nastuk, W. L., Ed., Academic Press, New York, 1964, chap. 5.
58. **Connor, G. I.,** Micropositioner in a closed-loop control system, *IEEE Trans. Biomed. Eng.,* 20, 114, 1973.
59. **Schreurs, A. W., Meijer, A. A., Bouman, L. N., and Bonke, F. I. M.,** Micromanipulator with an electrode driver used for microelectrode work, *Pflügers Arch.,* 346, 163, 1974.
60. **Donaldson, P. E. K.,** *Electronic Apparatus, for Biological Research,* Butterworth, London, 1958.
61. **Nastuk, W. L.,** *Physical Techniques for Biological Research,* Vols. 5 and 6, Academic Press, New York, 1964.
62. **Thomas, R. C.,** *Ion Sensitive Microelectrodes: How to Make and Use Them,* Academic Press, London, 1978.
63. **Sachs, F. and Auerbach, A.,** Single channel electrophysiology: use of patch clamp, *Meth. Enzymol.,* 103, 147, 1983.
64. **Purves, R. D.,** *Microelectrode Methods for Intracellular Recording and Ionophoresis,* Academic Press, London, 1981.
65. **Tauchi, M. and Kikuchi, R.,** A simple method for beveling micropipettes for intracellular recording and current injection, *Pflügers Arch.,* 368, 153, 1977.
66. **Fabiato, A. and Fabiato, F.,** Calcium induced release of calcium from the sarcoplasmic reticulum of skinned cells from adult human, dog, cat, rabbit, rat, and frog hearts and from fetal and newborn rat ventricles, *Ann. New York Acad. Sci.,* 307, 491, 1978.
67. **Wollenberger, A. and Will, H.,** Protein kinase catalyzed membrane phosphorylation and its possible relationship to the role of calcium in the adrenergic regulation of cardiac contraction, *Life Sci.,* 22, 1159, 1978.
68. **Bergveld, P.,** Alternative design of a unity gain follower with buffer, *IEEE Trans. Biomed. Eng.,* 25, 567, 1978.
69. **Muijser, H.,** A microelectrode amplifier with an infinite resistance current source for intracellular measurement of membrane potential and resistance changes under current clamp, *Experientia,* 35, 912, 1979.
70. **Peltoranta, M., Malmivuo, J., Nieminen, K., and Oja, S.,** Microelectrode amplifier for intracellular stimulation and recording, *Med. Biol. Eng. Comput.,* 21, 731, 1983.
71. **McGillivray, R. H. and Wald, R. W.,** Measurement of the maximal rate of rise of the cardiac action potential Vmax, *Med. Biol. Eng. Comput.,* 22, 275, 1984.
72. **Hondeghem, L. M. and Cotner, C. L.,** Measurement of Vmax of the cardiac action potential with a sample/hold peak detector, *Am. J. Physiol.,* 234, H312, 1978.
73. **Mabilde, C.,** Peak detector-memory system for electrophysiological signals, *Med. Biol. Eng. Comput.,* 18, 791, 1980.
74. **Tiedeman, J. S., Aronson, R. S., and Strauss, H. C.,** A circuit for measuring maximum negative or positive potentials, *J. Appl. Physiol.,* 38, 760, 1975.
75. **Lab, M. J. and Child, R. K.,** An automatic cardiac action potential duration meter, *Am. J. Physiol.,* 236, H183, 1979.
76. **Kentish, J. C. and Boyett, M. R.,** A simple electronic circuit for monitoring changes in the duration of the action potential, *Pflügers Arch.,* 398, 233, 1983.
77. **Elharrar, V. and Lovelace, D. E.,** On-line analysis of intracellular electrophysiological data using a microcomputer system, *Am. J. Physiol.,* 237, H400, 1979.
78. **Klein, D. L., Jenkins, J. M., and Ten Eick, R. E.,** Dual microcomputer analysis of cardiac transmembrane action potentials, *IEEE Trans. Biomed. Eng.,* 30, 819, 1983.
79. **Fusi, F., Piazzesi, G., Amerini, S., Mugelli, A., and Livi, S.,** A low cost microcomputer system for automated analysis of intracellular cardiac action potentials, *J. Pharmacol. Meth.,* 11, 61, 1984.

80. **Demjanenko, V. and Sachs, F.**, Computer interface for electrophysiological applications: simple modifications to a commercial (Datel) single board data-acquisition system, *Med. Biol. Eng. Comput.*, 20, 65, 1982.
81. **Moore, J. W. and Cole, K. S.**, Voltage clamp analysis, in *Physical Techniques in Biological Research*, Vol. 5, Nastuk, W. L., Ed., Academic Press, New York, 1964, chap. 5.
82. **Fozzard, H. A. and Beeler, G. W.**, The voltage clamp and cardiac electrophysiology, *Circ. Res.*, 37, 403, 1975.
83. **Smith, T. G., Barker, J. L., Smith, B. M., and Colburn, T. R.**, Voltage clamping with microelectrodes, *J. Neurosci. Meth.*, 3, 105, 1980.
84. **Finkel, A. S. and Redman, S.**, Theory and operation of a single microelectrode voltage clamp, *J. Neurosci. Meth.*, 11, 101, 1984.
85. **Kass, R. S., Siegelbaum, S. A., and Tsien, R. W.**, Three microelectrode voltage clamp experiments in calf cardiac Purkinje fibres: is slow inward current adequately measured?, *J. Physiol. London*, 290, 201, 1979.
86. **Brennecke, R. and Lindermann, B.**, Chopped-current clamp for current injection and recording of membrane polarization with single electrodes of changing resistance, *T.I.T. J. Life Sci.*, 1, 53, 1971.
87. **Hamill, O. P.**, Membrane ion channels, in *Topics in Molecular Pharmacology*, Burgen, A. S. V. and Roberts, G. C. K., Eds., Elsevier Science Publishers, Amsterdam, 1983, chap. 6.
88. **Auerbach, A. and Sachs, F.**, Patch clamp studies of single ionic channel, *Ann. Rev. Biophys. Bioeng.*, 13, 269, 1984.
89. **Sakmann, B. and Neher, E.**, Patch clamp techniques for studying ionic channels in excitable membranes, *Ann. Rev. Physiol.*, 46, 455, 1984.
90. **Schanne, O. F.**, Recording of contractile activity of cells in culture, *J. Appl. Physiol.*, 29, 892, 1970.
91. **Thompson, E. J., Wilson, S. H., Schuette, W. H., Whitehouse, W. C., and Nirenberg, M. W.**, Measurement of the rate and velocity of movement by single heart cells in culture, *Am. J. Cardiol.*, 32, 162, 1973.
92. **Curtis, D. R.**, Microelectrophoresis, in *Physical Techniques in Biological Research*, Vol. 5, Nastuk, W. L., Ed., Academic Press, New York, 1964, chap. 4.
93. **McDonald, R. L. and Barker, J. L.**, Neuropharmacology of spinal cord neurons in primary dissociated cell culture, in *Excitable Cells in Tissue Culture*, Nelson, P. G. and Lieberman, M., Eds., Plenum Press, New York, 1981, chap. 3.
94. **Kozhechkin, S. N.**, Microiontophoretic techniques in Pharmacology, *Trends Pharmacol. Sci.*, 1, 185, 1980.
95. **Athias, P., Groz, B., and Klepping, J.**, Determination of the ionic basis of spontaneous activity of cultured rat heart cells using microinjection techniques, *Biol. Cell.*, 37, 183, 1980.
96. **Clapham, D. E. and De Felice, L. J.**, Small signal impedance of heart cell membranes, *J. Membrane Biol.*, 67, 63, 1982.
97. **Sperelakis, N.**, Electrical properties of embryonic heart cells, in *Electrical Phenomena in the Heart*, De Mello, W. C., Ed., Academic Press, New York, 1972, chap. 1.
98. **Girardier, L., Forssman, W. G., and Matter, A.**, Conduction electrique intercellulaire dans des cultures de cellules myocardiques, *Helv. Physiol. Pharmacol. Acta*, 25, 167, 1967.
99. **De Haan, R. L. and Fozzard, H. A.**, Membrane response to current pulses in spheroidal aggregates of embryonic heart cells, *J. Gen. Physiol.*, 65, 207, 1975.
100. **De Haan, R. L. and De Felice, L. J.**, Electrical noise and rhythmic properties of embryonic heart cell aggregates, *Fed. Proc.*, 37, 2132, 1978.
101. **Clay, J. R., De Felice, L. J., and De Haan, R. L.**, Current noise parameters derived from voltage noise and impedance in embryonic heart cell aggregates, *Biophys. J.*, 28, 169, 1979.
102. **De Bruijne, J., Jongsma, H. J., and Van Ginneken, A. C. G.**, The passive electrical properties of spheroidal aggregates cultured from neonatal rat heart cells, *J. Physiol. (London)*, 355, 281, 1984.
103. **Pappano, A. J. and Sperelakis, N.**, Low K^+ conductance and low resting potentials of isolated single cultured heart cells, *Am. J. Physiol.*, 217, 1076, 1969.
104. **Lompre, A. M., Poggioli, J., and Vassort, G.**, Maintenance of fast Na channels during primary culture of embryonic chick heart cells, *J. Mol. Cell. Cardiol.*, 11, 813, 1979.
105. **Schanne, O. F., Ruiz-Ceretti, E., Rivard, C., and Chartier, D.**, Determinants of electrical activity in clusters of cultured cardiac cells from neonatal rats, *J. Mol. Cell. Cardiol.*, 9, 269, 1977.
106. **Schanne, O. F., St. Vincent, M., and Bkaily, G.**, Electrogenesis of freshly explanted spontaneously active clusters of neonatal rat ventricle cells, *Rev. Can. Biol. Exp.*, 42, 199, 1983.
107. **De Haan, R. L. and Gottlieb, S. H.**, The electrical activity of embryonic chick heart cells isolated in tissue culture singly or in interconnected cell sheets, *J. Gen. Physiol.*, 52, 643, 1968.
108. **McDonald, T. F., Sachs, H. G., and De Haan, R. L.**, Development of sensitivity to tetrodotoxin in beating chick embryo hearts, single cells, and aggregates, *Science*, 176, 1248, 1972.

109. **Matsuki, N. and Hermsmeyer, K.,** Tetrodotoxin-sensitive Na$^+$ channels in isolated single cultured rat myocardial cells, *Am. J. Physiol.*, 245, C381, 1983.

110. **McLean, M. J. and Sperelakis, N.,** Rapid loss of sensitivity to tetrodotoxin by chick ventricular myocardial cells after separation from the heart, *Exp. Cell Res.*, 86, 351, 1974.

111. **Renaud, J. F., Scana, A. M., Kazazoglou, T., Lombet, A., Romey, G., and Lazdunski, M.,** Normal serum and lipoprotein-deficient serum give different expression of excitability, corresponding to different stages of differenciation in chick cardiac cells in culture, *Proc. Natl. Acad. Sci. U.S.A.*, 79, 7768, 1982.

112. **Burt, J. M.,** Electrical and contractile consequences of Na$^+$ or Ca^{2+} gradient reduction in cultured heart cells, *J. Mol. Cell. Cardiol.*, 14, 99, 1982.

113. **Jacobson, S. L., Kennedy, C. B., and Mealing, G. A. R.,** Evidence for functional sodium and calcium ion channels in the membrane of cultured cardiomyocytes of adult rat, *Can. J. Physiol. Pharmacol.*, 61, 1312, 1983.

114. **Auclair, M. C.,** Activité électrique des cellules en culture de coeur de cobaye. Comparaison avec le Rat, *C.R. Soc. Biol.*, 170, 543, 1976.

115. **McDonald, T. F. and De Haan, R. L.,** Ion levels and membrane potential in chick heart tissue and cultured cells, *J. Gen. Physiol.*, 61, 89, 1973.

116. **Romey, G., Renaud, J. F., Fosset, M., and Lazdunski, M.,** Pharmacological properties of the interaction of sea anemone polypeptide toxin with cardiac cells in culture, *J. Pharmacol. Exp. Ther.*, 213, 607, 1980.

117. **McDonald, T. F. and Sachs, H. G.,** Electrical activity in embryonic heart cell aggregates: developmental aspects, *Pflügers Arch.*, 354, 151, 1975.

118. **Nathan, R. D. and De Haan, R. L.,** In vitro differenciation of fast Na$^+$ conductance in embryonic heart cell aggregates, *Proc. Natl. Acad. Sci. U.S.A.*, 75, 2776, 1978.

119. **Bernard, P. and Couraud, F.,** Electrophysiological studies on embryonic heart cells in culture. Scorpion toxin as a tool to reveal latent fast sodium channel, *Biochim. Biophys. Acta*, 553, 154, 1979.

120. **Bkaily, G., Sperelakis, N., Elishalom, Y., and Barenholz, Y.,** Effect of Na$^+$ or Ca^{2+}-filled liposomes on electrical activity of cultured heart cells, *Am. J. Physiol.*, 245, H756, 1983.

121. **McLean, M. J. and Sperelakis, N.,** Retention of fully differenciated electrophysiological properties of chick embryonic heart cells in culture, *Develop. Biol.*, 50, 134, 1976.

122. **McLean, M. J., Renaud, J. F., Niu, M. C., and Sperelakis, N.,** Membrane differenciation of cardiac myoblasts induced in vitro by an RNA-enriched fraction from adult heart, *Exp. Cell Res.*, 110, 1, 1977.

123. **Nathan, R. D.,** Aggregates of fetal rat heart cells: electrophysiology and tetrodotoxin sensitivity, *J. Mol. Cell. Cardiol.*, 13, 241, 1981.

124. **Jourdon, P. and Sperelakis, N.,** Electrical properties of cultured heart cell reaggregates from newborn rat ventricles: comparison with intact non cultured ventricles, *J. Mol. Cell. Cardiol.*, 12, 1441, 1980.

125. **Colquhoun, D., Neher, E., Reuter, H., and Stevens, C. F.,** Inward current channels activated by intracellular Ca in cultured cardiac cells, *Nature*, 294, 752, 1981.

126. **Fischmeister, R., De Felice, L. J., Ayer, R. K., Levi, R., and De Haan, R. L.,** Channel currents during spontaneous action potentials in embryonic chick heart cells. The action potential patch clamp, *Biophys. J.*, 46, 267, 1984.

127. **Sperelakis, N., Shigenobu, K., and McLean, M. J.,** Membrane cation channels — changes in developing heart, in cell culture, and in organ culture, in *Developmental and Physiological Correlates of Cardiac Muscle*, Lieberman, M. and Sano, T., Eds., Raven Press, New York, 1975, 209.

128. **Schanne, O. F., Ruiz-Ceretti, E., Payet, M. D., and Deslauriers, Y.,** Influence of varied (Ca^{2+})$_0$ and (Na$^+$)$_0$ on electrical activity of clusters of cultured cardiac cells from neonatal rats, *J. Mol. Cell. Cardiol.*, 11, 477, 1979.

129. **Renaud, J. F., Romey, G., Lombet, A., and Lazdunski, M.,** Differentiation of the fast Na$^+$ channel in embryonic heart cells: interaction of the channel with neurotoxins, *Proc. Natl. Acad. Sci. U.S.A.*, 78, 5348, 1981.

130. **Jacques, Y., Frelin, C., Vigne, P., Romey, G., Parjari, M., and Lazdunski, M.,** Neurotoxins specific for the sodium channel stimulate calcium entry in neuroblastoma cells, *Biochemistry*, 20, 6219, 1981.

131. **Pang, D. C. and Sperelakis, N.,** Veratridine stimulation of calcium uptake by chick embryonic heart cells in culture, *J. Mol. Cell. Cardiol.*, 14, 703, 1982.

132. **Jourdon, P.,** Effect of hyperpolarizing current on transmembrane potentials in newborn rat heart reaggregates, *Biol. Cell.*, 37, 189, 1980.

133. **Ebihara, L. and Johnson, E. A.,** Fast sodium current in cardiac muscle. A quantitative description, *Biophys. J.*, 32, 779, 1980.

134. **Cachelin, A. B., De Peyer, J. E., Kokubun, S., and Reuter, H.,** Sodium channels in cultured cardiac cells, *J. Physiol. (London)*, 340, 389, 1983.

135. **Kitzes, M. C. and Berns, M. W.,** Electrical activity of rat myocardial cells in culture: La^{3+}-induced alterations, *Am. J. Physiol.*, 237, C87, 1979.

136. **Nathan, R. D., Fung, S. J., Stocco, D. M., Barron, E. A., and Markwald, R. R.,** Sialic acid: regulation of electrogenesis in cultured heart cells, *Am. J. Physiol.*, 239, C197, 1980.

137. **Van Ginneken, A. C. G. and Jongsma, H. J.,** Slow inward current in aggregates of neonatal rat heart cells and its contribution to the steady-state current voltage relationship, *Pflügers Arch.,* 397, 265, 1983.
138. **Reuter, H.,** Properties of single calcium channels in cardiac cell culture, *Nature,* 297, 501, 1982.
139. **Surville, J. M., Athias, P., Groz, B., and Klepping, J.,** Etude de la relation fréquence-durée des potentiels d'action de cellules de coeur de rat maintenues en culture, *J. Physiol. Paris,* 75, 43, 1979.
140. **Robinson, R. B.,** Action potential characteristics of rat cardiac cells do not change with time in culture, *J. Mol. Cell. Cardiol.,* 14, 367, 1982.
141. **Koidl, B. and Tritthart, H. A.,** D-600 blocks spontaneous discharge, excitability and contraction of cultured embryonic chick heart cells, *J. Mol. Cell. Cardiol.,* 14, 251, 1982.
142. **Mead, R. H. and Clusin, W. T.,** Origin of the background sodium current and effects of sodium removal in cultured embryonic cardiac cells, *Circ. Res.,* 55, 67, 1984.
143. **Pelleg, A., Vogel, S., Belardinelli, L., and Sperelakis, N.,** Overdrive suppression of automaticity in cultured chick myocardial cells, *Am. J. Physiol.,* 238, H24, 1980.
144. **Ypey, D. L., Clapham, D. E., and De Haan, R. L.,** Development of electrical coupling and action potential synchrony between paired aggregates of embryonic heart cells, *Membrane Biol.,* 51, 75, 1979.
145. **McLean, M. J. and Sperelakis, N.,** Differences in degree of electrotonic interaction in highly differentiated and reverted cultured heart cell reaggregates, *J. Membrane Biol.,* 57, 37, 1980.
146. **Griepp, E. B., Peacock, J. H., Bernfield, M. R., and Revel, J. P.,** Morphological and functional correlation of synchronous beating between embryonic heart cell aggregates and layers, *Exp. Cell Res.,* 113, 273, 1978.
147. **Williams, E. H. and De Haan, R. L.,** Electrical coupling among heart cells in the absence of ultrastructurally defined gap junctions, *J. Membrane Biol.,* 60, 237, 1981.
148. **Lieberman, M., Kootsey, J. M., Johnson, E. A., and Sawanobori, T.,** Slow conduction in cardiac muscle. A biophysical model, *Biophys. J.,* 13, 37, 1973.
149. **Hasin, Y. and Barry, W. H.,** Myocardial metabolic inhibiton and membrane potential, contraction, and potassium uptake, *Am. J. Physiol.,* 247, H322, 1984.
150. **Athias, P., Variot, N., Grynberg, A., and Klepping, J.,** Application of the microperfusion of drugs interfering with calcium flux to the localization of calcium binding sites of isolated myocardial cells in culture, *J. Mol. Cell. Cardiol.,* 16-S3, 5, 1984.
151. **Halle, W. and Jentzch, K. D.,** Aktuelle Probleme der Wirkstofforschung, *Pharmazie,* 29, 433, 1974.
152. **Schanne, O. F. and Bkaily, G.,** Explanted cardiac cells: a model to study drug actions?, *Can. J. Physiol. Pharmacol.,* 59, 443, 1981.
153. **Sperelakis, N. and Lehmkuhl, D.,** Insensitivity of cultured chick heart cells to autonomic agents and tetrodotoxin, *Am. J. Physiol.,* 209, 693, 1965.
154. **Kaufmann, R., Tritthart, H., Rodenroth, S., and Rost, B.,** The mechanical and electrical activity of embryonic chick heart cells cultured in vitro, *Pflügers Arch.,* 311, 25, 1969.
155. **Josephson, I. and Sperelakis, N.,** 5'-guanylimidodiphosphate stimulation of slow Ca^{2+} current in myocardial cells, *J. Mol. Cell. Cardiol.,* 10, 1157, 1978.
156. **Bkaily, G. and Sperelakis, N.,** Injection of protein kinase inhibitor into cultured heart cells blocks calcium slow channels, *Am. J. Physiol.,* 246, H630, 1984.
157. **Clusin, W. T., Fischmeister, R., and De Haan, R. L.,** Caffeine-induced current in embryonic heart cells: time course and voltage dependence, *Am. J. Physiol.,* 245, H528, 1983.
158. **Cachelin, A. B., De Peyer, J. E., Kokubun, S., and Reuter, H.,** Ca^{2+}-channel modulation by 8-bromocyclic AMP in cultured heart cells, *Nature,* 304, 462, 1983.
159. **Li, T. and Sperelakis, N.,** Calcium antagonist blockade of slow action potentials in cultured chick heart cells, *Can. J. Physiol. Pharmacol.,* 61, 957, 1983.
160. **Renaud, J. F., Méaux, J. P., Romey, G., Schmid, A., and Lazdunski, M.,** Activation of the voltage dependent Ca^{2+} channel in rat heart cells by dihydropyridine derivatives, *Biochem. Biophys. Res. Commun.,* 125, 405, 1984.
161. **Kokubun, S. and Reuter, H.,** Dihydropyridine derivatives prolong the open state of Ca channels in cultured cardiac cells, *Proc. Natl. Acad. Sci. U.S.A.,* 81, 4824, 1984.
162. **Koidl, B. and Tritthart, H. A.,** The effects of ouabain on the electrical and mechanical activities of embryonic chick heart cells in culture, *J. Mol. Cell. Cardiol.,* 12, 663, 1980.
163. **Josephson, I. and Sperelakis, N.,** Ouabain blockade of inward slow current in cardiac muscle, *J. Mol. Cell. Cardiol.,* 9, 409, 1977.
164. **Josephson, I. and Sperelakis, N.,** Local anesthetic blockade of Ca^{2+}-mediated action potentials in cardiac muscle, *Eur. J. Pharmacol.,* 40, 201, 1976.
165. **Hering, S., Bodewei, R., and Wollenberger, A.,** Sodium current in freshly isolated and cultured single rat myocardial cells: frequency and voltage dependent block by mexiletine, *J. Mol. Cell. Cardiol.,* 15, 431, 1983.
166. **Goshima, K.,** Arrhythmic movements of myocardial cells in culture and their improvement with antiarrhythmic drugs, *J. Mol. Cell. Cardiol.,* 8, 217, 1976.

167. **Nathan, R. D. and Bhattacharyya, M. L.,** Perturbations in the membrane potential of cultured heart cells: role of calcium, *Am. J. Physiol.,* 247, H273, 1984.

168. **Guevara, M. R., Glass, L., and Shrier, A.,** Phase locking, period doubling bifurcations, and irregular dynamics in periodically stimulated cardiac cells, *Science,* 214, 1350, 1981.

169. **Clay, J. R., Guevara, M. R., and Shrier, A.,** Phase resetting of the rhythmic activity of embryonic heart cell aggregates, *Biophys. J.,* 45, 699, 1984.

170. **Van Meerwijk, W. P. M., Debruin, G., Van Ginneken, A. C. G., Vanhartevelt, J., Jongsma, H. J., Kruyt, E. W., Scott, S. S., and Ypey, D. L.,** Phase resetting properties of cardiac pacemaker cells, *J. Gen. Physiol.,* 83, 613, 1984.

171. **Jones, J. L., Lepeschkin, E., Jones, R. E., and Rush, S.,** Response of cultured myocardial cells to countershock-type electric field stimulation, *Am. J. Physiol.,* 235, H214, 1978.

172. **Shigeto, N., Shigeto, E., and Lieberman, M.,** Unidirectional conduction block and extrasystoles in synthetic strands of cultured heart cells, *Jap. Heart J.,* 23, 66, 1982.

173. **Hyde, A., Cheneval, J. P., Blondel, B., and Girardier, L.,** Electrophysiological correlates of energy metabolism in cultured heart cells, *J. Physiol. (Paris),* 64, 269, 1972.

174. **Auclair, M. C., Adolphe, M., Moreno, G., and Salet, C.,** Comparison of the effects of potassium cyanide and hypoxia on ultrastructure and electrical activity of cultured rat myoblasts, *Toxicol. Appl. Pharmacol.,* 37, 387, 1976.

175. **Seraydarian, M. W., Artaza, L., and Abbott, B. C.,** The effect of adenosine on cardiac cells in culture, *J. Mol. Cell. Cardiol.,* 4, 477, 1972.

176. **Shrier, A., Nahmias, A. J., and De Haan, R. L.,** Herpes virus infection alters electrical parameters of heart cell aggregates, *Am. J. Physiol.,* 234, C170, 1978.

177. **Spector, A. A., Mathur, S. N., Kaduce, T. L., and Hyman, B. T.,** Lipid nutrition and metabolism of cultured mammalian cells, *Prog. Lip. Res.,* 19, 155, 1981.

178. **Harary, I. and Farley, B.,** Studies in vitro on single beating rat heart cells. IX. The restoration of beating by serum lipids and fatty acids, *Biochim. Biophys. Acta,* 115, 15, 1966.

179. **Wenzel, D. G. and Kloeppel, J. W.,** Arrhythmogenicity of long chain fatty acids for cultured rat heart myocytes, *Toxicology,* 15, 105, 1980.

180. **Wenzel, D. G. and Kloeppel, J. W.,** Incorporation of saturated and cis- and trans-unsaturated long chain fatty acids in rat myocytes and increased susceptibility to arrhythmias, *Toxicology,* 18, 27, 1980

181. **Hasin, Y., Sapoznikov, D., Stein, O., and Stein, Y.,** Effect of fatty acid composition of rat heart myocytes on their electrical activity, *J. Mol. Cell. Cardiol.,* 14, 163, 1982.

182. **Grynberg, A., Athias, P., and Degois, M.,** Effect of change in growth environment on cultured myocardial cells investigated in a standardized medium, *In Vitro,* 22, 44, 1986.

Chapter 10

A BRIEF REVIEW OF SINGLE CHANNEL MEASUREMENTS FROM ISOLATED HEART CELLS

David E. Clapham

TABLE OF CONTENTS

I. INTRODUCTION

Ionic channels in the heart are rapidly being separated and defined by patch clamp techniques. Prior to the development of patch clamp techniques[1-3] and single whole-cell measurements,[2-6] separation of cardiac currents into individual components was difficult and beset by the technical problems of voltage-clamping a tissue (spatial homogeneity, ion accumulation, etc.). These problems were most severe in studying cardiac sodium and potassium currents.

This review summarizes recent electrophysiological findings in cardiac cells with emphasis on single-cell patch clamp experiments. Much of this work has been carried out on cells dissociated for less than 1 day rather than in long-term culture. Hopefully, freshly dissociated cells are more like native cells than cells cultured for many days. But, little is known about the effects of dissociation or culture conditions on the number of channels, appearance of new channels, or channel kinetics. In macrophages, for example, channels clearly develop in culture.[7] Loose patch clamp of native tissue[8,9] and intracellular dialysis of single whole cells will be major tools for defining these issues.

Other authors have reviewed cardiac electrophysiology more extensively.[10-12]

II. TECHNIQUES

A. Single Cell Isolation

One popular method of single cell isolation from heart tissue uses the Langendorff method of perfusion of the coronary arteries via the aorta.[13,14] Collagenase is usually the enzyme used in the perfusate with "Kraftbruhe"[15] or a similar solution serving as the recovery medium.[16] In heart tissue where these methods are not feasible, simple chopping of the tissue into small blocks followed by enzymatic dispersion also works well.[13,17-20]

The patch clamp technique requires obtaining an extremely high-resistance seal between the cell membrane and the glass pipette (gigaohm-seal or gigaseal). Generally, gigaseals are impossible to obtain on intact tissue, and enzymatic dispersion is thought to improve sealing by removal of the cell coat. Little is known about the conditions that enable seal formation, and most laboratories experience vagaries in seal success. In general, the condition of the cell is more important than technical factors such as the type of pipette glass in obtaining gigaseals.

B. Patch Clamp Techniques

Numerous detailed reviews of patch clamp techniques have been published. The most comprehensive overall source is Sakmann and Neher.[3] The general technique is summarized in Figure 1A, in which cell-attached, whole-cell, outside-out and inside-out variations of measurement are shown. These techniques allow one to measure single channel openings or whole-cell currents from single cells. Any portion of the cell membrane may be exposed to experimental solutions. Figure 1B shows a perfusion device[21,22] which can be used to switch the bathing solution rapidly (within 100 msec) to steady-state concentrations (whole-cell or detached patches). An improvement on this method may be used to switch solutions within 10 msec on detached patches.[23] Noma and co-workers[24] (Figure 1C) have developed a method of intraelectrode dialysis which enable them to perfuse the inside and outside face of the membrane simultaneously. Using this technique, solutions inside the pipette may be changed within 30 sec.

Almers et al.,[25] have used the principle of patch clamp to measure currents from large areas of membrane in intact tissue (Figure 1D). Since seals to intact tissues are not as tight (approximately 50 to 100 MΩ), this technique has been called "loose" patch clamp. Leak currents are subtracted electronically and accurate representations of net ionic currents through

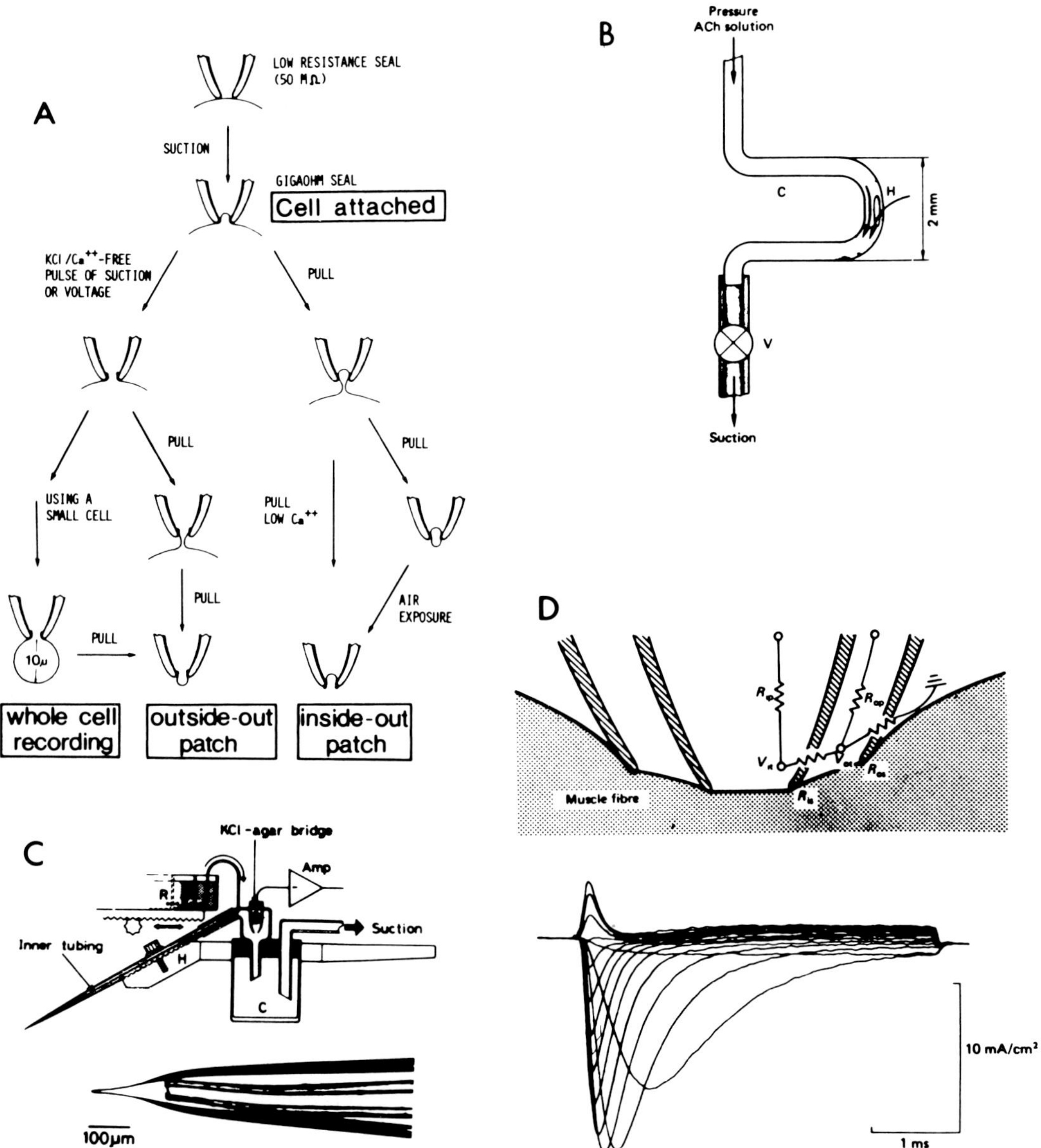

FIGURE 1.   (A) Variations of patch clamp techniques. After a gigaseal is obtained, single channels may be measured from the external face of the intact membrane or either face of the detached membrane. Alternatively, the whole membrane current may be recorded while the intracellular space is perfused. (From Hamill, O. P., Marty, A., Neher, E., Sakmann, B., and Sigworth, F. S., *Pflugers Arch.*, 391, 85, 1981. With permission.) (B) U-tube device for rapid (approx. 100 msec on rate) perfusion of the bath membrane face. Fluid flows rapidly from a test fluid reservoir through the U-tube until the exit valve is closed. The cell next to the hole is then perfused. (From Fenwick, E. M., Marty, A., and Neher, E., *J. Physiol.*, 331, 577, 1982. With permission.) (C) Intrapipette perfusion technique. A thin polyethylene tube extends to the tip of the pipette. Constant suction continually draws new solution into the tip (approx. 30 sec). (From Soejima, M. and Noma, A., *Pflugers Arch.*, 400, 424, 1984. With permission.) (D) Loose patch clamp. A double-walled electrode is used to isolate a large patch of membrane. The leakage across the seal is electronically compensated. (From Almers, W., Roberts, W. M., and Ruff, R. L., *J. Physiol.*, 347, 751, 1984. With permission.) Na currents from rat papillary muscle (courtesy W. Stühmer and R. Weiss).

Table 1
SINGLE IONIC CHANNELS IN CARDIAC CELL MEMBRANES

Name	Charge carrier	Activation mechanism	Modifier	$\gamma(pS)$[a]	Density (/um²)	Common blocker
			Inward Direction			
I_{Na}	Na^+	Voltage	—	15	1—16	TTX
$I_{Ca}(L)$	Ca	Voltage	cAMP, Catalytic sub-unit, dihydropyridines	0.5	—	Cd^{++}, Co^{++}, Mn^{++}, Ni^{++} D600, Dihydropyridines
I_{ca} (T)	Ca	Voltage	—	<0.5	—	$N:^{++}$,Co^{++}
I_f	Na,K	Voltage	cAMP	1	—	$[C_s]_o$
I_{TI} (?)	Na,K	$[Ca_i]$	—	35	—	—
			Outward Direction			
I_{K1}	K	Voltage	—	3.6	1.8	$[Ba]_o$ $[Cs]_i$
I_K (?)	K	Voltage	—	15 62	.02	Cs,TEA,4AP
I_A	K	Voltage	—	12(?)	—	4-AP
$I_{Ca\cdot k}$	K	Ca,voltage	—	200 ($[K]_o = 140$)	—	—
$I_{K\text{-}ACh}$	K	ACh (muscarinic)	GTP	3.3	—	$[Ba]_o$
$I_{K\text{-}Na}$	K	$[Na]_i$	—	207 ($[K]_i = 49$) ($[K]_o = 150$)	—	—
$I_{K\text{-}ATP}$	K	Loss of ATP	—	31—32	—	$[ATP]_i$ $[Ba]_o$
I_{Cl} (?)	Cl	Voltage	—	240 (symmetrical 140 Cl)	—	—

[a]   Unless otherwise noted, direct single channel measurements were made under physiologic conditions. See text for references. Data as of 1985.

Modified from Reuter, H., Ion channels in cardiac cell membranes, *Ann. Rev. Physiol.*, 46, 473, 1984.

the patch are obtained. The advantage of this technique is that ionic currents may be sampled under more physiologic conditions.

III. IONIC CURRENTS

A. Single Channel Currents

Table 1 lists the types of ionic channels identified in heart cell membranes at the level of the single channel. Recent developments concerning each type of net ionic current are summarized below.

1. Inward Currents
a. Na Channels

Cachelin et al.[26] demonstrated single sodium channels from cultured neonatal rat ventricular cells. Tetrodotoxin (TTX) blocked these channels from the extracellular face but not from the intracellular surface. Single channels averaged over many traces from cell-attached patches compared well with net I_{Na} measured in whole-cell recordings. As has been reported from other detached preparations, there was a large shift in kinetics toward more negative potentials.[27] Na channels have been recorded from rabbit ventricle,[28] rat ventricle,[29] and chick embryonic ventricle.[30] Whole-cell recording from isolated chick embryonic cells at three, seven and ten days of development reveals large (0.1 to 0.3 mA/cm^2), TTX-sensitive sodium currents similar to sodium currents in nerve.[64] Previous reports of TTX insensitivity described in these cells under nonclamped conditions may be due to sodium channel inactivation at low resting potentials.

The Na channel has been cloned and sequenced in *Electrophorus*[31] and reconstituted in lipid bilayers.[32] Experiments are underway to study the effects of altered gene sequence on gating and selectivity properties of the Na channel.

b. Ca Channels

At least two types of calcium channels have been measured in heart cells. A large body of work suggests that some calcium channels are phosphorylated via a cAMP-dependent pathway. Thus the effect of catecholamines on contractility and rate are mediated, at least in part, via a phosphorylation of the channel and subsequent increase in mean channel open time.[33-35] Recently Hess et al.[36] have proposed that Ca channels in heart cells have three different gating modes in the normal state. Calcium channel antagonists such as nitrendipine promote a mode with long closed times, while calcium channel agonists such as Bay K8644 promote a mode with long and frequent openings. A third mode consisting of short openings of intermediate frequency is the most common or native state. The mechanism of calcium channel ''rundown'' or disappearance of Ca channels from dialyzed cells or detached patches is still unclear. The disparity of results from different laboratories over rundown and Ca channel kinetics in part reflects different populations of Ca channels from tissue to tissue[12,37] (but see Brown et al.[38]).

Recently several groups have noted that in the absence of calcium, the calcium channel conducts sodium ions very well.[39,40] In fact, the conductance of the channel is higher in low [Ca]$_o$ than in normal [Ca]$_o$. This means that calcium ions themselves probably regulate entry into this channel.

c. Transient Inward Current Channel

The transient inward current in the presence of ouabain[42] may be composed of the Ca^{++}-activated nonspecific current[43] as has been described in the neonatal rat ventricle. Channel opening increases as a function of intracellular calcium above 0.5 μM. This channel does not have significant voltage dependence, but is temperature sensitive. The function of this channel in the normal heart is unknown, but it clearly can be arrhythmogenic under conditions where intracellular Ca is increased.

2. Pacemaker Channels

It now appears that I_f, a nonselective Na-K current, plays a role in pacemaker depolarization in some cells,[10] and that I_{K2} is a result of a potassium accumulation artifact.[43] In chick ventricular cell aggregates, I_{K2} has been described under conditions where potassium accumulation is not a problem,[44] but I_{K2} was not seen in whole-cell measurements from single chick ventricular cells.[65] Single channel measurements of I_f have now been reported.[45]

3. Outward Currents

a. I_{K1}

The inward rectifier channel is one of the most thoroughly studied heart cell potassium channel. It has been described in ventricle,[46-47] atrium, and in the node.[24] I_{K1} was originally thought to be a time-independent current, but more recent studies show a slow time-dependent inactivation.[46-47] Mild intracellular acidification increases net I_{K1}, but the conductance is eventually blocked in the pH 5 range.[48] Single channel conductance of I_{K1} is proportional to the square root of $[K]_o$, partially accounting for the profound effect of potassium on the beating rate. The density of I_{K1} channels is 10 to 100 times larger in ventricle than in sinoatrial (SA) or atrioventricular (AV) node.[24] The sparsity of potassium channels in nodal tissue accounts for the low resting membrane potential and must play a role in making these cells the dominant pacemakers.

b. I_{K-ACh}

The muscarinic ACh channel in the A-V and S-A nodes is similar in conductance and selectivity to I_{K1}, but displays different gating kinetics.[49,50] The ACh receptor complex *directly* increases the K channel open probability through a GTP-binding protein. Exogenous application of the purified G protein subunits has been shown to activate $I_k \cdot ACh$.[67,68]

c. I_{K-ATP}

Trube and Heschler[51] and Noma[52] reported a potassium current that was blocked by intracellular ATP in the 0.01 to 1 mM range. Ostensibly, the channel functions to hyperpolarize the cell when the cell is metabolically exhausted (normal $[ATP]_i$ = 3 to 4 mM). This channel was found in both atrial and ventricular cells, and is clearly distinct in conductance and kinetics from I_{K1}. Preliminary experiments by Trube and Heschler suggest that phosphorylation of the channel is not the mechanism by which ATP decreases conductance.

d. I_{K-Na}

In guinea pig ventricular cells a large conductance potassium channel is activated by $[Na]_i$ above 10 mM (K_D = 66 mM).[53] The channel is insensitive to $[Ca]_i$ and voltage, and is inwardly rectifying. A prolonged failure of the Na-K pump would activate this channel, repolarize the cell and perhaps favor increased Ca extrusion via Na-Ca exchange.

e. I_K

The delayed rectifier seen in whole-cell clamp and in intact tissue (i_{x1}, i_{x2})[54,55] has not yet been identified conclusively at the single channel level. Clapham and DeFelice[56] reported a 62 pS K-selective channel in chick embryonic ventricle which activated over the range -50 to $+10$ mV. The channels were relatively sparse (1/50 μm^2). However, it was clearly different in conductance from the known delayed rectifier channels (15 pS) recorded in nerve, muscle, and blood cells. Kakei and Noma[57] noted a 12 pS potassium channel in the rabbit atrioventricular node, but it was rare and has not been fully characterized. Recently a 15 pS channel potassium channel was measured in embryonic chick ventricle.[66] When voltage steps from -80 to 0 mV were made repetitively (50 to 100 times), the signal averaged single channel currents reproduced the whole cell current (I_K).

f. *Other Outward Currents*

Coulombe and Duclohier[58] reported an apparently chloride-selective channel of large conductance (400 to 450 pS) from newborn rat ventricular cultured cells. This is probably not the same channel as reported in reconstituted bilayers from bovine calf cardiac sarcolemma,[59] but resembles the chloride-selective channel in myotubes.[60]

4. *Intercellular Channels*

One gap junction channel is capable of synchronizing two cardiac cells as deduced from coupling between embryonic heart cell aggregates.[61] Kameyama[62] measured junctional conductance between two guinea pig ventricular cells and found no voltage dependence of transjunctional conductance. Raising $[Ca]_i$ reduces the junctional conductance. Spray et al.[63] report that the gap junctional channel in rat ventricle is voltage-independent, pH-dependent, and not very sensitive to calcium. Estimates set single channel conductance at approximately 100 pS.

IV. SUMMARY

At least three inward and six outward channels have been identified in various heart cell membranes. Of these, sodium, calcium, the transient inward, and the inward rectifier potassium currents were well known from previous macroscopic voltage clamp studies.

Intracellular regulation of large conductance potassium currents by ATP and sodium are probably ways of repolarizing the heart cell quickly to prevent metabolic exhaustion. Ca-activated potassium currents in heart have also been described.[69] The phosphorylation of the calcium channel via cAMP increases net calcium influx and contractile strength by prolonging the open time of calcium channels. Examples of *extracellular* modulation of channels are the control over calcium selectivity by $[Ca]_o$ and the activation of an inwardly rectifying K current by acetylcholine.

The membrane potential and thus the spontaneous beating of heart cells is exquisitely controlled by several ionic channels. The low density of potassium channels in nodal tissue creates low resting potentials and places them in a voltage range where they may be spontaneously active. On the other hand, ventricular cells have a relatively higher density of K channels, rest near E_K, and do not beat spontaneously. The overall low density of K channels in the heart allows for very fine control of the beating rate by activation of a very few channels. New channels will undoubtedly be found in heart tissue and more will be learned about the intracellular modulation of the known channels. The regulation of expression of channels by various heart tissues and the molecular mechanism of the protein "gate" are areas for future research.

ACKNOWLEDGMENTS

I express my appreciation to Drs. DeFelice, DeHaan, Fozzard, Irisawa, Neher, and Trube for reading the manuscript and for making helpful comments. This work was supported in part by NIH grant 7R01 HL34873.

REFERENCES

1. **Neher, E. and Sakmann, B.,** Single channel currents recorded from membrane of denervated frog muscle fibres, *Nature*, 260, 799, 1976.
2. **Hamill, O. P., Marty, A., Neher, E., Sakmann, B., and Sigworth, F. J.,** Improved patch-clamp techniques for high-resolution current recording from cells and cell-free membrane patches, *Pflügers Arch.*, 391, 85, 1981.
3. **Sakmann, B. and Neher, E., Eds.,** *Single Channel Recording,* Plenum Press, New York, 1983.
4. **Lee, K. S., Weeks, T. A., Kao, R. L., Akaike, N., and Brown, A. M.,** Properties of internally perfused, voltage-clamped, isolated nerve cell bodies, *J. Gen. Physiol.*, 71, 489, 1978.
5. **Lee, K. S., Weeks, T. A., Kao, R. L., Akaike, N., and Brown, A. M.,** Sodium currents in single heart muscle cells, *Nature*, 278, 269, 1979.
6. **Isenberg, G. and Klöckner, U.,** Glycocalyx is not required for slow inward current in isolated rat heart myocytes, *Nature*, 284, 358, 1980.
7. **Ypey, D. L. and Clapham, D. E.,** Development of a delayed outward-rectifying K^+ conductance in cultured mouse peritoneal macrophages, *Proc. Natl. Acad. Sci. U.S.A.*, 81, 3083, 1984.
8. **Stühmer, W. and Almers, W.,** Photobleaching through glass micropipettes: sodium channels without lateral mobility in the sarcolemma of frog skeletal muscle, *Proc. Natl. Acad. Sci. U.S.A.*, 79, 946, 1982.
9. **Stühmer, W., Roberts, W. M., and Almers, W.,** The loose patch clamp in single channel recordings, in *Single Channel Recording,* Sakmann, B. and Neher, E., Eds., Plenum Press, New York, 1983, 123.
10. **Reuter, H.,** Ion channels in cardiac cell membranes, *Ann. Rev. Physiol.*, 46, 473, 1984.
11. **Irisawa, H.,** Electrophysiology of single cardiac cells, *Jpn. J. Physiol.*, 34, 375, 1984.

12. **Noble, D.,** The surprising heart: a review of recent progress in cardiac electrophysiology, *J. Physiol.,* 353, 1, 1984.

13. **Taniguchi, J., Kokubun, S., Noma, A., and Irisawa, H.,** Spontaneously active cells isolated from the sinoatrial and atrioventricular nodes of the rabbit heart, *Jpn. J. Physiol.,* 31, 547, 1981.

14. **Dow, J. W., Harding, N. G. L., and Powell, T.,** Isolated cardiac myocytes. I. Preparation of adult myocytes and their homology with intact tissue, *Cardiovasc. Res.,* 15, 483, 1981.

15. **Isenberg, G. and Klöckner, U.,** Calcium-tolerant ventricular myocytes prepared by preincubation in a "KB Medium", *Pflügers Arch.,* 395, 6, 1982.

16. **Irisawa, H. and Nakayama, T.,** Isolation of a single pacemaker cell from rabbit S-A node, *Jikeikai Med. J.,* 30, 65, 1984.

17. **Hume, J. R. and Giles, W.,** Ionic currents in single isolated bullfrog atrial cells, *J. Gen. Physiol.,* 81, 153, 1983.

18. **Sheets, M. F., January, C. T., and Fozzard, H. A.,** Isolation and characterization of single canine Purkinje cells, *Circ. Res.,* 53, 544, 1983.

19. **Bustamente, J. O., Wantanabe, T., and McDonald, T. F.,** Single cells from adult mammalian heart: isolation procedure and preliminary electrophysiological studies, *Can. J. Physiol. Pharmacol.,* 59, 907, 1981.

20. **Marvin, W. J., Jr., Chittick, V. L., Rosenthal, J. K., Sandra, A., Atkins, D. L., and Hermsmeyer, K.,** The isolated sinoatrial node in primary culture from the newborn rat, *Circ. Res.,* 55, 253, 1984.

21. **Krishtal, O. A. and Pidoplichko, V. I.,** A receptor for protons in the nerve cell membrane, *Neuroscience,* 5, 2325, 1980.

22. **Fenwick, E. M., Marty, A., and Neher, E.,** A patch-clamp study of bovine chromaffin cells and of their sensitivity to acetylcholine, *J. Physiol.,* 331, 577, 1982.

23. **Brett, R. S., Dilger, J. P., and Adams, P. R.,** Improved concentration clamp for use with membrane patches, *Neurosci. Abstr.,* 241, 1984.

24. **Noma, A., Nakayama, T,. Kurachi, Y., and Irisawa, H.,** Resting K conductances in pacemaker and nonpacemaker heart cells of the rabbit, *Jpn. J. Physiol.,* 34, 245, 1984.

25. **Almers, W., Roberts, W. M., and Ruff, R. L.,** Voltage clamp of rat and human skeletal muscle: measurements with an improved loose-patch technique, *J. Physiol.,* 347, 751, 1984.

26. **Cachelin, A. B., de Peyer, J. E., Kokubun, S., and Reuter, H.,** Sodium channels in cultured cardiac cells, *J. Physiol.,* 340, 389, 1983.

27. **Fernandez, J. M., Fox, A. P., and Krasne, S.,** Membrane patches and whole cell membranes: a comparison of electrical properties in rat clonal pituitary (GH_3) cells, *J. Physiol.,* 356, 565, 1984.

28. **Grant, A. O., Starmer, C. F., and Strauss, H. C.,** Unitary sodium channels in isolated cardiac myocytes of rabbit, *Circ. Res.,* 53, 823, 1983.

29. **Kunze, D. L. and Brown, A. M.,** Unitary sodium currents in mammalian ventricular cells, *Physiologist,* 25, 300, 1982.

30. **Ten Eick, R., Matsuki, N., Quandt, F., and Yeh, J.,** Na channels in cultured embryonic chick ventricle: single channel conductance and open time: Effect of TTX, *Physiologist,* 25, 198, 1982.

31. **Noda, M., Shimizu, S., Tanabe, T., Takai, T., Kayano, T. Ikeda, Takahashi, H., Nakayama, H., Kanaoka, Y., Minamino, N., Kangawa, K., Matsuo, H., Raftery, R. A., Hirose, T., Inayama, S., Hayashida, H., Miyata, T., and Numa, S.,** Primary structure of *Electrophorus electricus* sodium channel deduced from cDNA sequence, *Nature,* 312, 121, 1984.

32. **Rosenberg, R. L., Tomiko, S. A., and Agnew, W. S.,** Reconstitution of neurotoxin-modulated ion transport by the voltage-regulated sodium channel isolated from the electroplax of *Electrophorus electricus,* *Proc. Natl. Acad. Sci. U.S.A.,* 81, 1239, 1984.

33. **Reuter, H.,** Calcium channel modulation by neurotransmitters, enzymes, and drugs, *Nature,* 301, 569, 1983.

34. **Cachelin, A. B., de Peyer, J. E., Kokubun, S., and Reuter, H.,** Ca^{2+} channel modulation by 8-bromocyclic AMP in cultured heart cells, *Nature,* 304, 462, 1983.

35. **Brum, G., Osterrieder, W., and Trautwein, W.,** Beta-adrenergic increase in the calcium conductance of cardiac myocytes studied with the patch clamp, *Pflügers Arch.,* 401, 111, 1984.

36. **Hess, P., Lansman, J. B., and Tsien, R. W.,** Different modes of Ca channel gating behaviour favoured by dihydropyridine Ca agonists and antagonists, *Nature,* 311, 538, 1984.

37. **Tsien, R. W.,** Calcium channels in excitable cell membranes, *Ann. Rev. Physiol.,* 45, 341, 1983.

38. **Brown, A. M., Camerer, H., Kunze, D. L., and Lux, H. D.,** Similarity of unitary Ca currents in three different species, *Nature,* 299, 156, 1982.

39. **Hess, P. and Tsien, R. W.,** Mechanism of ion permeation through calcium channels, *Nature,* 309, 453, 1984.

40. **Kostyuk, P. G.,** Intracellular perfusion of nerve cells and its effects on membrane currents, *Physiological Rev.,* 64, 435, 1984.

41. **Lederer, W. J. and Tsien, R. W.,** Transient inward current underlying arrhythmogenic effects of cardiotonic steroids in Purkinje fibres, *J. Physiol.,* 263, 73, 1976.

42. **Colquhoun, D., Neher, E., Reuter, H., and Stevens, C. F.,** Inward current channels activated by intracellular Ca in cultured cardiac cells, *Nature,* 294, 752, 1981.

43. **DiFrancesco, D.,** A new interpretation of the pacemaker current in calf Purkinje fibres, *J. Physiol.,* 314, 359, 1981.

44. **Clay, J. R. and Shrier, A.,** Analysis of subthreshold pacemaker currents in chick embryonic heart cells, *J. Physiol.,* 312, 471, 1981.

45. **DiFrancesco D.,** Characterization of single pacemaker channels in sino-atrial nodal cells, *Nature (London),* 324, 470, 1986.

46a. **Sakmann, B. and Trube, G.,** Conductance properties of single inwardly rectifying potassium channels in ventricular cells from guinea pig heart, *J. Physiol.,* 347, 641, 1984.

46b. **Sakmann, B. and Trube, G.,** Voltage-dependent inactivation of inward-rectifying single-channel currents in the guinea pig heart cell membrane, *J. Physiol.,* 347, 659, 1984.

47. **Kameyama, M., Kiyosue, T., and Soejima, M.,** Single channel analysis of the inward rectifier K current in the rabbit ventricular cells, *Jpn. J. Physiol.,* 33, 1039, 1983.

48. **Sato, R., Noma, A., Kurachi, Y., and Irisawa, H.,** Effects of intracellular acidification on membrane currents in ventricular cells of the guinea pig, *Circ. Res.,* 57, 553, 1985.

49. **Sakmann, B., Noma, A., and Trautwein, W.,** Acetylcholine activation of single muscarinic K^+ channels in isolated pacemaker cells of the mammalian heart, *Nature,* 303, 250, 1983.

50. **Soejima, M. and Noma, A.,** Mode of regulation of the ACh-sensitive K-channel by the muscarinic receptor in rabbit atrial cells, *Pflügers Arch.,* 400, 424, 1984.

51. **Trube, G. and Hescheler, J.,** Inward-rectifying channels in isolated patches of the heart cell membrane: ATP-dependence and comparison with cell-attached patches, *Pflügers Arch.,* 401, 178, 1984.

52. **Noma, A.,** ATP-regulated K^+ channels in cardiac muscle, *Nature,* 305, 147, 1983.

53. **Kameyama, M., Kakei, M., Sato, R., Shibasaki, T., Matsuda, H., and Irisawa, H.,** Intracellular Na^+ activates a K^+ channel in mammalian cardiac cells, *Nature,* 309, 354, 1984.

54. **Noble, D. and Tsien, R. W.,** Outward currents activated in the plateau range of potentials in cardiac Purkinje fibres, *J. Physiol.,* 195, 185, 1969.

55. **McDonald, T. F. and Trautwein, W.,** The potassium current underlying delayed rectification in cat ventricular muscle, *J. Physiol.,* 274, 217, 1978.

56. **Clapham, D. E. and DeFelice, L. J.,** Voltage-activated K channels in embryonic chick heart, *Biophys. J.,* 45, 40, 1984.

57. **Kakei, M. and Noma, A.,** Adenosine-5′-triphosphate-sensitive single potassium channel in the atrioventricular node cell of the rabbit heart, *J. Physiol.,* 352, 265, 1984.

58. **Coulombe, A. and Duclohier, H.,** A large unit conductance channel permeable to chloride ions in cultured rat heart cells, *J. Physiol.,* 350, 52P, 1984.

59. **Coronado, R. and Latorre, R.,** Detection of K^+ and Cl^- channels from calf cardiac sarcolemma in planar lipid bilayer membranes, *Nature,* 298, 849, 1982.

60. **Blatz, A. L. and Magleby, K. L.,** Single voltage-dependent chloride-selective channels of large conductance in cultured rat muscle, *Biophys. J.,* 43, 237, 1983.

61. **Clapham, D. E., Shrier, and DeHaan, R. L.,** Junctional resistance and action potential delay between embryonic heart cell aggregates, *J. Gen. Physiol.,* 75, 633, 1980.

62. **Kameyama, M.,** Electrical coupling between ventricular paired cells isolated from guinea pig heart, *J. Physiol.,* 336, 345, 1983.

63. **Spray, D. C., White, R. L., Mazet, F., and Bennett, M. V. L.,** Regulation of gap junctional conductance, *Am. J. Physiol.,* 248, H253, 1985.

64. **Clapham, D. E.,** Unpublished results.

65. **Clapham, D. E.,** Unpublished results.

66. **Clapham, D. E. and Logothetis, D. E.,** Delayed rectificer potassium currents in chick ventricle, *Biophys. J.,* 49, 540, 1986.

67. **Logothetis, D. E., Kurachi, Y., Galper, J., Neer, E., and Clapham, D. E.,** The Br subunits of GTP-Binding proteins activate the muscarinic K^+ channel in heart, *Nature (London),* 325, 321, 1987.

68. **Codina, J., Yuteni, A., Grenet, D., Brown, A. M., and Birnboumer, L.,** The α subunit of the GTP-binding protein, G_k, opens atrial potassium channels, *Science,* 236, 442, 1987.

69. **Callewaert, G., Vereecke, J., and Carmelict, E.,** Existence of a calcium-dependent potassium channel in the membrane of cow cardiac Purkinje cells, *Pflugers Arch.,* 406, 424, 1986.

Chapter 11

THE BEATING ACTIVITY OF CULTURED HEART CELLS

Kiyota Goshima, Eiji Kurimoto, Hiroyuki Kaneko, Shigeo Wakabayashi, and Syoyu Kobayasi

TABLE OF CONTENTS

I. INTRODUCTION

Cultured embryonic and postnatal myocardial cells show various interesting functions, resembling those of heart muscle tissue in vivo. Cultured myocardial cells are anatomically and functionally separate from nerves, connective tissue, and blood vessels, and moreover, they are in direct contact with the extracellular medium. Spontaneous beating is easily observed under a phase contrast microscope. The beating rate can also be monitored easily using photoelectric recording methods;[1-5] therefore, their use should greatly simplify studies on the direct effects of ions and other agents on the beating rate of myocardial cells. Unfortunately, the contractions in cultured cells have proved difficult to measure.[4] The striation pattern of sarcomeres in cultured myocardial cells cannot be detected under a phase contrast microscope, but this pattern can be observed clearly using a polarization microscope equipped with rectified lenses.[6] Researchers have successfully measured the changes of sarcomere lengths of cultured myocardial cells during rhythmic and arrhythmic beating, using a polarization microscope coupled to a video system (Figure 1).[5,6] Using this method, the effect of quinidine and procainamide on ouabain-induced arrhythmia[7] were studied, and changes in myofibril organization during cell fusion[8] and mitosis[9] were noted.

II. INVOLVEMENT OF A Na^+-Ca^{2+} EXCHANGE SYSTEM IN GENESIS OF OUABAIN-INDUCED ARRHYTHMIA IN CULTURED MOUSE AND QUAIL MYOCARDIAL CELLS[6]

It was attempted to determine whether toxic doses of ouabain cause transient or permanent increase in Ca^{2+}_i by using a polarization microscopy. It is well known that in the relaxed state, the concentration of Ca^{2+}_i in cardiac muscle is of the order of 10^{-7} M. It is also known that in the relaxed state, the sarcomere length of the myofibrils is about 2.2 to 2.3 μm. During excitation, the concentration of Ca^{2+}_i increases by one or two orders of magnitude, with concomitant shortening of the sarcomere length of myofibrils. Conversely, measurements of changes in sarcomere length during ouabain-induced arrhythmias (ouabain toxicity) can be used to measure change in the concentration of Ca^{2+}_i. Using the video method, we measured the changes of sarcomere length of cultured myocardial cells during ouabain-induced arrhythmia (Figure 2). During the normal beating cycle, the sarcomere length of cultured myocardial cells changed from 2.2 μm to 2.0 μm (Figure 3a). On the other hand, the sarcomere length remained in the shortened state (2.0 μm to 1.95 μm) throughout fibrillatory beating (Figure 3c,d). This observation suggests that Ca^{2+}_i is maintained at high concentration (about 10^{-6} M or more) throughout fibrillatory beating.

Next we considered the mechanism of the increase in Ca^{2+}_i on treatment with ouabain. Addition of ouabain caused gradual increases of both the Na content of myocardial cells and the rate of Ca uptake by myocardial cells, and fibrillatory beating appeared to develop when the Na content and the rate of Ca uptake exceeded the normal values by about 1.5 and 2.0×, respectively. The time course for the increase in the Na content was almost identical to the one for the increase in the rate of Ca uptake. Both the enhanced Na content and the increased rate of Ca uptake following ouabain treatment gradually decreased on washing out the ouabain. When the Na content and the rate of Ca uptake reached less than 1.3 and 1.5× the normal levels, respectively, fibrillatory beating ceased and normal beating restarted. Moreover, ouabain did not significantly affect the Ca efflux mechanism of myocardial cells. These observations suggest that inhibition of the Na pump by ouabain, and the resulting increase in the intracellular Na content, are related to increase in the rate of Ca uptake by the cells, which presumably accounts for the increase of Ca^{2+}_i during fibrillatory beating. Addition of ouabain did not affect the total Ca content (ionized Ca^{2+} plus bound or stored Ca) of myocardial cells. This may be because the total Ca content of myocardial

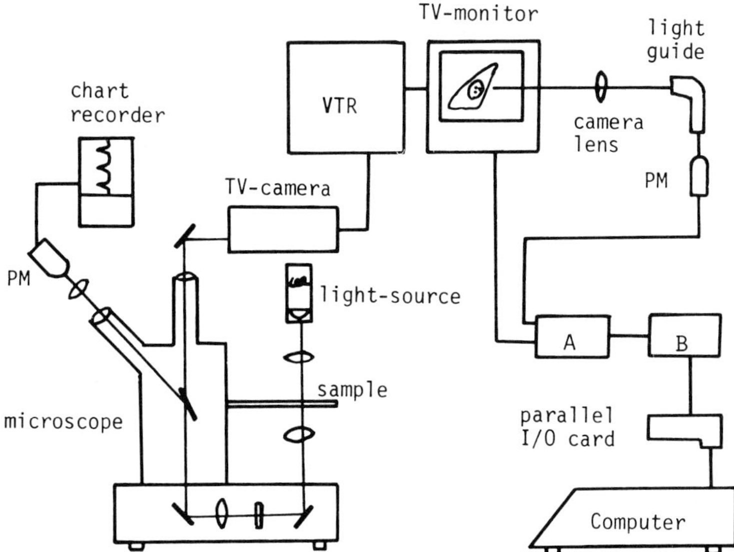

FIGURE 1.   Schematic diagram of the apparatus used for measuring beating rhythm of cultured myocardial cells. Images of cultured cells were recorded on videotape (National® NTR NV-8030) using a TV camera (Ikegami®, night vision camera, CTC-9000). The recorded images were displayed on a TV monitor. The motion of the beating cell was followed by recording the changes of light intensity at the edge of the beating cells. The edge of beating cell was brought into focus using a camera lens. The light signal was led through a light guide to a photomultiplier (PM) (RCA®, 1P21). The data on the time course of the beating rhythm were transferred into a microcomputer (Apple II® plus) through a circuit (A), a circuit (B) and a parallel I/O card (model Multibit Parallel I/O card, Toray, Japan) (described in detail in Reference 5). In some experiments, the light intensity variation due to the beating image was directly recorded with another light guide and a PM on a strip chart using a chart recorder (Nihonkoden®, RJG-4024).

cells is high, (presumably of the order of 10^{-3} mol/kg of cell water) and ouabain may have caused only small increases in the Ca content, which would not have been detectable in the experiments described above.

Incubation of myocardial cells in medium containing a low Na^+ and/or high Ca^{2+} concentration caused arrhythmias of the same type (fibrillatory beating) as those observed on addition of ouabain and increased the rate of Ca uptake by the cells without affecting the Na content. These observations suggest that increase in the rate of Ca uptake by myocardial cells may be partly responsible for the generation of arrhythmias.

We also studied the effect of the intracellular concentration of Na on the rate of Ca uptake using a model system. Myocardial cells were incubated in Ca^{2+}-free medium containing EGTA (0.1 mM) and various concentrations of Na^+ (0 to 140 mM) for 10 min, and then Na_i was measured (see Figure 4). The Na_i of myocardial cells was the same as that of the EGTA-medium. Thus, cells containing appropriate Na_i could be prepared. Then, we studied the effect of Na_i on ^{45}Ca-uptake by preincubating cells in EGTA medium (containing various concentrations of Na^+) and then incubating them in medium containing $^{45}Ca^{2+}$. A slight increase in the intracellular Na^+ concentration from the physiological concentration (20 mM) caused an appreciable increase in Ca uptake by the cells. Thus Ca uptake in myocardial cells is achieved by a carrier-mediated Na^+-Ca^{2+} exchange system, the stoichiometry of exchange being an Na^+:Ca^{2+} ratio of 3:1.[6,10,11] This is consistent with the model of electrogenic Na^+-Ca^{2+} exchange in the sarcolemma proposed by several groups of workers.[12-15]

(a)

(b)

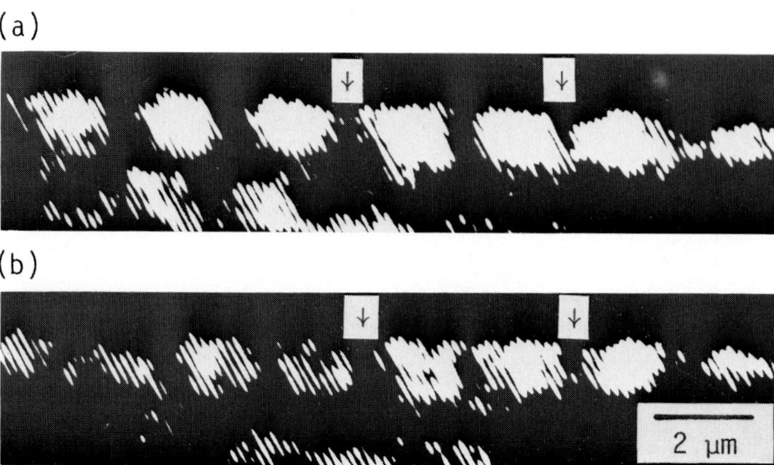

FIGURE 2.   Polarization micrography of a quail myocardial cell-sheet. The changes in sarcomere length of myocardial cells during normal and arrhythmic beating were recorded with a TV camera and a TV monitor. The images were recorded at a speed of 60 frames/ sec using a video tape-recorder. Then, each frame of the images was reproduced on a TV monitor in the ''stop-motion'' state. The stationary images observed on the TV monitor were photographed and changes in sarcomere length measured using a slide calliper. The myofibril I- and A-bands appear dark and bright, respectively: (a) before addition of ouabain; (b) 15 min after addition of $5 \times 10^{-6}\ M$ ouabain. The same myofibrils were observed in (a) and (b). The lengths of sarcomeres of (a) and (b) (the length between arrows) were plotted, (*) in Figure 3(a) and (**) in Figure 3(d), respectively.

Assuming that the Na^+-Ca^{2+} exchange system does not transport other ions, such as H^+, and that the Ca^{2+}_i is controlled by this system, the chemical reaction at equilibrium, on the basis of the model of electrogenic Na^+-Ca^{2+} exchange, can be written as follows:

$$\frac{Ca^{2+}_0}{Ca^{2+}_i} = \frac{Na_0^{+3}}{Na_i^{+3}} \exp(-E_m F/RT) \tag{1}$$

where E_m, F, R, T, Ca^{2+}_o and Na^+_o are the membrane potential, Faraday constant, universal gas constant, temperature, external Ca^{2+} concentration, and external Na^+ concentration, respectively. The resting potentials of myocardial cells with normal beating in the absence of ouabain and of cells showing fibrillatory beating in the presence of ouabain were almost equal (about -50 mV). However, it was difficult to know the precise value of the difference between resting potentials before and after addition of ouabain, because the values of resting potentials changed drastically from experiment to experiment. Based on calculation, an increase of Na^+_i to twice the normal level on treatment with ouabain, should lead to a Ca^{2+}_i of $8 \times$ the normal level. These observations suggest that the net increase in the steady-state level of Ca^{2+}_i and resultant shortening of sarcomere length throughout fibrillatory beating may, at least in part, be the consequence of the increased rate of Ca^{2+} uptake, which depends on the ouabain-induced increase in Na^+_i. But, our results do not exclude the possibility that the ouabain-induced increase in Ca^{2+}_i is due to enhancement of Na^+_i-induced efflux of Ca^{2+} from mitochondria.[15]

It is known that net increase in the steady-state level of Ca^{2+}_i suppresses generation of an action potential in muscle cells.[16] It is postulated that the increase in the steady-state level of Ca^{2+}_i also suppresses generation of an action potential in cultured myocardial cells. Thus, myocardial cells showing fibrillatory beating in the presence of ouabain generated an oscillatory potential of only small amplitude and high frequency.[6]

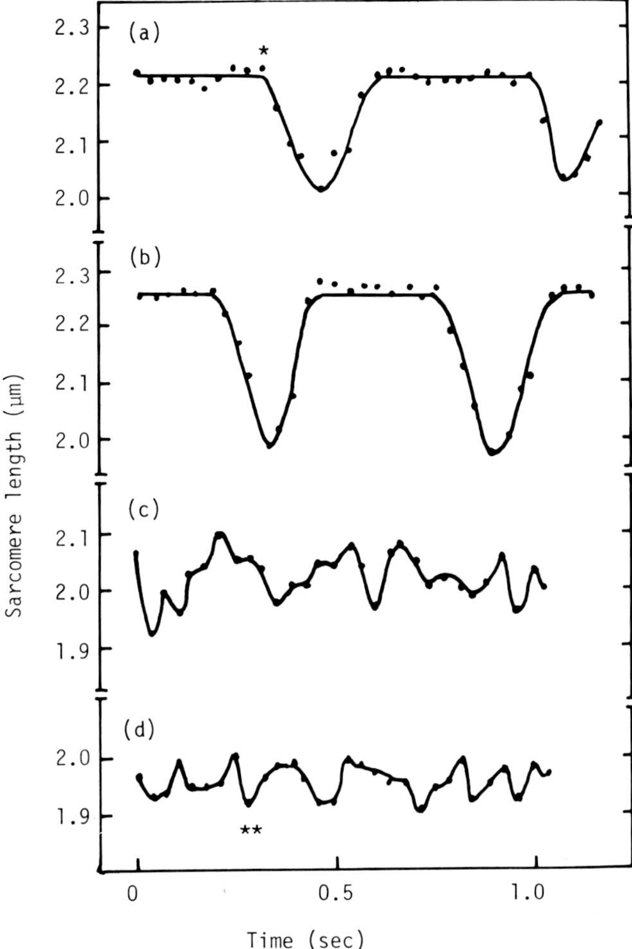

FIGURE 3.   Changes of sarcomere length of quail myocardial cells during normal and ouabain-induced arrhythmic beating. The two sarcomeres shown by arrows in Figure 2 were used for measurements of sarcomere length in Figures 3(a) to (d). The half-values of the lengths of the two sarcomeres are plotted. (*) Length of sarcomere shown in Figure 2(a), and (**) length of the sarcomere shown in Figure 2(b). Similar results were obtained in six independent experiments: (a) ouabain-free; (b) $5 \times 10^{-6}\,M$ Ouabain, 2 min; (c) $5 \times 10^{-6}\,M$ ouabain, 5 min; and (d) $5 \times 10^{-6}\,M$ ouabain, 15 min.

III. INHIBITION OF OUABAIN-INDUCED INCREASE IN Na_i OF CULTURED MOUSE AND QUAIL MYOCARDIAL CELLS BY QUINIDINE AND PROCAINAMIDE[7]

Quinidine and procainamide prevented ouabain-induced genesis of irregular beating (ouabain toxicity) (Figure 5). These antiarrhythmic drugs prevented ouabain-induced increases of Na^+_i (Figure 6) and Ca^{2+} uptake. The mechanism of these effects was then studied. These drugs did not affect the Na^+-Ca^{2+} exchange of the sarcolemma, Na^+-pumping or ouabain-binding in myocardial cells, but inhibited passive Na^+ influx by simple diffusion. These observations suggest that inhibition of passive Na^+ influx by quinidine or procainamide indirectly prevents ouabain-induced increase of the intracellular Na content, which presum-

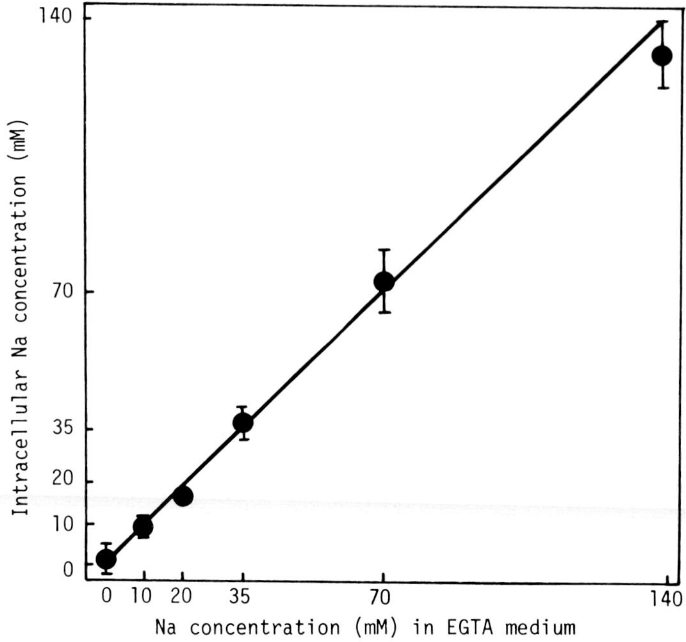

FIGURE 4.   Influence of Na^+ concentration (abscissa), in EGTA(Na-cho-line)-solution with Na^+ ranging from 0 to 140 mM (140 to 0 mM choline$^+$, substitution), on Na_i of quail myocardial cells. The cells were incubated in EGTA(Na-choline)-solution for 10 min at 25°C. Points and bars indicate mean S.D. of values in three experiments.

ably explains, at least in part, the inhibitory effect of both of these drugs on ouabain-induced irregular beating.[7] Eisner et al.[17] in their experiments with a Na^+-electrode in sheep Purkinje fibers, also suggested that lidocaine-induced decreased in Na^+_i is connected with its anti-arrhythmic action.

IV. STRUCTURAL CHANGES IN THE MYOFIBRILS DURING MITOSIS OF CULTURED NEWT MYOCARDIAL CELLS[9]

Myofibrils of embryonic myocardial cells in interphase and skeletal myotubes are mor-phologically stable structures which nevertheless exhibit turnover of myofibrillar proteins. Unlike skeletal myotubes, myocardial cells containing myofibrils synthesize DNA and undergo mitosis both in vivo and in culture. This observation raises the question: do myofibrils maintain their normal structure throughout mitosis? Many electron microscopic studies have been conducted to find the answer; however, conflicting results have been obtained. Some groups reported structural changes in the myofibrils, such as disintegration of the Z-band, occurring during mitosis, while others reported no such structural changes (see Reference 9). This may be due to the various different experimental animals used, or it may be that successive changes in myofibril structure during mitosis could not be observed continuously in these studies, because the preparations for electron microscopy were fixed. An attempt to resolve this question was made by observation of the myofibril striation patterns during mitosis without fixing the cells, using polarization microscopy. The mitotic index in cultures of embryonic newt myocardial cells was higher than in chick or rat myocardial cells. Moreover, in prophase, chromosomal condensation in newt embryonic myocardial cells could be detected easily, since these cells are larger than chick or rat cells. Therefore, we

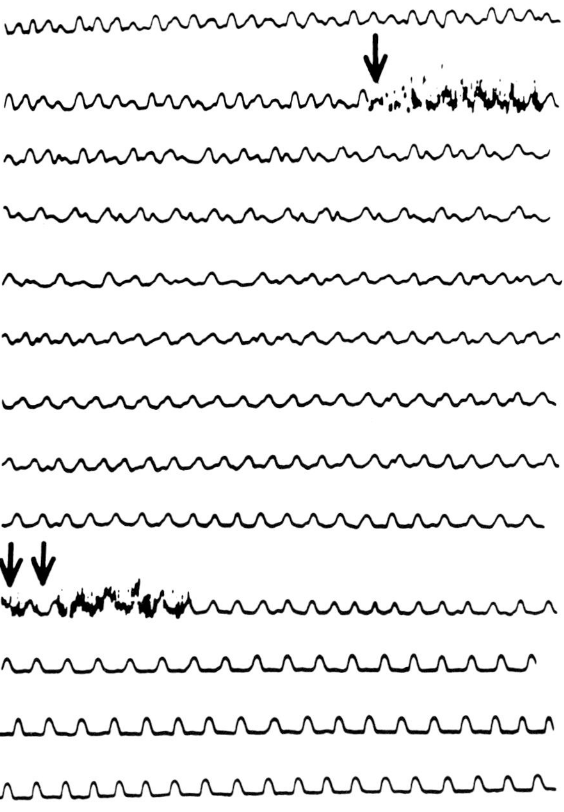

FIGURE 5. Photoelectric recording of recovery from ouabain-induced arrhythmia of a mouse myocardial cell sheet on addition of quinidine sulfate. The cell sheet was incubated in medium containing 0.2 mM ouabain for 20 min, and then 5 μM quinidine sulfate was added in the presence of ouabain at the time indicated by the single arrow. The final concentration of quinidine sulfate was increased to 10 μM, at the time indicated by the double arrows. Bar: 2 sec.

observed successive changes in the myofibril striation pattern of cultured newt myocardial cells using polarization microscopy and found that all myocardial cells undergo mitosis with disruptive changes in the striation pattern.

Two types of structural change were easily detected by polarization microscopy (Figure 7). In the first type, the striation pattern of the myofibrils became indistinct; i.e., the well-organized striation of A- (bright) and I- (dark) bands was lost in the region of the cell. We called this state *indistinct striation*. In the second type, the birefringence of sarcomeres completely disappeared. Even when the microscope stage was rotated to change the angle between the cell axis and the polarizing axis from 45° to 135°, no well-organized striation or birefringence, respectively, could be detected at any intermediate angle in the indistinct regions and the regions from which the birefringence had disappeared. Myofibril structural changes during mitosis that are observed using polarization microscopy typically progress as follows: in some regions, the striation pattern gradually became indistinct, regions containing indistinct striation exhibited weak contractions or no contractions at all, while other regions of the same cell, containing well-organized myofibrils with a clear striation pattern, exhibited normal contractions. The beating rhythm of the regions containing indistinct stria-

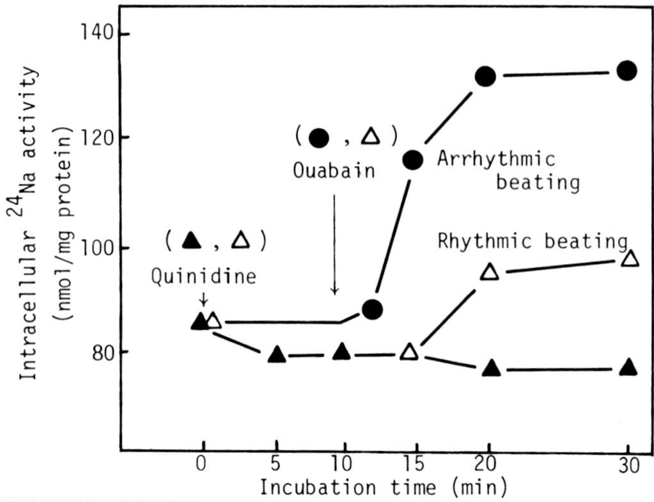

FIGURE 6.   Influence of quinidine on ouabain-induced increase in steady-state ^{24}Na activity of mouse myocardial cell-sheets. At the start of the experiment, $1 \times 10^{-5} M$ quinidine sulfate (▲,△) and after 10 min, $2 \times 10^{-4} M$ ouabain (●,○), was added. All experiments were performed at 25°C. Similar results were obtained in four independent experiments.

tion was synchronous with that of regions containing well-organized myofibrils. In some regions of indistinct striation, the birefringence of sarcomeres gradually decreased and finally disappeared, giving rise to loss of contractions. In the postmitotic state, the well-organized striation pattern of myofibrils reappeared.

Next, we examined the stages of mitosis at which the well-organized striation pattern of myofibrils started to become indistinct, which differed from cell to cell. In the micrographs shown in Figures 8 and 9, the striation patterns of cells started to become indistinct at metaphase and telophase, respectively. In most cases (19 of 21 cells), the striation pattern started to become indistinct at metaphase, but in two cells this did not occur until anaphase. In most cells (20 of 21 cells), the striation pattern of myofibrils disappeared during mitosis, but the stage at which this occurred differed from cell to cell — in most cells (18 of 21 cells), it began at metaphase and anaphase, but in two cells it started at telophase, and in one cell did not occur throughout the mitotic cycle.

Next, we compared the structural changes of myofibrils during mitosis using polarization microscopy with the ultrastructural changes observed by electron microscopy. Since the myofibril striation patterns observed by polarization microscopy were parallel to the coverslip and the cells were sectioned parallel to the coverslip, structural changes in the longitudinal plane of the myofibrils were only observed by electron microscopy. Figure 10 shows the regions of myofibrils where indistinct striation was observed by polarization microscopy and electron micrographs of the same regions. Z-bands had completely disappeared, and the arrangement of the disorganized sarcomeres was slightly disordered. In some cases, thin filaments had also disappeared from the regions where they had previously terminated at the Z-bands, although we could not verify whether thin filaments were absent in all cases.

Figure 11 shows the regions where the myofibril striation pattern had completely disappeared, using polarization microscopy. Only a few disorganized fascicles of myofilaments without Z-bands were seen. In small myofilament fascicles, thick filaments were easily identified by their typical diameter (about 15 nm), but it was not possible to identify thin filaments between the thick filaments.

The myofibril changes observed by electron microscopy in the present work are charac-

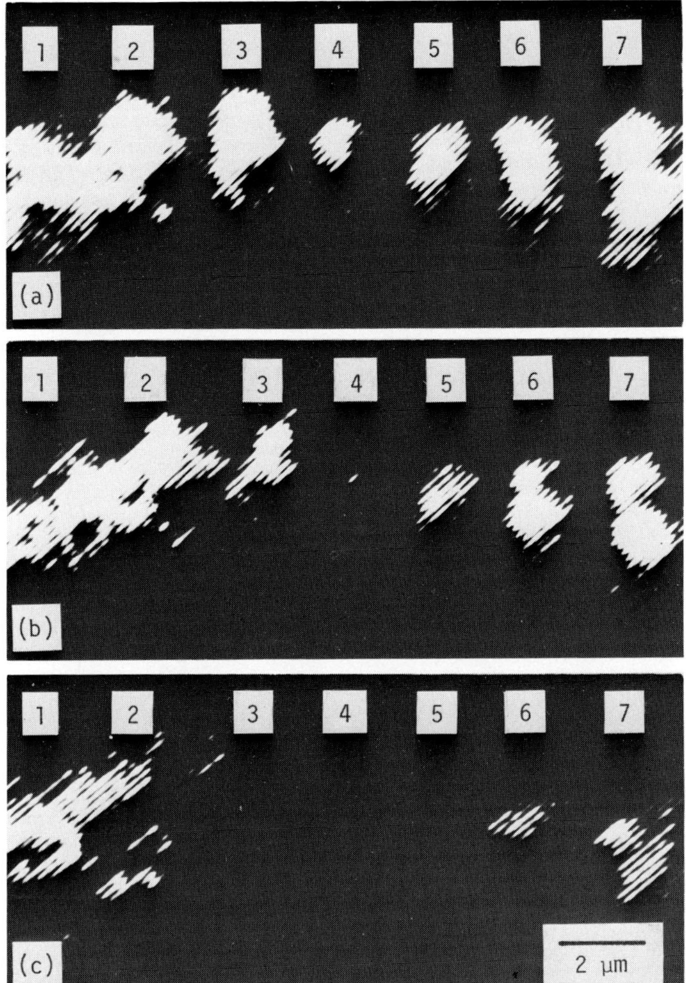

FIGURE 7.   Structural changes in the myofibril striation pattern of a cultured newt myocardial cell (8-day culture) as detected by polarization microscopy. (a) The well-organized striation pattern of A- (bright) and I- (dark) bands of five sarcomeres (Nos. 3 to 7) was clearly visible. However, two sarcomeres (Nos. 1 and 2) were indistinct. (b) Ten minutes later, the striation pattern of sarcomere (No. 3) became indistinct, while it became more indistinct in two sarcomeres (Nos. 1 and 2) and disappeared in one sarcomere (No. 4). (c) Twenty-two minutes after (a), the striation pattern (Nos. 6 and 7) became indistinct, and disappeared in two sarcomeres (Nos. 3 and 5). These photographs, taken in a cell during the mitotic stage of telophase, show the same region of the cell during the relaxed stage of beating.

terized by (1) disappearance of Z-bands and disarray of disorganized sarcomeres without Z-bands; (2) loss of thin myofilaments between thick myofilaments in some cases; (3) a tendency for myofibrils at the periphery of the cells to remain well-organized; and (4) reappearance of the well-organized myofibrils with Z-bands at the post-mitotic stage. These characteristics are consistent with those reported by Rumyantsev[18] who found that in in vivo rat heart (from 18-day-old embryos and 7-day-old rats), disintegration of the Z-bands in myofibrils at prometaphase was followed by progressive isolation of sarcomeres and scattering of myofilaments during subsequent stages of mitosis and then reconstitution of myofibrils in the early postmitotic stage.

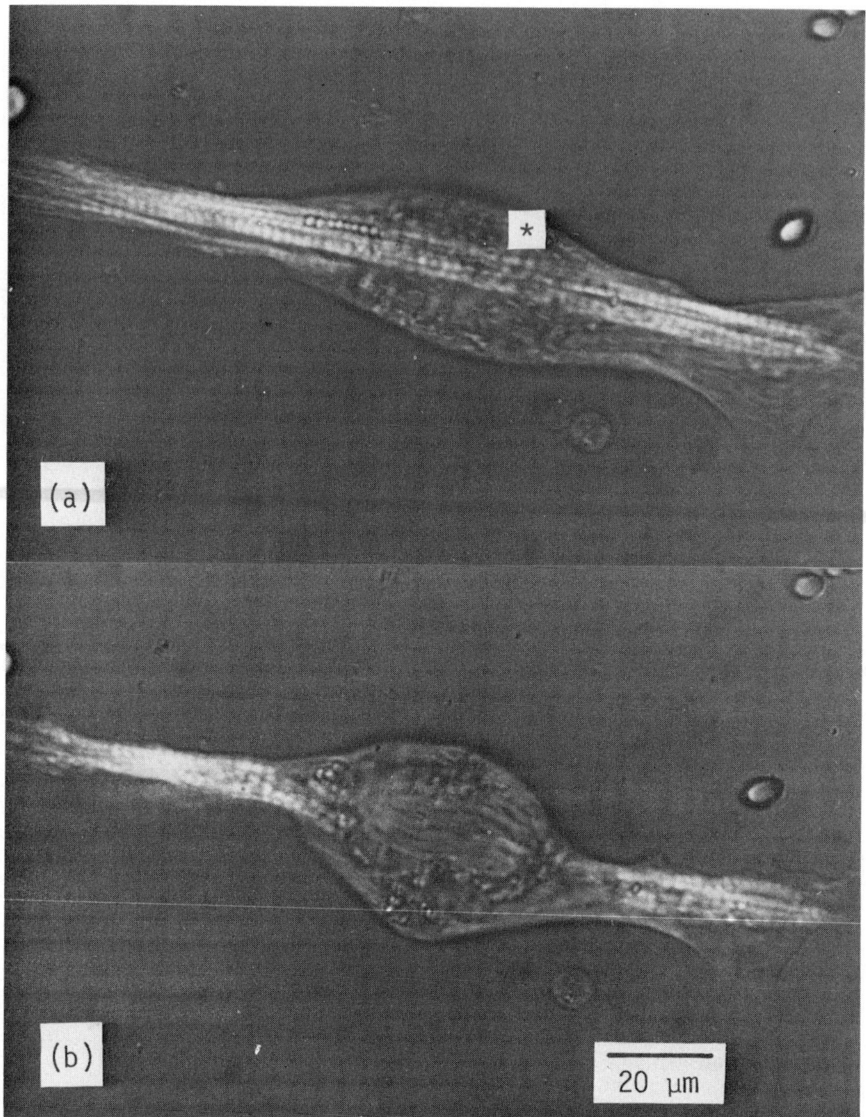

FIGURE 8.   Polarization microscopic observation of structural changes in cultured newt myocardial cell myofibrils (7-day culture). The striation pattern of myofibrils in the region indicated by asterisks (*) in (a) started to become indistinct at metaphase (a) and disappeared at anaphase (b) (21 min later).

As the disappearance of Z-bands was the first stage of the disruptive changes in myofibrils of the dividing myocardial cells, the mechanism of their disappearance during mitosis was considered in greater details. Rumyantsev[18] has suggested the involvement of an active Z-disk degenerating factor, such as an endogenous calcium-activated neutral protease, like the one isolated from heart muscles.[19] Immunofluorescence microscopy showed that this protease was localized at the Z-bands.[20] A protease of this nature might be involved in the disappearance of the myofibrils at the Z-bands in dividing myocardial cells. However, various aspects remain to be elucidated: (1) Do dividing myocardial cells contain more of this protease than non-dividing myocardial cells? (2) At what stage of mitosis is this protease activated and then inactivated? (3) How do calcium ions contribute to its action? and (4) How does

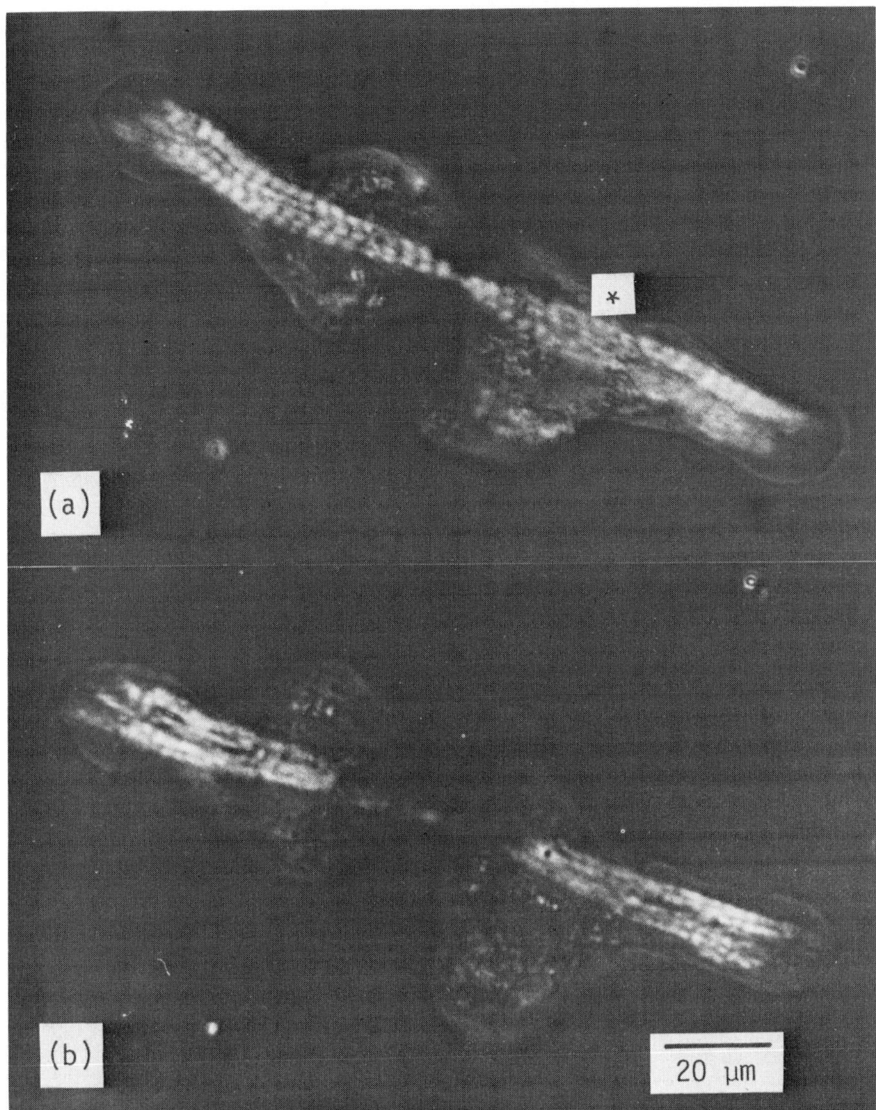

FIGURE 9.   Polarization microscopic observation of structural changes in myofibrils of a cultured newt myocardial cell (7-day culture). The striation pattern of myofibrils in the region indicated by asterisks (*) in (a) started to become indistinct at telophase (a) and disappeared at telophase (b) (20 min later).

this protease act locally to cause the disappearance of Z-bands?

Recent immunofluorescent and electron microscopic studies show that cytoskeletal proteins, such as desmin (skeletin) and vinculin, are localized at the periphery of Z-bands (see Reference 9). The studies also suggested that these proteins have important roles in the maintenance and/or formation of the well organized structure of myofibrils, although the details are as yet unknown. Mitosis of myocardial cells should provide a useful system for elucidating the functions of desmin, vinculin, and other cytoskeletal proteins in the maintenance and/or formation of well-organized myofibrils.

During mitosis, some single isolated myocardial cells (4 of 15 cells) maintained spontaneous beating, while others transiently ceased beating at the metaphase, anaphase, and/or

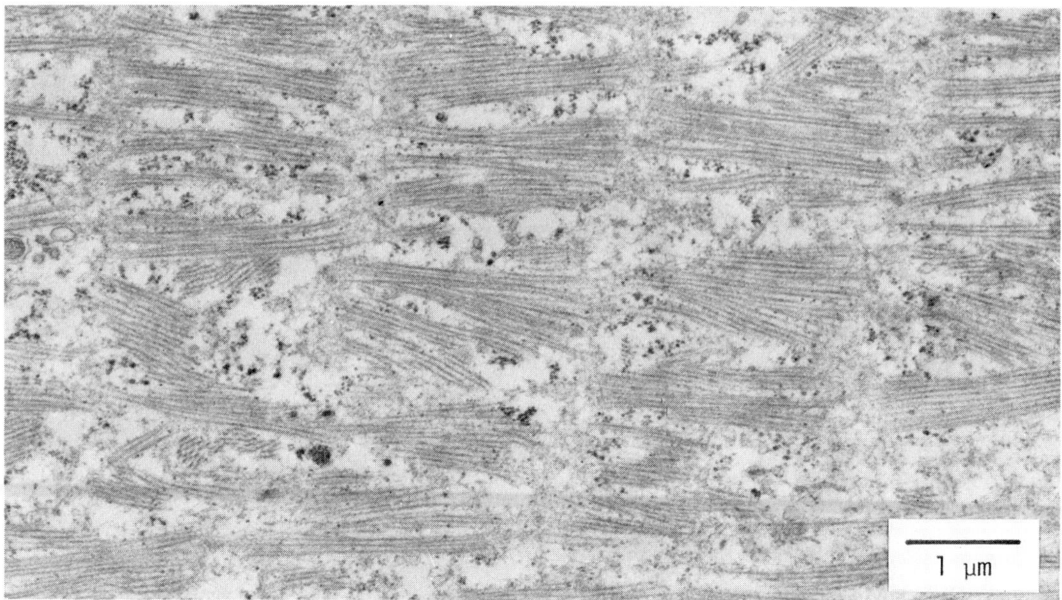

FIGURE 10.   An electron microscopic observation on the structural changes in myofibrils during mitosis corresponding with polarization microscopic observations. When the striation pattern of myofibrils in a beating myocardial cell (8-day culture) began to become indistinct at anaphase as observed by polarization microscopy, the cell was fixed and the region containing the indistinct striation was examined by electron microscopy. Complete disappearance of Z-bands and a slightly disorderly arrangement of disorganized sarcomeres without Z-bands are visible.

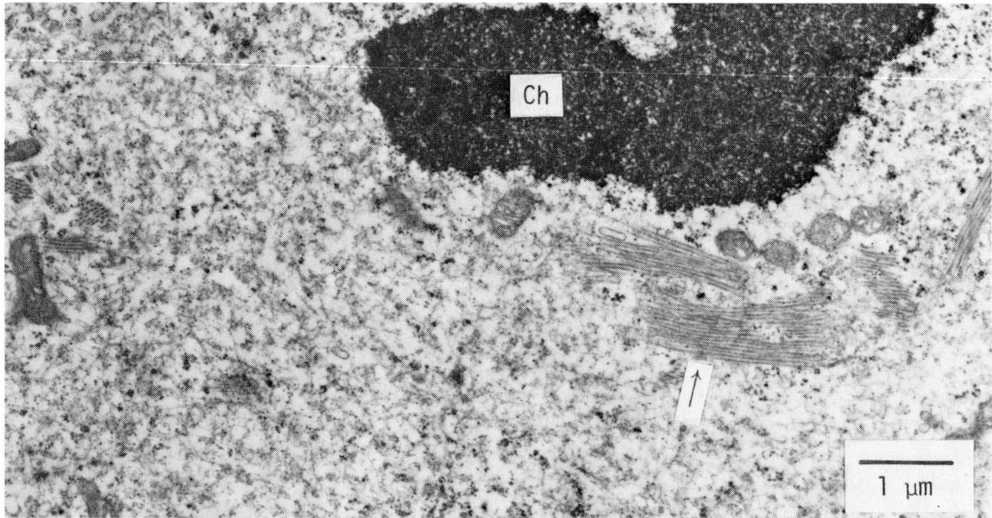

FIGURE 11.   An electron micrograph showing disorganization of myofibrils during mitosis consistent with polarization microscopic observations. When the striation pattern of myofibrils in a beating myocardial cell (9-day culture) partly disappeared at metaphase as observed by polarization microscopy, the cell was fixed and the region from which birefringence had disappeared was examined by electron microscopy. A few disorganized fascicles of myofilaments without Z-bands (arrow) may be seen. Ch are the chromosomes.

telophase, as previously described.[21] All the cells resumed spontaneous beating in the post-mitotic stage.

V. BEATING ACTIVITY OF HETEROKARYONS BETWEEN MYOCARDIAL AND NONMYOCARDIAL CELLS IN CULTURE[8]

Cultured mouse myocardial cells grown as monolayers may be fused by treatment with HVJ (Sendai virus). Myocardial cells may also be fused with quail myocardial cells, neuroblastoma cells, and nonexcitable cells, such as KB cells (a cell strain derived from a human epidermoid carcinoma). The beating activity of these heterokaryons was studied. Heterokaryons composed of myocardial cells from different species maintained spontaneous beating activity for 2 days or more. In one fused myocardial and neuroblastoma cell, activity was maintained for 22 to 26 hr, while in the fusion of a myocardial with a nonexcitable cell, such as the KB cell, activity was lost within 2 to 4 hr after addition of HVJ. Heterokaryons that had ceased to beat spontaneously did not contract on application of electrical field stimulation. The beating activity of the heterokaryons was inversely proportional to the ratio of nonmyocardial cells in the heterokaryons. Study of the rapid disappearance of beating activity in heterokaryons composed of a myocardial and a KB cell showed that both excitability of the cell membrane and myofibril organization were rapidly lost.

Before myocardial cells fused with KB cells, a well-organized striation pattern of myofibrils was observed by polarization microscopy, i.e., the bright A-bands and dark I-bands of the myofibrils were clearly visible (Figure 12a). But the organization was gradually disrupted as a function of time when the myocardial cell was fused with one or more cells of the KB, FL (a cell strain derived from a human amnion), or 3T3 cells (a cell strain derived from a whole mouse embryo). In the experiment shown in Figure 12, a heterokaryon containing a mouse myocardial and a KB cell started to lose the organized structure of myofibrils 2 hr after addition of HVJ (Figure 12b). On further incubation for 1 hr (Figure 12c) or 3 hr (Figure 12d), the well-organized striation pattern of myofibrils could no longer be seen by polarization microscopy. In preliminary observations by electron microscopy, we could also not detect the well-organized myofibrils in heterokaryons containing a mouse myocardial and a KB cell (2 hr after treatment with HVJ).

When myocardial cells fused to form giant myocardial cells, the structure of the well-organized myofibrils remained stable for 2 or more days after HVJ addition. However, when a myocardial cell was fused with one or more nonexcitable cells, such as KB cells, the myofibril organization was disrupted within 2 hr. The half-life of myofibrils of cultured myocardial cells is known to be of the order of several days.[22,23] Thus, the disruption of myofibrils by fusion with nonexcitable cells is very rapid compared with the half-life of myocardial cell myofibrils grown in normal medium. These results mean that the well-organized structure of myofibrils is unstable in the cytoplasm of myocardial plus KB cells. The observation that heterokaryons composed of two myocardial cells and one KB cell maintained beating activity for longer time periods than those composed of one myocardial and a KB cell, suggests that cytoplasmic components of KB cells must exceed a threshold concentration before myofibril disruption takes place. The observation that the heterokaryons composed of a myocardial and a neuroblastoma cell maintained their beating activity for a longer time period than those consisting of a myocardial and a KB cell, suggests that cytoplasmic components of neuroblastoma cells are less active than those of KB cells with respect to myofibril disruption.

FIGURE 12.   Successive observations of the disappearance of well-organized myofibrils in a mouse myocardial cell on fusion with a KB cell using a polarization microscope. (a) Typical appearance of well-organized striation of A- (bright) and I- (dark) bands observed in a single isolated mouse myocardial cell. (b) 2, (c) 3, (d) 5 hr after addition of HVJ, respectively. (b-d) The same region of the fused cell. Loss of the well-organized striation began within 2 hr after HVJ addition. No well-organized striation was detected in other regions of the heterokaryon, even when the microscope stage was rotated, changing the angle between the cell-axis and polarization-axis from 45° to 135°. (e) Phase contrast micrograph of the heterokaryon observed in (d). N_m and N_{KB} are the nuclei of myocardial and KB cells, respectively. Similar results were obtained in seven other experiments. Bar; (a-d) 5 μm, (e) 10 μm.

REFERENCES

1. **Okarma, T. B. and Kalman, S. M.,** Photoelectric monitoring of single beating heart cells in culture, *Exp. Cell Res.,* 69, 128, 1971.
2. **Goshima, K.,** Photoelectric demonstration of ouabain-induced arrhythmias in single isolated myocardial cells and in cell clusters in vitro, *Cell Struct. Funct.,* 2, 329, 1977.
3. **Jongsma, H. J., Tsjernina, L., and De Bruijne, J.,** The establishment of regular beating in populations of pacemaker heart cells. A study with tissue-cultured rat heart cells, *J. Mol. Cell. Cardiol.,* 15, 123, 1983.
4. **Harary, I., Wallace, G., and Bristol, G.,** A video-computer for the chronotropic and inotropic measurements of the beating of cultured heart cells, *Cytometry,* 3, 367, 1983.
5. **Kobayasi, S. and Goshima, K.,** Apparatus for measuring time intervals of the beating rhythm of cultured cardiac cells, *Rev. Sci. Instrum.,* in press.

6. **Goshima, K. and Wakabayashi, S.,** Involvement of an Na$^+$-Ca^{2+} exchange system in genesis of ouabain-induced arrhythmias of cultured myocardial cells, *J. Mol. Cell. Cardiol.,* 13, 489, 1981.
7. **Goshima, K. and Wakabayashi, S.,** Inhibition of ouabain-induced increase in Na content of cultured myocardial cells by quinidine and procainamide, *J. Pharm. Exp. Ther.,* 224, 239, 1983.
8. **Goshima, K., Kaneko, H., Wakabayashi, S., Masuda, A., and Matsui, Y.,** Beating activity of heterokaryons between myocardial and non-myocardial cells in culture, *Exp. Cell Res.,* 151, 148, 1984.
9. **Kaneko, H., Okamoto, M., and Goshima, K.,** Structural change of myofibrils during mitosis of newt embryonic myocardial cells in culture, *Exp. Cell Res.,* 153, 483, 1984.
10. **Wakabayashi, S. and Goshima, K.,** Kinetic studies on sodium-dependent calcium uptake by myocardial cells and neuroblastoma cells in culture, *Biochim. Biophys. Acta,* 642, 158, 1981.
11. **Wakabayashi, S. and Goshima, K.,** Partial purification of Na$^+$-Ca^{2+} antiporter from plasma membrane of chick heart, *Biochim. Biophys. Acta,* 693, 125, 1982.
12. **Mullins, L. J.,** Mechanism of Na/Ca transport, *J. Gen. Physiol.,* 70, 681, 1977.
13. **Horackova, M. and Vassort, G.,** Sodium-calcium exchange in regulation of cardiac contractility. Evidence for an electrogenic, voltage-dependent mechanism, *J. Gen. Physiol.,* 73, 403, 1979.
14. **Pitts, B. J. R.,** Stoichiometry of sodium-calcium exchange in cardiac sarcolemmal vesicles. Coupling to the sodium pump, *J. Biol. Chem.,* 254, 6232, 1979.
15. **Chapman, R. A.,** Control of cardiac contractility at the cellular level, *Am. J. Physiol.,* 245, H535, 1983.
16. **Hagiwara, S. and Nakajima, S.,** Effects of the intracellular Ca ion concentration upon the excitability of the muscle fiber membrane of a barnacle, *J. Gen. Physiol.,* 49, 807, 1966.
17. **Eisner, D. A., Lederer, W. J., and Sheu, S-S.,** The role of intracellular sodium activity in the antiarrhythmic action of local anaesthetics in sheep Purkinje fibres, *J. Physiol.,* 340, 239, 1983.
18. **Rumyantsev, P. P.,** Interrelations of the proliferation and differentiation processes during cardiac myogenesis and regeneration, *Int. Rev. Cytol.,* 51, 187, 1977.
19. **Mellgren, R. L.,** Canine cardiac calcium-dependent proteases: resolution of two forms with different requirements for calcium, *FEBS Lett.,* 109, 129, 1980.
20. **Ishiura, S., Sugita, H., Nonaka, I., and Imahori, K.,** Calcium-activated neutral protease. Its localization in the myofibril, especially at the Z-band, *J. Biochem.,* 87, 343, 1980.
21. **Kelly, A. M. and Chacko, S.,** Myofibril organization and mitosis in cultured cardiac muscle cells, *Dev. Biol.,* 48, 421, 1976.
22. **Desmond, W., Jr. and Harary, I.,** In vitro studies of beating heart cells in culture. XV. Myosin turnover and the effect of serum, *Arch. Biochem. Biophys.,* 151, 285, 1972.
23. **Clark, W. A., Jr. and Zak, R. J.,** Assessment of fractional rates of protein synthesis in cardiac muscle cultures after equilibrium labeling, *J. Biol. Chem.,* 256, 4863, 1981.

INDEX

Ectodermal cells, 61
EDTA, 144
EGF, see Epidermal growth factor
EGTA, 132, 171
Elastase, 8, 12
Electrical activity, 126, 138—140
  age of culture and, 145
  rate of, 144—145
  tissue origin and, 145
Electrical properties, passive, 137—138
Electrical stimulation, 26, 71
Electrogenic transport, 80
Electronic junction, 146
Electron microscopy, 176, 179—180
Electron probe microanalysis, 81
Electrophorus, 163
Electrophysiological studies, 125—152, see also
       Patch clamp method; Voltage clamp method
  advantages of cell culture for, 127
  age of culture, 145
  applications of, 147—150
  automaticity and coupling, 145—146
  cell culture mode, 127—129
  cell impalement, 133
  contraction transducing, 136
  electrical activity, 138—140
  electronic recording system, 133—135
  excitation-contraction coupling, 146—147
  growth medium for, 129
  handling of cultures for, 129—132
  impalement, 127, 132—133
  incubation chamber for, 130
  ionic determinants of membrane potential, 140—
    144
  of ion transport, 80
  microelectrodes for, 132—133
  microiontophoresis and microperfusion, 136—137
  micromanipulation in, 127, 131—132
  models of the pathological heart and, 148—150
  in monolayer cultures, 128
  new techniques in, 151—152
  nutritional studies, 149—150
  origin of tissue, 145
  passive electrical properties, 137—138
  perfusion versus static bath, 130—131
  in pharmacology, 147—148
  rate of activity, 144—145
  recording system for, 133—135
  single channel measurements, 159—165
  in sparse cultures, 127
  in spherical aggregates, 128
  in synthetic strands, 128
  tissue source for, 128—129
  visualization in, 131—132
Electrostatic recorder, 134
Embryoid body, 54—55, 60—61
  cystic, 56, 61
  formation in vitro, 55—56
Embryonal carcinoma cells
  nullipotent, 56
  pleuripotent, 53—65

Embryonic heart cells, 128
  culture techniques for, 9—11, 86
  calcium fluxes in, 119—120
  growth and development of, 34—37, 41—44
  ion fluxes in, 79
Endodermal cells, 56, 60—61
Endothelial cells, 13—14, 24, 35
Epidermal growth factor (EGF), 26
Epithelial cells, 149—150
Equipment, cell culture, 15
Erucic acid, 149—150
Eserine, 3
Ethylisopropylamiloride, 102
Excitation-contraction coupling, 110, 121—122,
    145—147
Extracellular matrix, 24
Extracellular space, 78—83, 96, 127

F

Fast cells, 141—144
Fast inward sodium current, 141—143
Fatty acids, 4, 12, 26, 39, 43—47
  dietary, 149—150
Fetal calf serum, 37
Fetuin, 24—25
Fibrillatory beating, 170—174
Fibroblasts, 13, 35
  ion fluxes in, 83—84, 88—89, 95—99
  metabolism in, 45
  in neuron culture, 69
Fibronectin, 24—25
Filtering of electrical signals, 134
Flame photometry, 80
Flattened cells, 38
FL cells, 181
Fluorescent indicator, 80
Flux chamber, 90—92
Fourier transform, 152
Frequency analysis, 152
Furosemide, 103

G

β-Galactosidase, 17
Galois, C.J.J., 1
Gap junction, 146, 164
Gestation, 35
Giant myocardial cells, 181
Gigaseal, 160—161
Girardi heart cells, 14, 98—99, 102
Glucocorticoids, 25—26, 71
Glucose, 12, 43—45
Glutamine, 26
Glycocalyx, 112, 121
Glycogen, 39
Glycolysis, 4, 43—44
Growth factors, 9—11, 37
Growth medium, 129
GTP, 162
5′-Guanylimidophosphate, 148

H

Ham's F10 medium, 15
Ham's F12 nutritional mixture, 25
Hatching, 35, 42
HEPES buffer, 130
Herpes virus, 149
Heterokaryons, 181—182
Hexokinase, 44
Histones, 40
Hormones, 24, 37
Hydrocortisone, 26—27
Hydrogen, sodium-hydrogen exchanger, 102
7β-Hydroxy cholesterol, 14
Hyperplasia, 37—38
Hypertrophy, 28, 37
Hypoxia, 149

I

Impalement, 127, 132—133
Incubation chamber, 130
Injury, 34
Ink-jet recorder, 134
Insulin, 13, 25—27
Intercalated disc, 56, 59
Intercellular channels, 164
Intermitotic period, 37
Interphase cells, 38
Interventricular septum, 116—117, 122
Intracellular compartment, calcium, 114—118
Intraelectrode dialysis, 160—161
Inward current, 162—163
Iodoacetate, 44
Ion channel, 159—165
Ionic currents, 162—164
Ionophore A23187, 14
Ion sensitive dye, 152
Ion-sensitive electrode, 80, 89, 151
Ion transport, 77—104, see also specific ions
    chemical analysis, 80—81
    chloride transport, 102—103
    computation of, 82—83
    contractility, 81
    electrophysiology, 80
    limitations of tissue culture preparations, 79
    methods of study of, 80—86
    naturally occurring cardiac muscle vs. cultured
        heart cells, 78—79
    potassium transport, 92—95
    preparation of cultured heart cells, 86—88
    radioisotopic flux, 81—86
    single channel, 159—165
    sodium transport, 96—102
    special techniques, 88—92
Iothalamic acid, 89
Isolated heart cells, 92
    preparation of, 160
    single channel measurements in, 159—165
Isoleucine, 12
Isoproterenol, 39, 45, 147—148

J

Junctional conductance, 164

K

KB cells, 181—182
Kraftbruhe, 160

L

Lactate, 45
Lactic dehydrogenase, 16—17, 19, 45—46
Lanthanum, 144, 146
Laurylacetate, 114
Leading cells, 145
Lidocaine, 148, 174
Linoleate, 150
Lipid-free medium, 4
Lipids, 149—150
Lipoprotein, 138
Liposomes, 148
Lithium, 103
Local anesthetics, 148
Lysosomal enzymes, 17

M

Manganese, 144, 146
Mass culture, 87
Maximal diastolic potential (MDP), 135, 140—141
MDP, see Maximal diastolic potential
Mechanism, 1
Medium, 9—12, 15, 18, 129—130
    basal, 24—25
    conditioned, 13, 65—72
    defined, 13, 24
    growth, 129
    hormonally supplemented, 24
    lipid-free, 4, 149—150
    serum-free, 14, 23—31
Medium 199, 25
Membrane conductance, background, 140
Membrane current, 87
Membrane potential, 103, 118, 122, 133, 165
    ionic determinants of, 140
Mesenchymal cells, 60
Mesodermal cells, 41, 56, 60—61
Messenger RNA, 39, 43
Metabolism
    in cultured cells, 44—45
    impaired, 149
    shift from anaerobic to aerobic, 43—47
Metallochromic indicator, 80
Metaphase, 38, 176, 178
Methanesulfonate, 103
Methylxanthines, 148
Mexiletine, 148
Microelectrode, 80, 127, 132—136
    connection to recording system, 133
Microiontophoresis, 136—137

The Heart Cell
In
Culture

Volume II

Editor

Arié Pinson, D.Sc.
Senior Investigator
Laboratory for Myocardial Research
Institute of Biochemistry
Hebrew University - Hadassah Medical School
Jerusalem, Israel

CRC Press, Inc.
Boca Raton, Florida

Library of Congress Cataloging-in-Publication Data

The Heart cell in culture.
  Includes index.
  1. Heart cells. 2. Cell culture. I. Pinson,
Arié, 1931-   [DNLM: 1. Cells, Cultured.
2. Myocardium--cytology. WG 280 H4365]
QP114.C44H43   1987   612'.17'0724   87-21864
ISBN 0-8493-4696-7 (set)

PREFACE

At the turn of the century, cell culture and other in vitro methods were developed and have since proven invaluable in research on differentiation, specific function, and metabolism of various tissues, which are often difficult to study in vivo due to the complexity of the interactions between different tissues and between tissues and body fluids.

In 1912, Burrows made the pioneering discovery that cardiac explants produced muscle cells that were capable of dividing and beating with a regular rhythm, thus demonstrating the myogenic nature of heart contraction. The tryptic method of dissociating cells was first described by Rous and Jones in 1916, and revived almost 40 years later by Moscona. In 1955, Cavanaugh isolated and grew heart cells from the chick embryo. Twenty-five years ago, Harary and Farley prepared the first cultures of neonatal rat heart cells. This method of culturing mammalian neonatal heart cells has proved to be a crucial step in the application of heart cell culture techniques to various research areas dealing with myocardial biology at the cellular level. As a direct consequence of this, since the early 1970s, there has been about a sixfold increase in the number of scientific papers per year dealing with heart cells in culture. Clearly, a summary of the progress in this field is long overdue.

My primary aim as the editor of *The Heart Cell in Culture* was not to provide a laboratory manual, although techniques are discussed when relevant. Rather, each chapter presents a critical review, demonstrating the progress made during the last 25 years as well as perspectives for the future.

The Heart Cell in Culture is presented in three multidisciplinary volumes, spanning a wide spectrum of topics and presenting state-of-the-art reviews. It is difficult to divide such a book into distinct parts, but it has been presented in three sections, one in each volume: the first on the heart cell culture system — historical background, the development of culturing techniques, and the function of such cells; the second on the basic biochemistry of the myocardial cell in culture; and the third on applied research in pharmacology and pathology of cardiac cells, including a chapter on cultured adult heart cells, which is becoming a popular tool in cardiac research. The minireview chapters in the series were selected with two main objectives in mind: to complement full-length comprehensive reviews in the same field by the addition of new dimensions and scope, in which case they immediately follow the main chapter, or to consider topics which are relatively new or show promise for the future development of research in cardiac biology. Indeed, I expect that these chapters should be extremely stimulating for this reason.

I hope that people working in the field of heart research will find useful information in these volumes, and that they will foster better understanding of cardiac biology and stimulate further progress in this important field of research.

I would like to thank Dr. R. Goldberg for her assistance with the thankless task of language editing, corrections to the text, and proofreading. I am grateful to all the contributors who made this book, *The Heart Cell in Culture*, possible.

Arié Pinson
Jerusalem, Israel

THE EDITOR

Dr. Arié Pinson, D.Sc., is a Senior Investigator at the Hebrew University-Hadassah Medical School and permanent visiting professor at the Faculty of Life and Environmental Sciences of the University of Dijon in France.

He studied Clinical Chemistry and Biology at the Univesity of Geneva in Switzerland. He earned his M.Sc. degree from the Hebrew University of Jerusalem in 1967 and his D.Sc. for a thesis on the Metabolism of Palmitic and Erucic Acid in Rat Heart Cell Cultures from the Univesity of Dijon in France in 1975.

Since 1975, the research carried out at Dr. Pinson's Laboratory for Myocardial Research has contributed toward establishing cultures of cardiac myocytes as a standard technique for studying the biochemistry of the heart in both health and disease. In a broader sense, these studies, often conducted in collaboration with leading scientists overseas, have advanced knowledge on cardiac cells in culture as a tool in the fields of development, differentiation, biochemistry, physiology, and medical research.

Dr. Pinson has been awarded two prizes for distinguished research by the Hebrew University. He has also held long-term fellowships from INSERM — Institut National de la Santé et de la Recherche Médicale (France), CNRS — Centre National de la Recherche Scientifique (France), DGRSRT — Délégation Générale à la Recherche Scientifique et Technique (France), ZWO — The Netherlands Society for the Advancement of Pure Research, and annual summer fellowships from INSERM.

A recognized authority in his field, he has frequently been invited to give plenary lectures at international conferences: at the Biochemical Society (U.K.) and at several ISHR (International Society for Heart Research) meetings and at other meetings focusing on cardiomyocytes.

Dr. Pinson has been invited to many seminars at universities and research institutes in Holland, Germany, France, and Israel. In 1984, he founded the IGHR (the Israeli Group for Heart Research), which is affiliated to the ISHR.

Dr. Pinson's current research interests include the biochemistry of cultured cardiomyocytes with particular emphasis on lipid and glucose metabolism, function and structure of sarcolemma, metabolism of high energy phosphates, and the use of this system as a model for elucidation of anoxic injury and iron overload in the heart.

CONTRIBUTORS

VOLUME II

John P. Bilezikian, M.D.
Associate Professor
Department of Medicine
Columbia University
New York, New York

Claudio M. Caldarera
Professor
Institute of Biochemistry
University of Bologna
Bologna, Italy

T. Chajek-Shaul
Professor
Department of Medicine and Lipid
  Research Laboratory
Hadassah University Hospital
Jerusalem, Israel

Carlo Clô, D.Sc.
Associate Professor
Department of Biochemistry
University of Bologna
Bologna, Italy

G. Friedman
Senior Lecturer
Department of Medicine and Lipid
  Research Laboratory
Hadassah University Hospital
Jerusalem, Israel

Jonas B. Galper, Ph.D.
Assistant Professor
Cardiovascular Division
Brigham and Women's Hospital
Boston, Massachusetts

Linda E. Kupfer, Ph.D.
Department of Pharmacology
Columbia University
New York, New York

Carla Pignatti
Assistant
Department of Biochemistry
University of Bologna
Bologna, Italy

Arié Pinson, D.Sc.
Senior Investigator
Department of Biochemistry
Hebrew University-
  Hadassah Medical School
Jerusalem, Israel

Lydie Rappaport
Senior Investigator, CNRS
Hôpital Lariboisière
INSERM
Paris, France

Richard B. Robinson, Ph.D.
Associate Professor
Department of Pharmacology
Columbia University
New York, New York

Ketty Schwartz
Directeur de Recherches
INSERM
Hôpital Lariboisière
Paris, France

Maria Seraydarian, Ph.D.
Professor
School of Nursing
University of California at Los Angeles
Los Angeles, California

Thomas W. Smith, M.D.
Professor
Cardiovascular Division
Brigham and Women's Hospital
Boston, Massachusetts

O. Stein
Department of Experimental Medicine
  and Cancer Research
Hebrew University-Hadassah Medical
  School
Jerusalem, Israel

Yechezkiel Stein
Professor
Chairman Department of Medicine and
 Lipid Research Laboratory
Hadassah University Hospital
Jerusalem, Israel

Christian Vial
Mâitre-Assistant
Laboratoire de Biologie
Université Lyon
Villeurbanne, France

Radovan Zak, Ph.D.
Professor
Department of Medicine
University of Chicago
Chicago, Illinois

In memory of my parents,
Zeev and Pessia Pinson

To my wife and children,
Yona, Gavriel, Halléli, and Shira

To Professor Prudent Padieu

TABLE OF CONTENTS

VOLUME I

TABLE OF CONTENTS

VOLUME II

TABLE OF CONTENTS

VOLUME III

Chapter 12

PROTEIN SYNTHESIS AND DEGRADATION IN CULTURED HEART CELLS

Radovan Zak

TABLE OF CONTENTS

I. INTRODUCTION

Methods used to measure protein synthesis and degradation can be divided into two categories: tracer kinetics and chemical assays of protein accumulation and/or degradation. Tracer kinetics is preferred by most investigators due to the need for high sensitivity while working with typical cultured cells. Hence, this chapter will cover tracer kinetics in more depth than the chemical methods.

All procedures that utilize radioactive tracers in order to obtain quantitative evaluations of the synthesis and degradation rates are based upon the same kinetic principle. The radiolabeled tracer flows from the precursor compartment (e.g., the extracellular pool of amino acids) into the product compartment (e.g., the amino acids of the protein pool). The rate of this flow, as well as its direction at any given time, depends on the difference between the specific radioactivities of the tracer in these two compartments. This difference in the specific radioactivities of the tracer is stated by the following differential equation:

$$dP^*/dt = k(F^* - P^*) \tag{1}$$

where P^* and F^* are the specific radioactivities of product and precursor, respectively, k is the rate constant of the radioactivity flux between the P and F compartments and t is time (length of the experiment). The above differential equation applies to in vivo as well as to in vitro studies.

Measurements of protein specific radioactivity are straightforward since an individual specific protein can be either immunoprecipitated prior to measurement of radioactivity or its specific radioactivity can be determined directly on SDS polyacrylamide gels after electrophoretic separation. In contrast, measurements of precursor specific radioactivity are rather difficult to obtain. This is due to the nature of the immediate precursor of protein aminoacyl-tRNA, which has a very small pool. Consequently, the measurement of its specific radioactivity not only requires sensitive methods, but also rapid and accurate sampling due to the rapid flux between this small precursor compartment and the large pool of proteins. Because of these problems, one has to frequently measure other pools and hope that a reasonable approximation of the true precursor pool will be achieved. One additional complication arises in that both the direction of fluxes and the interconnection between extracellular, intracellular, tRNA, and protein pools are not always known. This is illustrated in Figure 1. In addition to the direct pathways between free amino acid and protein pools, several other possibilities supported by some experimental evidence exist. For example, the intracellular compartment could be bypassed and the amino acid could move directly from extracellular space into the tRNA pool.[1] Alternatively, the amino acid derived from protein degradation might not equilibrate with the extracellular compartment rapidly enough and, thus, become preferentially reutilized for the synthesis of proteins.[2] Some of the methods used by various investigators to overcome the above problems will be discussed in Sections IV and V of this chapter. As far as the kinetic principles underlying the use of radiolabeled tracers are concerned, including their conceptual hazards, they have been addressed in many reviews published since the first application of radioisotopes in biological research in the early 1940s.[3-8]

As far as the chemical measurements are concerned, they are conceptually more simple than the kinetic studies. In order to evaluate the rate of synthesis, one has to measure the protein accumulation (growth) during the period of the experiment and then by independent means to estimate the rate of protein degradation, e.g., by measurement of amino acid release from the protein. The sum of the rates of growth (R_g) and degradation (R_d) gives the rate of synthesis (R_s):

$$R_s = R_g + R_d \tag{2}$$

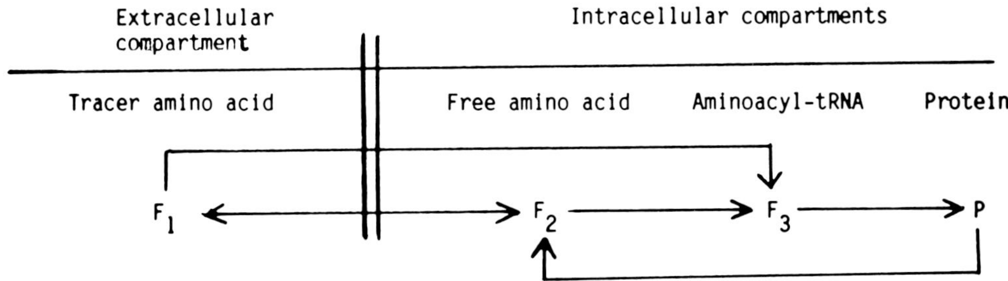

FIGURE 1.   Possible pathway of tracer flux through cellular compartments. Tracer amino acid added to culture medium (F_1) can enter the pool of free amino acids within the cell (F_2), form aminoacyl-tRNA complex (F_3), and finally be incorporated into product (P). It is conceivable, however, that the amino acid from compartment F_1 can bypass the compartment F_2 and enter directly into the pool of aminoacyl tRNA (F_3). Free amino acids derived from protein (P) degradation enter the pool of free intracellular amino acids (F_2). If the equilibrium between compartments F_2 and F_1 is not fast enough, it is conceivable that the amino acids derived from proteins are preferentially reutilized for protein synthesis.

The major drawback of this otherwise straightforward procedure is that the rate of growth is usually quite slow in vitro and is, therefore, difficult to estimate accurately over a short period of time when the cells are in culture. The measurement of protein degradation is also difficult, due to the fact that a portion of the amino acids derived from protein degradation is reutilized for protein synthesis. At present, only a few amino acids are not reutilized. Notable among these are those amino acids which undergo posttranslational modification such as hydroxyproline and the methyl derivatives of histidine and lysine. Some of the attempts used to minimize the problem of amino acids reutilization are discussed in Section V.

II. DEFINITION OF TERMS USED IN QUANTITATIVE EVALUATION OF PROTEIN SYNTHESIS AND DEGRADATION

The terminology and symbols used in studies of protein synthesis and degradation are not unified and, hence, are sometimes confusing and subject to dispute (e.g., References 9 to 11). In this chapter the definitions are given in quantitative terms, including units, so that their relationship to terms used by other investigators can be readily established. Koch in his review article[6] lists a variety of symbols and terms used by different investigators. The following terms and symbols are used in this chapter:

- R_g = rate of protein accumulation (growth); units, mass/time
- R_s = rate of protein synthesis, units, mass/time
- R_d = rate of protein degradation; unit, mass/time
- k_g = fractional growth rate; unit, time^{-1}
- k_s = fractional synthesis rate; unit, time^{-1}
- k_t = fractional turnover rate; unit, time^{-1}
- k_d = fractional degradation rate; unit, time^{-1}
- P^* = specific radioactivity of the tracer amino acid in the protein molecule; units, radioactivity/mass
- F^* = specific radioactivity of the tracer amino acid in the precursor pool. A subscript as in F^*_1, F^*_2, etc., indicates a sequential relationship between various compartments of the precursor; units, radioactivity/mass.

The difference between rate and fractional rate is frequently misunderstood. Therefore, this difference is worth elucidating. The rate represents the change of the entire pool per unit of time, hence, its units are mass/time. The larger the pool, i.e., more cells grown per dish, the larger the rate will be. Some investigators correct for the size of the pool by dividing the rate by the mass of the protein present per dish, e.g., one refers to the rate of incorporation as the mass of amino acid incorporated per gram of protein per hour. Although a correction for the size of the pool may be obtained in this manner, one cannot compare several proteins, since their relative amounts are likely to change in the course of the experiment, e.g., induction of some protein takes place.

Fractional rates are used to eliminate the ambiguities of rates. In this case the data represent a fraction (or percent) of the pool which underwent a given change. For example, $k_s = 0.1$ hr^{-1}, this means that the synthesis corresponded to 0.1 (or 10%) of the total protein pool per hour. The fractional rate is similar to the *corrected* rate in the sense that the mass of amino acid incorporated (or lost) per unit of time is expressed as the fraction of the total mass of the same amino acid in the protein pool (e.g., the dish). The dimensions of the fractional rate — reciprocal time — are that of frequency and hence, are independent of the size of a given pool, the amino acid composition and molecular weight of a given protein or protein mixture, the length of the experiment, and the species of amino acid used as a tracer. Most importantly, the computation is substantially simplified since the unified dimensions for the fractional rates allow for a direct comparison of the studied parameters such as k_g, k_s, k_t, etc.

III. CONCEPT OF THE STEADY AND NONSTEADY STATES IN STUDIES OF PROTEIN SYNTHESIS AND DEGRADATION

From the point of view of balance between the rates of protein synthesis and degradation, any biological system can exist in two states: (1) the steady state where there is no change in the amount of protein and (2) the nonsteady state where the pool of protein molecules changes, i.e., there is either negative or positive growth.

Dealing with the steady state is conceptually simple, since the flux of radioactivity between the precursor and product compartments in this case is solely the result of the degradation and simultaneous resynthesis of the protein molecules during the process referred to as turnover. The synthesis rate in this case equals that of degradation which, in turn, equals that of turnover. Hence, one can write that during a steady state situation:

$$k_s = k_d = k_t \tag{3}$$

The main objective of experiments in the steady state is to obtain an estimate of the rate at which a given protein turns over (k_t). This rate is often expressed in the more tangible term of protein half-life, T/2, which is related to k_t by the following relationship:

$$T/2 = \frac{\ln 2}{k_t} \tag{4}$$

The value of k_t can be obtained, as seen in Equation 3, by measurement of either the rate of synthesis or degradation. In most cases, however, the investigators have chosen the synthesis rate because of experimental convenience.

In the nonsteady state, when the total pool of protein P changes, additional factors influence the flux of radioactivity between precursor and product compartments which has to be evaluated. This is illustrated in Figure 2. It can be seen that when the growth is positive, i.e., the pool P expands, the radioactivity can be considered to enter the protein as a result of two processes: (1) addition of new protein molecules due to the expansion of the pool P (k_g) and (2) resynthesis of molecules P degraded during the process of turnover (k_t). Hence, the synthesis rate is equal to:

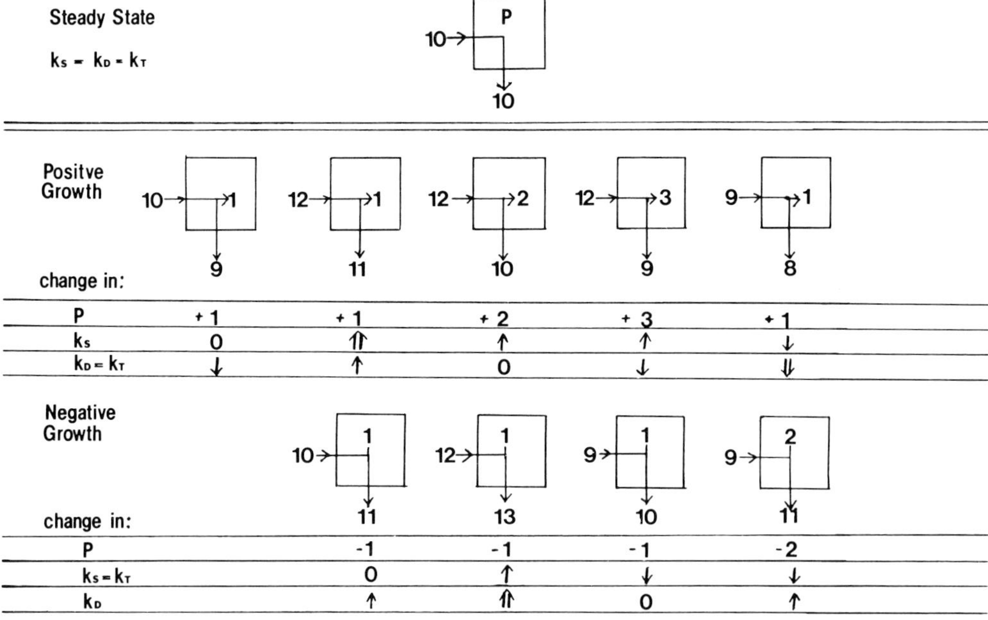

change in:					
P	+ 1	+ 1	+ 2	+ 3	+ 1
k_s	0	⇑	↑	↑	↓
$k_D = k_T$	↓	↑	0	↓	⇓

change in:				
P		-1	-1	-1	-2
$k_s = k_T$		0	↑	↓	↓
k_D		↑	⇑	0	↑

FIGURE 2. Contribution of synthesis and degradation to the size of the protein pool. In the *steady state*, 10 molecules enter the protein pool and 10 molecules are randomly degraded and leave the pool. There is no change in the total number of protein molecules. In the *nonsteady state of positive growth*, the number of molecules entering the pool exceeds those leaving it. The pool nevertheless continues to turn over. In the first example, 10 molecules enter (the same number as in the steady state, hence, there is no change in the rate of synthesis) and 9 molecules leave (this represents one molecule less than in the steady state). As a consequence of this imbalance, the pool expands by one molecule. The turnover of the pool decreased since only 9 molecules left and this loss was balanced by 9 of the total 10 molecules which entered the pool. In the *nonsteady state of negative growth*, the number of molecules leaving the pool exceeds those entering it. In the first example, 10 molecules entered (hence no change in the synthesis rate as compared to the steady state) and 11 molecules left the pool. (This represents an increase by 1 molecule as compared to the steady state). The size of the protein pool decreased by one but its turnover remained unchanged, since the 10 entering molecules were balanced by 10 molecules leaving just as in the steady state.

$$k_s = k_g + k_t \qquad (5)$$

In contrast, in the expanding system, an amino acid leaves the pool of protein P solely due to the degradation during turnover and, consequently, $k_d = k_t$. To define the growing system one has to determine k_s and k_g and to calculate the parameter, k_t, which also equals k_d by using Equation 5.

Figure 2 also demonstrates that the positive growth can be achieved by changes in the rates of both protein synthesis and degradation (which in this case equals the turnover as derived above). Thus, the pool P can expand by either increasing synthesis and not changing the turnover or by decreasing the turnover rate and not changing synthesis. The growth will also occur when the turnover rate declines more than the rate of synthesis. Alternatively, the pool P will expand when the rates of both synthesis and degradation increase, provided that the change in the first parameter is more than in the second one. The most rapid growth, however, will be achieved when the increase in the synthesis rate is accompanied by a decline in the degradation rate.

During negative growth the rate of protein degradation exceeds that of synthesis. In this case the incorporation of tracer amino acid occurs only due to the resynthesis of protein molecules which took place during turnover and, hence, $k_s = k_t$. In contrast, the amount of amino acid which leaves the protein pool is a sum of two processes: (1) the decline in the size of the pool P (i.e., negative growth, k_g) and (2) degradation which took place during turnover (k_t). Hence, the fractional degradation rate equals:

$$k_d = k_g + k_t \tag{6}$$

Similarly, as in the case of positive growth, both rates of synthesis (which in these cases equals the turnover) and degradation influence the rate at which the size of the protein pool declines. It can be seen in Figure 2 that the same loss occurs either when degradation increases in the absence of any change in turnover or when turnover (i.e., resynthesis) declines in the absence of any change in degradation, or when both rates increase but the magnitude of the change in degradation is larger than that of the turnover. Again, reciprocal change in turnover and degradation result in the fastest loss of protein mass.

IV. STUDIES OF PROTEIN SYNTHESIS

When tracers are used for measurements, the rate of protein synthesis is given by the rate of tracer incorporation into the protein with a consequent change in the protein specific radioactivity. As shown above, in the steady state and during negative growth, the incorporation of the tracer amino acid takes place solely during resynthesis of the protein degraded during turnover. During positive growth, however, both resynthesis (balanced by degradation during turnover) and expansion of the protein pool (i.e., addition of new protein molecules) contribute to tracer incorporation and, hence, to the synthesis rate.

Calculation of the synthesis rate by using Equation 1 is simplified when the precursor specific radioactivity does not change as could be expected in cell cultures. When a constant (C) is substituted for the precursor specific radioactivity F*, integration of Equation 1 yielded $P^* = C\,(1 - e^{-k_s t})$. This expression can be rearranged as follows:

$$P^*/C = (1 - e^{-k_s t}) \tag{7}$$

Evaluation of Equation 7 using selected values of k_s and the length of labeling (t) allows one to construct a simple graph or table relating the precursor/product ratio (P*/C) to k_s for each time t. Next, one measures the P*/F* ratio in cultured cells (F*/C) and the fractional synthesis rate is read directly from the graph or table.

For Equation 7 to provide an accurate estimate of k_s, a proper precursor compartment has to be measured. Since aminoacyl tRNA (rather than the specific radioactivity of the tracer amino acid in the medium) is the immediate precursor of protein, a certain time has to elapse for the tRNA to be charged. For experiments with cultured single cells, this process might be so rapid that no substantial error is introduced if the precursor is assumed to remain constant. However, it is advisable to verify this assumption by direct comparison of the specific radioactivities of the amino acids in the media and tRNA pool. The dansylation procedure of Airhart et al.[12] is well suited to this goal. In a case where the specific radioactivity in either precursor pool changes with time, one has to solve Equation 1 by using fitted curves for F* and P*. Equation 1 is then integrated graphically between the times t_1 and t_2. The estimate of k_s is obtained from:

$$k_s = \frac{P^*(t_2) - P^*(t_1)}{\displaystyle\int_{t_1}^{t_2} F^*(t)\,dt - \int_{t_1}^{t_2} P^*(t)\,dt} \tag{8}$$

When a specific mathematical expression for F* can be obtained, a computer assisted solution for Equation 1 can be used as described previously (e.g., Equation 3 in Reference 8).

An alternative method suitable for the assessment of fractional synthesis rates in cultured cells is the equilibrium approach.[13] The cells are incubated for a period of several days (depending on growth rates of a given type of cell) in radiolabeled medium and the protein specific radioactivity is determined as a function of time. If the precursor radioactivity in the medium remains constant over this period (which can be achieved by daily changes of the medium), the protein radioactivity increases asymptotically until a plateau is reached where the protein and media specific radioactivities become equal. (In a study of cultured chick embryonic cardiac myocytes, this occurred after 4 to 5 days of culture.[13]) The fractional synthesis rate can be calculated from the following equation:

$$P^*(t) = P^*(max)(1 - e^{-k_s t}), \tag{9}$$

where P* (max) is the plateau of the protein specific radioactivity. All the other terms are the same as those defined previously (Section II).

The main advantage of this approach is that the method is independent of precursor specific radioactivity. Moreover, any deviation of experimental data from Equation 9 during early incubation times can be used to evaluate the degree of amino acid reutilization. In a case where there is a slow equilibration between an amino acid in the medium and in the tRNA pool, or if the latter pool receives the amino acid preferentially from protein degradation, the experimental values will be smaller than predicted by Equation 9. The drawback of this method is that the cumulative value of k_s is only obtained over relatively long incubation times (usually several days).

Another method available combines the equilibrium approach and pulse labeling.[13] The cells are first cultured in the presence of one radioisotope of a given amino acid, e.g., ^{14}C-leucine, until a plateau of protein radioactivity is reached. At this point the amino acid in the protein and in the medium have equal specific radioactivities. Once the cells reach such an equilibrium, a second radioisotope is added (e.g., H[3]-leucine) and the cells are pulse labeled for a short period of time (e.g., 15 min to 4 hr). The protein is next isolated and the ratio of pulse/equilibrium label is determined. Since the ratio of the two radioisotopes is directly proportional to the specific readioactivities of F* and P*, Equation 8 can be used to calculate the fractional synthesis rate after substituting the isotope ratio for values of F* and P*.

The main advantage of the double isotope method is that no quantitation of the amino acid is necessary. The applicability of this method, of course, depends on the demonstration that the determined value of F* closely approximates to the true precursor of protein synthesis — the aminoacyl tRNA. In a case where the specific radioactivities of the amino acid in the media and in the aminoacyl-tRNA pools are grossly different, one has the option of "expanding" the extracellular pool by increasing the concentration of labeled amino acid. In studies of perfused organs,[16] it has been shown that the equilibrium between the two compartments is reached rapidly in such cases. The concentration of the labeled amino acid, however, must be higher than its physiological concentration in the serum and the investigator has to carefully consider whether or not this alters the rates of protein synthesis or degradation.

When the estimate of k_s is to be used to obtain additional parameters, namely that of the fractional turnover rate (see Equation 5), one also has to determine the fractional growth rate, k_g. This requires quantitation of specific protein as a function of time. The fractional growth rate represents a fraction of product, P, at time, t, by which the total number of P molecules have expanded:

$$k_g = \frac{[dP(t)/dt]}{P(t)} \tag{10}$$

To quantitate a specific protein, the investigator has several options in addition to the obvious one, such as spectrophotometric measurements of protein with defined spectra, e.g., cytochrome c. One option is to use radioimmunoassay,[14] since specific antibodies including monoclonal ones are becoming more and more available. The second option is to use the radioisotope dilution method.[14] In this case, standard protein (e.g., myosin heavy chain) is radiolabeled either in vivo or by covalent modification (e.g., iodination). The specific radioactivity of the standard is determined next and the known amount of the standard is added to the sample (unknown) in which the content of the same protein is to be determined. After thorough mixing, the protein is isolated, purified, and the fall in specific radioactivity is determined. The amount of unknown protein then equals:

$$\left(\frac{\text{Spec. radioac. of standard}}{\text{Spec. radioac. of standard + unknown}} - 1 \right) \times \text{amount of standard protein added}$$

This method is particularly suitable for protein separated by SDS-polyacrylamide gel electrophoresis, since the specific radioactivity can be determined by using excised bands of the desired protein. One has, however, to determine whether the harsh conditions of SDS denaturation did not result in the loss of label (as in the case of iodination).

When several proteins are to be studied, one can simplify the procedure because only one protein is quantitated and its ratio to the other protein is determined on SDS-gels, either by densitometric tracing of the stained proteins or from quantitative binding of [125]-labeled Coomassie blue to proteins.[14] It is important, however, to determine the stoichiometry of dye binding to each protein.

When working with cell suspension culture one can determine k_g by the replacement perfusion method.[15] In this case the cell density is maintained constant by the continuous removal and replacement of the media which remains at a constant volume. When the equilibrium of in- and outflow is reached, the following equation applies:

$$k_g = k_f = f/V \tag{11}$$

where k_f is the fractional flow rate in h^{-1}, f is flow rate in $m\ell/h$ and V is the volume of culture media in $m\ell$.

V. STUDIES OF PROTEIN DEGRADATION

Measurement of protein degradation is more difficult than estimating the synthesis rate, mostly because the degradation product, the free amino acid, is recycled back into the protein molecule. Consequently, the measurement of amino acid production underestimates the rate of protein degradation.

The production of labeled amino acid can be estimated using cultured cells by prelabeling the cells with the radioactive amino acid and then measuring the radioactivity in a combined deproteinized cell extract and in the medium before and after incubation.[17] A similar approach can be used to assay the production of free amino acids by chemical assay.

Several approaches have been used to minimize the reutilization of the amino acid produced during protein degradation. Some investigators have used chase with nonradioactive isotopes of tracer amino acid. In this case it is hoped that a large amount of nonradioactive amino acid added to the medium will dilute the radioisomer that is produced during degradation of intracellular proteins. Once reutilization of tracer is prevented, the rate of fractional degradation can be calculated using an equation for first order decay:

$$P^*(t) = P_0^* \, e^{-k_d t} \tag{12}$$

where P^*_o is protein specific radioactivity at time 0, i.e., before the chase has started.

The usefulness of the chase experiment, however, has to be evaluated by other means. High levels of some amino acids, such as leucine,[18,19] trypotophan,[20] and phenylalanine,[20] are known in some systems to reduce protein degradation. In cultures of cardiac myocytes derived from chicken embryonic heart, however, the chase methods have been shown to give identical results as pulse labeling with tracer leucine.[13]

The second alternative to prevent tracer reutilization is to block the protein synthesis with specific inhibitors such as puromycin[21] or cycloheximide.[22] The hazard of this approach is that in some cases, inhibition of protein synthesis has been shown to influence protein degradation (both stimulation and inhibition have been reported[23]). The second hazard is that blocking the protein synthesis results in the elimination of fast turning over proteins. Since the half-lives of some key regulatory enzymes are very short, they become rapidly eliminated, resulting in an artificial situation.

The procedures in which the probability of tracer recycling is the lowest are those in which the tracer is post-translationally modified — such as methylation of histidine, which takes place during myosin and actin biosynthesis. The modified amino acid is neither metabolized nor can it be incorporated into the nascent protein. Consequently, it is quantitatively released from the organism[24] and, presumably, the cells as well, although the method has yet not been used with cultured cells, probably because of its low sensitivity. (There are only three 3-methylhistidines per myosin molecule compared to 93 leucine residues.) The second drawback of this method is that it does not allow one to differentiate between the degradation of different proteins. Moreover, only some proteins are methylated — such as myosin heavy chains of fast skeletal muscles, while methylated amino acids are not present in cardiac myosin.

VI. CONCLUSION

The common kinetic principle underlying all tracer techniques in measurements of protein synthesis and degradation, is that the rate by which the radioactivity changes in the pool of protein molecules depends on the difference between the radioactivity entering due to protein synthesis and leaving due to protein degradation (Equation 1). The direction of this flux depends on the difference between the specific radioactivity of the tracer in the precursor (F*) and the product (protein, P*) pools. When F* exceeds P*, this specific radioactivity of the protein increases with time since its nonradioactive amino acid is being replaced by the radioactive one. On the other hand, when P* exceeds F* the radioactive amino acid within the protein molecule is replaced by that of the precursor pool which has a higher proportion of nonradioactive amino acids than the proteins do. The changes of protein specific radioactivity in each direction have to follow exponential functions, but over the short time interval of the experiment it can usually be fitted by a straight line. In cases where there is a preferential reutilization of the amino acid which enters the intracellular pool (due to protein degradation) for synthesis of proteins, the entry of radioactivity into the protein (when F* > P*) and the loss of radioactivity (when P* > F*) will both be delayed with a consequent error in estimates of protein synthesis and degradation. This can be avoided by measuring the immediate precursor of protein synthesis, the aminoacyl tRNA.

The most common confusion in interpretation of synthesis and degradation concerns the systems in which the size of the protein pool changes with time (nonsteady state), as is commonly the case with cultured cells. One way to avoid this confusion is to view the radioactivity entering the pool of protein molecules by two conceptually different, yet physically identical (formation of peptide bond) pathways. One is the formation of new peptide bonds during expansion (growth) of the protein pool. The other pathway is the resynthesis of protein molecules which were degraded due to the turnover. Turnover, defined as degradation balanced by resynthesis, takes place both in the absence of growth (steady state) as well as during growth (both positive and negative). The rate of radioactivity entry (synthesis rate) thus has a different meaning depending on the growth situation. In nongrowing and declining systems the synthesis reflects the rate of turnover. In an expanding system, however, it reflects turnover plus growth. Similarly, the rate of radioactivity lost (degradation rate) in nongrowing and expanding systems reflects the rate of turnover, while in the declining system it reflects turnover plus the rate of protein loss.

ACKNOWLEDGMENTS

This research was supported in part by the National Institute of Health grants HL-20592 and HL-16637 and by the Muscular Dystrophy Association.

REFERENCES

1. **Hod, Y. and Hershko, A.,** Relationship of the pool of intracellular valine to protein synthesis and degradation in cultured cells, *J. Biol. Chem.,* 251, 4458, 1976.
2. **Righetti, P., Little, E. P., and Wolf, G.,** Reutilization of amino acids in protein synthesis in HeLa cells, *J. Biol. Chem.,* 2416, 5724, 1971.
3. **Zilversmit, D. B.,** The design and analysis of isotope experiments, *Am. J. Med.,* 832, 848, 1960.
4. **Reiner, J.,** The study of metabolic turnover rates by means of isotopic tracers, *Arch. Biochem.,* 46, 53, 1953.
5. **Russell, J. A.,** The use of isotopic tracers in estimating rates of metabolic reactions, *Perspective in Biol. and Med.,* 1, 139, 1958.
6. **Koch, A. L.,** The evaluation of the rates of biological processes from tracer kinetic data, *J. Theor. Biol.,* 3, 283, 1962.
7. **Waterlow, J. C., Garlick, P. J., and Millward, D. J.,** *Protein Turnover in Mammalian Tissues and in the Whole Body,* Elsevier/North-Holland, New York, 1978.
8. **Zak, R., Martin, A. F., and Blough, R.,** Assessment of protein turnover by use of radioisotopic tracers, *Physiol. Rev.,* 59, 407, 1979.
9. **Kleiver, M.,** Meaning of ''turnover'' in biochemistry, *Nature,* 175, 342, 1955.
10. **Zilversmit, D. B.,** Meaning of ''turnover'' in biochemistry, *Nature,* 175, 863, 1955.
11. **Mawson, C. A.,** Meaning of ''turnover'' in biochemistry, *Nature,* 175, 317, 1955.
12. **Airhart, J., Vidrich, A., Khairallah, E. A.,** Compartmentation of free amino acids for protein synthesis in rat liver, *J. Biol. Chem.,* 140, 539, 1974.
13. **Clark, W. A. and Zak, R.,** Assessment of fractional rates of protein synthesis in cardiac muscle cultures after equilibrium labeling, *J. Biol. Chem.,* 256, 4863, 1981.
14. **Everett, A. W., Prior, G., Clark, W. A., and Zak, R.,** Quantitation of myosin in muscle, *Anal. Biochem.,* 130, 102, 1983.
15. **Spanier, A. M., Clark, W. A., and Zak, R.,** Replacement perfusion of cultured eucaryotic cells: a method for the accurate measurement of the rates of growth, protein synthesis and protein turnover, *J. Cell. Biochem.,* 26, 47, 1984.
16. **Mortimore, G. E., Woodside, K. H., and Henry, J. E.,** Compartmentation of free valine and its relation to protein turnover in perfused rat liver, *J. Biol. Chem.,* 247, 2776, 1972.
17. **Zak, R. and Drahota, Z.,** Release of methionine labeled with sulphur-35 from muscle tissue and mitochondria, *Nature,* 186, 973, 1960.
18. **Buse, M. G. and Weigand, D. A.,** Studies concerning the specificity of the effect of leucine on the turnover of proteins in muscles of control and diabetic rats, *Biochim. Biophys. Acta,* 475, 81, 1977.
19. **Fulks, R. M., Li, J. B., and Goldberg, A. L.,** Effect of insulin, glucose, and amino acids on protein turnover in rat diaphragm, *J. Biol. Chem.,* 250, 290, 1975.
20. **Hopgood, M. F., Clark, M. G., and Ballard, F. J.,** Inhibition of protein degradation in isolated rat hepatocytes, *Biochem., J.,* 164, 399, 1977.

21. **Schreiber, S. S., Oratz, M., Evans, C., Reff, F., Klein, I., and Rothschild, M. A.,** Cardiac protein degradation in acute overload in vitro: reutilization of amino acids, *Am. J. Physiol.,* 224, 1973.
22. **Woodside, K. H., Ward, W. F., and Mortimore, G. E.,** Effect of glucagon on general protein degradation and synthesis in perfused rat liver, *J. Biol. Chem.,* 249, 5458, 1984.
23. **Goldberg, A. L. and St. John, A. C.,** Intracellular protein degradation in mammalian and bacterial cells, part 2, *Ann. Rev. Biochem.,* 45, 747, 1976.
24. **Bates, P. C., Grimble, K. G., Sparrow, M. P., and Millward, D. J.,** Myofibrillar protein turnover. Synthesis of protein-bound 3-methylhistidine, actin, myosin heavy chain and aldolase in rat skeletal muscle in the fed and starved states, *Biochem. J.,* 214, 593, 1983.

Chapter 13

ISOMYOSINS, MICROTUBULES, AND INTERMEDIATE FILAMENTS

Ketty Schwartz and Lydie Rappaport

TABLE OF CONTENTS

I. INTRODUCTION

One of the characteristics of skeletal and cardiac muscles is their ability to adapt to the amount and type of work they are required to perform, both in physiological and pathological situations. The adaptative response of myocytes to new functional requirements involves several mechanisms including cell multiplication, cell growth and hypertrophy of pre-existing myocytes, changes in the pattern of expression of multigene families leading to a modified phenotype, and any combination of the above processes. The correlations between contractile activity and the biochemical changes induced by these mechanisms constitute an area of major interest to both biologists and clinicians. However, the search for the causal relationship between physiological function and the alterations in gene expression has only just started, and little is known as yet about the regulation of gene activity during muscle cell growth. The aim of this chapter is to describe and analyze two biochemical mechanisms that undergo changes during cardiac myocyte development: (1) isomyosin shifts and (2) cytoskeletal reorganization.

Myosin was the first contractile protein to be identified; it forms filaments and its role as the enzyme that catalyzes the transformation of chemical energy in the form of ATP into the mechanical work of muscle contraction is well established. It is a multisubunit protein, composed of two heavy chains (M.W. = 200,000), two light chains which may be phosphorylated (M.W. = 18,000 to 20,000), and two light chains which may not be phosphorylated (M.W. = 16,000 to 27,000). The existence of several isomyosins consisting of different heavy and light chains and different Ca^{2+}- and actin-stimulated ATPase activities was demonstrated more than 20 years ago when adult fast and slow skeletal muscles were compared to cardiac muscle. The differences in the ATPase activities of the various skeletal isomyosins account for certain differences in the physiological properties of muscles. For instance, in several types of skeletal muscle, the maximum velocity of shortening is correlated to myosin ATPase activity. Much less was known until recently about cardiac isomyosins, but advances in molecular and cell biology have enabled rapid development in this field during the past few years. Thus, the heart, in various animals, has been shown to contain several isomyosins whose relative amounts change according to developmental, hormonal, physiological, or pathological factors. Cardiac myocytes have now become a particularly attractive system for studying both the genetic expression of the myosin heavy chain multigene family, and the effect of isomyosin phenotypes on the mechanical properties of heart tissue.

The putative role of the cytoskeleton in controlling cardiac muscle cell development and myofibrillar assembly will also be considered in this chapter. The relationship between microtubules and microfilaments and contractile elements is now clear, and some aspects of muscular contraction seem to depend directly on the integrated dynamic actions of the plasma membrane, mechanical support of organelle distribution with respect to myofibril cross-striation and contraction-relaxation processes. Many of the proteins implicated in these processes have recently been characterized, including microtubular components, which probably play a key role. Much of the progress in this field in recent years may be attributed to improved immunological methods, allowing analysis of proteins present in only trace amounts.

Most of the research work to be discussed in this chapter was performed on intact muscles or isolated myocytes, with very few investigations on cardiac cell cultures. Since much evidence indicates heterogeneity and plasticity of cardiomyocytes, many future investigations of the environmental factors and mechanisms involved in muscle contraction will probably be performed in neonate or adult cell cultures, since the necessary techniques are already available.

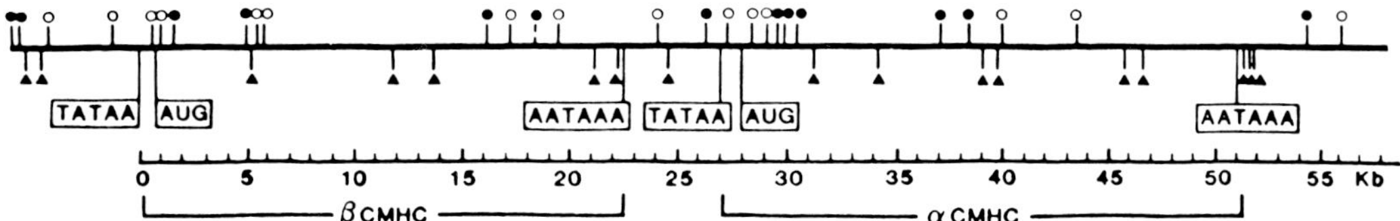

FIGURE 1.   Organization of α- and β-myosin heavy chain genes in the rat genome (5′ to 3′ orientation of the chromosomal map). Restriction endonuclease sites: ▲ BamHI; ● EcoRI; ○, HindIII; ⋯ ●, polymorphic EcoRI site in DNA. (From Mahdavi, V., Chambers, A. P., and Nadel-Ginard, B., Cardiac α- and β-myosin heavy chain genes are organized in tandem, *Proc. Natl. Acad. Sci. USA*, 81, 2626, 1984. With permission.)

II. ISOMYOSINS

A. Heavy and Light Chain Heterogeneity

Two classes of myosin heavy chains, referred to as α- and β-HC, respectively, are expressed in rabbit, rat, pig, beef, and human myocardium (Table 1). There are slight but well-defined differences in their peptide maps,[1-8] immunological properties,[3,4,9-14a] and mRNA sequences.[15-18] Comparison of their peptide maps, both in the absence and in the presence of specific immunoglobins, showed that the differences occurred at many sites in the molecule, within subfragments 1 and 2 (S1, S2), and in heavy meromyosin (HMM) and light meromyosin (LMM). This has been confirmed by nucleotide sequence analysis of the corresponding cDNA clones. In the rat, the transcripts of the light and heavy chain genes exhibit 95% homology in the coding region, but are clearly different in the untranslated 3′ ends,[15,16] where an additional untranslated sequence of 11 nucleotides is present in the α mRNA. The fragment corresponding to the LMM portion is very similar in both α- and β-chains, and almost completely identical stretches are interrupted by single nucleotide changes that occur at intervals of 25 to 100 bases. However, two regions are strikingly different: nucleotides 360 to 433 (with 30% divergence) and the 3′ end. The amino acid compositions derived from cDNA clones show more homology, since half the nucleotide substitutions occur in the third position and do not change the corresponding amino acid. The few changes in amino acids generally maintain the hydrophobicity and/or charge, and do not change the characteristic periodicity of LMM in the distribution of nonpolar groups and charged residues. Myosin heavy chains in the rabbit also show a high degree of homology.[16,17] As in the rat, the 3′ nontranslated sequences of the two rabbit isomyosin mRNAs are highly divergent. Clones corresponding to the center of the myosin heavy chain (subfragment 2 and the COOH-terminal portions of subfragment 1) exhibit 90% homology, both with respect to the nucleotide and derived amino acid sequences. Recent observations by Lompré et al.[19] and Sinha et al.[20] show the possible existence of two other myosin heavy chain mRNAs, α′ and β′. It is not clear, however, whether these new transcripts are the products of other genes that have not been identified with the probes used, or the result of an alternative splicing pathway of the gene primary transcripts. Therefore, although the existence of the α′- and β′-myosin HC genes cannot be ruled out, presently available data strongly suggest that the cardiac myosin heavy chain phenotype can be entirely accounted for by the expression of only two myosin heavy chain genes. These genes are linked in the genome, and span 50 kilobases of the chromosome, the α-gene being located 4 kilobases away from the β-gene (Figure 1).[18]

At the level of the light chain, several electrophoretically distinguishable isoforms exist, a characteristic feature of the adult heart being the difference in the apparent molecular weights of atrial and ventricular chains.[21-23] Two alkali light chains were found in more than seven animal species, one in the atria and one in the ventricles (ALC1 and VLC1, respectively, according to the terminology of Price et al.[23]). ALC1 cannot be distinguished from

Table 1
SCHEMATIC REPRESENTATION OF THE
SUBUNIT COMPOSITION OF CARDIAC
ISOMYOSINS

	Heavy chains	Light chain 1	Light chain 2
Ventricles			
V1	αα	VLC1	VLC2
V2	αβ	VLC1	VLC2
V3	ββ	VLC1	VLC2
Atria			
A1	αα	ALC1	ALC2
A2	ββ	ALC1	ALC2

LClemb, another alkali light chain present in embryonic skeletal muscles.[24] The phosphorylatable light chains ALC2 and VLC2 also differ in atrial and ventricular tissues.[23] In humans, each of these subunits can be resolved into several bands under specific electrophoretic conditions. The nature (proteolysis, isoforms, and genetic differences) of these bands is not clear.[23] Three types of microheterogeneity were also found in human VLC2.[25]

B. Isomyosin Structure

Each myosin molecule is composed of two heavy chains and four light chains, two alkaline ones and two phosphorylatable ones, which could give rise to a large number of different isomyosins by combinations of the various types of chain described above. So far, five isomyosins have been identified, three in the ventricles and two in the atria (Table 1).

Hoh et al.[26] were the first to achieve electrophoretic separation of the ventricular isomyosins under nondissociating conditions. The terminology of these authors is now widely accepted: these isomyosins are designated as V1, V2, and V3, depending on their electrophoretic mobility. They are heavy chain isoforms — V1 and V3 being composed of α-α and β-β homodimers, respectively, whereas V2 is an α-β heterodimer.[2,27] All three contain the same light chains, VLC1 and VLC2. V1, V2, and V3 exhibit clearly different Ca^{2+}-dependent and actin-dependent ATPase activities, in the order V1 > V2 > V3.[8,9,26,28-31] It should, however, be pointed out that precise comparison has only been made between V1 and V3, which can be obtained in pure form from 3-week-old and adult hypophysectomized rats or from hypothyroid and hyperthyroid rabbits (see below). The intermediate activities of V2 were deduced from gel analysis, which is not very rigorous, but it has not yet been possible to isolate this isozyme. The activities of V1 and V3 differ by a factor of 3.5 to 4, in both rat and rabbit.[8,28,31]

Two isozymes from atrial tissue, A1 and A2, have been electrophoretically resolved. Both of these differ from V1, V2, and V3.[26] The nature of these two bands is not clear, but there is increasing evidence that the atria contain α- and β-ventricular HC. Atrial and ventricular isomyosins show strong immunological cross-reactions and very similar enzymatic properties.[13,32-34] Moreover, in rats and rabbits, the ventricular myosin heavy chain mRNA sequences are expressed in the atrial tissue.[19,20] The most likely hypothesis at the present time is that the atrial isomyosins are composed of α- and/or β-heavy chains (HC) which are associated in different ways with the atrial light chains ALC1 and ALC2, which would account for the different electrophoretic mobilities of atrial and ventricular isomyosins.

C. The Nature and Regulation of Isomyosin Shifts

Cardiac myosin isozyme composition has been observed to change under a number of physiological and pathological conditions, such as ontogenic development, increased cardiac work, exercise, and hormonal disturbances (Table 2).

Table 2
MYOSIN HEAVY CHAIN DISTRIBUTION IN HEART UNDER VARIOUS PHYSIOLOGICAL AND PATHOLOGICAL CONDITIONS

	Rat		Rabbit		Man, pig, and beef	
	Atria	Ventricles	Atria	Ventricles	Atria	Ventricles
Ontogenic development						
In-utero	αβ	αβ	—	β	—	αβ
Birth	αβ	αβ	—	αβ	—	αβ
Young	α	α	α	αβ	αβ	αβ
Adults	α	αβ	α	β	αβ	αβ
Hypothyroidism	αβ	β	α	β	—	—
Hyperthyroidism	α	α	α	α	—	—
Diabetes	—	αβ	—	—	—	—
Diabetes + Insulin	—	αβ	—	—	—	—
Hemodynamic overload of the ventricle	α	αβ	—	β	—	β
Hemodynamic overload of the atrium	α	—	—	—	αβ	—

Note: The characters in small type (α and β) indicate small amounts of the corresponding heavy chains.

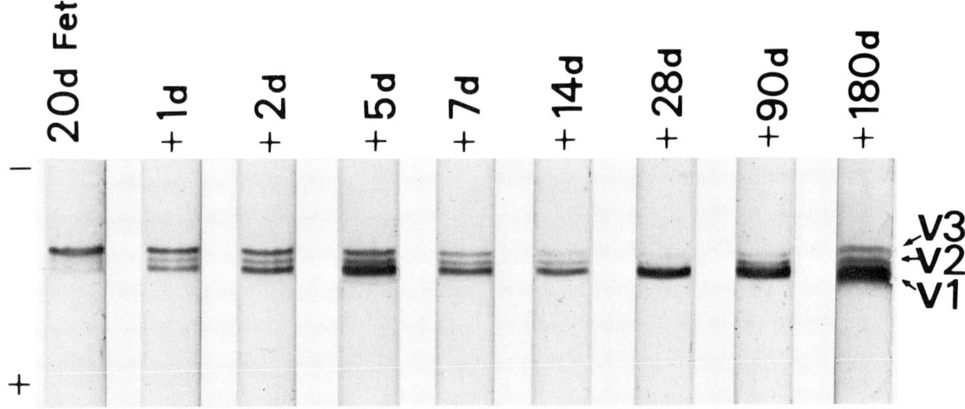

FIGURE 2.   Age-dependent evolution of cardiac isomyosins in the rat. Typical electrophoretic patterns obtained from both ventricles of fetuses and from left ventricles of neonates and adults. (From Lompré A. M., Nadal-Ginard, B., and Mahdavi, V., Expression of the cardiac α- and β-myosin heavy chain genes is developmentally and hormonally regulated, *J. Biol. Chem.*, 259, 6437, 1984. With permission.)

1. Ontogenic Development

The developmental distribution of the various cardiac myosin chains depends on the animal species and their tissue distribution, in the ventricle or atrium. Mainly the β HC occurs in the ventricles of the fetuses of all the species tested so far; however, during late gestation, both heavy chains are present, but the β chain is always more abundant.[4,29] In pig, guinea pig, beef, and man, no major changes occur at or immediately after birth, and the β chain is always markedly predominant. In adult man, there is little regional variation in the proportion of fibers containing α chains in the right and left ventricles within the same heart.[13,14a] In contrast, specific regions in the bovine ventricular myocardium contain the α chain, including the free wall of the right ventricle and some Purkinje fibers.[11] In rats (Figure 2) and rabbits, the relative amount of heavy chains (or V1) markedly increases after birth and subsequently declines in adult rabbits, whereas it remains predominant in adult rats.[26,29] A specific and rather similar regional distribution of α and β chains is also found in these two animal species.[10,11] Thus, they display the same trend as the bovine heart, in as far as the right ventricle contains more α chains than the left ventricle, and small bundles corresponding to the Purkinje strands also contain α chains. In contrast to bovine heart, fibers containing α chains are present in both the right and left ventricles of rat and rabbit, mostly in the subepicardial layers. The distribution of α and β heavy chains at the cellular level was elucidated by double immunolabeling of isolated caridac myocytes and cultured heart cells (Figure 3).[36] Three types of myocytes exist, containing α chains only, β chains only and both types of chain, respectively. In the latter case, α and β chains are identically distributed, since the same myofibrils react with both antimyosin immumoglobulins. These data are consistent with earlier histochemical observations of cryostat sections, showing that these two isoenzymes are uniformly distributed within each ventricular cell.[37] A striking feature of heavy chain isomyosin distribution in ventricular tissue is its heterogeneity in different myocytes, as compared to its homogeneity within the same myocyte.

In the atria, the relative proportions of α and β HC also vary considerably in the tissues of different animal species. In small animals such as rats and rabbits, atrial tissues are almost completely homogeneous, since practically all the fibers contain α chains, and only a few β chains.[34] In bovine and human heart, greater heterogeneity is observed, since most fibers of the left atrium contain α chains while most right atrium fibers contain both types of chain.[14,34] The interatrial septum and specific regions of the right atrium such as the crista terminalis and the main conduction pathways between the sinus and atrioventricular nodes only include β chains (or at least heavy chains which cross-react with anti-β immunoglobulins).

There are very few experimental studies on the developmental regulation of myosin light chains, and rather surprisingly, the evidence of changes comes from man and beef.[35,38] Fetal and adult atria contain the same characteristic light chains, ALC1 and ALC2. The fetal ventricles also contain ALC1, which is a major component throughout the latter half of gestation and subsequently slowly disappears in the postnatal period. Adult ventricles contain only VLC1 and VLC2, whereas adult atria, particularly the left one, contain small amounts of VLC1 and VLC2 in addition to ALC1 and ALC2. The precise distribution of these light subunits within different fiber types is not known, probably because specific immunoglobulins have not yet been isolated.

2. Hormonal Imbalances

In the ventricles, administration of thyroid hormone (T3 or T4) stimulates the synthesis of the α heavy chain and represses that of the β chains, whereas thyroidectomy or hypophysectomy has the reverse effect.[26] The extent of the change depends on the initial phenotype: for example, administration of thyroid hormone has very little effect on the rat in which α chains already predominate, but markedly affects the rabbit, pig, and guinea pig, in which β chains predominate.[1,3,4,10-12,31,33] After thyroxine treatment, all ventricular fibers in the rabbit react with anti-α myosin HC immunoglobulins,[11] and the myosin pattern on native gels is the same in the papillary muscle and the free wall, showing that all myocytes respond simultaneously to the thyroxine trigger.[39] The nature of this trigger is unknown, but recent experiments show that naturally occurring and synthetic thyroid analogs both have the same effect, suggesting that the mechanism for cardiac isomyosin regulation might involve a primary signal, possibly related to dietary carbohydrate and modulated by thyroid hormone and glucocorticoids.[40] It is of interest that the burst of α-myosin HC synthesis, normally occurring after birth, is correlated to a rapid increase in the serum level of thyroid hormone; however, during later periods of development, the changes become independent of thyroid hormone, and during ontogenic development the regulation of isomyosin shifts is not directly linked to thyroid hormone levels.[27] Differentiation in vitro differs from that

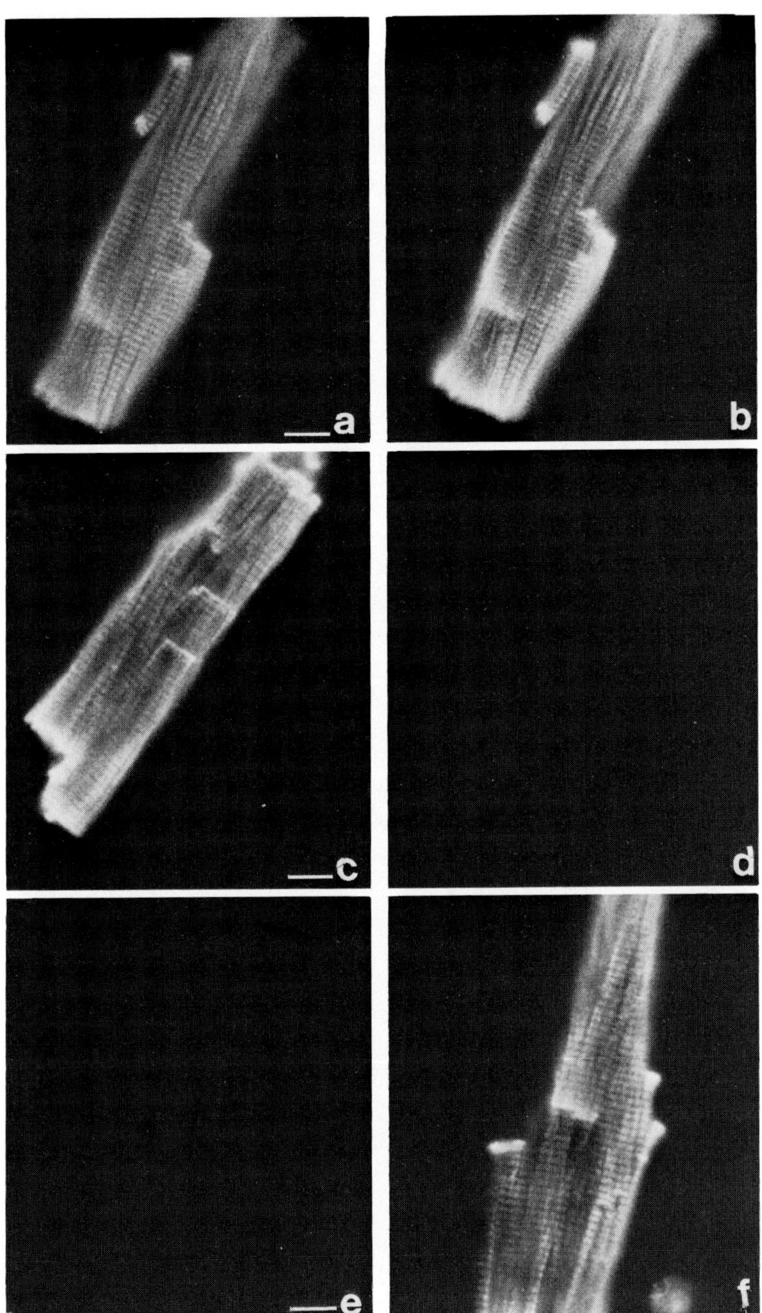

FIGURE 3.   Double indirect immunofluorescence micrograph of myocytes isolated from an 8-week-old rat. Cells were incubated with anti-VI myosin (a, c, and e) and anti-V3 myosin (b,d,f). Figures a and b show an example of a myocyte stained with both antibodies. Note that the reactive fibers exhibit the same overall morphology. In c and d, cells were stained with anti-V3 myosin only, whereas in e and f, cells were only stained with anti-VI myosin. (Bar = 10 μm).

in vivo, since embryonic rat cardiac myocytes after 7 days in culture contain predominantly V3. In a medium containing a physiological concentration of T4, isomyosin profiles shift to the in vivo pattern, with V1 predominating after day 4 of culture. This would support the hypothesis that the changes occurring at birth in vivo are greatly influenced by thyroid hormones.[41]

Differences in the ventricular light chain pattern or in the electrophoretic mobility and enzymatic activities of the atrial isomyosins due to thyroid hormone imbalance have not been observed.[33] However, recent data show that trace amounts of β chains are synthetized in the atria of hypothyroid neonate, young and adult rats, indicating that atria and ventricles respond similarly to thyroxine from a qualitative, but differently from a quantitative view point.[42]

Insulin has the same effect as thyroxine on isomyosin expression, since streptozocin-induced diabetes in rats is accompanied by a shift from the α to the β form.[43] In this model, identical results were obtained, whether the myosin heavy chains were isolated from purified myosin synthetized in the intact heart or in a cell-free system.[44] Furthermore, in diabetic rats, the reduced T4 and T3 levels do not explain the predominance of β myosin, suggesting that insulin is also able to modulate α- and β-gene transcription antithetically.

Finally, shifts from α to β-chains have been reported in gonadectomized male and female rats, suggesting that sex hormones might also exert a regulatory role.[45]

3. Hemodynamic Overload of the Heart

In contrast to thyroxine-induced hypertrophy, pressure overload hypertrophy results in a high proportion of β chains, and a low proportion of α chains. In this case too, the extent of change depends on the initial phenotype: thus in ventricles, the changeover is almost complete in the rat,[9,10,46,47] fairly extensive in the rabbit,[8] and only slight in pig and man.[13,14,48] In atria, which mainly have α-HC, recent observations show a dramatic change in patients with atrial hypertrophy due to mitral stenosis and/or mitral insufficiencies.[14,49] In rats, however, atrial hypertrophy secondary to ventricular overload does not change the electrophoretic mobilities of atrial isomyosins.[50] The different responses of man and rat could either be due to technical factors or to species variation, but the most likely explanation is that in man, the overload is directly imposed on the atria, while in rats it is only secondary.

Also, in contrast to thyroxine-induced hypertrophy, the isomyosin redistribution in the rat is asynchronous within the same heart, and varies depending on the location of the myocytes. In pressure overload of the left ventricle, the time-course of β isomyosin synthesis begins in the left papillary muscle, and then shifts to the left free wall and finally to the right ventricle.[51] Similar differences between the right and left ventricles were observed in the rat during renal hypertension, in addition to differences between the subendocardial region containing only β chains, and the subepicardial regions, which exhibit a mixed pattern of α and β chains.[10]

The response to volume overload varies according to the pathological state, and may either cause a slight increase in β chains or no change at all.[46] On the other hand, exercise (swimming or running) increases α chain expression.[52,53]

Shifts in light chain isoforms in response to cardiac pressure-overload hypertrophy were also found in human atria, where the normal adult isoforms, ALC1 and ALC2, are replaced by the ventricular isoforms, VLC1 and VLC2 (particularly VLC2 and a related form of this light chain).[38] In human hypertrophied ventricles, the reappearance of ALC1 has been reported.[54]

4. Regulation

Little is known at present about the molecular basis of the regulation of cardiac myosin heavy chain genes. In pioneering experiments, Hoh and Egerton[55] showed that the increased

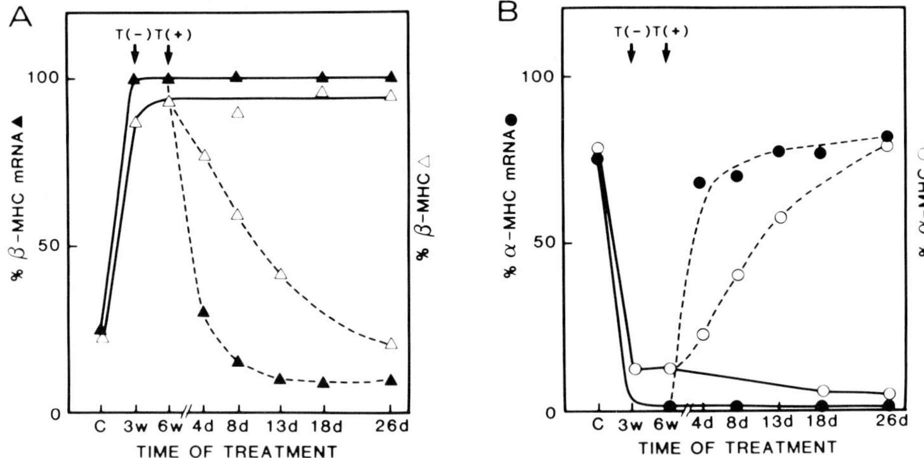

FIGURE 4.   Effect of thyroid hormone on the relative levels of ventricular α and β myosin heavy chains mRNA's and isozymes. Solid line, thyroidectomized rats; dashed lines, thyroidectomized rats injected daily with L-thyroxin α - MHC mRNA (●) and isozyme (○); β-MHC mRNA (▲) and isozyme (△). (From Lompré, A. M., Nadel-Girard, B., and Mahdavi, V., Expression of the cardiac ventricular α- and β-myosin heavy chain genes is developmentally and hormonally regulated, *J. Biol. Chem.*, 259, 6437, 1984. With permission.)

incorporation of labeled amino acids into V1 myosin in vitro, observed after administration of T3 to rats, was prevented by pretreatment with actinomycin, which suggested pretranslational control. A close correlation was indeed found several years later between the relative abundance of myosin heavy chain mRNAs and the corresponding isozymes[19,56] (Figure 4). Furthermore, mRNA levels are also correlated to the synthesis rates of the corresponding proteins.[56] The time lag observed between mRNA and protein decay and accumulation is due to the differences between the apparent half-lives of various mRNAs, approximately 3 days, and of proteins, which is 7 days. An important point is that the $t_{1/2}$ values for proteins and mRNAs are similar during the induction and deinduction processes, suggesting that changes in heavy chain stability do not play a significant role in inducing isomyosin shifts. Taken together, these data indicate that cardiac myosin heavy chain isoforms are antithetically regulated at the pretranslational level. It is not yet clear whether this is due to a change in the transcription rate or a change in mRNA processing.

D. Functional Implications of Isomyosin Shifts

The isomyosin shifts and the segregation of different forms of myosin in different types of cardiac muscle fibres might greatly affect heart physiology, and also have clinical implications. The contractile properties of cardiac fibers are largely determined by the type of myosin they contain. Thus, the velocity of shortening in rat and rabbit papillary muscles at zero load and 50% isometric tension is directly proportional to myosin ATPase activity and to heavy chain isomyosin content, the presence of V1 isomyosin corresponding to the highest contraction velocity.[57,58] This correlation is independent of muscle length and weight, and the resting tension present at 95% L_{max}. Over the entire force-velocity curve, papillary muscles containing 100% V1 isomyosin shorten more quickly than those containing 100% V3.[58] It is not clear why the ratio of unloaded shortening velocities is approximately six, whereas that of Ca^{2+}-stimulated ATPase activities is between three and four. This discrepancy could be due to differences in experimental conditions or to regulatory factors that are lost during myosin isolation and purification. The recent findings of Winegrad et al.[59] indeed show that a regulatory system sensitive to β-adrenergic stimulation does not survive the procedures used in isolating myosin.

Since V1 and V3 myosins regulate, at least partially, muscle shortening velocity, the homogeneity of isomyosin distribution within individual myocytes is rather puzzling as different regions of the cell should not contract at different velocities. α and β chains are structurally linked in both intact hearts and cultured cells — a strong indication that this linkage is a general feature of the sarcomeric organization of cardiac tissue.[36] One possibility is that the slower myosins might act as a drag on the force, in which case the consequence of the close coexistence of α and β chains would be an intermediate velocity between the high velocity contraction of pure α chains and the slow one of pure β chains.[59]

On the other hand, there are indications that cardiac isomyosins are regulated by cyclic AMP, and that the α heavy chain is preferentially stimulated by β-adrenergic activity.[60] A very good correlation exists between the relative amount of V1 myosin and the maximum Ca^{++}-activated force of contraction obtained by stimulation of the cAMP-dependent adrenergic system. The changes in Ca^{++}-activated ATPase occurs within as short a time as 30 sec; this might constitute a form of thick filament regulation of contraction, in addition to the calcium regulation of troponin on the thin filaments, and might provide a means of modulating the maximum contraction force.

In addition to force and velocity, a third parameter of contraction, the energy cost is also clearly related to the α/β chain ratio. The presence of V1 isomyosin is associated with increased tension-dependent heat, probably because cross-bridges with the more rapidly cycling V1 break down at a faster rate while tension is maintained.[61,62] In that case, hearts containing V1 would sacrifice economy to the ability to develop tension, and, in contrast, those containing V3 would develop tension more economically, but at the expense of contraction velocity.

The physiological implications of a mechanism by which the heart can modulate velocity, force, and energy cost are, of course, considerable. It should, however, be pointed out that, due to species differences, this mechanism seems to be effective only in the ventricles of small animals, and not in those of large ones such as man.

III. CYTOSKELETON

This term is commonly used to define several types of cytoplasmic filaments. Among the cytoskeletal elements so far identified in heart myocytes (Table 3), some are ubiquitous in cells, and others are specific for muscle tissue. They are defined either according to their size and appearance under the electron microscope, e.g., intermediate filaments and microtubules, or on the basis of their properties, e.g., actin-binding proteins. Purification of the proteins constituting these filaments has made the production of monospecific antibodies and their use in immunomicroscopy possible. Cytoskeletal elements appear in isolated myocytes as a pattern which may either be loosely distributed inside the sarcoplasm (Figure 5), or coincide with the myofibrillar Z-discs, or be linked to the sarcoplasmic membranes (Figure 6) or to the intercalated discs (Figure 6). Immunolabeling combined with either optical or electron microscopy has recently made it possible to establish that most of the Z-discs are composed of α-actinin and several other proteins.[63-69] The intermediate desmin filaments surrounding the Z-disc structures, form attachments between the myofibrils,[67-76] and are connected to the desmosomal plaques of the intercalated discs.[77] Several proteins, which play a key role in anchoring the actin filament bundles to the plasma membrane in nonmuscle cells,[78,79] also anchor the desmin filaments to the sarcolemma and the actin filaments at the level of the fascia adherens in heart.[71,78,80-82]

Other proteins, good candidates for having an organizational role in the maintenance of cellular structure, include titin and nebulin, that stretch over the whole length of the sarcomere and join the Z-discs together,[83] and myomesin present in the M-band, which may function as a kind of nucleation site for myosin filaments and constitute an anchoring site for thick

Table 3
MAJOR CYTOSKELETON PROTEINS IDENTIFIED IN HEART MUSCLE

Type	Mr × 10⁻³	Location	Presumed role	Ref.
Actin-binding proteins				
Vinculin	130	F. adherens	Sarcolemma-myofibr. binding	63, 65, 66
α-Actinin	105	F. adherens Z. disc	Promotes attachment of new actin filaments to the Z band	68
Filamin	250	Z. disc	Cross linker	64, 69, 88
Myosin-binding proteins				
Myomesin	165	M-band	Assembly and alignment of thick filaments	84, 85
Titin	10^4 (T_1, T_2)	A band and A-I junctions	Connects Z lines and lines longitudinally	83
Nebulin	5.10^3	N_2 lines		
Intermediate filaments (10 nm diam)				
Desmin	55	Intermyofibril lattice Z disc level	Transversal linkage of adjacent myofibrils	67-77
Desmin-associated proteins				
Synemin	240	Desmoplaques	Attachment of desmin filaments	70, 80
Spectrin	230	Subplasmalemmal region	to the sarcolemma	82
Tubulin	53 52	Intermyofibrillar network-nuclear memb.	Nuclear shape Distribution of organelles in single cells	106—108
MAPS 1—2	270—340		Binding of microtubules to organelles, tubulin polymerization	100—104

filaments.[84,85] The links between some of these cytoskeletal components with the myofibrillar structures are schematically represented in Figure 7. The description that follows chiefly concerns the intermediate filaments and microtubules in isolated rat heart myocytes and their distribution in growing myocytes together with the expression of the α- and β-HC isomyosins, intracellular markers of myocyte contractile activity.

A. Intermediate Filaments

Intermediate filaments (diameter: 10 nm, situated between actin filaments (6 nm in diameter) and myosin filaments (18 nm in diameter)), exhibit remarkable tissue specificity in their composition, even if they are indistinguishable by ultrastructural criteria. Two types have been recognized in heart, desmin and vimentin filaments, which differ in their biochemical composition and immunological properties (for reviews, see References 67 to 72 and 85). Desmin (55 kDaltons) was first identified in Purkinje cells[70,87] and is characteristic of adult muscle,[88-93] while vimentin, typical of mesenchymally derived cells,[72,73] coexists with desmin in embryo muscle cells,[92,93] and is a component of nonmuscle cell intermediate filaments in postnatal muscle (Figure 8).

Intermediate filaments have the common feature of possessing relatively low solubility. They are purified by tissue extraction, first with detergents and then high-salt buffers, which remove most of the proteins. After two-dimensional polyacrylamide gel electrophoresis desmin is resolved into three spots which have different charges, reflecting different states of phosphorylation of the molecule.[72] The concentration of desmin in adult heart muscle,

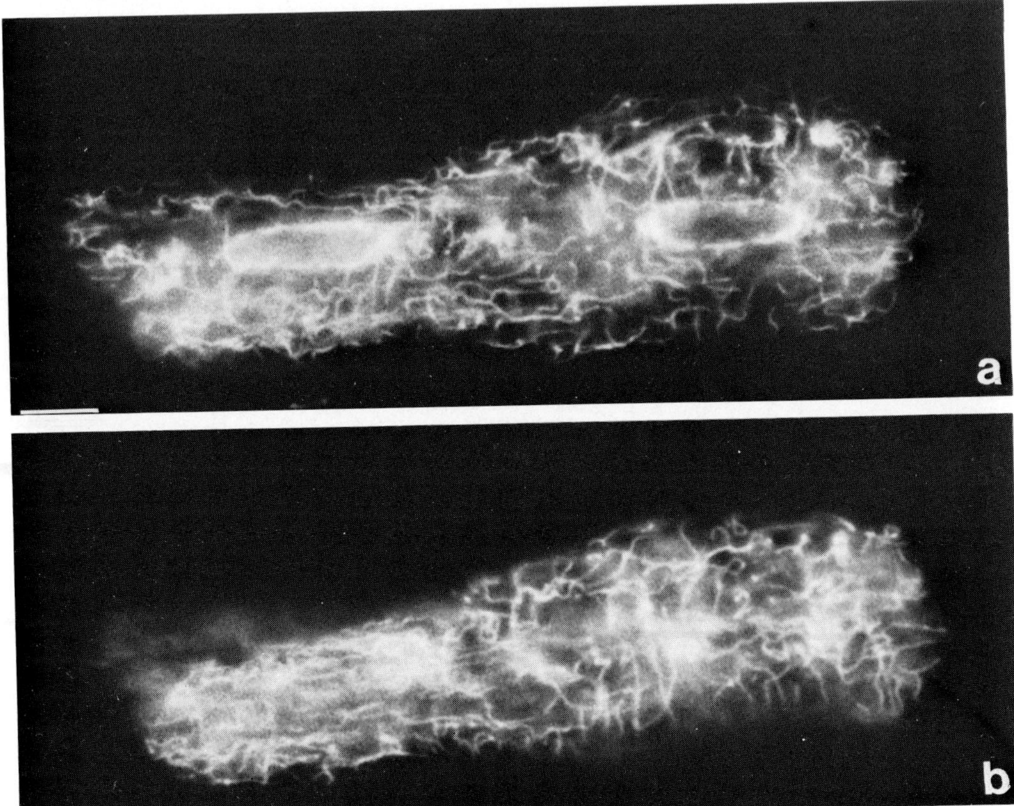

FIGURE 5.　Indirect immunofluorescence labeling of tubulin in adult rat heart myocytes. Tubulin antiserum was raised against purified brain tubulin, and purified on an affinity column. The focus was at the level of the nuclei (a) and sarcolemma (b) in the same cell. Note that staining is concentrated around the nuclei and scattered throughout the sarcoplasm (bar = 10 μm).

evaluated by immunoquantitation, is relatively high compared to its concentration in skeletal muscle (4 and 0.4 mg/100 mg total proteins, respectively).[94] Despite the stability of the intermediate filaments and their resistance to extraction, their organization in living cells is actively controlled inside those cells.[90-98] In muscle cells with rudimentary assemblies of contractile filaments, the intermediate filaments are scattered throughout the cytoplasm or located in cell junction areas but are seldom present near the Z-discs. In the ensuing period of development when myofibrils are rapidly assembled, intermediate filaments mostly run transversely at the level of the Z-discs and are located at the end junctions of the intercalated discs.[67-76] Some of these filaments are longitudinally oriented along the myofibrils and near the sarcolemma and nuclear membranes.[67,97] The accuracy of desmin detection at the ultrastructural level was recently improved by immuno-electron microscopy, making it possible to show that in adult heart desmin labeling on the intercalated discs is confined to the interfibrillar spaces[63,76] (Figure 9) and concentrated near the desmosomes.[77]

Using cultured 3-day-old heart myocytes and immunofluorescence labeling of both intermediate desmin filaments and Z-disc structures, we observed that desmin location on the Z-discs is closely correlated with myofibril alignment,[74] as first suggested by Lazarides.[71] Thus, desmin is clearly associated with the Z-discs in growing cells in vivo when the forces of tension induce myofibril alignment in register (Figure 10a and b). In contrast, in cultured myocytes with myofibrils randomly oriented,[99] the distribution of desmin in the cell is dramatically modified: desmin is less frequently localized close to the Z-discs, and appears

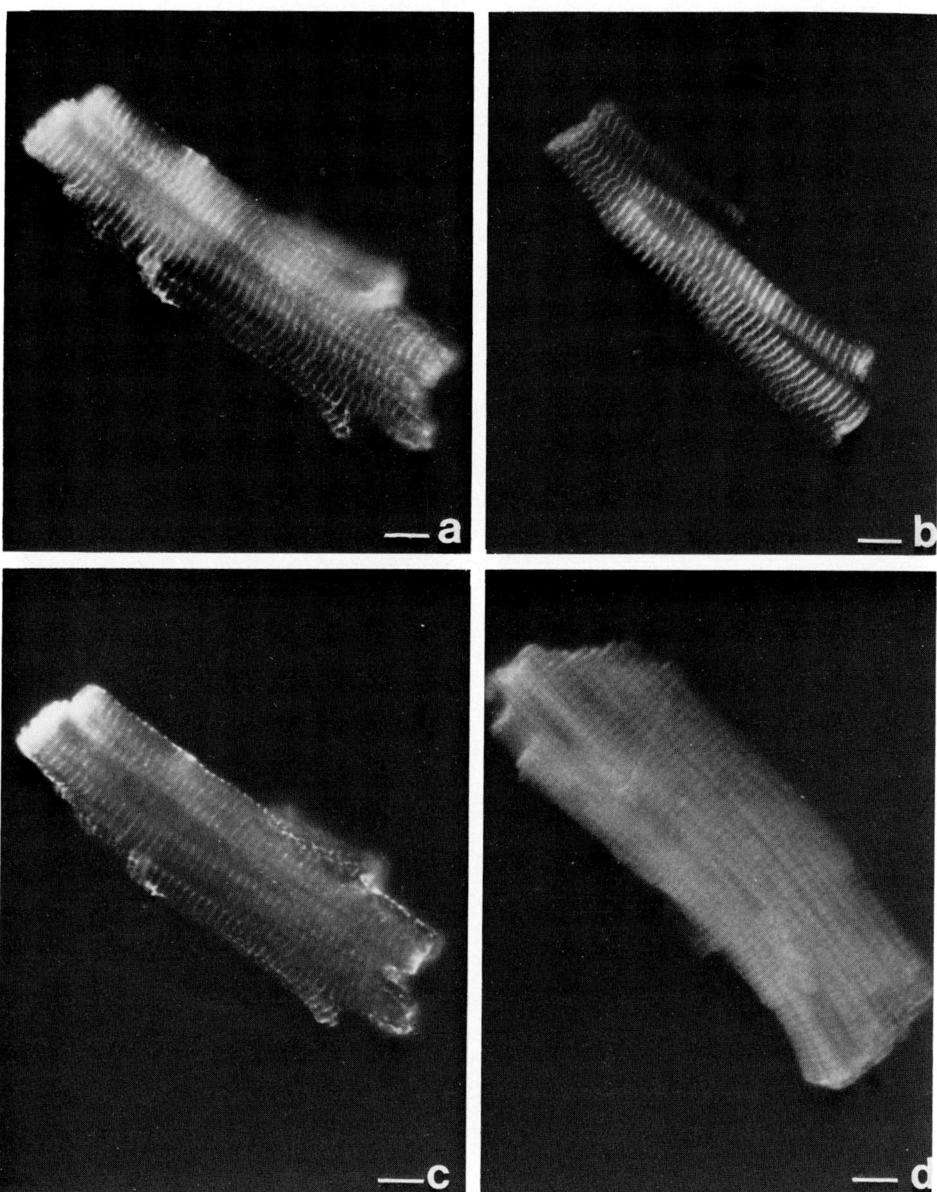

FIGURE 6. Immunofluorescence staining of isolated myocytes obtained from a 3-week-old rat. Cells were labeled with guinea pig antibodies to desmin (a) or with rabbit antibodies to α-actinin (b), vinculin (c), or filamin (d). Guinea pig antibodies were visualized in the rhodamine channel and rabbit antibodies in the FITC channel. Double staining of a cell (a,c) showed the distribution of vinculin along the sarcolemma, while desmin was visible on the Z-discs and intercalated discs. The α-actinin (b) and filamin (d) show a similar distribution to desmin. Note that staining with anti-tubulin (Figure 5) is completely different from these labelings (bar = 10 μm). The anti-α ackinin and anti-vinculin were gifts from B. Jockusch and anti-filamin was kindly given by V. Koteliansky.

as foci dispersed in the sarcoplasm or in long cytoplasmic processes. However, desmin is still present on the intercalated discs formed *de novo* during culture (Figure 10c and d). Electrophoretic and autoradiographic analyses of ^{35}S-methionine incorporation into newly synthesized proteins reveals that in cultured cells, desmin is neosynthetized and that its electrophoretic properties are unchanged. Thus the disappearance of desmin from Z-discs, but not from intercalated discs, is probably related to the change in myofibril organization.

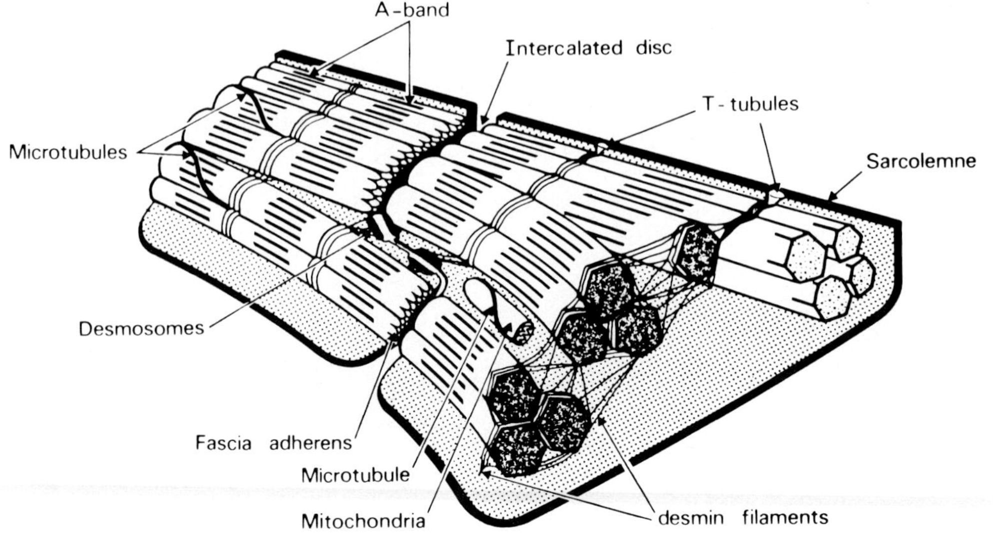

FIGURE 7.   Simplified schematic representation of the distribution of intermediate desmin filaments and microtubules along the myofibrils.

It is worth noting that the evolution of desmin binding to the Z-discs that we observe in cultured myocytes, exhibiting randomly oriented myofibrils,[74] is the reverse of that described during development.[90-98] In addition, it emerges from these experiments that the binding of desmin to membranous structures (intercalated discs) and to myofibrillar structures (Z-discs) is of different types, since desmin disappears from the Z-discs and actively binds to the newly formed cell junctions, simultaneously.

B. Microtubules

Although microtubules have been extensively studied in recent years (for review see Reference 100), they have been little investigated in muscle. They are obviously very conservative structures, consisting mainly of tubulin. The latter protein is a heterodimer (110 kDaltons), containing one α and one β chain which, depending on various factors, assemble to form microtubules, 25 nm in diameter. Several proteins (microtubule-associated proteins, MAPs), copurified with tubulin during assembly-disassembly cycles, were shown to affect microtubule nucleation and elongation in in vitro assays.[100,101] Some of these proteins might have a bridging role in microtubule interactions with other cellular structures, including cytoskeletal filaments and cytoplasmic organelles or membranes.[100-104]

The concentration of tubulin in adult heart myocytes is a thousand times lower than in brain (10 μg/100 mg total proteins).[105] Nevertheless immunofluorescent labeling of tubulin appears to be concentrated around the nuclei, and as scattered bright spots and rods in the cytoplasm (Figure 5). Stainings of tubulin are however never superimposable from one cell to another, thus reflecting the constant state of flux of microtubules within the cell.[105]

In electron microscopy, microtubules in heart myocytes appear as a network that surrounds the nuclei (Figure 11), and as a helicoidal pattern running across myofibrils at the I-band and closely linked to organelles, including the mitochondria, sarcoplasmic reticulum and Golgi apparatus.[106-109]

C. Relationship of Microtubules and Intermediate Filaments

There are several indications that intermediate filaments and microtubules are somehow functionally associated in different types of cells.[109,110] Their distribution is correlated in

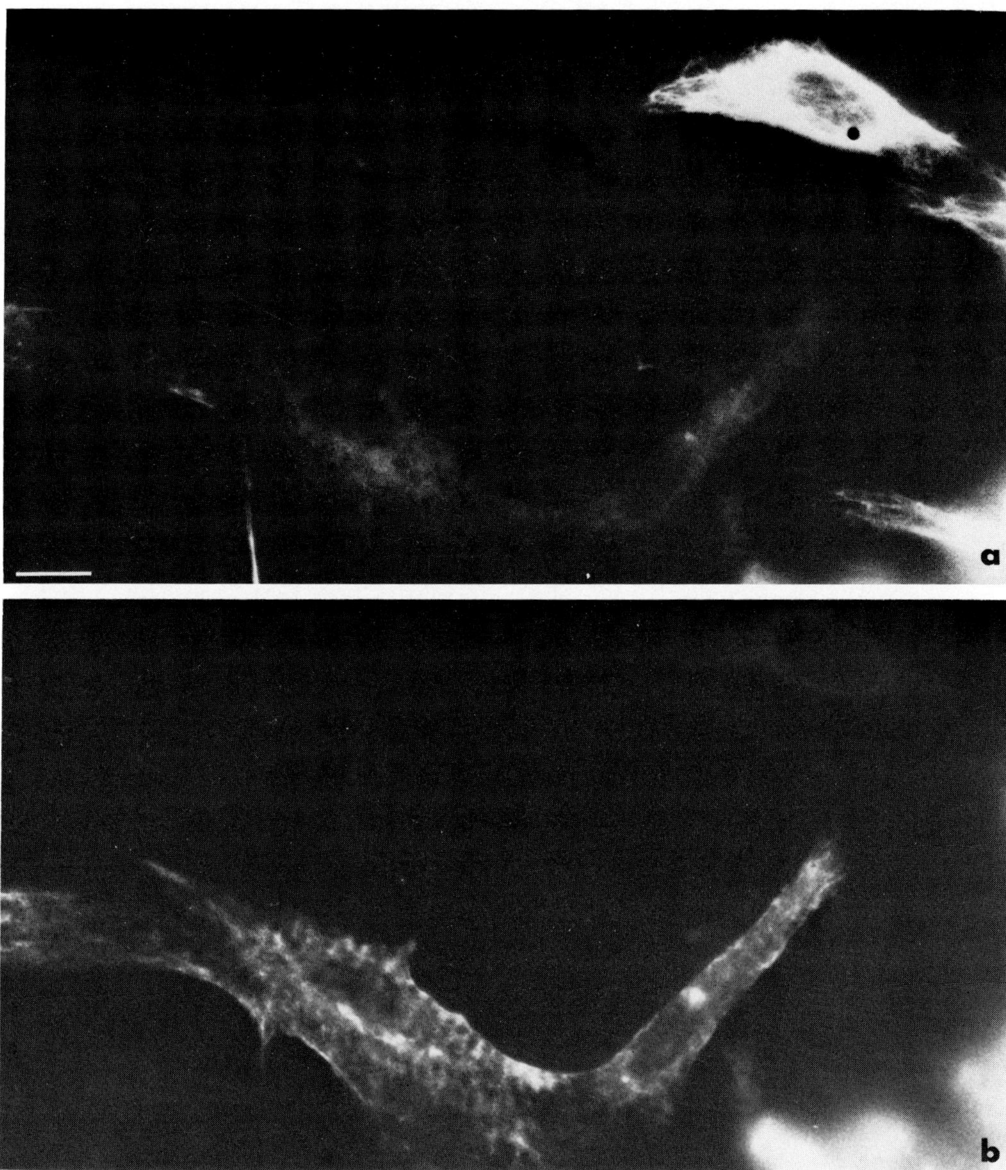

FIGURE 8.   Cultured neonate rat heart myocytes stained with both antibodies to vimentin (a) and desmin (b). Note that myocytes only stained with anti-desmin antibodies, while nonmuscle cells only stained with anti-vimentin (bar = 10 μm).

such a way that in most cases they can only be distinguished after disruption of the microtubule network by antimitotic agents that induce the collapse of the intermediate filaments around the nuclei.[111,112] The relationship of the microtubules and intermediate filaments was investigated in the heart using double immunofluorescence labeling and culture of neonate myocytes.

We have already pointed out the differences between the respective distributions of microtubules and intermediate filaments in rod-shaped heart myocytes, in which desmin is arranged in the typical pattern striated fibers, while tubulin is mostly concentrated around the nuclei and loosely scattered throughout the sarcoplasm (Figures 5 and 6). Moreover, while 1-day cultured rod-shaped myocytes still exhibit a clearly striated pattern of desmin, a modified microtubule pattern appears in the form of a meshwork running from the nuclei

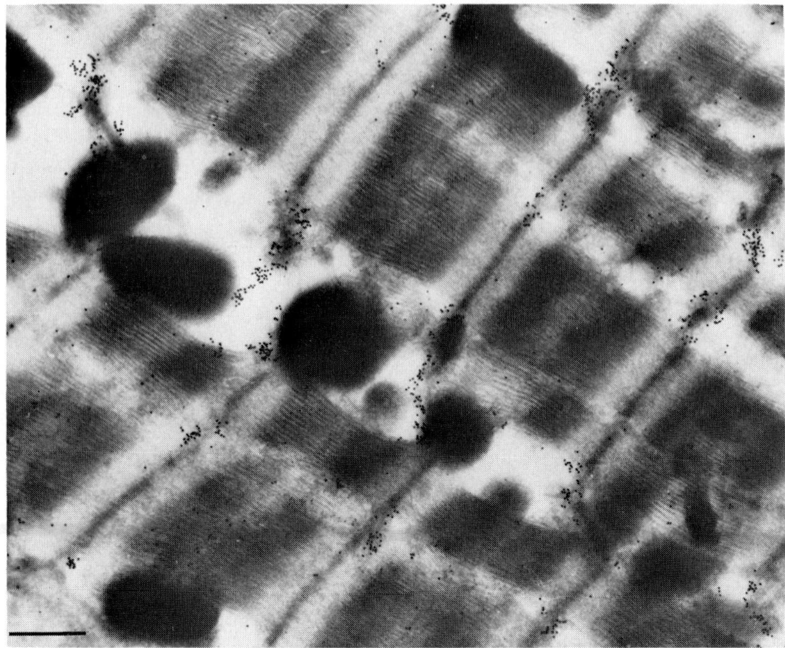

FIGURE 9.   Electron micrograph of a longitudinal, ultrathin frozen section of rat ventricular papillary muscle, immunolabeled with anti-desmin antibodies and with gold-labeled anti-rabbit IgGs. Labeling is mostly confined to the intermyofibrillar spaces between the Z-discs and is sometimes visible near mitochondria. No obvious labeling is present on the Z-discs themselves (bar = 1 μm). (Photograph by courtesy of Dr. S. Watkins.)

towards the sarcoplasm (Figure 12). This pattern remains unchanged with time during culture, whereas as shown in Figure 10, desmin disappears from the Z-discs when myofibrils become progressively randomly distributed throughout the cell.[74] Addition of the antimitotic drug, colchicine, which disrupts microtubules,[111] has minor effects on desmin distribution in rod-shaped cells.[105] However, as in nonmuscle cells, addition of this drug after 3 days of culture induces the collapse of the desmin filaments into a perinuclear mass (Figure 13). Consequently, it may be concluded that microtubules and intermediate filaments are distributed in a different manner in rod-shaped and spread cultured myocytes. However, the effect of colchicine in cultured myocytes suggests that the two structures are in some way associated, but that their binding is weaker than that of desmin to the Z-discs and thus to the bundles of registered myofibrils.

D. Functions of Microtubules and Intermediate Filaments

We shall now consider some of the roles of these filaments in heart myocytes which seem to be quite distinct.

1. Shape of Myocytes

The shape of nonmuscle cells greatly depends on the microtubule network.[100,111,112] In certain embryonic muscle cells, disruption of the microtubule network by colchicine induces spherical cells in which the myofibrils are arranged in a concentric pattern around centrally located nuclei[114] but does not affect the shape of isolated mature myocytes.[105] The arrangement of various combinations of cytoskeletal components among the myofibrils, either transverse as intermediate filaments, longitudinal as titin, and helicoidal as microtubules together with the compacted registered collection of myofilaments, is probably responsible

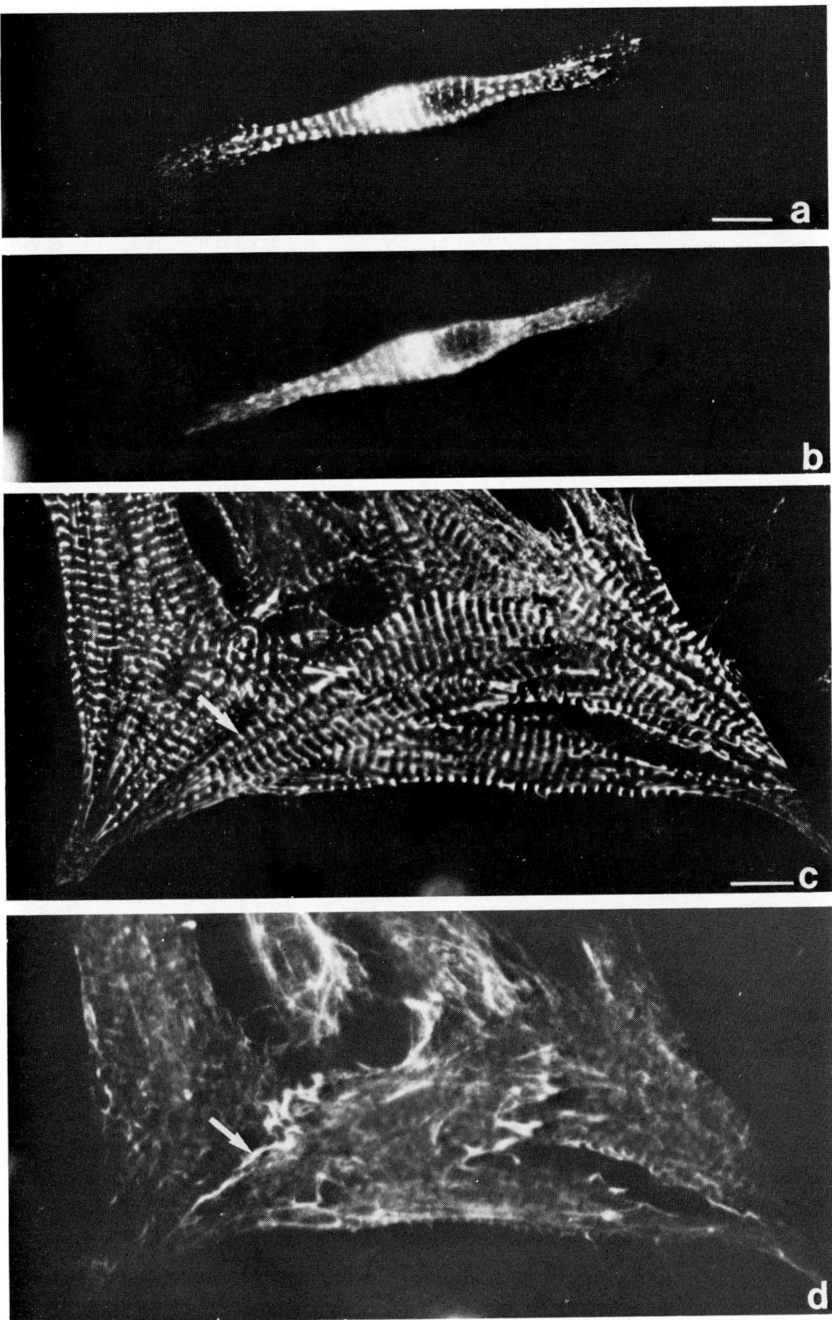

FIGURE 10.   Double immunolabeling of cardiac myocytes with antibodies to α-actinin (a,c) and desmin (b,d) after 1 day (a,b) and 3 days (c,d) of culture. Note (1) The random organization of myofibrils, identified by the anti α-actinin labeling, that correlates with the absence of desmin at the level of the Z discs, and (2) The desmin staining on the newly formed cell junctions (arrows) (bar = 10 μm).

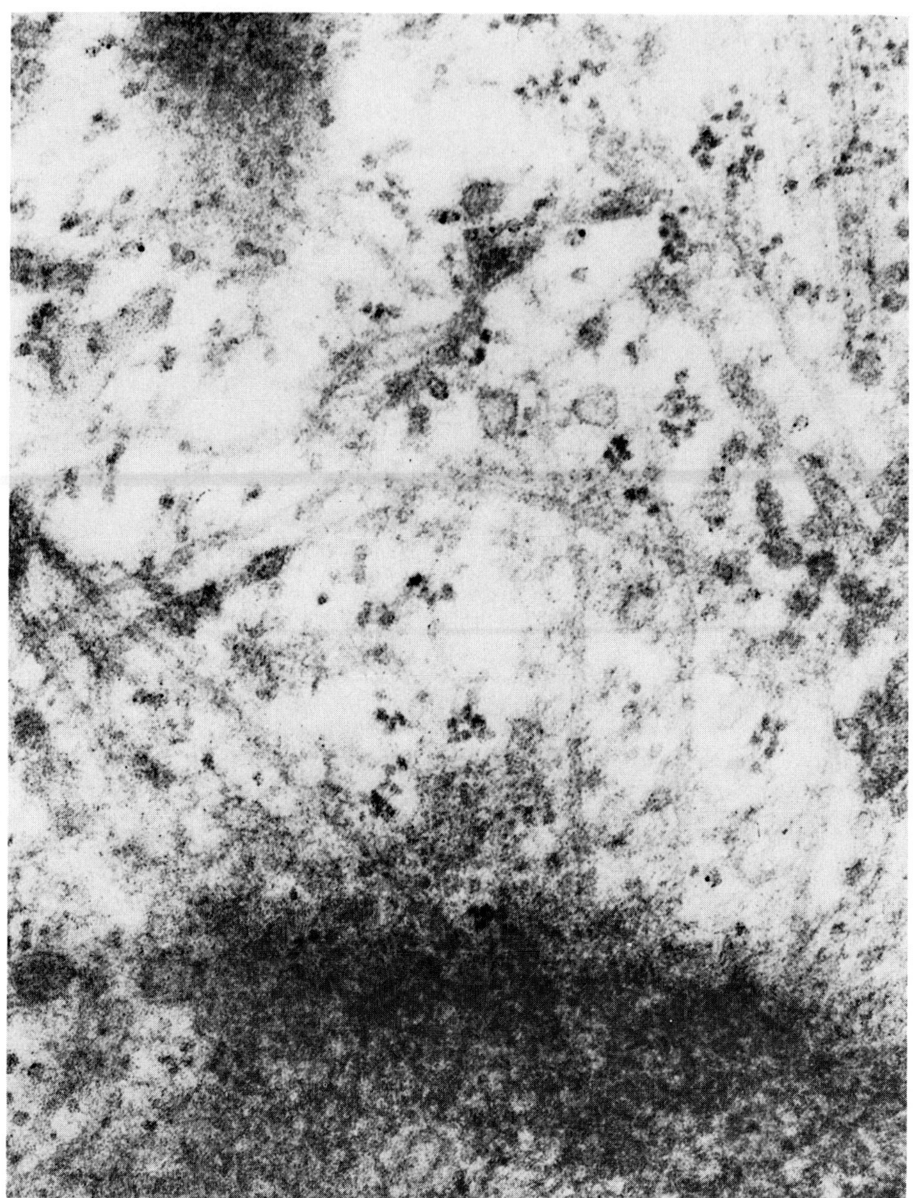

FIGURE 11.   Electron microscopic visualization of perinuclear microtubules in rabbit heart, 7 days after experimental aortic stenosis (original magnification × 54,000) (photograph by courtesy of Pr. P. Y. Hatt).

for maintaining the overall shape of the isolated cardiac adult myocytes. Microtubules might play a specific role in maintaining cellular structures in position during contraction-relaxation processes, and sarcoplasmic transport.[115]

2. Growth of Myocytes

Studies of striated muscle indicate that longitudinally oriented microtubules are prominent during early muscle development but decrease in number as the myofibrils develop and fill the sarcoplasm.[114,116,117] Both microtubules and intermediate filaments are believed to be involved in the elongation of muscle cells and in the assembly of sarcomeres during my-

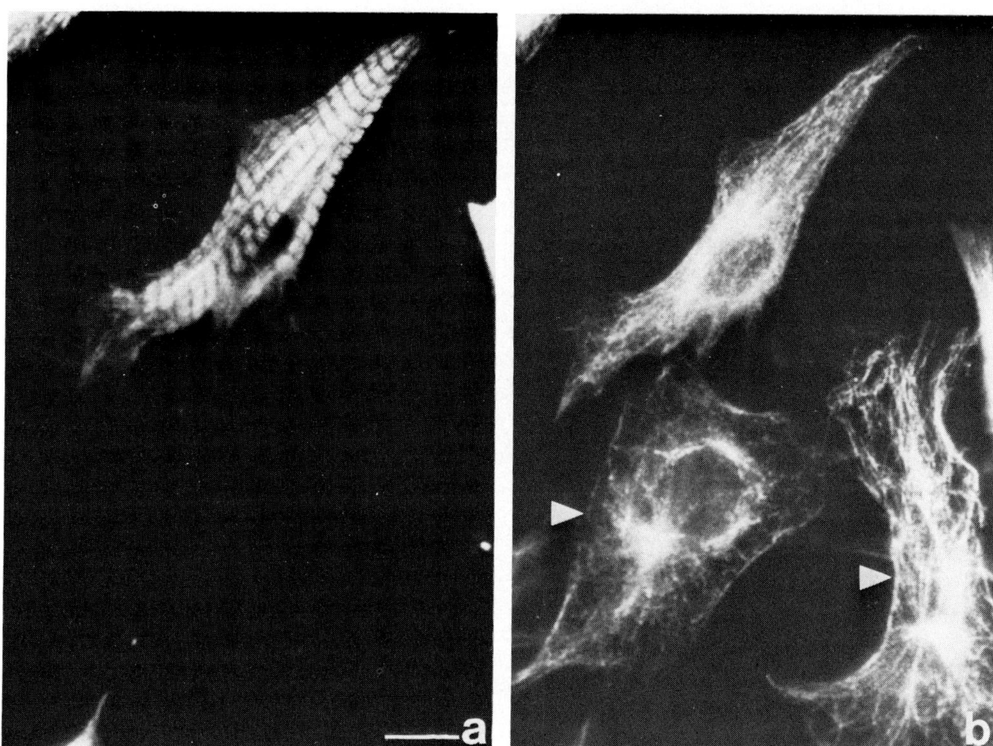

FIGURE 12.   Double immunostaining of 3-day-old rat heart cultured myocytes with antibodies raised against myosin (a) and tubulin (b). Note that the microtubule pattern in cultured myocytes (cell reacting with myosin) resembles the corresponding pattern in nonmuscle cells (head of arrow) more closely than that of adult myocytes, shown in Figure 5 (bar = 10 μm).

ogenesis.[114,117-120] They therefore seem to exert great influence upon the final arrangement of the contractile apparatus in muscle tissue.

We attempted to establish whether these structures are also involved when a rapid increase in myofibrillar volume occurs in myocytes, during adult heart growth. For this investigation, heart growth was mechanically stimulated by aortic stenosis inducing a pressure overload.[5] As stated above, a shift in the expression of myosin occurs under these conditions: α-HC myosin, which is the only form present in 3-week old rat, is replaced by the fetal isoform β-HC myosin (see above). Thus, the β form was used as a biochemical marker to evaluate cardiac adaptation to overloading in single myocytes.[121,122] Tubulin, desmin, and β-HC isomyosin distributions were examined by double immunofluorescence in isolated myocytes at various times after overload-induced growth. No significant changes in the desmin pattern were observed. The distribution and synthesis of desmin and newly synthesized sarcomere proteins seem to be harmoniously correlated. It seems that it is essentially only under pathological conditions when myofibrillar structures are that altered, that intermediate filaments become particularly numerous and disorganized in heart[70] as well as skeletal muscle.[123] It is, however, possible that the reduction in the number of myofibrils facilitates the observation of intermediate filaments. In contrast, a rearrangement of microtubules in parallel arrays occurs soon after aortic stenosis and coincides with heart growth (Figure 14).[121,122] This alteration in the microtubule pattern does not occur simultaneously in all the myocytes, which exhibit various types of microtubule labeling at any given time. Furthermore, these alterations are transitory, since 15 days after growth stimulation, the microtubule pattern in all the myocytes resembles the one observed in control cells. Double immunofluorescence

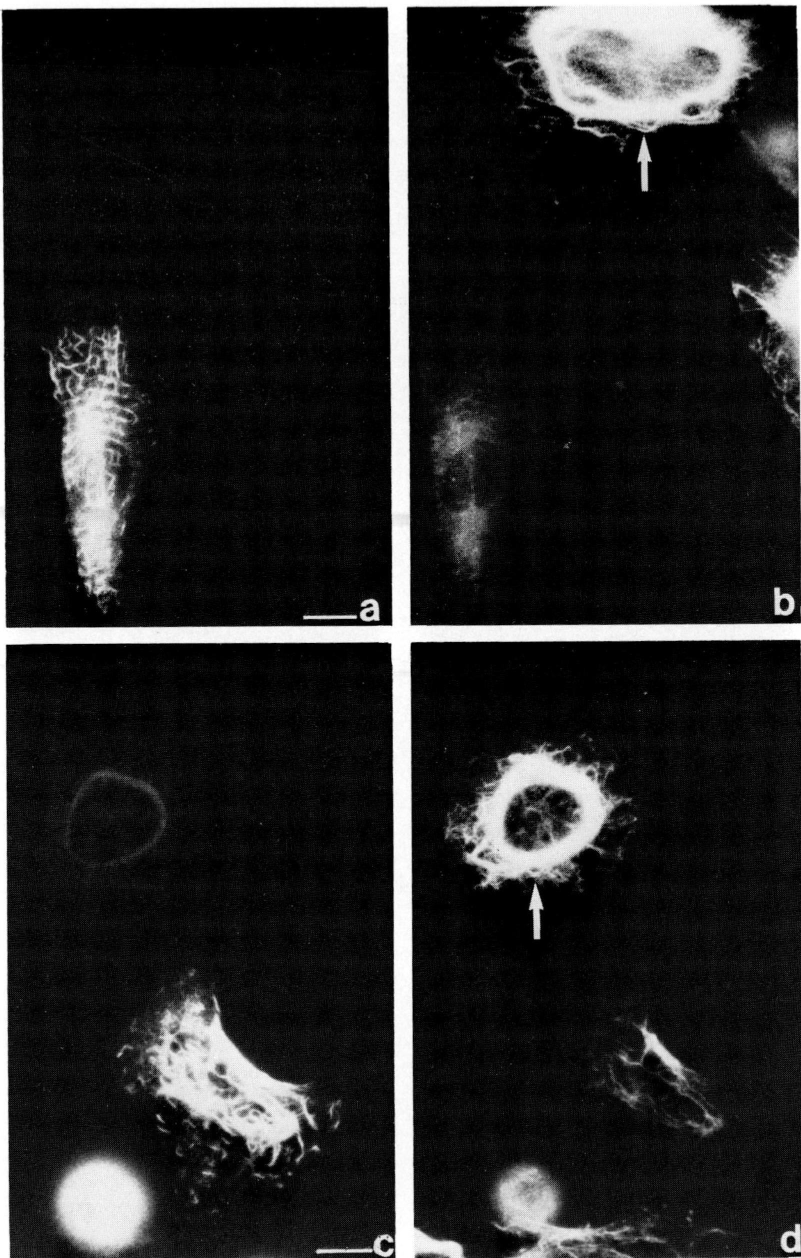

FIGURE 13.   Effects of the tubulin-specific drug, colchicine, on the distribution of desmin and vimentin in cultured heart cells. Colchicine ($5 \times 10^{-5}\,M$) was added for 24 hr either 1 day (a,b) or 2 days (c,d) after plating. Cells were doubly labeled with antibodies against desmin (a,c) and vimentin (b,d) as in Figure 8. Note after drug treatment, the specific collapse of vimentin filaments in nonmuscle cells (arrows). The collapse of desmin in muscle cells was only significant 2 days after culture (bar = 10 μm).

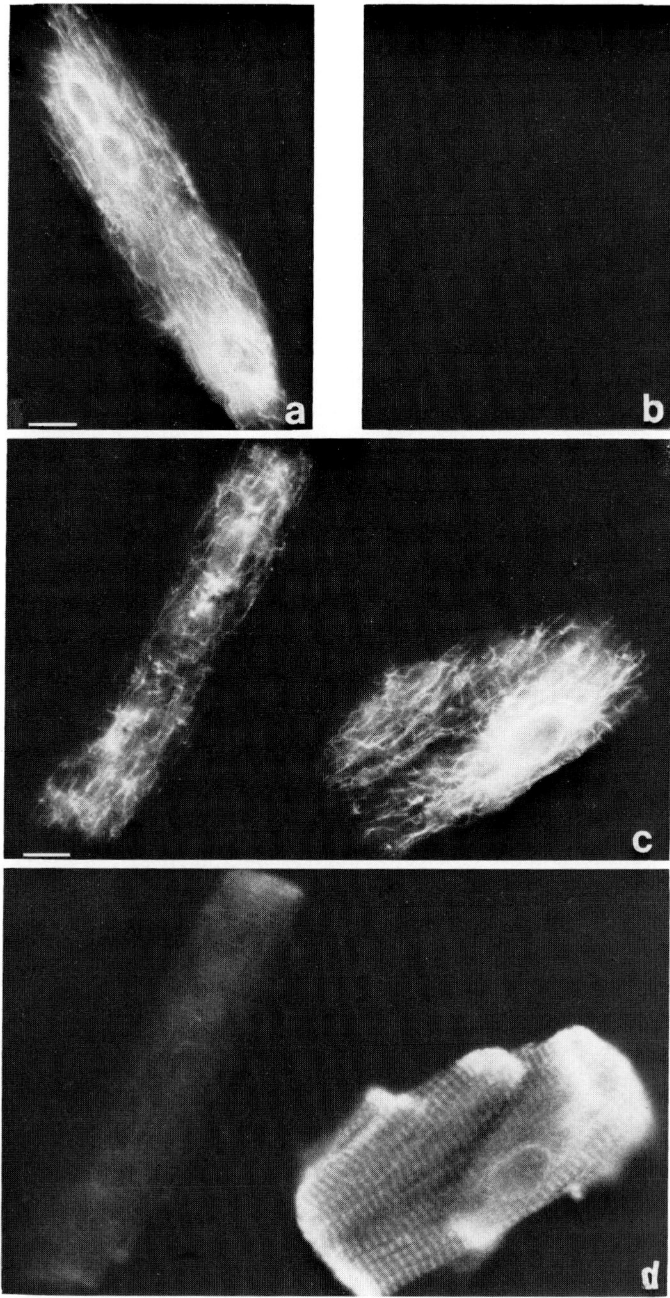

FIGURE 14. Various patterns of double indirect immunofluorescence labeling of myocytes isolated at various times after aortic stenosis. Isolated myocytes 4 days (a,b) and 10 days (c,d) after stenosis were labeled with antitubulin (a,c) and anti-V3 myosin (b,d). In (c) and (d), two myocytes with different immunolabelings were viewed in the same field. Normal (c) and altered (a,c) microtubule patterns were not related to the presence (c) or absence (b,d) of V3-myosin, (bar = 10 μm).

myocyte labeling with anti-tubulin and anti-βHC myosin shows that the alteration in the microtubule pattern precedes the appearance of β-HC myosin in the myocytes.[121] Similar reorganization of the microtubule pattern is also observed when the growth of hypothyroid rat myocytes is stimulated by administration of thyroxine.[124] Under these conditions, the newly expressed α-HC myosin may be visualized in all the myocytes as early as 12 hr after growth stimulation by thyroxine.

Thus microtubules appear to be associated with the series of cellular events involved at the onset of growth, whether stimulation is hormonal or mechanical, and regardless of the type of the newly expressed isomyosins.[126] Dynamic instability is a general property of microtubules and may be fundamental in explaining cellular microtubule organization and its role in myocyte growth. Altered microtubule orientation is also observed when fetal rat heart myocytes are cultured in the presence of dibutyryl-cyclic AMP (DBcAMP).[125] However, the biochemical events inducing variations in the patterns of microtubules and intermediate filaments have not as yet been defined. Microtubule-associated proteins (MAP2) and desmin are phosphorylated by different types of kinases.[100,127,128] The relationship of these enzymes with the filaments and their regulation and role remain to be elucidated. There is also evidence that in nonmuscle cells Ca^{2+} is involved in controlling the assembly of microtubules.[129]

To sum up, cytoskeletal elements may constitute a control system that processes stimuli and regulates cellular responses during acute growth. Microtubules might be actively involved in certain nonspecific cellular events that occur in all growing cells such as protein synthesis and distribution of organelles. Intermediate filaments may have a more passive role in maintaining the myofibrillar register.

IV. FUTURE LINES OF RESEARCH

In this chapter, we have reviewed data on isomyosin expression and the distribution of microtubules and intermediate filaments in isolated adult heart myocytes. Great strides have recently been made in our knowledge of the biochemical properties of myofibrillar proteins and their expression in heart. It would now be of interest to learn more about the turnover, intracellular location, and interactions of myofibrillar and cytoskeletal components in living cells. For example, the close relationship between the two isomyosins in myocytes has been clearly demonstrated by immunofluorescence, but it is not yet clear exactly how two proteins with different ATPase activities interact during contraction processes. Similarly, proteins such as titin and nebulin are biochemically well-defined in heart, and their concentrations in heart cells have been reported as being about 50% of the actin concentration.[83] However, their interaction with myofibrillar structures and their role in contraction are not yet known. Although cytoskeletal elements are obviously essential in determining the organization of organelles inside the living myocyte, their functions in this cell are as yet mostly deduced from electron microscopy studies. Finally, the nature and importance of the factors mediating myocyte adaptational processes, induced either mechanically or hormonally, are as yet not clear. According to recent results, they could be energetic,[40] secreted in serum[130] or depend on cyclic nucleotides.[59,60] They may interfere with the preferential expression of isoforms as well as with variations in the organization of cytoskeleton elements. Various Ca^{2+}/ Calmodulin and cAMP-dependent protein kinases that phosphorylate protein components of either myofibrils or the cytoskeleton may have an important role in regulating the relationship between the latter structures.

The accuracy of the ultrastructural location of these proteins and factors could be greatly improved by experiments on living cells by using either micro-injections,[131] or protein labeling in the presence of various agonists, or molecular hybridization *in situ*. Neonatal cultured myocytes are easy to prepare, react well in response to a number of myocardial stimuli but possess certain characteristics of immature cells. Techniques for isolating Ca^{2+}-resistant adult myocytes have recently much improved. They display well-differentiated adult characteristics, but are still difficult to maintain in long-term culture. Both neonatal and adult cells provide an opportunity to study the cardiac muscle cell without the constrains common to intact animals or isolated organs and, therefore, make it possible to gain insight into the effect of extrinsic or cellular factors on the molecular mechanisms in cardiac myocytes.

ACKNOWLEDGMENTS

This research was supported by INSERM. The authors are indebted to J. L. Samuel for collaboration in research, comments and suggestions during the course of these studies and to P. Cagnac for expert assistance in typing and editing the manuscript.

REFERENCES

1. **Flink, I. L., Rader, J. H., and Morkin, E.,** Thyroid hormone stimulates synthesis of a cardiac myosin isozyme, *J. Biol. Chem.,* 254, 3105, 1979.
2. **Hoh, J. F. Y., Yeoh, G. P. S., Thomas, M. A. W., and Higginbottom, L.,** Structural differences in the heavy chains of rat ventricular myosin isozymes, *FEBS Lett.,* 97, 330, 1979.
3. **Schwartz, K., Lompré, A. M., Bouveret, P., Wisnewsky, C., and Whalen, R. G.,** Comparisons of rat cardiac myosins at fetal stages, in young animals and in hypothyroid adults, *J. Biol. Chem.,* 257, 14412, 1982.
4. **Chizzonite, R. A., Everett, A. W., Clark, W. A., Jakovcic, S., Rabinowitz, M., and Zak, R.,** Isolation and characterization of two molecular variants of myosin heavy-chain from rabbit ventricle. Change in their content during normal growth and after treatment with thyroid hormone, *J. Biol. Chem.,* 257, 2056, 1982.
5. **Klotz, C., Swynghedauw, B., Mendes, H., Marotte, F., and Léger, J. J.,** Evidence for new forms of cardiac myosin heavy chains in mechanical overloading and in ageing, *Eur. J. Biochem.,* 115, 415, 1981.
6. **Lompré, A. M., Han, K., Bouveret, P., Richard, C., and Schwartz, K.,** Comparison of the tryptic digestion pattern of subfragments 1 from V1 and V3 rat cardiac isomyosins, *Eur. J. Biochem.,* 139, 459, 1984.
7. **Schier, J. J. and Adelstein, R. S.,** Structural and enzymatic comparison of human cardiac muscle myosins isolated from infants, adults and patients with hypertrophic cardiomyopathy, *J. Clin. Invest.,* 69, 816, 1982.
8. **Litten, R. Z., Martin, B. J., Low, R. B., and Alpert, N. R.,** Altered myosin isozyme patterns from pressure-overloaded and thyrotoxic hypertrophied rabbit hearts, *Circ. Res.,* 50, 846, 1982.
9. **Lompré, A. M., Schwartz, K., d'Albis, A., Lacombe, G., Thiem, N. V., and Swynghedauw, B.,** Myosin isoenzyme redistribution in chronic heart overload, *Nature,* 282, 105, 1979.
10. **Gorza, L., Pauletto, P., Pessina, A. L., Sartore, S., and Schiaffino, S.,** Isomyosin distribution in normal and pressure overloaded rat ventricular myocardium. An immunohistochemical study, *Circ. Res.,* 49, 1003, 1981.
11. **Sartore, S., Gorza, L., Pierobon-Bormioli, S., Dalla Libera, L., and Schiaffino, S.,** Myosin types and fiber types in cardiac muscle. I. Ventricular myocardium, *J. Cell Biol.,* 88, 226, 1981.
12. **Clark, W. A., Chizzonite, R. A., Everett, A. W., Rabinowitz, M., and Zak, R.,** Species correlations between cardiac isomyosins, *J. Biol. Chem.,* 257, 5449, 1982.
13. **Mercadier, J. J., Bouveret, P., Gorza, L., Schiaffino, S., Clark, W. A., Zak, R., Swynghedauw, B., and Schwartz, K.,** Myosin isoenzymes in normal and hypertrophied human ventricular myocardium, *Circ. Res.,* 53, 52, 1983.
14. **Gorza, L., Mercadier, J. J., Schwartz, K., Thornell, L. E., Sartore, S., and Schiaffino, S.,** Myosin types in the human heart: an immunofluorescence study of normal and hypertrophied atrial and ventricular myocardium, *Circ. Res.,* 54, 694, 1984.
14a. **Bouvagnet, P., Leger, J., Pons, F., Dechesne, C., and Leger, J. J.,** Myosin types and fiber types in human hearts: an anatomical description, *Circ. Res.,* 55, 794, 1984.
15. **Mahdavi, V., Periasamy, M., and Nadal-Ginard, B.,** Molecular characterization of two myosin heavy chain genes expressed in the adult heart, *Nature,* 297, 659, 1982.

16. **Sinha, A. M., Umeda, P. K., Kavinsky, C. J., Rajamanickman, C., Hsu, H. J., Jakovcic, S., and Rabinowitz, M.,** Molecular cloning of mRNA sequences for cardiac α and β form myosin heavy chain: expression in ventricles of normal, hypothyroid and thyrotoxic ventricles, *Proc. Natl. Acad. Sci. USA,* 79, 5847, 1982.

17. **Kavinsky, C. J., Umeda, P. K., James, E. L., Sinha, A. M., Nigro, J. M., Jakovcic, S., and Rabinowitz, M.,** Analysis of cloned mRNA sequences encoding subfragment 2 and part of subfragment 1 of α- and β- myosin heavy chains of rabbit heart, *J. Biol. Chem.,* 259, 2775, 1984.

18. **Mahdavi, V., Chambers, A. P., and Nadal-Ginard, B.,** Cardiac α- and β-myosin heavy chain genes are organized in tandem, *Proc. Natl. Acad. Sci. USA,* 81, 2626, 1984.

19. **Lompré, A. M., Nadal-Ginard, B., and Mahdavi, V.,** Expression of the cardiac ventricular α- and β-myosin heavy chain genes is developmentally and hormonally regulated, *J. Biol. Chem.,* 259, 6437, 1984.

20. **Sinha, A. M., Friedman, D., Nigro, J. M., Jakovcic, S., Rabinowitz, M., and Umeda, P. K.,** Expression of rabbit ventricular α-myosin heavy chain messenger RNA sequences in atrial muscle, *J. Biol. Chem.,* 259, 6674, 1984.

21. **Long, L., Fabian, F., Mason, D. T., Wikman-Coffelt, J.,** A new cardiac myosin characterized from the canine atria, *Biochem. Biophys. Res. Comm.,* 76, 626, 1977.

22. **Syrovy, I., Delcayre, C., and Swynghedauw, B.,** Comparison of ATPase activity and light subunits in myosins from left and right ventricles and atria in seven mammalian species, *J. Mol. Cell. Cardiol.,* 11, 1129, 1979.

23. **Price, K. M., Littler, W. A., and Cummins, P.,** Human atrial and ventricular myosin light-chain subunits in the adult and during development, *Biochem. J.,* 191, 571, 1980.

24. **Whalen, R. G., Butler-Browne, G. S., and Gros, F.,** Identification of a novel form of myosin light chain present in embryonic light chain, *J. Molec. Biol.,* 126, 415, 1978.

25. **Klotz, C., Léger, J. J., and Elzinga, M.,** Comparative sequence of myosin light chains from normal and hypertrophied human hearts, *Circ. Res.,* 50, 201, 1982.

26. **Hoh, J. H., McGrath, P.A., and Hale, P. T.,** Electrophoretic analysis of multiple forms of rat cardiac myosin: Effect of hypophysectomy and thyroxine replacement, *J. Mol. Cell. Cardiol.,* 10, 1053, 1978.

27. **Everett, A. W., Chizzonite, R. A., Clark, W. A., and Zak, R.,** Relationship of changes in molecular forms of myosin heavy chains to endogenous level of thyroid hormone during postnatal growth. In: perspectives in cardiovascular research, Vol. 8 Tarazi, R. C. and Dunbar, J. B., Eds., Raven Press, New York, 1983, 83.

28. **Pope, B., Hoh, J. F. Y., and Weeds, A.,** The ATPase activity of rat cardiac myosin isoenzymes, *FEBS Lett.,* 118, 205, 1980.

29. **Lompré, A. M., Mercadier, J. J., Wisnewsky, C., Bouveret, P., Pantaloni, C., d'Albis, A., and Schwartz, K.,** Species and age-dependent changes in the relative amounts of cardiac myosin isoenzymes in mammals, *Develop. Biol.,* 84, 286, 1981.

30. **Malhotra, A., Penpargkul, S., Fein, F. S., Sonnenblick, E. H., and Scheuer, J.,** The effect of streptozotocin-induced diabetes in rats on cardiac contractile proteins, *Circ. Res.,* 49, 1243, 1981.

31. **Martin, A. F., Pagani, E. D., and Solaro, R. J.,** Thyroxine-induced redistribution of isoenzymes of rabbit ventricular myosin, *Circ. Res.,* 50, 117, 1982.

32. **Dalla Libera, L. and Sartore, S.,** Immunological and biochemical evidence for atrial-like isomyosin in thyrotoxic rabbit ventricle, *Biochim. Biophys. Acta,* 669, 84, 1981.

33. **Banerjee, S. K.,** Comparative studies of atrial and ventricular myosin from normal, thyrotoxic and thyroidectomized rats, *Circ. Res.,* 52, 131, 1983.

34. **Gorza, L., Sartore, S., and Schiaffino, S.,** Myosin types and fiber types in cardiac muscle. II. Atrial myocardium, *J. Cell. Biol.,* 95, 838, 1982.

35. **Whalen, R. G., Sell, S. M.,** Myosin from fetal hearts contains the skeletal muscle embryonic light chain, *Nature,* 286, 731, 1980.

36. **Samuel, J. L., Rappaport, L., Mercadier, J. J., Lompré, A. M., Sartore, S., Triban, C., Schiaffino, S., and Schwartz, K.,** Distribution of myosin isozymes within single cardiac cells. An immunohistochemical study, *Circ. Res.,* 52, 200, 1983.

37. **Weisberg, A., Winegrad, S., Tucker, M., and Mc Clellan, G.,** Histochemical detection of specific isozymes of myosin in rat ventricular cells, *Circ. Res.,* 51, 802, 1982.

38. **Cummins, P.,** Transitions in human atrial and ventricular myosin light-chain isoenzymes in response to cardiac-pressure-overload-induced hypertrophy, *Biochem. J.,* 205, 195, 1982.

39. **Loiselle, D. S., Wendt, I. R., and Hoh, J. F. Y.,** Energetic consequences of thyroid-modulated shifts in ventricular isomyosin distribution in the rat, *J. Musc. Res. Cell. Motil.,* 3, 5, 1982.

40. **Sheer, D. and Morkin, E.,** Myosin isoenzyme expression in rat ventricles: effects of thyroid hormone analogs, catecholamines, glucocorticoids and high carbohydrate diet, *J. Phamacol. Exp. Ther.,* 229, 872, 1984.

41. **Nag, A. C. and Cheng, M.,** Expression of myosin isoenzymes in cardiac muscle cells in culture, *Biochem. J.,* 221, 21, 1984.

42. **Schwartz, K., Samuel, J. L., Syrovy, I., Marotte, F., Wisnewsky, C., and Rappaport, L.,** Effects of ontogenic development, thyroid hormone and hemodynamic overload on rat atrial isomyosins; 13th European Conference on Muscle and Motility, Gwatt, September 23-28, 1984 (abstr.).

43. **Dillmann, W. H.,** Diabetes mellitus induces changes in cardiac myosin of the rat, *Diabetes*, 29, 579, 1980.

44. **Dillmann, W. H., Barrieux, A., and Reese, G.,** Effect of diabetes and hypothyroidism on the predominance of cardiac myosin heavy chains synthesized in vivo or in a cell-free system, *J. Biol. Chem.*, 259, 2035, 1984.

45. **Schaible, T. F., Malhotra, A., Ciambrone, G., and Scheuer, J.,** The effects of gonadectomy on left ventricular function and cardiac contractile proteins in male and female rats, *Circ. Res.*, 54, 38, 1984.

46. **Mercadier, J. J., Lompré, A. M., Wisnewsky, C., Samuel, J. L., Bercovici, J., Swynghedauw, B., and Schwartz, K.,** Myosin isoenzymic changes in several models of rat cardiac hypertophy, *Circ. Res.*, 49, 525, 1981.

47. **Rupp, H.,** The adaptative changes in the isoenzyme pattern of myosin from hypertrophied rat myocardium as a result of pressure overload and physical training, *Basic, Res. Cardiol.*, 76, 79, 1981.

48. **Wisenbaugh, T., Allen, P., Cooper, IV G., Holzgrefe, H., Beller, G., and Carabello, B.,** Contractile function, myosin ATPase activity and isozymes in the hypertrophied pig left ventricle after chronic progressive pressure overload, *Circ. Res.*, 53, 332, 1983.

49. **Mercadier, J. J., Menasché, P., de la Bastie, D., Wisnewsky, C., Piwnica, A., and Schwartz, K.,** Left atrial isomyosin changes in patients with mitral valve disease, *Circulation*, 70(II), 197, 1984.

50. **Bugaisky, L. B., Siegel, E. L., and Whalen, R. G.,** Myosin isozyme changes in the heart following constriction of the aorta of 25-day-old rat, *FEBS Lett.*, 161, 230, 1983.

51. **Mercadier, J. J., Lecarpentier, Y., Delcayre, C., Lompré, A. M., Swynghedauw, B., and Schwartz, K.,** Mécanismes biochimiques de l'adaptation myocardique dans l'hypertrophie cardiaque, *Arch. Mal. Coeur*, 75, 1179, 1982.

52. **Penpargkul, S., Malhotra, A., Shaible, T., and Scheuer, J.,** Cardiac contractile proteins and sarcoplasmic reticulum in hearts of rats trained by running, *J. Appl. Physiol., Respirat.Environ Exer. Physiol.*, 48, 409, 1980.

53. **Pagani, E. D. and Solaro, R. J.,** Swimming exercise, thyroid state, and the distribution of myosin isoenzymes in rat heart, *Am. J. Physiol.*, 245, H713, 1983.

54. **Tuchschmid, C. R., Srihari, T., Hirzel, H. O., and Schaub, M. C.,** Structural variants of heavy chains of atrial and ventricular myosins in hypertrophied human hearts, in *Cardiac Adaptation to Hemodynamic Overload, Training and Stress*, Jacob, R., Gülch, R. W., and Kissling, G., Eds., Steinkopff Verlag, Darnstadt, 1983, 123.

55. **Hoh, J. F. Y. and Egerton, L. G.,** Action of triiodothyronine on the synthesis of rat ventricular myosin isoenzymes, *FEBS Lett.*, 101, 143, 1979.

56. **Everett, A. W., Sinha, A. M., Umeda, P. K., Jakovcic, S., Rabinowtiz, M., and Zak, R.,** Regulation of myosin synthesis by thyroid hormone: relative change in the α- and β-myosin heavy chain mRNA levels in rabbit heart, *Biochemistry*, 23, 1596, 1984.

57. **Schwartz, K., Lecarpentier, Y., Martin, J. L., Lompré, A. M., Mercadier, J. J., and Swynghedauw, B.,** Myosin isoenzymic distribution correlates with speed of myocardial contraction, *J. Mol. Cell. Cardiol.*, 13, 1071, 1981.

58. **Pagani, E. D. and Julian, F. J.,** Rabbit papillary muscle myosin isozymes and the velocity of muscle shortening, *Circ. Res.*, 54, 586, 1984.

59. **Winegrad, S.,** Regulation of cardiac contractile proteins. Correlations between physiology and biochemistry, *Circ. Res.*, 55, 565, 1984.

60. **Winegrad, S., McClelan, G., Tucker, M., and Lin, L.,** Cyclic AMP regulation of myosin isozymes in mammalian cardiac muscle, *J. Gen. Physiol.*, 81, 749, 1983.

61. **Alpert, N. R. and Mulieri, L. A.,** Heat, mechanics, and myosin ATPase in normal and hypertrophied heart muscle, *Fed. Proc.*, 41, 192, 1981.

62. **Alpert, N. R. and Mulieri, L. A.,** Increased myothermal economy of isometric force generation in compensated cardiac hypertrophy induced by pulmonary artery constriction in the rabbit, *Circ. Res.*, 50, 491, 1982.

63. **Tokuyasu, K. T., Dutton, A. H., Geiger, B., and Singer, S. Y.,** Ultrastructure of chicken cardiac muscle as studied by double immunolabeling in electron microscopy, *Proc. Natl. Acad. Sci. USA*, 78, 7619, 1981.

64. **Koteliansky, V. E., Glukhova, M. A., Shirinsky, V. P., Bahaev, V. R., Kandalenko, V. F., Rukosuev, V. S., and Smirnov, V. N.,** Identification of a filamin-like protein muscle, *FEBS Lett.*, 125, 44, 1981.

65. **Koteliansky, V. E. and Gneusher, G. N.,** Vinculin localization in cardiac muscle, *FEBS Lett.*, 159, 158, 1983.

66. **Pardo, J. V., Angelo Siliciano, J. D., and Craig, S. W.,** Vinculin is a component of an extensive network of myofibril-sarcolemma attachment regions in cardiac muscle fibers, *J. Cell Biol.*, 97, 1081, 1983.

67. **Eriksson, A. and Thornell, L. E.**, Intermediate (skeletin) filaments in heart Purkinje fibers. A correlative morphological and biochemical identification with evidence of a cytoskeletal function, *J. Cell Biol.*, 80, 231, 1979.

68. **Price, M. G. and Sanger, J. W.**, Intermediate filaments in striated muscle: a review of structural studies in embryonic and adult skeletal and cardiac muscle, in *Cell and Muscle Motility*, Vol. 3, Dowben, R. M. and Shay, J. W., Eds., Plenum Press, New York, 1983, 1.

69. **Sanger, J. W., Mittall, B., and Sanger, J. M.**, Analysis of myofibrillar structure and assembly using fluorescently labeled contractile proteins, *J. Cell Biol.*, 98, 825, 1984.

70. **Thornell, L. E., Johansson, B., Ericksson, A., Lehto, V. P., and Virtanen, I.**, Intermediate filament and associated proteins in the human heart: an immunofluorescence study of normal and pathological hearts, *Eur. Heart J.*, 5(Supp. F), 231, 1984.

71. **Lazarides, E.**, Intermediate filaments as mechanical integration of cellular space, *Nature*, 283, 249, 1980.

72. **Lazarides, E.**, Intermediate filaments: a chemically heterogeneous, developmentally regulated class of proteins, *Ann. Rev. Biochem.*, 51, 219, 1982.

73. **Osborn, M., Geisler, N., Shaw, G.,Sharp, G., and Weber, K.**, Intermediate filaments, *Cold Spring Harbor Symp. Quant. Biol.*, 46, 413, 1981.

74. **Samuel, J. L., Jockusch, B. M., Bertier-Savalle, B., Escoubet, B., Marotte, F., Swynghedauw, B., and Rappaport, L.**, Myofibrillar organization and desmin in rat heart myocytes, *Bas. Res. Cardiol.*, 80 (Suppl. 1), 119, 1985.

75. **Tokuyasu, K. T., Dutton, A. H., and Singer, S. Y.**, Immunoelectron microscopic studies of desmin (skeletin) localization and intermediate filament organization in chicken cardiac muscle, *J. Cell Biol.*, 97, 1736, 1983.

76. **Tokuyasu, K. T.**, Visualization of longitudinally-oriented intermediate filaments in frozen sections of chicken cardiac muscle by a new staining method, *J. Cell. Biol.*, 97, 562, 1983.

77. **Kartenbeck, J., Franke, W. W., Moser, J. G., and Stoffels, V.**, Specific attachment of desmin filaments to desmosomal plaques in cardiac myocytes, *EMBO J.*, 2, 735, 1983.

78. **Jockusch, B. M.**, Pattern of microfilaments organization in animal cells, *Mol. Cell. Endocrin.*, 29, 1, 1983.

79. **Weeds, A.**, Actin-binding proteins regulators of cell architecture and motility, *Nature*, 296, 811, 1982.

80. **Granger, B. L. and Lazarides, E.**, Synemin: a new high molecular weight protein associated with desmin and vimentin filaments in muscle, *Cell*, 22, 727, 1980.

81. **Goodman, S. R., Zagon, P. S., and Kulikowski, R.**, Identification of a spectrin-like protein in nonerythroid cells, *Proc. Natl. Acad. Sci. USA*, 78, 7570, 1981.

82. **Price, M. G.**, Lazarides, E., Expression of intermediate filament-associated proteins, paranemin and synemin, in chicken development, *J. Cell Biol.*, 97, 1860, 1983.

83. **Wang, K.**, Myofilamentous and myofibrillar connections: role of nebulin and intermediate filaments, in *Muscle Development, Molecular and Cellular Control*, Pearson, M. Z. and Epstein, H. F., Eds. Cold Spring Harbor Lab., New York, 1982, 439.

84. **Eppenberger, H. M., Bahler, M., Doetschman, T. C., Eppenberger, M., Perriard, J. C., Studer, D., Wallimann, T., Strehler, E.**, The protein M myomesin in cross-striated muscle cells during myofibrillogenis, in *Muscle Development, Molecular and Cellular Control*, Pearson, M. Z. and Epstein, H. F., Eds. Cold Spring harbor Lab, New York, 1982, 429.

85. **Strehler, E. E., Pelloni, G., Heizmann, C. W., and Eppenberger, H. W.**, Biochemical and ultrastructural aspects of 165,000 mol wt protein in cross-straited chicken muscle, *J. Cell Biol.*, 86, 775, 1980.

86. **Anderton, B. H.**, Intermediate filaments: a family of homologous structures, *J. Muscle Res. Cell Motil.*, 2, 141, 1981.

87. **Thornell, L. E. and Ericksson, A.**, Filament systems in the Purkinje fibers of the heart, *Am. J. Physiol.*, 241, H291, 1981.

88. **Lazarides, E. and Hubbard, B. D.**, Immunological characterization of the subunit of the 100 Å filaments from muscle cells, *Proc. Natl. Acad. Sci. USA*, 73, 4344, 1976.

89. **Gard, D. L. and Lazarides, E.**, The synthesis and distribution of desmin and vimentin during myogenesis in vitro, *Cell*, 19, 263, 1930.

90. **Carlsson, E., Kjorell, U., Thornell, L. E., Lambertsson, A., and Strehler, E.**, Differentiation of the myofibrils and the intermediate filament system during postnatal development of the rat heart, *Eur. J. Cell. Biol.*, 27, 67, 1982.

91. **Danton, S. I. and Fischman, D. A.**, Immunocytochemical analysis of intermediate filaments in embryonic heart cells with monoclonal antibodies to desmin, *J. Cell Biol.*, 98, 2179, 1984.

92. **Tokuyasu, K. T., Maher, P. A., and Singer, S. J.**, Distributions of vimentin and desmin in developing chick myotubes in vivo. Immunofluorescence study, *J. Cell Biol.*, 98, 1961, 1984.

93. **Lazarides, E.**, The distribution of desmin (100 Å) filaments in primary cultures of embryonic chick cardiac cells, *Exp. Cell Res.*, 112, 265, 1978.

94. **Bertier-Savalle, B.**, Ph.D. thesis, Clermont-Ferrand University, Clermont-Ferrand, France, 1985.

95. **Forbes, M. S. and Sperelakis, N.**, The membrane systems and cytoskeletal elements of mammalian myocardial cells, in *Cell and Muscle Motility* Vol. 3, Dowben, R. M. and Shay, J. W., Eds., Plenum Press, New York, 1983, 89.

96. **Fuseler, J. M., Shay, J. W., and Feit, H.**, The role of intermediate (10 nm) filament in the development and integration of the myofibrillar contractile apparatus in the embryonic mammalian heart: in *Cell and Muscle Motility*, Vol. 1, Dowben, M. and Shay, J. W., Eds., Plenum Press, New York, 1983, 205.

97. **Fuseler, J. W. and Shay, J. W.**, The association of desmin with the developing myofibrils of cultured embryonic rat heart myocytes, *Dev. Biol.*, 91, 448, 1982.

98. **Bennet, G., Fellini, S. A., Toyama, Y., and Holtzer, H.**, Redistribution of intermediate filament subunits during skeletal myogenesis and maturation in vitro, *J. Cell Biol.*, 82, 577, 1979.

99. **Kulikowski, R. R.**, Myofibrillogenesis in vitro. Implications for early cardiac morphogenesis in *Perspectives in Cardiovascular Research*, Vol. 5, Pexieder, T., Ed., Raven Press, New York, 1981, 367.

100. **Dustin, P.**, *Microtubules*, Springer-Verlag, Berlin, 1984.

101. **Mitchison, T. and Kirschner, M.**, Dynamic instability of microtubule growth, *Nature*, 312, 237, 1984.

102. **Pollard, T. D., Selden, S. C., and Griffith, L. M.**, Actin-Microtubule interactions, in *Biological Functions of Microtubules and Related Structures*, Sakai, H., Muhri, H., and Borisy G. G., Eds., Academic Press, 1982, 311.

103. **Nakamura, Y. and Ueda, K.**, Connection between microtubules and mitochondria, *Cytologia*, 47, 713, 1983.

104. **Vallee, R. B.**, MAP$_2$ (Microtubule Associated Protein 2), in *Cell and Muscle Motility:* Vol. 5, Shay, J. W. Ed., Plenum Press, New York, 1984, 289.

105. **Samuel, J. L., Schwartz, K., Lompré, A. M., Delcayre, C., Marotte, F., Swynghedauw, B., and Rappaport, L.**, Immunological quantitation and localization of tubulin in adult rat heart isolated myocytes, *Eur. J. Cell Biol.*, 31, 99, 1983.

106. **Ferrans, V. Y. and Roberts, S. C.**, Intermyofibrillar nuclear myofibrillar connections in human and canine myocardium. An ultrastructural study, *J. Mol. Cell. Cardiol.*, 5, 247, 1973.

107. **Hatt, P. Y., Bejal, G., Moravec, J., and Swynghedauw, B.**, Heart failure. An Electron Microscopic study of the left ventricular papillary muscle in aortic insufficiency in the rabbit, *J. Mol. Cell. Cardiol.*, 1, 235, 1970.

108. **Golstein, M. A. and Entman, M. L.**, Microtubules in mammalian heart muscle, *J. Cell Biol.*, 80, 183, 1979.

109. **Geiger, B. and Singer, S. J.**, Association of microtubules and intermediate filaments in chicken gizzard cells as detected by double immunofluorescence, *Proc. Natl. Acad. Sci. USA*, 77, 4769, 1980.

110. **Summerhayes, I. C., Wong, D., and Bochen, L.**, Effect of microtubules and intermediate filaments on mitochondrial distribution, *J. Cell Sci.*, 61, 87, 1983.

111. **De Brabander, M., Geuens, G., Buydens, R., Willebords, R., and De Mey, J.**, Microtubule stability and assembly in living cells: the influence of metabolic inhibitors, taxology and pH, *Cold Spring Harbor Symp. Quant. Biol.*, 46, 227, 1982.

112. **Weber, K. and Osborn, M.**, Microtubule and intermediate filament networks in cells viewed by immunofluorescence microscopy, in *Cytoskeletal Elements and Plasma Membrane Organization*, Poste, G. and Nicolson, G. L., Eds., North Holland, 1981, 1.

113. **Tucker, R. W.**, Role of microtubules and centrioles in growth regulation of mammalian cells, in *Cell and Muscle Motility*, Dowben, R. M. and Shay, J. W., Eds., Vol. 3, 1983, 259.

114. **Warren, R. H.**, The effect of colchicine on myogenesis in vivo in Rana pipiens and Rhodnius prolixus (Hemiptera), *J. Cell Biol.*, 39, 544, 1968.

115. **Eckel, J. and Reinaver, H.**, Effects of microtubule-disrupting agents on insulin binding and degradation in isolated cardiocytes from adult rat, *Hoppe-Seyler's Z. Physiol. Chem.*, 364, 845, 1983.

116. **Auber, J.**, La myofibrillogenese du muscle strié II Vertebrés, *J. Microsc. (Paris)*, 8, 367, 1969.

117. **Cartwright, J., Jr. and Goldstein, M. A.**, Microtubules in soleus muscles of the postnatal and adult rat, *J. Ultrastr. Res.*, 79, 74, 1982.

118. **Toyama, Y., Forry-Schaudies, S., Hoffman, B., and Holtzer, H.**, Effects of taxol and colcemid on myofibrillogenesis, *Proc. Natl. Acad. USA*, 79, 6556, 1982.

119. **Behrendt, H.**, Effect of anabolic steroids on rat heart muscle cells. I Intermediate Filaments, *Cell Tiss. Res.*, 180, 303, 1977.

120. **Berner, P. A., Somlyo, A., and Somlyo, A. P.**, Hypertrophy-induced increase of intermediate filaments in vascular smooth muscle, *J. Cell Biol.*, 88, 96, 1981.

121. **Samuel, J. L., Bertier, B., Bugaisky, L., Marotte, F., Swynghedauw, B., Schwartz, K., and Rappaport, L.**, Different distributions of microtubules, desmin filaments and isomyosins during the onset of cardiac hypertrophy in the rat, *Eur. J. Cell Biol.*, 34, 300, 1984.

122. **Rappaport, L., Samuel, J. L., Bertier, B., Bugaisky, L., Marotte, F., Mercadier, A., and Schwartz, K.**, Isomyosins microtubules and desmin during the onset of cardiac hypertrophy in the rat, *Eur. Heart J.*, 5(Suppl. F), 243, 1984.

123. **Stoeckel, M. E., Osborn, M. Porte, A., Sacrez, A., Batzenschlager, A., and Weber, K.,** An unusual familial cardiomyopathy characterized by aberrant accumulations of desmin-type intermediate filaments, Virchows Arch. *Pathol. Anat. Physiol., 393,* 53, 1984.

124. **Rappaport, L., Samuel, J. L., Bertier-Savalle, B., Marotte, F., and Schwartz, K.,** Microtubules and desmin filaments during the onset of heart growth in the rat, in *Adult Heart Muscle Cells,* Piper, H. M. and Spieckermann, P. G., Eds., Springer Verlag, New York, 1985, 129.

125. **Nath, K., Shay, J. W., and Bollon, A. P.,** Relationship between dibutyryl cyclic AMP and microtubule organization in contracting heart muscle cells, *Proc. Natl. Acad. Sci. USA,* 75, 319, 1978.

126. **O'Connor, C. M., Gard, D. L., and Lazarides E.,** Phosphorylation of intermediate filament proteins by cAMP dependent protein kinases, *Cell,* 23, 135, 1981.

127. **Schulman, H.,** Phosphorylation of microtubule-associated proteins by a Ca^{2+} calmodulin-dependent protein kinase, *J. Cell Biol.,* 99, 11, 1984.

128. **Rappaport, L., Leterrier, J. F., Virion, A., and Nunez, J.,** Phosphorylation of microtubule associated proteins, *Eur. J. Biochem.,* 62, 539, 1976.

129. **Perry, G., Brinkley, B. R., and Bryan, J.,** Interaction of calcium-calmodulin in microtubule assembly in vitro, in *Cell and Muscle Motility,* Vol. 2, Dowben, M. and Shay, J. W., Eds., 1982, 73.

130. **Kira, Y., Ebisawa, K., Koizumi, T., Ogata, E., and Ito, Y.,** Evidence for a hormonal factor mediating the effect of a pressure load on lysine incorporation in rabbit heart, *Biochem. Biophys. Res. Commun.,* 107, 492, 1982.

131. **Kreis, T. E. and Birchmeier, W.,** Microinjection of fluorescently labeled proteins into living cells with emphasis on cytoskeletal proteins, *Int. Rev. Cytol.,* 75, 209, 1982.

Chapter 14

INTRACELLULAR ENERGY TRANSPORT IN HEART CELL CULTURES

Maria W. Seraydarian and Christian Vial

TABLE OF CONTENTS

I. INTRODUCTION

The interest in the chemical energy associated with mechanical work and tension developed by a muscle has a long and exciting history and has often served as a major vehicle in the development of modern physiology and biochemistry.[1] Only relatively recently has this topic been considered in terms of the regulation of energy metabolism and of the intracellular energy transport mechanism involving creatine (Cr), phosphocreatine (PCr), and creatine kinase (CK), particularly in the heart muscle.[2-5] The proposed energy shuttle from the site of energy production, the mitochondria, to the site of energy utilization, the myofibrils, (Figure 1) satisfies the myocardial functional requirements of a high energy resting metabolism with precise coupling of energy production to utilization and efficient access to the available energy for the contraction process. The importance of the shuttle mechanism is still controversial and continues to stimulate a great deal of discussion and experimentation, especially with regard to the cytosolic signal for mitochondrial energy production. This question is not merely of academic interest, since an understanding of the energy metabolism and regulation in the myocardium is a *sine qua non* condition for a rational approach to prevention and treatment of cardiac disorders.

In 1907 it was discovered[6] that muscles were capable of contracting in the absence of oxygen. Under anaerobic conditions, lactic acid was produced which was then thought to be linked directly to the process of contraction, until Lundsgaard[7] demonstrated that contraction (and contracture) could occur in iodoacetate-treated muscles without the formation of lactic acid. At about the same time, the work of the Eggletons[8] and Fiske and Subbarow[9] showed that muscles contained a large amount of PCr which is degraded during the course of contractions. This marked the end of the "lactic acid era" and the beginning of the "melting pot era"[10] with either PCr or ATP[11] being proposed as the immediate source of chemical energy for muscle contraction. PCr was assigned a prominent role based on the observations that during muscle contraction, "phosphagen" (PCr) disappearance was concomitant with liberation of free Cr and the PCr hydrolysis correlated to the energetics of muscle contraction. This was a pioneering step in muscle physiology. It has already been demonstrated that the decrease of PCr was accompanied by an increase in respiration; moreover, addition of Cr and inorganic phosphate (Pi) to the muscle mince resulted in net synthesis of PCr and in increased oxygen consumption lasting 10 to 20 min.[1] The role of ATP as the immediate energy source began to emerge with the observation that ADP was rephosphorylated by PCr in the CK-catalyzed reaction[12] and the discovery of myosin ATP-ase by Engelhart and Ljubimova.[13] The dependence of various "contractile models" on ATP[14] gave further credibility to ATP as the source of chemical energy for the mechanical events in muscle. However, there was no direct proof that ATP underwent hydrolysis during contraction and that the enthalpy production was accounted for by the ATP hydrolysis.

Biochemists, in response to the challenge to provide direct evidence for ATP utilization,[15] developed rapid freezing techniques, but the results showed no significant breakdown of ATP during bursts of muscle activity, either in a single muscle twitch or in short tetanic contraction.[16] The amount of PCr hydrolyzed was roughly as expected for single muscle twitches and the rate was proportional to the tension maintained and to the work performed during short tetani. Both single twitches and tetanic contractions occurred concomitantly with PCr breakdown but not with that of ATP, and with the appearance of Cr but not of ADP. It was thus assumed, quite correctly, that ADP produced in the process of contraction is rephosphorylated to ATP as soon as it is formed. Finally, a direct relationship between ATP utilized and the work done was demonstrated in 1962 by Cain and Davies[17] when CK was inhibited by 2, 4-fluorodinitrobenzene. The CK-catalyzed transphosphorylation, first described by Lohman,[12] indicated that PCr acted as a reservoir for high-energy phosphates.

The mechanism of the transformation of chemical into mechanical energy is still not

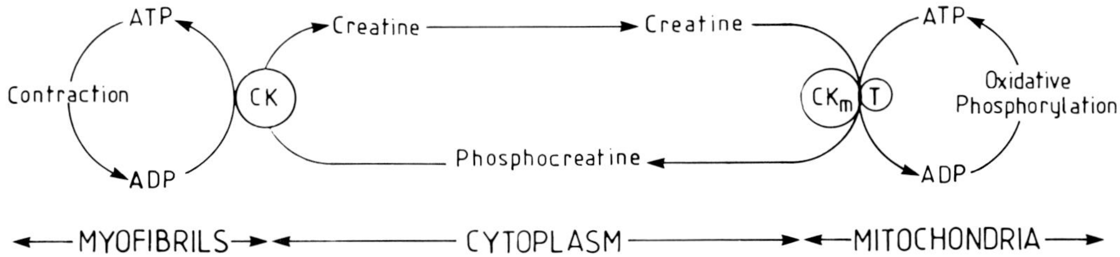

FIGURE 1.   Current model of intracellular energy transport. CKm, mitochondrial creatine kinase; CK, myofibrilar creatine kinase; T, translocase.

clear.[18] Experimental evidence suggests that at rest, ATP is bound to myosin mainly as a complex of myosin with ATP hydrolysis products. The binding of Ca^{2+} to troponin C activates the actin-myosin interaction, with concurrent release of ADP and Pi. Presumably this causes tilting of the mobile myosin cross-bridge, with relative sliding of the myosin and actin filaments past each other producing motion and force. One ATP molecule is split for each make-and-break cycle of the actin-myosin interaction. There is no evidence that the Cr-PCr system is directly involved in any steps in the contraction process other than the CK-catalyzed rephosphorylation of ADP. Yet, there is undoubtedly a close correlation between muscle function and PCr, the function being impaired when PCr is not replenished or CK is inhibited.[5]

Thus "obligatory" ATP rephosphorylation in muscle seems to operate at the site of ATP utilization. Nevertheless, the rate of mitochondrial oxidative phosphorylation is determined by the rate of ATP hydrolysis in extramitochondrial reactions, so that an increased flux through energy utilizing reactions results in increased oxygen uptake. The energy production by the mitochondria is coupled to oxygen uptake reflecting the activity of the mitochondrial respiratory chain in response to a cytosolic signal.[19] Several models for the regulation of energy metabolism essentially explain how the ADP produced outside the mitochondria enters the mitochondria and therefore effectively stimulates oxidative phosphorylation.[20] Recent work suggests that the Cr-PCr-CK system in contractile tissues is involved in the regulation of energy metabolism in the mitochondria and in the intracellular transport.[2-5] PCr subserves the well established role as an "energy buffer", providing high-energy phosphate in the reverse CK reaction, ADP + PCr → ATP + Cr. It has been suggested that Cr is rephosphorylated by mitochondrial ATP in the forward CK reaction, ATP + Cr → ADP + PCr. This was supported by the discovery of CK bound to myofibrils[21] and to mitochondrial membranes.[22] Thus ATP used by the myofibrils in the process of contraction is rephosphorylated by PCr at the site of utilization, Cr being the de facto end product of this process. Cr acts as an effector molecule in the coupling of energy utilization at the myofibrils and energy production in the mitochondria. The energy shuttle from the mitochondria to the myofibrils maintains a high steady-state ATP level, favoring ATP binding to myosin and creating a concentration gradient for Cr between the myofibrils and the mitochondria.

In this chapter, intracellular energy transport will be discussed, with special emphasis on the role of the mitochondrial CK (CKm) in the heart, in general, and in the heart cell cultures, in particular.

II. CREATINE KINASE ISOZYMES IN CARDIAC AND SKELETAL MUSCLE

A. General Considerations

In 1943, Banga[23] separated a component of crude muscle extract which catalyzed the forward CK reaction. The enzyme derived from rabbit muscle was subsequently crystallized

Table 1
SUBCELLULAR DISTRIBUTION OF MUSCLE CK

% of total CK activity

Source	Mito	Myofibrils	Microsomes	Soluble	Ref.
Rabbit heart	6	—	1	70	34
Rat heart	48	—	—	—	22
Rat M. quadriceps	7	—	—	—	22
Pigeon M.pectoralis	25	—	—	—	22
Rat heart	22	8	4	35	35
Rat heart	17	4	2.5	77	36
Human heart	8.5	3.5	2.5	88	36
Guinea pig heart	6	4	1	89	37
Human heart	10	2	3	85	38

Note: Results obtained by differential centrifugation.

Table 2
HEART CK ISOENZYME DISTRIBUTION

% of total CK activity

Source	CKm	MM	MB	BB	Ref.
Dog heart	9	89	—[a]	—[a]	39
Rat heart	33	57	11	11	39
Rat heart	35	40	21.2	3.5	4
Rat heart	20	66	13	—	40
Rat heart	20	61	18	1	41
Rat myocytes	24	60	15	1	41
Rat myocytes	22	53	20	5	42
Human heart	13	65	20	2	38
Mouse heart	9	86	5	—	43

Note: Results obtained from electrophoretic separation and integration of the peak areas.

[a]   Trace.

by Kuby et al.,[24] and the molecular weight determined as about 82,000 daltons.[25] Kinetic studies[26] led to the conclusion that Mg·ADP and Mg·ATP are the substrates of CK. More details on the kinetics may be found in Watts' review.[27] In 1965, Dawson et al.[28] presented evidence for the dimeric structure of CK and for the existence of two different subunits: the M and B monomers. Consequently the enzyme exists in three forms: the MM-type predominates in muscles, the BB-type is mainly found in nervous tissue, and the MB hybrid form is present in small amounts in heart muscle, lungs, stomach, and spleen.[29-31] A mixture of MM and BB dimers can be denatured by urea and guanidinium chloride, leading to the formation of the MB form following dialysis.[32] A similar result is observed after repeated freezing and thawing of MM and BB mixtures.[33] The three isozymes may be separated electrophoretically: the BB isozyme migrates most rapidly towards the anode; MM migration is either very slow towards cathode or does not occur at all; and MB is found in an intermediate position. A fourth distinct type of isozyme migrating cathodically, was isolated from heart mitochondria by Jacobs.[22] The subcellular distribution of CK isozymes in the heart (Table 1 and References 22, 34 to 38) and the precentage of each isozyme form (Table 2, References 4, 38 to 43) show species differences. The latter also depends upon the stage of development as will be seen in Section V.

B. Extramitochondrial Creatine Kinase

Since large amounts of the MM and MB isoenzymes can be extracted from myocardium in isotonic media, they are considered to be soluble cytosolic enzymes; while another fraction of the enzyme is bound to particulate components of the heart cell. It was later established that the MB isozyme is freely soluble in the cytosol,[39] and that it can leak out of the cells when the plasma membrane is damaged, e.g., during ischemia.[44,45] In addition, some of the MM isozyme is bound to the sarcoplasmic reticulum (e.g., References 29, 46) and some associated with the myofibrils.[21,47] The MM isoenzyme has also been detected bound to sarcolemmal and nuclear membranes.[48-50] In skeletal muscle, MM isoenzyme is an integral component of the M-line,[21,51,52] the possible sites of interaction on the myosin molecule being in subfragments 1 and 2.[53-57] Curiously, chicken heart contains almost exclusively the BB isoenzyme which is located within the Z-line region of sarcomeres.[58]

Using a specific antiserum to the M subunit, the cell-free translation products of mRNA from canine myocardium have been immuno-precipitated and shown to be translated as the mature polypeptide.[59] The cDNA sequences and translated polypeptide sequences of rabbit, rat, chick, and Torpedo M subunits have been determined, recently.[60-63] There is a high degree of homology in the amino acid sequence of the M form in all these species and the region around the active site, cysteine, appears to be highly conserved.

It was recognized very early on that the function of the CK isozymes bound either to myofibrils or to membranes might be to phosphorylate ADP, thus providing ATP in the immediate vicinity of the various ATPase activities associated with these structures.[64-67]

It has been shown that at the myofibrillar level:

1. P_i might be preferentially liberated by myosin ATPase from PCr-generated ATP rather than from ATP added to the medium.[68]
2. Part of the MgADP produced by Mg^{2+} ATPase is not made available to pyruvate kinase added to the medium.[69]
3. In the presence of MgADP, PCr prevents rigor tension development and decreases rigor tension more than externally added ATP and;[70-71]
4. There is enough CK activity bound to the M line to rephosphorylate the ATP hydrolyzed by the Mg^{2+}-ATPase.[72]

It has been claimed that ATP produced by sarcolemmal-bound CK is preferentially used by the Na^+ pump;[49] however, this conclusion has been disputed by Philipson and Nishimoto.[73] The nuclear CK might be involved in some processes occurring during mitosis such as spindle elongation.[74,75]

Experiments performed with frog heart have stressed the importance of Cr and PCr levels and of myofibrillar- or sarcolemmal-bound CK isozymes in the control of contractility and electrical activity in this muscle; this topic has been previously reviewed.[76]

C. Mitochondrial Creatine Kinase

1. Identification of CKm

The occurrence of a mitochondrial isozyme of CK was first reported in 1964 by Jacobs and Heldt[77] who observed that[32] P_i was incorporated into PCr when isolated mitochondria (from pigeon or rat skeletal muscle as well as heart muscle) were incubated aerobically with Krebs cycle substrates. This incorporation was stimulated by the addition of Cr, P_i, or ADP. The high levels of CK activity responsible for the PCr synthesis were attributable to a special form of the enzyme (CKm) with cathodic mobility in cellulose acetate electrophoresis at pH 8.8.[22] This was in contrast to the mobility of the cytoplasmic MM, MB, or BB isozymes. CKm belonged to the same constant proportion group as the mitochondrial respiratory enzymes, and it was postulated that the function of the enzyme was to synthetize PCr from mitochondrially generated ATP. High energy phosphate might thus be transferred between the intra- and extra-mitochondrial compartments. The existence of the CKm isozyme was confirmed by several investigators[4,78] and demonstrated by histochemical staining for enzymes activity *in situ*.[79,80] The activity of this isozyme was found to be between 10 and 35% of the total CK activity of the muscle, depending upon the analytical methods used and the species studied (Table 2). Using a rather indirect method, Kupriyanov et al.[81] estimated the molar content of CKm in heart mitochondria as 1 mol of active site of CKm per mole of adenine nucleotide translocase; however, this has not been otherwise confirmed.

CKm is immunologically distinct from the cytoplasmic isozymes, an antibody against CKm does not cross-react with cytosolic CK and conversely, anti-serum against cytosolic CK exhibits no reaction with CKm.[33]

Based on fractionation experiments conducted in phosphate buffers, CKm was believed to be present in the intermembrane space,[82] and Scholte et al.[83] later demonstrated that CKm was in fact bound to the outer surface of the inner mitochondrial membrane. After disruption of the outer membrane by hypotonic swelling or digitonin treatment, CKm can be easily solubilized by phosphate buffer in a concentration and pH-dependent manner.[13-15] This phenomenon is reversible.[84,86] Adriamycin prevents CKm reassociation to the mitochondrial membrane,[88,89] and it has recently been claimed that cardiolipin is involved in the binding.[90] CKm can also be solubilized by adenine nucleotides[85,91] or negatively charged organomercurial reagents.[92] In the latter case, solubilization may be brought about by the titration of a small number of thiol groups probably located on the enzyme but different from the so-called "essential" thiol groups. Two forms of solubilized CKm were observed in cellulose acetate electrophoresis of enzyme from beef, rat, and human[4,93,94] but not in canine heart;[95] the two forms have been called m_1 and m_2 by Hall et al.[93] A high concentration of 2-mercaptoethanol can partially convert m_2, the form which moves rapidly towards the cathode, into m_1, the more slowly moving form; this reaction is partially reversible after concentration by ultrafiltration. The physiological significance of these forms, if any, is not presently understood.

Perryman et al.[59] have shown that canine CKm is encoded by the nuclear genome and that the in vitro translated CKm subunit has a mol wt 6,000 daltons greater than that of the mature form. They assume that, like several other mitochondrial enzymes, premitochondrial CKm is subjected to a proteolytic cleavage during translocation from the cytoplasm into mitochondria.

Until now the cathodically migrating CK isozyme has been assumed to be mitochondrial CK; recently, however, it has been reported that the cell nucleus might also be a possible source of a cathodically migrating CK.[80]

2. CKm in the Regulation of Mitochondrial Respiration

Studies using isolated mitochondria have provided considerable experimental support for a role of CKm in the regulation of mitochondrial respiration. The first investigators who discovered the CK activity associated with heart mitochondria,, observed that Cr was able to stimulate the rates of substrate oxidation and of state 4 oxygen consumption and also to increase the rate of PCr synthesis.[2,3,34,77,97] In the presence of Cr, CKm is able to continuously regenerate ADP, which is the primary acceptor in oxidative phosphorylation. On the other hand, PCr addition results in a decrease of the amount of oxygen taken up during State 3, due to a competition between CKm and oxidative phosphorylation for ADP utilization.[97] This aspect has been recently reviewed by Jacobus.[98]

3. Kinetic Studies of CKm

The kinetic parameters of CKm do not basically differ from those of the MM isozyme from various sources. The mechanism of the reaction is believed to be a rapid-equilibrium random reaction with synergism in binding of the substrates.

Table 3
APPARENT MICHAELIS CONSTANTS FOR HEART CKm

| | Estimated rate | ATP production | ATP | Apparent Km | | | Ref. |
				Cr	ADP	PCr	
Rat	PCr	—	0.1	6			3
	NADPH	—	—	—	0.035	0.72	
Rat	NADH	PEP/PK	0.7	5			99
	NADPH	—	—	—	0.05	0.5	
Beef	pHstat	—	0.056	4.5	0.015	0.31	100
Purified							
Rat	PCr	OP	0.037	—	—	—	101
	PCr	PEP/PK	0.200	—	—	—	
	O_2	OP	0.034	—	—	—	
Soluble	PCr	PEP/PK	0.200	—	—	—	
Rabbit	PCr	OP	0.09	—	—	—	102
		PEP/PK	0.13	—	—	—	
Mitoplasts	PCr	OP	0.084	—	—	—	
		PEP/PK	0.090	—	—	—	
Rabbit	PCr	OP	0.11	—	—	—	103
	PCr	PEP/PK	0.09	—	—	—	
	NADH	PEP/PK	0.78	—	—	—	
Mitoplasts	PCr	OP	0.21	—	—	—	
Soluble	PCr	PEP/PK	0.11	—	—	—	
	NADH	PEP/PK	0.63	—	—	—	
Rat	NADH/NADPH	—	0.24 (sucrose)	—	0.048	—	104
			0.51 (KCl)	—	0.15	—	

Notes: The rates of reaction were estimated either by NADH/NADPH or PCr production or by O_2 consumption. ATP was generated either by mitochondrial oxidative phosphorylation (OP) or by the phosphoenolpyruvate pyruvate kinase system (PEP/PK). Km values are mM.

Table 3[3,99-104] presents a summary of the apparent Michaelis constants for CKm, the methods of measuring reaction rates, and the sources of ATP used by several investigators. The Km values for ATP and Cr are always higher than those for ADP and PCr, making the reverse reaction (ATP synthesis) more favorable than the forward reaction (PCr synthesis). Furthermore, the rate of the reverse reaction has been found to be higher than, or equal to that of the forward reaction. Table 4[101,103,105-108] shows that the intrinsic kinetic constants for CKm are not significantly different in mitochondrial and solubilized CKm. Therefore, the binding of the enzyme to the mitochondrial membrane does not seem to result in a higher affinity for ATP (i.e., favoring PCr synthesis). The conditions used to determine CKm kinetic parameters are important. Different results were obtained when: (1) the rates of reaction of PCr were determined directly or in a coupled enzyme assay and (2) intra- or extra-mitochondrially generated ATP was used.

It is apparent from Table 4 that each time the reaction has been assessed spectrophotometrically by the pyruvate kinase-lactate dehydrogenase coupled enzyme assay, the Km for ATP was higher than that obtained either by direct estimation of PCr synthesis or by measurement of the increase in oxygen consumption, with ATP derived either from oxidative phosphorylation or from the phosphoenolpyruvate/pyruvate kinase regenerating system. The apparent Km for ATP in the two latter cases was smaller than the true Km obtained by the spectrophotometric method (Table 3). The reason for this discrepancy is not clear. The efficiency of the phosphoenolpyruvate pyruvate kinase system in regenerating ATP and

Table 4
KINETIC CONSTANTS FOR BOUND OR SOLUBLE KCm AND THE MM ISOZYME

Source	Method	pH	K_iATP	KATP	K_iCR	KCr	Ref.
Rabbit MM	pH-stat	8	1.2	0.48	15.6	6.1	105
	pH-stat	9	0.7	0.25	24.4	8.6	106
Calf MM	pH-stat	8.8	0.97	0.78	53	21	107
Pig CKm							
Bound	pH-stat	8.8	0.56	0.30	15.7	9.7	103
Soluble	pH-stat	8.8	0.6	0.2	11.2	4.5	
Rabbit CKm							
Bound	pH-stat	8.8	0.7	0.5	19.6	16.2	103
Bound	PEP/CK	7.4	1.4	0.36	34.1	4.4	
Soluble	PEP/CK	7.4	2.4	0.16	39.8	3.95	
Rat CKm							
Bound	PEP/CK	7.4	0.73	0.73	5	5	101
Soluble	PEP/CK	7.4	0.6	0.35	8.5	5	
Rat CKm							
Bound	PEP/CK	7.4	0.75	0.15	28.8	5.2	108
	OP	7.4	0.29	0.014	29.4	5.2	

maintaining a low level of ADP might be the critical factor.[109] Another possibility might be that CKm is inhibited either by phosphoenolpyruvate or by impurities found in the commercial preparations of this compound.[110,111]

Jacobus and Saks[108] have demonstrated (by using two different ways of estimating the rate of PCr synthesis) that with respiration intact a tenfold decrease of the dissociation constant of MgATP from the ternary complex CKm-Cr-MgATP occurred, as compared to the value obtained with rotenone and oligomycin-treated mitochondria with an external ATP-regenerating system. The stability of the ternary complex is enhanced by an unknown mechanism and this is interpreted by the authors as a possible evidence for the direct channeling of mitochondrial ATP to CKm, mediated by the adenine nucleotide translocase.

Experiments using labeled Pi or ATP were conducted in an effort to elucidate whether ATP coming from oxidative phosphorylation has preferred access to CKm (over external ATP). There is still no agreement between different investigators.[112-114] The relative contribution of the *de novo* synthesized and the external ATP pools to the synthesis of PCr might depend upon the total amount of nucleotides with the mitochondrial contribution decreasing when external ATP concentration increases. The outer mitochondrial membrane might constitute a micro-environment for CKm by creating a partial diffusion barrier to the efflux of newly synthetized ATP and to the influx of ATP from the medium.[115]

Possible "coupling" between CKm and adenine nucleotide translocase has been studied. It has been demonstrated that ADP generated by CKm was able to overcome the atractyloside-induced inhibition of respiration, but that CKm activity itself is not affected by atractyloside.[104] When ADP was produced by CKm, a fivefold increase of atractyloside concentration was required to inhibit respiratory stimulation to the same degree as that obtained with added ADP.[116] Recently it was demonstrated that the forward CKm reaction inhibits translocation of external ADP into the mitochondrial matrix, and the reverse CKm reaction inhibits translocation of ATP in a similar way. The apparent inhibition of nucleotide exchange is neither due to Cr nor to PCr. CKm reacts with the newly exported nucleotides and the latter effectively competes with nucleotides added to the medium for the transport back into the matrix. These results support functional coupling between PCr synthesis by CKm, adenine nucleotide translocase, and oxidative phosphorylation.[117]

On the other hand, a specific channeling of high-energy phosphate by PCr due to a slower diffusion in the cytoplasm of adenylate than of Cr has been proposed but not proven. Also, a facilitated diffusion mechanism for PCr was suggested, especially in myocytes where the distance between ATP-utilizing and ATP-generating sites is very small. The only functions of CK would be the classical energy storage and the potential activation of glycolysis by release of Pi.[118] This latter role, regulation of glycolysis by removal or release of Pi, has been previously suggested for muscle.[119-120]

The proposed function of CKm in the intracellular energy transport is primarily based on experiments with isolated mitochondria. Investigations at the cellular and organ levels of biological complexity have provided indirect experimental support for the energy shuttle. The results from kinetic studies remain equivocal.

III. CREATINE KINASE ISOZYMES IN HEART CELL CULTURES

The rat myocardial cell culture is a stable preparation characterized by spontaneous rhythmic contractions uncomplicated by either innervation or hormonal influences and essentially with no diffusion barriers.[121] Such a model system permits investigations at the cellular level and has been utilized in the study of the energy metabolism of the myocardium. With no evidence for any direct participation of PCr in the process of contraction, we have observed that the rate of spontaneous contractions correlated with the intracellular concentration of PCr. In the presence of oligomycin, an inhibitor of oxidative phosphorylation, the cessation of contractions within 30 to 60 min following the addition of the drug correlated with the depletion of PCr while ATP remained practically unchanged.[122] In another series of experiments, when the cells were grown in medium either containing 5 m*M* creatine or pulsed with 50 m*M* creatine, there was an increase of intracellular PCr, with no change in ATP, these changes being independent of ATP concentration. The increase in PCr was dependent on aerobic metabolism and was inhibited by oligomycin. Since the only metabolic path known for the aerobic synthesis of PCr is via CK, the results suggested that CKm was present in the cultured cells. Since CKm was not found in fetal hearts of several species[43] its presence in the myocardial cell culture remained to be demonstrated.

Using electrophoresis on cellulose acetate strips, four CK isozymes were identified in the cultured rat myocardial cells (Figure 2A): the CKm, MM, MB, and BB. The bands were located by comparing their electrophoretic mobility to that of purified CKm, MM, and BB on accompanying strips. CKm was absent in nonmuscle cells derived from the same neonatal rat hearts and from a muscle cell line, L6 (Figures 2B and 2C).

The CKm was present in all the cultures derived from neonatal rats (mixed, 1- to 6-days old). Figure 3 shows the percentage increase in CKm in myocardial cells with time in culture. The changes in CKm are masked by the increase of the nonmuscle cell population with time in culture, as confirmed by the increase in the percentage of BB and decrease of the total CK mU/mg protein. The estimated population of myocardial cells after 96 hr in culture is only 50%.

The percentage of mitochondrial CK increased with the age of the newborn rats from which the culture was derived, suggesting that the majority of cells might be derived from myocytes and not from fetal cells. Figure 4 shows the electrophoretogram of CK isoenzymes of myocardial cell cultures prepared from fetal, 18 days after gestation, 1-day-old newborn and 6-day-old newborn rat hearts, and the corresponding fresh tissue. The adult rat heart CK isoenzyme pattern is also shown. The MM isozyme is dominant at all the developmental stages of myocardial cells, while BB decreases with progressive development. Figure 4 also demonstrates that the CK isoenzymes of cultured cells tend to correspond to the isozyme patterns typical of the developmental stage of tissue from which the cultures were derived; the percentage of isoenzyme distribution approximates the fresh tissue, although total CK in the fresh tissue is much higher than the activity in cultured cells at 96 hr (1.5 U/mg

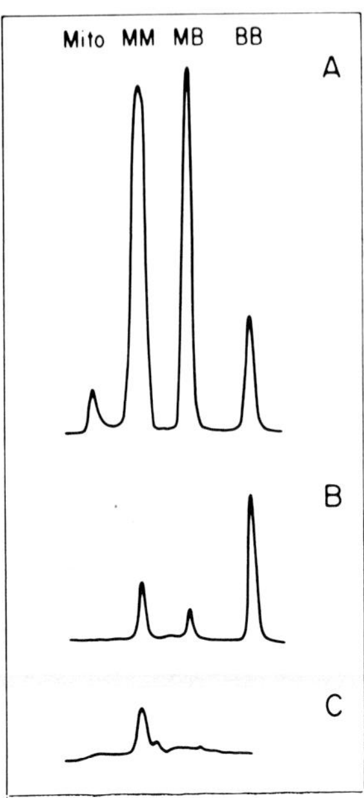

FIGURE 2.   Electrophoretogram of CK isoenzymes on cellulose acetate strips from 96 hr cultures. A. Rat myocardial cells, 18 μg protein. B. Rat non-muscle cells derived from the same hearts as in (A), 17 μg protein. C. L6 muscle cells, 100 μg protein. (From Van Brussel, E., Yang, J. J., and Seraydarian, H. W., Isozymes of creatine kinase in mammalian cell cultures, *J. Cell. Physiol.*, 116, 221, 1983. With permission.)

protein in culture, 6.2 U/mg protein in fetal heart, 7.3 U/mg protein in 1- to 6-day-old rat heart, and 21.4 U/mg protein in 5-month-old rat heart). Similar results were obtained with myosin isozymes separated by electrophoresis on polyacrylamide gels, i.e., the postnatal myocardial myosin pattern of the predominant V_1 isozyme was also characteristic of cultured cells derived from neonatal hearts; rat fetal myocardial cells are characterized by predominant V_3 isozyme.[42,123]

The absence of CKm in the fetal hearts[43] indicates that the regulation of energy production in the mitochondria by Cr and energy transport by PCr develop postnatally. The possibility that Cr (and/or oxygen tension) acts as the trigger for the development of the regulatory mechanisms requires comprehensive study, but preliminary results showed that when cells were cultured in the presence of 20 mM Cr the CKm levels increased. In a series of 20 experiments the mean increase of CKm was 44% (after 48 to 192 hr in culture) with p < 0.005 in the paired T-test. Rat myocardial cell cultures could serve as an appropriate model for the study of the trigger mechanism for the synthesis of CKm.

Indeed, the increase of CKm in cells grown in medium enriched with 20 mM Cr indicate that Cr might be the postnatal trigger for CKm synthesis. It is conceivable that the regulation of energy production by Cr at the mitochondria and the intracellular energy shuttle via the Cr-PCr-CKm system develops concomitantly with the increased myocardial energy demand, perhaps coupled to the increased arterial oxygen tension.

Although it is very likely that in the heart and in the skeletal muscle CKm catalyzes the resynthesis of PCr at the mitochondria and thus plays a crucial role in intracellular energy transport, and that Cr is an effector molecule in the coupling of energy production to energy utilization, the concept remains controversial (e.g., Reference 118).

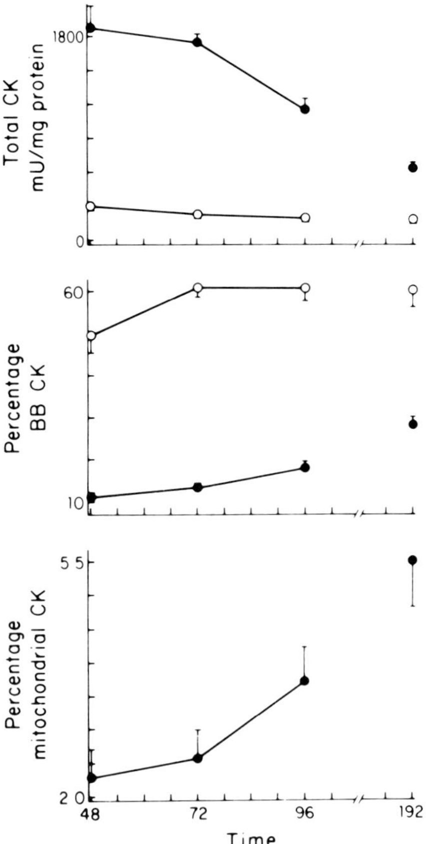

FIGURE 3.   Total CK and percentage distribution of BB and mitochondrial CK in cultured myocardial (●) and non-muscle cells (○) derived from the same rat heart. Each point is a mean value of 10 experiments ± SEM. (From Van Brussel, E., Yang, J. J., and Seraydarian, H. W., Isozymes of creatine kinase in mammalian cell cultures, *J. Cell. Physiol.*, 116, 221, 1983. With permission.)

IV. OXYGEN MEASUREMENT IN HEART CELL CULTURES

In the study of energy metabolism it eventually becomes essential to measure the cellular oxygen consumption. Until now, the basic measurement of oxygen consumption in cultured cells called for manipulations of the cells with potential damage to the cell membrane. Most commonly, in order to study oxygen consumption, the cultured myocardial cells were removed from the substratum in the culture dish by protease treatment, and the oxygen consumption was measured in cell suspensions using a commercial oxygraph. The number of viable cells actually decreased during the course of the measurements of oxygen uptake and the myocardial cells were always quiescent. Measurements performed on myocardial cells still attached to the substratum in the dish and contracting spontaneously could provide important data on the rate of oxygen consumption. A Lucite attachment which permits the measurement of oxygen consumption in cells in culture without manipulating the cells was constructed. The attachment fits over commercially available culture dishes and has an oxygen electrode built into it. Oxygen uptake of cells was thus measured while the cells were attached to the substratum in the culture dish and could be observed using an inverted phase microscope.[124]

A schematic drawing of the Lucite attachment is shown in Figure 5. Figure 6A shows

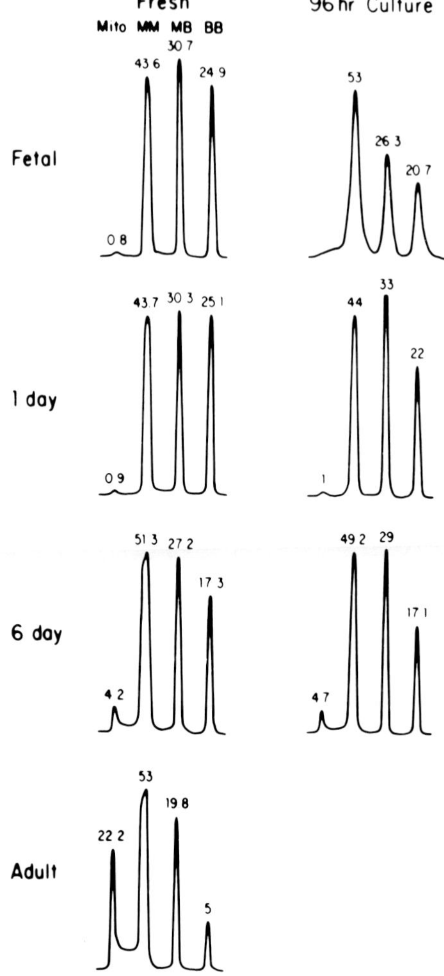

FIGURE 4.   Electrophoretogram of CK isozymes on cellulose acetate strips from fresh tissues and corresponding 96 hr cultures. Fetal rat, 18 days gestation, fresh heart and cultured myocardial cells; 1-day-old newborn rat, fresh heart and cultured myocardial cells; 6-day-old rat, fresh heart and cultured myocardial cells; adult rats 5 months old, fresh heart. Approximately 10 mU CK per strip. (Fron Van Brussel, E., Yang, J. J., and Seraydarian, H. W., Isozymes of creatine kinase in mammalian cell cultures, *J. Cell. Physiol.*, 116, 221, 1983. With permission.)

the stability of oxygen level in air-saturated F-10 medium in a culture dish without cells and the stability of oxygen level when nitrogen-saturated F-10 medium has been introduced (Figure 6C). The uptake of oxygen by cultured myocardial cells in air-saturated F-10 is shown in Figure 6B (up to the arrow). The addition of 2 mM KCN to the medium (at the location of the arrow) completely stopped oxygen uptake by the cells.

Figure 7 shows that the addition of 3 mM EGTA (which chelates free calcium) to the medium resulted in the cessation of beating and in a decrease in the rate of oxygen uptake (Figure 7B). As expected the addition of 3 mM EGTA [Ethylene Glycol bis(Beta-Amino-ethylether) N-N′-tetracetic acid] to the medium prior to the measurement produced a de-creased rate of oxygen consumption from the start of the measurement (Figure 7A). The rates of oxygen consumption of beating and quiescent myocardial cells were thus obtained: 40.5 nmol/min/mg protein and 27.6 nmol/min/mg protein, respectively. Other mammalian cells studied did not show significant changes of oxygen consumption rate in the presence

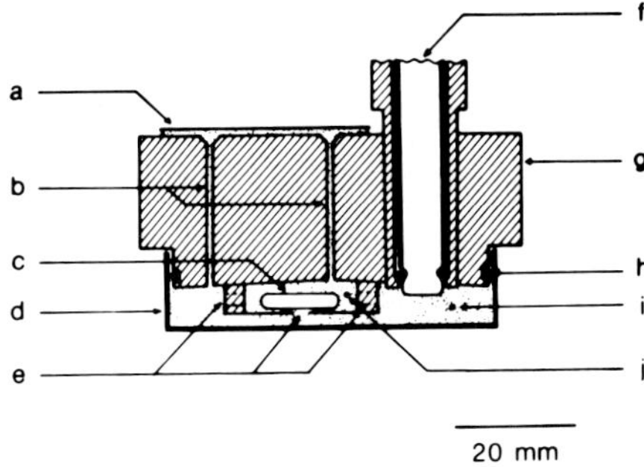

FIGURE 5.   Attachment for measuring oxygen uptake of cells in culture (cross sectional view). (a) Parafilm, (b) holes, 1mm diameter, (c) magnetic stirring bar, (d) culture dish (Falcon #3002), (e) holes, 2.5 mm diameter, (f) polarographic oxygen electrode, (g) Lucite attachment, (h) O-ring, (i) medium in culture dish, and (j) circular stirring chamber. (From Takenori, Y., Yang, J. J., Ricchiuti, N. V., and Seraydarian, M. W., Oxygen consumption of mammalian myocardial cells in culture: measurements in beating cells attached to the substrate of the culture dish, *Anal. Biochem.*, 145, 302, 1985. With permission.)

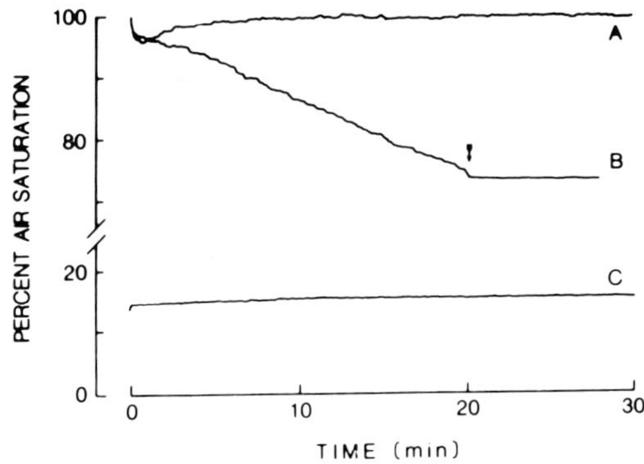

FIGURE 6.   Typical traces of the changes of oxygen level in the medium at 37°C. (A) air-saturated F-10 medium without cells; (B) air-saturated F-10 medium with myocardial cells 72 hr in culture (646 μg total protein). 2 mM KCN added at arrow. (C) nitrogen-saturated F-10 medium without cells. (From Takenori, Y., Yang, J. J., Ricchiuti, N. V., and Seraydarian, M. W., Oxygen consumption of mammalian myocardial cells in culture: measurements in beating cells attached to the substrate of the culture dish, *Anal. Biochem.*, 145, 302, 1985. With permission.)

and absence of EGTA. Therefore, the change in the rate of oxygen consumption of myocardial cells in the presence and absence of EGTA is attributed to the higher (by about 47%) rate of oxygen uptake of beating myocardial cells in culture. In adult heart the rate of oxygen uptake is known to be about 35% greater in beating as compared to the quiescent heart.[125] Thus the relative rates of oxygen consumption of beating and quiescent myocardium and of beating and quiescent cultured neonatal myocardial cells are comparable.

The reported oxygen uptake of non-beating whole heart[125] is about 2.5 ml/100 g wet wt/min or 15.5 nmol/min/mg protein. The oxygen uptake rate of adult cardiac myocytes has been reported to be 69.5 nmol/min/mg protein[126] in conditions roughly comparable to those in neonatal culture. The difference in the oxygen uptake of heart cells in culture derived from newborn rat and from adult myocardium might be explained in the light of the difference in the content of CKm in the two cell populations (with a higher level of CKm in adult heart[4] than in the heart cells from newborn rat[42]). These results are consistent with the view that CKm plays an important role in the coupling of energy utilization and energy production

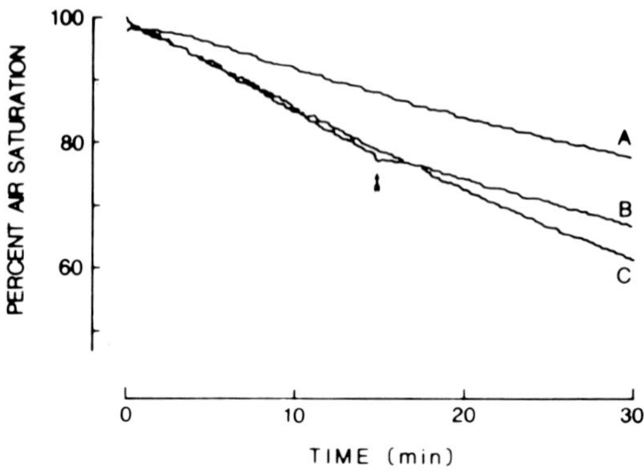

FIGURE 7. Typical traces of oxygen uptake of myocardial cells in culture at 37°C in the presence and absence of EGTA. A, B, and C are measurements in the different culture dishes, each about 72 hr in culture, with 689 g, 710 g, and 798 g total protein, respectively. (A) myocardial cells in F-10 + 3 m*M* EGTA; (B) myocardial cells in F-10, 3 mM EGTA added at arrow; (C) myocardial cells in F-10. (From Takenori, Y., Yang, J. J., Ricchiuti, N. V., and Seraydarian, M. W., Oxygen consumption of mammalian myocardial cells in culture: measurements in beating cells attached to the substrate of the culture dish, *Anal. Biochem.*, 145, 302, 1985. With permission.)

in heart cells and its low concentration in the neonatal cell in culture might limit the respiratory activity. It is expected that the rates of oxygen consumption will change in parallel with various concentrations of Cr in myocardial cells but not in either nonmuscle heart cells or in the L6 muscle cell line.

Oxygen consumption of quiescent myocardial cells was about twice that of the heart non-muscle cells, HeLa or L-6 cells, consistent with the data that the adult myocardium has a high resting level of energy metabolism.[125] This characteristic of myocardial cell culture further demonstrates its usefulness as a model system in the studies of the myocardial energy metabolism. It is reasonable to assume that specific regulatory mechanisms are needed to prevent a possible depletion of ATP resulting from further energy demand during activity surperimposed on the high resting level of metabolism in the heart.

V. DEVELOPMENTAL ASPECTS OF CREATINE KINASE ISOZYMES

A. Changes in the Heart During the Pre- and Postnatal Periods

Analysis of the heart CK isozyme composition at various times before and after birth reveals a complex picture of the appearance and disappearance of isozymes which have been studied in detail in some species. Striking changes occur in the fetal mouse before birth: the 14-day-old fetal heart contains only BB isozymes; three days later about 80% of the activity is due to the presence of the M form, containing isozymes MB and MM. CKm is not detectable at birth, it amounts to 1% of total CK activity in 6-day-old mice and it reaches the adult level of 9% on day 25 after birth. During the same period, there is a decrease in BB-isozyme content from around 3 to 4 to 0%, a decrease in MB-isozyme from 24 to 5% and an increase in total CK activity.[29,11,127] In contrast, in a large animal, such as the lamb, MM is already present at birth, suggesting that the events which take place towards the end of the fetal life of the mouse (decrease of BB, transient increase of MB, and increase of MM) occur well before birth in the lamb. A low concentration of CKm is also present at birth which increases threefold during the following months to attain the adult level.[128] The lamb heart Cr content increased simultaneously with the accumulation of the MM isozyme and the CKm isozyme.

It has been suggested that the neonatal increase of CK activity in the rat heart was concomitant with a marked rise in the respiratory enzymes, ATP synthesis and/or the myo-fibrillar ATPase.[129] However, it should be noted that neither CKm nor MM isozymes are ubiquitous, they are not found, for instance, in frog or chick heart, so it appears that at least in these species, CKm is not essential for energy metabolism.[127]

B. Striated Muscle Cells in Culture

CK activity or the transition from BB to MM are often used as biochemical markers for muscle differentiation. The expression of the isozymes in myogenic cell cultures has been studied extensively.[130-132] The changes which are observed in chick myogenic cell culture undergoing differentiation are similar to those observed in embryogenic muscle. There is, however, an incomplete transition in vitro as compared to the in vivo situation: a significant amount of MB and BB isozymes remains in cultured cells. So far, this has not been satisfactorily explained.[133-135]

The regulation of muscle-specific gene expression with a focus on CK isozymes accu-mulation and biosynthesis, in particular, has been studied by Eppenberger's group. It seems that the BB to MM transition is principally governed by changing rates of protein synthesis of the two types of CK subunits which depends upon the levels of mRNA for the two forms rather than by differential degradation rates.[136-138] According to these authors, B-CK synthesis is a property of myogenic cells whereas a high rate of M-CK synthesis could be characteristic for myotubes. The results suggest that transcriptional control is most likely to trigger the isozyme switch.

Finally, it should be recalled that Cr, which can be regarded as an end product of muscle activity, has also been shown to play a role in myogenesis: addition of Cr stimulates the rate and extent of fusion of cultured skeletal muscle myoblasts into myotubes.[122,139] In addition, Cr has been reported to enhance the rate of synthesis of contractile proteins.[140] Immunofluorescence staining allows the detection of M-line-associated MM-CK very early in the development of myogenic cells. This association of MM-CK with myofibrils may represent an important step in the improvement of their contractile properties and in their physiological maturation.[141-142] The specific case of mammalian heart cells in culture has been discussed in Chapter 3.

C. Cardiac Hypertrophy

The subcellular distribution and tissue content of CK isozymes is altered by cardiac hypertrophy. Ingwall's results have shown that in two models of canine hypertrophy, the extent of hypertrophy correlated with the accumulation of the "fetal-type" isozymes MB and BB. During stable hypertrophy MM and CKm levels do not vary; however, the transition from compensated to uncompensated hypertrophy is accompanied by a 30 and 50% decrease in activities, respectively. Simultaneous NMR analysis of high energy phosphate levels and of the flux through the CK reaction has shown that these two parameters were also reduced in hypertrophy.[143-144]

VI. CONCLUSIONS

The intracellular energy transport in the myocardium involving the Cr-PCr-CKm system is consistent with the experimental results obtained from a number of laboratories. As shown in Figure 1, the proposed energy shuttle depends on the presence of CKm bound to mito-chondria and CK bound to myofibrils; thus PCr phosphorylated at mitochondria by CKm diffuses to the myofibrils where myofibrillar CK maintains the ATP level. The system operates efficiently because of the spatial displacement between the two events. It is tempting to speculate that the function of PCr in the energy transport has the advantage of PCr not being susceptible to nonspecific enzymatic reactions. Furthermore, since ATP is directly involved in the process of contraction and PCr is not, some regulation of muscle activity prior to a potential depletion of ATP is allowed for. Finally, the regulation of energy metabolism and the energy shuttle by the Cr-PCr-CKm system is concomitant with the continuous energy demand of the myocardium superimposed upon a high resting metabolism. This system is not ubiquitous and its development in myocardial cells is usually triggered postnatally.

ACKNOWLEDGMENTS

The authors' work reported in this paper was supported by grants from the Public Health Service NHLBI # 21080, the American Heart Association #81940, and the American Heart Association, Greater Los Angeles Affiliate (to M.W.S.). We thank the CNRS for financial support LP 5421 (to C.V.).

REFERENCES

1. **Needham, D. M.,** *Machina Carnis,* Cambridge University Press, New York, 1981.
2. **Bessman, S. P. and Fonyo, A.,** The possible role of mitochondrial bound creatine kinase in regulation of mitochondrial respiration, *Biochem. Biophys. Res. Commun.,* 22, 597, 1966.
3. **Jacobus, W. E. and Lehninger, A. L.,** Creatine kinase of rat heart mitochondria, *J. Biol. Chem.,* 248, 4803, 1973.
4. **Saks, V. A., Chernousova, G. B., Voronkov, I. I., Smirnov, J. N., and Chazov, E. I.,** Study of energy transport mechanism in myocardial cells, *Circ. Res.,* 34 (Suppl. III), 138, 1974.
5. **Seraydarian, M. W. and Abbott, B. C.,** The role of the creatine-phosphocreatine system in muscle, *J. Mol. Cell. Cardiol.,* 8, 741, 1976.
6. **Fletcher, W. N. and Hopkins, F. G.,** *J. Physiol. (London),* 35, 247, 1907.
7. **Lundsgaard, E.,** Untersuchungen uber Muskelkontraktionen ohne Milchsaurebildung, *Biochem. Z.,* 217, 162, 1930.
8. **Eggleton, P. and Eggleton, G. P.,** The physiological significance of "phosphagen", *J. Physiol.,* 63, 155, 1927.
9. **Fiske, C. H. and Subbarow, Y.,** The nature of the "inorganic phosphate" in voluntary muscle, *Science,* 65, 401, 1927.
10. **Carlsson, F. D. and Wilkie, D. R.,** Muscle Physiology, *Prentice-Hall Biol. Sci. Series,* Prentice-Hall, Inc., Englewood Cliffs, N. J., 1974.
11. **Fiske, C. H. and Subbarow, Y.,** *Science,* 70, 381, 1929.
12. **Lohman, K.,** Uber die enzymatische Aufspaltung der Creatinphosphosaure; zugleich ein Beitrag zum Chemismus der Muskelkontraktion, *Biochem. Z.,* 271, 264, 1934.
13. **Engelhart, V. A. and Ljubimova, M. N.,** Myosine and adenosine-triphosphatase, *Nature,* 144, 668, 1939.
14. **Szent-Gyorgyi, A.,** *Chemistry of Muscular Contraction,* Academic Press, New York, 1947.
15. **Hill, A. V.,** A challenge to biochemists, *Biochim. Biophys. Acta,* 4, 4, 1950.
16. **Mommaerts, W. F. H. M.,** Energetics of contraction, *Physiol. Rev.,* 40, 427, 1969.
17. **Cain, D. F. and Davies, R. E.,** Breakdown of adenosine triphosphate during a single contraction of working muscle, *Biochem. Biophys. Res. Commun.,* 8, 361, 1962.
18. **Homsher, E. and Kean, C. J.,** Skeletal muscle energetics and metabolism, *Ann. Rev. Physiol.,* 40, 93, 1978.
19. **Hansford, R. G.,** *Curr. Top. Bioenerg.,* 10, 297, 1980.
20. **Jacobus, W. E.,** Respiratory control and the integration of heart high energy phosphate metabolism by mitochondrial creatine kinase, *Ann. Rev. Physiol.,* 47, 707, 1985.
21. **Turner, D. C., Walliman, T., and Eppenberger, H. M.,** A protein that binds specifically to the M-line of skeletal muscle is identified as the muscle form of creatine kinase, *Proc. Natl. Acad. Sci. USA,* 70, 702, 1973.
22. **Jacobs, H. K., Heldt, H. W., and Klingerberg, M.,** High activity of creatine kinase in mitochondria from muscle and brain and evidence for a separate mitochondria isoenzyme of creatine kinase, *Biochem. Biophys. Res. Commun.,* 16, 516, 1964.

23. **Banga, I.,** *Studies from the Institute of Medical Chemistry,* Vol. 3, Szent-Györgyi, A., Ed., S. Karger, Basel, 1943, 59.
24. **Kuby, S. A., Noda, L., and Lardy, H. A.,** Adenosinetriphosphate-creatine transphosphorylase. Isolation of the crystalline enzyme from rabbit muscle, *J. Biol. Chem.,* 209, 191, 1954.
25. **Noda, L., Kuby, S. A., and Lardy, H. A.,** Adenosinetriphosphate-creatine transphosphorylase. Homogeneity and physicochemical properties, *J. Biol. Chem.,* 209, 203, 1954.
26. **Kuby, S. A., Noda, L., and Lardy, H. A.,** Adenosinetriphosphate-creatine transphosphorylase. Kinetics studies, *J. Biol. Chem.,* 210, 65, 1954.
27. **Watts, D. C.,** Creatine kinase, in *The Enzymes.,* Vol. 8 (part A), Boyer, P., Ed., Academic Press, New York, 1973, 383.
28. **Dawson, D. M., Eppenberger, H. M., and Kaplan, N. O.,** Creatine kinase: evidence for dimeric structure, *Biochem. Biophys. Res. Commun.,* 21, 346, 1965.
29. **Eppenberger, H. M., Eppenberger, M., Richterich, R., and Aebi, H.,** The ontogeny of creatine kinase isozymes, *Develop. Biol.,* 10, 1, 1964.
30. **Eppenberger, H. M., Dawson, D. M., and Kaplan, N. O.,** The comparative enzymology of creatine kinases: isolation and characterization from chicken and rabbit tissue, *J. Biol. Chem.,* 242, 204, 1967.
31. **Roberts, R., Henry, P. D., and Sobel, B. E.,** An improved basis for enzymatic estimation of infarction size, *Circulation,* 52, 743, 1975.
32. **Dawson, D. M., Eppenberger, H. M., and Kaplan, N. O.,** The comparative enzymology of creatine kinases: physical and chemical properties, *J. Biol. Chem.,* 242, 210, 1967.
33. **Roberts, R. and Grace, A. M.,** Purification of mitochondrial creatine kinase, *J. Biol. Chem.,* 255, 2870, 1980.
34. **Sobel, B. E., Shell, W. E., and Klein, M. S.,** An isoenzyme of creatine phosphokinase associated with rabbit heart mitochondria, *J. Molec. Cell. Cardiol.,* 4, 367, 1972.
35. **Scholte, H. R.,** On the triple localization of creatine kinase in heart and skeletal muscle cells of the rat: evidence for the existence of myofibrillar and mitochondrial isozymes, *Biochem. Biophys. Acta,* 305, 413, 1973.
36. **Desjardins, P. R. and Pesclovitch, R.,** Subcellular localization of human atypical creatine kinase, *Clin. Chim. Acta.,* 135, 35, 1983.
37. **Ogunro, E. A., Peters, T. J., and Hearse, D. J.,** Subcellular compartmentation of creatine kinase isoenzymes in guinea pig heart, *Cardiovasc. Res.,* 11, 250, 1977.
38. **Blum, H. E., Weber, B., Deus, B., and Gerok, W.,** The mitochondrial izozyme from human heart muscle, in *Creatine Kinase Isoenzymes,* Lang, H., Ed., Springer-Verlag, New York, 1981, 19.
39. **Ingwall, J. S.,** The Creatine kinase system in hypertrophied heart, *J. Molec. Cell. Cardiol.,* 15 (Suppl. II), 38, 1983.
40. **Vatner, D. E. and Ingwall, J. S.,** Effects of moderate pressure overload cardiac hypertrophy on the distribution of creatine kinase isozymes, *Proc. Soc. Exp. Biol. Med.,* 175, 5, 1984.
41. **Murphy, M. P., Hohl, C., Brierly, G. P., and Altschuld, R. A.,** Release of enzymes from adult rat heart myocytes, *Circ. Res.,* 51, 560, 1982.
42. **Van Brussel, E., Yang, J. J., and Seraydarian, H. W.,** Isozymes of creatine kinase in mammalian cell cultures, *J. Cell. Physiol.,* 116, 221, 1983.
43. **Hall, N. and De Luca, M.,** Developmental changes in creatine kinase isozymes in neonatal mouse hearts, *Biochem. Biophys. Res. Commun.,* 66, 988, 1975.
44. **Konttinen, A. and Sommer, H.,** Determination of serum creatine kinase isoenzymes in mitochondrial infarction, *Am. J. Cardiol.,* 29, 817, 1972.
45. **Wagner, G. S., Roe, C. R. Limbird, L. E., Rosati, R. A., and Wallace, A. G.,** The importance of identification of the myocardial specific isoenzyme of creatine kinase (MB form) in the diagnosis of acute myocardial infarction, *Circulation,* 47, 263, 1973.
46. **Baskin, R. J. and Deamer, D. W.,** A membrane bound creatine phosphokinase in fragmentated sarcoplasmic reticulum, *J. Biol. Chem.,* 245, 1345, 1970.
47. **Ottaway, J. H.,** Evidence for binding of cytoplasmic creatine kinase to structured elements in heart muscle, *Nature,* 215, 521, 1967.
48. **Saks, V. A., Lipina, N. V., Sharov, V. G., Smirnov, V. N., Chazov, E. I., and Grosse, R.,** The localization of the MM isoenzyme of creatine kinase on the surface membrane of myocardial cells and its functional coupling to ouabain-inhibited (Na, K) ATPase, *Biochim. Bophys. Acta.,* 465, 550, 1977.
49. **Grosse, R., Spitzer, E., Kupriyanov, V. V., Saks, V. A., and Repke, K. R. H.,** Coordinate interplay between (Na^+/K^+) antiport across the membrane of vesicles formed from the plasmic membrane of cardiac muscle cell, *Biochim. Biophys. Acta,* 603, 143, 1980.
50. **Erashova, N. S., Saks, V. A., Sharov, V. G., and Lyzlova, S. N.,** Creatine kinase bound to cardiac cells nuclei, *Biochem. Biophys. Res. Commun.,* 82, 1217, 1978.
51. **Walliman, T., Turner, D. C., and Eppenberger, H. M.,** Localization of creatine kinase isoenzymes in myofibrils. I. Chicken skeletal muscle, *J. Cell. Biol.,* 75, 297, 1977.

52. **Walliman, T., Pellon, G., Turner, D. C. and Eppenberger, H. M.,** Monovalent antibodies against MM-creatine kinase remove the M-line from myofibrils, *Proc. Natl. Acad. Sci. USA,* 75, 4296, 1978.
53. **Botts, J., Stone, B. D., Wang, A. T. L., and Mendelson, R. A.,** EPR resonance and nanosecond fluorescence depolarization studies on creatine kinase interaction with myosin and its fragments, *J. Supramol. Struct.,* 3, 141, 1975.
54. **Mani, R. S. and Kay, C. M.,** Physiochemical studies on the creatine kinase M-line protein and its interaction with myosin fragments, *Biochim. Biophys. Acta,* 45, 391, 1976.
55. **Herasymowych, O. S., Mani, R. S., Kay, C. M., Bradley, R. D., and Scraba, D. G.,** Ultrastructure studies on the binding of creatine kinase and the 165,000 molecular weight component to the M-band of muscle, *J. Mol. Biol.,* 136, 193, 1980.
56. **Mani, R. S. and Kay, C. M.,** Fluorescence studies on the interaction of muscle M-line proteins, creatine kinase and the 165,000 dalton component with each other and with myosin and myosin subfragments, *Int. J. Biochem.,* 13, 1197, 1981.
57. **Woodhead, J. L. and Lowey, S.,** An in vitro study of the interactions of skeletal muscle M-protein and creatine kinase with myosin and its subfragments, *J. Mol. Biol.,* 168, 831, 1983.
58. **Walliman, T., Kuhn, J. J., Pelloni, G., Turner, D. C., and Eppenberger, H. M.,** Localization of creatine kinase isoenzymes in myofibrils. II. Chicken heart muscle, *J. Cell Biol.,* 75, 318, 1977.
59. **Perryman, M. B., Strauss, A. W., Olson, J., and Roberts, R.,** In vitro translation of canine mitochondrial creatine kinase messenger RNA, *Biochem. Biophys. Res. Commun.,* 110, 967, 1983.
60. **Putney, S., Herlihy, W., Royal, N., Pang H., Aposhian, H. V., Pickering, L., Belagaje, R., Biemann, K., Page, D., Kuby, S., and Schimmel, P.,** Rabbit muscle creatine kinase cDNA cloning primary structure and detection of human homologues, *J. Biol. Chem.,* 259, 14,317, 1984.
61. **Benfield, P. A., Zivin, R. A., Miller, L. S., Sowder, R., Smythers, G. W., Henderson, L., Oroszlan, S., and Pearson, M. L.,** Isolation and sequence of cDNA clones coding for rat skeletal muscle creatine kinase, *J. Biol. Chem.,* 259, 14979, 1984.
62. **Ordahl, C. P., Evans, G. L., Copper, T. A., Kunz, G.,and Perriard, J. C.,** Complete cDNA derived amino acid sequence of chick muscle creatine kinase, *J. Biol. Chem.,* 259, 15,224 1984.
63. **Giraudat, J., Devillers-Thiery, A., Pierrard, J. C., and Changuex, J. P.,** Complete nucleotide sequence of Torpedomarmorata in RNA coding for the 43000 daltons protein: muscle specific creatine kinase, *Proc. Natl. Acad. Sci. USA,* 81, 7313, 1984.
64. **Perry, S. V.,** Creatine phosphokinase and the enzymic and contractile properties of the isolated myofibrils, *Biochem. J.,* 57, 427, 1954.
65. **Carvahlo, A. P. and Motta, A. M.,** The role of ATP and of a bound phosphoryl group acceptor on Ca binding and exchangeability in SR, *Arch. Biochem. Biophys.,* 142, 201, 1971.
66. **Yagi, K. and Noda, L.,** Phosphate transfer to myofibrils by creatine kinase, *Biochim. Biophys. Acta,* 43, 249, 1960.
67. **Yagi, K. and Mase, R.,** Coupled reaction of creatine kinase and myosin A. ATPase, *J. Biol. Chem.,* 237, 397, 1962.
68. **Bessman, A. P. and Yang, Y. C. T.,** Intimate coupling creatine phosphokinase and myofibrillar adenosine triphosphatase, *Biochem. Biophys. Res. Commun.,* 96, 1414, 1980.
69. **Saks, V. A., Venturaclapier, R., Huchua, Z. A., Preobradzhensky, A. N., and Emelin, I. N.,** Further evidence for compartmentation of adenosine nucleotides in cardiac myofibrillar and sarcolemmal coupled ATPase creatine kinase systems, *Biochim. Biophys. Acta,* 803, 254, 1984.
70. **McClellan, G., Welsberg, A., and Winegrad, S.,** Energy transport from mitochondria to myofibrils by a creatine phosphate shuttle in cardiac cells, *Am. J. Physiol.,* 14, C423, 1983.
71. **Veksler, U. I., and Kapelko, V. I.,** The effect of phosphocreatine on the rigor tensions of EGTA-treated rat mitochondrial fibers, *Biochem. Biophys. Acta,* 803, 264, 1984.
72. **Walliman, T., Schlosser, T., and Eppenberger, H. M.,** Function of M-line-bound creatine kinase as intramyofibrillar ATP regenerator at the receiving end of the phosphorylcreatine shuttle in muscle, *J. Biol. Chem.,* 259, 5238, 1984.
73. **Philipson, K. D. and Nishimoto, A. Y.,** ATP produced by myocardial sarcolemmal-bound creatine kinase is not preferentially used by the Na$^+$ pump, *Biochem. Biophys. Res. Commun.,* 124, 696, 1984.
74. **Cande, W. Z.,** Creatine kinase role in anaphase chromosome movement, *Nature,* 304, 557, 1983.
75. **Koons, S. J., Eckert, B. S., and Zobel, C. R.,** Immunofluorescence and inhibitor studies on creatine kinase and mitosis, *Exp. Cell Res.,* 140, 401, 1982.
76. **Saks, V. A., Rosenshtraukh, L. V., Smirnov, V. N., and Chazov, E. I.,** Role of creatine phosphokinase in cellular function and metabolism, *Can. J. Physiol. Pharmacol.,* 56, 691, 1978.
77. **Jacobs, H. K. and Heldt, H. W.,** Sur la creatine kinase mitochondriale *Bull. Soc. Chim. Biol.,* 46, 188, 1964.
78. **Kleine, T. O.,** Localization of creatine kinase in microsomes and mitochondria of human heart, skeletal muscle and cerebral cortex, *Nature,* 207, 1393, 1965.

79. **Baba, N., Kim, S., and Farrell, E. C.,** Histochemistry of creatine kinase, *J. Mol. Cell. Cardiol.,* 8, 599, 1976.

80. **Sharov, V. A., Saks, V. A., Smirnov, V. N., and Chazov, E. I.,** An electron microscopic histochemical investigation of the localization of creatine kinase in heart cells, *Biochim. Biophys. Acta,* 468, 495, 1977.

81. **Kupriyanov, V. V., Elizarov, G. U., and Saks, V. A.,** Determination of the molar content of creatine kinase in heart mitochondria using SH reagents, *Biokhimya.,* 46, 930, 1981.

82. **Klingenberg, M. and Pfaff, E.,** In regulation of metabolic processes in mitochondria, *BBA Library,* Vol. 7, 180, 1966. Elsevier.

83. **Scholte, H. R., Weijers, P. J., and Wit-Peters, E. M.,** The localization of mitochondrial creatine kinase and its use for the determination of the sidedness of submitochondrial particles, *Biochem. Biophys. Acta,* 291, 764, 1973.

84. **Vial, C., Font, B., Goldschmidt, D., and Gautheron, D. C.,** Dissociation and reassociation of creatine kinase with heart mitochondria: pH and phosphate dependence, *Biochem. Biophys. Res. Commun.,* 88, 1352, 1979.

85. **Lipskaya, T. Y., Temple, V. D., Belousova, L. V., Molokova, E. V., and Rybink, I. V.,** Investigation of the interaction of mitochondrial creatine kinase with the membranes of the mitochondria, *Biokhimya.,* 45, 877, 1980.

86. **Font, B., Vial, C., Goldschmidt, D., Eichenberger, D., and Gautheron D. C.,** Heart mitochondrial creatine kinase solubilization. Effect of mitochondrial swelling and SH groups reagents, *Arch. Biochem. Biophys.,* 212, 195, 1981.

87. **Hall, N. and De Luca, M.,** Binding of creatine kinase to heart and liver mitochondria in vitro, *Arch. Biochem. Biophys.,* 201, 674, 1980.

88. **Newman, R. A., Hacker, M. P., and Fagan, M. A.,** Adriamycin-mediated inhibition of creatine kinase binding to heart mitochondrial membrane, *Biochem. Pharmacol.,* 31, 109, 1982.

89. **Vial, C., Goldschmidt, D., and Font, B.,** Effect of anthracyclines on the interaction of mitochondrial creatine kinase, *J. Mol. Cell. Cardiol.,* 15 (Suppl. 1), 153, 1983.

90. **Muller, M., Moser, R., Cheneval, D., and Carafoli, E.,** Cardiolipin is the membrane receptor for mitochondrial creatine kinase, *J. Biol. Chem.,* 260, 3839, 1985.

91. **Marcillat, O., Goldschmidt, D., Font, B., and Vial, C.,** Effect of creatine kinase association with heart mitochondria on its apparent kinetic properties, J. Mol. Cell. Cardiol., 16, suppl. 2, 128, 1984.

92. **Font, B., Vial, C., Goldschmidt, D., Eichenberger, D., and Gautheron, D. C.,** Effects of SH groups reagents on creatine kinase interaction with the mitochondrial membrane, *Arch. Biochem. Biophys.,* 220, 541, 1983.

93. **Hall, N., Addis, P., and De Luca, M.,** Purification of mitochondrial creating kinase: two interconvertible forms of the active enzyme, *Biochem. Biophys. Res. Commun.,* 76, 950, 1977.

94. **Wevers, R. A., Mull-Steinbusch, M. W. F. J., and Soons, J. B. J.,** Mitochondrial creatine kinase in the human heart, *Clin. Chim. Acta,* 101, 103, 1980.

95. **Roberts, R.,** Purification and characterization of mitochondrial creatine kinase in *Heart Creatine Kinase,* Jacobus, W. W. and Ingwall, J. S., Eds., Williams and Wilkins, Baltimore, 1980, 31.

96. **Jockers-Wretou, E.,** The cell nucleus as a possible source of cathodally migrating serum creatine kinase, *Clin. Chem.,* 30, 1268, 1985.

98. **Jacobus, W. E.,** Respiratory control and the integration of heart high energy phosphate metabolism by mitochondrial creatine kinase, *Ann. Rev. Physiol.,* 47, 707, 1985.

99. **Saks, V. A., Chernousova, G. B., Gukowsky, D. E., Smirnov, V. N., and Chazov, E. I.,** Studies of energy transport in heart cells. Mitochondrial isoenzyme of creatine phosphokinase: kinetic properties and regulatory action of Mg^{2+} ions, *Eur. J. Biochem.,* 57, 273, 1975.

100. **Hall, N., Addis, P., and De Lica, M.,** Mitochondrial creatine kinase. Physical and kinetic properties of the purified enzyme from beef heart, *Biochemistry,* 18, 1745, 1979.

101. **Saks, V. A., Kupriyanov, V. V., Elizarova, G. V., and Jacobus, W. E.,** The importance of creatine kinase localization for the coupling of mitochondrial phosphorylcreatine production to oxidative phosphorylation, *J. Biol. Chem.,* 255, 755, 1980.

102. **Erickson-Viitanen, S., Geiger, P. J., Viitanen, P., and Bessman, S. P.,** Compartmentation of mitochondrial creatine phosphokinase. II. The importance of the outer mitochondrial membrane for mitochondrial compartmentation, *J. Biol. Chem.,* 257, 14,405, 1982.

103. **Marcillat, O. and Vial, C.,** unpublished results, 1985.

104. **Vandegaer, K. M. and Jacobus, W. E.,** Evidence against direct transfer of the adenine nucleotides by the heart mitochondrial creatine kinase adenine nucleotide translocase complex, *Biochem. Biophys. Res. Commun.,* 109, 442, 1982.

105. **Morrison, J. F. and James, E.,** The mechanism of the reaction catalyzed by adenine triphosphate-creatine phosphotransferase, *Biochem. J.,* 97, 37, 1965.

106. **Maggio, E. T., Kenyon, G. L., Markam, G. D., and Reed, G. M.,** Properties of a CH_3S-blocked creatine kinase with the altered catalytic activity, *J. Biol. Chem.,* 252, 1202, 1977.

107. **Jacobs, H. K. and Kuby, S. A.,** Kinetic properties of the crystalline adenosine triphosphate-creatine transphosphorylase from calf brain, *J. Biol. Chem.,* 245, 3305, 1970.

108. **Jacobus, W. E. and Saks, V. A.,** Creatine kinase of heart mitochondria: changes in its kinetic properties induced by coupling to oxidative phosphorylation, *Arch. Biochem. Biophys.,* 219, 167, 1982.

109. **Altschuld, R.,** Interaction between mitochondrial creatine kinase and oxidative phosphorylation, in *Heart Creatine Kinase,* Jacobus, W. E. and Ingwall, J. S., Eds., Williams and Wilkins, Baltimore, 1980, 27.

110. **Fitch, C. D., Chevli, R., and Jellinek, M.,** Phosphocreatine does not inhibit rabbit muscle phosphofructokinase or pyruvate kinase, *J. Biol. Chem.,* 254, 11357, 1979.

111. **Tornheim, K. and Lowenstein, J. M.,** Creatine phosphate inhibition of heart lactate deshydrogenase and muscle pyruvate kinase is due to a contaminant, *J. Biol. Chem.,* 254, 10,586, 1979.

112. **Altschuld, R. A. and Brierley, G. P.,** Interaction between the creatine kinase of heart mitochondria and oxidative phosphorylation, *J. Molec. Cell. Cardiol.,* 9, 875, 1977.

113. **Lipskaya, T. Y., Temple, V. I., Belousova, L. V., and Molokova, E. V.,** Interaction between heart mitochondrial creatine kinase and oxidative phosphorylation, *Biokhimiya,* 45, 1015, 1980.

114. **Yang, W. C. T., Geiger, P. J., and Bessman, S. P.,** Formation of creatine phosphate from creatine on ^{32}p-labeled ATP by isolated rabbit heart mitochondria, *Biochem. Biophys. Res. Commun.,* 7, 882, 1977.

115. **Erickson-Viitanen, S., Viitanen, P., Geiger, P. J., Yang, W. C. T., and Bessman, S. P.,** Compartmentation of mitochondria creatine phosphokinase. I. Direct demonstration of compartmentation with the use of labeled precursors, *J. Biol. Chem.,* 257, 14,395, 1982.

116. **Moreadith, R. W. and Jacobus, W. E.,** Creatine kinase of heart mitochondria. Functional coupling of ADP transfer to the adenine nucleotide translocase, *J. Biol. Chem.,* 257, 899, 1982.

117. **Barbour, R. L., Ribaudo, J., and Chan, S. H. P.,** Effect of creatine kinase activity on mitochondrial ADP/ATP transport. Evidence for a functional interaction, *J. Biol. Chem.,* 259, 8246, 1984.

118. **Meyer, R. A., Sweeney, H. L., and Kushmerick, N. J.,** A simple analysis of the "phosphocreatine shuttle", *Am. J. Physiol.,* 246, C365, 1984.

119. **Davuluri, S. P., Hird, F. J. R., and McLean, R. M.,** A reappraisal of the function and synthesis of phosphoarginine and phosphocreatine in muscle, *Comp. Biochem. Physiol.,* 69B, 329, 1981.

120. **Hird, F. J. R. and McLean, R. M.,** Synthesis of phosphocreatine and phosphoarginine by mitochondria from various sources, *Comp. Biochem. Physiol.,* 76B, 1, 1983.

121. **Harary, I. and Farley, B.,** In vitro studies of beating heart cells in culture. I. Growth and organization, *Exp. Cell Res.,* 29, 451, 1963.

122. **Seraydarian, M. W., Artaza, L., and Abbot, B. C.,** Creatine and control of energy metabolismin cardiac and skeletal muscle cells in culture, *J. Mol. Cell. Cardiol.,* 6, 405, 1974.

123. **Lompre, A. M., Mercadier, J. J., Wisnewsky, C., Bouveret, P., Pantaloni, C., D'Albis, A., and Schwartz, K.,** Species and age-dependent changes in the relative amounts of cardiac myosin isozymes in mammals, *Dev. Biol.,* 84, 286, 1981.

124. **Takenori, Y., Yang, J. J., Ricchiuti, N. V., and Seraydarian, M. W.,** Oxygen consumption of mammalian myocardial cells in culture: measurements in beating cells attached to the substrate of the culture dish, *Anal. Biochem.,* 145, 302, 1985.

125. **Gibbs, C. L.,** Cardiac energetics, *Physiol. Rev.,* 58, 174, 1978.

126. **Burns, A. H. and Ready, W. J.,** Amino acid stimulation of oxygen and substrate utilization by cardiac myocytes, *Am. J. Physiol.,* 235, E461, 1978.

127. **Ingwall, J. S., Kramer, M. F., and Friedman, W. F.,** Developmental changes in heart creatine kinase, in *Heart Creatine Kinase,* Jacobus, W. E. and Ingwall, J. S., Eds., Williams and Wilkins, Baltimore, 1980, 9.

128. **Ingwall, J. S., Kramer, M. F., Woodman, D., and Friedman, W. F.,** Maturation of energy metabolsim in the lamb: changes in myosin ATPase and creatine kinase activities, *Pediatr. Res.,* 15, 1128, 1981.

129. **Baldwin, K. M., Cooke, D. A., and Cheadle, W. G.,** Enzyme alterations in neonatal heart muscle during development, *J. Molec. Cell. Cardiol.,* 9, 651, 1977.

130. **Turner, D. C., Maier, V., and Eppenberger, H. M.,** Creatine kinase and aldolase isoenzyme transition in cultures of chick skeletal muscle cells, *Develop. Biol.,* 27, 63, 1964.

131. **Mooris, G. E., Cook, A., Cole, R. J.,** Isozymes of creatine phosphokinase during myogenis in vitro, *Exp. Cell Res.,* 74, 582, 1972.

132. **Dym, H., Turner, D. C., Eppenberger, H. M., and Yaffe, D.,** Creatine kinase isoenzyme transition in actinomycin D-treated differentiating muscle cultures, *Exp. Cell Res.,* 113, 15, 1978.

133. **Lough, J. and Bischoff, R.,** Differentiation of creatine phosphokinase during myogenesis: quantitative fractionation of isozymes, *Develop. Biol.,* 29, 410, 1977.

134. **Perriard, J. C., Caravatti, M., Perruard, E., and Eppenberger, H. M.,** Quantitation of creatine kinase isoenzyme transitions in differentiating chick embryonic breast muscle and myogenic cell cultures by immunoadsorption, *Arch. Biochem. Biophys.,* 191, 90, 1978.

135. **Perriard, J. C., Perruard, E., and Eppenberger, H. M.,** Detection and relative quantitation of mRNA for creatine kinase isozymes in RNA from myogenic cell cultures and embryonic chicken tissues, *J. Biol. Chem.,* 252, 6529, 1978.
136. **Caravatti, M. and Perriard, J. C.,** Turnover of the creatine kinase subunits in chicken myogenic cell cultures and in fibroblasts, *Biochem. J.,* 196, 377, 1981.
137. **Perriard, J. C.,** Developmental regulation of creatine kinase isozymes in myogenic cell culture from chicken. Levels of mRNA for creatine kinase subunit M and B, *J. Biol. Chem.,* 254, 7036, 1979.
138. **Caravatti, M., Perriard, J. C., and Eppenberger, H. M.,** Developmental regulation of creatine kinase in myogenic cell cultures from chicken. Biosynthesis of creatine kinase subunits M and B, *J. Biol. Chem.,* 254, 1388, 1979.
139. **Seraydarian, M. W. and Artaza, L.,** Regulation of energy metabolism by creatine in cardiac and skeletal muscle cells in culture, *J. Mol. Cell. Cardiol.,* 8, 669, 1976.
140. **Ingwall, J. S. and Wildenthal. K.,** Role of creatine in the regulation of cardiac protein synthesis, *J. Cell Biol.,* 68, 159, 1976.
141. **Eppenberger, H. M., Perriard, J. C., and Walliman, T.,** Analysis of creatine kinase isozymes during muscle differentiation, *Isozymes: Current Topics in Biological and Medical Research,* 7, 19, 1983.
142. **Walliman, T., Moser, H., and Eppenberger, H. M.,** Isoenzyme-specific localization of M-line bound creatine kinase in myogenic cells, *J. Muscle Res. Mot.,* 4, 429, 1983.
143. **Ingwall, J. S. and Fossel, E. T.,** Changes in the creatine kinase system in the hypertrophied myocardium of the dog and rat, *Persp. Cardiovasc. Res.,* 7, 610, 1983.
144. **Ingwall, J. S.,** Changes in creatine kinase system during the transition from compensated to uncompensated hypertrophy in the spontaneously hypertensive rat, *Persp. Cardiovasc. Res.,* 8, 145, 1983.

Chapter 15

STUDIES ON LIPOPROTEIN LIPASE IN HEART CELL CULTURES

O. Stein, Y. Stein, G. Friedman, and T. Chajek-Shaul

TABLE OF CONTENTS

I. INTRODUCTION

Heart cell cultures have been used to study various aspects of lipid metabolism. Those studies were concerned mainly with the interaction of free fatty acids added to the culture medium on metabolic and contractile properties of the cultured cells. Many of these studies have been covered in several recent reviews,[1,2] and this chapter will focus mainly on lipoprotein lipase, which plays a pivotal role in lipoprotein metabolism in vivo. The heart is one of the richest sources of lipoprotein lipase and, therefore, heart cell cultures provide a very convenient model for studying various aspects of this very important enzyme. For further information on LPL activity and its action in lipoprotein metabolism, the reader is referred to several excellent recently published reviews on this subject.[3-5]

A. Cell Type Responsible for Lipoprotein Lipase (LPL) Secretion

Lipoprotein lipase, the key enzyme in the catabolism of chylomicron and very low density lipoprotein, triacylglycerol, is present in the heart in two compartments, one intracellular, the other situated on the cell surface.[6] The surface compartment has been localized to the plasma membrane of endothelial cells, and it is also the site of action of the enzyme.[7] Studies in vivo have shown that lipoprotein lipase in the rat heart is delivered to its site of action by a vesicular transport, which can be inhibited by colchicine.[8] However, the cellular origin of the enzyme has been in dispute so far, i.e., whether it is synthesized in the myocardial muscle cells, or in the mesenchymal cells of the interstitium, which comprises the capillary endothelium, pericytes, and fibroblasts. Methods have been developed to culture cells derived from newborn rat heart,[9] and it has been shown that the cultured cells are capable of synthesizing lipoprotein lipase.[10] With the help of microscopy and cytochemistry it has been established that such cultures contain two types of cells, one which represents the cardiac myocytes, the other which comprises the mesenchymal cells.[11] It has become possible to separate the two cell types with the aid of a ''preplating'' technique[12] and to obtain cultures with a preponderance of one cell type.

In the studies which will be reviewed here, the standard procedure of cell isolation and preplating involved the following steps: sequential trypsinization followed by two periods of preplating during which the nonbeating cells were allowed to attach to the culture dish and as a result, the final suspension was maximally depleted of mesenchymal cells. The cells which had attached during the first period were designated as F_1, those in the second as F_2, and the remaining beating cells as M. The ultrastructural features of the F_1, F_2, and M cultures allowed the characterization of the cells in each group. Thus the F_1 and F_2 cultures consisted mainly of mesenchymal and endothelial cells. The cells in the M cultures were primarily cardiac muscle cells, but with longer time in culture, the number of mesenchymal cells increased.[13]

Lipoprotein lipase activity was determined in cells which had been cultured for 3 to 11 days. The enzyme activity was inhibited $92.7 \pm 2.6\%$ by the addition of 1 M NaCl and $77.1 \pm 2.7\%$ protamine sulfate. In the absence of serum the loss of activity was 85.7 ± 4.9.[14] As shown in Table 1, enzyme activity increased with time in culture in the three preparations studied (F_1, F_2, and M), being highest in the F_1 preparation for all incubation times. The steep increase in enzyme activity in the F cultures between days 3 and 5 could either be due to the recovery of the cells following trypsinization, or to an increase in the number of those cells responsible for production of the enzyme in the F_1 cultures. The presence of lipoprotein lipase on the cell surface was also demonstrated via the use of its native substrate, i.e., triacylglycerol present in very low density lipoproteins (VLDL).[15] The hydrolysis of the labeled VLDL triacylglycerol was compared in the F_1 and M cultures and was up to 10 times faster in the F_1 cultures than in M cultures (Table 2).

Henson et al.[16] have reported on lipoprotein lipase activity in cultured heart cells. In order

Table 1
LIPOPROTEIN LIPASE ACTIVITY IN CULTURED MESENCHYMAL (F) AND MYOGENIC (M) CELLS OF RAT HEARTS

	Lipoprotein lipase activity, nmol fatty acid released/mg cell protein/hr		
Days in culture	F_1	F_2	M
3	136.3 ± 16.4	55.9 ± 16.5	34.3 ± 2.4
5	441.0 ± 18.9	171.5 ± 25.8	29.0 ± 3.7
8	650.0 ± 84.3	262.5 ± 35.0	107.0 ± 8.2
11	799.5 ± 87.9		98.9 ± 16.4

Note: F_1 and F_2 cultures consisted mainly of mesenchymal cells with features common to fibroblasts, smooth muscle cells, and endothelial cells. The cells in the M cultures consisted mainly of cardiac muscle cells.

Adapted from Chajek, T. et al., *Biochim. Biophys. Acta,* 528, 456, 1978. With permission.

Table 2
COMPARISON OF HYDROLYSIS OF VERY LOW DENSITY LIPOPROTEIN (VLDL) TRIACYLGLYCEROL BY F_1 AND M CULTURES

	Hydrolysis of VLDL triaelglycerol, nmol fatty acid/mg cell protein	
Time (min)	F_1	M
20	110	—
30	220	20
60	430	40
120	910	100

Note: Very low density lipoprotein, 220 nmol triaclyglycerol/mℓ, was added to culture medium devoid of serum, containing fatty acid-poor bovine serum albumin, 4 g/dℓ. Values are means of 4 to 14 dishes.

Adapted from Chajek, T. et al., *Biochim. Biophys. Acta,* 528, 466, 1978. With permission.

to determine cellular location of the enzyme, myosin ATPase was used as an index of muscle content of the cultures. They observed no change in LPL specific activity when muscle content was either decreased or increased; therefore, their conclusion was that both the beating cardiac myocytes and the mesenchymal cells could be the source of the enzyme. It seems, however, that our M cultures did contain less mesenchymal cells, because of the introduction of a second preplating step (F_2), which permitted a sharper separation between the two cell populations. Moreover, the lipoprotein lipase activity determined in our studies was measured in the presence of heparin and was about one order of magnitude higher than that reported by Henson et al.[16] In spite of the differences in experimental procedure, it seems of interest to point out that there was a steep rise in lipoprotein lipase activity in the muscle-enriched preparation between days 5 and 8 in both studies.

B. Mechanism of LPL Secretion

In analogy to the findings in the perfused heart, LPL could be released from cultured heart cells into the medium by a brief exposure to heparin. While a very small release of LPL was observed in heparin-free medium, addition of heparin resulted in a sharp increase

Table 3
EFFECT OF 2-DEOXYGLUCOSE AND TUNICAMYCIN ON HEPARIN-RELEASABLE AND RESIDUAL LIPOPROTEIN LIPASE ACTIVITY IN F_1 HEART CELL CULTURES

	Lipoprotein lipase activity, nmol fatty acid released/mg cell protein/hr	
Inhibitor	**Heparin releasable**	**Residual**
None	280 ± 20	562 ± 12
2-Deoxyglucose, 2.5 mM	95 ± 8	172 ± 2
None	225 ± 13	663 ± 33
Tunicamycin, 5 μg/mℓ	96 ± 3	224 ± 24

Note: The cells were incubated with 2-deoxyglucose for 3 hr and with tunicamycin for 24 hr. Heparin, 5 U/mℓ, was added to the culture medium 3 min prior to the termination of the experiment. Values are means ± S.E. of triplicate dishes.

Adapted from Friedman, G. et al., *Atherosclerosis,* 36, 289, 1980. With permission.

in enzyme release within the first 3 to 5 min. The heparin-releasable activity amounted to 20 to 30% of total cellular LPL and was 90% inhibited by 1 M NaCl. Exposure of the heart cells to heparin resulted in a continuous release of enzyme, so that 30 min after addition of heparin more than 80% of the enzyme activity was found in the medium; at 3 hr total enzyme activity doubled, and about 95% was found in the medium.[14]

The finding of a continuous release of lipoprotein lipase activity into post-heparin medium raised the question as to whether it reflects enzyme activation or *de novo* synthesis. The half-life of lipoprotein lipase in the cultured cells was determined after addition of cyclo-heximide to the culture medium. The heparin releasable and residual enzyme activity as well as the total activity, determined independently in cells not exposed to heparin, had a $t_{1/2}$ of 35 min. Following almost complete enzyme depletion the cells were able to resynthesize the enzyme. When cycloheximide was removed from the system, full restoration of enzyme activity was achieved after 1 hr.[14]

Use was made of colchicine in order to determine whether the transport of lipoprotein lipase in the cultured heart cells is affected by this drug, in analogy to the findings in the intact rat heart.[8] Pretreatment of the cells with colchicine for 4 hr resulted in a pronounced decrease in the fraction of heparin-releasable activity. At the same time, the residual activity remained high in the colchicine-treated cells. Thus, while enzyme synthesis was not impaired, its transport to the cell surface (from which it was released by heparin) was affected by the treatment with the drug.[14]

Reduction in LPL activity was observed when 2-deoxyglucose was added to the culture medium[17] at a relatively low concentration (2.5 mM). The inhibition was apparent after 2 hr of incubation and progressed more rapidly at longer time intervals (Table 3). Protein and glycoprotein synthesis were measured and, while no interference with the incorporation of [³H]leucine into total cellular protein was observed, a marked and progressive fall occurred in the incorporation of labeled glucosamine, mannose, and galactose into glycoproteins.[17] In rat preadipocytes, a 30% decrease in protein synthesis occurred when 5 mM 2-deoxy-glucose was included in a glucose-containing medium.[18] At the same time there appeared to be a complete disappearance of lipoprotein lipase activity, which was interpreted as being the result of the inhibition of enzyme synthesis.[18] Whether the decrease in lipoprotein lipase activity upon addition of 2-deoxyglucose to the culture medium of the heart cells was due to inhibition of enzyme synthesis, or to the formation of a protein lacking an oligosaccharide

Table 4
**EFFECT OF ANION AND CATION EXCHANGE
RESINS ON THE EFFECTIVENESS OF HORSE
SERUM DIALYSATE IN PREVENTING THE
DECREASE IN LIPOPROTEIN LIPASE ACTIVITY
IN F_1 HEART CULTURES DURING INCUBATION
WITH DIALYZED SERUM**

Incubation medium		Lipoprotein lipase activity, nmol fatty acid released/mg cell protein/hr	
Serum	Dialysate	Exp I	Exp II
Nondialyzed	None	1108 ± 17	1270 ± 11
Dialyzed	None	391 ± 13	314 ± 4
Dialyzed	Untreated	1076 ± 22	1034 ± 61
Dialyzed	post-DEAE cellulose	928 ± 25	932 ± 7
Dialyzed	post-CMcellulose	379 ± 14	366 ± 10

Adapted from Friedman, G. et al., *Biochim. Biophys. Acta,* 619, 650, 1980. With permission.

side chain which might be essential for the expression of enzyme activity, remains to be determined.

Tunicamycin has been shown to interfere selectively with the formation of dolichol-bound N-acetylglucosamine derivatives and, hence, exposure to tunicamycin results in the synthesis of glycoproteins deficient in asparagine-linked oligosaccharides.[19,20] When tunicamycin was added to the heart cell cultures, a 3 hr pretreatment period was required in order to be able to demonstrate the fall in LPL activity.[17] This was probably due to a slow intracellular penetration of the antibiotic.[20] However, after this lag period a 60 to 70% fall in enzyme activity and a 50% reduction in the incorporation of labeled glucosamine and mannose into glycoproteins occured, while protein synthesis decreased by only 9% (Table 3). It is plausible that the protein core of lipoprotein lipase may be formed even when its glycosylation is impaired by tunicamycin or 2-deoxyglucose. The reduction in enzyme activity observed would, therefore, support the possibility that sugar residues are indeed important for the activity of the enzyme. In order to resolve the above-mentioned possibilities in a more definitive manner, the actual formation of the protein should be confirmed with the aid of a specific antibody and the mode of degradation of the enzyme should be elucidated.

C. Regulation of LPL Synthesis
1. Serum Factors

For full expression of LPL activity, the culture medium requires the presence of relatively high concentrations of serum. A mixture of 10% fetal calf serum and 10% horse serum was found to be optimal.[10] Reduction of serum concentration to 5% caused marked impairment in the synthesis of LPL.[21] Inadvertently, it was found that dialyzed serum, although used at optimal concentrations, was not able to support LPL synthesis, even though it did not interfere with synthesis of cellular glycoproteins.[22] Addition of the dialysate to the culture medium containing the dialyzed serum completely restored LPL activity. The effectiveness of the dialysate was not influenced by trichloroacetic acid precipitation, ether extraction, exposure to pronase, or heating to 80°C for 10 min; it was retained after chromatography on DEAE cellulose, but was lost after elution from CM cellulose columns (Table 4). These results suggested that the factor(s) in the dialysate are positively charged, low molecular weight compounds. A series of basic polyamines was added to the culture medium containing the dialyzed serum and a partial prevention of the fall in enzyme activity was achieved.[22]

Table 5
VLDL INDUCED DECREASE IN
LIPOPROTEIN LIPASE ACTIVITY IN F_1
HEART CELL CULTURES

	Lipoprotein lipase activity, nmol fatty acid released/mg cell protein/hr	
Additions to medium	**Exp I**	**Exp II**
None	1218	1263
Albumin	1179	1176
VLDL 0.75 mg triacylglycerol/mℓ	460	509
VLDL + albumin	500	484
VLDL + glucose + insulin	436	651
LDL 0.18 mg triacylglycerol/mℓ	948	1005

Note: Cells were grown in culture medium containing 20% fetal calf and horse serum. All additions were done to complete culture medium. Incubation was carried out for 3 hr. The concentration of albumin was 30 mg, glucose 1 mg, and insulin 5 μg/mℓ.

Adapted from Friedman, G. et al., *Biochim. Biophys. Acta,* 573, 521, 1979. With permission.

These findings suggested that the polyamines might affect the residence time of lipoprotein lipase by preventing its inactivation. Indeed, loss of lipoprotein lipase activity of heart homogenates incubated at 37°C could be retarded by the addition of spermine and spermidine.[22]

2. Substrate

In man and in experimental animals, the nutritional state has been shown to affect lipoprotein lipase in adipose tissue, heart and skeletal muscle.[23] As the enzymic activity of heart and skeletal muscle was found to be higher in the fasting than in the fed state, it was postulated that increase of lipoprotein lipase activity was due to a rise in glucagon to insulin ratio, which is known to occur during fasting.[24] However, no correlation could be found between the lipoprotein lipase activity of rat heart and plasma glucagon levels.[25] An alternative explanation could be the LPL in the heart is regulated by the serum levels of its substrate, i.e., the triacylglycerol-rich lipoproteins. This hypothesis seemed amenable to investigation in a simple model system of heart cell cultures which were shown to synthesize the enzyme very actively.[13]

Lipoprotein lipase activity was studied in rat heart cell cultures grown in the presence of 20% fetal calf and horse serum; the final concentration of triacylglycerol was 0.03 mg/mℓ.[26] When the enzyme activity had reached high levels, the cells were incubated for 24 hr in a medium containing 20% serum derived from fasted or fed rats. No change in enzyme activity occurred in the presence of serum from fasted rats, but a 50% fall was observed using serum from fed rats. When the complete culture medium was supplemented with rat plasma VLDL (0.075 to 0.75 mg triacylglycerol per milliliter) a pronounced decrease in lipoprotein lipase activity occurred after 3 to 5 hr of incubation (Table 5). As the addition of VLDL to the culture medium resulted in a smaller decrease of heparin releasable as compared to the residual activity, it seems that there was no direct inhibition of surface bound enzyme activity and that the transport of the enzyme to the cell surface was not affected. Thus, addition of VLDL to the culture medium probably resulted in a fall in enzyme synthesis, even though total protein synthesis remained unchanged. This inhibition could be reproduced by increasing

Table 6
EFFECT OF ADDITION OF HYDROCORTISONE TO COMPLETE CULTURE MEDIUM AT DIFFERENT STAGES OF GROWTH ON LIPOPROTEIN LIPASE IN F_1 HEART CELL CULTURES

Hydrocortisone added on day	Duration of culture (days)	Lipoprotein lipase activity, nmol fatty acid released/mg cell protein/hr	
		Control	Treated
2	3	82.6 ± 6.2	116.0 ± 4.1
2	4	187.1 ± 4.8	245.5 ± 9.0
2	5	367.3 ± 19.0	973.0 ± 45.0
—	6	619.3 ± 10.2	—
7	8	694.1 ± 28.0	1119.8 ± 41.0

Notes: The cells were grown in complete culture medium containing 20% serum for up to 8 days and in the presence of hydrocortisone, 0.05 μg/mℓ added to this medium from day 2 to 7 till the termination of the experiment. Values are means ± S.E. of triplicate determinations.

Adapted from Friedman, G. et al., *Biochim. Biophys. Acta*, 531, 222, 1978. With permission.

free fatty acid concentration of the medium; however, addition of excess albumin to VLDL-containing medium did not prevent the fall in enzyme activity.[26] These results obtained with cultured rat heart cells suggest that plasma levels of triacylglycerol-rich lipoproteins might modulate the lipoprotein lipase activity in the heart in vivo.

3. Hormonal Regulation

Hormonal regulation of lipoprotein lipase was studied extensively in the intact animal and was shown to vary with the tissue studied.[27] Thus while lipoprotein lipase of adipose tissue is increased after insulin administration,[28] conflicting evidence was presented with respect to the heart.[29] The culture system seemed to provide a convenient tool for studying hormonal regulation of lipoprotein lipase synthesis.

Lipoprotein lipase activity in F_1 heart cell cultures increased markedly between day 3 and 5 and to a lesser degree at later time intervals.[21] Addition of hydrocortisone to the culture medium resulted in an increase in lipoprotein lipase activity at all stages of culture (Table 6). Lipoprotein lipase activity did not increase after addition of insulin to the complete culture medium. When the cells were cultured in the presence of medium containing 5% serum (serum poor), the increase in lipoprotein lipase activity between days 3 and 6, was much lower. A significant rise in lipoprotein lipase activity was encountered by addition of hydrocortisone and insulin to the serum-poor medium. When serum-poor medium was replaced by 20% serum-containing medium, between day 6 and 7 of culture a fall in enzyme activity occurred, which could be prevented by hydrocortisone with or without addition of insulin. Estradiol, growth hormone, or glucagon when added to serum-containing medium, or serum-poor medium did not affect LPL activity. Thus, lipoprotein lipase of heart is controlled by glucocorticoids.[21]

The reciprocal control of lipoprotein lipase of adipose tissue and heart by fasting and feeding was referred to in Section b and the mechanism underlying these nutritional changes has been attributed in part to the effects of insulin and catecholamines.[3] In a review on regulation of lipoprotein lipase activity,[30] it was suggested that: "if cyclic AMP is a mediator of hormonal action in both cases, either the hormones concerned cause opposing changes

Table 7

EFFECT OF CYCLIC AMP EFFECTORS ON LIPOPROTEIN LIPASE ACTIVITY IN THE MESENCHYMAL RAT F_1 HEART CELL CULTURES

	Lipoprotein lipase activity nmol fatty acid released/ mg cell protein/hr		
Additions to cultures	**Control**	**+ 1 M NaCl**	**− Serum**
None	851 ± 12[a]	2 ± 0.9	68 ± 3
435 μM dibutyryl cyclic AMP	2271 (2256, 2286)	0	175 (182, 168)
100 μM 3-isobutyl-1-methylxanthine	1271 ± 101[b]	26 ± 15	112 ± 20
1 $\mu g/m\ell$ each of cholera toxin + ganglioside GM$_1$	1413 ± 54[c]	7 ± 3.9	109 ± 4

Note: All incubations were carried out for 24 hr. Values are mean ± S.D. of triplicate dishes. Control determination of enzyme activity under standard conditions. + 1 M NaCl, assay is in the presence of 1 M NaCl; − Serum, assay without rat serum. A vs. b and a vs. c, P < 0.001. Values in brackets are duplicates.

Adapted from Friedman, G. et al., *Biochim. Biophys. Acta*, 752, 106, 1983. With permission.

in cyclic AMP concentrations at the two sites (heart and adipose tissue), or a change in the concentration of the nucleotide in a particular direction produces opposite effects on the activity of the enzyme in the two tissues''. Cultured cells derived from neonatal rat hearts[14] as well as preadipocytes isolated from rat epididymal fat pads[31] which synthesize lipoprotein lipase were used as models to examine these possibilities.

F_1 heart cells were grown under optimal conditions till they reached peak values of LPL activity. At that stage, they were exposed to agents known to modulate cyclic AMP.[32] The response of the heart cell cultures to cholera toxin, 3-isobutyl-1-methylxanthine (IBMX), and dibutyryl cAMP during the first 3 to 6 hr was quite variable. No decrease in LPL activity was encountered after incubation with cholera toxin, while with IBMX and with dibutyryl cAMP the reduction in enzyme activity was not more than 25 to 30%.[32] However, after 24 hr incubation with all three effectors, an increase in lipoprotein lipase activity was seen in the heart cell cultures, which ranged between 100 to 250% (Table 7). In cholera toxin-treated cultures, the rise in LPL was preceded by a fourfold increase in cellular concentration of cAMP, but cAMP returned to basal levels at the time of maximal increase of LPL.[32] Incubation of heart cell cultures with dibutyryl cAMP for 24 hr resulted in a sixfold increase of heparin-releasable lipoprotein lipase activity, while residual activity was doubled. The rise in surface-bound lipoprotein lipase was evidenced also by an increase in the lipolysis of chylomicron triacylglycerol. The dibutyryl cAMP-induced heparin-releasable and residual lipoprotein lipase activity had the same $t_{1/2}$ as the basal activity, indicating that the rise in LPL was due to an increase in enzyme synthesis.[32]

The up-regulation of lipoprotein lipase in heart cell cultures by modulation of cAMP may be compared to other cellular systems in which induction of different enzymes was studied. Thus, in cultured fibroblasts the addition of dibutyryl cAMP[33] resulted, after 10 hr, in the induction of phosphodiesterase activity, which remained elevated during the 48 hr of the experiment. Cyclic AMP (or dibutyryl cAMP) was shown to induce hepatic enzymes, such as tyrosine aminotransferase or phosphoenolpyruvate carboxykinase, in culture.[34] The temporal sequence of molecular events following stimulation of a hormonal membrane receptor in the adrenal medulla which cause the rise in activity of tyrosine hydroxylase has been summarized in a recent review.[35] It appears that the rise in cyclic AMP is a transient event, lasting no longer than 2 hr; it is followed by activation of cytosol cAMP-dependent protein kinase, translocation of the catalytic subunits of the protein kinase into the nucleus, and enhancement of messenger RNA synthesis.[35] In hamster DMK cells, arylhydrocarbon (benzo[a]pyrene) hydroxylate, which can be induced by dibutyryl cAMP, had a K_m of 0.15 μM, while substrate-induced enzyme had a K_m of 4.3 μM.[36] The apparent K_m of basal and dibutyryl cAMP-induced lipoprotein lipase in the heart cell culture were 0.124 and 0.093 mM, respectively (i.e., not statistically different).[32] Thus, it seems that dibutyryl cAMP-induced lipoprotein lipase activity resulted from an accelerated synthesis of the enzyme protein, rather than from a change in the affinity of the enzyme for its substrate.

Addition of dibutyryl cAMP or 3-isobutyl-1-methylxanthine to rat preadipocytes grown in serum-containing culture medium resulted in a progressive decrease in lipoprotein lipase activity released into the culture medium, so that at 6 to 8 hr, enzyme activity ranged between 20 and 30% of that recovered in the control dishes. However, after 24 hr, LPL activity in preadipocytes increased by 100 to 250% in analogy to the heart cell cultures.[32]

The reason for the difference in response of cultured preadipocytes and heart cells to the effectors during the first 8 hr of incubation has not been elucidated, but could be related to a possible absence of a hormone-sensitive lipase in the heart cells, and hence to a difference in intracellular metabolism of triacylglycerol. On the other hand, a common mechanism can be postulated for the long-term effect of cAMP on the induction of lipoprotein lipase activity in both types of cultures. It probably involves mRNA and protein synthesis, which culminates in increased enzyme activity.

D. The Role of LPL in the Uptake of Chylomicron Cholesteryl Ester

Perfused rat hearts have been shown to take up esterified cholesterol from cholesterol-rich chylomicrons — this uptake being independent of protein endocytosis.[37] Since removal of lipoprotein lipase from the heart by heparin resulted in a 95% reduction in the clearance of the cholesteryl ester from the perfusate,[37] it seemed that lipoprotein lipase might play a role in the uptake of esterified cholesterol by nonhepatic tissues. The relationship between lipoprotein lipase activity and the uptake of chylomicron cholesteryl ester was then studied in rat heart cell cultures.[38] A saturable uptake of ^{3}H-cholesteryl linoleyl ether was encountered which was accompanied by almost complete hydrolysis of the triacylglycerol portion of the chylomicrons (Table 8).

The uptake of cholesteryl linoleyl ether (a nonhydrolyzable analogue of cholesteryl ester), was much higher than that of iodinated chylomicron proteins, suggesting that these may be independent processes.[38] The heart cell cultures were used to study a possible relationship between lipolysis of chylomicron triacylglycerol and the uptake of cholesteryl ester. In this culture system, chylomicron triacylglycerol is hydrolyzed by a membrane-supported lipoprotein lipase, synthesized by the cells,[14,15] the activity of which can be lowered by exposure to a medium containing dialyzed serum. A decrease in the uptake of cholesteryl linoleyl ether, when labeled chylomicrons were added to heart cell cultures pretreated with dialyzed serum, was related to the lower lipoprotein lipase activity of these cultures. To determine whether the uptake of cholesteryl linoleyl ether was related to the process of triacylglycerol lipolysis per se, or to the presence of lipoprotein lipase, hydrolysis of triacylglycerol and the uptake of cholesteryl ether were dissociated. Lipolysis of chylomicron triacylglycerol was achieved by exposure of chylomicrons (labeled with ^{14}C-fatty acid in their triacylglycerol portion and with ^{3}H-cholesteryl linoleyl ether) to membrane-supported lipoprotein lipase in three different systems. The chylomicrons were injected into functionally eviscerated rats, or were perfused through rat hearts or exposed to heart cell cultures. The chylomicron "remnants" produced in this manner had lost most of their triacylglcerol when they added to heart cell cultures in which the uptake of cholesteryl ether was to be measured. As the same amount of labeled cholesteryl linoleyl ether was added to the cultures in the form of intact chylomicrons or "remnants", the finding of a similar uptake of the label indicated that concurrent lipolysis of triacylglycerol was not mandatory for the uptake process. A comparable decrease in the uptake of cholesteryl linoleyl ether occurred when intact chy-

Table 8
UPTAKE OF [³H]CHOLESTERYL LINOLEYL ETHER BY CULTURED F₁ RAT HEART CELLS INCUBATED WITH MESENTERIC DUCT RAT CHYLOMICRONS

Incubation medium		Cellular uptake/6 hr			
		[³H]Cholesteryl ether		¹⁴C-Labeled lipid	
Chylomicron triacylglycerol (µg/mℓ)	Lipolysis of triacylglycerol (%)	dpm	as % of medium ³H	dpm	as % of medium ¹⁴C
10	94.9 ± 9.7	214 ± 28	5.6 ± 1.4	—	—
50	91.1 ± 5.8	446 ± 15	2.2 ± 0.1	179 ± 17	15.6 ± 1.8
100	87.1 ± 1.2	567 ± 46	1.4 ± 0.1	302 ± 51	13.4 ± 2.0
200	86.6 ± 5.7	770 ± 31	1.0 ± 0.1	617 ± 75	13.1 ± 1.6
500	78.7 ± 3.5	960 ± 40	0.5 ± 0.1	879 ± 199	8.0 ± 1.4

Notes: The cells were cultured in complete medium for 8 days. On the day of the experiment, the medium was removed and replaced with 1 mℓ medium without serum containing 4% bovine serum albumin and varying concentrations of rat chylomicrons which had been labeled biosynthetically with [¹⁴C]palmitic acid and by transfer in vitro with [³H]cholesteryl linoleyl ether. After 6 hr of incubation at 37°C, the medium was collected and the degree of lipolysis was determined. The percentage lipolysis was calculated from the amount of label found in medium fatty acid plus label in cells, divided by the total radioactivity added to medium. The cell layer was washed 3× with 0.2% albumin, 3× with phosphate buffered saline, and removed with 0.05% trypsin. Following inactivation of the trypsin by addition of serum-containing medium, the cells were pelleted, the pellet washed twice by resuspension in 4 mℓ phosphate-buffered saline and radioactivity was determined after extraction of the lipids according to the method of Folch et al. Values are means ±S.D. of triplicate dishes.

Adapted from Chajek-Shaul, T. et al., *Biochim. Biophys. Acta,* 666, 147, 1981. With permission.

lomicrons or chylomicron remnants were added to cultures with lowered lipoprotein lipase activity.[38] It appears, therefore, that the presence of lipoprotein lipase rather than concomitant triacylglycerol lipolysis is important for the uptake of cholesteryl linoleyl ether.

This aspect was studied further in the heart cell cultures using exogenous lipoprotein lipase isolated from bovine milk and chylomicrons labeled with [³H]cholesteryl linoleyl ether.[39] The direct involvement of the enzyme molecule in the enhancement of cholesteryl ester uptake by the heart cells was also suggested by the finding of a much higher uptake of the labeled analogue from chylomicron remnants (prepared using membrane-supported lipoprotein lipase), when milk lipoprotein lipase was added to the incubation medium.

Additional evidence to support the thesis that binding and internalization of chylomicron cholesteryl linoleyl ether requires the presence of lipoprotein lipase attached to the cell surface, was derived from the observation that in F₁ heart cell cultures, colchicine interferes with the transport of lipoprotein lipase to the cell surface.[14] After exposure to colchicine, total lipoprotein lipase activity of the culture remains unchanged, but there is a pronounced fall in surface-bound enzyme, determined as heparin-releasable activity. A correlation was found between the decrease of heparin-releasable surface lipoprotein lipase and the reduction in the uptake of chylomicron cholesteryl ether.[40] In order to further test the hypothesis that binding of LPL to the cell surface is mandatory for the uptake of chylomicron cholesteryl ester, heparin was added to the incubation medium. Addition of heparin to the culture medium reduced the lipoprotein lipase-catalyzed uptake of the labeled cholesteryl linoleyl ether by 90%.[40] This was also true when milk LPL was added to the culture medium in the presence of heparin (Table 9).

Table 9
EFFECT OF HEPARIN ON THE UPTAKE OF [³H]CHOLESTERYL LINOLEYL ETHER BY RAT F₁ HEART CELLS AND PREADIPOCYTES IN CULTURE

	F$_1$ rat heart cells in culture			Rat preadipocytes	
	Cellular [³H]cholesteryl linoleyl ether (% of medium label)		**Lipolysis of ¹⁴C-labeled triacyl-glycerol (%)**	**Cellular [³H]cholesteryl linoleyl ether (% of medium label)**	**Lipolysis of ¹⁴C-labeled triacyl-glycerol (%)**
Additions	**Exp I**	**Exp II**			
None	2.8 ± 0.35	4 ± 0.5	85 ± 3.8	3.4	89
Heparin (3 min)	0.8	1.4 ± 0.01	97 ± 2.8	—	—
Heparin (6 hr)	0.6 ± 0.01	0.2 ± 0.03	86 ± 2.3	0.6	96
Milk lipoprotein lipase	11 ± 0.55	19 ± 0.50	99 ± 1.0	20.2	99.4
Heparin (3 min) + milk lipoprotein lipase	1.2	2.5 ± 0.43	98 ± 1.5	—	—
Heparin (6 hr) + milk lipoprotein lipase	0.8 ± 0.28	0.5 ± 0.09	97 ± 1.0	2.2	99.4

Notes: F$_1$ rat heart cells were cultured for 8 days and preadipocytes for 10 days in their respective complete media. Lipoprotein lipase activity was 936 and 998 nmol fatty acid released per mg cell protein per hr in heart cells and 246 and 560 nmol fatty acid released per mg cell protein per hr in preadipocyte cells and in the culture medium, respectively. At the beginning of the experiment the culture medium was removed and the cells were incubated in 1 mℓ of F$_{10}$ medium without serum containing 4% bovine serum albumin and rat chylomicrons (50 µg triacylglycerol) labeled with [¹⁴C]palmitic acid (over 85% in triacylglycerol) and with [³H]cholesteryl linoleyl ether. Heparin, 5 units/mℓ, was present in some dishes throughout the 6 hr incubation period; to some dishes, heparin was added for 3 min, the heparin-containing medium was removed, the cell layer washed 3× with phosphate-buffered saline and the labeled chylomicrons and milk lipoprotein lipase, 5 µg protein/mℓ (where indicated), were added to 1 mℓ fresh F$_{10}$ medium containing 4% bovine serum albumin. Values are means ± S.D. of triplicate dishes.

Adapted from Chajek-Shaul, T. et al., *Biochim. Biophys. Acta,* 666, 216, 1981. With permission.

One may speculate that different active sites on the lipoprotein lipase molecule could be responsible for the catalysis of triacylglycerol hydrolysis and for the binding of the enzyme to the cell surface. The effect of the enzyme on cholesteryl linoleyl ether uptake is apparently related to the latter category.

The mechanism of cholesteryl ester uptake and the role of lipoprotein lipase in this process are not known. Two possible modes of internalization of the labeled lipid could be envisaged, i.e., by adsorptive or by membrane endocytosis. Thus the role of lipoprotein lipase could be one of a ligand between the membrane and the esterified cholesterol. On the other hand, the presence of the enzyme might have perturbed the lipid domains of the plasma membrane, allowing the incorporation of the esterified cholesterol into the membrane proper.

REFERENCES

1. **Spitzer, J. A.,** TI studies of substrate metabolism in isolated myocytes, *Adv. Exp. Biol. Med.,* 161, 217, 1983.
2. **Dow, J. W., Harding, N. G. L., and Powell, T.,** Isolated cardiac myocytes. II. Functional aspects of mature cells, *Cardiovasc. Res.,* 15, 549, 1981.
3. **Cryer, A.,** Tissue lipoprotein lipase activity and its action in lipoprotein metabolism, *Int. J. Biochem.,* 13, 525, 1981.

4. **Quinn, D., Shirai, K., and Jackson, R. L.,** Lipoprotein lipase: mechanism of action and role in lipoprotein metabolism, *Proc. Lipid Res.,* 22, 35, 1982.

5. **Hamosh, M. and Hamosh, P.,** Lipoprotein lipase, its physiological and clinical significance, *Mol. Aspects Med.,* 6, 199, 1983.

6. **Chajek, T., Stein, O., and Stein, Y.,** Interaction of concanavalin A with membrane-bound and solubilized lipoprotein lipase of rat heart, *Biochim. Biophys. Acta,* 431, 507, 1976.

7. **Blanchette-Mackie, E. J. and Scow, R. O.,** Sites of lipoprotein lipase activity in adipose tissue perfused with chylomicrons, *J. Cell Biol.,* 51, 1, 1971.

8. **Chajek, T., Stein, O., and Stein, Y.,** Interference with the transport of heparin-releasable lipoprotein lipase in the perfused rat heart by colchicine and vinblastine, *Biochim. Biophys. Acta,* 388, 260, 1975.

9. **Harary, I. and Farley, B.,** In vitro studies on single beating rat heart cells. Part 1 (Growth and organization), *Exp. Cell Res.,* 29, 451, 1963.

10. **Pinson, A., Frelin, C., and Padieu, P.,** The lipoprotein lipase activity in cultured beating heart cells of the postnatal rat, *Biochimie,* 55, 1261, 1973.

11. **Pollinger, I. S.,** Identification of cardiac myocytes in vivo and in vitro by the presence of glycogen and myofibrils, *Exp. Cell Res.,* 76, 243, 1973.

12. **Pollinger, I. S.,** Separation of cell types in embryonic heart cell cultures, *Exp. Cell Res.,* 63, 78, 1970.

13. **Chajek, T., Stein, O., and Stein, Y.,** Rat heart in culture as a tool to elucidate the cellular origin of lipoprotein lipase, *Biochim. Biophys. Acta,* 488, 140, 1977.

14. **Chajek, T., Stein, O., and Stein, Y.,** Lipoprotein lipase of cultured mesenchymal rat heart cells. I. Synthesis, secretion and releasability by heparin, *Biochim. Biophys. Acta,* 528, 456, 1978.

15. **Chajek, T., Stein, O., and Stein, Y.,** Lipoprotein lipase of cultured mesenchymal rat heart cells. II. Hydrolysis of labeled very low density lipoprotein triacylglycerol by membrane-supported enzyme, *Biochim. Biophys. Acta,* 528, 466, 1978.

16. **Henson, L. C., Schotz, M. C., and Harary, I.,** Lipoprotein lipase in cultured heart cells. Characteristics and cellular location, *Biochim. Biophys. Acta,* 487, 212, 1977.

17. **Friedman, G., Stein, O., and Stein, Y.,** Lipoprotein lipase activity in cultured mesenchymal rat heart cells. Part 5. Effect on enzyme activity of the glycosylation inhibitors, 2-deoxyglucose and tunicamycin, *Atherosclerosis,* 36, 289, 1980.

18. **Glick, J. M., McGlynn, A. M., and Rothblat, G. H.,** Inhibition of lipoprotein lipase synthesis and release in cultured preadipocytes, in *Proc. 5th Int. Symp. Atherosclerosis, Oral Presentation 473,* Houston, Texas, November, 1979.

19. **Struck, D. K. and Lennarz, W. J.,** Evidence for the participation of saccharide-lipids in the synthesis of the oligosaccharide chain of ovalbumin, *J. Biol. Chem.,* 252, 1007, 1977.

20. **Speake, B. K. and White, D. A.,** The effect of tunicamycin on the glycosylation of lactating-rabbit mammary glycoproteins, *Biochem. J.,* 180, 481, 1979.

21. **Friedman, G., Stein, O., and Stein, Y.,** Lipoprotein lipase of cultured mesenchymal rat heart cells. III. Effect of glucocorticouds and insulin on enzyme formation, *Biochim. Biophys. Acta,* 531, 222, 1978.

22. **Friedman, G., Stein, O., and Stein, Y.,** Lipoprotein lipase activity in F_1 heart cell cultures. Effect of dialyzable serum factors on enzyme stability and enzyme synthesis, *Biochim. Biophys. Acta,* 619, 650, 1980.

23. **Broensztajn, J., Otway, S., and Robinson, D. S.,** Effect of fasting on the clearing factor lipase (lipoprotein lipase) activity of fresh and defatted preparations of rat heart muscle, *J. Lipid Res.,* 11, 102, 1970.

24. **Lithell, H., Boberg, J., Hellsing, K., Lundqvist, G., and Vessby, B.,** Lipoprotein-lipase activity in human skeletal muscle and adipose tissue in the fasting and fed states, *Atherosclerosis,* 30, 89, 1978.

25. **Kotlar, T. J. and Borensztajn, J.,** Oscillatory changes in muscle lipoprotein lipase of fed and starved rats, *Am. J. Physiol.,* 233, E316, 1977.

26. **Friedman, G., Stein, O., and Stein, Y.,** Lipoprotein lipase of cultured mesenchymal rat heart cells. IV. Modulation of enzyme activity by VLDL added to the culture medium, *Biochim. Biophys. Acta,* 573, 521, 1979.

27. **Robinson, D. S.,** *Comprehensive Biochemistry,* Vol. 18, Florkin M. and Stotz, E. H., Eds., Elsevier, Amsterdam, 1970, 51.

28. **Cryer, A., Riley, S. E., Williams, E. R., and Robinson, D. S.,** Effect of nutritional status on rat adipose tissue, muscle and post heparin plasma clearing-factor lipase activities: their relationship to triglyceride fatty acid uptake by fat cells and to plasma insulin concentrations, *Clin. Sci. Mol. Med.,* 50, 213, 1976.

29. **Borensztajn, S., Samols, D. R., and Rubenstein, A. H.,** Effects of insulin on lipoprotein lipase activity in the rat heart and adipose tissue, *Am. J. Physiol.,* 223, 1271, 1972.

30. **Robinson, D. S. and Wing, D. R.,** Regulation of adipose tissue clearing factor lipase activity, in *Adipose Tissue Regulation and Metabolic Functions,* Jeaurenaud, B. and Hepp, D., Eds., Thieme, Stuttgart, 1970, 41.

31. **Bjorntorp, P., Karsson, M., Pertoft, H., Pettersson, P., Sjostrom, L., and Smith, L. L.,** Isolation and characterization of cells from rat adipose tissue developing into adipocytes, *J. Lipid Res.,* 19, 316, 1978.
32. **Friedman, G., Chajek-Shaul, T., Stein, O., and Stein, Y.,** Modulation of lipoprotein lipase activity in cultured rat mesenchymal heart cells and preadipocytes by dibutyryl cyclic AMP, cholera toxin and 3-isobutyl-1-methylxanthine, *Biochim. Biophys. Acta,* 752, 106, 1983.
33. **D'Armiento, M., Johnson, G. S., and Pastan, I.,** Regulation of adenosine 3′:5′-cyclic monophosphate phosphodiesterase activity in fibroblasts by intracellular concentrations of cyclic adenosine monophosphate, *Proc. Natl. Acad. Sci. USA,* 69, 459, 1979.
34. **Wicks, W. D., Leichtling, B. H., Wimalasena, J., Roper, M. D., Su, J.-L., Su, Y.-F., Howell, S., Harden, T. K., and Wolfe, B. B.,** *Advances in Nucleotide Research,* Vol. 9, George, W. J. and Ignarro, L. J., Eds., Raven Press, New York, 1978, 411.
35. **Guidotti, A., Chuang, D. M., Hollenbeck, R., and Costa, E.,** *Advances in Nucleotide Research,* Vol. 9, George, W. J. and Ignarro, L. J., Eds., Raven Press, New York, 1978, 185.
36. **Yamasaki, H., Huberman, E., and Sacks, L.,** Regulation of aryl hydrocarbon (benzo(a)pyrene hydroxylase) activity in mammalian cells, *J. Biol. Chem.,* 250, 7766, 1975.
37. **Fielding, C. J.,** Metabolism of cholesterol-rich chylomicrons, *J. Clin. Invest.,* 62, 141, 1978.
38. **Chajek-Shaul, T., Friedman, G., Halperin, G., Stein, O., and Stein, Y.,** Uptake of chylomicron [³H]cholesteryl linoleyl ether by mesenchymal rat heart cell cultures, *Biochim. Biophys. Acta,* 666, 147, 1981.
39. **Freidman, G., Chajek-Shaul, T., Stein, O., Olivecrona, T., and Stein, Y.,** The role of lipoprotein lipase in the assimilation of cholesteryl linoleyl ether by cultured cells incubated with labeled chylomicrons, *Biochim. Biophys. Acta,* 666, 156, 1981.
40. **Chajek-Shaul, T., Friedman, G., Stein, O., Olivecrona, T., and Stein, Y.,** Binding of lipoprotein lipase to the cell surface is essential for the transmembrane transport of chylomicron cholesteryl ester, *Biochim. Biophys. Acta,* 712, 200, 1982.

Chapter 16

METABOLSIM OF GLUCOSE AND LIPIDS IN CULTURED HEART CELLS

Arié Pinson

TABLE OF CONTENTS

I. INTRODUCTION

Carbohydrate and lipid metabolism in the heart including the role they play during the performance of cardiac work have been extensively reviewed.[1-3] This chapter will be limited to those aspects of carbohydrate and lipid metabolism which have been studied in heart cells in culture, pinpointing findings which have extended our knowledge of the metabolism of the heart. Pioneering studies with cardiomyocytes giving similar results to those described in other systems (e.g., in vivo or in the perfused heart) and those which are in conflict with other data will also be discussed.

Heart cell cultures would seem to be an ideal system for studying metabolic events at a cellular level, especially for kinetic investigations, since the extracellular environment may be controlled. However, there are several limitations: the cells in culture must have adapted to their new substratum and, since the culture medium is usually changed every 48 hr, the substrate is effectively supplied in a "batchwise" fashion. Consequently, on the one hand, cellular metabolism is continually adapting to the changing environment, while on the other hand, the catabolic products are never completely removed from the vicinity of the cells. In addition, in contrast to the in vivo heart, where ATP almost exclusively derived by oxidative phosphorylation is the energy source for its sole function as a pump, cultured heart cells, which clearly resemble those *in situ* in their ability to contract spontaneously and rhythmically, are not required to act against the frictional resistance of blood vessels; therefore, their energy requirements correspond to those for isotonic contractions. One consequence of this is that there are less mitochondria in cultured cells than in those in vivo. In spite of the reservations mentioned above, the basic metabolism, structure, and electrophysiological properties of cardiomyocytes resemble those of the intact heart. Thus studies of the metabolism of cultured heart cells, in which the fate of a substrate can easily be followed from its uptake to the utilization of its metabolites, have certainly contributed to extending our knowledge of the physiology and biochemistry of the heart, even though this system has not as yet been fully exploited.

II. GLUCOSE METABOLISM

Glucose-6-phosphate is at the "crossroads" of glucose metabolism, with several different pathways being available from that point: (1) glycolysis yielding pyruvate (and lactate) usually followed by complete oxidation via the citric acid cycle, (2) glycogen synthesis via uridine diphosphoglucose (UDP glucose), and (3) the pentose phosphate shunt.

In the heart, under normal aerobic conditions, glucose does not constitute an important energy source, and even in conditions of extremely high work load with glucose as the only available substrate in the extracellular fluid, glucose oxidation only accounts for 40% of the energy required for contraction, the rest being derived from intracellular triglycerides (TGs).[4-6]

A. Glucose Uptake

Early studies in the perfused heart showed that several sugars, including glucose, share a common transport pathway[7] — the relative amount of sugar (75%) taken up at various different sugar concentrations remains constant. Thus, the use of nonmetabolizable sugars facilitates the study of glucose uptake mechanisms. Transport was shown to be regulated primarily by insulin. In addition, the only rate-limiting step for uptake at high extracellular glucose concentrations is phosphorylation by hexokinase.[8-10]

In the only extensive study on glucose transport to date in cultured heart cells, Paris et al.[11] compared the metabolism of 3-O-methyl glucose, which enters the cell but is not metabolized, and 2-deoxyglucose, which undergoes phosphorylation by hexokinase but may not be further utilized. They showed that the rate of transport is several fold higher than the phosphorylation rate. The latter is limited by the rate of transport into the cell, reaching a steady state within a few minutes and is dependent on the extracellular carbohydrate concentration. The concentration of the intracellular pool of sugar is below the K_m value for hexokinase and determines the rate of phosphorylation, which is increased by countertransport and decreased by cytochalasin B, so that any changes in carbohydrate metabolism are reflected in the rate of 2-deoxyglucose phosphorylation. Glucose transport was not affected by the presence of fatty acids in the medium or by preloading the cells with fatty acids.[11]

These findings conflict with those reported by Nealy et al.,[12] who found that in the working perfused heart, palmitate inhibited glucose uptake. This discrepancy is possibly due to the contribution of a regulation system in cultured heart cells which responds to the concentration of extracellular hexoses.[11] The decrease in transport in cultured cells following incubation with cycloheximide suggests that the synthesis of a carrier protein is involved and that its activity may be modulated under appropriate conditions.

Other studies[13,14] have pointed to the importance of the nature of the extracellular medium for the glucose uptake process. Supplementation of the medium with 20% serum resulted in glucose uptake at a constant rate of 3 μmol/hr/mg protein. Thus the medium became depleted of glucose after 10 to 12 hr, well before the next medium change. In the absence of serum, however, glucose was taken up at about half this rate.[14] The presence of serum in the medium stimulated both glucose transport and hexokinase activity.[15] These rates of glucose uptake are an order of magnitude higher than those found in both perfused[8] (30 μmol/hr/g of tissue) and in fetal heart[16] (66 μmole/hr/g of tissue). Other investigators have reported similar findings (about 1.5 μmol/hr/mg protein) with cultured cells.[17-19] However, the results of Ross and McCarl[20] (0.5 μmol/h/g of tissue) were very similar to the findings with perfused hearts. These differences may be put down to variations in the populations of the different cell types, since cardiomyocytes take up glucose at a much higher rate than nonmuscle cells.[20]

B. Glycogen Metabolism

As mentioned previously, glycogen is synthesized via uridine diphosphoglucose (UDG) by two glycogen synthases — a (dephosphorylated) and b (phosphorylated) — which are regulated by synthase kinase. Glycogen synthase transfers α-glycosyl groups to the 4-glucosyl position. A branching enzyme is also involved in glycogen formation.

A debranching enzyme and a phosphorylase participate in glycogen degradation.[21]

The enzymatic regulation of glycogen synthesis and breakdown, which is different from that in the liver, has as yet to be studied in heart cell cultures.

On incubation with labeled glucose, the labeled glycogen peak occurs within 2 to 4 hr of renewing the medium, at which point the glucose concentration in the medium is 75% of the initial level. Following this initial increase, glycogen labeling decreases in two phases: (1) a rapid one with a half-life of 10 hr, and (2) one with a half life of 100 hr. However, these values do not reflect the true glycogen turnover, since labeled glucose incorporation into glycogen is still taking place during the first phase. The true half-life is probably less than 10 hr. Furthermore, the glucose incorporated into glycogen from the medium never exceeds more than 5% of the total.[13-20]

Despite sharp fluctuations in glucose incorporation, glycogen content is more or less constant in such cultures (40 to 60 μg/mg protein)[17,20,22] and is similar to the values found in vivo, with glygocen deposits in the heart being much lower than in skeletal muscle.[23] The total glycogen content in cultured heart cells increases with the age of the culture — from 5 μg/mg protein immediately after plating to 50 μg/mg of protein after 2 to 3 weeks in culture.[22] Glycogen accumulation in older cultures has also been demonstrated by microscopic examination following staining by the Schiff method.[24]

C. Lactic Acid Production

Glycolysis in the heart is regulated via the activities of phosphofructokinase and glyceraldehyde-3-phosphate dehydrogenase. The supply of glucose-6-phosphate, either following glucose uptake and phosphorylation by hexokinase or via glycogen degradation, is a crucial step in the pathway from glucose to pyruvate and lactate.[1,3]

Under conditions of normal oxygen supply the heart does not produce lactic acid — glucose is either being utilized for glycogen synthesis or undergoing complete oxidation. The LDH H_4 isoenzyme pattern is present, and consequently, lactate is oxidized to pyruvate. Lactic acid is only the end product of glucose degradation during oxygen deprivation, when the energy is supplied to the ischemic area via glycolysis or in the partially anaerobic fetal heart.

Studies with cardiomyocytes have shown that most of the glucose incorporated is rapidly converted to lactate which is then released into the extracellular space.[13-15,20] However, lactate production is not a result of oxygen restriction in the culture medium.[14] It has been shown that the released lactate is not derived from endogenous glycogen, but from newly incorporated glucose and is not dependent on the concentration of lactate in the medium.[13,14] However, the presence of serum in the medium (at an optimum concentration of 20%) promotes glucose uptake and lactate production.[14] According to Ross and McCarl, nonmuscle cells in culture only oxidize glucose to lactate, while in cardiomyocytes, some glucose is also metabolized via the citric acid cycle.[20]

The amount of lactate produced also varies with the initial seeding density.[25] Indeed, it is difficult to compare the amount of glucose converted to lactate in various experiments, since the methods for preparing cell cultures and the initial seeding densities were different in each research group, affecting the proportion of myocytes and nonmuscle cells in the culture. Both our group[13] and Ross and McCarl[20] have reported similar values for the lactate production, 40 μmol/mg protein (more than 65% of the incorporated glucose) and 0.88 μmol/hr/mg protein (representing 75% of the glucose taken up from the medium), respectively. These values are 4-fold higher than the data of Orloff and McCarl.[18]

It has been anticipated that lactate production in cultured heart cells might serve as a mechanism for removing excess glucose from the cells since only a small amount is used in energy production and a mere 5% is stored in the form of glycogen.[13,14] The shift in the LDH isoenzyme pattern from the heart (H_4) to the muscular (M_4) type may be a reflection of this,[26] although it might also be related to the changes in the relative proportion of the different types of cells in the culture.[27] It should be noted that as far back as 1966, Randle et al. reported that in the nonworking perfused heart, significant amounts of glucose and pyruvate were converted to lactate.[28] These findings are possibly related to the fact, already mentioned in the introduction, that the cells in culture are not performing work against resistance, a direct contrast to the in vivo situation.

Only after the medium has been completely depleted of glucose and over 65% has been transformed to lactate, does the lactate concentration in the medium begin to decrease in a linear fashion, as it either undergoes oxidation or transamination.[13,14,20]

D. Glucose Oxidation Pathways

The mechanisms of control of glycolysis and of complete oxidation in the heart have been discussed in detail by Randle and Tubbs.[3]

The growth of heart cells in culture occurs in two distinct phases: an initial phase of cell division, until confluency is reached,[29] and a nondividing phase with continued cell growth. These two phases are clearly reflected in the glucose oxidation via two pathways in cultured cells: (1) The pentose phosphate pathway, and (2) Glucose oxidation via the citric acid cycle and oxidative phosphorylation. The importance of both of these pathways has been studied in cultured cells — the oxidation of glucose (and also fatty acids) has been estimated by measuring the release of $[^{14}C]\text{-}CO_2$ from an appropriately labeled substrate, either directly in the culture flask,[25] or by means of a simple multi-channel trapping device[30,31] or in more recent work by the rather sophisticated "radiorespirometer" working along similar lines.[32]

1. The Pentose Phosphate Pathway

This pathway is very active in the early embryonic heart, with decreasing activity as the heart approaches the adult form, when the division index decreases.[33,34] The end product of this pathway, ribose-phosphate, is a precursor for nucleoside and subsequent DNA synthesis. As ribose is formed from carbons 2 to 6 of the glucose molecule, studies on the relative importance of each pathway have been performed by comparing the amount of $^{14}C\text{-}CO_2$ released from $1\text{-}^{14}C$-glucose obtained via the pentose phosphate shunt and the citric acid cycle to that produced from $6\text{-}^{14}C$-glucose via the citric acid cycle alone. Warshaw and Rosenthal[25] reported enhanced glucose oxidation in conditions of low density plating, presumably reflecting increased flow through the pentose phosphate pathway, which is inhibited when the cells reach confluency.[25,35] We have reported[13,14] similar findings, as well as a link between this pathway and ^3H-thymidine incorporation into DNA (see Figure 2 in Reference 13). Thus, DNA synthesis, which comes after the pentose phosphate pathway, increases in parallel with the flow through the pentose shunt, after a delay of several hours, a finding consistent with the increased requirement for ribose phosphate in proliferating cultures.

2. Glucose Oxidation

The level of glucose oxidation via the tricarboxylic acid cycle has been shown to remain constant during all the phases of heart cell culture.[13,25] However, conflicting results have been reported by Schroedl et al.,[36] who found that the rates of glucose utilization varied in cultures of different ages, which they interpreted as a postnatally induced decline of the utilization of glucose as an energy source.

E. Pyruvate and Lactate Oxidation

As mentioned above, we have reported that after glucose is taken up from the medium, lactate released back into the medium (formed by glycolysis of glucose) is reutilized within the cell, either in oxidation or in further transformations. Rosenthal and Warshaw[35] found a lag phase in the time course of $^{14}CO_2$ formation from $[2\text{-}^{14}C]$-pyruvate as compared to that produced from $[1\text{-}^{14}C]$-pyruvate. Following this lag, $^{14}CO_2$ production from both substrates increased linearly, although $^{14}CO_2$ formation from $[1\text{-}^{14}C]$-pyruvate occurred at a higher rate. One possible explanation for this is that the CO_2 derived from carbon-1 is produced directly by the action of pyruvate dehydrogenase, while CO_2 from carbons 2 and 3 is formed via incorporation into acetyl-CoA followed by the citric acid cycle reactions.

Ross and McCarl[20] found that pyruvate was taken up at a rate of 4.79 ± 0.92 nmol/min/mg protein. About 50% of the incorporated pyruvate is reexported as lactate, the remainder being oxidized. They also confirmed an earlier finding[35] that $^{14}CO_2$ release from $[2\text{-}^{14}C]$-pyruvate lagged behind that formed from $[1\text{-}^{14}C]$-pyruvate by about 10 min, and found that the steady state for oxidation was reached much faster with the former (within 20 min) than with the latter substrate (within 80 to 100 min). They also showed that with lactate oxidation $^{14}CO_2$ release from $[1\text{-}^{14}C]$-lactate increased linearly over a period of about 2 hr and then fell.[20] More $^{14}CO_2$ was also produced from $[1\text{-}^{14}C]$-lactate than from $[1\text{-}^{14}C]$-pyruvate under similar conditions. As yet no satisfactory explanation has been given for this finding. Furthermore, it was also found that 15% of the total lactate is incorporated into cellular components — two-thirds of this into water soluble molecules and one-third into lipids and proteins.

In conclusion, the above results show that in cultured cells, pyruvate is either oxidized or converted to lactate, but is not transformed into other cellular components, while lactate undergoes either oxidation or transformation to other compounds. Similar findings have been reported in the perfused heart,[2,28] indicating that the LDH isoenzyme profile does not play a significant role in the regulation of pyruvate-lactate formation under low work-load conditions.

III. LIPID METABOLISM

There is a well established dual role for lipids in the heart. On the one hand, they are the major functional-structural component of the sarcolemma, while on the other hand, they serve as the major energy source of the myocardium. For this reason storage of lipids is much more important than that of glycogen, since they provide a much more efficient energy source for the heart.[37,38]

Quantitative studies of heart metabolism *in situ* with regard to substrate utilization were greatly advanced when cardiac catheterization techniques, allowing blood samples to be taken from the intact animal, were devised. By this means, Bing et al.[39,40] found the cardiac oxygen consumption to be much too high to be accounted for by the oxidation of glucose, lactate, and pyruvate, alone. Furthermore, when they extracted the fatty acids (FAs) and ketone bodies,[41] they discovered that 16 and 67% of the oxygen taken up was utilized in glucose oxidation and lipid oxidation, respectively. Taking into consideration the state of the art at the time it was difficult to reach the definite conclusion that nonesterified fatty acids (NEFAs) are the major energy-providing substrate in the heart. This was confirmed later by studying the extraction of NEFAs from the plasma by the human myocardium.[42] Since that time, lipid metabolism in the heart has been extensively studied.[1-3] In the following sections, certain aspects of these studies will be reviewed, concentrating on the contribution of the cultured heart cell system to this field.

A. "Nonspecific" Effects of FAs

It has been shown that if heart cells are grown in a lipid-deficient medium, the resulting loss of beating ability is accompanied by changes of the levels of some of the enzymes involved in oxidation pathways and the pentose shunt.[43,44] Addition of serum lipids to the medium restored both beating and enzymatic activities.[45,46] It was found that whereas lipids were obligatory for maintenance of the beating function, cell growth was also possible in lipid-depleted serum. The absence of lipids apparently had a long-term effect on the beating function, since beating was not maintained even if ATP, either derived from glycolysis or oxidative phosphorylation, was kept at a high level.[47] Moreover, restoration of beating lagged behind the addition of FA, suggesting that the relationship was not one of a simple cause and effect.[45] Indeed the maintenance of beating function is clearly not dependent on energy charge alone since different FAs affect it in various ways, i.e., palmitic acid restored beating whereas linoleic acid did not, although they behave in a similar manner as far as oxidation goes.[48] Thus, in a serum free medium, palmitate was more efficient in restoring cardio-myocyte beating than linoleate, although the latter supported mitochondrial metabolism.[45,49,50]

Fatty acids (FAs) are not the only compounds capable of restoring the beating to quiescent heart cells. Cortisol acetate was capable of reinitiating beating in such cultures with an accompanying increase of cellular energy charge.[22] Recently, it has been reported that increasing the phosphatidyl choline/sphingomyelin (PC/SM) ratio has a similar effect, also causing the restoration of beating.[51] Although this was explained by the authors as being due to changes in the sarcolemmal composition to a state which is similar to that of younger cells — a "turning back of the clock", a veritable rejuvenation! — it is more likely that it is the result of a "nonspecific", as yet undefined effect of lipids. As has already been shown, changes in the sarcolemmal lipid composition influence the viscosity[52] and action potential duration[53] and the beating function in turn, in these cells. FAs also have a complementary arrhythmogenic effect. Indeed, increased plasma NEFA or extracellular FA accumulation during oxygen restriction and deprivation, are apparently related to cardiac arrhythmias;[54] although this cause and effect relationship remains controversial,[55] a similar effect in cultured heart cells has been reported.[56] Glucose reduced the arrhythmogenic effect of NEFAs, possibly by increasing the rate of TG synthesis.

B. Lipid Composition

Many systematic studies of FA composition in cultured heart cells have been aimed at determining how it differes from that in the in vivo heart and the variations that occur with aging of these cultures.[50,57,58] Although it is difficult to compare these reports directly, the findings are generally similar — any differences seem to be a result of the various culture media used, mainly with respect to the type and quantity of serum present. However, the FA profile in 3-day-old cultured cells was essentially similar to that in newborns,[58] as increased levels of linoleic acid (18:2) and a fall in arachidonic acid (20:4) were the main age-related changes.[50,58] Although in cells of this type FA transformations (including chain elongation, shortening, saturation, and desaturation) may occur,[49,59,60] the above-mentioned age-related changes were apparently due to the use of different sera containing fatty acids[58] and to overgrowth by nonmuscle cells. The molar ratio of cholesterol/phospholipids is also high (1.66) in cultured heart cells, with phosphatidyl ethanolamine and phosphatidyl choline constituting 34.8 and 39% of the total phospholipids (PLs), respectively.[61] Sarcolemmal viscosity is dependent on the content of saturated and unsaturated PLs present in the membrane[52,62] and, as already mentioned, affects its function.[52,53,62] In conclusion, studies of the cellular lipid composition and of changes in the lipid profile are a useful tool for studying cardiomyocyte function only when they are coupled with studies of other factors affecting cellular activity.

C. Fatty Acid Uptake and Distribution

Fatty acids (FAs) are transported complexed to plasma albumins which have several high affinity FA binding sites.[63] Less than 1% of the FAs circulate either in the free state or in a loosely bound form. It has been reported that the rate of FA uptake depends on the molar ratio of FA to albumin — since the concentration of the latter is usually constant, the FA level (generally around 0.5 mM, but which can vary within a fairly large range) is the critical factor.[64]

Although FA transport across cell membranes has been studied in many systems, the understanding of this topic is far from complete, particularly since it is very difficult to dissociate transport and subsequent utilization. Indeed, many data represent the sum of uptake and utilization. Stein and Stein[65] found that more than 90% of the incorporated FA is converted to the di- or triglyceride or phospholipid form within 1 min following a 15 sec ^{3}H-oleic acid pulse. There is no direct confirmation of a specific FA binding site on the cell membrane and the exact nature of the carrier molecule (if one exists) is as yet unknown. It has been suggested that the first step is a passive transport process, during which the hydrophobic FA methyl residues become solubilized within the hydrophobic components of the outermost membrane of the bilayer, possibly at specific FA or FA-albumin binding sites.[66-69] The next step is the translocation of the FA from the outer to the inner of the membrane layers, involving an acyl-coenzyme A intermediate[70] — diffusion only seems to play a role at very high free fatty acid (FFA) concentrations.

In a series of studies on FA uptake and metabolism in the cultured heart cell system,[71-74] two phases have been shown to occur in the uptake of labeled FA:s (1) passive diffusion, predominating at FA levels of greater than 200 μM and (2) a saturatable specific binding process with a K_m of about 10 μM. Furthermore, FA uptake in cardiomyocytes, in agreement with findings in liver cells,[69] is dependent on the total FA (both free and albumin-bound FA).[71] The various long chain FAs competed for uptake, and since the existence of a specific binding protein could not be demonstrated, it was assumed that this specific transport system is located on the cell surface.[71] Maximal uptake of palmitate occurred at palmitate:albumin ratios ranging from 7 to 10, far different from the in vivo value of approximately 2.

Albumin enhanced both uptake and subsequent esterification of palmitate but reduced that of oleate.[73] This was due to binding of most of the palmitate molecules to the low- and medium-affinity binding sites in albumin; while in the case of oleate, an unsaturated FA, albumin acts as an alternative site to the cardiac cells for its uptake, thus lowering oleate uptake.[73]

This uptake process was shown not to be energy dependent, but is probably a facilitated one.[74] In calcium-sensitive isolated adult myocardial cells (in albumin free incubation medium) FA uptake occurred in two stages: (1) A nonsaturatable phase corresponding to FA accumulation in the FFA fraction, which was directly proportional to chain length, and (2) A saturatable component, independent of chain length, which included FAs that had either been further esterified or converted to CO_2.[75,76] Thus there are apparently two pools of fatty acids: (1) A minor FA pool, which rapidly equilibrates with the extracellular FAs and where almost immediate conversion takes place; transport into this pool is possibly carrier mediated, and (2) The major FFA pool into which FAs are taken up by simple diffusion.

It is unlikely that the differences between these two culture systems are of physiological significance, but rather that they represent an adaptation in calcium intolerant cardiac myocytes.

As mentioned above, after uptake FAs are immediately metabolized in the intracellular channels of the sarcoplasmic reticulum.[65] The existence of a cytosolic FA binding protein, with a mol wt of 12,000 for binding to FA-acyl CoA, thus suppressing its detergent-like properties which could damage the membranes and impair enzymatic reactions, has also been demonstrated.[77,78]

In the first step, a FA-acyl CoA thioester is formed by the action of acyl CoA synthetases, which may be specific for FAs. This is subsequently further metabolized by one of two routes: transport to the mitochondrion or esterification to form TGs and PLs. Although several research groups have studied FA transformation and incorporation into the various classes of lipids,[49,59,60] these did not include kinetic studies. Detailed kinetic investigations have recently been carried out by our group.[79] In experiments in which palmitate incorporation was followed during a 90 min period, it was found that most of the FAs were immediately esterified, FFA only accounting for 5 to 10% of the total FA incorporated. Diglycerides do not accumulate in this system, indicating that the FA conversion rate is greater than the uptake. On depletion of the cells of energy-providing substrates and treatment with an uncoupler, such as dinitrophenol, an 80% inhibition of FA esterification occurred, but FFA did not accumulate within the cells. Impairment of esterification has been shown to lead to export of both diglycerides and FAs. Under normal conditions FA incorporation into TGs and PLs is approximately equal. Most of the TGs were located in the 100,000 g supernatant fraction, whereas PLs were generally associated with the particulate and membranous fractions.

Fatty Acid Oxidation

In most schemes representing cellular FA metabolism, the mitochondrial route is shown as the major one for the metabolism of exogenous FA, indicating that exogenous and not endogenous FAs are preferentially utilized for energy supply. However, some reports including Stein and Stein, and our own work, seem to suggest that this is not in fact the case, and that most FAs are in the esterified form shortly after uptake.[65,79] Moreover, as far back as 1964, Spitzer and Gold[80] showed that in the hind limbs of dogs at rest, FA extracted from the blood was not directly utilized, but was esterified and stored. The FAs oxidized were derived by hydrolysis of TGs. Furthermore, the turnover of stored FAs was accelerated

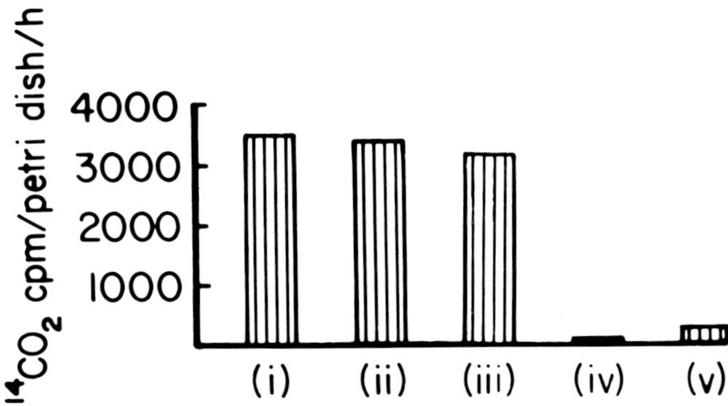

FIGURE 1.   Release of $^{14}CO_2$ either from endogenous (i to iii) or from newly incorporated (iv to v) 16-^{14}C palmitate by 6-day-old cardiac cells. (i to iii) Preincubation of cells for 13 hr with 16-^{14}C palmitate. Media were then replaced with the following: either (i) no added FA or (ii) with added unlabeled palmitate, or (iii) with added labeled palmitate. In (iv) preincubation with unlabeled palmitate for 13 hr preceding 1 hr incubation with labeled palmitate. In (v) incubation was for 1 hr with labeled palmitate without preincubation.

during exercise. Zierler et al.[81] proposed that the muscle used endogenous FAs during exercise, which were replenished on recovery. Dagenais et al.[82] found a delay of 30 min between labeled oleate uptake and the appearance of $^{14}CO_2$. In resting muscle, FFA uptake reached a steady state within 10 min, while the plateau in $^{14}CO_2$ only occurred after 2.5 to 3 hr of continuous [^{14}C] oleate perfusion, during which time only 5 to 10% of the extracted oleate was oxidized to CO_2, although $^{14}CO_2$ release continued at the maximal rate for several hours after the cessation of perfusion.[82] These findings demonstrate that in skeletal muscle, FAs are not oxidized directly, but first converted to the esterified form and stored in lipid droplets, generally in the vicinity of the mitochondria.

Based on early experiments showing that perfused hearts were depleted of glycogen in less than 30 min in the absence of exogenous substrate, Shipp et al.[83] suggested that endogenous lipids were being utilized to provide energy, although it was not clear whether PLs[83] or TGs[84,85] were preferred.

In cultured heart cells, it was demonstrated by a simple experiment that endogenous lipids were perferentially utilized over FAs that had just entered the cell[60,86] (see Figure 1). Preincubation of the cultured cells with [16-^{14}C] palmitate until it was completely depleted from the medium (13 hr) was followed by measurement of $^{14}CO_2$ release after replacement with media containing: (1) No added FAs, and (2) unlabeled palmitate supplementation, or (3) labeled palmitate supplementation; or either (4) with, or (5) without preincubation for 13 hr with unlabeled palmitate, preceding incubation with labeled palmitate for 1 hr. Therefore, the utilization of endogenous substrate only (in (1) and (2)) of both endogenous and newly incorporated palmitate (in(3)), and from newly incorporated palmitate alone (in (4) and (5)) were measured. No significant differences in the released $^{14}CO_2$ were noted in experiments (1) to (3), while in experiment (4) it was only 3% as compared with about 8% in experiment (5) of the total $^{14}CO_2$ released from cells which had been preincubated with labeled palmitate (in experiments 1 to 3). The above data support the preferential oxidation of endogenous lipids over those which have just entered the cell, but do not indicate whether most of the energy-supplying FAs at the mitochondrial sites are TGs or PLs.

In order to undergo oxidation, acyl CoA thioesters are transported into the mitochondria after having been converted to acyl carnitine at the outer mitochondrial membrane. On reaching the inner membrane, the carnitine is exchanged with the inner mitochondrial membrane carnitine by a transferase enxyme, the acyl group is then transferred to the mitochondrial matrix CoA and finally, β-oxidation occurs.[87] (A detailed review of these reactions may be found in Reference 3). In cardiomyocytes there are conflicting data regarding the importance of FA oxidation as compared to that of other substrates. Ross and McCarl[20] concluded that glucose, pyruvate, and lactate contribute more towards energy production than palmitate (see Table 6 in Reference 20). Stanisz et al.[88] reported that more energy is derived from glutamine metabolism (55%) than from either glucose (40%) or FAs (5%). Fujimoto and Harary[43] demonstrated a shift from lipid to carbohydrate metabolism in aging cultures, which they thought may be due to low intracellular levels of the enzymes involved in fatty acid metabolism[45,47] or in a postulated hypoxic glycolytic pathway in such cells.[17] However, according to our data,[14,60] FAs are the major energy source for cardiac myocytes and the decrease in FA oxidation on aging could be eliminated by preincubating the cells with palmitate so that palmitate oxidation is maintained at a constant level throughout. In addition, the cell populations in culture vary as a function of age, and this may also be of some significance.[29] Clearly, this topic warrants reexamination in the light of these findings.

The degradation of fatty acids occurs primarily via β-oxidation involving four sequential reactions.[89] FA-acyl CoAs or chain lengths from C_4 to C_{18} undergo oxidation via the "oxidase", and only under certain conditions, i.e., under oxygen restriction, have short-chain intermediates of palmitate and stearate oxidation been isolated.[90,91] There were no differences in the oxidation rates (i.e., the amount of $^{14}CO_2$ released) from [^{14}C]palmitate labeled at either at carbon 1, 6 or 11, either when fed to intact rats[92] or in perfused hearts.[93,94]

Samuel et al.[71] showed that long chain FAs are completely oxidized to CO_2 in cultured heart cells. However, we have reported a striking difference in the production of $^{14}CO_2$ from either [1-^{14}C]- and [16-^{14}C]-palmitate in heart cells in culture,[31,60,86] the release being several fold higher from carboxy- than from ω-labeled palmitate. This implied that the two termini of the palmitate molecule had different catabolic fates. A similar phenomenon was demonstrated in fat cells from both normal and starved rats, but no explanation was offered.[95]

In a study using rat hepatocytes, although total oxidation of [1-^{14}C] or [16-^{14}C]-palmitate, expressed as the sum of [^{14}C]CO_2 and [^{14}C]-acid soluble products, was almost identical during a 30 min period, considerably less [^{14}C]CO_2 was produced from [16-^{14}C]palmitate, being matched by a similar increase in acid soluble radioactivity.[96] These data were interpreted, in accordance with the early work of Brown et al.[97] as being due to the preferential incorporation of ω-labeled C_2 units into ketone bodies. If the same were true in heart cells, reutilization of the ketone bodies would be expected, as they are the energy source of choice in the myocardium[98]

Our studies revealed considerable release into the medium of water-soluble products derived from the methyl terminus of the palmitate molecule.[31,60,86] This extracellular radioactivity did not disappear even after long incubation periods, indicating that the released metabolite could not be further metabolized. This product was identified as 3-hydroxybutyrate by column chromatography, GLC, and mass spectrometry.[31,86] We also showed that 80% of the radioactivity was associated with the L(+)-isomer by using McCann and Greville's method[99] for separating D(−)-and L(+)-isomers. Therefore, the major product from the ω-terminal of [16-^{14}C]palmitate was undoubtedly L(+)3-hydroxy butyrate. In addition, the disappearance of radioactivity from the medium when cells were incubated with DL-[U^{14}C]3-hydroxybutyrate occurred in two phases (see Figure 5 in Reference 31): a rapid decline with a half-life of about 7 hr and a slower one with a half-life of 48 hr, corresponding to the utilization of the D(−) and the L(+)-isomers, respectively.

The coenzyme A derivative of L(+)-3-hydroxy butyrate is an intermediate in long chain FA oxidation.[100] It is believed that the L(+)-isomer of 3-hydroxy butyrate only participates in β-oxidation while attached to the enzyme complex as the coenzyme A derivative. The free 3-hydroxy butyrate released into the body fluids is in the D(−) form. Although some early studies indicated that mammalian tissues were able to metabolize L(+)-3-hydroxy butyrate both in vivo and in vitro,[101] it was thought that its metabolism was of no physiological importance.[101,102] Furthermore, mitochondrial 3-hydroxy butyrate dehydrogenase, which oxidizes 3-hydroxy butyrate to acetoacetate, has an absolute sterospecificity for the D(—)-isomer.[103] Based on the uneven distribution of radioactivity in both the [^{14}C]CO$_2$ and in the water soluble 3-hydroxybutyrate, we suggested that L(+)hydroxybutyryl CoA is detached and deacylated to form the free acid, which is transported across the membrane, contributing to the water-soluble, radioactive 3-hydroxy butyrate in the extracellular medium. It may, however, be metabolized at a very slow rate, as indicated by the [^{14}C]DL-3-hydroxybutyrate disappearance from the medium.[31] Furthermore, other radioactive metabolites were also secreted into the medium, particularly from [1-^{14}C]palmitate metabolism. However, under such conditions, the total excreted radioactivity (distributed among several compounds, one of which was identified as 3-hydroxy butyrate) was considerably lower. We propose that this is the D(−) isomer, in keeping with previous reports on ketone body production in the heart.[104] This is consistent with earlier reports by Opie[5] and by Evans et al.[93] of high concentrations of 3-hydroxy butyrate in the extracellular compartment in perfused hearts. It is also possible that some of the early reports in which 3-hydroxy butyrate was chemically determined, in fact unknowingly dealt with the L(+)-isomer.

E. Abnormal Fatty Acid Metabolism

Erucic acid (EA, Δ-13 docosenoic acid; C$_{22:1}$) is the major fatty acid contained in rapeseed oils. An EA-rich diet causes an accumulation of EA-rich triglycerides in heart muscle cells, which leads to cardiac lipidosis if fed to rats and other animals. This occurs in two distinct phases: (1) a rapid increase in cardiac lipid content in the first three days following an EA-rich diet, and (2) a subsequent continuous slow decline in lipid content. These phenomena are accompanied by an inflammatory response of the myocardium and necrosis. Other organs neither show any significant accumulation of fat nor pathological changes on exposure to EA (for a review on this topic see Reference 105). Since the metabolism of EA is different from that of the lower homologues, it may serve as a convenient tool for studying regulation of FA metabolism in the heart. There are three possible causes of EA accumulation: (1) an enhanced rate of uptake, (2) a decreased oxidation rate, and (3) slow lipolysis of intracellular EA-containing triglycerides which could hinder EA recycling.

We have shown in biochemical studies and by electron microscopy that EA accumulates in cultured heart cells,[59,60] as well as EA-induced changes in the physical and electrophysiological properties of the cultured cardiomyocyte sarcolemma. On the one hand, sarcolemmal fluidity was increased, thereby changing the exposure of the membrane moieties to the extracellular environment, as manifested in lactoperoxidase-catalyzed radioiodination;[52] on the other hand, these erucic acid induced changes increased the duration of the cardiac cell action potential and depressed the overshoot potential.[53] Our work also showed that EA uptake occurs at a similar rate to that of other FAs,[59,60] and is subsequently incorporated into different classes of lipids (although its esterification is somewhat delayed).[59,60,106]

The second possibility, that cardiac cells have a decreased ability to oxidize EA, has been studied by using both erucate labeled at C-14 (on double bond) and that carrying label on the carboxyl carbon. Experiments showed[59,60,107,108] that ^{14}CO$_2$ originating from the carboxyl group appears rapidly, whereas that originating from the carbon-14 lagged behind for about 12 hr (Figure 1 in Reference 108). It has also been shown that EA chain shortening occurs in two phases: first to C$_{20:1}$ and then to C$_{18:1}$ (Figure 2 in Reference 108). Thus, apparently EA cannot be directly oxidized in the mitochondrion but first undergoes chain-shortening in two steps to form oleic acid (C$_{18:1}$ which may then be cleaved further by the β-oxidation enzyme complex). Lazarrow's finding[109] that chain shortening occurs in the peroxisomes is consistent with the delay in EA oxidation. This phenomenon of chain shortening occurs with other FAs, albeit at low rates, as lower-labeled homologues, such as $C_{14:0}$, $C_{12:0}$ and $C_{10:0}$, derived from [16-^{14}C]palmitate could be detected in heart cell cultures.[110] However, delayed oxidation cannot explain EA accumulation in myocardial cells since: (1) Other cells, such as cultured hepatocytes or cultured skeletal muscle cells, with a similar pattern of EA oxidation did not accumulate trierucin,[60] and (2) Less than 10% of the total incorporated FAs are oxidized within the time period considered.

The third possibility, that lipid accumulation might be due to impaired cleavage of EA esterified TGs, was investigated by comparing intracellular lipase activities in homogenates from the heart (where EA accumulates) and from skeletal muscle (where it does not) employing a homologous series of triglycerides, from trilaurin to trierucin. It has been clearly demonstrated[111] that for all FAs from C_{12} to C_{18}, lipolytic activity was unrelated to FA chain length. In homogenates from both sources, cleavage of the unsaturated FA was higher than that of the homologous saturated TG, and the rate of lipolysis of tri-Δ-11-eicosenoin ($C_{20:1}$) was similar in both tissues, but much lower than lipolysis rates of the lower TGs. However, although trierucin cleavage was low in skeletal muscle homogenates, it was undetectable in those from the heart, even on increasing the enzyme concentration. In mixtures of TGs there was no preferential hydrolysis of any particular TG, except that in heart homogenates, trierucin did not compete for lipolytic activity. Thus the lipases it contains clearly do not cleave trierucin, which may well explain why EA-rich TGs are selectively accumulated in the hearts of animals fed on high EA-containing diets.

As mentioned above, following the dramatic accumulation of EA-rich lipids in the heart, cardiac lipid content slowly returns to normal, even when the EA-rich diet is continued. This may be due to adaptation of the cardiomyocyte to EA involving the synthesis of new lipases. This possibility was investigated by comparing the kinetics of [14-^{14}C]erucate and [10-^{14}C]oleate.[112] As would be expected these two FAs were rapidly incorporated into TGs and PLs. On short pulse labeling, cellular recycling of oleic acid rich TGs was rapid, while that of EA-containing ones lagged far behind. However, it was proven conclusively that the synthesis of new lipases could not explain the phenomena described above, i.e., the formation of EA-containing TGs and the subsequent return to normal lipid composition, since protein synthesis inhibitors had no effect on this system. However, chloroquine, a known inhibitor of lysosomal activity, further delayed cleavage of EA-rich TGs and also that of oleic acid rich TGs, but only following incorporation of large amounts of the latter. Therefore, it has been concluded that EA-rich TGs are cleaved by lysosomal lipases, which are only active during conditions of lipid accumulation, such as EA uptake or excessive uptake of a "regular" FA, such as oleic acid.

IV. CONCLUSIONS AND PERSPECTIVES

The glucose metabolism in cultured heart cells has been termed an impaired or hypoxic one, since lactate is produced from glucose in cells which are normally oxidative muscle cells. On careful examination of the available literature on both cultured cardiomyocytes and perfused hearts, it becomes evident that lactate production should not be considered as an abnormal response but, on the contrary, as an expression of regulation of the cellular energy requirements (see Section II.C). Therefore, in this respect, cultured heart cells are more akin to the heart at rest or under low work loads in which lactate production also occurs.

A review of the literature shows that much remains to be learned with respect to the mechanisms of carbohydrate metabolism. Clearly, the shift from the H_4 to the M_4 forms of the LDH isozyme is an inadequate explanation for the lactate production. Other "crossroads" of enzymatic control, such as the active and inactive forms of pyruvate dehydrogenase, governing the flow of pyruvate into the citric acid cycle, and the enzymes activating glycogen synthesis and degradation, deserve further study.

Fatty acids constitute the main source of metabolic energy in cultured cardiomyocytes, just as in the in vivo heart. Furthermore, the use of these cultures has allowed progress to be made in our understanding of FA transport across membranes. In other studies the fact that preferential oxidation of FAs derived by the cleavage of intracellular reserves over newly incorporated FAs takes place, was confirmed.

Another highly significant achievement made by the use of cultured heart cells was the discovery of the pathway for incomplete oxidation of FAs leading to the release of L(+)3-hydroxy butyrate, an intermediate in β-oxidation. It will be interesting to examine the physiological role of this pathway, bearing in mind that L(+)3-hydroxy butyrate, when injected in to suckling rats, may be utilized in the synthesis of cholesterol and other lipids, chiefly in the brain.[113]

Heart cell cultures have enabled a feasible proposal to be put forward, explaining EA-induced cardiac lipidosis and the subsequent decrease in intracellular lipid content, involving lysosomal lipases. The finding that EA, a very long chain FA, has to undergo two cycles of chain shortening (yielding oleic acid) before oxidation, gives some indication of the role of peroxisomal oxidation. It would be interesting to determine whether EA is unique in this respect or whether other long chain FAs, such as $C_{24:1}$, are metabolized in a similar manner.

Little mention has been made in this review of interactions between carbohydrate and lipid metabolism in cultured heart cells — surprisingly, this field has sufferred from almost complete neglect. Even though the relationship of FA utilization to glucose content is frequently mentioned in many of the works cited in this chapter, a deep study of this aspect has not been attempted. Cells in culture should provide an ideal system, considering the ease of manipulation and control, for carrying out such studies on a cellular level.

At the present time, only a tentative scheme can be drawn up for the metabolic pathways of lipids within the cell — the flow of substrates between the various organelles and the sites where they are stored or metabolized, and control by hormonal as well as other mechanisms is far from clear.

Finally, since the cultured cardiomyocyte is a rhythmically contracting cell, it may be expected that ATP utilization and some enzymic activities will fluctuate during cycles of cardiac contraction and relaxation. The cultured cell system should offer an ideal system for such studies, as the technical difficulties of freeze-clamping[114,115] should be easier to overcome than in the perfused heart.

ACKNOWLEDGMENTS

The author's work reported in this chapter was supported by fellowships from the Institut National de la Santé et de la Recherche Médicale (INSERM), France; the Centre National de la Recherche Scientifique (CNRS), France; and the Délégation Générale à la Recherche Scientifique et Technique (DGRST), France; and grants from INSERM, DGRST, the Chief Scientist, Ministry of Health, the State of Israel, and the Hebrew University and Hadassah Joint Grant Committee. In addition Travel Fellowship were granted each year by INSERM, France.

I would like to thank Dr. L. Opie for his helpful comments on reading the text.

REFERENCES

1. **Neely, J. R. and Morgan, H. E.**, Substrate and energy metabolism of Heart, *Annu. Rev. Physiol.*, 36, 413, 1974.
2. **Neely, J. R., Whitmer, K. M., and Mochizuki, S.**, Effects of mechanical activity and hormones on myocardial glucose and fatty acid utilization, *Circ. Res.*, 37, 733, 1975.
3. **Randle, P. J. and Tubbs, P. K.**, Carbohydrate and fatty acid metabolism, in *Handbook of Physiology, The Cardiovascular System, Section 2*, Vol. 1, American Physiological Society, Bethesda, Md., 1979, 805.
4. **Neely, J. R., Rovetto, M. J., and Oran, J. F.**, Myocardial utilizaiton of carbohydrates and lipids, *Prog. Cardiovasc. Dis.*, 15, 289, 1972.
5. **Opie, L. H.**, Myocardial energy metabolism, *Adv. Cardiol.*, 12, 70, 1974.
6. **Illingworth, J. A., Ford, W. C. L., Kobayashi, K., and Williamson, J. R.**, *Rec. Adv. Stud. Card. Struct. Metabol.*, 8, 271, 1975.
7. **Park, C. R., Morgan, H. E., Henderson, M. J., Regen, D. M., Cadenas, E., and Post, R. L.**, The regulation of glucose uptake in muscle as studied in the perfused rat heart, *Rec. Prog. Hormone Res.*, 17. 49, 1961.
8. **Morgan, H. E., Henderson, M. J., Regen, D. M., and Park, C. R.**, Regulation of glucose uptake in muscle. I, *J. Biol. Chem.*, 236, 253, 1961.
9. **Morgan, H. E., Cadenas, E., Regen, D. M., and Park, C. R.**, Regulation of glucose uptake in muscle. II, *J. Biol. Chem.*, 236, 262, 1961.
10. **Post, R. L., Morgan, H. E., and Park, C. R.**, Regulation of glucose uptake in muscle. III, *J. Biol. Chem.*, 236, 269, 1961.
11. **Paris, S., Pouysségur, J., and Ailhaud, G.**, Sugar Transport in chick empbryo cardiac cells in cultures. Analysis by countertransport, relationship to phosphorylation and effect of glucose starvation, *Biochim. Biophys. Acta*, 602, 644, 1980.
12. **Neely, J. R., Bowman, R. H., and Morgan, H. E.**, Interaction of fatty acid and glucose oxidation by cultured heart cells, *Am. J. Physiol.*, 216, 804, 1969.
13. **Frelin, C., Pinson, A., Moalic, J. M., and Padieu, P.**, Energy metabolism of beating rat heart cell cultures. II. Glucoes metabolism, *Biochimie*, 56, 1597, 1974.
14. **Frelin, C., Pinson, A., Athias, P., Surville, J. M., and Padieu, P.**, Glucose and Palmitate metabolism by beating rat heart cells in culture, *Pathol. Biol.*, 27, 45, 1979.
15. **Frelin, C.**, The growth of heart cells in cuture. Evidence for a multiple activation of the pleiotypic program, *Biochimie*, 60, 627, 1978.
16. **Clark, C. M.**, Carbohydrate metabolism in isolated foetal rat heart, *Am. J. Physiol.*, 220, 583, 1971.
17. **Rosenthal, M. D. and Warshaw, J. B.**, Interaction of fatty acid and glucose oxidation by cultured heart cells, *J. Cell Biol.*, 58, 332, 1973.
18. **Orloff, K. G. and McCarl, R. L.**, The effect of metabolic inhibitors on cultured rat heart cells, *J. Cell Biol.*, 57, 225, 1973.
19. **Ziegler, B., Lippmann, H. G., Mehling, R., and Jutzi, E.**, Characterization of growth and glucose utilization of a cell culture of chicken embryo heart, *Biol. Med. Ger.*, 25, 555, 1970.
20. **Ross, P. D. and McCarl, R. L.**, Oxidation of carbohydrates and palmitate by intact cultured neonatal rat heart cells, *Am. J. Physiol.*, 246, H 389, 1984.
21. **Hers, H. G.**, The control of glycogen metabolism in the liver, *Annu. Rev. Biochem.*, 45, 167, 1976.
22. **Anastasia, J. V. and McCarl, R. L.**, Effects of cortisol on cultured rat heart cells. Lipase activity, fatty acid oxidation, glycogen metabolism and ATP levels as related to the beating phenomenon, *J. Cell Biol.*, 57, 109, 1973.
23. **Jolley, R. L., Cheldelin, V. H., and Newburgh, R. W.**, Glucose catabolism in foetal and adult heart, *J. Biol. Chem.*, 233, 1289, 1958.
24. **Pinson, A. and Padieu, P.**, Les cultures des cellules isolées: technique croissance et application, *Cah. Nutr. Diet.* 8, 237, 1975.
25. **Warshaw, J. B. and Rosenthal, M. D.**, Changes in glucose oxidation during growth of embryonic heart cells in culture, *J. Cell Biol.*, 52, 283, 1972.
26. **Cahn, R. D.**, Developmental changes in embryonic enzyme patterns: The effect of oxidative substrates on lactic dehydrogenase in beating chick embryonic heart cell cultures, *Dev. Biol.*, 9, 327, 1964.
27. **Van Der Laarse, A., Hollaar, L., Kokshoorn, L. J. M., and Witteveenm, A. G. J.**, The activity of cardio-specific isoenzymes of creatine phosphokinase and lactate dehydrogenase in monolayer cultures of neonatal rat heart cells, *J. Mol. Cell. Cardiol.*, 11, 501, 1979.
28. **Randle, P. J., Garland, P. B., Hales, C. N., Newsholme, E. A., Denton, R. M., and Pogson, C. I.**, Interactions of metabolism and the physiological role of insulin, *Rec. Prog. Horm. Res.*, 22, 1, 1966.

29. **Yagev, S., Heller, M., and Pinson, A.,** Changes in cytoplasmic and lysosomal enzyme activities in cultured rat heart cells: the relationship to cell differentiation and cell population in cultures, *In Vitro,* 20, 893, 1984.
30. **Pinson, A. and Padieu, P.,** Quelques aspects du métabolisme des acides gras à longue chaine par les cellules bathmotropes de coeur de rat en culture, *Biochimie,* 56, 1587, 1974.
31. **Pinson, A., Degrès, J., and Heller, M.,** Partial and incomplete oxidation of palmitate by cultured beating cardiac cells from neonatal rats, *J. Biol. Chem.,* 254, 8331, 1979.
32. **Ross, P. D., McCarl, R. L., and Hartzell, C. R.,** Radiorespirometry and metabolism of cultured cells attached to petri dishes, *Anal. Biochem.,* 112, 378, 1981.
33. **Coffey, R. G., Cheldelin, V. H., and Newburgh, R. W.,** Glucose utilization by chick embryo heart homogenates, *J. Gen. Physiol.,* 48, 105, 1964.
34. **Green, M. H. and Landau, B. R.,** Contribution of the pentose cycle to glucose metabolism in muscle, *Arch. Biochem. Biophys.,* 111, 569, 1965.
35. **Rosenthal, M. D. and Warshaw, J. B.,** Fatty acid and glucose oxidation by cultured rat heart cells, *J. Cell Physiol.,* 93, 31, 1977.
36. **Schroedl, N. A., Hartzell, C. R., Ross, P. D., and McCarl, R. L.,** Glucose metabolism, insulin effects, and developmental age of cultured neonatal rat heart cells, *J. Cell Physiol.,* 113, 132, 1982.
37. **Hochachka, P. W., Neely, J. R., and Driedzic, W. R.,** Integration of lipid utilization with Krebs cycle activity in muscle, *Fed. Proc.,* 36, 2009, 1977.
38. **Christiansen, K.,** Membrane-bound lipid particles from beef heart acylglycerol synthesis, *Biochim. Biophys. Acta,* 380, 390, 1975.
39. **Bing, R. J., Hammond, M. M., Handelsman, J. C., Powers, S. R., Spencer, F. C., Eckenhoff, J. E., Goodale, W. T., Hafkenshiel, J. H., and Kety, S. S.,** Measurement of coronary blood flow, oxygen consumption, and efficiency of left ventricle in man, *Am. Heart J.,* 38, 1, 1949.
40. **Bing, R. J., Siegel, A., Vitale, A., Balboni, P., Sparks, E., Taeschler, E., Klapper, M., and Edwards, S.,** Metabolic studies on the human heart in vitro. I. Studies on carbohydrate metabolism of the human heart, *Am. J. Med.,* 15, 284, 1953.
41. **Bing, R. J., Siegel, A., Ungar, I., and Gilbert, M.,** Metabolism of the human heart. II. Studies on fat, ketone and amino acid metabolism, *Am. J. Med.,* 16, 504, 1954.
42. **Gordon, R. S., Jr.,** Unesterified fatty acid in human blood plasma. II. The transport function of unesterified fatty acid, *J. Clin. Invest.,* 36, 810, 1957.
43. **Fujimoto, A. and Harary, I.,** Studies in vitro on single beating rat-heart cells. IV. The shift from fat to carbohydrate metabolism in culture, *Biochim. Biophys. Acta,* 86, 74, 1964.
44. **Kuramitsu, H. and Harary, I.,** Studies in vitro on single beating rat-heart cells. III. Enzyme changes and loss of specific function in culture, *Biochim. Biophys. Acta,* 86, 65, 1964.
45. **Harary, I., McCarl, R., and Farley, B.,** Studies in vitro on single beating rat heart cells. IX. Restoration of beating by serum lipids and fatty acids, *Biochim. Biophys. Acta,* 115, 22, 1966.
46. **Fujimoto, A. and Harary, I.,** The effect of lipids on enzyme levels in beating rat heart cells, *Biochem. Biophys. Res. Commun.,* 20, 456, 1965.
47. **Harary, I. and Slater, E. C.,** Studies in vitro on single beating rat heart cells. VIII. The effect of Oligomycin, Dinitrophenol and Ouabain on the beating rate, *Biochim. Biophys. Acta,* 99, 127, 1965.
48. **Stein, O. and Stein, Y.,** Metabolism in vitro of palimitic and linoleic acid in the heart and diaphragm of essential fatty acid deficient rats, *Biochim. Biophys. Acta,* 84, 621, 1964.
49. **Haggerty, D. F., Jr., Gershenson, L. E., Harary, I., and Mead, J. F.,** The metabolism of linoleic acid in mammalian cells in culture, *Biochem. Biophys. Res. Commun.,* 21, 568, 1965.
50. **Gershenson, L. E., Harary, I., and Mead, J. F.,** Studies in vitro on single beating heart cells. X. The effect of linoleic and palmitic acids on beating and mitochondrial phosphorylaiton, *Biochim. Blophys. Acta,* 131, 50, 1967.
51. **Yechiel, E. and Barenholtz, Y.,** Relationship between membrane lipid composition and biological properties of rat myocytes. Effects of aging and manipulation of lipid composition, *J. Biol. Chem.,* 260, 9123, 1985.
52. **Mersel, M., Amar, A., Benenson, A., Heller, M., Hallaq, H., Padieu, P., and Pinson, A.,** Membrane fluidity and accessibility of sarcolemmal proteolipids to enzymatic radio-iodination in cultured cardiac myocytes, in *Advances in Studies on Heart Metabolism,* Caldarera, C. M. and Harris, P., Eds., Clueb, Bologna, 1982, 88.
53. **Athias, P., Pinson, A., Frelin, C., Padieu, P., and Klepping, J.,** Comparative study on the effects of exogenous palmitate and erucate on intracellular electric properties of cultured beating heart cells, *J. Mol. Cell. Cardiol.,* 11, 755, 1979.
54. **Gmeiner, R., Apstein, C. S., and Brachfeld, N.,** Effect of palmitate on hypoxic cardiac performance, *J. Mol. Cell. Cardiol.,* 7, 227, 1975.
55. **Opie, L. H. and Lubbe, W. F.,** Are fatty acids arrhythmogenic?, *J. Mol. Cell Cardiol.,* 7, 155, 1975.

56. **Wenzel, D. G. and Kleoppel, I. W.,** Arrhythmogenicity of long-chain fatty acids for cultured rat heart myocytes, *Toxicology,* 15, 105, 1980.

57. **Szuhaj, B. F. and McCarl, R. L.,** Fatty acid composition of rat hearts as influenced by age and dietary fatty acids, *Lipids,* 8, 241, 1973.

58. **Rogers, C. G.,** Fatty acids and Phospholipids of adult and new born rat hearts and of cultured beating neonatal rats, *Lipids,* 9, 541, 1974.

59. **Pinson, A. and Padieu, P.,** Erucic acid metabolism by cultured beating heart cells of the postnatal rat, *Rec. Adv. Stud. Card. Struct. Metab.,* 7, 29, 1976.

60. **Pinson, A.,** Métabolisme de l'acide palmitique et de l'acide érucique par les cellules cardiaques bathmotrpes de rat en culture, D. Sc. Thesis in Biochemistry, University of Dijon, France, January, 1975.

61. **St. Geme Jr., J. W., Martin, H. L., Davis, C. W. C., and Mead, J. F.,** A biochemical concept of cellular production of paramyxoviruses, *Pediat. Res.,* 11, 174, 1977.

62. **Benenson, A., Mersel, M., Heller, M., and Pinson, A.,** Radioiodination technique for localization of lipids and proteolipids in membranes: application to heart cells in cultures, in *Methods in Studying Cardiac Membranes, II,* Dhalla, N. S., Ed., CRC Press, Inc., Boca Raton, Fla., 1984, 59.

63. **Wosilait, W. D., Soler-Argilaga, C., and Nagy, P.,** A theoretical analysis of the binding of palmitate by human serum albumin, *Biochem. Biophys. Res. Commun.,* 71, 419, 1976.

64. **Eaton, R. P., Berman, M., and Steinberg, D.,** Kinetic studies of plasma free fatty acid and triglyceride metabolism in man, *J. Clin. Invest.,* 48, 1560, 1969.

65. **Stein, O. and Stein, Y.,** Lipid synthesis intracellular transport and storage. III. Electron microscopic radioanthographic study of the rat heart perfused with tritiated oleic acid, *J. Cell Biol.,* 36, 63, 1968.

66. **Spector, A. A.,** Transport and utilization of free fatty acid, *Ann. N. Y. Acad. Sci.,* 149, 768, 1968.

67. **Spector, A. A., Ashbrook, J. D., Santos, E. C., and Fletchter, J. E.,** Quantitative analysis of uptake of free fatty acid by mammalian cells: lauric acid and human erythrocytes, *J. Lip. Res.,* 13, 445, 1972.

68. **Mahadevan, S. and Sauer, F.,** Effect of trypsin, phospholipases and membrane-impermeable reagents on the uptake of palmitic acid by rat liver cells, *Arch. Biochem. Biophys.,* 164, 185, 1974.

69. **Weisiger, R., Gollan, I., and Ockner, R.,** Receptors for albumin on the liver cell surface may mediate uptake of fatty acids and other albumin-bound substances, *Science,* 211, 1048, 1981.

70. **Maloy, S. R., Ginsburgh, C. H., Simons, R. W., and Nunn, W. D.,** Transport of long and medium chain fatty acids by Escherichia coli K12, *J. Biol. Chem.,* 256, 3735, 1981.

71. **Samuel, D., Paris, S., and Ailhaud, G.,** Uptake and metabolism of fatty acids cardiac cells from chick embryo, *Eur. J. Biochem.,* 74, 583, 1976.

72. **Franchi, A. and Ailhaud, G.,** Incorporation of a fatty acid containing a photosensitive group into lipids of cultured cardiac cells from chick embryo, *Biochimie,* 59, 813, 1977.

73. **Paris, S., Samuel, D., Jacques, Y., Gache, C., Franchi, A., and Ailhaud, G.,** The role of serum albumin in the uptake of fatty acids by cultured cardiac cells from chick embryo, *Eur. J. Biochem.,* 83, 235, 1978.

74. **Paris, S., Samuel, D., Romey, G., and Ailhaud, G.,** Uptake of fatty acids by cultured cardiac cells from chick embryo. Evidence for a facilitation process without energy dependence, *Biochimie,* 61, 361, 1979.

75. **DeGrella, R. F. and Robley, J. L.,** Uptake and metabolism of fatty acids by dispersed adult rat heart myocytes, *J. Biol. Chem.,* 255, 9731, 1980.

76. **DeGrella, R. F. and Robley, J. L.,** Uptake and metabolism of fatty acids by dispersed adult rat heart myocytes, *J. Biol. Chem.,* 255, 9739, 1980.

77. **Mishkin, S., Stein, L., Gatmaitan, Z., and Arias, I. M.,** The binding of fatty acids to cytoplasmic proteins: binding to Z protein in liver and other tissues of the rat, *Biochem. Biophys. Res. Commun.,* 47, 997, 1972.

78. **Mishkin, S. and Trucotte, R.,** The binding of long chain fatty acid CoA to Z, a cytoplasmic protein present in liver and other tissues of the rat, *Biochem. Biophys. Res. Commun.,* 57, 918, 1974.

79. **Brandes, R., Pinson, A., and Heller, M.,** Transport and metabolism of free fatty acids in cultured heart cells, in *Enzymes of LIpid Metabolism,* Freysz, L., Gatt, S., Dreyfus, H., and Massareli, R., Eds., Plenum Press, New York, in press.

80. **Spitzer, J. J. and Gold, M.,** Free fatty acid metabolism by skeletal muscle, *Am. J. Physiol.,* 206, 159, 1964.

81. **Zierler, K. L., Maseri, A., Klassen, G., Rabinowitz, D., and Burgess, J.,** Muscle metabolism during exercise in man, *Trans. Assoc. Am. Physicians,* 81, 266, 1968.

82. **Dagenais, G. R., Tancredi, R. G., and Zierler, K. L.,** Free fatty acid oxidation by forearm muscle at rest, and evidence for an intramuscular lipid pool in human forearm, *J. Clin. Invest.,* 58, 421, 1976.

83. **Shipp, J. C., Thomas, J., and Crevasses, L. E.,** Oxidation of carbon-14-labelled endogenous lipids by isolated perfused rat heart, *Science,* 143, 371, 1964.

84. **Olson, R. E. and Hoeschen, R. J.,** Utilization of endogenous lipids by the isolated perfused rat heart, *Biochem. J.,* 103, 796, 1967.

85. **Denton, R. M. and Randle, P. J.,** Concentration of glycerids and phospholipids in rat heart and gastrocnemius muscles, *Biochem. J.,* 104, 416, 1967.
86. **Pinson, A., Frelin, C., and Padieu, P.,** Palmitate oxidation by beating heart cells, *Rec. Adv. Stud. Card. Struct. Metabl.,* 12, 667, 1978.
87. **Ramsay, R. R. and Tubbs, P. K.,** The mechanism of fatty acid uptake by heart mitochondria: an acylcarnitine-carnitine exchange, *FEBS Lett.,* 54, 21, 1975.
88. **Stanisz, J., Wice, B. M., and Kennell, D. E.,** Comparative energy metabolism in cultured heart muscle and Hela cells, *J. Cell Physiol.,* 115, 320, 1983.
89. **Huxtable, R. J. and Wakil, S. J.,** Comparative mitochondrial oxidation of fatty acid, *Biochim. Biophys. Acta,* 239, 168, 1971.
90. **Lynen, F.,** Biosynthesis of saturated fatty acids, *Fed. Proc.,* 20, 941, 1961.
91. **Stewart, H. B., Tubbs, P. K., and Stanley, K. K.,** Intermediates of fatty acid oxidation, *Biochem. J.,* 132, 61, 1973.
92. **Weinman, E. O., Chaikoff, I. L., Dauben, W. G., Gee, M., and Entenman, C.,** Relative rates of conversion of the various atoms of palmitic acid to carbon dioxide by the intact rat, *J. Biol. Chem.,* 184, 735, 1950.
93. **Evans, J. R., Opie, L. H., and Shipp, J. C.,** Metabolism of palmitic acid in perfused rat heart, *Am. J. Physiol.,* 205, 766, 1963.
94. **Posner, B., Matsuzaki, F., and Raben, M. S.,** Differential oxidation of C-1, C-11 and C-16 of palmitate by rat tissues, *Fed. Proc.,* 23, 269, 1964.
95. **Harper, R. D. and Saggerson, E. D.,** Factors affecting fatty acid oxidation in fat cells isolated from rat white adipose tissue, *J. Lipid Res.,* 17, 516, 1976.
96. **Christiansen, R. Z.,** The effect of clofibrate feeding on hepatic fatty acid metabolism, *Biochim. Biophys. Acta,* 530, 314, 1978.
97. **Brown, G. W., Jr., Chapman, D. D., Matheson, H. R., Chaikoff, I. L., and Dauben, W. G.,** Acetoacetate formation in the liver. III. On the mechanism of acetoacetate formation from palmitic acid, *J. Biol. Chem.,* 209, 537, 1954.
98. **Williamson, J. R. and Krebs, H. A.,** Acetoacetate as fuel of respiration in the perfused rat heart, *Biochem. J.,* 80, 540, 1961.
99. **McCann, W. P. and Greville, G. D.,** D(−) and l(+)-β-Hydroxy-butyric acid, *Biochem. Prep.,* 9, 63, 1962.
100. **Wakil, S. J. and Mahler, H. R.,** Studies on the fatty acid oxidizing system of animal tissues. V. Unsaturated fatty acyl coenzyme A hydrolase, *J. Biol. Chem.,* 207, 125, 1954.
101. **Lehninger, A. L. and Greville, G. D.,** The enzymatic oxidation of d-and l-β-hydroxybutyrate, *Biochim. Biophys. Acta,* 12, 188, 1953.
102. **Klee, C. B. and Sokoloff, L.,** Changes in the D(−)-β-hydroxy-butyric dehydrogenase activity during brain maturation in the rat, *J. Biol. Chem.,* 242, 3880, 1967.
103. **Lenninger, A. L., Sudduth, H. C., and Wise, J. B.,** D-β-Hydroxybutyric dehydrogenase of mitochondria, *J. Biol. Chem.,* 235, 2450, 1960.
104. **Krebs, H. A.,** Rate control of the tricarboxylic acid cycle, *Adv. Enzyme Regul.,* 8, 335, 1970.
105. **Vles, R. D.,** Nutritional aspects of rapeseed oil, in *The Role of Fat in Human Nutrition,* Vergroesen, A. J., Ed., Academic Press, London, 1975, 433.
106. **Rogers, C. G.,** Erucic acid and phospholipids of newborn rat heart cells in culture, *Lipids,* 12, 375, 1977.
107. **Pinson, A. and Padieu, P.,** Erucic acid metabolism by cultured beating heart cells of the postnatal rat, in *Myocardial Cell Damage,* Proc. 6th annual meeting of the International study group for research in cardiac metabolism, Freiburg i. Br., abstract 91, 1973.
108. **Pinson, A. and Padieu, P.,** Erucic acid oxidation by beating heart cells in culture, *FEBS Lett.,* 39, 88, 1974.
109. **Lazarrow, P. B.,** Rat liver peroxisomes oxidize Palmitoyl coenzyme-A, *J. Cell Biol.,* 70, A87, 1976.
110. **Pinson, A., Frelin, C., and Padieu, P.,** unpublished data, 1975.
111. **Mersel, M., Heller, M., and Pinson, A.,** Intracellular lipase activities in heart and skeletal muscle homogenates. The absence of trierucin cleavage by the heart: a possible biochemical basis for erucic acid lipidosis, *Biochim. Biophys. Acta,* 572, 218, 1979.
112. **Aupècle, P. and Pinson, A.,** The role of cardiac lysosomal lipases in triacylglycerol cleavage, *FEBS Lett.,* 144, 93, 1982.
113. **Webber, R. J. and Edmond, J.,** Utilization of L(+)-3-hydroxy-butyrate, acetoacetate, and glucose for respiration and lipid synthesis in the 18-day old rat, *J. Biol. Chem.,* 252, 5222, 1977.
114. **Thompson, C. I., Rubio, R., and Berne, R. H.,** Changes in adenosine and glycogen phosphorylase activity during the cardiac cycle, *Am. J. Physiol.,* 238, H389, 1980.
115. **Wollenberger, A., Babskii, E. B., Krause, E. G., Genz, S., Blohm, D., and Bogdanova, E. V.,** Cyclic changes in levels of cyclic AMP and cyclic GMP in frog myocardium during the cardiac cycle, *Biochem. Biophys. Res. Commun.,* 55, 446, 1973.

Chapter 17

POLYAMINES IN CULTURED HEART CELLS

C. M. Caldarera, C. Pignatti, and C. Clô

I. INTRODUCTION

In spite of the great number of papers in the literature dealing with the role of polyamines in the heart muscle, knowledge on the involvement of these polycations in the metabolic and functional activity of heart cells in culture is as yet very limited. We hope that this minireview will stimulate interest in this area.

II. BIOLOGICAL FUNCTIONS OF POLYAMINES

The naturally occurring polyamines, spermine (SPM), spermidine (SPD) and the precursor putrescine (PTC) are widely distributed in the cells of higher eukaryotes.

In the vast literature dealing with polyamines as biological effectors, their ability to stimulate DNA, RNA, and protein synthesis,[1-4] to enhance the fidelity of translation,[5] to stabilize membranes against degradative reactions[6] by inhibiting phospholipase activity[7] and peroxidation,[8] and to exert protective and restorative effects on mitochondrial energy-linked processes,[9] have been accorded particular relevance.

Even though most polyamine functions can be explained in terms of their nature of polybasic cations capable of interacting with biological polyanions[10] such as nucleic acids and phospholipids, their specific involvement in the regulation of cell cycle, growth, division, differentiation, and function of various cell types[1-4] continues to warrant intensive investigation with respect to their site and mechanism of action. Significant progress in elucidating the physiological role of polyamines has been accomplished by using potent and specific inhibitors[11] of the enzymes involved in their biosynthesis: ornithine decarboxylase (ODC), S-adenosylmethionine decarboxylase (AdoMetDC) and propylamine transferases.[6]

Both the activity of ODC, a short-lived enzyme ($t_{1/2} =$ 10 to 20 min) which initiates the polyamine synthesis chain by catalyzing PTC formation, and the intracellular content of polyamines, are particularly elevated in tissues undergoing rapid division or rapidly synthesizing RNA and proteins, such as regenerating liver,[12] tumor cells,[13] embryonic tissues,[14] and hypertrophied heart.[15]

The positive influence of polyamines on cell growth and proliferation is supported by results obtained using tissue cultures:

1. In a variety of cultured cells, ODC and AdoMetDC activities rapidly increase shortly after exposure to trophic stimuli,[16-18] or after transformation of normal cells with oncogenic viruses or chemical carcinogens;[13,19]
2. ODC, AdoMetDC, and polyamine concentrations are cell-cycle dependent, with a first peak in late G_1, preceding the initiation of DNA synthesis, and a second peak in G_2, preceding cell division;[20,21]
3. Inhibition of polyamine synthesis depresses DNA, RNA, and protein synthesis, resulting in a reduction or suppression of cell proliferation;[11,22,23]
4. Exogenously administered polyamines, at relatively low concentrations, can stimulate cell growth[18,24,25] and reverse the antiproliferative effect of polyamine synthesis inhibitors.[11,22,23]

Besides affecting cell growth and proliferation, polyamines have been shown to play an important role in the differentiation of many types of cultured cells, such as neuroblastomas,[26] chondrocytes,[27] myoblasts,[28] fibroblasts,[29] promyelocytic cells,[30] and erythroleukemic cells.[31]

However, a unifying hypothesis on the molecular mechanism whereby polyamines influence cellular activity has yet to be defined. Possibly such a hypothesis should evoke interactions between these amines and one or more intracellular factors closely involved in the regulation of macromolecular synthesis, division, and differentiation, such Ca^{2+} and the cyclic nucleotides (cAMP and cGMP).[32-34] All these factors are considered the main components of an internal signaling system which regulates the activity of most cells. In particular, it appears that in most types of cultured cells, including heart cells, cAMP preferentially promotes cell-specific patterns of differentiation and may act as a negative signal for the induction of cell proliferation, while Ca^{2+} and cGMP may be the primary signals responsible for switching on the division program.[35-38] A characteristic feature of mitogen-stimulated cultured cells is a rapid influx of Ca^{2+}, a depletion of cAMP and an accumulation of cGMP.[33,34,39] Furthermore, it appears that Ca^{2+} itself is involved in the regulation of both basal level and mitogen-stimulated changes of cAMP[36,40] and cGMP,[37] via inhibition of adenylate cyclase[41] and stimulation of cAMP-dependent phosphodiesterase (cAMP-PDE)[42] and guanylate cyclase[43] activities. In addition, cGMP strongly reduces the cAMP level by increasing both soluble and particulate cAMP-PDE, as observed by our group in quiescent heart cell cultures.[44] It has recently been reported that in the heart, as well as in kidney cortex, polyamines mediate rapid increase of free cytosolic Ca^{2+} (by enhancing Ca^{2+} influx and mobilizing intracellular Ca^{2+}) induced by hormonal stimuli.[45-47] On the other hand, Ca^{2+} is required for the activation of ODC in many cell types, such as epithelial cells,[48] hepatoma cells,[49] lymphocytes,[50] and astrocytoma cells.[51]

These findings, together with the evidence of the involvement of polyamines in the control of cellular cAMP and cGMP levels, discussed below, support the hypothesis proposing close mutual interactions between all these factors. Greater knowledge of these would be very useful in studying the influence of polyamines on the regulation of cellular activity, particularly in the case of heart cells, whose metabolism and function are closely dependent on Ca^{2+} and cyclic nucleotides.

III. POLYAMINES AND THE HEART

The involvement of polyamines in the development, metabolism, and functional activity of the heart has been demonstrated. Thus, in developing heart, the activity of ODC is high at birth, and falls progressively to a lower level in adult tissue,[52] concomitantly with decreased responsiveness to various effectors.[53] In perfused heart, exogenous polyamines increase amino acid uptake and their incorporation into myocardial proteins[54] as well as [³H]-ribose incorporation into RNA.[55] Furthermore, factors or conditions which lead to cardiac hypertrophy, such as aortic constriction, hypoxia, exercise, catecholamines, and tyroxine, cause an early and rapid increase in ODC activity and polyamine content.[15,56-59] The increase in ODC activity by β-adrenergic agonists is a cAMP-mediated event[60] which requires Ca^{2+}.[61] By using specific inhibitors of ODC, such as α-difluoromethylornithine (α-DFMO)[59] or 1-3 diaminopropane,[62] it has been observed that, in contrast to tyroxine-induced hypertrophy,[63] the increase of polyamine content plays a crucial role in the genesis of heart hypertrophy via activation of the noradrenergic cardiac system. Furthermore, since polyamines have been found to stimulate the acetylation of histones in the perfused heart,[55,58] representing an essential gene derepression mechanism closely related to RNA synthesis, it has been suggested that the increase of polyamine synthesis following β-adrenergic stimulation is responsible for the increased acetylation of histones observed during heart hypertrophy.[64]

IV. EXPERIMENTS ON HEART CELL CULTURES

Cultured cells are usually regarded as an appropriate model system in studying the role that naturally occurring compounds play in specific metabolic processes as well as in growth, division and differentiation, under controlled conditions. The use of heart cell cultures provides a further advantage, since they retain the specialized characteristics of differentiation of the organ of origin, including myosin-synthesizing activity,[65] spontaneous beating ability,[66] and sensitivity to many drugs or hormones,[67-69] as well as to changes in the availability of crucial elements, such as oxygen[70,71] and calcium.[72] Furthermore, differentiation and division are not mutually exclusive events in heart cells in culture.[73,74]

In our experiments, primary beating heart cell cultures, obtained from 8 to 10-day-old chick embryos according to the repeated trypsinization procedure of DeHaan,[75] were routinely used. Cells were grown as monolayers in Eagle's minimum essential medium (MEM F-15, GIBCO), supplemented with 10% fetal calf serum, 10% triptose phosphate broth, antibiotics, and buffered with 2.5 mM Hepes, pH 7.4. In order to prevent multiplication of fibroblast-like cells,[76] the seeding density was at least 7 to 8 $\times$ 10^5 cells per mℓ of medium.

A. Polyamines and Synthesis of Macromolecules

Experimental evidence shows that in cultured cells, the transition from a resting to a growing state is accompanied by an early and net increase of ODC and AdoMetDC activities, leading to polyamine accumulation.[77-79] This appears to be a general phenomenon (independent of the type of cell and nature of the mitogenic signal) which most likely could be included in the series of biochemical events referred to as "the pleiotypic response" of the cell by Hershko et al.[80]

1. Polyamines and Serum

The proliferative activity of heart cells in culture,[81] as well as of other cell types,[82,83] is mainly ensured by the presence of and depends on the amount of serum in the incubating medium. Serum, in addition to providing macromolecular factors and low molecular weight nutrients,[83] is the sole source of lipids,[84] whose central role for the long-term maintenance of cultured heart cell functions has been well established.[85,86] Re-addition of serum to confluent and serum-starved heart cells causes a 20-fold increase in the activity of ODC within 6 hr,[87] followed by accumulation of polyamines, particularly PTC and SPD.[87-89] AdoMetDC activity is also stimulated by serum readdition.[90] Polyamine accumulation is not dependent on the source of serum, horse,[91] or fetal calf,[87-89] but it is proportional to the amount of serumused[89] and accompanied by increases in protein, RNA, and DNA synthesis.[88,91] The presence of specific inhibitors of polyamine biosynthesis, such as α-DFMO,[88] α-methylornithine,[91] a reversible inhibitor of ODC, and methylglyoxal bis (guanylhydrazone),[91,92] which blocks SPD and SPM formation by inhibiting AdoMetDC, prevents the stimulatory effect of serum on macromolecular synthesis, suggesting that, as in other cell types,[11,23] polyamine accumulation is a requirement for and not simply a result of the action of serum on heart cells in culture.

2. Polyamines and Phospholipids

The importance of lipids in the action of serum on cultured heart cells is supported by the fact that cells fed with lipid-depleted serum lose the ability to beat[85] and to proliferate,[93] and that optimal growth and beating can be restored by treatment with serum lipids.[94] The effect of phospholipids (PL), as normal components of the lipidic fraction of the serum, on macromolecular synthesis and on polyamine metabolism has been studied in primary heart cell cultures. In growing cells, the addition of a complex of purified PL from calf heart (at low concentration) potentiates the stimulatory effect of serum (5%) on protein, RNA, and DNA specific activity as well as on polyamine accumulation.[88,95] This effect is particularly evident when the cells approach confluency or are completely confluent, when their ability to synthesize lipid *de novo* or to take up precursors or nutrients being greatly reduced.[96] Furthermore, in quiescent or serum-starved cells, exogenous PLs, such as phosphatidylcholine or phosphatidylethanolamine, partially substitute for serum in causing the onset of DNA synthesis.[88] Similarly to serum readdition, the addition of each individual PL leads to a significant increase of PTC, SPD, and SPM content by rapidly stimulating ODC and

AdoMetDC activities.[88,97] The increase in ODC activity is prevented by actinomycin D or cycloheximide, thus suggesting *de novo* synthesis of the enzyme. The failure to observe any increase in PL-induced DNA synthesis in the presence of α-DFMO,[88] supports the idea of a mediating role for polyamines in PL action.

3. Polyamines and Oxygen

The rate of macromolecular synthesis and division of cultured heart cells is increased progressively as the oxygen tension (pO_2) is reduced from high (80%) to low (5 to 2 %) values.[70,98,99] The same phenomenon has been observed with other cell types, including fibroblasts[100] and HeLa cells.[101] Although the exact mechanism of the action of oxygen remains to be elucidated, experimental evidence suggests that its earliest effect occurs either at, or before, the transcriptional level and may involve factor(s) closely related to cell growth and proliferation.[98,101] If the onset of ODC activity and the polyamine content are events which follow mitogenic signals, then they would be expected to occur when the environmental pO_2 falls. Indeed, as compared with normoxic (20% O_2) condition, in heart cells at lower pO_2s (5%, 10%), a significant increase in ODC activity[102] and in polyamine accumulation,[99] which parallels the onset of RNA and DNA synthesis, may be observed. The opposite pattern of ODC and polyamine levels occurs in cells exposed to higher pO_2s (40%, 80%), as well as in the hearts of rats exposed to absolute hyperoxia (100% O_2).[102] The dependence of ODC on the environmental oxygen is further supported by the observation that the net increase in ODC activity, usually induced by isoproterenol via cAMP accumulation (see below), is prevented when cells are maintained under high pO_2 (80%).[102] The failure of exogenous administered cAMP to elevate ODC activity when cells are shifted from 20 to 80% O_2, suggests that high pO_2 levels do not affect the β-receptor system, but most likely reduce the activity of the enzyme.[102] Since ODC requires SH groups for optimal catalytic activity,[103] it is conceivable that the inactivation of ODC at high pO_2 may be due to changes in the redox state of the enzyme thiol-groups, either as a result of direct oxidation and/or secondary to perturbation of the cellular protective apparatus for thiol groups.[102,104] Based on these considerations and in view of the reported protective role of polyamines against peroxidative processes and in membrane stabilization, one may speculate that polyamine deficiency, secondary to reduced ODC activity, contributes to increase the cell damage under hyperoxic conditions.

Furthermore, shifting the cells from high to low pO_2, leads to a significant increase in the rate of histone acetylation,[99] particularly of arginine-rich fractions, suggesting that the cellular gene activity is modulated by the environmental oxygen availability. Since at different pO_2s the changes in histone acetylation vary in a similar manner to the intracellular polyamine content, which have been shown to stimulate histone acetylation[64] as well as other gene derepression mechanisms,[105,106] it seems that, as with serum and PL, polyamines play an important role in the oxygen-mediated control of growth and proliferation of cultured heart cells.

4. Exogenous Polyamines

A direct proof of the ability of polyamines to positively affect the gene activity of heart cells, comes from experiments in which polyamines were added to cultures in a serum-free medium,[18] in order to prevent the formation of toxic oxidation products by serum amin-oxidases.[1] At extremely low doses (0.1 μM PTC, 1 μM SPM or 10 μM SPD), i.e., about 100-fold less than those present in heart cells and in the range of those found in physiological fluids,[17] polyamines can substitute for serum in stimulating thymidine incorporation into DNA.

B. Polyamines and Cyclic Nucleotide Metabolism

There is a considerable body of experimental data supporting the generalization that, in different biological systems, many of the effects of polyamines are fundamentally opposite to the known biological effects of cAMP and qualitatively similar to those elicited by cGMP. This is particularly true as far as the regulation of cell growth is concerned, both in normal and in pathological conditions. The opposite effects of these two nucleotides are supported by the fact that cGMP (like polyamines) accumulates while cAMP decreases in cell cultures stimulated to proliferate,[36,107,108] and that cGMP or cAMP congeners, respectively, induce or inhibit macromolecular synthesis and cell division.[36,109] In view of the antagonistic effects elicited by the two nucleotides and since a concomitant change in their intracellular concentrations is not a general rule in cells undergoing division, it has been proposed that a fall in cAMP/cGMP molar ratio may be an adequate stimulus for the induction of cell proliferation. Moreover, cAMP and cGMP have been recognized as possible final mediators of cardiac function during various physiological and pathological conditions.[110-112] While cAMP appears to be associated with a positive inotropic effect and is mainly related to β-adrenergic receptors, cGMP has closer connections with the muscarinic receptors and seems to be a negative signal for the rate and intensity of contraction in beating heart cell cultures and in the heart muscle. It therefore follows that detailed knowledge of the inter-relationship existing between polyamines and cyclic nucleotides would provide useful information on their specific role in the control of cellular activity, and would be of particular interest in the study of cardiac performance.

1. Intracellular Polyamines

As observed with polyamines, the intracellular cAMP and cGMP levels respond specifically to growth conditions, and display strict dependence on the presence and on the amount of serum in the incubating medium.[36,107] Serum-mediated changes in content of both cAMP and cGMP also occur in primary quiescent heart cell cultures. Indeed, with respect to serum-starved cells, incubation with increasing doses of fetal calf serum leads to a progressive depletion of cAMP and accumulation of cGMP (as well as of polyamines), resulting in a progressive decrease in cAMP/cGMP ratio.[113,114] The effect of serum is mediated via the stimulation of both guanylate cyclase and cAMP-PDE and the inhibition of both adenylate cyclase and cGMP-PDE activities.[114] Qualitatively similar data were obtained when phospholipids, which stimulate polyamine accumulation, were used in place of serum.[88,115]

The possibility that in cultured heart cells, endogenous polyamines may influence the steady-state levels of cAMP and cGMP is based on the observation that selective inhibition of polyamine synthesis leads to accumulation of cAMP and depletion of cGMP by affecting the specific cyclase and phosphodiesterase activities.[91,92,113,114,116] Furthermore, in the presence of polyamine inhibitors, both serum[91,114] and phospholipids[88] have no effect on cyclic nucleotide metabolism, providing further evidence that polyamine accumulation is an important step in the action of both serum and phospholipids. Based on these data, and taking into consideration the opposite effects of cAMP and cGMP on cell growth, it may well be that the antiproliferative effect of polyamine inhibitors is also due to their ability to influence cyclic nucleotide metabolism.

Besides affecting cellular cyclic nucleotide content, endogenous polyamines appear to influence cyclic nucleotide responses of heart cells to different effectors, either hormonal or nonhormonal.[91,117-119] In contrast to polyamine-deficient cells, polyamine-rich ones, which show higher guanylate cyclase and cAMP-PDE but lower adenylate cyclase and cGMP-PDE activities, exhibit an impaired responsiveness to cAMP-mediated effectors, such as PGE$_1$, isoproterenol, noradrenaline, glucagon, or isobutylmethylxanthine. Conversely, polyamine accumulation in heart cells is associated with a much higher sensitivity to serum, insulin, acetylcholine, and morphine, whose action involves an early and rapid decline of cAMP and an increase of cGMP content. The probability that polyamines may be important in defining the responsiveness of heart cells is supported by the fact that inhibition of polyamine synthesis is accompanied by either improved or impaired sensitivity to cAMP- or to cGMP-mediated effectors, respectively.[91,117,118]

2. Exogenous Polyamines

A further substantial indication that polyamines play a role in the control of cyclic nucleotide metabolism has been furnished by studies on the effects of exogenously administered polyamines. The addition of each amine alone, at the same concentrations as those which affect DNA synthesis (0.1 μM PTC, 1 μM SPM, 10 μM SPD, see above), to confluent and serum-starved cultures, causes a significant decrease of cAMP and an increase of cGMP levels, resulting in a dramatic fall of the cAMP/cGMP ratio.[113,120,121] Higher concentrations of the amines are less effective in causing cAMP depletion,[120] but can induce increased cGMP accumulation.[119] The effects of polyamines have been observed maximally during short-term (minutes) incubation,[121] but are still evident even after a 2 hr exposure of the cells.[119] In the latter case, polyamines accumulate significantly inside the cells, probably by an energy-dependent system, such as in the Girardi heart cell cultures in work done by Alfheim and Hongslo,[122] and in human fibroblasts in the studies of Pohjanpelto.[123] The ability of polyamines to reduce cAMP content in a dose-inverse manner has also been observed with other cultured cell types, such as fibroblast, glioma, neuroblastoma, and neuroblastoma-glioma hybrid cells.[120]

Moreover, in cultured heart cells, as well as in the other cell types mentioned, the presence of SPM, SPD, or PTC in the incubating medium, strongly counteracts the effects of the various cAMP-mediated hormones or chemicals causing cAMP to accumulate inside the cells.[120] Once more, polyamine efficiency is inversely proportional to the concentration utilized. The generality of this phenomenon together with the evidence that polyamines can affect the basal level of cAMP, apparently make the cellular membrane receptors very unlikely as a primary site of action. In addition, the antagonistic effect of exogenous polyamines towards cAMP-mediated effectors, is consistent with the decreased responsiveness of polyamine-rich cells to the same effectors (see above).

Indeed, experimental evidence indicates that the effects of exogenous polyamines on cellular cyclic nucleotide levels occurs via rapid changes in the specific cyclase and phosphodiesterase activities, and seems to exclude changes in the rate of efflux from the cells, at least as far as cAMP is concerned.[124] The addition of each amine alone to the cells leads to an early and significant increase of both particulate and especially soluble guanylate cyclase activities, by increasing the affinities of the enzymes for the substrate as well as their catalytic activity.[121] Furthermore, in vitro experiments have shown that polyamines can substitute for Mn^{2+} as cofactors for guanylate cyclase probably by interacting with its cation site.[121] Besides stimulating guanylate cyclase, exogenous polyamines inhibit cellular cGMP-PDE within a few minutes.[113,121] Once again the soluble enzyme is preferentially affected by each of the amines, particularly by SPD and SPM. This inhibitory effect is supported by the results of Kincaid et al.,[125] which showed that polyamines significantly decrease the activity of a specific soluble cGMP-PDE purified from bovine brain.

As observed in cultured fibroblasts by Wright et al.,[126] the membrane-bound adenylate cyclase activity is strongly decreased in heart cells exposed to each of these amines and especially by SPD.[119,127] Higher concentrations of each amine were less effective in reducing the enzymic activity,[129] as also indicated by in vitro experiments.[127] On the other hand, shortly after the addition of polyamines to the medium, a net increase may be observed in cAMP-PDE activity in both the soluble and notably in the particulate cellular fractions.[87,119] In both cases, the low affinity form of the enzyme, which is known to be stimulated by Ca^{2+}-calmodulin complex,[128] is preferentially affected by each amine, PTC being particularly effective. Again, higher concentrations of all three amines are less effective[119] or may bring about a decrease in cAMP-PDE activity.[87] Kinetic analyses have shown that a net increase in the affinity of the enzymes for cAMP occurs after the addition of polyamines to the cultures.[87]

Table 1 summarizes the relationships between exogenous or endogenous polyamines and cyclic nucleotide metabolism, based on the results to date.

3. Polyamines and the Self-Regulatory Mechanism of cAMP Content

An increase in cAMP-PDE activities due to factors inducing cAMP accumulation (activators of adenylate cyclase, inhibitors of cAMP-PDE or cAMP congeners), has been shown to occur in several cell types.[128-130] Cyclic AMP-activated protein kinase is involved in this regulation, which represents a "servomechanism" by which cAMP can control its own accumulation inside the cells. Furthermore it is of particular relevance that ODC activity also increases following cellular cAMP accumulation,[60,103,131,132] suggesting that alternative mechanisms for ODC regulation operate inside the cells, including a cAMP-dependent (catecholamines) and a cAMP-independent (serum or insulin) mechanism.

As in the case of cAMP-PDE, the induction of ODC by cAMP involves cAMP-dependent protein kinase activation. It also requires Ca^{2+} and other components, possibly amino acids (asparagine), present in the culture medium and is mediated at a post-transcriptional level.[103,133] On the other hand, the ability of polyamines to inhibit cAMP-dependent protein kinase activity, as reported by other groups,[134,135] may be the basis of another self-regulatory mechanism of polyamine synthesis and consequently of cell growth.

A close temporal relationship between cellular cAMP accumulation and the subsequent rise in ODC activity and in polyamine content has been observed in cultured heart cells exposed to noradrenaline, isoproterenol or dibutyryl-cAMP.[87,116] The increase in polyamine levels is then followed by a significant increase in the activities of both high and low affinity components of cAMP-PDE. The presence of an inhibitor of polyamine synthesis in the incubating medium, strongly reduces or completely prevents the onset of cAMP-PDE activities, suggesting that polyamine accumulation is a requirement for the cAMP-mediated induction of cAMP-PDE. This might provide a rationale for the stimulatory effect of cAMP on ODC.

However, cAMP-PDE activity can also be increased by factors, such as serum and insulin,[128] which lower cAMP levels and therefore operate via a cAMP-independent mechanism. Readdition of fetal calf serum to quiescent and serum-starved heart cells,[116] while decreasing cAMP content, stimulates ODC activity and polyamine accumulation, which in turn is paralleled by a significant increase in cAMP-PDE activities. The last mentioned effect is prevented when polyamine accumulation is blocked by specific inhibitors of their synthesis.[114,116] It is quite interesting to note that, as observed with exogenous polyamines, endogenous polyamines appear to be preferentially involved in the regulation of the low affinity form of cAMP-PDE, both in the case of cAMP-dependent and of the cAMP-independent mechanism of induction. Figure 1 summarizes our present working hypothesis.

The fact that β-adrenergic agonists and serum, which only have in common the ability to stimulate ODC activity and polyamine accumulation, produce a common response, suggests that these polycations may mediate their effects in cultured heart cells.

V. CONCLUSIONS AND FUTURE TRENDS

The foregoing review provides evidence that, as well as in many other types of cells tested so far, polyamines seem to play a very important role in the mechanism of regulation of the metabolic and functional activity of heart cells in culture. The ability of ODC activity and polyamine content to fluctuate rapidly in response to several kinds of stimuli, including oxygen, serum, phospholipids, and β-adrenergic agonists, gives support to this view. The

Table 1
RELATIONSHIPS BETWEEN POLYAMINES AND CYCLIC NUCLEOTIDES IN CULTURED HEART CELLS

| | Exogenous polyamines | | Endogenous polyamine content | | |
	Effect	Ref.	High	Low	Ref.
cAMP content	Depletion	92,113,119—121,124	Low	High	89,91,92,114,116—118
cGMP content	Accumulation	113,119—121	High	Low	89,114,117,118
Adenylate cyclase	Inhibition	119,127	Low	High	114,117
cAMP-PDE	Stimulation	87,119,124,127	High	Low	87,114,116,117
Guanylate Cyclase	Stimulation	119,121,127	High	Low	114,117
cGMP-PDE	Inhibition	113,119,121,127	Low	High	114,117
Responsiveness to cAMP-mediated effectors	Decreased	120	Decreased	Increased	91,116—119
Responsiveness to cGMP-mediated effectors	Increased	120	Increased	Decreased	117—119

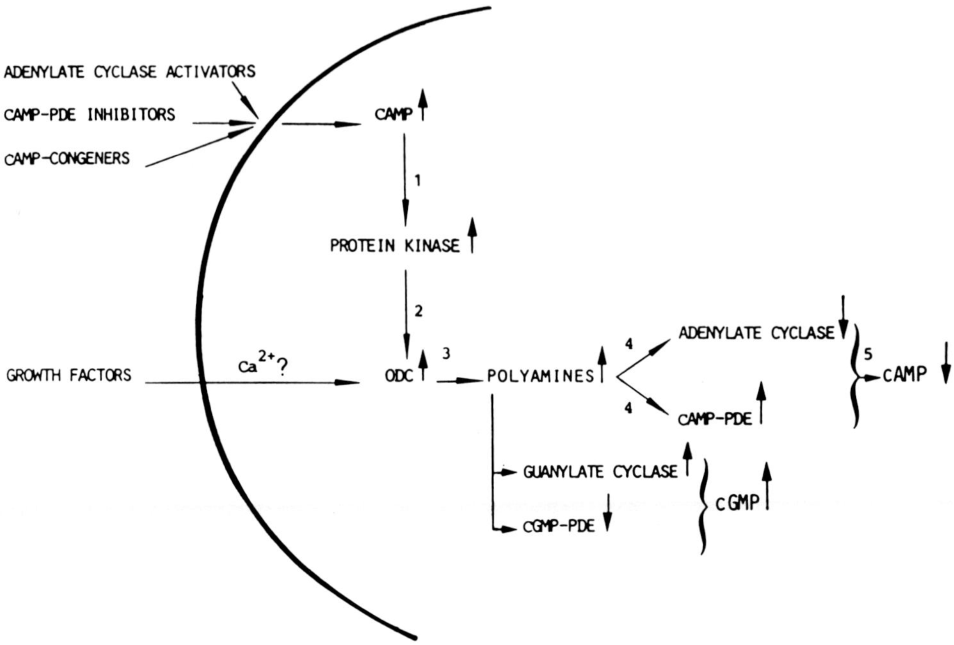

FIGURE 1. Hypothesis for an involvement of endogenous polyamines in cyclic nucleotide regulation. Numbers 1 to 5 indicate the hypothetical sequence of events occurring in the self-regulation of cAMP levels.

close dependence of ODC activity on environmental oxygen as well as on many other components of the culture medium or drugs, must be taken into account when growth and proliferation of cultured heart cells are investigated under controlled conditions, since ODC activity is correlated with cell division.

Another interesting aspect is the relationship between exogenous phospholipids and cellular polyamines. Lipids have been shown to ensure optimal growth and beating ability of cultured heart cells, and evidence that exogenously administered phospholipids stimulate polyamine accumulation has been presented in this review. Obviously, further studies are necessary to elucidate this phenomenon, but it may be that the newly synthesized and free polycations partly neutralize the negative phosphate groups of phospholipids and protect them against peroxidative reactions, as observed in other model systems.

Furthermore, in heart cells polyamines are closely involved in the regulation of cAMP and cGMP metabolism — a phenomenon certainly worthy of more detailed investigation. Ca^{2+} and calmodulin might be required for this action. Changes in the levels of the two nucleotides are related to hormone or drug-induced changes in beating frequency and, therefore, changes in the intracellular or extracellular polyamine concentration might have important implications with respect to the electrophysiological properties of both heart cells in culture and heart muscle. In this regard it is relevant to note that the sensitivity of heart cells towards different cyclic nucleotide-mediated effectors, which are known to affect contractility, is most likely influenced by the concentration of endogenous polyamines or by the presence of extremely low doses of these polycations in the culture medium. Based on these considerations, it would be worthwhile investigating whether polyamine accumulation which occurs during heart hypertrophy might account in part not only for the impaired cardiac performance, but also for the decreased responsiveness to catecholamines in the hypertrophied heart.[136] Also a possible involvement of polyamines in the organization and integrity of the cytoskeleton of heart cells may be a very exciting trend in future research.

Heart cells in culture offer a very useful experimental system for evaluating the cellular interactions of polyamines with other effectors, mainly Ca^{2+}, cAMP and cGMP, with their precise interaction being of fundamental importance for the optimal performance of beating heart cells.

ACKNOWLEDGMENTS

The invaluable editorial assistance of Dr. B. Tantini and Dr. S. Marmiroli is gratefully acknowledged. Part of the work was supported by a grant from the C.N.R., Rome, Progetto Finalizzato "Oncologia" (n° 84.00520.44).

REFERENCES

1. **Bachrach, U.**, *Function of Naturally Occurring Polyamines*, Academic Press, New York, 1973.
2. **Raina, A. and Jänne, J.**, Physiology of the natural polyamines, putrescine, spermidine and spermine, *Med. Biol.*, 53, 121, 1975.
3. **Pegg, A. E. and McCann, P. P.**, Polyamine metabolism and function, *Am. J. Physiol.*, 243 (Cell Physiol. 12), C212, 1982.
4. **Tabor, C. W. and Tabor, H.**, Polyamines, *Ann. Rev. Biochem.*, 53, 749, 1984.
5. **Atkins, J. F., Lewis, J. B., Anderson, C. W., and Gesteland, R. F.**, Enhanced differential synthesis of proteins in a mammalian cell-free system by addition of polyamines, *J. Biol. Chem.*, 250, 5688, 1975.
6. **Tabor, H. and Tabor, C. W.**, Biosynthesis and metabolism of 1,4 diaminobutane, spermidine, spermine, and related amines, *Adv. Enzymol.*, 36, 203, 1972.
7. **Sechi, A. M., Cabrini, L., Landi, L., Pasquali, P., and Lenaz, G.**, Inhibition of phospholipase A_2 and phospholipase C by polyamines, *Arch. Biochem. Biophys.*, 186, 248, 1978.
8. **Tadolini, B., Cabrini, L., Landi, L., Varani, E., and Pasquali, P.**, Polyamine binding to phospholipid vesicles and inhibition of lipid peroxidation, *Biochem. Biophys. Res. Commun.*, 122, 550, 1984.
9. **Toninello, A., Di Lisa, F., Siliprandi, D., and Siliprandi, N.**, Protective and restorative effects of spermine on the membrane potential of rat liver mitochondria, in *Advances in Polyamines in Biomedical Science*, Caldarera, C. M. and Bachrach, U., Eds., CLUEB, Bologna, 1984, 31.
10. **Tadolini, B., Cabrini, L., and Sechi, A. M.**, Polyamine interaction with biological polyanions, in *Advances in Polyamines in Biomedical Science*, Caldarera, C. M. and Bachrach, U., Eds., CLUEB, Bologna, 1984, 37.
11. **Heby, O. and Jänne, J.**, Polyamine antimetabolites: biochemistry, specificity, and biological effects of inhibitors of polyamine synthesis, in *Polyamines in Biology and Medicine*, Morris, D. R. and Marton, L. J., Eds., Marcel Dekker, New York, 1981, 243.
12. **Russell, D. H. and Snyder, S. H.**, Amine synthesis in rapidly growing tissues: ornithine decarboxylase activity in regenerating rat liver, chick embryo, and various tumors, *Proc. Natl. Acad. Sci. USA*, 60, 1420, 1968.
13. **Scalabrino, G. and Ferioli, M. E.**, Polyamines in mammalian tumors, Part I, *Adv. Cancer Res.*, 35, 151, 1982.
14. **Caldarera, C. M., Barbiroli, B., and Moruzzi, G.**, Polyamines and nucleic acids during development of the chick embryo, *Biochem. J.*, 97, 84, 1965.
15. **Pegg, A. E. and Hibasami, H.**, Polyamine metabolism during cardiac hypertrophy, *Am. J. Physiol.*, 239, E372, 1980.
16. **Russell, D. H.**, *Polyamines in Normal and Neoplastic Growth*, Russell, D. H., Ed., Raven Press, New York, 1973, 1.
17. **Jänne, J., Pösö, H., and Raina, A.**, Polyamines in rapid growth and cancer, *Biochim. Biophys. Acta*, 473, 241, 1978.
18. **Clô, C., Orlandini, G. C., Casti, A., and Guarnieri, C.**, Polyamines as growth stimulating factors in eukariotic cells, *Ital. J. Biochem.*, 25, 94, 1976.
19. **Bachrach, U., Menashe, M., Faber, J., Desser, H., and Seiler, N.**, Polyamine biosynthesis and metabolism in transformed human lymphocytes, *Adv. Polyamine Res.*, 3, 259, 1981.
20. **Heby, O., Gray, J. W., Lindl, P. A., Marton, L. J., and Wilson, C. B.**, Changes in L-ornithine decarboxylase activity during the cell cycle, *Biochem. Biophys. Res. Commun.*, 71, 99, 1976.
21. **Rupniak, H. T. and Paul, D.**, Polyamines and the control of the cell cycle in animal cells, in *Polyamines in Biology and Medicine*, Morris, D. R. and Marton, L. J., Eds., Marcel Dekker, New York, 1981, 315.

22. **Mamont, P. S., Duchesne, M. C., Grove, J., and Bey, P.,** Anti-proliferative properties of DL-α-difluoromethylornithine in cultured cells. A consequence of the irreversible inhibition of ornithine decarboxylase, *Biochem. Biophys. Res. Commun.*, 81, 58, 1978.

23. **Mamont, P. S., Duchesne, M. C., Ohlenbusch, A. M. J., and Grove, J.,** Effects of ornithine decarboxylase inhibitors on cultured cells, in *Enzyme-Activated Irreversible Inhibitors*, Seiler, N., Jung, M. J., and Koch-Weser, J., Eds., Elsevier/North-Holland Biomedical Press, Amsterdam, 1978, 43.

24. **Ham, R. G.,** Putrescine and related amines as growth factors for a mammalian cell line, *Biochem. Biophys. Res. Commun.*, 14, 34, 1964.

25. **Pohjanpelto, P.,** Relationship between putrescine and the proliferation of human fibroblasts in vitro, *Exp. Cell Res.*, 80, 137, 1973.

26. **Chen, K. Y. and Liu, A. Y. C.,** The role of polyamines in the differentiation of mouse neuroblastoma cells, *Adv. Polyamine Res.*, 4, 743, 1983.

27. **Takigawa, M., Takano, T., and Suzuki, F.,** Effects of parathyroid hormone and cyclic AMP analogues on the activity of ornithine decarboxylase and expression of the differentiated phenotype of chondrocytes in culture, *J. Cell. Physiol.*, 106, 259, 1981.

28. **Stoscheck, C. M., Florini, J. R., and Richman, R. A.,** The relationship of ornithine decarboxylase activity to proliferation and differentiation of L6 muscle cells, *J. Cell. Physiol.*, 102, 11, 1980.

29. **Bethell, D. R. and Pegg, A. E.,** Polyamines are needed for the differentiation of 3T3-L1 fibroblasts into adipose cells, *Biochem. Biophys. Res. Commun.*, 102, 272, 1981.

30. **Huberman, E., Weeks, C., Heirmann, A., Callihan, M., and Slaga, T.,** Alterations in polyamine levels induced by phorbol diesters and other agents that promote differentiation in human promyelocytic leukemia cells, *Proc. Natl. Acad. Sci. USA*, 78, 1062, 1981.

31. **Gazitt, Y. and Bachrach, U.,** Turnover and analysis of polyamines and their acetyl derivatives in differentiating Friend erythroleukemia cells, *Adv. Polyamine Res.*, 4, 757, 1983.

32. **Berridge, M. J.,** The interaction of cyclic nucleotides and calcium in the control of cellular activity, *Adv. Cyclic Nucleotide Res.*, 6, 1, 1975.

33. **Chlapowski, F. J., Kelly, L. A., and Butcher, R. W.,** Cyclic nucleotides in cultured cells, *Adv. Cyclic Nucleotide Res.*, 6, 245, 1975.

34. **Goldberg, N. D. and Haddox, M. K.,** Cyclic GMP metabolism and involvement in biological regulation, *Ann. Rev. Biochem.*, 46, 823, 1977.

35. **Swierenga, S. H. H., MacManus, J. P., and Whitfield, J. F.,** Regulation by calcium of the proliferation of heart cells from young adult rats, *In Vitro*, 12, 31, 1976.

36. **Schönhöfer, P. S. and Peters, D. H.,** Role of cyclic nucleotide in cultured cells, in *Cyclic 3'5'-Nucleotides: Mechanisms of Action*, Cramer, H. and Shultz, J., Eds., John Wiley & Sons, London, 1977, 107.

37. **Coffey, R. G., Hadden, E. M., Lopez, C., and Hadden, J. W.,** cGMP and calcium in the initiation of cellular proliferation, *Adv. Cyclic Nucleotide Res.*, 9, 661, 1978.

38. **Ralph, R. K.,** Cyclic AMP, calcium and control of cell growth, *FEBS Lett.*, 161, 1, 1983.

39. **Pastan, I. H., Johnson, G. S., and Anderson, W. B.,** Role of cyclic nucleotides in growth control, *Ann. Rev. Biochem.*, 44, 491, 1975.

40. **Harary, I., Renaud, J. F., Sato, E., and Wallace, G. A.,** Calcium ions regulate cyclic AMP and beating in cultured cells, *Nature*, 261, 60, 1976.

41. **Robinson, G. A., Butcher, R. W., and Sutherland, E. W.,** The catecholamines, in *Cyclic AMP*, Robinson, G. A., Butcher, R. W., and Sutherland, E. W., Eds., Academic Press, New York, 1971, 146.

42. **Teo, T. S. and Wang, J. H.,** Mechanism of activation of a cyclic adenosine 3', 5' monophosphate phosphodiesterase from bovine heart by calcium ions, *J. Biol. Chem.*, 248, 5950, 1973.

43. **Wallach, D. and Pastan, T.,** Stimulation of membranous guanylate cyclase by concentrations of calcium that are in the physiological range, *Biochem. Biophys. Res. Commun.*, 72, 860, 1976.

44. **Clô, C., Pignatti, C., Marmiroli, S., Tantini, B., and Caldarera, C. M.,** Regulation of cyclic AMP-phosphodiesterase activity of quiescent heart cell cultures by cyclic GMP, *Ital. J. Biochem.*, in press.

45. **Koenig, H., Goldstone, A. D., and Lu, C. Y.,** β-adrenergic stimulation of Ca^{2+} fluxes, endocytosis, hexose transport, and amino acid transport in mouse kidney cortex is mediated by polyamines, *Proc. Natl. Acad. Sci. USA*, 80, 7210, 1983.

46. **Koenig, H., Goldstone, A., and Lu, C. Y.,** Polyamines regulate calcium fluxes in a rapid plasma membrane response, *Nature*, 305, 530, 1983.

47. **Fan, C. C. and Koenig, H.,** β-adrenergic stimulation of calcium fluxes and membranes transport processes in rat heart is mediated by polyamines, *Fed. Proc.*, 43, 735, 1984.

48. **Langdon, R. C., Fleckman, P., and McGuire, J.,** Calcium stimulates ornithine decarboxylase activity in cultured mammalian epithelial cells, *J. Cell. Physiol.*, 118, 39, 1984.

49. **Canellakis, Z., Theoharides, T. C., Bondy, P. K., and Triarhos, E. T.,** The role of calcium in the induction of ornithine decarboxylase in rat HTC cells, *Life Sci.*, 29, 707, 1981.

50. **Otani, S., Matsui, I., Nakajima, S., Masutani, M., Mizoguchi, Y., and Morisawa, S.,** Induction of ornithine decarboxylase in guinea pig lymphocytes by the divalent cation ionophore A23187 and phytohemagglutinin, *J. Biochem.*, 88, 77, 1980.
51. **Gibbs, J. B., Hsu, C. Y., Terasaki, W. L., and Brooker, G.,** Calcium and microtubule dependence for increased ornithine decarboxylase activity stimulated by β-adrenergic agonists, dibutyryl cyclic AMP, or serum in a rat astrocytoma cell line, *Proc. Natl. Acad. Sci. USA*, 77, 995, 1980.
52. **Lau, C. and Slotkin, T. A.,** Regulation of ornithine decarboxylase activity in the developing heart of euthyroid or hyperthyroid rats, *Mol. Pharmacol.*, 18, 247, 1980.
53. **Das, R. and Kanungo, M. S.,** Activity and modulation of ornithine decarboxylase and concentrations of polyamines in various tissues of rats as a function of age, *Exp. Gerontol.*, 17, 95, 1982.
54. **Gibson, K. and Harris, P.,** The in vitro and in vivo effects of polyamines on cardiac protein biosynthesis, *Cardiovasc. Res.*, 8, 668, 1974.
55. **Caldarera, C. M., Rossoni, C., and Casti, A.,** Involvement of polyamines in ribonucleic acid synthesis as a possible biological function, *Ital. J. Biochem.*, 25, 33, 1976.
56. **Russell, D. H., Shiverick, K. T., Hamrell, B. B., and Alpert, N. R.,** Polyamine synthesis during initial phases of stress-induced cardiac hypertrophy, *Am. J. Physiol.*, 221, 1287, 1971.
57. **Feldman, M. J. and Russell, D. H.,** Polyamine biogenesis in left ventricle of the rat heart after aortic constriction, *Am. J. Physiol.*, 222, 1199, 1972.
58. **Caldarera, C. M., Orlandini, G., Casti, A., and Moruzzi, G.,** Polyamines and nucleic acid metabolism in myocardial hypertrophy of the overloaded heart, *J. Mol. Cell. Cardiol.*, 6, 95, 1974.
59. **Bartolomé, J., Huguenard, J., and Slotkin, T. A.,** Role of ornithine decarboxylase in cardiac growth and hypertrophy, *Science*, 210, 793, 1980.
60. **Russell, D. J., Byus, C. V., and Manen, C. A.,** Proposed model of major sequential biochemical events of a trophic response, *Life Sci.*, 19, 1297, 1976.
61. **Guarnieri, C., Flamigni, F., Muscari, C., and Caldarera, C. M.,** Involvement of calcium ions in the activation of ornithine decarboxylase by isoprenaline evaluated *in situ* in the perfused heart, *Biochem. J.*, 212, 241, 1982.
62. **Caldarera, C. M., Davalli, P., Flamigni, F., Clô, C., Toni, R., and Guarnieri, C.,** Inhibition of rat heart hypertrophy by 1-3 diaminopropane, in *Advances in Studies on Heart Metabolism*, Caldarera, C. M. and Harris, P., Eds., CLUEB, Bologna, 1982, 519.
63. **Pegg, A. E.,** Effect of α-difluoromethylornithine on cardiac polyamine content and hypertrophy, *J. Mol. Cell. Cardiol.*, 13, 881, 1981.
64. **Casti, A., Guarnieri, C., Dall'Asta, R., and Clô, C.,** Effect of spermine on acetylation of histones in rabbit heart, *J. Mol. Cell. Cardiol.*, 9, 63, 1977.
65. **Desmond, W. Jr. and Harary, I.,** In vitro studies of beating heart cells in culture. XV. Myosin turnover and the effect of serum, *Arch. Biochem. Biophys.*, 151, 285, 1972.
66. **McLean, M. J. and Sperelakis, N.,** Retention of fully differentiated electro-physiological properties of chick embryonic heart cells in culture, *Dev. Biol.*, 50, 134, 1976.
67. **Moura, A. M. and Simpkins, H.,** Cyclic AMP levels in cultured myocardial cells under the influence of chronotropic and inotropic agents, *J. Mol. Cell. Cardiol.*, 7, 71, 1975.
68. **Boder, G. B., Harley, R. J., and Johnson, J. S.,** Recording system for monitoring automaticity of heart cells in culture, *Nature*, 231, 531, 1971.
69. **Ghanbari, H. and McCarl, R. L.,** Involvement of cyclic nucleotides in the beating response of rat heart cells in culture, *J. Mol. Cell. Cardiol.*, 8, 481, 1976.
70. **Karsten, U., Kössler, A., Janiszewski, E., and Wollenberger, A.,** Influence of variations in pericellular oxygen tension on individual cell growth, muscle-characteristic proteins, and lactate dehydrogenase isoenzyme pattern in cultures of beating rat heart cells, *In Vitro*, 9, 139, 1973.
71. **DeLuca, M. A., Ingwall, J. S., and Bittl, J. A.,** Biochemical responses of myocardial cells in culture to oxygen and glucose deprivation, *Biochem. Biophys. Res. Commun.*, 59, 749, 1974.
72. **Harary, I., Renaud, J. F., Sato, E., and Wallace, G. A.,** Calcium ions regulate cyclic AMP and beating in cultured heart cells, *Nature*, 261, 60, 1976.
73. **Polinger, I. S.,** Growth and DNA synthesis in embryonic chick heart cells, in vivo and in vitro, *Exp. Cell Res.*, 76, 253, 1973.
74. **Weinstein, R. B. and Hay, E. D.,** Deoxyribonucleic acid synthesis and mitosis in differentiated cardiac muscle cells of chick embryos, *J. Cell Biol.*, 47, 310, 1970.
75. **DeHann, R. L.,** Spontaneous activity of cultured heart cells, in *Factors Influencing Myocardial Contractility*, Tanz, R. D., Kaverland, F., and Jay, R., Eds., Academic Press, New York, 1967, 217.
76. **Speicher, D. W., Peace, J. N., and McCarl, R. L.,** Effects of plating density and age in culture on growth and cell division of neonatal rat heart primary cultures, *In Vitro*, 17, 863, 1981.
77. **Lembach, K. J.,** Regulation of growth in vitro. I. Control of ornithine decarboxylase levels in untransformed and transformed mouse fibroblasts by serum, *Biochim. Biophys. Acta*, 354, 88, 1974.

78. **Jänne, J., Pösö, H., and Raina, A.**, Polyamines in rapid growth and cancer, *Biochim. Biophys. Acta,* 473, 241, 1978.

79. **Tomita, Y., Nakamura, T., and Ichihara, A.**, Control of DNA synthesis and ornithine decarboxylase activity by hormones and amino acids in primary cultures of adult rat hepatocytes, *Exp. Cell Res.,* 135, 363, 1981.

80. **Hershko, A., Mamont, P., Shields, R., and Tomkins, G. M.**, Pleiotypic response, *Nature New Biol.,* 232, 206, 1971.

81. **Frelin, C. and Padieu, P.**, Pleiotypic response of rat heart cells in culture to serum stimulation, *Biochimie,* 58, 953, 1976.

82. **Dulbecco, R.**, Topoinhibition and serum requirement of transformed and untransformed cells, *Nature,* 227, 802, 1970.

83. **Holley, R. W.**, Control of growth of mammalian cells in cell cultures, *Nature,* 258, 487, 1975.

84. **Bailey, J. M.**, Lipid metabolism in cultured cells. VI. Lipid biosynthesis in serum and synthetic growth media, *Biochim. Biophys. Acta,* 125, 226, 1966.

85. **Harary, I., Seraydarian, M., and Gerschenson, L. E.**, Effect of lipids on contractility of cultured heart cells, in *Factors Influencing Myocardial Contractility,* Tanz, R. D., Kaverland, F., and Jay, R., Eds., Academic Press, New York, 1967, 231.

86. **Renaud, J. F., Scana, A. M., Kazazoglou, T., Lombet, A., Romez, G., and Lazdunski, M.**, Normal serum and lipoprotein-deficient serum give different expression of excitability, corresponding to different stages of differentiation, in chick cardiac cells in culture, *Proc. Natl. Acad. Sci. USA,* 79, 7768, 1982.

87. **Clô, C., Tantini, B., Coccolini, M. N., and Caldarera, C. M.**, Involvement of polyamines in cyclic AMP metabolism in heart cell cultures, *Adv. Polyamine Res.,* 3, 333, 1981.

88. **Pignatti, C., Tantini, B., Rossoni Caldarera, C., Turchetto, E., and Clô, C.**, Involvement of polyamines in phospholipid-induced changes of cyclic nucleotide contents and macromolecular synthesis in cultured heart cells, *Adv. Polyamine Res.,* 4, 331, 1983.

89. **Clô, C., Pignatti, C., Tantini, B., and Caldarera, C. M.**, A possible involvement of endogenous polyamines in the regulation of the steady-state levels of cyclic nucleotides in heart cell cultures, *Ital. J. Biochem.,* 31, 437, 1982.

90. **Pignatti, C., Caldarera, C. M., Tantini, B., and Clô, C.**, Polyamine mediate the effects of phosphatidylcholine and phosphatidylethanolamine on myocardial cell cultures, Bat Sheva Seminar on Polyamines in Growth and Differentiation Processes, Jerusalem, July 4 to 9, 1982.

91. **Clô, C., Coccolini, M. N., Tantini, B., and Caldarera, C. M.**, Increased sensitivity of heart cell cultures to norepinephrine after exposure to polyamine synthesis inhibitors, *Life Sci.,* 27, 67, 1980.

92. **Clô, C., Orlandini, G. C., Guarnieri, C., and Caldarera, C. M.**, Effect of serum or polyamines on cyclic AMP levels in myocardial cell cultures, *Bull. Mol. Biol. Med.,* 2, 48, 1977.

93. **Fujimoto, A. and Harary, I.**, The effect of lipids on enzyme levels in beating rat heart cells, *Biochem. Biophys. Res. Commun.,* 20, 456, 1965.

94. **Harary, I., McCarl, R., and Farley, B.**, Studies in vitro on single beating rat heart cells, *Biochim. Biophys. Acta,* 115, 15, 1966.

95. **Clô, C., Pignatti, C., Coccolini, M. N., Tantini, B., and Turchetto, E.**, Effect of exogenous phospholipids on protein synthesis in confluent heart cell cultures, *Ital. J. Biochem.,* 30, 271, 1981.

96. **Howard, B. V. and Howard, W. J.**, Lipid metabolism in cultured cells, *Adv. Lipid Res.,* 12, 51, 1974.

97. **Caldarera, C. M., Pignatti, C., Tantini, B., Marmiroli, S., and Clô, C.**, Phospholipid-induced increase of polyamine synthesis in heart cell cultures, *J. Mol. Cell. Cardiol.,* 15 (Suppl. 3) 2, 1983.

98. **Hollenberg, M.**, Effect of oxygen on growth of cultured myocardial cells, *Circ. Res.,* 28, 148, 1971.

99. **Clô, C., Orlandini, G. C., Guarnieri, C., and Caldarera, C. M.**, Role of oxygen on growth rate and gene activity in cultured chick-embryo heart cells, *Biochem. J.,* 154, 253, 1976.

100. **Brosemer, R. W. and Rutter, W. J.**, Effect of oxygen tensions on the growth and metabolism of a mammalian cell, *Exp. Cell Res.,* 25, 101, 1961.

101. **Rueckert, R. and Muller, G.C.**, Effect of oxygen tension on Hela cell growth, *Cancer Res.,* 20, 944, 1960.

102. **Caldarera, C. M., Flamigni, F., Muscari, C., Clô, C., and Guarnieri, C.**, Evidence for sulphydrylic involvement in the inactivation of rat heart ornithine decarboxylase by oxygen radicals and hyperoxia, *Adv. Polyamine Res.,* 4, 525, 1983.

103. **Canellakis, E. S., Viceps-Madore, D., Kyriakidis, D. A., and Heller, J. S.**, The regulation and function of ornithine decarboxylase and of the polyamines, *Curr. Top. Cell. Regul.,* 15, 155, 1979.

104. **Guarnieri, C., Flamigni, F., Davalli, P., Clô, C., and Caldarera, C. M.**, Inhibition of rat heart ornithine decarboxylase by superoxide radicals, *Ital. J. Biochem.,* 31, 63, 1982.

105. **Cory, M., Henry, D. W., Taylor, D. L., and Koskela, K. J.**, Inhibitors of histone methylation, *Chemico-Biological Interactions,* 9, 253, 1974.

106. **Das, R. and Kanungo, M. S.**, Effects of polyamines on in vitro phosphorylation and acetylation of histones of the cerebral cortex of rats of various age, *Biochem. Biophys. Res. Commun.,* 90, 708, 1979.

107. **Moens, W., Vokaer, A., and Kram, R.,** Cyclic AMP and cyclic GMP concentrations in serum and density-restricted fibroblast cultures, *Proc. Natl. Acad. Sci. USA,* 72, 1063, 1975.
108. **Rudland, P. S., Seeley, M., and Seifert, W. E.,** Cyclic GMP and cyclic AMP levels in normal and transformed fibroblasts, *Nature,* 251, 417, 1974.
109. **Pastan, I. H., Johnson, G. S., and Anderson, W. B.,** Role of cyclic nucleotides in growth control, *Ann. Rev. Biochem.,* 44, 491, 1975.
110. **George, W. J., Busuttil, R. W., Paddock, R. J., White, L. A., and Ignarro, L. J.,** Opposing regulatory influences of cyclic guanosine monophosphate and cyclic adenosine monophosphate in the control of cardiac muscle contraction, in *Recent Advances in Studies on Cardiac Structure and Metabolism,* Roy, P. E. and Harris, P., Eds., University Park Press, Baltimore, 1975, 243.
111. **Krause, E. G. and Wollenberger, A.,** Cyclic nucleotides and heart, in *Cyclic 3':5'-Nucleotides: Mechanism of Action,* Cramer, H. and Schultz, J., Eds., John Wiley & Sons, London, 1977, 229.
112. **Vapaatalo, H., Metsä-Keletä, T., Parantainen, J., Palo-Oja, T., Kangasaho, M., and Laustiola, K.,** The role of cyclic nucleotides and prostaglandins in heart function, *Acta Biol. Med. Germ.,* 37, 785, 1978.
113. **Clô, C., Tantini, B., Pignatti, C., and Caldarera, C. M.,** Increased cyclic GMP content in confluent and serum-restricted heart cell cultures exposed to polyamines, *J. Mol. Cell. Cardiol.,* 15, 139, 1983.
114. **Clô, C., Pignatti, C., Tantini, B., Marmiroli, S., and Caldarera, C. M.,** Cyclic nucleotide metabolism in quiescent chick embryo heart cell cultures treated with α-difluoromethylornithine, in *Advances in Polyamines in Biomedical Science,* Caldarera, C. M. and Bachrach, U., Eds., CLUEB, Bologna, 1984, 11.
115. **Clô, C., Tantini, B., Turchetto, E., Manfroni, S., and Pignatti, C.,** Modulation of cyclic nucleotide metabolism by exogenous phospholipids in heart cell cultures, *J. Mol. Cell. Cardiol.,* 15 (Suppl. 3), 3, 1983.
116. **Clô, C., Tantini, B., Pignatti, C., and Rossoni Caldarera, C.,** A possible mediating role of polyamines in cyclic AMP-phosphodiesterase stimulation in confluent heart cell cultures, in *Advances in Studies on Heart Metabolism,* Caldarera, C. M. and Harris, P., Eds., CLUEB, Bologna, 1982, 139.
117. **Clô, C., Tantini, B., Pignatti, C., Marmiroli, S., and Caldarera, C. M.,** Polyamines influence heart cell (HC) sensitivity to cyclic nucleotide-mediated effectors, *J. Mol. Cell. Cardiol.,* 16, (Suppl. 2), 7, 1984.
118. **Clô, C., Pignatti, C., Tantini, B., Marmiroli, S., and Caldarera, C. M.,** Relevance of cellular polyamines in the response of heart cells to cyclic nucleotide-mediated effectors, in *Peptide Hormones, Biomembranes and Cell Growth,* Bolis, C. G., Frati, L., and Verna, R., Eds., Plenum Press, London, 1984, 207.
119. **Clô, C., Pignatti, C., Tantini, B., Marmiroli, S., and Caldarera, C. M.,** Cyclic nucleotide response of chick embryo heart cell cultures after preincubation with polyamines, *Ital. J. Biochem.,* 34, 203A, 1985.
120. **Clô, C., Caldarera, C. M., Tantini, B., Benalal, D., and Bachrach, U.,** Polyamines and cellular adenosine 3':5'-cyclic monophosphate, *Biochem. J.,* 182, 641, 1979.
121. **Clô, C., Tantini, B., Pignatti, C., Guarnieri, C., and Caldarera, C. M.,** Regulation of cyclic nucleotide metabolism by polyamines in heart cell cultures, *Adv. Polyamines Res.,* 4, 667, 1983.
122. **Alfheim, I. and Hongslo, J. K.,** Studies on the in vitro uptake of polyamines by Girardi heart cells, *Acta Pharmacol. Toxicol.,* 46, 171, 1980.
123. **Pohjanpelto, P.,** Putrescine transport is greatly increased in human fibroblasts initiated to proliferate, *J. Cell. Biol.,* 68, 512, 1976.
124. **Clô, C., Tantini, B., Coccolini, M. N., and Caldarera, C. M.,** Mediation of polyamine-induced decrease of cyclic AMP content by cyclic AMP-phosphodiesterase in chick heart cell cultures, *J. Mol. Cell. Cardiol.,* 13, 773, 1981.
125. **Kincaid, R. L., Manganiello, V. C., and Vaughan, M.,** Effect of spermine on activity and stability of calcium-dependent guanosine 3':5'-monophosphate phosphodiesterase, *J. Biol. Chem.,* 254, 4970, 1979.
126. **Wright, R., Buehler, B. A., and Rennert, O. N.,** Polyamines and adenylate cyclase in normal and cystic fibrosis (CF) cultured fibroblast, *Pediatric Res.,* 10, 373, 1976.
127. **Clô, C., Tantini, B., Pignatti, C., Marmiroli, S., and Caldarera, C. M.,** Polyamine regulation of cyclic nucleotide enzymes, *Ital. J. Biochem.,* 33, 43A, 1984.
128. **Strada, S. J. and Thompson, W. J.,** Multiple forms of cyclic nucleotide phosphodiesterases: anomalies or biologic regulators?, *Adv. Cyclic Nucleotide Res.,* 9, 265, 1978.
129. **Pledger, W. J., Thompson, W. J., Epstein, P. M., and Strada, S. J.,** Regulation of cyclic nucleotide phosphodiesterase forms by serum and insulin in cultured fibroblasts, *J. Cell. Physiol.,* 100, 497, 1979.
130. **Ball, E. H., Seth, P. K., and Sanwal, B. D.,** Regulatory mechanisms involved in the control of cyclic adenosine 3':5'-monophosphate phosphodiesterases in myoblasts, *J. Biol. Chem.,* 255, 2962, 1980.
131. **Byus, C. V. and Russell, D. H.,** Ornithine decarboxylase activity. Control by cyclic nucleotides, *Science,* 187, 650, 1975.
132. **Russell, D. H. and Haddox, M. K.,** Cyclic AMP-mediated induction of ornithine decarboxylase in normal and neoplastic growth, *Adv. Enzyme Regul.,* 17, 61, 1979.

133. **Costa, M. and Nye, J. S.,** Calcium, asparagine-and cAMP are required for ornithine decarboxylase activation in intact Chinese hamster ovary cells, *Biochem. Biophys. Res. Commun.,* 85, 1156, 1978.
134. **Bachrach, U., Katz, A., and Hochman, J.,** Polyamines and protein kinase. I. Induction of ornithine decarboxylase and activation of protein kinase in rat glioma cells, *Life Sci.,* 22, 817, 1978.
135. **Murray, A. W., Froscio, M., and Rogers, A.,** Effect of polyamines on cyclic AMP-dependent and independent protein kinases from mouse epidermis, *Biochem. Biophys. Res. Commun.,* 71, 1175, 1976.
136. **Bristow, M. R., Ginsburg, R., Minobe, W., Cubicciotti, R. S., Sageman, W. S., Lurie, K., Billingham, M. E., Harrison, D. C., and Stinson, E. B.,** Decreased catecholamine sensitivity and β-adrenergic-receptor density in failing human hearts, *New. Eng. J. Med.,* 307, 205, 1982.

Chapter 18

MUSCARINIC CHOLINERGIC RECEPTORS IN CULTURED HEART CELLS

Jonas B. Galper and Thomas W. Smith

TABLE OF CONTENTS

I. INTRODUCTION

In this chapter, we summarize new insights obtained from studies of muscarinic cholinergic receptors in cardiac tissue, with emphasis on studies of cultured heart cells. The muscarinic cholinergic receptor comprises an essential link in the pathway by which the parasympathetic nervous system modulates the function of the heart, blood vessels, and visceral organs. In the heart, vagal stimulation leads to the release of acetylcholine at nerve endings in the heart, leading to a decrease in the beating rate and in the force of contraction. Many of the functional consequences of muscarinic stimulation stem from the ability of muscarinic stimulation to antagonize the effects of adrenergic (sympathetic) stimulation of the heart. A major factor in the postsynaptic component of muscarinic inhibition of sympathetic action has been attributed to the reduction in accumulation of cyclic AMP following beta-adrenergic stimulation.[1] In addition, the response of the heart to muscarinic stimulation is associated with increased K^+ permeability of the heart cell membrane[2] a decrease in the slow inward current carried by Ca^{++} ions[3] increased levels of cyclic GMP[4] increased turnover of phosphatidylinositol in the cell membrane,[5] and changes in the phosphorylation pattern of specific proteins in the cell.[6] Many, if not most, of these effects are presumed to be interactive and not independent of one another.

Hormones and neurotransmitters are well known as being able to regulate the number or affinity of plasma membrane receptors in excitable cells as well as other tissues under hormonal control.[7] Binding of agonist to cell surface receptors in a number of instances induces endocytosis of receptors. Since a critical number of muscarinic receptors may be necessary to mediate a given physiologic response, it might be expected that the cardiac cell could modulate the level of muscarinic responsiveness through agonist-induced changes in receptor number or affinity. Such fine tuning of the response to muscarinic stimulation may be of substantial importance in modulating cardiac function. We shall review in this chapter evidence derived from studies of cultured heart cells indicating that muscarinic agonists do indeed play an essential part in controlling responsiveness of the heart to subsequent muscarinic stimulation.

II. VAGAL INNERVATION OF THE HEART

By way of further background, the autonomic nervous system is characterized in general by a two-neuron chain. Cell bodies of primary preganglionic neurons are located in the central nervous system and send out axons to synapses with the secondary postganglionic neuron, whose cell body is located in a peripheral autonomic ganglion. The postganglionic axon then passes to its distribution in the innervated organ. The parasympathetic input to the vagus nerve originates in the dorsal motor and salivatory nuclei of the medulla oblongata. The dorsal motor nucleus is the principal source of preganglionic parasympathetic fibers destined to innervate the heart. Fibers from the cardiac plexus synapse with ganglia located primarily in the atria, with a particularly rich collection of ganglia in the region of the sinoatrial and atrioventricular nodes.

Both preganglionic and postganglionic nerve terminals release acetycholine. Preganglionic cholinergic receptors are nicotinic receptors, while postganglionic receptors in the heart are muscarinic receptors. These distinctions are based on observations that drugs such as nicotine and dimethylphenylpiperazinium, which are selective nicotinic agonists, suppress sinoatrial pacemaker activity in isolated chick embryo hearts.[8] However, these effects are inhibited by atropine, a specific muscarinic antagonist, and hexamethonium, a specific nicotinic antagonist. Thus, vagal stimulation by nicotine can be inhibited either directly at the preganglionic nicotinic receptor or at the postganglionic muscarinic receptor by specific antagonists.

Pappano has provided an excellent review of the development of parasympathetic innervation of the heart.[9] Development from fertilization to hatching in the chicken lasts 21 days. In the chick embryo, vagal fibers appear in the region of the truncus arteriosus during the third day *in ovo*. Vagal fibers reach the atria toward the end of the fourth day, and the interatrial septum on the fifth day. Innervation of sinoatrial node and atrial wall has been demonstrated on days six and seven, respectively.[10] The atrioventricular groove is reached by vagal fibers by the ninth day. Far less is known regarding the development of parasympathetic innervation of the human heart. Nerve cells are known to be present at about 5 weeks of gestation, while ganglia appear to develop between the fifth and twelfth weeks of gestation.[11]

In the chick heart, inhibitory transmission between postganglionic neurons and sinoatrial pacemaker cells appears on the twelfth day *in ovo*.[12] Available evidence indicates that muscarinic receptors are present in the embryonic chick heart for a substantial period prior to ingrowth of the vagus nerve. Exogenously added acetylcholine mediates only a very modest decrease in atrial beating rate of chick hearts 2 to 3 days *in ovo*; however, an increased sensitivity of beating to acetylcholine does not appear until the seventh to tenth day *in ovo*, following vagal innervation.[13,14]

Galper and colleagues demonstrated that beating rate in embryonic chick hearts 2 to 4 days *in ovo* decreased by 20 to 30% in response to 1 mM carbamylcholine, while beating rates of embryos 5 to 6 days *in ovo* decreased by 90% at the same carbamylcholine concentration with a half-maximal effect at $10^{-5}\,M$ (see Figure 1). In 7- to 12-day embryonic chick hearts sensitivity continues to increase rapidly, with beating rates being totally inhibited at $10^{-6}\,M$ carbamylcholine (half-maximal effect at $3 \times 10^{-8}\,M$).

Interestingly, at all ages from 2 1/2 to 18 days *in ovo*, the number of muscarinic receptors per milligram of cell protein remains unchanged as measured by the binding of the labeled muscarinic antagonist [^{3}H] quinuclidinylbenzilate ([^{3}H]QNB)[15] (see Figure 2). Thus, muscarinic receptors are present prior to ingrowth of the vagus nerve, but these receptors are relatively insensitive to muscarinic agonists. The development of sensitivity following vagal innervation does not appear to be due to an alteration in receptor numbers as measured by radioligand binding, but rather to events subsequent to binding of agonist to the receptor.

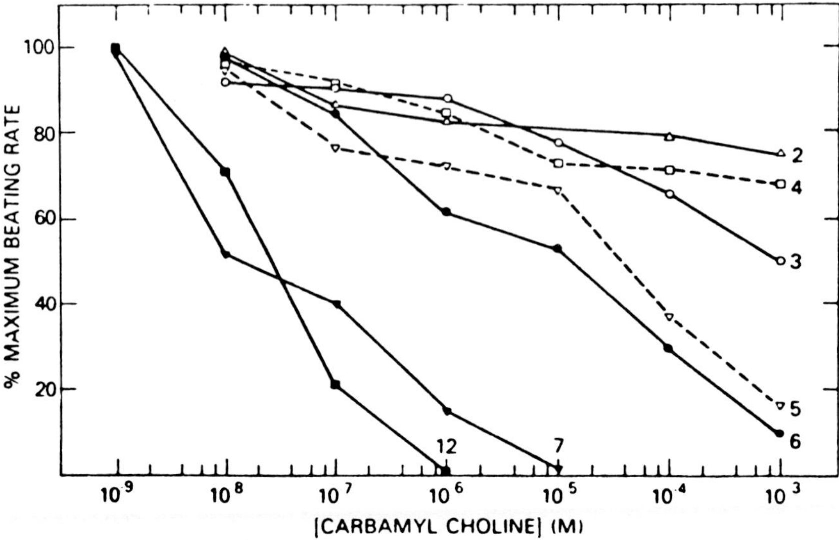

FIGURE 1.   Reduction of heart rate by carbamylcholine. Heart beating rates were determined at 37° C using a video motion detector. Hearts were treated with isoproterenol (0.1 μM) and incubated until a stable rate was achieved (approximately 5 min). Then, successive doses of carbamylcholine were added to the preparation and the rate of beating was determined every 30 sec. Time was allowed for recovery between successive doses. The minimum beating rate at a given dose was plotted as percentage of the isoproterenol control rate. Each point represents an average of 15 to 20 hearts. Day 12 represents 5 hearts studied using a perfusion technique. $\triangle$, Day 2; $\bigcirc$, Day 3; $\square$, Day 4; $\triangle$, Day 5; $\bullet$, Day 6; $\blacktriangle$, Day 7; $\blacksquare$, Day 12 hearts (From Galper, J. B., Klein, W., Catterall, W. A., *J. Biol. Chem.*, 252, 8692, 1977. With permission.)

III. PHYSIOLOGY OF THE MUSCARINIC RESPONSE

A. The Expression of Muscarinic Function in Heart Cell Cultures

As noted above, parasympathetic stimulation of the heart mediates a decrease in the rate and force of contraction. These effects are due at least in part to antagonism of the effects of the sympathetic nervous system on the heart. Cultured heart cells differ in a number of important ways from the intact heart. Heart cell cultures are essentially devoid of neuronal elements and hence, in the absence of exogenous effectors, neuronally activated functions may be below basal levels observed in intact, innervated hearts. Heart cell cultures prepared in the conventional fashion are usually a mixture of heart cell types derived from a mixture of atrium and ventricle, pacemaking tissues and specialized conduction tissue. Furthermore, the state of aggregation of heart cells affects heart cell function in culture. Hence the properties of monolayer cultures differ significantly from aggregates. During growth of heart cells in culture, the relative distribution of growth factors, lipoproteins, cholesterol, and phospholipids in the growth medium does not generally reflect in vivo conditions. Consequently, certain properties of heart cells in culture do not accurately reflect responsiveness of the heart in vivo. Early studies suggested that chicken heart cells in culture were insensitive to the effects of beta-adrenergic and cholinergic agonists.[16] Nevertheless, other groups demonstrated negative chronotropic effects of cholinergic agonists after 15 to 30 min of incubation,[17,18] and a more recent study using a rapid flow technique suggested that cultured heart cells are capable of an immediate response to as little as 1 nM carbamylcholine.[19] In our studies,[20] we routinely find a 15 to 20% carbamylcholine-mediated decrease in beating rate with a maximal effect at $10^{-3}M$ carbamylcholine. Recently, Reynaud et al.[21] demonstrated that growth conditions markedly affected the muscarinic responsiveness of cultured

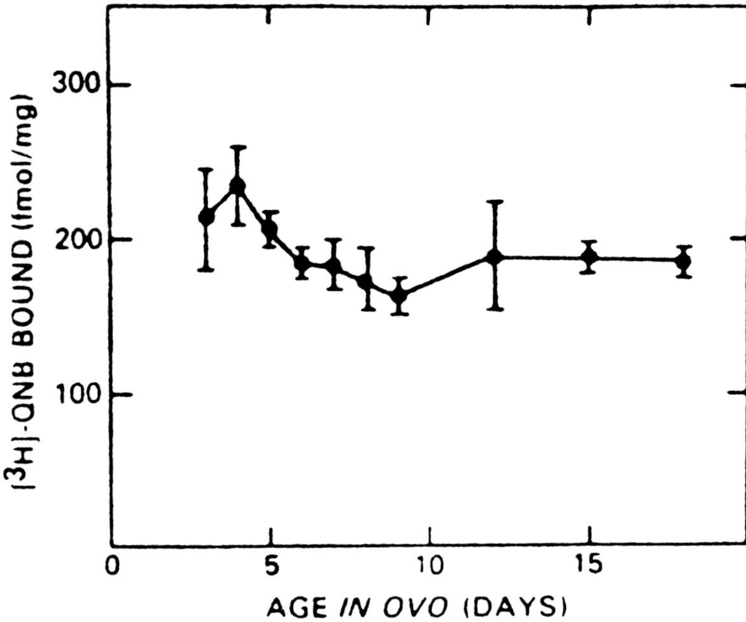

FIGURE 2.   Developmental changes in the number of muscarinic receptors in embryonic chick hearts. Hearts were dissected from 5 to 10 embryos of each age, cut open, and washed thoroughly. Homogenates were prepared and QNB binding was measured after a 60 min incubation at room temperature with 2 nM[^{3}H]QNB. Parallel incubations in the presence of 100 μM oxotremorine were carried out and the [^{3}H]QNB bound nonspecifically under these conditions was subtracted. Each point represents the mean ($\pm$ S.E.) of three to five such experiments. (From Galper, J. B., Klein, W., Catterall, W. A., *J. Biol. Chem.*, 252, 8692, 1977. With permission.)

heart cells. Growth of embryonic chick heart cell in medium supplemented with lipoprotein-depleted fetal calf serum (FCS) resulted in a marked increase in responsiveness to muscarinic agonists. Inhibition of beating was complete at 10 μM oxotremorine with a half-maximal effect at 6nM oxotremorine. Comparison with cultures grown with control serum demonstrated that cultures grown in lipoprotein-depleted serum contained 2.5 fold more [^{3}H]QNB binding sites, and total cholesterol per mg protein was increased twofold. We have confirmed these studies in our laboratory and have further demonstrated that the effects of lipoprotein depletion appear to be reversed by reconstitution of the LDL fraction in the lipoprotein-depleted medium. These studies suggest an important role for cholesterol and cholesterol metabolism in the control of expression of muscarinic function in cultured heart cells.

B. Electrophysiology of the Muscarinic Response

Vagal stimulation or acetylcholine exposure produces characteristic changes in sinus venosus and atrial muscle that include hyperpolarization of the resting membrane potential and shortening of action potential duration. These changes are associated with increased K$^+$ permeability and decreased Ca^{2+} permeability. Early studies by Hutter and Trautwein documented hyperpolarization in response to vagal stimulation and also accelerated repolarization of the action potential in atrial tissue and sinoatrial node, resulting in decreased duration of the action potential.[22] These effects were found to be sensitive to extracellular K$^+$ concentration, and Harris and Hutter[23] documented an increased rate of ^{42}K$^+$ efflux from a sinus venosus preparation loaded with ^{42}K$^+$ following vagal stimulation or acetylcholine exposure. It is now widely accepted that the response to muscarinic stimulation involves an increase in K$^+$ conductance of the heart cell membrane. This has been confirmed and more extensively quantified by Glitsch and Pott[24] in studies of guinea pig atria using voltage clamp techniques.[25] Despite further detailed studies,[26] it remains uncertain whether acetylcholine acts by inducing an increased conductance of the inward rectifying time-independent K^+ channel (I_K), or by altering the conductance of a different specific K^+ channel.

1. Evidence for the Presence of a Single Muscarinic K^+ Channel in Isolated Pacemaker Cells

Recently, Sakmann et all.[27] using patch clamp analysis of enzymatically dispersed rabbit sinoatrial and atrioventricular nodal cells presented data that support the presence of an acetylcholine-dependent K^+ conductance that differs in gating and conductance properties from the inward rectifying resting K^+ channels in atrial and ventricular cells. Two K^+ currents with pulse durations having time constants of 1 to 4 msec and 48 msec were shown. While the current with a time constant of 48 msec was insensitive to acetylcholine, the 1 to 4 msec current showed a biphasic distribution of both burst duration and closed time duration in the presence of low concentrations of acetylcholine. From these data, a two-step model for channel activation has been proposed in which an initial rapidly reversible channel opening is followed by more prolonged activation.

2. Acetylcholine Effects on Ca^{2+} Current

The effect of acetylcholine on Ca^{2+} movement into cardiac cells is controversial. An acetylcholine-mediated increase in outward K^+ current shortens the action potential duration. Thus, the time course of the slow inward current (I_{si}) is abbreviated. This almost certainly contributes to the negative inotropic effect of vagal stimulation on the heart. In addition, there is evidence of a more direct inhibitory effect of muscarinic agonists on Ca^{2+} entry. Voltage clamp studies in frog atrial trabeculae[28] have documented an acetylcholine-mediated decrease in I_{si} at acetylcholine concentrations of $2 \times 10^{-8}M$ or above. While the relative importance of these effects is dependent on species and details of experimental conditions, it seems clear that alterations in action potential duration mediated by changes in K^+ conductance and more direct effects on I_{si} mediate the negative inotropic effects of muscarinic stimulation.

3. Effects of Acetylcholine on Ventricle and Purkinje Fibers

Hino and Ochi[29] studied acetylcholine effects on membrane currents in guinea pig papillary muscle. I_{si} was substantially decreased in the absence of any apparent effect on K^+ currents or action potential duration. Carmeliet and Ramon[30-32] have observed yet another pattern in sheep Purkinje fibers, which demonstrated prolongation of the action potential, hyperpolarization of the maximum diastolic potential, and an increase in the rate of diastolic (phase 4) depolarization. Voltage clamp studies showed a decrease in I_{si} and a shift in the activation curve for the pacemaker current (I_{K2}) by a few millivolts in the depolarization direction in response to acetylcholine. While no uniform picture emerges from these studies, taken together, they indicate that a combination of acetylcholine-mediated changes in K^+ and Ca^{2+} permeabilities accounts for observed changes in membrane potential, beating rate, and contractility. The relative importance of each effect is species dependent.

4. Comparison Between Electrophysiologic Response to Muscarinic Stimulation in Atrium and Ventricle

Several groups have suggested that in the absence of beta-adrenergic stimulation, acetylcholine produced negligible effects on contractile state of ventricles from embryonic chick heart as well as in certain mammalian species.[33] Josephson and Sperelakis[34] used a two-microelectrode voltage clamp method to study the effects of acetylcholine on I_{si} in aggregate cultures of embryonic chick ventricular cells 16 to 20 days *in ovo*. They demonstrated that isoproterenol ($10^{-6}M$) increased I_{si} at all clamp potentials, while subsequent addition of acetylcholine decreased I_{si} to control levels seen in the absence of isoproterenol. They further found no effect of acetylcholine on late outward K^+ current as would have been expected if acetylcholine had altered K^+ permeability.[34]

Inoue et al.[35] demonstrated that in both atrial and ventricular strips, acetylcholine reduced the overshoot and the duration of action potentials. However, the effect in the atrium was associated with both a membrane hyperpolarization and a reduction of membrane resistance, while the effects in the ventricle were associated with neither. Our own recent studies of cultured atrial and ventricular cells from chick hearts 14 days *in ovo* grown in media supplemented with lipoprotein-depleted serum demonstrated inhibition of beating in atrial cultures with an IC_{50} of $2 \times 10^{-7}M$ carbamylcholine, compared to a maximum effect of only 10 to 15% inhibition of beating rate at 10^{-5} M in ventricular cultures. However, in the presence of 10^{-5} M isoproterenol, both atria and ventricles demonstrated a similar negative chronotropic response to carbamylcholine. We further demonstrated that while carbamylcholine increased the half-time for $^{42}K^+$ efflux from atrial cultures by 40%, no effect of carbamylcholine on the half-time of efflux of $^{42}K^+$ from ventricular cultures could be demonstrated.[36] One interpretation of these data is that muscarinic agonists may alter I_{si} either by an indirect effect on K^+ permeability or a direct effect on Ca^{2+} permeability in atrial cultures, but in cultured ventricular cells, only the effect Ca^{2+} permeability is evident.

C. Ion Flux Studies
1. Potassium Fluxes

Hutter described a 2- to 3-fold increase in $^{42}K^+$ efflux from the turtle sinus venosus in response to vagal stimulation. Cassalle[37,38] found a 12 to 20% increase in the rate of $^{42}K^+$ uptake by guinea pig sinoatrial node in response to acetylcholine. Cultured heart cells offer a useful system for studying acetylcholine-induced changes in K^+ flux due to the absence of appreciable diffusion barriers. However, these cultures tend to have decreased sensitivity to muscarinic agonists when compared to the intact hearts from which they are derived. For example, half-maximal inhibition of beating in intact chick embryo hearts 12 days *in ovo* occurs at $10^{-7}M$ carbamylcholine,[15] while beating in cultures from these hearts requires $7 \times 10^{-5}M$ carbamylcholine for half-maximal inhibition. The cause of this altered sensitivity may be related to the aggregation state of the cultured heart cells, or to other neurohumoral or developmental factors present in the intact heart but not expressed in cultured heart cells.

Despite this difference in muscarinic sensitivity, cultured heart cells respond in a predictable and characteristic manner to muscarinic stimulation. We find a half-time of efflux of $^{42}K^+$ from cultured chick embryo ventricular cells loaded to asymptote with $^{42}K^+$ of 13.8 min in control cells, decreasing to 9.0 min in the presence of carbamylcholine (see Figure 3). Similarly, the half-time of uptake of $^{42}K^+$ decreased from 13.8 to 8.8 min in the presence of carbamylcholine. Thus, altered potassium permeability was apparent in the presence of carbamylcholine although total intracellular K^+ content remained constant.[20] Since steady-state intracellular K^+ content is maintained in the presence of acetylcholine-induced increases in K^+ permeability, the passive movement of K^+ out of the cell down its concentration gradient must be balanced by an increase in K^+ movement into the cell caused by the carbamylcholine-induced hyperpolarization and/or by an increase in active transport of K^+ into the cell via Na^+,K^+-ATPase, both of which would tend to cause hyperpolarization.

2. Calcium Fluxes

In electrically paced guinea pig atria, acetylcholine (2 to $4 \times 10^{-6}M$) decreased contraction amplitude by 80% concomitant with a 40% decrease in $^{45}Ca^{2+}$ uptake over 5 min.[39] Additional studies suggest decreased calcium influx in atrial preparations in response to muscarinic stimulation.[40,41] Definitve studies of the effects of muscarinic agonists on Ca^{2+} movements

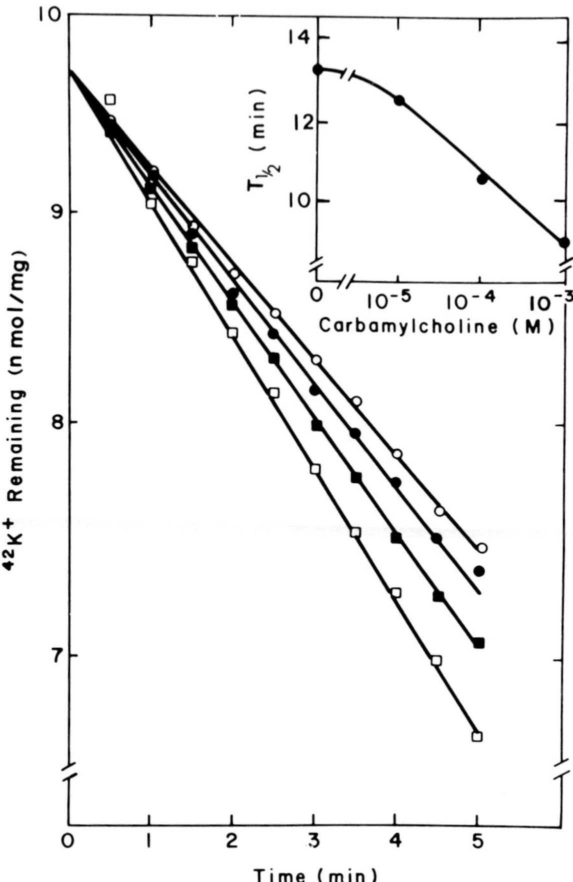

FIGURE 3.   Effect of carbamylcholine on $^{42}K^+$ efflux from cultured heart cells. Dishes containing heart cells cultured from chicken embryos 10 days *in ovo* and grown on glass coverslips were labeled overnight with 0.2 μCi/mℓ leucine [4,5-^{3}H], incubated for 3 hr in growth medium 5 μCi/mℓ in $^{42}K^+$ (total K^+ = 4.5 mM), rinsed, and placed in an efflux chamber. Cells were perfused with growth medium containing unlabeled K^+ and the indicated concentrations of carbamylcholine; $\bigcirc$, no carbamylcholine; $\bullet$, $10^{-5}M$; $\blacksquare$, $10^{-4}M$; $\square$, $10^{-3}M$; $\triangle$, $10^{-3}M$ carbamylcholine plus $10^{-6}M$ atropine. $^{42}K^+$ efflux was determined at the times indicated plotted as $^{42}K^+$ remaining in the cells at time t, calculated by subtracting total efflux $^{42}K^+$ from the cell during time t from the initial $^{42}K^+$ content of the cell at time zero. $^{42}K^+$ at time zero was determined as the mean K^+ content/mg protein of 14 coverslips exposed to 5 μCi/mℓ $^{42}K^+$ for 3 hr and rinsed, and $^{42}K^+$ was determined. Each curve is the mean of two sets of seven replicate determinations each. The lines are the least-squares fit of the $\log_{10}$ of $^{42}K^+$ remaining at time t to a straight line and are plotted on a logarithmic scale vs time. Correlation coefficients were at least 0.99 for all plots. Insert: The effect of increasing carbamylcholine concentration on the half-time of $^{42}K^+$ efflux. The half-time of efflux is calculated from the relationship $T_{1/2}$ = 1n 2/k_1. (From Galper, J. B., Dziekan, L. C., Miura, D. S., and Smith, T. W., *J. Gen. Physiol.*, 80, 231, 1982. With permisssion.)

in cultured heart cells, and their relationship to altered contractile state, have not yet been published.

D. The Role of Cyclic Nucleotides in Mediating the Muscarinic Response

Altered intracellular cyclic GMP and cyclic AMP in response to muscarinic stimulation suggest that cyclic nucleotides may serve as second messengers in mediating a response to muscarinic stimulation. Cyclic AMP is well established as a second messenger in the chronotropic and inotropic responses of the heart to beta-adrenergic stimulation. Cyclic AMP levels have been shown to correlate with increases in slow inward current (I_{si}) and in enhanced contractility. In addition, agents including histamine and phosphodiesterase inhibitors that increase intracellular cAMP levels independent of beta-adrenergic effects also increase I_{si} and increase contractility.[42,43] These and other studies demonstrate that cAMP, presumably acting through the stimulation of protein phosphorylation via cAMP-activated protein kinase, is capable of increasing Ca^{2+} influx.

It is now well established that the antagonist effects of parasympathetic stimuli on responses to sympathetic stimulation are mediated, at least in part, by the ability of muscarinic agonists to lower intracellular cyclic AMP levels. Hormonal stimulation of adenylate cyclase activity involves interactions of three separate components: the hormone receptor, a catalytic subunit that converts ATP to cyclic AMP, and at least two guanine nucleotide regulatory proteins which, in the presence of GTP, couple hormone binding to the stimulation or inhibition of the catalytic unit.[44] GTP and GTP analogues increase adenylate cyclase activity by shifting the equilibrium point of adenylate cyclase from a low activity state, in which GDP is bound to the stimulatory guanine nucleotide regulatory protein, to a high activity state in which GTP or a nonhydrolyzable GTP analogue is bound. A large number of studies[44-47] document the role of the stimulatory guanine nucleotide regulatory protein (G_s) as an intermediary in coupling the interaction of hormone or neurotransmitter with receptor to the stimulation of adenylate cyclase.

More recently, a number of hormones and neurotransmitters have been found to lower cAMP levels or to attenuate the increase due to stimulatory agents.[48] Pioneering studies by Murad and colleagues[49] demonstrated that acetylcholine decreased the synthesis of cAMP in cardiac tissue. Independent inhibitory and stimulatory control of adenylate cyclase activity is supported by studies of the development of beta-adrenergic stimulation and muscarinic cholinergic inhibition of adenylate cyclase in the embryonic chick heart. We observed that while adenylate cyclase in embryonic chick hearts of all ages studied (2.5 to 18 days *in ovo*) can be stimulated half-maximally by $10^{-6}M$ isoproterenol, prior to ingrowth of the vagus nerve (day 3 *in ovo*), the concentration of carbamylcholine required for half-maximal inhibition of adenylate cyclase was 40 times greater than that required in the hatched chick.[50,51] In addition, inhibition of adenylate cyclase by muscarinic agonists in 3-day hearts was only 10% of that observed in the heart of the hatched chick. No difference in muscarinic receptor number could be demonstrated by binding of [³H]QNB in chick hearts from 3 to 12 days *in ovo*.[15] Thus, during embryologic development in the chick, the beta-adrenergic stimulatory effect on adenylate cyclase and the muscarinic inhibitory effect do not develop in a parallel fashion. This suggests that prior to vagal innervation, a factor (possibly an inhibitory guanine nucleotide binding protein) that couples muscarinic receptor occupancy to cyclase inhibition is not yet normally expressed.

The studies of Biegon and Pappano,[52] as well as our own studies,[50] indicate that muscarinic inhibition of both basal and isoproterenol-stimulated adenylate cyclase develop in parallel in the chick embryo heart and hence may occur via a single mechanism. At any rate, it seems safe to conclude that inhibition of adenylate cyclase activity is an important physiologic mechanism by which parasympathetic stimulation of the heart tends to offset both basal and beta-adrenergic stimulated contractile state.

The role of cGMP in the heart continues to be a subject of debate. In the isolated perfused heart, acetylcholine is known to raise cGMP levels while cAMP levels decrease somewhat or remain constant. The increase in cGMP tends to parallel the negative inotropic effect of acetylcholine, but the decrease in beating rate in response to acetylcholine does not demonstrate the same correlation with cyclic GMP levels.[4] On the basis of these and other experiments, it has been suggested that beta-adrenergic agonists stimulate cAMP formation, which is responsible for mediating the increased rate and force of contraction in the heart, while acetylcholine stimulates cGMP formation, which opposes the action of cAMP and decreases the rate and force of contraction. Permeant cGMP analogues, such as the dibutyryl derivative, slowed the beating rate in cultures of rat heart ventricular cells by 15%, while similar concentrations of dibutyryl cAMP increased beating rate by 15%. Butyrate alone had no effect.[53] Another permeant derivative, 8-bromo-cyclic GMP decreases both contractility and inward Ca^{2+} current in guinea pig atrium.[54] Isoproterenol partially inhibits these effects of 8-bromo-cyclic GMP. Watanabe and Besch[55] found only a small effect of acetylcholine on myocardial contractility in the guinea pig atrium, although acetylcholine increased levels significantly. In the presence of isoproterenol, however, the same acetylcholine concentration decreased the inotropic response to isoproterenol substantially. These effects of cGMP on contractility have not been demonstrated, however, by all investigators.[56-58] Differing species under study as well as differing calcium concentrations in physiologic media may account for some of the discrepancies in published work. At the very least, it appears necessary to invoke compartmentation if cGMP is to be viewed as an important modulator of myocardial contractile state. This conclusion comes from studies of sodium nitroprusside, an agent that elevates cyclic GMP levels dramatically (nearly 20-fold in cat atria), yet has no significant effect on contractile state. In the same preparation, acetylcholine decreased contractile force development without an effect that could be measured on cGMP levels.[59]

It is of interest that guanylate cyclase, the enzyme that catlyzes the conversion of GTP to cGMP, exists in heart in both soluble and membrane bound forms,[60] and only the soluble form is activated by sodium nitroprusside.[61] This finding lends some credibility to the possibility of highly localized effects of cGMP, e.g., in the sarcolemmal membrane.

IV. BIOCHEMICAL ASPECTS OF MUSCARINIC RESPONSIVENESS IN THE HEART

Interaction of muscarinic cholinergic agonists with their receptors in the heart sets in motion a complex series of events that include inhibition of adenylate cyclase,[48] changes in levels of cGMP,[4] phosphorylation of specific proteins by cyclic GMP-activated protein kinases,[62] reduced levels of phosphorylation of specific proteins due to reduced cAMP-dependent protein kinase activation, and changes in the turnover of phosphatidylinositol.[63] As indicated earlier, these events presumably underly alterations in Ca^{2+} entry and/or K^+ conductance. Although the relationship of these physiologic events to the underlying biochemical mechanisms remains for the most part unclear, the discussions that follow will attempt to provide a framework for considering these relationships.

A. Ligand Binding to the Muscarinic Receptor
1. Labeled Antagonist Binding
Data reviewed in this section support the view that interaction of muscarinic agonist (A) with its receptor (R) forms an agonist-receptor complex (AR); the interactions of AR with the guanine nucleotide regulatory subunit and GTP are important steps in the coupling of agonist binding to the biochemical and physiological events that follow.

The availability of the potent muscarinic antagonist quinuclindinyl-benzilate in tritiated form ([³H]QNB) has facilitated the study of binding interactions between muscarinic ligands and receptors. Although this radioligand is relatively hydrophobic and tends to bind to lipid bilayers in a nonspecific manner, careful selection of assay conditions makes possible up to 98% specific binding, usually defined as radioactivity bound that can be displaced by unlabeled QNB or other ligand. This permits the indirect assessment of agonist binding to the receptor by measurement of displacement of [³H]QNB. Potency series for the relative binding affinity of agonists and antagonists have been determined for embryonic chick heart[15] or rat heart[64] that correlate well with pharmacological potency for muscarinic stimulation or inhibition. Studies of Galper and colleagues[15] indicate that the binding of [³H]QNB occurs via a two-step process, with initial formation of a rapidly reversible, low-affinity complex followed by conversion to a more slowly reversible, high affinity complex with the receptor.

In parallel with observations of β-adrenergic agonist-receptor interactions, we have found that competition between carbamylcholine and 1 nM [³H]QNB for binding to receptors from cultured heart cells shows a complex pattern in which at least two agonist binding sites can be identified: (1) high affinity binding sites (R_H) constituting 26% of total sites with an IC_{50} for inhibition of [³H]QNB binding of 3.9 × 10⁻⁷M; and (2) low affinity sites (R_L) with an IC_{50} of 4.5 × 10⁻⁵M for inhibition of antagonist binding.[65] The analysis of these complex binding curves is illustrated in Figure 4.

Using analogous competitive binding methods, Rosenberger and colleagues[66] found that guanine nucleotides and Na⁺ decreased the affinity of the muscarinic receptor for agonist. Oxotremorine, a potent muscarinic agonist, shifted the IC_{50} for [³H]QNB by about 1 log unit in the direction of lower affinity. Also in analogy with the beta-adrenergic receptor, despite these pronounced changes in agonist affinity in response to Na⁺ and guanine nucleotides, only negligible effects on antagonist binding were evident. Similar findings were reported by Berrie et al.[67] for rat heart receptors.

In homogenates of cultures of embryonic chick heart cells, we have demonstrated that addition of guanine nucleotides had no effect on the total number of receptors, but resulted in the conversion of a subset of 26% of high-affinity receptors to a low-affinity form.[65] Intact cells exposed briefly (1 to 15 min) to muscarinic agonists and washed carefully to remove unbound agonist and then homogenized showed a decrease in the apparent number of receptors by 26% as judged by [³H]QNB binding. Binding to the high-affinity receptor subset in these homogenates could no longer be demonstrated, and all of the residual binding was of the low affinity type. These findings suggest that the loss of receptor following brief agonist exposure may involve persistent binding of agonist to the high affinity form of the receptor (R_H), leaving these high affinity receptors unavailable for labeled antagonist binding. Guanine nucleotide exposure appears to mediate the conversion of R_H to R_L. Hulme et al.[68] reported a GTP-mediated conversion of receptors from intact rat heart from a super high affinity form (K_d 5 × 10⁻⁸M) and high-affinity form (K_d 2 × 10⁻⁶ M) to a predominant low-affinity state (K_d 10⁻⁴ M).

Thus, guanine nucleotides decrease the affinity of both beta-adrenergic receptors and muscarinic receptors for their respective agonists. Important differences, however, have been demonstrated between these two effects of guanine nucleotides. Harden and colleagues[69] found that N-ethylmaleimide, a sulfhydryl reagent, caused decreased affinity of oxotremorine for the muscarinic receptor. This effect was similar to the alteration in oxotremorine affinity in the presence of GTP. However, pretreatment of membranes with N-ethylmaleimide at concentrations as high as 1 mM had no effect on beta-adrenergic binding to receptors, and the affinity of the beta receptor for agonists maintained its GTP sensitivity after *N*-ethylmaleimide treatment.

We have compared the affinity of muscarinic receptors for agonists in intact cultured heart cells and homogenates from these cells using competitive binding of agonists with [³H]QNB. While the IC_{50} for oxotremorine competition with [³H]QNB was 4 × 10⁻⁷ M in homogenates,

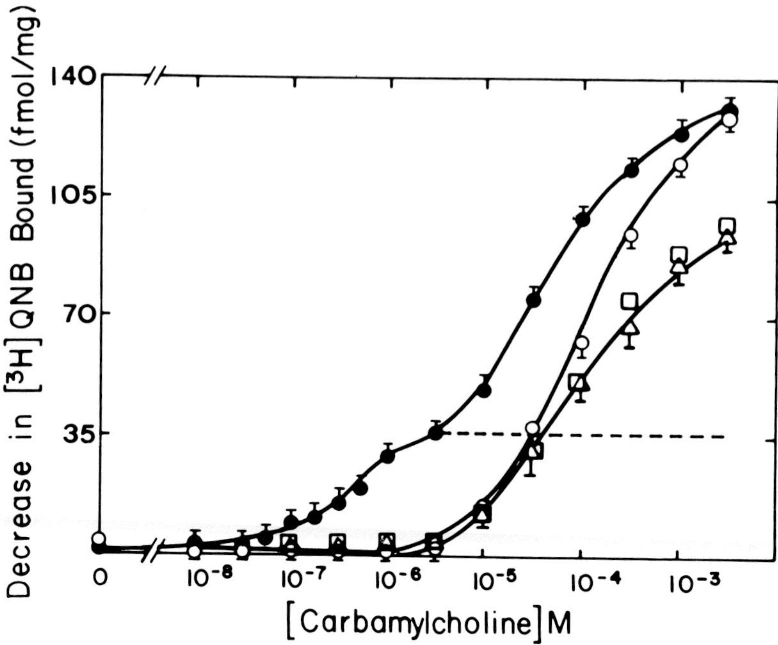

FIGURE 4.   Effect of carbamylcholine and guanine nucleotides on apparent receptor affinity for agonists. Replicate culture dishes were incubated with ($\triangle$) and without ($\bullet$) 10^{-3} carbamylchoine for 15 min; cells were harvested, homogenized, and incubated for 1 hr rat room temperature in medium containing 1 nM [^{3}H]QNB at the concentration of agonist indicated and [^{3}H]QNB binding determined. In a similar manner, an aliquot of homogenate from control cells not previously exposed to agonist were incubated with $10^{-4}M$ Gpp(NH)p ($\bigcirc$) and [^{3}H]QNB binding determined in the presence of the indicated concentrations of agonists. Inhibition of [^{3}H]QNB binding at each concentration of agonist is plotted in fmol/mg. 100% inhibition corresponds to a decrease in [^{3}H]QNB bound of 140 fmol/mg. Each point represents the mean of three replicate determinations repeated n times; error bars represent S.E. The solid lines are drawn by eye. $\bullet$, control (n = 25); $\bigcirc$, control plus $10^{-4}M$Gpp(NH)p (n = 20); $\triangle$, carbamylcholine pretreatment for 15 min (n = 20); $\square$, data derived by subtracting the sub class of high affinity receptors from the control curve; ---, extrapolation of high affinity shoulder. (From Galper, J. B., Dziekan, L. C., O'Hara, D. S., and Smith, T. W., *J. Biol. Chem.*, 257, 10344, 1982. With permission.)

binding to intact cells gave an IC$_{50}$ of 4 $\times$ $10^{-6}M$ and an increase in the Hill coefficient from 0.58 to 0.82 respectively.[65,70] These data indicate a significant decrease in receptor heterogeneity and an apparent loss of high-affinity binding sites in intact cells. In a similar experiment, Nathanson[71] found an increase in the Hill coefficient from 0.50 to 0.92 in cell homogenates and intact cells, respectively. Computer analysis of these data showed a single class of low-affinity receptors in intact cells with K$_d$ of 1.2 $\times$ $10^{-5}M$.[71] These findings are consistent with the hypothesis that high-affinity muscarinic receptors are either markedly decreased or absent in intact cells. We consider it likely that the high affinity receptor in the intact cells has a very short half-time and is converted to a low affinity form in the presence of endogenous GTP.

2. Labeled Agonist Binding

Two radiolabeled muscarinic agonists have become available commercially: [^{3}H]oxotremorine-M and [^{3}H]cis-methyldioxalane. Both of these radioligands tend to bind to filters and give high levels of nonspecific binding to membrane preparations in filtration assays.

In addition, oxotremorine-M tends to oxidize readily to a form that does not yield specific binding. Due to these problems, fewer data are available regarding direct assessment of agonist binding than displacement of labeled antagonist by unlabeled agonist. Studies of central nervous system tissue[72,73] suggest that both guanine nucleotides and Na^+ decrease agonist binding to a significant extent. Studies in our own laboratory of [³H]cis-methyldioxalane binding to embryonic chick heart homogenates have been possible through the development of improved techniques for separation of bound and free radioligand. We can obtain 92% specific binding as determined by competition with unlabeled cis-methyldioxalane or atropine. Studies with this radiolabeled agonist demonstrate a subclass of high affinity receptors (k_d 16 × $10^{-9}M$) and a low affinity site (K_d 4 × $10^{-7}M$). Kinetics of formation and dissociation of the agonist-receptor complex are complicated, with both association and dissociation phases demonstrating a biphasic pattern. The fraction of [³H]cis-methyldioxalane with the more rapid dissociation rate increased in the presence of guanine nucleotides, with a half-maximal effect at 10^{-6} M Gpp(NH)p. These findings are consistent with the view that muscarinic receptor exists in at least two forms, and that both Na^+ and guanine nucleotides mediate the interconversion of the high affinity receptor to a low-affinity form. The data further support the view that agonist persistently bound to the high-affinity receptor is released during guanine nucleotide-mediated conversion to the low-affinity form.

B. Developmental Aspects of Muscarinic Responsiveness — Evidence for Interaction of GTP with Agonist-High Affinity Receptor Complex in Mediation of Physiologic Response

Developmental studies of physiologic responsiveness, and the cellular and subcellular changes associated with development, serve as useful tools to elucidate mechanisms of receptor function. Earlier, we reviewed the evidence that before ingrowth of the vagus nerve, the response of the embryonic chick heart to muscarinic agonists is markedly reduced compared to the state that exists after vagal innervation. Muscarinic agonists had a very limited effect on either beating rate or K^+ permeability in cells cultured from hearts prior to vagal innervation (days 3 and 4 *in ovo*).[74] At 10 days *in ovo*, however, high concentrations of carbamylcholine produced a 33% increase in the rate of $^{42}K^+$ efflux and a 15% decrease in beating rate of cultured heart cells.[20] These data are summarized in Table 1 and Table 2.

Since radiolabeled antagonist binding revealed no difference in the number of muscarinic receptors in chick hearts 2.5 to 18 days *in ovo*, we performed additional experiments to examine the affinity state of receptors from hearts 3.5 days *in ovo* (see Figure 5a). Studies of carbamylcholine competition with [³H]QNB binding to homogenates of cells cultured from hearts 3 days *in ovo* demonstrated equal numbers of high and low affinity receptors for carbamylcholine. However, guanine nucleotides had no effect on the affinity of these receptors for agonist. Interestingly, we found two lots of horse serum that induced the development of both a beating rate response and a K^+ permeability response to muscarinic stimulation in cells cultured from hearts 3 days *in ovo* (see Table 2). Furthermore, homogenates of cultured cells from these young embryonic hearts contained a 1.5-fold higher amount of high affinity receptors than was apparent in control cells, and demonstrated conversion of R_H to R_L in the presence of guanine nucleotides[74] (Figure 5b). The coexistence of the induction of high-affinity receptors and responsiveness to guanine nucleotides by these lots of serum with the development of a physiologic response to muscarinic stimulation suggests that the high-affinity receptor-guanine nucleotide interaction is important in the mediation of the physiologic response. We postulated that prior to vagal innervation of the heart, a factor (possibly a guanine nucleotide regulatory protein) responsible for coupling receptor occupancy to physiologic response was not present, or was present in an inactive form.

Table 1
EFFECT OF CARBAMYLCHOLINE
ON BEATING RATE IN HEART
CELL CULTURES FROM CHICK
EMBRYOS $3\frac{1}{2}$ and 10 days *IN OVO*

Beating rate/min ($\pm$ S.E.)

Days *in ovo*[a]	Control	Carbamylcholine $(10^{-3}M)$
10 (n = 20)	140 $\pm$ 5	118 $\pm$ 7
$3\frac{1}{2}$ (n = 6)	130 $\pm$ 8	137 $\pm$ 6

Note: Cells grown on glass coverslips were perfused in Sykes Moore chambers, and the beating rate was determined. For each determination, cells were first perfused with growth medium. After establishment of a stable baseline, perfusion was switched to medium $10^{-3}M$ in cholinergic agonist for 5 min and the effect on beating rate determined. All observations in a given series were performed on the same group of cells in the microscopic field.

[a]   Age of the embryo from which hearts were taken.

C. Inhibitory Guanine Nucleotide Regulatory Protein

A protein initially referred to as islet-activating protein (IAP) has proven to be a useful tool in examining the mechanisms underlying muscarinic inhibition of adenylate cyclase. Substantial data now supports the view that this protein isolated from cultures of *Bordetella pertussis* interacts with an inhibitory guanine nucleotide binding protein commonly referred to as G_i or N_i. This islet-activating protein was originally shown to potentiate the effects of various insulin secretagogues.[75] Hazeki and Ui[76] demonstrated that levels of cAMP in the presence of isoproterenol in heart cells were significantly higher in the presence of islet-activating protein. Stimulation of cultures from adult rat heart with muscarinic agonists in the presence of isoproterenol resulted in a marked inhibition of cAMP accumulation. Muscarinic agonists had no effect, however, on cAMP levels in the presence of isoproterenol in cells pretreated with islet-activating protein, indicating that this substance interferes with the action of agonists that inhibit adenylate cyclase. Katada and Ui[77] extended these studies to show that IAP enhanced the effect of GTP or GTP plus isoproterenol on cAMP production in membranes from cultured neural cells. Nicotinamide adenine dinucleotide (NAD) and ATP were required for this response, and the reaction was shown to involve the transfer of an ADP ribose moiety from NAD to a 41,000 dalton protein. The parallel between ADP ribosylation of the IAP substrate clearly bears a striking resemblance to the ADP ribosylation of the 42,000 dalton cholera toxin substrate that is known to be the stimulatory guanine nucleotide regulatory subunit (G_s) that is essential in the activation of adenylate cyclase by stimulatory hormones such as catecholamines.

Recently, G_i has been shown to be composed of at least three subunits designated alpha, beta, and gamma.[78-81] Alpha$_i$ has been shown to be a 41,000 dalton protein that binds guanine nucleotides and is the substrate for IAP. The beta subunit is a 35,000 dalton protein that has been shown to inhibit G_s-stimulated adenylate cyclase activity and appears to be identical to the 35,000 dalton beta subunit of G_s. The gamma subunit has been identified as a 5,000 dalton protein occurring in association with G_s and G_i.[81] Based on the finding that binding

Table 2
EFFLUX RATE OF $^{42}K^+$ FROM CELLS
GROWN IN CONTROL MEDIUM AND
MEDIUM SUPPLEMENTED WITH
HORSE SERUM 9938

	$t_{1/2}$ of $^{42}K^+$ efflux ($\pm$ S.E., n = 7)	
Days *in ovo*[a]	**Control**	**Carbamylcholine ($10^{-3}M$)**
10	13.2 ± 0.2	8.8 ± 0.3
$3_{1/2}$ (control serum)	9.8 ± 0.4[b]	8.7 ± 0.3
$3_{1/2}$ (serum 9938)	10.5 ± 0.2[b]	7.5 ± 0.4

Note: Experiments were carried out as described in Figure 3. After labeling to equilibrium (3 hr) with $^{42}K^+$, cells were incubated for the indicated times in the continuing presence of $5\mu Ci/m\ell$ of $^{42}K^+$ with or without $10^{-3}M$ carbamylcholine. $^{42}K^+$ efflux was measured as described in Figure 3. Half-times for efflux were derived from slopes of regression lines obtained from the efflux data as described in Figure 3. The correlation coefficients for these regression lines were at least 0.99. The data for each curve represent the mean of seven replicate determinations. Values for Control and Carbamylcholine are in minutes.

[a] Age of the embryo from which hearts were taken for culture.

[b] The differences of control values for $t_{1/2}$ of efflux for cultures of hearts $3^{1}/_{2}$ days *in ovo* grown in control serum or in serum 9938 are significant within 98% confidence limits. The difference between the control rate of efflux and the rate of efflux in the presence of $10^{-3}\ M$ carbamylcholine is significant within 99% confidence limits.

of beta subunit to G_s results in inhibition of cyclase activity, a model has been proposed for inhibition of adenylate cyclase. In this model, binding of inhibitory agonists (such as acetylcholine) in the presence of guanine nucleotide results in release of the beta subunit of G_i and subsequent binding of this subunit to G_s with consequent inhibition of adenylate cyclase activity.[78-81]

D. A Model of Agonist-Receptor-Guanine Nucleotide Regulatory Protein Interaction

The data cited above suggest the following model for the interaction of muscarinic agonist, high-affinity receptor, and guanine nucleotide regulatory protein. This parallels the scheme presented by DeLean et al.[82] for the β-adrenergic receptor.

$$\text{step 1} \qquad \text{step 2} \qquad \text{step 3}$$

$$R_HG_i + A \rightarrow R_HG_iA + GTP \rightarrow (R_HG_i \cdot GTP) + A \rightarrow R_L + (G_i \cdot GTP)$$

In this scheme, R_H is a high-affinity receptor, R_L is a low-affinity receptor, G_i is an inhibitory guanine nucleotide regulatory protein, and A is the agonist. Gpp(NH)p, a nonhydrolyzable analogue of GTP, is often substituted for GTP in experimental usage. Only

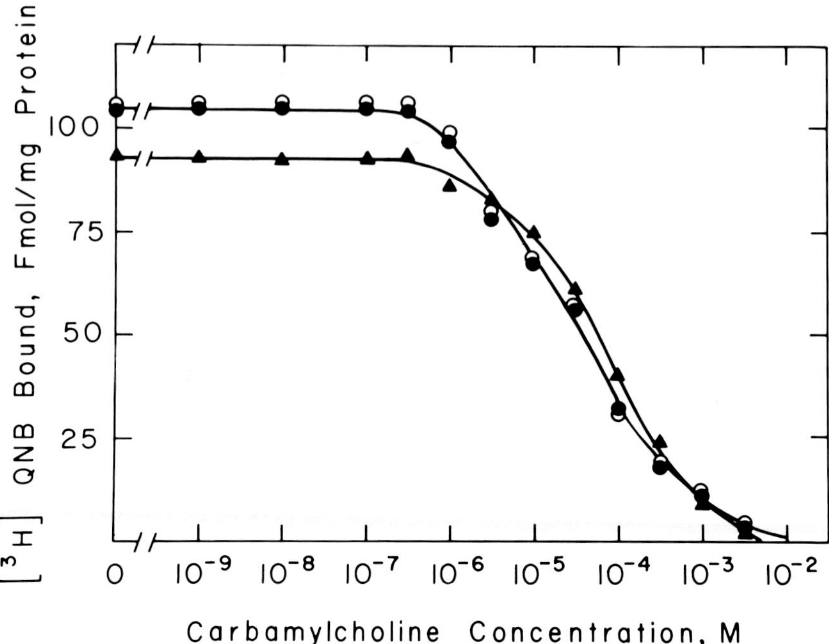

FIGURE 5. (A) Effect of carbamylcholine and guanine nucleotides on apparent receptor affinity for agonist. Replicate culture dishes of heart cells from embryos $3\frac{1}{2}$ days *in ovo* were labeled overnight with 0.01 Ci/mℓ of [U-^{14}C] leucine. Cells were washed and incubated with ($\blacktriangle$) and without ($\bullet$) $10^{-3}M$ carbamylcholine for 15 min; cells were harvested, homogenized and incubated for 1 hr rat room temperature in medium containing 1 nM[^{3}H]QNB at the concentrations of the agonist indicated, and [^{3}H]QNB binding determined. In a similar manner, aliquots of homogenates from control cells not previously exposed to agonist were incubated with $10^{-4}M$ Gpp(NH)p ($\bigcirc$) and [^{3}H]QNB binding determined in the presence of the indicated concentration of agonist. Each point represents the mean of three replicate determinations repeated 12 times for each group. The curves are drawn by eye. (B) Effect of carbamylcholine and guanine nucleotides on apparent affinity for agonist in cells cultured from hearts of embryos $3\frac{1}{2}$ days *in ovo* in medium supplemented with 2% horse serum, lot 9938. Cells were grown for 3 days in culture with medium M-199 supplemented with 4% fetal calf serum and 2% horse serum lot 9938. $\bigcirc$, control (n = 12); $\bullet$, control plus $10^{-4}\,M$ Gpp(NH)p (n = 12); $\blacktriangle$, after carbamylcholine pretreatment for 15 min (n = 12). (From Galper, J. B., Dziekan, L. C., and Smith, T. W., *J. Biol. Chem.*, 259, 7382, 1984. With permission.)

indirect evidence is available at present for the species in parentheses. According to this model, in the absence of GTP, agonist binds persistently to $R_H G_i$ to form $R_H G_i A$ (step 1), thus causing these R_H sites to be unavailable for binding of labeled antagonist. This explains why brief exposure to agonist results in the loss of 26% of [^{3}H]QNB binding sites. According to this scheme, binding of A and formation of $R_H G_i A$ permits interaction of $R_H G_i A$ with GTP, with release of agonist (step 2) and regeneration of the receptor in low-affinity form (step 3). These receptors would now be available for binding of labeled antagonist. Available data do not exclude the possibility that the conversion of R_H to R_L following the binding of GTP to $R_H G_i A$ (step 2 and step 3) constitutes a single-step process with immediate conversion of R_H to R_L and simultaneous dissociation of A, R_L, and $G_i \cdot$GTP from the complex.

By analogy with the β-adrenergic receptor, it is likely that a guanine nucleotide binding protein must be present in order for the muscarinic receptor to be influenced by guanine nucleotides. Hence, the absence of a significant effect of guanine nucleotides on the affinity of R_L for agonist indicates that G_i may only be associated with the high-affinity state of the receptor. It may be that conversion of R_H to R_L in the presence of guanine nucleotides is

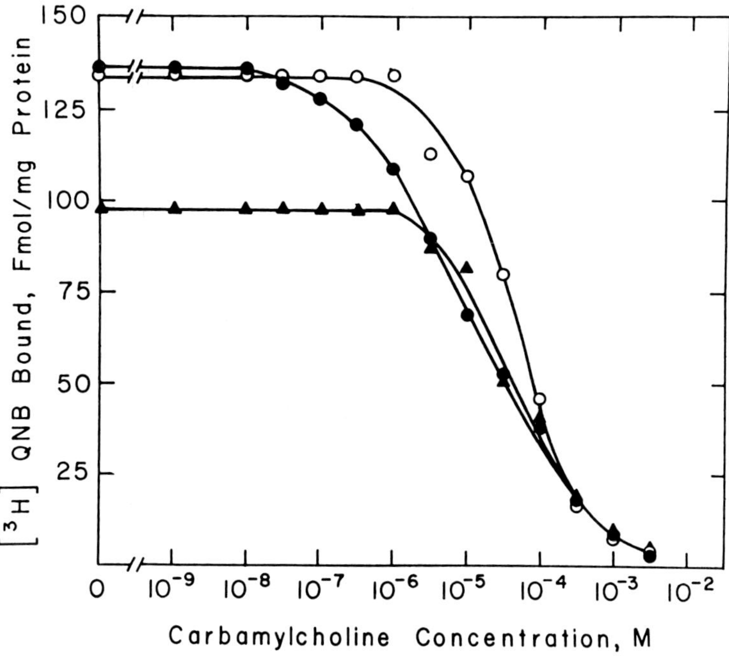

FIGURE 5B.

associated with the release of the guanine nucleotide regulatory protein (step 3). On the other hand, regeneration of R_H from R_L may be associated with binding of an inhibitory guanine nucleotide regulatory protein to the low-affinity form of the receptor. Thus, conversion of R_L to R_H might be as follows:

$$R_L + G_i \rightarrow R_H G_i$$

If interaction of R_L and G_i is required to generate R_H, then the absence of a functional inhibitory guanine nucleotide regulatory protein in heart cells cultured from embryos prior to vagal innervation could account for the inability to demonstrate R_H in these cells. Absence of G_i or the existence of G_i in an inactive form could also account for the inability of guanine nucleotides to alter affinity of muscarinic receptors for agonist in cells cultured from hearts of embryos prior to vagal innervation. Recent data from our laboratory suggest that levels of the α subunit of G_i increase during embryonic development of the chick heart parallel with increased coupling of the muscarinic receptor to aphysiological response.[50]

Previous work from our laboratory demonstrates that there is no difference in the response of adenylate cyclase to GTP and isoproterenol in homogenates from hearts in embryos taken prior to or following ingrowth of the vagus nerve.[51] Stimulatory guanine nucleotide regulatory protein must therefore be present and functionally coupled to beta receptor and adenylate cyclase both before and after vagal innervation. This leads to the speculation that the stimulatory guanine nucleotide regulatory protein (G_s) is present and active at a time prior to functional representation of the inhibitory guanine nucleotide regulatory protein, G_i.

We discussed previously the studies of Wantanabe et al.[46] demonstrating that muscarinic agonists not only result in adenylate cyclase inhibition in homogenates of myocardium in the presence of GTP, but also antagonize the guanine nucleotide-mediated decrease in affinity of β-adrenergic receptors for isoproterenol. Muscarinic agonists, therefore, appear to regulate both the affinity of beta-adrenergic receptors and the activity of adenylate cyclase by a GTP-dependent mechanism. In the scheme put forward above, the species $G_i \cdot$GTP might mediate adenylate cyclase inhibition and interfere with the GTP-stimulated decrease in the affinity of the β-adrenergic receptor for agonist. This would result in the effective uncoupling of G_s from both the β-adrenergic receptor and the catalytic unit of adenylate cyclase. Release of the beta subunit of G_i from the G_i·GTP complex would be consistent with the model discussed in the previous section for muscarinic inhibition of adenylate cyclase.

E. Muscarinic Stimulation of Protein Kinases and Phosphorylated Intermediates

As noted above, it seems unlikely that all of the physiologic effects of muscarinic agonists could result from the inhibitory effect on adenylate cyclase. Changes in cAMP levels have been shown to modulate Ca^{2+} permeability, for example, but have not been shown to affect K^+ permeability directly. It should be noted, however, that both Clusin[83] and Isenberg[84] have suggested that changes in Ca^{++} permeability might control levels of K^+ permeability. In any event, several studies have demonstrated that muscarinic agonists have a positive effect on the phosphorylation of specific cellular proteins, and these effects cannot be explained fully by muscarinic agonist-induced alteration in cAMP levels.

Studies of Hartzell and Titus[85] indicate that ^{32}P incorporation into a 165,000 dalton protein from frog atrium was increased by β-adrenergic agonists and decreased by muscarinic cholinergic agonists. Isoproterenol-induced changes in phosphrylation of this protein were correlated well with altered contractile state in terms of both concentration and time. In contrast, however, carbamylcholine decreased tension earlier and at lower concentrations than were required for reduced ^{32}P incorporation. Carbamylcholine had a direct inhibitory effect on cAMP function distal to its effect on adenylate cyclase activation and cAMP production, and it is possible that the muscarinic effect on cAMP production and the direct inhibitory effect on cAMP function both play a role in the decreased ^{32}P incorporation into the 165,000 dalton protein. This latter protein appears to be the C-protein component of the myofibril.

Although space constraints do not permit a detailed review, numerous cyclic GMP-dependent protein kinases are known to exist, and muscarinic agonists have been shown to stimulate at least some of these kinases. Thus, schemes for muscarinic regulation of myocardial function must take into account effects of cGMP on phosphorylation of myocardial proteins involved in Ca^{2+} movement, myofibrillar function, and K^+ permeability as well as indirect effects of muscarinic agonists on cAMP levels by inhibition of adenylate cyclase and direct inhibitory effects on cAMP activity.

F. Phosphatidylinositol Turnover: A Possible Mediator of Physiologic Response to Muscarinic Stimulation

It is known that muscarinic receptor activation increases phosphatidylinositol turnover in a number of tissues. As noted previously, this effect has been postulated to be responsible for changes in Ca^{2+} permeability. Carbamylcholine also stimulated phosphatidyl inositol turnover in dissociated embryonic chick heart cells. Using [³H]inositol to prelabel phosphatidylinositol, Brown and Brown[86] determined phosphatidylinositol turnover by measuring release of [³H]inositol-1-phosphate in the presence of lithium which inhibited dephosphorylation of the inositol. Using these methods they demonstrated a threefold carbamylcholine-stimulated increase in turnover of phosphatidylinositol (PI) in rat atria.[86]

In order to determine whether separate classes of muscarinic receptors were responsible for muscarinic stimulation of PI turnover and inhibition of adenylate cyclase activity,[87] Brown and Brown compared the time course and concentration-dependence of muscarinic agonist-mediated stimulation of PI turnover and inhibition of adenylate cyclase activity in suspensions of embryonic chick heart cells. They demonstrated that the time course of PI turnover lagged significantly behind the decrease in cAMP levels, and that the IC_{50} for a carbamylcholine-mediated decrease in cAMP levels was 100-fold less than that for the stimulation of PI hydrolysis. Furthermore, although both carbamylcholine and oxotremorine inhibited cAMP formation, oxotremorine had no significant effect on PI turnover.[86] These data suggest that these two physiologic responses to muscarinic stimulation may be coupled to separate classes of muscarinic receptors.

G. Differences in the Biochemical Response to Muscarinic Stimulation in Atrium and Ventricle

The marked difference between the physiologic response of the embryonic chick atrium and ventricle to muscarinic stimulation could be due to differences in: (1) muscarinic receptor number; (2) distribution of muscarinic receptors among high- and low-affinity states; (3) the ability of guanine nucleotides to mediate conversion of R_H to R_L; or (4) the ability to couple agonist binding to inhibition of adenylate cyclase.

1. Comparison of Receptor Number in Cultures of Atrium and Ventricle

Siegel and Fischbach[88] could not demonstrate a significant difference in the binding of [³H]QNB and [³H]methylscopolamine to myocytes from cultures of atria and ventricle from chick embryo hearts 8 days *in ovo*. However, microelectrode studies demonstrated that many more atrial myocytes (89%) became hyperpolarized in the presence of $10^{-4}M$ carbamylcholine than ventricular cells (26%). In studies designed to determine the effect of innervation on cultures of atrium and ventricle, they found that growth of cultures in the presence of brain extracts produced a 2.6-fold increase in the number of carbamylcholine-sensitive ventricular cells while [³H]methylscopolamine binding sites remained essentially constant.[88] In recent studies in our laboratory, we have demonstrated that [³H]QNB binding was 15 to 20% higher in homogenates of atria of hearts 14 days *in ovo* than in homogenates of ventricles. However, studies of the binding of [³H]cis-methyldiozalane to heart homogenates demonstrated that there were 1.5-fold more high affinity muscarinic receptors in atria than in ventricle, and that atrial high-affinity receptors were tenfold more sensitive to guanine nucleotide conversion to a low-affinity form than those in the ventricle.

2. Comparision of Sensitivity of Adenylate Cyclase in Atrium and Ventricle to Muscarinic Inhibition

Finally, the ability of a given concentration of the muscarinic agonist carbamylcholine to inhibit isoproterenol-stimulated adenylate cyclase activity was fourfold higher in ventricle than in atrium. Hence, as discussed previously, muscarinic stimulation of atrial cells appears to be associated with both changes in K^+ permeability and inhibition of adenylate cyclase activity. However, in ventricular cells, inhibition of β-adrenergic agonist-stimulated adenylate cyclase activity may be the primary mechanism of muscarinic function. Since muscarinic stimulation of PI turnover appears to be independent of muscarinic inhibition of adenylate cyclase activity,[87] these studies suggest that at least three separate responses to muscarinic stimulation can be detected in the heart. The presence of increased numbers of R_H and increased responsiveness of R_H to GTP in atrial tissue further supports the hypothesis that different affinity states (or different gene product populations) of the muscarinic receptor may be coupled to specific physiologic functions.

V. REGULATION OF MUSCARINIC RESPONSIVENESS BY AGONISTS

A general property of cells sensitive to hormonal and neurotransmitter influences is the ability to respond to changes in agonist concentration by regulating the number or affinity of cell surface receptors.[89] Such changes in receptor number or affinity frequently result in changes in the physiologic responsiveness of the cell to hormone or neurotransmitter stimulation.[7] In the cultured chick heart cell preparation, we have noted a decrease in the ability

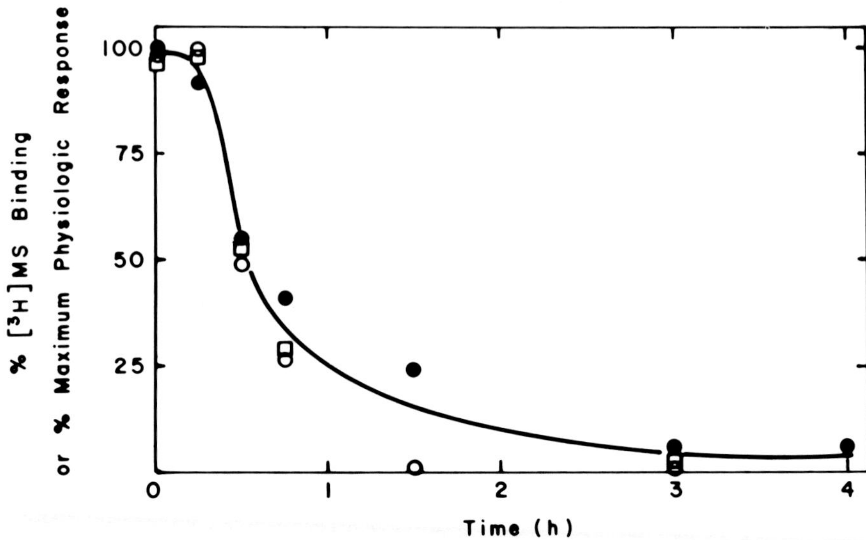

FIGURE 6.   Correlation between the time course of agonist-mediated receptor loss and the agonist-mediated decrease in K^+ permeability and beating rate response to muscarinic agonist. The fraction of [³H]MS binding sites subject to agonist control remaining at each time (●) was derived by subtracting the subclass of receptors not subject to agonist control (49 ± 6 fmol/mg protein) from total [³H]MS bound for each point and determining the percent of maximum [³H]MS binding remaining at each time point. The percent of maximal K^+ efflux response (■) and percent maximal beating rate response (○) at each time are shown. (From Galper, J. B., Dziekan, L. C., Miura, D. S., and Smith, T. W., *J. Gen. Physiol.*, 80, 231, 1982. With permission.)

of carbamylcholine to decrease beating rate following preexposure of cells to muscarinic agonists. While no effect of agonist exposure was apparent during the first 12 min, the beating rate response to carbamylcholine gradually disappeared and was completely absent after a 90 min preexposure, with a loss of half of the response after an exposure time of 30 min.[20]

Electrophysiologic techniques have been useful in studies if muscarinic responsiveness of intact cardiac tissue. Prolonged carbamylcholine exposure results in a decay in the membrane hyperpolarization and K^+ conductance responses.[90,91] Evidence available from these and other studies indicates that it is unlikely that alterations in K^+ distribution during muscarinic stimulation can explain the attenuation of the electrophysiologic response to a subsequent application of muscarinic agonist. Evidence from our own and other laboratories supports the view that the decreased response following prolonged agonist exposure is related to alteration in receptor number or affinity for agonist.

A. Ion Flux Studies

As noted previously, altered $^{42}K^+$ efflux in response to muscarinic agonist exposure probably reflects altered K^+ permeability in chick heart cell cultures. We have examined the effect of various durations of exposure to carbamylcholine on the ability of subsequent exposure to increase the $^{42}K^+$ efflux rate. The extent of increase in $^{42}K^+$ efflux decreased by 50% after a 30 min carbamylcholine exposure, and the typical muscarinic response disappeared completely after a 90 min preexposure. The time course of this loss of K^+ efflux effect was closely related to the loss of ability of carbamylcholine to decreased beating rate in cells preexposed to carbamylcholine.[20] These findings are summarized in Figure 6. Cells preexposed to carbamylcholine for 3 hr, then washed and incubated in fresh medium, became fully responsive again to carbamylcholine after 12 hr. This response was blocked by the protein synthesis inhibitor cycloheximide.

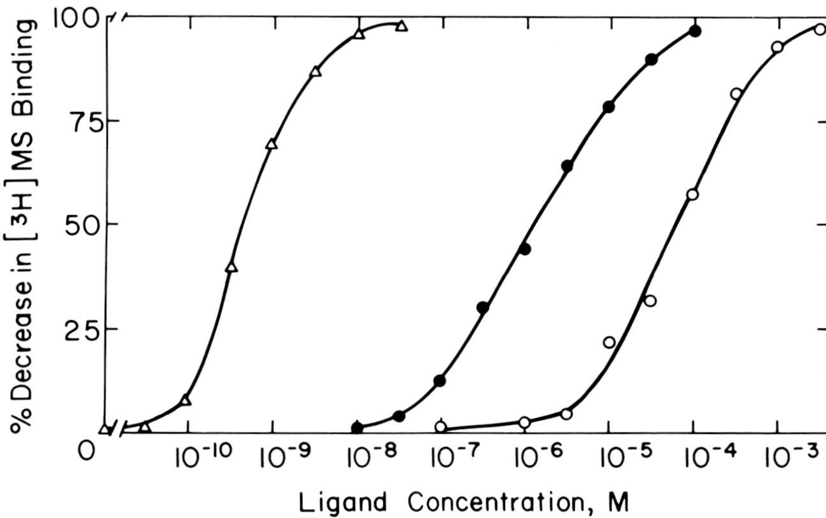

FIGURE 7.   Inhibition of [³H]MS binding to intact cultured heart cells by muscarinic agonists and antagonists. Replicate cultures grown in multiwell dishes were labeled overnight with [U-¹⁴C]leucine, washed, and incubated at 37°C with 2 n*M*[³H]MS plus the indicated concentrations of muscarinic ligand. Each point represents the mean of four replicate determinations. Hill coefficients were derived by the method of Brown and Hill. △, unlabeled methylscopolamine; ●, oxotremorine; ○, carbamylcholine. (From Galper, J. B., Dziekan, L. C., O'Hara, D. S., and Smith, T. W., *J. Biol. Chem.*, 257, 10344, 1982. With permission.)

B. Agonist-Induced Changes in Receptor Number

Cultured chick heart cells respond to a 3-hr exposure to muscarinic agonist with a decrease of up to 70% in the number of muscarinic receptor sites as judged by [³H]QNB binding to homogenates of heart cell cultures. The remaining receptors show an unchanged affinity for [³H]QNB, indicating that the reduced number of receptors reflected a loss of binding sites rather than a change in affinity for the antagonist.

Physiologic response to muscarinic stimulation in the heart requires agonist interaction with receptors on the cell surface. Thus, numbers of receptors measured in homogenates of cells may not reflect physiologic receptors in the intact cell available for agonist binding and modulation of physiologic state. We therefore developed an assay for the binding of the muscarinic antagonist [³H]methylscopolamine ([³H]MS) to intact cultured heart cells. Binding of [³H]MS was saturable and of high affinity; the relative potency of muscarinic ligands to compete with [³H]MS for binding to intact cells followed the order expected for binding to a muscarinic receptor,[65] as illustrated in Figure 7.

The cultured heart cell preparation provided a suitable model system for the study of changes in the number of [³H]MS binding sites in the intact cell after preexposure to agonist for various times. A brief lag phase occurred, after which 72% of [³H]MS binding sites disappeared during a 6 hr agonist exposure.[20] Half of these receptors were lost over the initial 30 min, and binding sites recovered over a 12-hr incubation period in the presence of fresh media. This recovery was inhibited by the protein synthesis inhibitor cycloheximide. The time course of agonist-induced decrease in receptor number demonstrated two subclasses of [³H]MS binding sites. After 70% of the receptors were lost, a residual subclass of 30% of receptors continued to bind [³H]MS after periods of agonist exposure up to 6 hr. This residual 30% of sites, however, was not sufficient to mediate a physiologic response to agonist binding since a 3-hr exposure to agonist removed essentially all physiologic responsiveness (see Figure 8). Thus, 70% of the muscarinic receptor sites in the cultured heart cell appeared to be subject to agonist control, and this same population appears to be critical

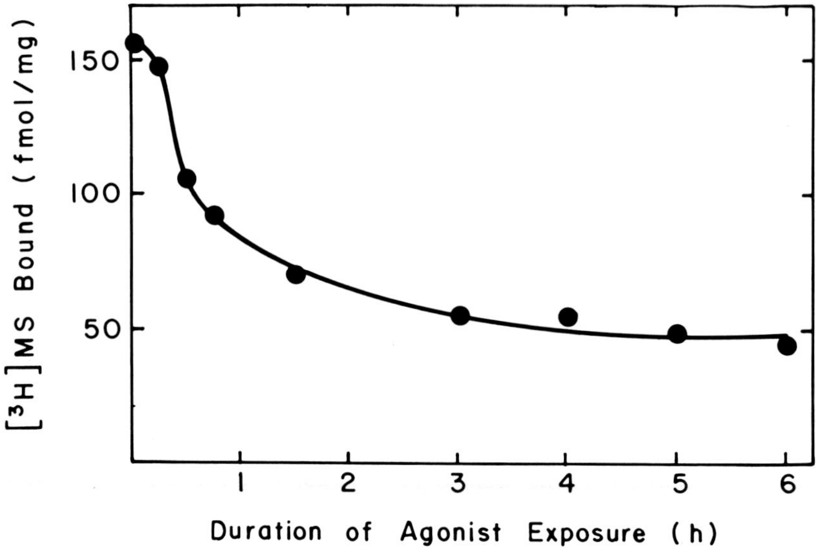

FIGURE 8.   Time dependence of agonist-induced receptor loss measured by the binding of [³H]MS to intact cells. Replicate cultures in multiwell dishes were incubated overnight in medium containing 0.01 μCi/mℓ leucine [¹⁴C (U)]. Growth medium was replaced by fresh medium containing $10^{-3}M$ carbamylcholine incubated for the times indicated, rinsed, and incubated for 1 hr with 2 nM[³H]MS, and washed. Each point represents the mean of four replicate determinations repeated six times and is corrected for nonspecific binding. (From Galper, J. B., Dziekan, L. C., Miura, D. S., and Smith, T. W., *J. Gen. Physiol.*, 80, 231, 1982. With permission.)

for the mediation of a physiologic response. A close correlation existed between the time course of the agonist-mediated decrease in binding of [³H]MS and the time course of the agonist-induced decrease in K^+ permeability and beating rate responses. All three of these parameters showed a 15-min lag period followed by decreases that were maximal after 6 hr and half-maximal at about 30 min[20] (see Figure 6). This close relationship between the number of receptors evident in the intact cell and the physiologic responsiveness of the cell suggests that agonist modulation of receptor number constitutes a sensitive mechanism for the control of muscarinic responsiveness of the cultured heart cell.

The relative potency of muscarinic agonists to mediate downregulation of the muscarinic receptor closely parallels their pharmacologic potency. These findings differ from those of Nathanson et al.,[92] who studied the relative potency of various muscarinic agonists to mediate downregulation as measured by the loss of [³H]QNB binding sites in heart cell cultures. These authors demonstrated that carbamylcholine was more potent than oxotremorine in mediating receptor loss. Since oxotremorine has been shown to be a more potent agonist than carbamylcholine as measured by competitive binding with [³H]QNB, they concluded that oxotremorine is a partial agonist.

C. Muscarinic Receptor Processing During Agonist-Induced Receptor Loss

In several well-studied systems, binding of a physiologic ligand to a receptor results in a process known as receptor-mediated endocytosis. Agents including colchicine and vinblastine that inhibit microtubule function interfere with the process of endocytosis. As summarized in Figure 9, we found that intact cultured heart cells preexposed to muscarinic agonists in the presence of colchicine and vinblastine showed a marked decrease in the loss of receptors compared to cells exposed to agonist alone.[93] These findings are consistent with the hypothesis that agonist exposure causes a change in the configuration of the receptor in

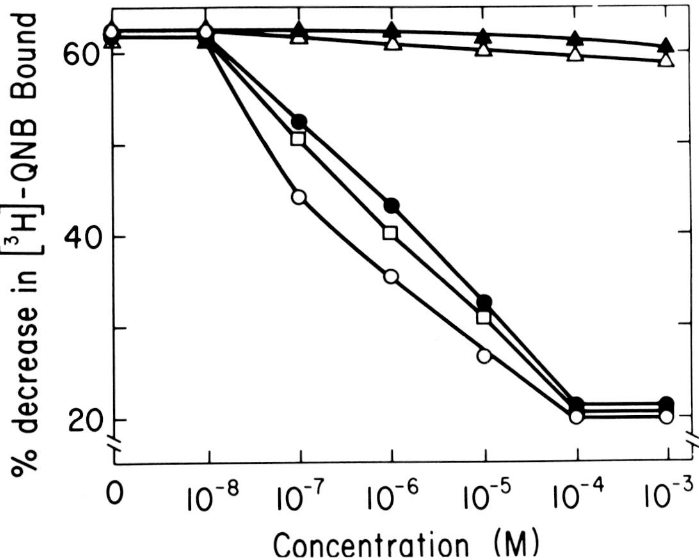

FIGURE 9.   Effect of inhibitors of microtubule and microfilament function on agonist modulaton of receptor number. Groups of three replicate cultures were preincubated with the indicated concentration of agent for 2 hr. Fresh medium containing $10^{-3}M$ carbamylcholine and the indicated concentration of each agent was then added. After further incubation for 3 hr, cells were harvested and [³H]QNB binding to homogenates determined. Each point is the mean of three replicate determinations. ▲, cytochalasin B; △, lumicolchicine; ●, cholchicine; □, podophyllotoxin; ○, vinblastine. (From Galper, J. B. and Smith, T. W., *J. Biol. Chem.*, 255, 9571, 1980. With permission.)

the cell membrane, or some other alteration in receptor state, which renders the receptor inaccessible to ligand binding and subject to endocytosis.

1. Evidence for an Intermediate State of the Receptor in the Membrane after Agonist Binding but Before Receptor Degradation

As noted above, agonist exposure, especially when prolonged, may result in an altered receptor configuration that no longer permits productive association of agonist with the receptor. Although such an altered form of the receptor may not yield a physiologic response upon agonist binding, it remains possible that a hydrophobic radioligand might continue to bind specifically to such an altered form of the receptor. To test this hypothesis, we compared the binding of [³H]MS and the more hydrophobic antagonist [³H]QNB to cells subjected to prior agonist exposure for graded lengths of time. The results of these experiments, summarized in Figure 10, show that [³H]QNB continued to bind specifically to muscarinic receptors for 90 min after [³H]MS binding had essentially disappeared. Analysis of these data suggests a sequential conversion of the receptor from a form *A* that gave physiologic response to agonist and bound both [³H]QNB and [³H]MS to form *B* that bound only [³H]QNB and did not give a physiologic response to agonist. The next step in processing involves a form *C*, presumably a degraded form of the receptor, that was able to bind neither [³H]MS nor [³H]QNB. Thus, *B* may constitute an intermediate state of the receptor formed during endocytosis that is more intimately associated with the cell membrane, but is inaccessible to agonist and has not yet progressed to a degraded state.

D. Recovery of Muscarinic Receptors Following Agonist Preexposure

Recently, Hunter and Nathanson[94] studied the recovery of muscarinic responsiveness in

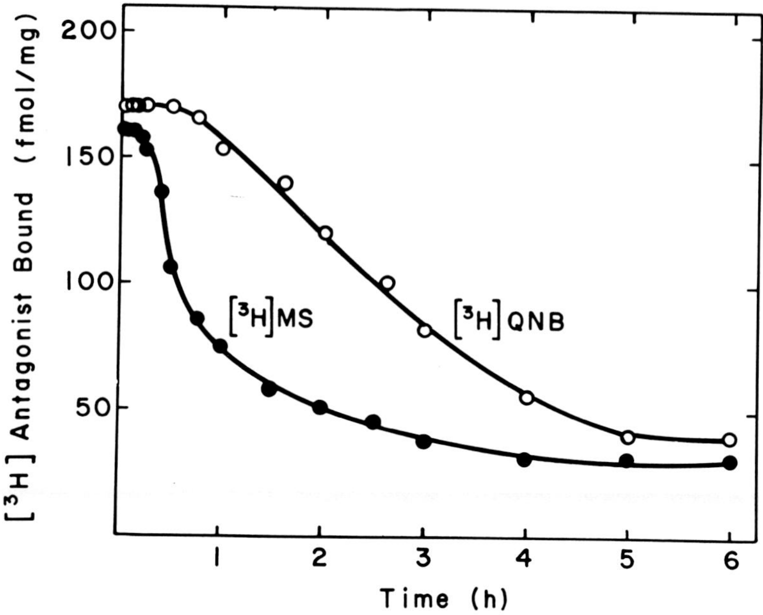

FIGURE 10.   Time course of agonist-induced receptor loss in intact cells measured by the binding of [³H]MS or [³H]QNB. (A) cells from replicate cultures grown in multiwell dishes were grown overnight in medium containing 0.10 Ci/mℓ of [U-¹⁴C] leucine. Cells were washed and incubated for the indicated times at 37°C in HEPES-buffered M-199 containing $10^{-3}M$ carbamylcholine, rinsed with 3 mℓ of fresh medium, and incubated with medium 1 nM in [³H]QNB or 2 nM[³H]MS for 1 hr at 37°C. Cells were then washed 3 × with 3 mℓ each of wash solution and specific ³H-antagonist bound determined. Each point represents the mean of six sets of four replicate determinations each. Nonspecific binding was determined independently as the mean of four replicate determinations for each point. Solid lines are drawn by eye. ○, [³H]QNB bound; ●, [³H]MS bound. (B) derived from Figure 10A. ●(A), residual [³H]MS binding sites plotted as a function of duration of agonist exposure; ○ (B), the number of receptors at a given time t that bind [³H]QNB but not [³H]MS, derived by subtracting A, the residual [³H]MS binding sites at times t, from the total sites that bind [³H]QNB at the same time; □ (C), the number of receptor sites that no longer bind [³H]QNB, derived by subtracting the total number of [³H]QNB binding sites at time t (Figure 10A) from total [³H]QNB binding sites present at time 0. The residual ligand binding after 6-hr exposure to agonist has been subtracted from each curve. Solid lines are drawn by eye. (From Galper, J. B., Dziekan, L. C., O'hara, D. S., and Smith, T. W., *J. Biol. Chem.*, 257, 10344, 1982. With permission.)

atria of chick embryos exposed for 8 hr to carbamylcholine. Comparison of the time course of recovery of [³H]QNB binding sites in atrial homogenates and the recovery of sensitivity of beating rate to carbamylcholine demonstrated that while [³H]QNB binding reached control levels 20 hr after washout of carbamylcholine, the physiologic response as measured by the IC$_{50}$ for inhibition of beating continued to recover for up to 36 hr. Using a competitive binding assay between [³H]QNB and various concentrations of carbamylcholine as a measure of affinity for agonist, they could not detect any significant difference in receptor affinity for agonist or in responsiveness of receptor affinity to guanine nucleotides in homogenates of control atria and atria that had recovered for 20 hr following 8 hr of agonist exposure. They did, however, demonstrate an 11-fold decrease in the sensitivity of isoproterenol-stimulated adenylate cyclase activity to inhibition by carbamylcholine in atria following a 20 hr recovery from agonist exposure. These data suggest that coupling of muscarinic receptors to a physiologic response is part of the recovery process following agonist-induced

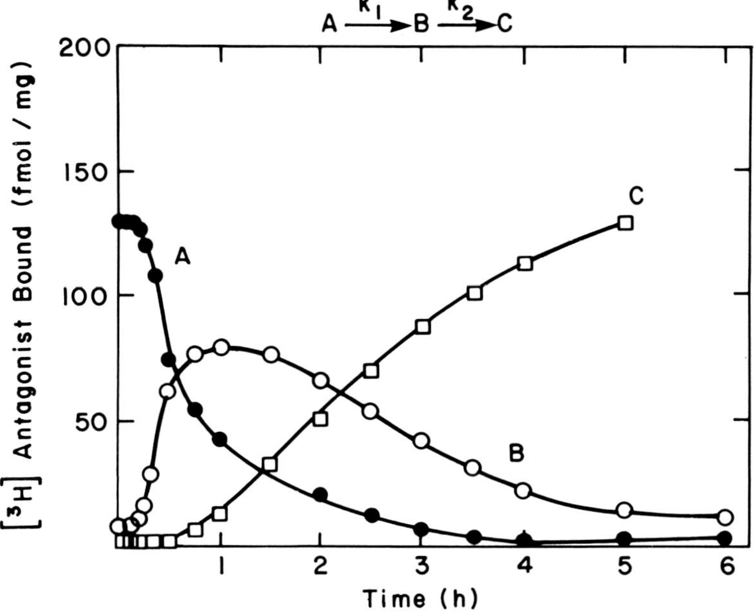

FIGURE 10B.

loss of receptor function. Furthermore, the coupling of receptors to physiologic function lags behind the reappearance of muscarinic receptors. Comparison of these findings with the apparent absence of coupling of muscarinic receptors to a physiologic response in the embryonic chick heart prior to ingrowth of the vagus nerve (see prior discussion) suggests an interesting parallel between recovery from downregulation and embryonic development of a muscarinic response. Hence, coupling factors and not receptor number appear to play a critical role in the appearance of a physiologic response to muscarinic stimulation during both embryonic development and recovery from receptor downregulation.

VI. THE INTERACTION OF MUSCARINIC RECEPTORS AND K⁺ CHANNELS: THE ROLE OF GUANINE NUCLEOTIDE REGULATORY PROTEINS

Recently, whole cell recordings from embryonic chick heart atrial cells in culture have demonstrated that acetylcholine and/or GTP stimulated K^+ current could be inhibited by pretreatment of cells with pertussis toxin.[95] Reconstitution of the patches from these cells with purified β subunits of G_i resulted in the GTP independent and/or pertussis toxin insensitive activation of this same K^+ current. This activation of the muscarinic induced K^+ current was not due to the presence of contaminating α_i in the preparation of β. These data suggest that activation of the muscarinic receptor might directly activate a K^+ channel by releasing the β subunit of α_i.[97] However, work from a second group suggested that a specific α protein, "α_{40}" with molecular weight of 40,000 was capable of reconstituting the stimulation of muscarinic K^+ current in these patches.[97] The resolution of these two observations is pending.

VII. SUMMARY AND CONCLUSIONS

Vagal stimulation leads to the release of acetylcholine and stimulation of muscarinic cholinergic receptors in the heart. In the atrium, this leads to hyperpolarization of the membrane potential and shortening of action potential duration. Both of these phenomena are caused by an increase in K^+ permeability and the development of a time independent inwardly rectifying outward current carried by K^+ ion. Decreased contractility in response to acetylcholine is related at least in part to a decrease in the slow inward Ca^{2+} current, due either to direct inhibition of I_{si} or indirectly by decreasing action potential duration and hence the time during which Ca^{2+} can flow into the cell.

Pathways by which binding of muscarinic agonist to its receptor alter K^+ permeability and I_{si} are complex and incompletely understood, but may involve interaction of the K^+ channel with the β subunit of G_i or an α subunit of G_i.

Cultured heart cells have been particularly useful in studies of the development of muscarinic responsiveness and hence for studies of the factors involved in a physiologic response to muscarinic stimulation. Muscarinic receptors in the intact heart and heart cell cultures exist in both high- and low-affinity states. After agonist exposure, guanine nucleotides mediate the conversion of high-affinity receptors (R_H) to a low-affinity form (R_L), associated with the release of bound agonist. It appears that both high affinity receptors and guanine nucleotide sensitivity must be present to permit a physiologic muscarinic response. A model is presented in which binding of muscarinic agonist to R_H stimulates binding of GTP to an agonist-receptor complex followed by the conversion of R_H to R_L with release of agonist. A receptor-inhibitory guanine nucleotide binding protein complex is postulated to be involved.

Binding of muscarinic agonists to the receptor is coupled to a GTP-dependent adenylate cyclase inhibition via the inhibitory guanine nucleotide regulatory protein G_i. Compelling evidence indicates that G_i is independent of the stimulatory guanine nucleotide regulatory protein G_s that couples β-adrenergic agonist binding to GTP-dependent stimulation of adenylate cyclase. Binding of muscarinic agonists to their receptors also interferes with GTP-stimulated decrease in affinity of the β-adrenergic receptor for β-adrenergic agonists. Thus, inhibition of adenylate cyclase appears to involve interference with, or uncoupling of, β-adrenergic receptor interactions with G_s as well as uncoupling of the interaction of G_s with the catalytic subunit of adenylate cyclase.

Adenylate cyclase inhibiton is an important mechanism by which muscarinic stimulation mediates functional changes in cardiac cells. Resulting decreases in cAMP levels are associated with decreased phosphorylation of specific proteins that appear to be related to control of membrane permeability and contractile state. It is likely that muscarinic stimulation of cyclic GMP (cGMP) formation as well as altered turnover of phosphatidylinositol are also involved in the muscarinic response.

Studies in the cultured heart cell preparation indicate that the number of muscarinic receptors available at the cell surface is critical in the control of muscarinic responsiveness. Agonist exposure leads to decreasing numbers of receptors that closely parallels the decrease in physiologic responsiveness with respect to both agonist concentration and duration of exposure. Altered receptor number appears to involve a change in the configuration of the receptor within the membrane, followed by endocytosis and ultimate degradation of the receptor. Figure 11 summarizes the receptor populations defined in our studies of cultured chick embryo heart cells. In the cultured heart cell, ''spare'' receptors do not appear to be available and alteration in the numbers of receptors in the sarcolemmal membrane therefore constitutes a sensitive mechanism for modulation of muscarinic responsiveness, and hence the balance between sympathetic and parasympathetic tone in the heart.

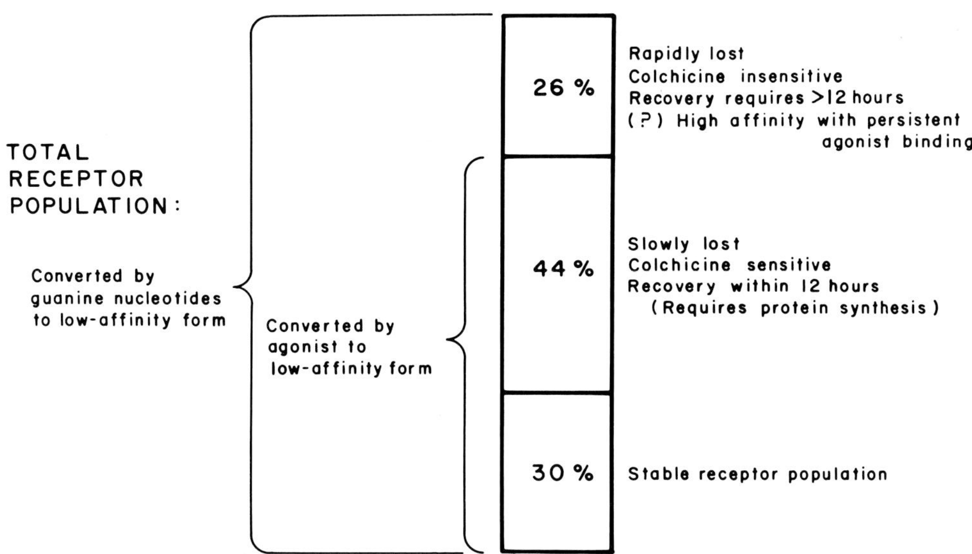

FIGURE 11.   Schematic representation of response of muscarinic cholinergic receptor population to agonist, guanine nucleotide, or colchicine exposure. (From Galper, J. B. and Smith, T. W., *J. Biol. Chem.*, 255, 9571, 1980. With permission.)

REFERENCES

1. **Levy, M. N.**, Sympathetic-parasympathetic interactions in the heart, *Circ. Res.*, 29, 437, 1971.
2. **Hutter, O. F.**, in *Nervous Inhibition*, Florey, E., Ed., Pergamon Press, Elmsford, N.Y., 1961, 114.
3. **Ten Eick, R., Nawrath, H., McDonald, T. F., and Trautwein, W.**, On the mechanism of the negative inotropic effect of acetylcholine, *Pflugers Arch.*, 361, 207, 1976.
4. **George, W. J., Polson, J. B., O'Toole, A. G., and Goldberg, N. D.**, Elevation of guanosine 3'-5'-cyclic phosphate in rat heart after perfusion with acetyl-choline, *Proc. Natl. Acad. Sci. USA*, 66, 398, 1970.
5. **Michell, R. H., Jafferji, S. S., and Jones, L. M.**, in *Advances in Experimental Medicine and Biology*, Vol. 83, Bazan, H., Brenner, R., and Giusto, H., Eds., Plenum Press, New York, 1976, 447.
6. **Lindemann, J. P., Jones, L. R., and Watanabe, A. M.**, Effect of muscarinic cholinergic agonist on beta receptor induced increases in membrane protein phosphorylation in intact ventricles, *Clin. Res.*, 28, 471A, 1980.
7. **Catt, K. J., Harwood, J. P., Aguilera, A., and Dufau, M. L.**, Hormonal regulation of peptide receptors and target cell responses, *Nature*, 280, 109, 1979.
8. **Pappano, A. J.**, Onset of chronotropic effects of nicotinic drugs and tyramine on the sino-atrial pacemaker in chick embryo heart: relationship to the development of autonomic neuroeffector transmission, *J. Pharmacol. Exp. Ther.*, 196, 76, 1976.
9. **Pappano, A. J.**, Ontogenetic development of autonomic neuroeffector transmission and transmitter reactivity in embryonic and fetal hearts, *Pharmacol. Rev.*, 29, 3, 1977.
10. **Abel, W.**, Further observations on the development of the sympathetic nervous system in the chick, *J. Anat. Physiol.*, 47, 35, 1912.
11. **Gardner, E. and O'Rahilly, R.**, The nerve supply and conducting system of the human heart at the end of the embryonic period proper, *J. Anat.*, 121, 571, 1976.
12. **Pappano, A. J. and Loffelholz, K.**, Ontogenesis of adrenergic and cholinergic neuroeffector transmission in chick embryo heart, *J. Pharmacol. Exp. Ther.*, 191, 468, 1974.
13. **Pappano, A. J. and Skowronek, C. A.**, Reactivity of chick embryo hearts to cholinergic agonists during ontogenesis: decline in desensitization at the onset of cholinergic transmission, *J. Pharmacol. Exp. Ther.*, 191, 109, 1974.
14. **DuFour, J. J. and Pasternak, J. M.**, Effects chronotropes de l'acetylcholine sur le coeur de l' embryon du poulet, *Helv. Physiol. Pharmacol. Acta*, 18, 563, 1960.

15. **Galper, J. B., Klein, W., and Catterall, W. A.,** Muscarinic acetylcholine receptors in developing chick heart, *J. Biol. Chem.,* 252, 8692, 1977.

16. **Sperelakis, N. and Lehmkul, D.,** Insensitivity of cultured chick heart cells to autonomic agents and tetrodotoxin, *Am. J. Physiol.,* 209, 693, 1965.

17. **Boder, G. B. and Johnson, J. S.,** Comparative effects of some cardioactive agents on automaticity of cultured heart cells, *J. Mol. Cell. Cardiol.,* 4, 453, 1972.

18. **Ertel, R. J., Clarke, D. E., Chao, J. C., and Franke, F. R.,** Autonomic receptor mechanisms in embryonic chick myocardial cell cultures, *J. Pharmacol. Exp. Ther.,* 178, 73, 1971.

19. **Hermsmeyer, K. and Robinson, R. B.,** High sensitivity of cultured cardiac muscle cells to autonomic agents, *Am. J. Physiol.,* 233, c172, 1977.

20. **Galper, J. B., Dziekan, L. C., Miura, D. S., and Smith, T. W.,** Agonist-induced changes in modulation of K^+ permeability and beating rate by muscarinic agonists in cultured heart cells, *J. Gen. Physiol.,* 80, 231, 1982.

21. **Renaud, J. F., Scanu, A. M., Kazazoglou, T., Lombet, A., Romey, G., and Luzdunski, M.,** Normal serum and lipoprotein-deficient serum give different expression of excitability, corresponding to different stages of differentiation, in chicken cardiac cells in culture, *Proc. Natl. Acad. Sci. USA,* 79, 7768, 1982.

22. **Hutter, D. F. and Trautwein, W.,** Vagal and sympathetic effects on the pacemaker fibers in the sinus venosus of the heart, *J. Gen. Physiol.,* 89, 715, 1956.

23. **Harris, E. J. and Hutter, O. F.,** The action of acetylcholine on the movement of potassium ions in the sinus venosus of the heart, *J. Physiol.,* 133, 58, 1956.

24. **Glitsch, H. G. and Pott, L.,** Effects of acetylcholine and parasympathetic nerve stimulation on membrane potential in quiescent guinea pig atria, *J. Physiol.,* 279, 655, 1978.

25. **Garnier, D., Nargeot, J., Ojeda, C., and Rougier, O.,** The action of acetylcholine on background conductance in frog atrial trabeculae, *J. Physiol.,* 274, 381, 1978.

26. **Osterreider, W. A., Noma, A., and Trautwein, W.,** On the kinetics of the potassium channel activated by acetylcholine in the S-A node of the rabbit heart, *Pflugers Arch. Eur. J. Physiol.,* 306, 101, 1980.

27. **Sakmann, B., Noma, A., and Trautwein, W.,** Acetylcholine activities of single muscarinic K^+ channels in isolated pacemaker cells of the mammalian heart, *Nature,* 303, 250, 1983.

28. **Giles, W. and Noble, S.,** Changes in membrane currents in bullfrog atrium produced by acetylcholine, *J. Physiol.,* 261, 103, 1976.

29. **Hino, N. and Ochi, R.,** Effect of acetylcholine on membrane currents in guinea pig papillary muscle, *J. Physiol.,* 307, 183, 1980.

30. **Carmeliet, E. and Ramon, J.,** Electrophysiological effects of acetylcholine in sheep cardiac Purkinje fibers, *Pflugers Arch.,* 387, 197, 1980.

31. **Carmeliet, E. and Ramon, J.,** Effect of acetylcholine on time-independent currents in sheep cardiac Purkinje fibers, *Pflügers Arch.,* 387, 207, 1980.

32. **Carmeliet, E. and Ramon, J.,** Effects of acetycholine on time-dependent currents in sheep cardiac Purkinje fibers, *Pflugers Arch.,* 387, 217, 1980.

33. **Biegon, R. L. and Pappano, A.,** Dual mechanism for inhibition of calcium-dependent action potentials by acetylcholine in avian ventricular muscle, *Circ. Res.,* 46, 353, 1980.

34. **Josephson, I. and Sperelakis, N.,** On the ionic mechanism underlying adrenergic-cholinergic antagonism in ventricular muscle, *J. Gen. Physiol.,* 79, 69, 1982.

35. **Inoue, D., Hachisu, M., and Pappano, A. J.,** Acetylcholine increases resting membrane potassium conductance in atrial but not in ventricular muscle during muscarinic inhibition of Ca^{2+}-dependent action potentials in chick heart, *Circ. Res.,* 53, 158, 1983.

36. **Galper, J. B. and Smith, T. W.,** unpublished data.

37. **Musso, E. and Vassalle, M.,** Inhibitory actions of acetylcholine on potassium uptake of the sinus node, *Cardiovasc. Res.,* 9, 490, 1975.

38. **Lipsius, S. L. and Vassalle, M.,** Effects of acetylcholine on potassium movements in the guinea pig sinus node, *J. Pharmacol. Exp. Ther.,* 201, 669, 1977.

39. **Grossman, A. and Furchgott, R. F.,** The effects of various drugs on calcium exchange in the isolated guinea-pig left auricle, *J. Pharmacol. Exp. Ther.,* 145, 162, 1964.

40. **Hoditz, H. and Lullmann, H.,** Der Enfluss von acetylcholin auf den calcium satz ruhsender und kontrahierender vorhofnuskulatur in vitro, *Experimentia,* 20, 279, 1964.

41. **Prokopczuk, A., Pytkowski, B., and Lawartowski, B.,** Effect of acetylcholine on calcium efflux from atrial myocardium, *Eur. J. Pharmacol.,* 70, 1, 1981.

42. **Watanabe, A. M. and Besch, H. R., Jr.,** Cyclic adenosine monophosphate modulation of slow calcium influx channels in guinea pig hearts, *Circ. Res.,* 35, 316, 1974.

43. **Tsien, R. W.,** Cyclic AMP and contractile activity in the heart, *Advances in Cyclic Nucleotide Research,* Greengard, P. and Robison, G. A., Eds., Academic Press, New York, 1977, 363.

44. **Ross, E. M. and Gilman, A. G.,** Biochemical properties of hormone sensitive cyclase, *Ann. Rev. Biochem.,* 49, 533, 1980.

45. **Cassel, D. and Selinger, F.,** Mechanism of adenylate cyclase activation placement of bound GDP by GTP, *Proc. Natl. Acad. Sci. USA*, 75, 4155, 1978.
46. **Watanabe, A. M., McConnaughey, M. M., Strawbridge, R. A., Fleming, J. W., Jones, L. R., and Besch, H. R.,** Muscarinic cholinergic receptor modulation of β-adrenergic receptor affinity for catecholamines, *J. Biol. Chem.*, 253, 4833, 1978.
47. **Stadel, J. M., DeLean, A., and Lefkowitz, R. J.,** A high affinity agonist β-adrenergic receptor complex is an intermediate for catecholamine stimulation of adenylate cyclase in turkey and frog erythrocyte membranes, *J. Biol. Chem.*, 255, 1436, 1980.
48. **Jakobs, K. H., Aktories, K., and Schultz, G.,** GTP-dependent inhibition of cardiac adenylate cyclase by muscarinic cholinergic agonists, *Naunyn-Schmiedeberg's Arch. Pharmacol.*, 310, 113, 1979.
49. **Murad, F., Chi, Y. M., Rall, T. W., and Sutherland, E. W.,** Effects of catecholamines and choline esters on the formation of adenosine 3′,5′-Phosphate by preparations from cardiac muscle and liver, *J. Biol. Chem.*, 237, 1233, 1962.
50. **Liang, B. T., Hellmitch, M. R., Neer, E. J., and Galper, J. B.,** Development of muscarinic cholinergic inhibition of adenylate cyclase in embryonic chick heart: its relationship to changes in the inhibitory guanine nucleotide regulatory protein, *J. Biol. Chem.*, 261, 9011, 1986.
51. **Alexander, R. W., Galper, J. B., Neer, E. J., and Smith, T. W.,** Non-coordinate development of β-adrenergic receptors and adenylate cyclase in chick heart, *Biochem. J.*, 204, 825, 1982.
52. **Biegon, R., and Pappano, A. J.,** Dual mechanism for inhibition of calcium dependent action potentials by acetylcholine in avian ventricular muscle-relationship to cyclic AMP, *Circ. Res.*, 46, 353, 1980.
53. **Krause, E. G., Halle, W., and Wollenberger, A.,** Effect of dibutyryl cyclic GMP on cultured beating rat heart cells, in *Advances in Cyclic Nucleotide Research*, Vol. 1, Greengard, P. and Robison, G. A., Eds., Raven Press, New York, 1972, 301.
54. **Kohlhardt, M. and Haap, K.,** 8-bromo-guanosine 3,5′-monophosphate mimics the effect of acetylcholine on slow response action potential and contractile force in mammalian atrial myocardium, *Adv. Cyclic Nucleotide Res.*, 5, 473, 1972.
55. **Watanabe, A. M. and Besch, H. R., Jr.,** Interaction between cyclic adenosine monophosphate and cyclic guanosine monophosphate in guinea pig ventricular myocardium, *Circ. Res.*, 37, 309, 1975.
56. **Endo, M. and Shimizu, T.,** Failure of dibutyryl and 8-bromo-cyclic GMP to mimic the antagonistic action of carbachol on the positive inotropic effects of sympathomimetic amines in the canine isolated ventricular myocardium, *Jpn. J. Pharmacol.*, 29, 423, 1979.
57. **Brooker, G.,** Dissociation of cyclic GMP from the negative inotropic action of carbachol in guinea pig atria, *J. Cyclic Nucleotide Res.*, 3, 407, 1977.
58. **Mirro, M. J., Bailey, J. C., and Watanabe, A. M.,** Dissociation between the electrophysiological properties and total tissue cyclic guanosine monophosphate content of guinea pig atria, *Circ. Res.*, 45, 225, 1979.
59. **Diamond, J., Ten Eick, R. E., and Trapani, A. J.,** Are increases in cyclic GMP levels responsible for the negative inotropic effects of acetylcholine in the heart?, *Biochem. Biophys. Res. Commun.*, 79, 912, 1977.
60. **Goldberg, N. D., O'Dea, R. F., and Haddox, M. K.,** in *Advances in Cyclic Nucleotide Research*, Greengard, P. and Robison, G. A., Eds., Raven Press, New York, 1978, 13.
61. **Revtyak, G., Jones, L. R., Besch, H. R., Jr., and Watanabe, A. M.,** Characterization of particulate and soluble guanylate cyclase from canine myocardium, *Circulation*, 58 (Suppl. II), 439, 1978.
62. **Kuo, J. R., Wyatt, G. R., and Greenguard, P.,** Cyclic nucleotide dependent protein kinases: Partial purification and some properties of guanosine 3′,5′-monophosphate-dependent and adenosine 3′,5′-monophosphate-dependent protein kinases from various tissues and species of arthropoda, *J. Biol. Chem.*, 246, 7159, 1971.
63. **Schacht, J. and Agranoff, B. W.,** Stimulation of hydrolysis of phosphatidic acid by cholinergic agents in guinea pig synaptosomes, *J. Biol. Chem.*, 249, 1551, 1974.
64. **Fields, J. Z., Roeske, W. R., and Morkin, E.,** Cardiac muscarinic cholinergic receptors. Biochemical identification and characterization, *J. Biol. Chem.*, 253, 3251, 1978.
65. **Galper, J. B., Dziekan, L. C., O'Hara, D. S., and Smith, T. W.,** The biphasic response of muscarinic cholinergic receptors in cultured heart cells to agonists, *J. Biol. Chem.*, 257, 10344, 1982.
66. **Rosenberger, L. B., Yamamura, H., and Roeske, W. R.,** Cardiac muscarinic cholinergic receptor binding is regulated by Na$^+$ and guanine nucleotides, *J. Biol. Chem.*, 255, 820, 1980.
67. **Berrie, C. D., Birdsall, N. J. M., Burgen, A. S. V., and Hulme, E. C.,** Guanine nucleotides modulate muscarinic receptor binding in the heart, *Biochem. Biophys. Res. Commun.*, 87, 1000, 1979.
68. **Hulme, E. C., Berrie, C. P., Birdsall, N. J. M., and Burgen, A. S. V.,** Interactions of muscarinic receptors with guanine nucleotides and adenylate cyclase, in *Drug Receptors and their Effectors*, Birdsall, N. J. M., Ed., MacMillan Publishers, London, 1981, 23.

69. **Harden, T. K., Scheer, A. G., and Smith, M. M.,** Differential modification of the interaction of cardiac muscarinic cholinergic and β-adrenergic receptors with a guanine nucleotide binding component, *Mol. Pharmacol.*, 21, 570, 1982.

70. **Galper, J. B. and Smith, T. W.,** Properties of muscarinic acetylcholine receptors in heart cell cultures, *Proc. Natl. Acad. Sci. USA*, 75, 5851, 1978.

71. **Nathanson, N. M.,** Binding of agonists and antagonists to muscarinic acetylcholine receptors on intact cultured heart cells, *J. Neurochem.*, 41, 1545, 1983.

72. **Birdsall, N. J. M., Berrie, C. P., Burgern, A. S. V., and Hulme, E. C.,** Modulation of binding properties of muscarinic receptors. Evidence for receptor effector coupling, in *Receptors for Neurotransmitters and Peptide Hormones*, Pepeu, G., Kuhar, M. J., and Enna, S., Eds., Raven Press, New York, 1980, 107.

73. **Ehlert, F. J., Yamamura, H. I., Triggle, D., and Roeske, W. R.,** The influence of guanyl-5′-yl-imidodiphosphate and sodium chloride on the binding of the muscarinic agonist, ^{3}H cis methyldeoxalane, *Eur. J. Pharmacol.*, 61, 317, 1980.

74. **Galper, J. B., Dziekan, L. C., and Smith, T. W.,** The development of physiologic responsiveness to muscarinic agonists in chick embryo heart cell cultures: role of high affinity receptors and sensitivity to guanine nucleotides, *J. Biol. Chem.*, 259, 3777, 1984.

75. **Katada, T. and Ui, M.,** Perfusion of the pancreas isolated from pertussis-sensitized rats: potentiation of insulin secretory responses due to β-adrenergic stimulation, *Endocrinology*, 101, 1247, 1977.

76. **Hazeki, O. and Ui, M.,** Modification by islet-activating protein of receptor mediated regulation of cyclic AMP accumulation in isolated rat heart cells, *J. Biol. Chem.*, 256, 2856, 1981.

77. **Katada, T. and Ui, M.,** ADP ribosylation of the specific membrane protein of C_6 cells by islet-activating protein associated with modification of adenylate cyclase activity, *J. Biol. Chem.*, 257, 7210, 1982.

78. **Katada, T., Bokoch, G. M., Northup, J., Ui, M., and Gilman, A. G.,** The inhibitory guanine nucleotide-binding regulatory component of adenylate cyclase: properties and function of the purified porotein, *J. Biol. Chem.*, 259, 3568, 1984.

79. **Katada, T., Northup, J. K., Bokoch, G. M., Ui, M., and Gilman, A. G.,** The inhibitory guanine nucleotide-binding regulatory component of adenylate cyclase: subunit dissociation and guanine nucleotide-dependent hormonal inhibition, *J. Biol. Chem.*, 259, 3578, 1984.

80. **Katada, T., Bokoch, G. M., Smigel, M. D., Ui, M., and Gilman, A. G.,** The inhibitory guanine nucleotide-binding regulatory component of aenylate cyclase: Subunit dissociation and the inhibition of adenylate cyclase in S_{49} lymphoma cyc- and wild type membranes, *J. Biol. Chem.*, 259, 3586, 1984.

81. **Hildebrandt, J. D., Codina, J., Risinger, R., and Birnbaumer, L.,** Identification of a subunit associated with the adenyl cyclase regulatory proteins N_s and H_i, *J. Biol. Chem.*, 259, 2039, 1984.

82. **DeLean, A., Stadel, J. M., and Lefkowitz, R. J.,** A ternary complex model explains the agonist-specific binding properties of the adenylate cyclase-coupled β-adrenergic receptor, *J. Biol. Chem.*, 255, 7108, 1980.

83. **Clusin, W. T.,** Correlation between relaxation and automaticity in embryonic heart cell aggregates, *Proc. Natl. Acad. Sci. USA*, 77, 679, 1980.

84. **Isenberg, G.,** Cardiac Purkinje fibers: $Ca^{2+}{}_i$ controls potassium permeability via the conductance components g_{k1} and g_{k2}, *Pflügers Arch.*, 371, 77, 1977.

85. **Hartzell, H. C. and Titus, L.,** Effect of cholinergic and adrenergic agonists on phosphorylation of a 165,000 dalton myofibrillar protein in intact cardiac muscle, *J. Biol. Chem.*, 257, 2111, 1982.

86. **Brown, S. L. and Brown, J. H.,** Muscarinic stimulation of phosphatidylinositol metabolism in atria, *Mol. Pharmacol.*, 24, 351, 1983.

87. **Brown, J. H. and Brown, S. L.,** Agonists differentiate muscarinic receptors that inhibit cyclic AMP formation from those that stimulate phosphoinositide metabolism, *J. Biol. Chem.*, 259, 3777, 1984.

88. **Siegel, R. E. and Fischbach, G. D.,** Muscarinic receptors and responses in intact embryonic chick atrial and ventricular heart cells, *Dev. Biol.*, 101, 346, 1984.

89. **Raff, M.,** Self-regulation of membrane receptors, *Nature*, 259, 265, 1976.

90. **Tokimasa, T., Hasuo, H., and Kobetsu, K.,** Desensitization of the muscarinic acetylcholine receptor of atrium in bullfrogs, *Jpn. J. Physiol.*, 31, 83, 1981.

91. **Jalife, J., Hamilton, A. J., and Moe, G. K.,** Desensitization of the cholinergic receptor at the sinoatrial cell of the kitten, *Am. J. Physiol.*, 238, H439, 1980.

92. **Nathanson, N. M., Holttum, J., Hunter, D. D., and Halvorsen, S. W.,** Partial agonist activity of oxotremorine at muscarinic acetylcholine receptors in the embryonic chick heart, *J. Pharmacol. Exp. Ther.*, 229, 455, 1984.

93. **Galper, J. B. and Smith, T. W.,** Agonist and guanine nucleotide modulation of muscarinic cholinergic receptors in cultured heart cells, *J. Biol. Chem.*, 255, 9571, 1980.

94. **Hunter, D. D. and Nathanson, N. M.,** Decreased physiological sensitivity mediated by newly synthesized muscarinic acetylcholine receptors in embryonic chicken heart, *Proc. Natl. Acad. Sci. USA*, 81, 3582, 1984.

95. **Pfaffinger, P. J., Martin, J. M., Hunter, D. D., Nathanson, N. M., and Hille, B.,** GTP-binding proteins couple cardiac muscarinic receptors to a K channel, *Nature,* 317, 536, 1985.
96. **Logothetis, D. E., Kurachi, Y., Galper, J., Neer, E. J., and Clapham, D. E.,** The $\beta\gamma$ subunits of GTP-binding proteins activate the muscarinic K^+ channel in heart, *Nature,* 325, 321, 1987.
97. **Codina, J., Yatani, A., Grenet, D., Brown, A. M., and Birnbaumer, L.,** The α subunit of the GTP binding protein G_k opens atrial potassium channels, *Science,* 236, 442, 1987.

Chapter 19

ALPHA AND BETA ADRENERGIC RECEPTORS IN CULTURED HEART CELLS

Richard B. Robinson, Linda E. Kupfer, and John P. Bilezikian

TABLE OF CONTENTS

I. INTRODUCTION

Cardiac tissue was successfully maintained in culture in the form of explants as early as 1912,[1] but enzyme-dissociated cultures of single myocardial cells were not produced until 1955.[2] Another decade passed before the autonomic responsiveness of cardiac cultures was studied. Research activity in this area has increased considerably since these early studies, and the culture system has proven useful in elucidating the regulation and expression of autonomic responsiveness. In this chapter, we will describe what has been learned about the adrenergic system in myocardial cultures and the extent to which this has expanded our knowledge of the comparable system in the intact myocardium. The chapter will confine itself to a discussion of monolayer cultures of enzyme-dispersed embryonic or neonatal heart cells and will not discuss those studies involving organ or explant culture, or those employing acutely dispersed adult myocardial tissue.

II. ADRENERGIC RESPONSIVENESS

Most studies of adrenergic responsiveness of cardiac cells in culture have taken advantage of the fact that these cultures tend to beat spontaneously, and have, therefore, measured the chronotropic response to adrenergic agonists. In addition, the majority of early studies employed cultures derived from the embryonic chick. The earliest reference to such a study was a review article by Wollenberger in 1964[3] citing earlier unpublished experiments. It was reported that epinephrine, at a threshold concentration of 2×10^{-9} M, elicited an increase in spontaneous rate. It was stated the effect could be antagonized by β- but not by α-adrenergic blockers. This positive result was overshadowed a year later by a study that found monolayer cultures to be totally unresponsive to either epinephrine or norepinephrine.[4] Later researchers were successful in demonstrating a positive chrontropic response to β-adrenergic agonists,[5,6] but the results were seldom consistent with respect to the effective concentration range. This apparent irreproducibility of autonomic responsiveness between laboratories, coupled with other reported discrepancies between intact and cultured myocardium,[4] led to the general perception that heart cells did not maintain their functional characteristics in vitro.

However, numerous subsequent studies have clearly demonstrated that, under appropriate culture and experimental conditions, heart cultures are highly responsive to catecholamines. Hermsmeyer and Robinson[7] studied the possible basis for the variability in reported results and concluded that observed adrenergic responsiveness depended critically on experimental conditions. In the embryonic chick ventricle cultures, they demonstrated a rapid and profound tachyphylaxis that contributed to the attenuation of the response when certain protocols were employed. In particular, use of a superfusion system decreased the ED_{50} concentration by more than 3000× relative to that observed using a static test chamber. When a static chamber was employed, the magnitude of the ED_{50} concentration was depressed, and it was shifted to the right when the drug was added further away from the test cell, resulting in a more gradual equilbration period (Figure 1). Moreover, if the experiment was conducted in serum-containing medium, the response to catecholamines was further attenuated. It can therefore be concluded that, at least in the case of monolayer cultures derived from embryonic chick ventricle, functional adrenergic receptors persist in vitro, but the response can be very labile.

At the same time as these studies of embryonic chick myocytes were being conducted, mammalian cardiac tissue was being cultured by other investigators beginning with Harary and Farley.[8] By far the most common alternatives to the embryonic chick have been the neonatal rat or mouse myocardium as a culture source. In 1974, Goshima demonstrated that both atrial and ventricular cells from fetal or neonatal mice responded to norepinephrine $(10^{-6}M)$ with an increase in spontaneous rate.[9] Lane et al.[10] extended these findings by

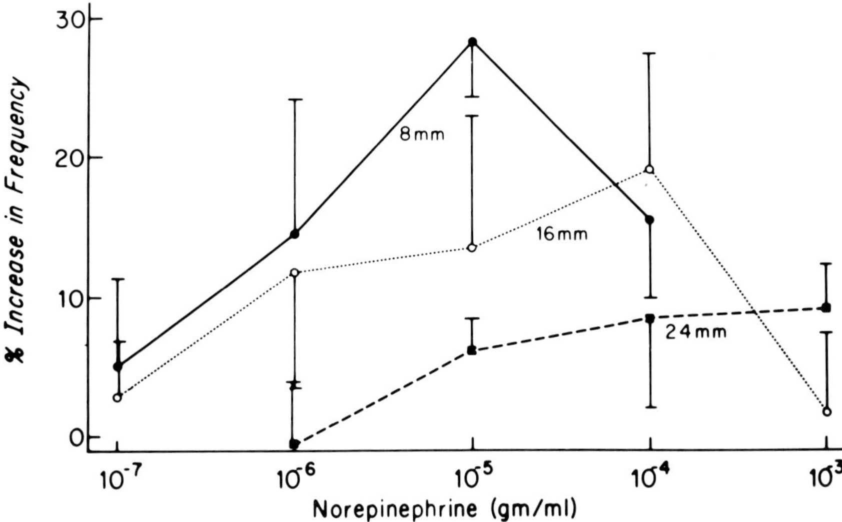

FIGURE 1.   Increase in frequency of spontaneous contraction as a function of norepinephrine concentration depended on distance from which norepinephrine was applied in the static-chamber method. Here, curves are shown with application at the three distances indicated — 8, 16, and 24 mm. These experiments were carried out on cells cultured from 3 to 5 days. At a distance of 8 mm, sensitivity was much less than either of previous methods, ED_{50}, being approximately 3 μM (10^{-6} g/mℓ) with no dilution factor introduced. Distances less than 8 mm could not be used because of responses to control applications. Curves from applications at 16- and 24-mm distances were difficult to quantitate in terms of ED^{50}, and curve at 24 mm could be taken as no response whatsoever. Depressed maxima at greater distances occurred even at high concentrations and, thus, insensitivity could not be due to dilution differences. Values plotted represent means $\pm$ SE (n = 2 to 7 for each dose.) (From Hermsmeyer, K. and Robinson, R. B., High sensitivity of cultured cardiac chick cells cells to autonomic agents, *Am. J. Physiol.*, 233, C172, 1977. With permission.)

constructing complete concentration-response curves for a series of adrenergic agonists.[10] They reported responses of single myocytes to epinephrine and norepinephrine in the range of 10^{-9} to $10^{-5}M$. In addition, they demonstrated that α- and β-adrenergic agonists caused a positive chronotropic response, and that each response was blocked by the appropriate antagonist. This was the first evidence that the same cell could possess both α- and β-adrenergic (as well as cholinergic) receptors.

Using neonatal rat cultures, Kaumann and Wittmann reported a positive chronotropic response to as little as $4 \times 10^{-11}M$ isoproterenol.[11] Furthermore, they conducted a detailed pharmacological analysis of the apparent equilibrium constant of the adrenergic receptor for a β-adrenergic antagonist, obtaining a value from the cultures that was comparable to data from the intact heart. In 1977, Karsten et al.[12] found that neonatal rat cultures responded selectively to α- and β-adrenergic agonists. Activation of either class of receptor resulted in a positive chronotropic response, but only β-receptor activation also resulted in an increase in cAMP levels. Marvin et al.[13] confirmed that neonatal rat cardiac cultures were highly sensitive to catecholamines, and collected the first complete concentration-response curve to norepinephrine in this species (Figure 2). Similar studies have been performed on cultures derived from the newborn cat[11] and the fetal dog.[14] Kaumann compared the chronotropic response of neonatal rat cultures to isoproterenol and to the partial agonist, dichloroisoproterenol, and found evidence for spare β-adrenergic receptors for isoproterenol, but not for dichloroisoproterenol.[15] Hermsmeyer et al.[16] in confirming the high sensitivity of single neonatal rat ventricle cells to both α- and β-adrenergic agonists, also demonstrated some partial agonist activity for the α- and β- antagonists phentolamine and propranolol. Freyss-

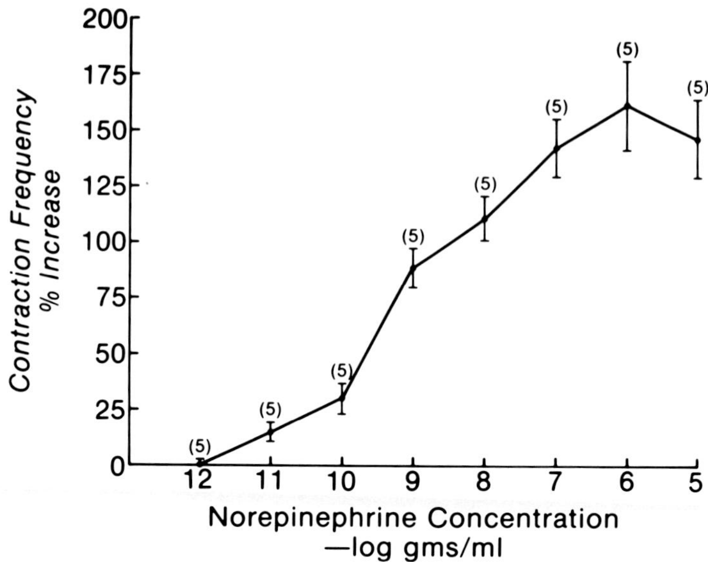

FIGURE 2.   High sensitivity of spontaneously contracting isolated rat myocardial cells in culture shown by the positive chronotropic effect of norepinephrine (ED_{50} = 2.3 nM). The cells were exposed to 10-$\mu\ell$ pulses of norepinphrine in the concentrations plotted in a noncirculating chamber at 37°C, and the peak response was recorded. The number of cells is indicated at each datum point. Each cell included was exposed to the neurotransmitter for 30 sec for each of five or less doses and was always observed to return to baseline rate ± 20% in control solution. Source hearts (n = 3) exposed to norepinephrine suffusion had an ED_{50} of 100 nM. (From Marvin, W., Jr., Robinson, R. B., and Hermsmeyer, K., Correlation of function and morphology of neonatal rat and embryonic chick cultured cardiac and vascular muscle cells, *Circ. Res.*, 45, 528, 1979. With permission.)

Beguin et al.[17] demonstrated that beta selective agonists caused a positive chronotropic effect in culture with an order of potency comparable to their β_1 subselectivity.[17]

In 1983, Fisher et al.[18] extended the earlier observation of rapid β receptor desensitization in the embryonic chick to rat cultures, and demonstrated that the same phenomenon does not occur with respect to the α-adrenergic receptor. Marsh et al.[19] studied the inotropic response to isoproterenol in embryonic chick cultures. Although they were unable to demonstrate a chronotropic response, they used an optoelectric measuring apparatus[20] to show an inotropic response between 10^{-9} and 10^{-6} M. As with the chronotropic studies of others, the inotropic response to β receptor activation was found to desensitize rapidly in the embryonic chick cultures.

III. MECHANISM OF ACTION OF THE ADRENERGIC RESPONSE

Many studies have attempted to correlate the actions of catecholamines on the heart cultures with activation of the adenylate cyclase/cyclic AMP system. Moura and Simpkins[21] reported a dose-dependent increase in cellular cAMP upon incubation of neonatal rat monolayer cultures with epinephrine or norepinephrine. The effective concentration range appeared roughly comparable to that reported by others to cause a positive chronotropic response. Harary et al.[22] confirmed these results and reported that the effect of norepinephrine on cAMP level was increased and more prolonged in low calcium solutions. They speculated that the dependence of the effect on calcium concentration could arise from the effect of calcium on phosphodiesterase activity. Renaud et al.[23] reported that basal cAMP levels in embryonic chick cultures were low compared to those in intact source tissue, but could be greatly increased by isoproterenol. In an interesting series of experiments, Lawrence et al.[24] first demonstrated that dibutyryl cAMP in the millimolar concentration range could mimic the effect of norepinephrine on both the spontaneous rate and the action potential configuration (amplitude and duration) in mouse myocardial cultures. They then prepared co-cultures of mouse heart cells and rat ovarian granulosa cells and demonstrated that follicle-stimulating hormone (FSH) could mimic the chronotropic and action potential effects of norepinephrine. Presumably, the FSH was activating a receptor on the granulosa cell to form an intracellular intermediate which was able to cross the gap junction between the granulosa and the myocardial cell. It was proposed that the intracellular substance was cAMP. Freyss-Beguin et al.[17] carried out a detailed comparison of the effect of a series of β-adrenergic agonists on both beating rate and cAMP levels. They found a good correlation between the two parameters, as well as a correlation between response and degree of β_1 potency of the agonists.

Recently the patch clamp technique[25] has begun to be employed to investigate the mechanism of action of catecholamines. Cell culture is well suited to this approach since it provides clean membranes required to attain the necessary tight seal between electrode and cell. Reuter et al.[26] have used this technique with neonatal rat cultures for studying the single channel characteristics of the calcium channel. They concluded that β-adrenergic agonists increased the calcium current by increasing the probability that the channel would open, rather than by increasing the magnitude of the current through the channel. They also considered an increase in the total number of available channels as being a likely contributing effect to the action of catecholamines. Comparable results to those with isoproterenol were obtained using 8-bromo-cAMP. However, analysis of the 8-bromo-cAMP effect suggested that it was due to a shortening of the interval between openings rather than to an increase in the mean ''open times'' of the channel, whereas the effect of isoproterenol was thought to result from an increase in the mean ''open time''.

IV. THE IDENTIFICATION OF ADRENERGIC RECEPTORS IN CULTURED HEART CELLS

The data reviewed in the first part of this chapter provides pharmacological evidence for the existence of both α- and β-adrenergic receptors in cultured heart preparations. In order to demonstrate directly the presence of α- and β-adrenergic receptors, radioligands of appropriate adrenergic selectivity and specific activity are necessary. These radioligands have become available over the past ten years.[27,28] They allow analysis of adrenergic receptors per se and a correlation with certain biochemical and pharmacological properties of the heart. Application of the inherent advantages of radioligand techniques to the cultured heart cell allows one to make use of the inherent advantages of cell culture as opposed to in vivo and cardiac membrane preparations. These include knowing rather precisely the specific type of cell being studied. With primary cardiac cell culture, the complicating feature of neuronal input is also avoided. Furthermore, the use of cell culture enables the study of adrenergic receptors in an environment that can be precisely controlled. In addition to knowing the composition of the culture medium, exposure of the cells to hormones or drugs can be controlled in a way that is not possible with other experimental models. This latter advantage will be illustrated in our studies investigating the effect of thyroid hormone on cardiac cells.

V. CHOICE OF RADIOLIGAND

The specific radioligand used to detect α- and β-adrenergic receptors presents a particular problem with cultured cells. Cultured cells are only available in limited quantities, thus making it difficult to utilize a radioligand of relatively low specific activity. Thus, the fairly generous variety of tritiated radioligands that have been developed for the identification of α- and β-adrenergic receptors in cardiac membrane preparations are not generally useful in cardiac cell cultures. High specific activity can be achieved by the use of iodinated radioligands which have become available. For β-adrenergic receptors, the most useful iodinated radioligands are iodohydroxybenzylpindolol ([125I]IHYP), iodopindolol ([125I]IP), and iodocyanopindolol ([125I]ICYP). With regard to primary cardiac cell culture, [125I]IHYP and [125I]ICYP have both been used with success. However, the low specific activity tritiated radioligands ([3H] propranolol and [3H]dihydroalprenolol) have been used as well. In the case of the tritiated radioligands for the detection of β-adrenergic receptors, it has been necessary to use larger amounts of cellular material and higher concentrations of radioligand, complicating true appreciation of putative high affinity beta-receptor sites.

For the detection of α-adrenergic receptors, only recently has an iodinated radioligand become available. This ligand is an α_1 specific α-adrenergic inhibitor, [125I]-I-2-[b-(4-hydroxyphenyl) ethylaminomethyl]tetralone ([125I]IBE 2254.[29] To our knowledge, the many tritiated α-adrenergic antagonists that have been used to identify α-adrenergic receptors in cardiac membrane preparations ([3H]WB4101, [3H]prazosine, ([3H]phentolamine, [3H]clonidine and [3H]dihydroergocryptine) have not been used successfully in cultured cardiac cells. It is noteworthy that with the exception of [3H]clonidine, radioligands used for detecting α- or β-adrenergic receptors have been of the antagonist form. It has been possible to utilize α- or β-adrenergic inhibitors as radioligands because these compounds bind competitively at the true receptor sites of interest.

VI. IDENTIFICATION OF BETA-ADRENERGIC RECEPTORS

Beta adrenergic receptors were first identified in cultured cardiac cells by Tsai and Chen.[30] The tritiated ligand, [3H]dihydroalprenolol ([3H]DHA), was employed. Although the major thrust of that work was to evaluate the effects of thyroid hormone on the β-adrenergic receptor, partial characterization of the beta receptor was accomplished. In that study, the binding affinity (K_d) was 2.7×10^{-7} M and the binding capacity (B_{max}) was approximately 2 pmol/mg protein. These values are only estimates because formal equilibrium or kinetic binding analyses were not performed.

More recently, Kaumann has also successfully used a tritiated radioligand, [3H] (−) propranolol, to identify β-adrenergic receptors in cultured rat myocardial cells.[15] In this study, binding was rapid ($t_{1/2} = 28 \pm 2$s) and was easily dissociable ($t_{-1/2} = 84 \pm 17$s). Other binding characteristics best fitted a single site model with an estimated K_d of 4.2 nM and a B_{max} of 228 fmol/mg protein. Binding was stereoselective for the (−)isomer of the agonist or antagonist. A particular feature of this study was the ability to correlate maximum chronotropic effects of (−)isoproterenol with occupation of β receptors. It was shown that only 15% of β receptors need be occupied by the agonist in order to achieve maximum chronotropy in this system.

A more complete assessment of β adrenergic receptors in cultured rat cardiac myoblasts was provided in 1980 by Lau et al.[31] using [125I]IHYP, and in 1984 by Kupfer et al.[45] using [125I]ICYP. It was important to utilize a cell culture methodology that provided a generally homogeneous population of cardiac myoblasts after 4 days of culture. With the differential plating procedure of Blondel et al.,[33] it was possible to achieve enrichment with cardiac myoblasts of greater than 80% at the time of assay. The β-adrenergic receptor on myoblasts was characterized by a single very high affinity binding site with K_d of 88 ± 33 pM and a B_{max} of $7,600 \pm 2,100$ sites per cell (Figure 3). Using [125I]ICYP, a ligand with greater affinity, the K_d was found to be 24 ± 3 pM; the B_{max} was $12,000 \pm 2,600$ sites per cell. With either radioligand, binding was shown to be kinetically appropriate with rapid asso-

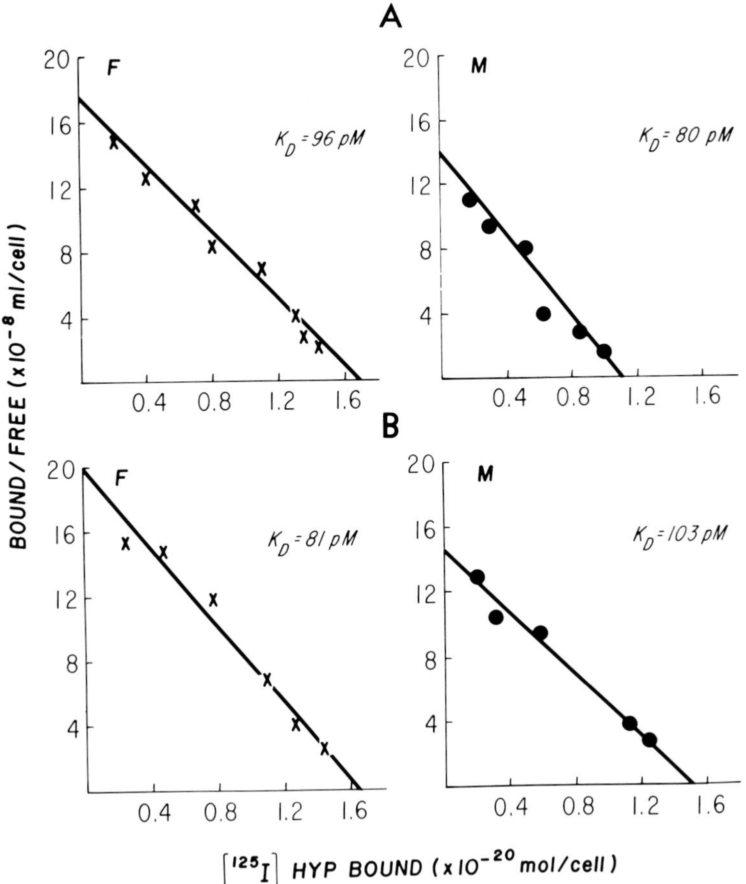

FIGURE 3.   Scatchard analyses of the equilibrium binding of [^{125}I]IHYP to myocardial (M) and fibroblast (F) cells. A and B represent two separate sets of experiments. In each experiment, the M and F cultures were derived from a single preparation. Each data point is the mean of three to four determinations. The line in each panel was drawn by linear regression analysis. (From Lau, Y. H., Robinson, R. B., Rosen, M. R., and Bilzikian, J. B., subclassification of beta-adrenergic receptors in cultured rat cardiac myoblasts and fibroblasts, *Circ. Res.*, 47, 41, 1980. With permission.)

ciation and dissociation constants. In addition, it could be shown that binding of these iodinated radioligands to cultured myoblasts cells was stereoselective for the ($-$)isomer of the antagonist, propranolol, and the agonist, isoproterenol. A series of adrenergic agonists indicated β_1 subselectivity with regard to the order of potency of adrenergic agonists in inhibiting binding (isoproterenol > epinephrine = norepinephrine; Figure 4). In addition, the β_1 selective adrenergic inhibitor, practolol, was very effective in inhibiting binding (Figure 5).

These data may be compared with those obtained for relatively pure cultures of fibroblasts (obtained from ventricular tissue) which could conceivably introduce erroneous results if the population of myocardial cells contained a significant admixture of fibroblasts. Fibroblast cells also contain β adrenergic receptors with similar affinity ($K_d = 71 \pm 23$ pM) and capacity ($B_{max} = 9,000 \pm 2,400$ sites per cell; Figure 3). The major differentiation, however, between the β receptor of the fibroblast and the β receptor of the mycardial cell is that the fibroblast has β_2 subselectivity. The order of binding potency in the fibroblasts thus contrasts with the myoblasts in that isoproterenol > epinephrine > norepinephrine (Figure 4). More-

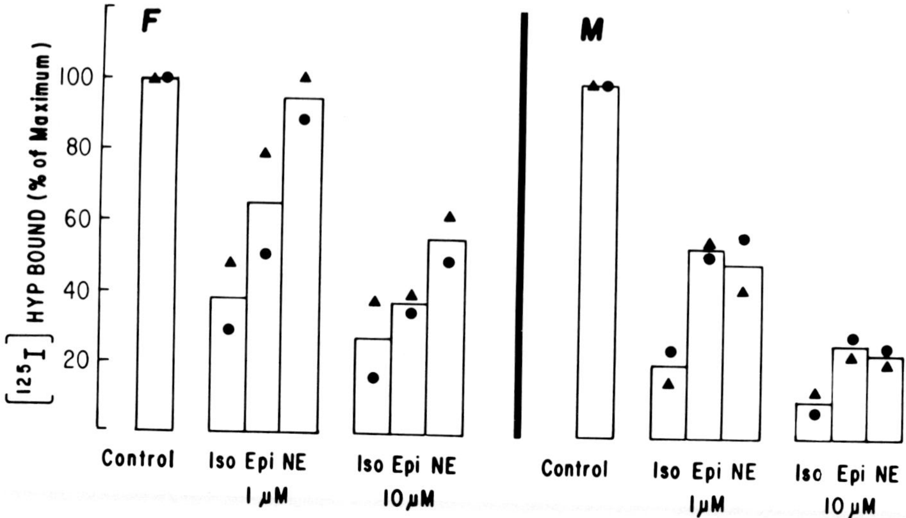

FIGURE 4. The relative potencies of beta-adrenergic agonists in inhibiting the specific binding of [^{125}I]IHYP to M and F cells. 1-Isoproterenol (Iso), 1-epinephrine (Epi) and 1-norepinephrine (NE), each 1 μM or 10 μM were tested as inhibitors of maximal specific [^{125}I]IHYP binding to F and M cells. The results shown are the mean ± SD of three experiments for which data points were obtained in triplicate. At each concentration, the differences between inhibition by epinephrine and norepinephrine were significant for F cells (p < 0.05), but not for M cells (p > 0.05). (From Lau, Y. H., Robinson, R. B., Rosen, M. R., and Bilezikian, S. P., Subclassification of beta-adrenergic receptors in cultured rat cardiac myoblasts and fibroblasts, *Circ. Res.*, 47, 41, 1980. With permission.)

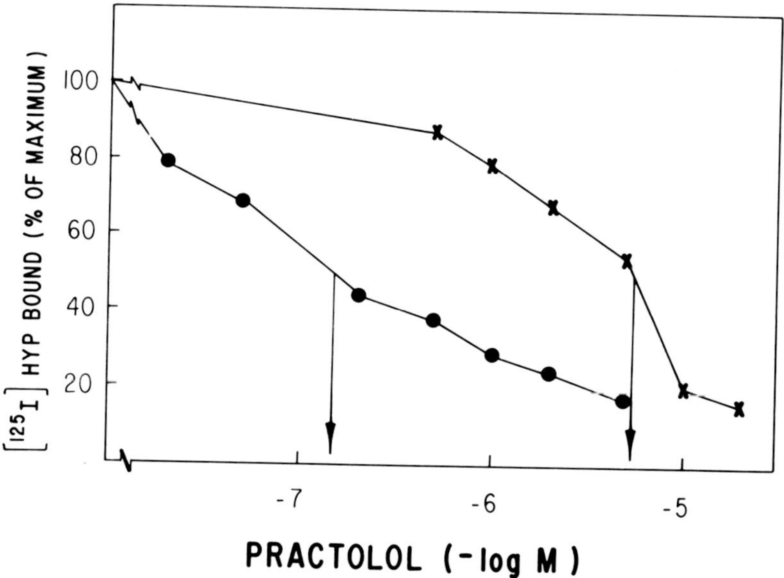

FIGURE 5. The binding affinity of practolol for beta receptor sites in M and F cells. Practolol at the indicated concentrations was tested as an inhibitor of maximal specific binding of [^{125}I]IHYP to M (●) or F (X) cells. For both curves, the data points shown in the most sensitive portion of the dose response curve (0.5 to 5 μM) are the average of three separate experiments on two different M and F cultures. The data points for each experiment were obtained in triplicate. (From Lau, Y. H., Robinson, R. B., Rosen, M. R., Bilezikian, J. P., Subclassification of beta-adrenergic receptors in cultured rat cardiac myoblasts and fibroblasts, *Circ. Res.*, 47, 41, 1980. With permission.)

over, the β-selective antagonist, practolol, is a weak inhibitor of binding in the fibroblast cell (Figure 5). In these studies characterizing the β-adrenergic receptor of the cultured myocardial cell, it was shown that the biochemical marker of beta adrenergic function, adenylate cyclase activity, correlated very well with the binding characteristics. Thus, the degree to which the adrenergic agonists, isoproterenol, epinephrine, and norepinephrine, stimulated adenylate cyclase activity in these preparations mirrored their respective abilities to compete in binding.

VII. IDENTIFICATION OF ALPHA-ADRENERGIC RECEPTORS

The identification of α-adrenergic receptors in primary cultures of myocardial cells lagged behind the demonstration of β-adrenergic receptors by several years. This remained so until the development of [^{125}I]IBE 2254 by Engel et al.,[29] that enabled direct identification of adrenergic receptors in myocardial cells. It was anticipated, however, that such receptors would be present on the basis of pharmacological studies in cultured cardiac cells as previously reviewed. In our system, when spontaneously contracting ventricular myocardial cells were superfused with the α-adrenergic agonist, phenylephrine, the rate of contraction increased in a concentration-dependent manner. The enhanced chronotropic effect was completely reversed when medium lacking the drug was subsequently used. At low concentrations of phenylephrine (1 to 100 n*M*), the positive chronotropic response was totally abolished by the α-adrenergic antagonist, prazosine (Figure 6). This observation suggested that the chronotropic response to phenylephrine was, at least in part, an α_1 mediated phenomenon. It was shown that [^{125}I]IBE 2254 binding to rat heart was rapid, saturable, stable, and readily reversible. Binding reached equilibrium within 12 to 15 min. The rate constant for association (k_1) was 2.8×10^8 min^{-1}M^{-1} and for dissociation (k_i) was 0.173 min^{-1} (Figure 7). An estimate of the equilibrium dissociation constant (K_d) was 690 pM. This value agreed fairly closely with the K_d estimated by equilibrium binding and analyzed according to the method of Scatchard (i.e., 324 ± 42, mean + SEM, Figure 8).[34] The B_{max} calculated according to this method was 33,000 ± 4,000 sites per cell. Further studies indicated that the binding reaction was stereoselective for the (−)isomer of norepinephrine. In experiments comparing the ability of a series of unlabelled α-adrenergic catecholamines to compete with [^{125}I]IBE 2254 for binding sites, it was shown that the binding site could best be described as an α_1 receptor (BE 2254 = prazosine > phentolamine > yohimbine; Figure 9).

VIII. REGULATION OF ALPHA- AND BETA-ADRENERGIC RECEPTORS BY THYROID HORMONE IN CULTURED RAT MYOCARDIAL CELLS

These observations establish directly the existence of β_1- and α_1-adrenergic receptors in primary cultures of neonatal rat myocardial cells. The facile demonstration of these receptors with appropriate high specific activity radioligands has made it possible to begin to address questions of clinical relevance in a very easily controllable setting. One area of particular importance is the influence of thyroid hormone on the heart. Great interest in the mechanism by which thyroid hormone influences the adrenergic system relates to clinical observations associating thyroid hormone and the catecholamines. It has been appreciated for years that many of the clinical features of excess thyroid hormone appear to reflect increased adrenergic activity.[35,36] With respect to the heart, numerous experiments have shown that cardiac tissue from hyperthyroid rats demonstrate consistent increases in the number of β receptors, with no significant changes in β-receptor affinity.[37,38] In contrast, experimental models of hypothyroidism have been associated with a decrease in the number of β-adrenergic receptors, once again associated with no apparent change in affinity. These studies were extended to provide an analysis of α-adrenergic receptors in the setting of altered thyroid hormone levels.

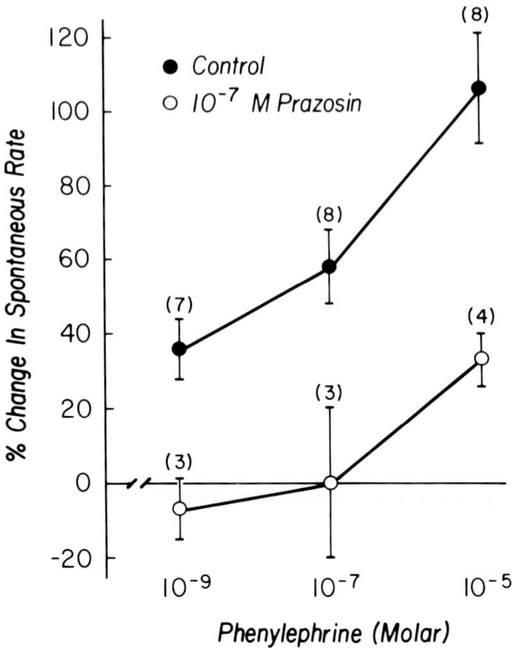

FIGURE 6.   The chronotropic response of cultured rat ventricular cells to phenylephrine. The response plotted is the mean percent increase in spontaneous contractile rate over baseline: (●) Response to phenylephrine alone; (○) response to phenylephrine in the presence of prazosine. The prazosin was added prior to and maintained throughout the superfusion of phenylephrine. The concentration of prazosin used did not, by itself, significantly alter the spontaneous rate. Values in parentheses above the data points indicate the number of separate experiments at each concentration. Mean ± SEM. (From Kupfer, L. S., Robinson, R. B., and Bilezikian, J. P., Identification of Alpha$_1$ adrenegic receptors in cultured rat myocardial cells with a new iodinated alpha$_1$ adrenergic antagonist [^{125}I]IBE 2254, *Circ. Res.*, 51, 250, 1982. With permission.)

In general, α-adrenergic receptors in membrane preparations from rat heart are decreased in hyperthyroidism.[38] In hypothyroidism, on the other hand, the number of α-adrenergic receptors has been reported to be increased,[39] decreased,[40,41] or unaltered.[42] According to some reports, α-receptor affinity for the radiolabeled antagonist also change in hypothyroidism and hyperthyroidism,[43,44] while many other studies have not detected any changes in the K_d.[38]

The methodology employed to obtain the above information has generally employed the model of the animal first rendered hyperthyroid or hypothyroid and then studied with respect to the resulting in vitro cardiac membrane preparation. Although this approach to the problem has yielded informative data, it is potentially limited in interpretive value because of the pervasive effects of thyroid hormone on many different organs. Thus it is not possible to conclude whether the changes observed in these studies reflect direct effects of thyroid hormone on the heart or secondary effects due to the actions of thyroid hormone on other tissues. It is thus particularly valuable to use the cultured myocardial cell to study this question because it offers an environment that is precisely controlled. The direct effects of thyroid hormone on cardiac adrenergic receptors can be observed in the absence of neuronal

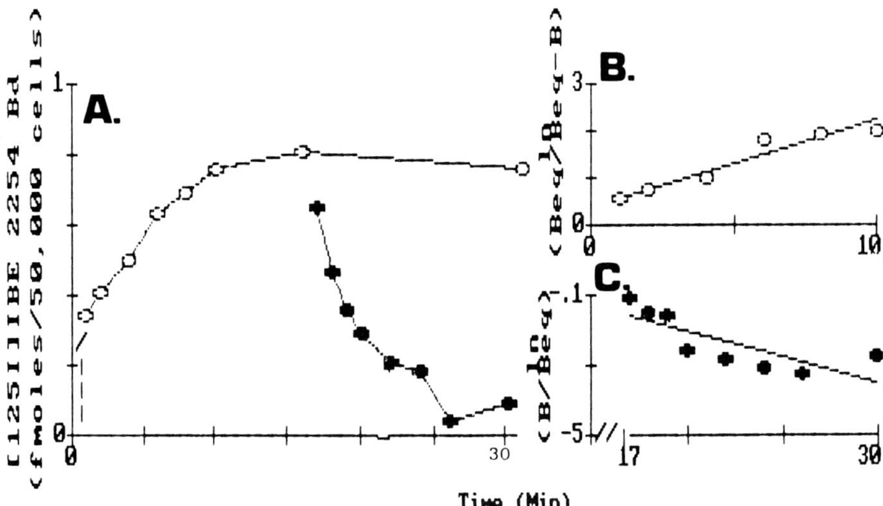

FIGURE 7.   Kinetics of [125I]IBE 2254 binding to cultured rat heart cells. A. Association kinetics (open symbols): Cultured rat heart cells (250,000/mℓ) were incubated with [125I]IBE 2254 (25 pM) at 37°C and specific binding was determined at the indicated time points. Dissociation kinetics (closed symbols): After binding reached equilibrium (17 min), excess unlabelled BE 2254 (50 μM) was added to the reaction mixture. Aliquots were removed at designated times. B. The association rate constant (k_1) was determined according to the equation $k_1 = k_{ob} - k_{-1}/[L]$, where k_{ob} is the pseudofirst order rate constant for association calculated from the slope of the plot ln $(B_{eq}/B_{eq} - B)$ vs. time (B_{eq} equals specific binding at a given time point): k_{-1} is the rate constant for dissociation; and [L] is the concentration of [125I]IBE 2254 employed in these experiments. C. The first order rate constant for dissociation (k_{-1}) was calculated from the slope of the plot, ln (B/B_{eq}) vs. time. For both association and dissociation kinetics, slopes were determined by linear regression analysis. The experiment shown is representative of three separate experiments.

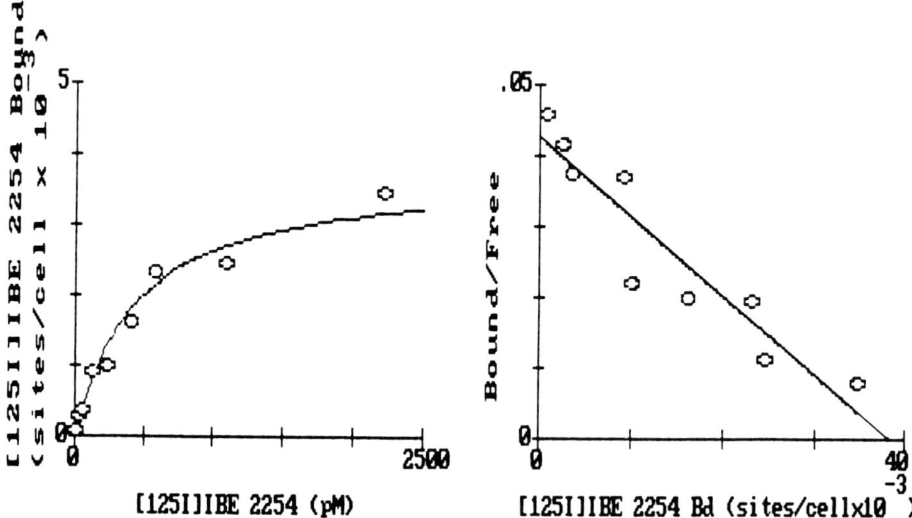

FIGURE 8.   Equilibrium binding of [125I]IBE 2254 to cultured rat heart cells. Cardiac cells were incubated with increasing concentrations of [125I]IBE 2254 and specific binding was determined. The line describing the relationship, Bound/Free vs. amount bound was determined by linear regression analysis and is shown according to the method of Scatchard.[34] The experiment shown is representative of three separate experiments.

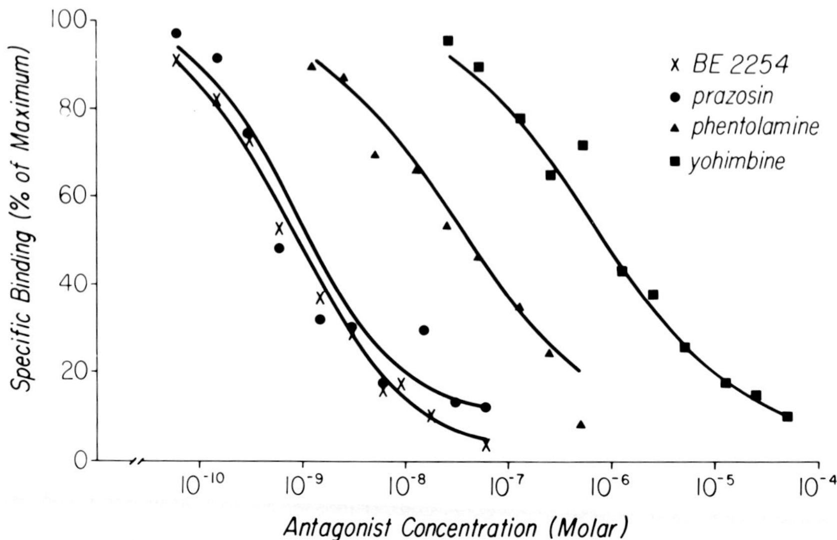

FIGURE 9.   Competition of [^{125}I]IBE 2254 binding to myocardial cells by alpha-adrenergic antagonists. The data are expressed as a percentage of maximum specific binding of [^{125}I]IBE at 25 pm. Each value is the mean of three separate experiments performed in triplicate. The curves drawn represent the best fit of the data when analyzed by log probit transformation and by least squares linear regression analysis. (From Kupfer, L. S., Robinson, R. B., and Bilezikian, J. P., Identification of alpha$_1$ adrenergic receptors in cultured rat myocardial cells with a new iodinated alpha$_1$ adrenergic antagonist [^{125}I]IBE 2254, *Circ. Res.*, 51, 250, 1982. With permission.)

input, the vasculature and other hormones. Moreover, the cultured neonatal rat heart is particularly well suited to this kind of investigation because myocardial cells obtained from neonatal rats have been exposed in vivo to negligible levels of thyroid hormone.

Cultured neonatal rat heart cells were grown in the presence of 10 nM triiodothyronine (T$_3$) and compared to cells grown in the absence of additional thyroid hormone.[45] A single class of high affinity α-adrenergic receptors was observed in both situations. The B$_{max}$ for α receptors was significantly decreased in the presence of T$_3$ (B$_{max}$ = 18,000 ± 4,300 sites per cell) as compared to control cells (B$_{max}$ = 37,000 ± 4,500 sites per cell). The affinity for [^{125}I]IBE 2254 was significantly increased in the cells exposed to T$_3$ (K$_d$ = 150 ± 37 pM) as compared to control cells (K$_d$ = 420 ± 24 pM). Using [^{125}I]ICYP under similar experimental conditions, the data show that excess thyroid hormone was associated with an increase in B$_{max}$ (17,000 ± 4,000 per cell) as compared to control cells (12,000 ± 2,600 sites per cell). The affinity for the radioligand was similar in both control conditions and under thyroid hormone treatment (K$_d$ = 24 ± 3.3 pM vs. 29 ± 4.0 pM, respectively). It is noteworthy that the thyroid hormone associated increase in the number β-adrenergic receptors contrasts with the decrease in number of α-adrenergic receptors. Quantitatively, however, the decrease in α receptor number, both in terms of sites per cell and the percentage decrease, is much greater than the increase in β receptor number. A summary of these data is shown in Table 1.

The subselectivity of α- and β-adrenergic receptors under conditions of excess thyroid hormone in cultured myocardial cells remained unchanged as compared to control conditions. The receptors could still be clearly designated as α_1 or β_1. Finally, it was possible to study the enzyme, adenylate cyclase, as a function of thyroid hormone concentration. The isoproterenol-associated increase in adenylate cyclase activity was greater in thyroid hormone treated cells than in the control (500 vs. 375 pM cAMP per mg per 10 min). The 33% increase in maximum isoproterenol responsiveness is close to the 42% increase in β-adrenergic receptor number measured under the same conditions. The concentration at which

Table 1
**THYROID HORMONE INDUCED CHANGES IN
ALPHA AND BETA ADRENERGIC RECEPTORS**

	Control	Thyroid treated
Alpha adrenergic receptor		
K_d (pM)[a]	420 ± 24	150 ± 37[b]
B_{max}(sites/cell)[a]	$37,000 \pm 4500$	$18,000 \pm 4300$[b]
Beta adrenergic receptor		
K_d (pM)[c]	24 ± 3.3	29 ± 4.0[b]
B_{max} (sites/cell)[c]	$12,000 \pm 2600$	$17,000 \pm 4000$[b]

Note: Cells were grown in serum-containing medium.

[a] n = 4.
[b] Values calculated using the Student's paired *t*-test, p < 0.05; values
    are mean ± SEM.
[c] n = 5.

isoproterenol half-maximally stimulated adenylate cyclase activity was not altered by excess thyroid hormone.

There is very little information in the literature with which to compare these results. However, Tsai and Chen[30] have shown that in cultured neonatal rat heart cells a similar increase in binding capacity as a function of thyroid hormone occurs. These investigators also found an increase in isoproterenol-associated cAMP accumulation as a function of thyroid hormone. Karliner has recently studied the effect of T_3 upon α_1 receptors in cultured neonatal ventricular myocytes.[46] Although in many respects their brief report does not provide enough details for meaningful comparison with results reported above, it is noteworthy, nevertheless, that T_3 administration was associated with a decrease in α_1-adrenergic receptors when the data were expressed on the basis of cell surface area.

Although these experiments clearly document a direct effect of thyroid hormone on the α- and β-adrenergic receptor system of the heart cell, they were limited by the use of culture medium that contains physiological concentrations of thyroid hormone. Additional preliminary experiments, therefore, have been performed with a serum free medium that allowed us to explore the effect of thyroid hormone in a setting that was initially devoid of any hormone.[47] Thus, one could analyze the effects of thyroid hormone over a range of concentrations from athyroid to hyperthyroid. The range of T_3 used was equivalent to 0.01 pM to 1 nM free T_3. The α receptor number decreased as a function of thyroid hormone concentration. The affinity of [^{125}I]IBE 2254 showed a dose related increase in binding affinity from 161 pM ($T_3 = 10^{-12}$) to 121 pM ($T_3 = 10^{-10}$) to 40 pM ($T_3 = 10^{-8}$). The binding capacity decreased at these increasing concentrations from 22,000 to 16,000 to 13,200 sites per cell. The correlation between the increase in thyroid hormone concentration and the decrease in K_d and B_{max} were both significant (p < .05). Thus, in a similar fashion to experiments with serum-containing culture media, but in a somewhat more sophisticated manner, the studies in serum free medium indicate clearly that thyroid hormone directly affects both alpha receptor affinity and binding capacity.

Experiments with [^{125}I]ICYP in serum free medium indicate that T_3 in increasing concentrations leads to an increase in β receptor sites from 1,700 sites per cell ($T_3 = 10^{-11}$) to 3,000 sites per cell ($T_3 = 10^{-9}$) to 4,000 per cell ($T_3 = 10^{-7}$). There was no consistent change in affinity for [^{125}I]ICYP as a function of thyroid hormone concentration. These results for the β receptor, therefore, also confirm the results of experiments in which serum-containing medium is used.

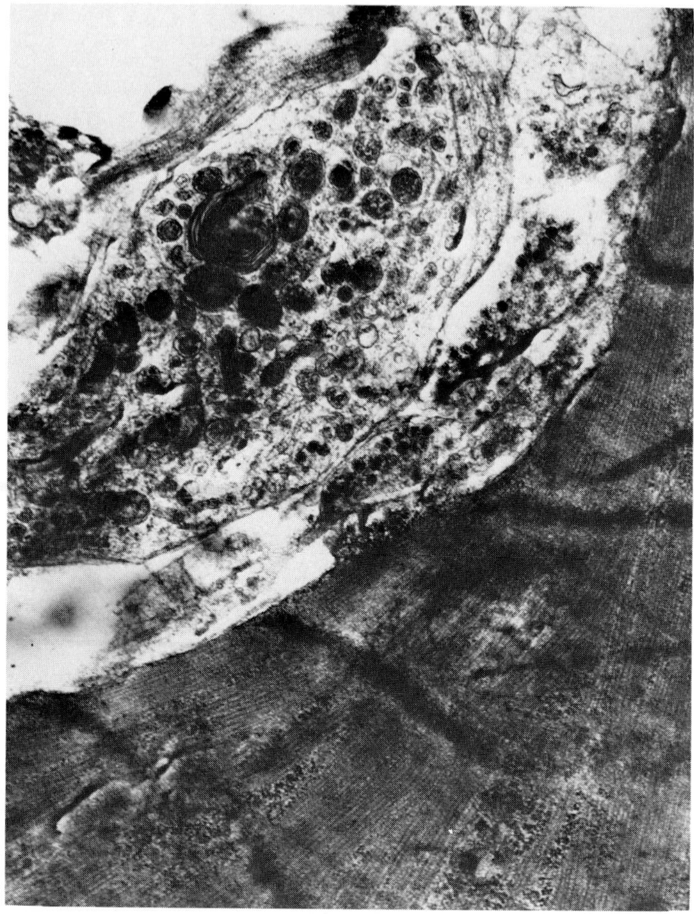

FIGURE 10.   Electron micrograph of a close contact between a cardiac cell and neuron in culture showing typical synaptic vesicle-like structures within the nerve cell and the high degree of myofibril organization of the muscle cell. The original magnification was 23,000 ×. (From Robinson, R. B., Models of cardiac development: transplants, organ culture, cell dispersion, and cell culture, in *The Developing Heart*, Legato M. J., Ed., Martinus Nijhoff, Boston, Mass., 1984, 69. With permission.)

IX. CONTROL OF ADRENERGIC FUNCTION IN CULTURED MYOCARDIAL CELLS VIA INNERVATION

The vast majority of cardiac culture preparations have been derived from fetal or neonatal rather than adult tissue. This would be expected to affect the observed adrenergic responsiveness, since it is known to change with development.[48] Since it has been proposed that developmental changes in the adrenergic system are related to the ontogeny of sympathetic innervation, it is worth considering the effect of innervation on the adrenergic responsiveness of myocardial cultures. Cultures have been reported to contain a catecholamine-like substance, but no adrenergic neurons.[49] When sympathetic ganglia are added to heart cultures, long-lasting specific associations are made between the neurons and the muscle cells of the heart cultures.[50] These nerve-muscle associations are still formed in the presence of β-adrenergic antagonists[51,52] and are functional.[52,53] Close associations (Figure 10) and functional innervation[54] may also be demonstrated in co-cultures of myocardial cells and enzyme dispersed sympathetic neurones. When the chronotropic response of these co-cultures is compared to that of noninnervated myocardial cultures an interesting result is obtained — the positive chronotropic response to α-adrenergic agonists, typically observed in culture, is changed to a negative chronotropic response following in vitro innervation.[55] This change in α-adrenergic response upon innervation in culture is similar to that observed in the intact myocardium during neonatal development,[56,57] at a time when the heart is becoming sympathetically innervated.

X. SUMMARY

The information in this chapter indicates that the culture of myocardial cells can be used advantageously to explore physiological and pharmacological properties of the adrenergic catecholamines in the heart, allowing direct analysis, relatively free of contaminating cells and neuronal input. Although information regarding the effects of the catecholamines on isolated heart cells is still rather limited, the techniques have been worked out and should provide additional useful information in the future. The availability of iodinated radioligands for the detection of α- and β-adrenergic receptors has begun also to provide useful information regarding their presence and characteristics in heart cells. Protocols exploring the influence of thyroid hormone on the cultured myocardial cell are the first examples illustrating how receptor binding techniques and the cultured cell model can be used advantageously to study the interaction between the adrenergic catecholamines and other hormone systems. It is expected that additional outstanding questions in molecular cardiology may be addressed through the use of heart cell cultures.

ACKNOWLEDGMENTS

Some of the studies reported in this chapter were supported, in part, by NIH Grants HL 28958 and HL 20859. Dr. Robinson was a Senior Investigator of the New York Heart Association, at the time of those studies.

REFERENCES

1. **Burrows, M. T.,** Rhythmische kontraktionen der isolierten herzmuskelzellen ausserhalb des organismus, *Muench. Med. Wochenschr.,* 59, 1473, 1912.
2. **Cavanaugh, M. W.,** Pulsation, migration and division in dissociated chick embryo heart cells in vitro, *J. Exp. Zool.,* 128, 573, 1955.
3. **Wollenberger, A.,** Rhythmic and arrhythmic contactile activity of single myocardial cells cultured in vitro, *Circ. Res.,* 15, II184, 1964.
4. **Sperelakis, N. and Lehmkuhl, D.,** Insensitivity of cultured chick heart cells to autonomic agents and tetrodotoxin, *Am. J. Physiol.,* 209, 693, 1965.
5. **Harrison, D. C., Kleiger, R. E., and Merigan, T. C.,** Action of isoproterenol on heart cells in tissue culture, *Proc. Soc. Exp. Biol. Med.,* 124, 122, 1967.
6. **Ertel, R. J., Clarke, D. E., Chao, J. C., and Franke, F. R.,** Autonomic receptor mechanisms in embryonic chick embryonic cell cultures, *J. Pharmacol. Exp. Ther.,* 178, 73, 1971.
7. **Hermsmeyer, K. and Robinson, R. B.,** High sensitivity of cultured cardiac muscle cells to autonomic agents, *Am. J. Physiol.,* 233, C172, 1977.
8. **Harary, I. and Farley, B.,** In vitro studies of single beating heart cells, *Science,* 131, 1674, 1960.
9. **Goshima, K.,** Initiation of beating in quiescent myocardial cells by norepinephrine, by contact with beating cells and by electrical stimulation of adjacent FL cells, *Exp. Cell Res.,* 84, 223, 1974.
10. **Lane, M.-A., Sastre, A., Law, M., and Salpeter, M. M.,** Cholinergic and adrenergic receptors on mouse cardiocytes in vitro, *Dev. Biol.,* 57, 254, 1977.

11. **Kaumann, A. J. and Wittmann, R.,** Apparent equilibrium constant between beta-adrenoceptors and a competitive antagonist on a cultured pacemaker cell of mammalian heart, *Naunyn-Shmiederberg's Arch. Pharmacol.,* 287, 23, 1975.

12. **Karsten, U., Wallukat, G., and Wollenberger, A.,** Alpha- and beta-adrenergic receptor stimulation in cultures of beating rat heart cells and its effects on automaticity and cyclic AMP levels, *Acta Biol. Med. Ger.,* 36, K63, 1977.

13. **Marvin, W., Jr., Robinson, R. B., and Hermsmeyer, K.,** Correlation of function and morphology of neonatal rat and embryonic chick cultured cardiac and vascular muscle cells, *Circ. Res.,* 45, 528, 1979.

14. **Robinson, R. B., Reder, R. F., and Danilo, P., Jr.,** Preparation and electrophysiological characterization of cardiac cell cultures derived from the fetal canine ventricle, *Dev. Pharmacol. Ther.,* 7, 307, 1984.

15. **Kaumann, A. J.,** Cultured heart cells as a model for beta-adrenoceptors in a heart pacemaker: chronotropic spare beta-adrenoceptors and spare adenyl cyclase for (−)-ISO but not for (−)-DCI in rat cardiocytes, *Naunyn-Schmiedeberg's Arch. Pharmacol.,* 320, 119, 1982.

16. **Hermsmeyer, K., Mason, R., Griffen, S. H., and Becker, P.,** Rat cardiac muscle single cell automaticity responses to alpha- and beta-adrenergic agonists and antagonists, *Circ. Res.,* 51, 532, 1982.

17. **Freyss-Beguin, M., Griffaton, G., Lechat, P., Picken, D., Quennedey, M. C., Rouot, B., and Schwartz, J.,** Comparison of the chronotropic effect and the cyclic AMP accumulation induced by beta$_2$-agonists in rat heart cell culture, *Br. J. Pharmac.,* 78, 717, 1983.

18. **Fisher, R. A., Harary, I., and Thomas, J. A.,** Homologous desensitization of rat heart cells to beta-adrenergic stimulation and the absence of alpha-adrenergic or prostaglandin E$_1$ effects, *J. Cyclic. Nucl. Protein Phosphor. Res.,* 9, 449, 1983.

19. **Marsh, J. D., Barry, W. H., and Smith, T. W.,** Desensitization to the inotropic effect of isoproterenol in cultured ventricular cells, *J. Pharmacol. Exp. Ther.,* 223, 60, 1982.

20. **Sinclair, A. J., Miller, H. A., and Harrison, D. C.,** An electrooptical monitoring technique for heart cells in tissue culture, *J. Appl. Physiol.,* 29, 747, 1970.

21. **Moura, A.-M. and Simpkins, H.,** Cyclic AMP levels in cultured myocardial cells under the influence of chronotropic and inotropic agents, *J. Mol. Cell. Cardiol.,* 7, 71, 1975.

22. **Harary, I., Renaud, J.-F., Sato, E., and Wallace, G. A.,** Calcium ions regulate cyclic AMP and beating in cultured heart cells, *Nature,* 261, 60, 1976.

23. **Renaud, J.-F., Sperelakis, N., and Le Douarin, G.,** Increase of cyclic AMP levels induced by isoproterenol in cultured and non-cultured chick embryonic hearts, *J. Mol. Cell. Cardiol.,* 10, 281, 1978.

24. **Lawrence, T. S., Beers, W. H., and Gilula, N. B.,** Transmission of hormonal stimulation by cell-to-cell communication, *Nature,* 272, 501, 1978.

25. **Hamill, O. P., Marty, A., Neher, E., Sakmann, B., and Sigworth, F. J.,** Improved patch-clamp techniques for high-resolution current recording from cells and cell-free membrane patches, *Pflügers Arch.,* 391, 85, 1981.

26. **Reuter, H., Cachelin, A. B., De Peyer, J. E., and Kokubun, S.,** Modulation of calcium channels in cultured cardiac cells by isoproterenol an 8-bromo-cAMP, *Cold Spring Harbor Symp. Quant. Biol.,* 48, 193, 1983.

27. **Watanabe, A. M., Jones. L. R., Manalan, A. S., and Besch, H. R.,** Cardiac autonomic receptors: recent concepts from radiolabeled ligand binding studies, *Circ. Res.,* 50, 161, 1982.

28. **Collucci, W. S.,** Alpha-adrenergic receptors in cardiovascular medicine, in *Receptor Science in Cardiology,* Haft, J. I. and Karliner, J. S., Eds., Futura, New York, 1984.

29. **Engel, G. and Hoyer, D.,** [^{125}I]IBE 2254, a new high affinity radioligand for alpha one adrenoceptors, *Eur. J. Pharmacol.,* 73, 221, 1981.

30. **Tsai, J. S. and Chen, A.,** Effect of L-triiodothyronine on [^{3}H]dihydroalprenolol binding and cyclic AMP response to adrenaline in cultured heart cells, *Nature,* 275, 138, 1978.

31. **Lau, Y. H., Robinson, R. B., Rosen, M. R., and Bilezikian, J. P.,** Subclassification of beta-adrenergic receptors in cultured rat cardiac myoblasts and fibroblasts, *Circ. Res.,* 47, 41, 1980.

32. **Kupfer, L. E., Robinson, R. B., and Bilezikian, J. P.,** Identification of alpha$_1$ adrenergic receptors in cultured rat myocardial cells with a new iodinated alpha$_1$ adrenergic antagonist [^{125}I]IBE 2254, *Circ. Res.,* 51, 250, 1982.

33. **Blondel, B., Roijen, I., and Cheneval, J. P., Jr.,** Heart cells in culture: a simple method for increasing the proportion of myoblasts, *Experentia,* 27, 356, 1971.

34. **Scatchard, G.,** The attractions of proteins for small molecules and ions, *Ann. N.Y. Acad. Sci.,* 51, 660, 1949.

35. **Werner, S. C.,** Hyperthyroidism, in *The Thyroid: A Fundamental and Clinical Text,* 4th ed., Werner, S. C. and Ingbar, S. H., Eds., Harper & Rowe, New York, 1978, 591.

36. **Irvine, W. J. and Toft, A. D.,** The diagnosis and treatment of thyrotoxicosis, *Clin. Endocrinol.,* 5, 687, 1976.

37. **Williams, R. S. and Lefkowitz, R. J.**, The effects of thyroid hormone on adrenergic receptors, in *Molecular Basis of Thyroid Hormone Action*, Oppenheimer, J. H. and Samuels, H. H., Eds., Academic Press, New York, 1983, 325.
38. **Bilezikian, J. P. and Loeb, J. N.**, The influence of hyperthyroidism and hypothyroidism on alpha and beta-adrenergic receptor systems and adrenergic responsiveness, *Endocr. Rev.*, 4, 378, 1983.
39. **Sharma, V. K. and Bannerjee, S. P.**, Alpha adrenergic receptors in rat heart, *J. Biol. Chem.*, 253, 5277, 1978.
40. **McConnaughey, M. M., Jones, L. R., Watanabe, A. M., Besch, H. R., Williams, L. T., and Lefkowitz, R. J.**, Thyroxine and propylthiouracil effects on alpha and beta adrenergic receptor number, ATPase activities and sialic acid content of rat cardiac membrane vesicles, *J. Cardiovasc. Pharmacol.*, 1, 609, 1979.
41. **Noguchi, A. and Whitsett, J. A.**, Ontogeny of alpha$_1$ adrenergic receptors in the rat myocardium: effects on hypothyroidism, *Eur. J. Pharmacol.*, 86, 43, 1983.
42. **Williams, R. S. and Lefkowitz, R. J.**, Thyroid hormone regulation of alpha adrenergic receptors: studies in rat myocardium, *J. Cardiovasc. Pharmacol.*, 1, 181, 1979.
43. **Ciaraldi, T. and Marinetti, G. V.**, Thyroxine and propylthiouracil effects in vivo in alpha and beta-adrenergic receptors in rat heart, *Biochem. Biopys. Res. Commun.*, 74, 984, 1977.
44. **Kunos, G., Mucci, L., and O'Regan, S.**, The influence of hormonal and neuronal factors on rat heart adrenoceptors, *Br. J. Pharmacol.*, 71, 371, 1980.
45. **Kupfer, L. E., Bilezikian, J. P., and Robinson, R. B.** Regulation of α and β adrenergic receptors by triiodothyronine in cultured rat myocardial cells, *Naunyn-Schmiedeberg's Arch. Parmacol.*, 334, 275, 1986.
46. **Karliner, J., Simpsin, P., Moazed, D., and Park, B.**, Norepinephrine and T$_3$-induced myocardial cell hypertrophy is associated with down regulation of alpha$_1$-adrenergic receptors, *Circ. Res.*, 70, II-10A, 1984.
47. **Kupfer, L. E.**, Identification, characterization and regulation by triiodothyronine of adrenergic receptors in cultured rat heart cells, Ph.D. thesis, Columbia University, 1984.
48. **Pappano, A. J.**, The development of postsynaptic cardiac autonomic receptors and their regulation of cardiac function during embryonic, fetal and neonatal life, in *Physiology and Pathophysiology of the Heart*, Speralakis, N., Ed., Martinus Nijhoff, Boston, 1984, 355.
49. **Jellinek, M., Sperelakis, N., Napolitano, L. M., and Cooper, T.**, 3,4-dihydroxyphenylalanine in cultured ventricular cells from chick embryo heart, *J. Neurochem.*, 15, 959, 1968.
50. **Mark, G. E., Chamley, J. H., and Burnstock, G.**, Interactions between autonomic nerves and smooth and cardiac muscle cells in tissue culture, *Dev. Biol.*, 32, 194, 1973.
51. **Campbell, G. R., Chamley, J. H., and Burnstock, G.**, Lack of effect of receptor blockers on the formaiton of long-lasting associations between sympathetic nerves and cardiac muscle cells in vitro, *Cell Tiss. v. Res.*, 187, 551, 1978.
52. **Purves, R. D., Hill, C. E., Chamley, J. H., Mark, G. E., Fry, D. M., and Burnstock, G.**, Functional autonomic neuromascular junctions in tissue culture, *Pflügers Arch.*, 350, 1, 1974.
53. **Marvin, W. J., Jr., Atkins, D. L., Chittick, V. L., Lund, D. D., and Hermsmeyer, K.**, In vitro adrenergic and cholinergic innervation of the developing rat myocyte, *Circ. Res.*, 55, 49, 1984.
54. **Furshpan, E. J., MacLeisch, P. R., O'Lague, P. H., and Potter, D. D.**, Chemical transmission between rat sympathetic neurons and cardiac myocytes developing in microcultures: evidence for cholinergic, adrenergic and dual-function neurons, *Proc. Natl. Acad. Sci.*, 73, 4225, 1976.
55. **Drugge, E., Rosen, M. R., and Robinson, R. B.**, Neuronal regulation of the development of the α adrenergic chronotropic response in the rat heart, *Circ. Res.*, 57, 415, 1985.
56. **Rosen, M. R., Hordof, A. J., Ilvento, J. P., and Danilo, P., Jr.**, Effects of adrenergic amines on electrophysiological properties and automaticity of neonatal and adult canine purkinje fibers: evidence for alpha- and beta-adrenergic actions, *Circ. Res.*, 40, 390, 1977.
57. **Reder, R. F., Danilo, P., Jr., and Rosen, M. R.**, Developmental changes in alpha adrenergic effects on canine purkinje fiber automaticity, *Dev. Pharmacol. Ther.*, 7, 94, 1984.
58. **Robinson, R. B.**, Models of cardiac development: transplants, organ culture, cell dispersion, and cell culture, in *The Developing Heart*, Legato, M. J., Ed., Martinus Nijhoff, Boston, 1985, 69.

INDEX

A

Acetoacetate, 87
Acetylcholine, 100, 112—113, 116—120
Actin-binding protein, 22—23
α-Actinin, 23, 25, 29
Actinomycin D, 99
Action potential, 83, 87, 115—117
Active transport, 117
Acyl CoA synthetase, 84
Acyl CoA thioester, 85
Adenine nucleotide translocase, 43, 48
S-Adenosylmethionine decarboxylase (AdoMetDC), 96—99
Adenylate cyclase, 97, 151
  in adrenergic response, 146
  in atrium, 129
  in muscarinic response, 119, 124—128, 134, 136
  polyamines and, 100—104
  thyroid hormone and, 154—155
  in ventricle, 129
AdoMetDC, see S-Adenosylmethionine decarboxylase
α-Adrenergic agonist, 145
β-Adrenergic agonist, 102, 114, 121, 144—145
β-Adrenergic antagonist, 145
Adrenergic neuron, 156
α-Adrenergic receptor, 143—157
  identification of, 147, 151—154
    radioligands for, 147—148
  regulation by thyroid hormone, 151—155
β-Adrenergic receptor, 100, 121, 125, 136, 143—157
  desensitization of, 146
  identification of, 147—151
    radioligands for, 147—148
  regulation by thyroid hormone, 151—155
Adrenergic response, 144—146
  control by innervation, 156—157
  mechanism of action of, 146—147
β-Adrenergic stimulation, 119
Adriamycin, 46
Albumin, 83—84
Amino acids
  pools of, 2—3, 6
  post-translationally modified, 9
  reutilization of, 3, 7—9
  uptake of, 97
Aminoacyl-tRNA, 2—3, 6—7
Antimyosin immunoglobulin, 18
Aortic constriction, 97
Aortic stenosis, 30—31, 33
Arachidonic acid, 83
Arrhythmia, 83
Arylhydrocarbon hydroxylate, 70—71
ATP, 42—43, 124
ATPase, 34
  calcium-activated, 22
  magnesium, 45

  myosin, 14, 21, 42, 45
  sodium-potassium, 117
Atractyloside, 48
Atrioventricular node, 113, 116
Atrium
  adenylate cyclase in, 129
  hypertrophy of, 20
  muscarinic response in, 116—117, 129
Atropine, 113, 123
Axon, postganglionic, 113

B

Beating function
  fatty acids and, 82—83
  oxygen consumption by, 52—53
Beating rate
  adrenergic response and, 147
  muscarinic response and, 112—116, 120, 123—124, 130—134
8-Bromo-cyclic AMP, 147
8-Bromo-cyclic GMP, 120

C

Calcium, 96, 102, 104
Calcium channel, 147
Calcium fluxes, 112, 116—120, 136
Calcium permeability, 128
Calmodulin, 34, 101, 104
cAMP, see Cyclic AMP
Carbamylcholine, 113—114, 117—121, 123—125, 129—134
Carbohydrates, metabolism of, 86
Cardiolipin, 46
Carnitine, 85—86
Carrier protein, 79
Catecholamines, 69, 97, 102, 144
Cell culture, see Heart cell culture
Cell differentiation, 96—97
Cell division, 99, 104
Cell growth, 30—34, 96, 99—100
  negative, 5—6
  positive, 5—6
  protein metabolism during, 4—6
Cell junction, 26, 29
Cell proliferation, 96—97, 99—100
Cell shape, 28—30
cGMP, see Cyclic GMP
Chloroquine, 88
Cholera toxin, 70
Cholesterol, 83, 89, 115
Cholesteryl ester, chylomicron, 71—73
Cholesteryl linoleyl ether, 71—73
Cholinergic agonist, 114
Cholinergic receptor
  muscarinic, see Muscarinic receptor; Muscarinic response

glucose uptake and, 79
  metabolism of, 82—89
    abnormal, 87—88
  nonesterified, 82
  nonspecific effects of, 82—83
  oxidation of, 84—89
  pools of, 84
  uptake of, 83—84
Fetal heart
  creatine kinase in, 54
  isomyosins in, 17—18
Fibroblast, 149—151
Filaments, see Intermediate filaments
Filamin, 23, 25
2,4-Fluorodinitrobenzene, 42
Follicle-stimulating hormone (FSH), 147
Force-velocity curve, 21
FSH, see Follicle-stimulating hormone

G

Glucagon, 68—69, 100
Glucocorticoids, 18, 69
Glucose
  metabolism of, 78—82, 88
  oxidation of, 80—82
  uptake of, 78—79
Glutamine, 86
Glyceraldehyde-3-phosphate dehydrogenase, 80
Glycogen, 78—79, 85
Glycogen synthase, 79
Glycolysis, 49, 78, 81—82
Golgi apparatus, 26
Growth factors, 104
Growth hormone, 69
GTP, 120, 127
Guanine nucleotide regulatory protein, 119—120,
    124—125, 135—136
Guanine nucleotides, 123
Guanylate cyclase
  in muscarinic response, 120
  polyamines and, 100—101, 103—104

H

Heart cell culture, see also specific topics
  creatine kinase in, 49—52
  glucose metabolism in, 78—82
  intracellular energy transport in, 41—56
  lipid metabolism in, 82—89
  lipoprotein lipase in, 63—73
  muscarinic cholinergic receptors in, 111—137
  oxygen consumption in, 51—54
  polyamines in, 95—105
  protein synthesis and degradation, 1—10
Heavy meromyosin (HMM), 15
Hemodynamic overload, isomyosins and, 17, 20
Heparin, 65—67, 72—73
Hexamethonium, 113
Hexokinase, 78—79
Histamine, 119

Histones, acetylation of, 98
HMM, see Heavy meromyosin
Hormones
  cyclic AMP-mediated, 101
  isomyosins and, 18—20
  regulation of lipoprotein lipase by, 69—71
Hydrocortisone, 69
3-Hydroxybutyrate, 86—87, 89
3-Hydroxybutyrate dehydrogenase, 87
Hyperoxia, 99
Hyperpolarization, 117
Hyperthyroidism, 17, 151—155
Hypertrophy, cardiac, 55, 97, 104
  pressure overload, 20
Hypophysectomy, 18
Hypothyroidism, 17, 20, 151—155
Hypoxia, 97

I

IAP, see Islet-activating protein
I-band, 26
IBE 2254, 148, 151, 153—155
IBMX, see 3-Isobutyl-1-methylxanthine
ICYP, see Iodocyanopindolol
IHYP, see Iodohydroxybenzylpindolol
Immuno-electron microscopy, 24
Immunofluorescence, 34
Immunofluorescence labeling, 27
Inflammatory response, 87
Inotropic response, 116, 119—120, 146
Insulin, 68, 78, 100
  isomyosins and, 17, 20
  regulation of lipoprotein lipase by, 69
Interatrial septum, 18
Intercalated disc, 22, 24—26
Interfibrillar space, 24, 28
Intermediate filament, 22—29, 34—35
  function of, 28, 30—31, 33—34
  relationship to microtubules, 26—28, 31—32
Intracellular energy transport, 41—56
  creatine kinase isozymes
    in cardiac and skeletal muscle, 43—49
    in heart cell cultures, 49—52
  developmental aspects of myocardial creatine
      kinase, 54—55
  oxygen measurement, 51—54
Iodination, 8
Iodocyanopindolol (ICYP), 148, 154—155
Iodohydroxybenzylpindolol (IHYP), 148—150
Iodopindolol (IP), 148
Ion fluxes, muscarinic response and, 117—119, 130
IP, see Iodopindolol
Ischemia, 45, 80
Islet-activating protein (IAP), 124
3-Isobutyl-1-methylxanthine (IBMX), 70—71, 100
Isometric tension, 21
Isomyosins, 14—22, 34—35, 50
  age-dependent evolution of, 17—18
  diabetes and, 17, 20
  heavy and light chain heterogeneity, 15—16

The Heart Cell In Culture

Volume III

Editor

Arié Pinson, D.Sc.
Senior Investigator
Laboratory for Myocardial Research
Institute of Biochemistry
Hebrew University - Hadassah Medical School
Jerusalem, Israel

CRC Press, Inc.
Boca Raton, Florida

Library of Congress Cataloging-in-Publication Data

The Heart cell in culture.
  Includes index.
  1. Heart cells. 2. Cell culture. I. Pinson,
Arié, 1931-   [DNLM: 1. Cells, Cultured.
2. Myocardium--cytology. WG 280 H4365]
QP114.C44H43   1987   612'.17'0724   87-21864
ISBN 0-8493-4696-7 (set)

International Standard Book Number 0-8483-4696-7 (set)
International Standard Book Number 0-8493-4697-5 (Volume I)
International Standard Book Number 0-8493-4698-3 (Volume II)
International Standard Book Number 0-8493-4699-1 (Volume III)

Library of Congress Card Number 87-21864
Printed in the United States

PREFACE

At the turn of the century, cell culture and other in vitro methods were developed and have since proven invaluable in research on differentiation, specific function, and metabolism of various tissues, which are often difficult to study in vivo due to the complexity of the interactions between different tissues and between tissues and body fluids.

In 1912, Burrows made the pioneering discovery that cardiac explants produced muscle cells that were capable of dividing and beating with a regular rhythm, thus demonstrating the myogenic nature of heart contraction. The tryptic method of dissociating cells was first described by Rous and Jones in 1916, and revived almost 40 years later by Moscona. In 1955, Cavanaugh isolated and grew heart cells from the chick embryo. Twenty-five years ago, Harary and Farley prepared the first cultures of neonatal rat heart cells. This method of culturing mammalian neonatal heart cells has proved to be a crucial step in the application of heart cell culture techniques to various research areas dealing with myocardial biology at the cellular level. As a direct consequence of this, since the early 1970s, there has been about a sixfold increase in the number of scientific papers per year dealing with heart cells in culture. Clearly, a summary of the progress in this field is long overdue.

My primary aim as the editor of *The Heart Cell in Culture* was not to provide a laboratory manual, although techniques are discussed when relevant. Rather, each chapter presents a critical review, demonstrating the progress made during the last 25 years as well as perspectives for the future.

The Heart Cell in Culture is presented in three multidisciplinary volumes, spanning a wide spectrum of topics and presenting state-of-the-art reviews. It is difficult to divide such a book into distinct parts, but it has been presented in three sections, one in each volume: the first on the heart cell culture system — historical background, the development of culturing techniques, and the function of such cells; the second on the basic biochemistry of the myocardial cell in culture; and the third on applied research in pharmacology and pathology of cardiac cells, including a chapter on cultured adult heart cells, which is becoming a popular tool in cardiac research. The minireview chapters in the series were selected with two main objectives in mind: to complement full-length comprehensive reviews in the same field by the addition of new dimensions and scope, in which case they immediately follow the main chapter, or to consider topics which are relatively new or show promise for the future development of research in cardiac biology. Indeed, I expect that these chapters should be extremely stimulating for this reason.

I hope that people working in the field of heart research will find useful information in these volumes, and that they will foster better understanding of cardiac biology and stimulate further progress in this important field of research.

I would like to thank Dr. R. Goldberg for her assistance with the thankless task of language editing, corrections to the text, and proofreading. I am grateful to all the contributors who made this book, *The Heart Cell in Culture,* possible.

Arié Pinson
Jerusalem, Israel

THE EDITOR

Dr. Arié Pinson, D.Sc., is a Senior Investigator at the Hebrew University-Hadassah Medical School and permanent visiting professor at the Faculty of Life and Environmental Sciences of the University of Dijon in France.

He studied Clinical Chemistry and Biology at the Univesity of Geneva in Switzerland. He earned his M.Sc. degree from the Hebrew University of Jerusalem in 1967 and his D.Sc. for a thesis on the Metabolism of Palmitic and Erucic Acid in Rat Heart Cell Cultures from the Univesity of Dijon in France in 1975.

Since 1975, the research carried out at Dr. Pinson's Laboratory for Myocardial Research has contributed toward establishing cultures of cardiac myocytes as a standard technique for studying the biochemistry of the heart in both health and disease. In a broader sense, these studies, often conducted in collaboration with leading scientists overseas, have advanced knowledge on cardiac cells in culture as a tool in the fields of development, differentiation, biochemistry, physiology, and medical research.

Dr. Pinson has been awarded two prizes for distinguished research by the Hebrew University. He has also held long-term fellowships from INSERM — Institut National de la Santé et de la Recherche Médicale (France), CNRS — Centre National de la Recherche Scientifique (France), DGRSRT — Délégation Générale à la Recherche Scientifique et Technique (France), ZWO — The Netherlands Society for the Advancement of Pure Research, and annual summer fellowships from INSERM.

A recognized authority in his field, he has frequently been invited to give plenary lectures at international conferences: at the Biochemical Society (U.K.) and at several ISHR (International Society for Heart Research) meetings and at other meetings focusing on cardiomyocytes.

Dr. Pinson has been invited to many seminars at universities and research institutes in Holland, Germany, France, and Israel. In 1984, he founded the IGHR (the Israeli Group for Heart Research), which is affiliated to the ISHR.

Dr. Pinson's current research interests include the biochemistry of cultured cardiomyocytes with particular emphasis on lipid and glucose metabolism, function and structure of sarcolemma, metabolism of high energy phosphates, and the use of this system as a model for elucidation of anoxic injury and iron overload in the heart.

CONTRIBUTORS

VOLUME III

Michael W. Berns, Ph.D.
Professor
Department of Development and Cell
  Biology
Department of Surgery
University of California
Irvine, California

Hans H. Eppenberger, Ph.D.
Professor
Department of Cell Biology
Swiss Federal Institute of Technology
Zurich, Switzerland

Wieland Gevers, Ph.D.
Professor
Department of Medical Biochemistry
University of Cape Town
Cape Town, South Africa

Amirav Gordon, M.D.
Associate Professor
The Hubert H. Humphrey Center for
  Experimental Medicine and Cancer
  Research
Hebrew University-Hadassah Medical
  School
Jerusalem, Israel

Michael Heller, Ph.D.
Associate Professor
Department of Biochemistry
Hebrew University-
  Hadassah Medical School
Jerusalem, Israel

Chaim Hershko, M.D.
Professor
Department of Medicine
Shaare Zedek Medical Center
Hebrew University-
  Hadassah Medical School
Jerusalem, Israel

H. Hietter
Centre de Neurochimie
C.N.R.S.
Strasbourg, France

Kiyoshi Hosono, M.D.
Vice President and Professor
Department of Medicine
National Defense Medical College
Saitama, Japan

Hideyuki Ishida
1st Department of Medicine
National Defense Medical College
Saitama, Japan

Margarita C. Kitzes, Ph.D.
Department of Anatomy and Physiology
California State University
Long Beach, California

Gabriella Link
Assistant
Department of Nutrition
Hebrew University-
  Hadassah Medical School
Jerusalem, Israel

B. Luu, Ph.D.
Directeur de Recherches
Centre de Neurochimie
C.N.R.S.
Strasbourg, France

Marcel Mersel, Ph.D.
Chargé de Recherches
Centre de Neurochimie du C.N.R.S.
Strasbourg, France

Chandrasekharm N. Nagineni, Ph.D.
Assistant Research Physiologist
Division of Nephrology
VA Medical Center
Sepulveda, California

Tetsuya Nakamura, Ph.D.
Tsukuba Research Laboratories
Eisai Co., Ltd.
Ibaraki, Japan

Jean-Claude Perriard, Ph.D.
Professor
Department of Cell Biology
Swiss Federal Institute of Technology
Zurich, Switzerland

Arié Pinson, D.Sc.
Senior Investigator
Department of Biochemistry
Hebrew University-
 Hadassah Medical School
Jerusalem, Israel

Hans Michael Piper, M.D., Ph.D.
Professor
Department of Physiology
University of Düsseldorf
Düsseldorf, West Germany

Irmelin Probst, Ph.D.
Institute of Biochemistry
University of Goettingen
Goettingen, West Germany

Beat Schäfer, Ph.D.
Department of Cell Biology
Swiss Federal Institute of Technology
Zurich, Switzerland

Peter Schwartz, Ph.D.
Institute of Anatomy
University of Goettingen
Goettingen, West Germany

T. Scott-Burden, Ph.D.
Senior Lecturer
Department of Medical Biochemistry
Cape Town University
Cape Town, South Africa

Maria Seraydarian, Ph.D.
Professor
School of Nursing
UCLA
Los Angeles, California

Rolf Spahr, M.A.
Department of Physiology
University of Düsseldorf
Düsseldorf, West Germany

P. G. Spieckermann, M.D.
Professor
Institute of Physiology
University of Vienna
Vienna, Austria

Noboru Suzuki, Ph.D.
Tsukuba Research Laboratories
Eisai Co., Ltd
Ibaraki, Japan

Beatrice A. Wittenberg, Ph.D.
Professor
Department of Physiology and
 Biophysics
Albert Einstein College of Medicine
Bronx, New York

In memory of my parents,
Zeev and Pessia Pinson

To my wife and children,
Yona, Gavriel, Halléli, and Shira

To Professor Prudent Padieu

TABLE OF CONTENTS

VOLUME I

TABLE OF CONTENTS

VOLUME II

TABLE OF CONTENTS

VOLUME III

Chapter 20

INTERACTIONS AND MODES OF ACTION OF CARDIAC GLYCOSIDES WITH CULTURED HEART CELLS

Michael Heller

TABLE OF CONTENTS

I. INTRODUCTION

The bicentennial anniversary of Withering's discovery of the therapeutic powers of the foxglove cannot be allowed to pass without quoting from his famous monograph "An Account of the Foxglove and Some of its Medical Uses" which was published in 1785:[1]

"In the year 1775, my opinion was asked concerning a family *receipt* for the cure of dropsy. I was told that it long had been kept a secret by an old woman in Shropshire, who had sometimes made cures after the more regular practitioners had failed . . . this medicine was composed of twenty or more different herbs but it was not very difficult for one conversant in these subjects to perceive that the active herb could be no other than *foxglove.*"

Ever since Withering's time, an explanation for the way in which cardiac glycosides exert their effects has been sought. The modern era or the "molecular" approach to the mode of action of cardiac glycosides began over 30 years ago when Schatzmann described the specificity of cardiac glycosides in inhibiting the sodium and potassium active transport in erythrocytes for the first time.[2]

This was shortly followed by Skou's demonstration that the transport process is an enzymatic one involving a Na-K-ATPase in the plasma membrane which is specifically inhibited by cardiac glycosides.[3] Later, Repke identified the ATPase responsible for Na^+ and K^+ transport ATPase as the digitalis receptor.[4]

In extensive experimental work, almost all the known metabolic processes which may partly explain the cellular basis of the positive inotropic action of cardioactive steroids have been investigated. These studies are aimed at establishing a link between the biochemical effects of these drugs and the enhanced contractility of the myocardium.

Difficulties in interpreting the data obtained from experiments with intact myocardial preparations, led several research groups to employ cultured, single beating cardiac muscle cells which yielded more straightforward results.[5,6] Indeed, for the last 30 years this technique has proven invaluable for investigating the biochemistry, pharmacology, electrophysiology, and cellular biology of the heart.

The earliest studies on the response of explanted or cultured heart cells to chronotropic and inotropic agents such as cardiac glycosides were done by Wollenberger and his associates during the early 1960s.[7] Cultured, single cells from embryonic chick ventricles appeared to be particularly sensitive to digitalis. Digitoxin at a concentration of 1 nM caused a measurable increase in the rate of beating of the cells.[8] A detailed study was done on arrhythmias produced in single cells at digitoxin concentration of 0.1 μM or more, using microcinematography.[9]

Since then, methodology has improved. Chronotropic and inotropic measurements of contractility can now be monitored and recorded using a TV video system. A camera attached to a computerized, digitalized device enables determination, with a high degree of precision and reproducibility, of contractility, rate of beating, velocity of relaxation, the time taken to reach the peak contraction, relaxation, and twitch times.[10-13]

Quantitative correlation between the binding of cardiac glycosides to the receptors and the induction of the physiologic consequences can conveniently be studied in cultured cells. The beating muscle cell of the myocardium constitutes a simple model for the study of drug action in general, and for the action of cardiac glycosides, in particular.

There are several reasons for the suitability of this system:[14,15]

1.   The sensitivity of these cells to various of the pharmacologically active group of cardiac glycosides, eliciting increases in both amplitude and in the velocity of contraction, serves as a model for inotropic effects. Furthermore, the active transport of Na^+ and K^+ vary as a function of the concentration of the drug.
2.   The receptor(s) for the drug have been shown to be located on the surface of the

sarcolemma facing the extracellular space, allowing studies of the interaction with the drug to be made. Such processes may be characterized without disturbing the integrity of the membrane.

3. Much simpler interpretation of the results is possible in this system, since certain complicating factors, primarily diffusional and intercellular barriers as well as neural effects, are absent.

At present, no review has been published so far specifically covering this topic. Therefore, in this review, the intention is to consider and discuss various aspects with direct bearing on the application of cultured cardiac cells in elucidating the mechanism of action of cardiac glycosides. These include studies on binding of the cardiac glycosides and their effects on transport, flux, and contractility as well as on receptor regulation, but hardly any electrophysiologic studies. Certain aspects have received more attention than others. The pertinent questions are (1)"What have the cultured heart cells contributed to our understanding of how cardiac glycosides act?" or (2)"Have the actions of cardiac glycosides advanced our knowledge and understanding of the cation fluxes in cardiac cells and how these regulate intracellular processes?" It seems that these two questions describe two sides of the same coin and both of them will be considered in this review.

II. SOURCE OF CELLS

It is surprising to find that only a very limited number of animal species have been used as sources of hearts for such studies. The most common ones are hearts of chick embryos and postnatal rats. There are a few reports of studies on cells from adult animals such as dog, rat, and rabbit.

The reasons why the animal species employed for this purpose are so limited most probably stem from methodological difficulties. The two types of cell systems from immature animals mentioned above have received more intense and thorough investigation than expected. Furthermore, many aspects of the biochemistry, physiology, and pharmacology of the heart considered pertinent to human heart, have also been demonstrated in postnatal and embryonic cells. In contrast to cells from adult animals, primary cultures of these cells can be relatively easily prepared and lend themselves to various manipulations. The well-characterized properties of these two types of cells, accumulated from numerous studies over the years, also explain in part their frequent use. As far as interactions with digitalis are concerned, in contrast to other tissues or cells from adult rats, postnatal heart cells that are "digitalis-sensitive" tissues, respond with many characteristics similar to those of several other species commonly used because of certain similarities in their responses to those in humans.

Furthermore, procedures are available which enable separation of cardiac muscle cells from other types of cells present in the heart, without altering the native "conformation" of the sarcolemma — the site of digitalis action.

It should, however, be stressed that considerable progress has recently been made concerning the preparation and characterization of stable myocytes from adult species, including human hearts.[16] It is to be anticipated that further reports of pharmacological and biochemical studies employing adult cardiomyocytes will be published soon.

III. METHODOLOGY

The two types of cells commonly employed in studies with cardiac glycosides are prepared from hearts from 11-day-old chick embryos or 1- to 2-day-old rats.

Experiments employed to determine the effects of cardiac glycosides on cellular activities fall into three categories:

1.   Binding of ^{3}H-labeled cardiac glycoside: (a) under equilibrium conditions; (b) the kinetics of association and dissociation; and (c) competition in binding between labeled and unlabeled drugs and other effectors.
2.   Fluxes — transport of monovalent and divalent cations, rates, absolute concentrations, and contractility of cells in the presence of cardiac glycosides.
3.   Regulation of receptors with respect to their number and properties.

The advantage of using the synchronously beating monolayers of cells can be appreciated especially in the following contexts:

1.   Definition of unidirectional and net fluxes of Na^+, K^+, and Ca^{2+}, which are very difficult to interpret due to considerable diffusional barriers and uncertainty with respect to the volume of the interstitial space, both in the intact animal and tissue models. This has to be done with good temporal resolution under conditions identical to those in which the inotropic state is determined.
2.   There is a substantial positive inotropic response to cardiac glycosides which may be recorded and monitored using the appropriate equipment.
3.   Rapid onset of the inotropic response on exposure to cardiac glycosides and rapid offset following washout of the drug may be obtained.
4.   They are particularly useful for distinguishing and resolving components of cation fluxes, particularly those of Ca^{2+}, in beating preparations which are also under the influence of cardiac glycosides.
5.   Neural influences or any other factors that could alter the contractile state or ion transport, e.g., endogenous catecholamines, are absent in these preparations.

The parameters mentioned above could or, perhaps, should be measured simultaneously, and should yield relevant information provided that the methodology is reliable and sensitive enough.

Basic physical characteristics of cultured rat myocardial cells indicate that they are oblate spheroid in shape (apparent surface area (A) = 2,250 μm^2; cell volume (V) = 2,600 μm^3);[17] those for embryonic cardiac cells indicate a spherical shape (A = 670 μm^2 and V = 1,630 μm^3),[18] and in cells from adult rat ventricles, A = 1,400 μm^2/cell.[19]

IV. PROPERTIES OF THE CARDIAC GLYCOSIDE BINDING SITE

The interaction of any drug with its target tissue is expected to have biological consequences. If a "specific" biological site is involved, it may be classified as a "receptor" when the interactions, frequency of interactions, or occupation of these site(s) initiate physiological or pharmacological effects.

The binding of radioactively labeled cardiac glycosides (^{3}H or ^{14}C) to cells may either be followed by "destructive" or "nondestructive" methods. This may either be done by scraping the labeled cells and counting them directly in a detergent-containing fluorophor, or after extraction of the labeled digitalis.[20] The alternative "nondestructive" method for determining the cell surface receptors is by using condensed phase radioluminescence (CPR) by recording the very low intensity photons of light emitted when tritiated drugs are located close to the natural fluorophor (e.g., tryptophan) in the membrane proteins.[21] This method has been used recently to measure binding of various ^{3}H-labeled drugs to myocytes, but has unfortunately not yet been applied to cardiac glycosides.

The binding may be described in terms of "total" binding and "nonspecific" binding. Two sets of experimental conditions are employed in the case of cardiac cells (and probably with other preparations too). Total binding is estimated by exposure to increasing concentrations of the drug and measurement of the amount remaining bound, either in an isotonic medium devoid of K^+ or at concentrations below 1 mM until equilibrium is attained. Values for nonspecific binding are obtained by using either (1) heat-denatured cells (e.g., 60°C, 60 min) or (2) in the presence of high concentrations (>0.1 mM) of unlabeled ouabain. The "specific" binding is then calculated by subtraction of the "nonspecific" from the "total" binding.[20] Binding to intact cells is very sensitive to changes in the extracellular concentration of potassium.[20,22,23]

Detailed description and systematic analysis of properties of cardiac glycoside receptors in cells has only recently been performed, although there were a few earlier reports. If there is only one population of sites (one class of receptors) a simple saturation isotherm yielding a linear "Scatchard" plot would be expected.[24] Nonhomogenous populations of binding sites (multiple classes of receptors) however, would give rise to a curvilinear Scatchard plot. This plot may also be analyzed according to the methods of Feldman,[25] Weidemann et al.[26] or Fehlmann et al.[27] Other curve-fitting programs may be used to give the binding parameters, e.g., the equilibrium dissociation constant (K_D) and the maximal binding capacity (B_{max}). These values may be expressed as the number of binding sites per cell or per unit surface area.

For cardiac glycosides, many studies have been done which showed the existence of only one population of sites. However, recently, several reports have appeared describing additional sites in noncardiac and cardiac tissues as well as in cultured cells.[28]

In cells from both rat and chick embryos, some results support a homogeneous population of receptors, whilst others indicate that there are heterogenous populations.

A. Chick Embryo Cells

However, careful studies of the binding characteristics under equilibrium conditions (K_D) and measurements of association and dissociation rate constants (k_a and k_d, respectively) have led to different conclusions.

Erdmann et al.[29] showed the existence of a single class of receptors with the following parameters:

1. Maximal binding sites (B_{max}) of 2.6 pmol/mg protein which corresponds to 860,000 sites per cell and a density of 1,280 molecules per μm^2 (this value was calculated taking into account data for the chick cell size since the reported value of 430 molecules per μm^2 is based on data from the rat which apparently has larger cells).
2. Binding constant (K_D) of 0.15 to 0.19 μM were determined with association rate constants (k_a) = 3.2×10^4 m^{-1}s^{-1} and dissociation rate constants (k_d) = 4.5×10^{-3}s^{-1} at 37°C.
3. K^+ ions induce a decrease in the association rate without affecting the process of dissociation.
4. The dissociation is more sensitive to temperature, the rates of association being less affected; this results in an increase in the equilibrium constant (K_D) with increasing temperature in the range from 9 to 37°C.

The consequences of the interaction and occupation of sites with ouabain, e.g., inhibition of ($^{86}Rb^+$ + K^+) influx, decrease of intracellular K^+ and increase of intracellular Na^+, occurred with EC_{50} values very close to the K_D value, i.e., 0.58 μM, 0.67 μM, and 0.74 μM, respectively.

Also, the inhibition of the sarcolemmal Na-K ATPase occurred with an IC_{50} of 1.4 μM, which correlated well with the K_D value of 0.43 μM for ouabain binding to the single class of sites in the isolated sarcolemma of chick embryo heart.

There is, however, a discrepancy between the degree of occupancy of the binding sites by ouabain and inhibition of active transport of $^{86}Rb^+$.

Up to 40% of the sites bind ouabain without causing a stoichiometric inhibition of the pump.[29]

A single class of receptors was also reported by Smith's group:[10,23,30]

1. In the K^+-free medium, maximum binding values, represent the total number of receptors with B_{max} = 5.6 to 5.9 pmol ouabain per mg protein at equilibrium, which correspond to 910,000 to 958,000 sites per cell or a receptor density of 1,360 to 1,450 molecules of pump/μm^2.

2. At physiological concentrations of K^+ (i.e., 4 mM), only a fraction of the sites (i.e., 38%) were occupied so that the K_D values in K^+-free medium and 4 mM K^+ containing medium were 0.12 to 0.14 and 0.49 μM, respectively. K^+ only affected the association rate constants (k_a) which were 0.17 $\pm$ 0.04 and 0.045 $\pm$ 0.007 pM^{-1} min^{-1} in 0 and 4 mM K^+, respectively. Dissociation rate constants (k_d) were not affected, being in the 0.22 to 0.26 min^{-1} range. At lower concentrations of ouabain (e.g., 100 pM), only 10% of the sites were occupied at physiological $[K^+]_o$ concentration of 4 mM.

Ouabain at 0.1 μM did not produce an increase in contractile force (amplitude of cell motion). Only at 0.6 to 1 μM was a significant elevation observed after a delay of 30 sec, reaching a maximum of 50% within 5 to 7 min. No effect on beating rate was detectable. Indeed, more than 10% of total binding sites must be occupied before a significant change in contractile state occurs.

However, the same cells may apparently also display features corresponding to a heterogeneous population of sites:

1. Lazdunski's group[12] has described two classes of sites with high affinity (K_D = 33 nM), low capacity (B_{max} = 1.2 pmol/mg protein or 1.5 $\times$ 10^5 sites per cell, corresponding to a receptor density of 220 molecules per μm^2) and a second class of low-affinity sites (K_D = 3.9 μM) with a high capacity (B_{max} = 11.8 pmol/mg protein or 1.4 $\times$ 10^6 sites per cell corresponding to a receptor density of 2,090 molecules per μm^2 surface area).

- The high affinity sites were detected by equilibrium binding only and no measurable consequences of this binding could be assigned to them.
- Both the low and high affinity sites already exist at a very early stage of development; both retain the same affinities throughout embryonic life cycle (5 to 20 days *in ovo*). The number of low affinity receptor sites/cell also remains constant during embryonic development. The amount of high affinity sites, however, decreases after 14 days *in ovo* from 1.3 to 0.4 pmol/mg protein.

2. All the measured physiological parameters affected by ouabain (e.g., ^{86}Rb$^+$ influx [Na$^+$]$_i$, ^{45}Ca^{2+} flux and inotropic effects) were inhibited at an EC$_{50}$ = 1 to 6 μM, a value which correlates well with the K_D values of 2 to 7 μM for the low affinity receptor.

The reasons for the conflicting results are not clear but may be related to differences in experimental procedures.

B. Rat Postnatal Cells

High and low affinity binding sites for cardiac glycosides in these cells have been described:

1. Different but close values for binding constants were reported by Erdmann's and Heller's groups, for the high affinity sites, 32 nM[31] and 89 nM, respectively.[22]

2.   In one study, kinetics of binding showed a single association rate constant (k_a) of 1.2 to 1.5 $\times$ 10^4 $m^{-1}s^{-1}$ with two dissociation rate constants of k_{d1} = 9 $\times$ $10^{-4}s^{-1}$ and k_{d2} = 2 $\times$ $10^{-5}s^{-1}$.[31] Reported values for the K_D of the low affinity site were: 1 μM or 7.1 μM or even higher values.[22,31] The binding capacities (B) for the high affinity sites were 0.2 pmole/mg protein, corresponding to 80,000 sites per cell[31] or 147,000 sites per cell,[22] representing only about 8% of low affinity sites having a maximal binding capacity (B_{max} = 2.6 pmol/mg protein or 10^6 sites per cell).[22]
3.   Both classes of sites are sensitive to K^+, which affects the association velocity only, and not the dissociation rates.[31]
4.   The binding and dissociation of ouabain are strongly temperature-dependent in the range between 9 to 37°C, with a Q_{10} value of ca. 3.[31]
5.   Ouabain dissociates from the high affinity sites rapidly, with a $t_{1/2}$ = 12 min, whereas this process is much slower at the low affinity sites, with $t_{1/2}$ = 575 min.[31]
6.   Due to these differences, the ratio of ouabain molecules bound to the high and low affinity sites at different ouabain concentrations may be estimated. At 10 nM, 0.1 μM, 1 μM, and 10 μM ouabain, the ratio is 13, 4.21, 0.6, and 0.13, respectively.[31]
7.   Thus, at concentrations of below 0.5 μM ouabain, binding to the high affinity sites predominates, whereas at higher ouabain concentrations the low affinity sites also become occupied. This range of concentrations was termed the *critical digitalis concentrations.*[20,22]

Systematic studies have revealed that only with the low affinity site are the binding characteristics correlated to inhibition of the pump. This was reflected in decreased active ($^{86}Rb^+$ + K^+) influx, a decrease in cellular K^+ and an increase in cellular Na^+ with EC_{50} values corresponding to 13 μM, 19 μM, and 10 to 100 μM ouabain, respectively.[31]

In another report by McCall's group,[17] specific ouabain binding was measured for a range of ouabain concentrations. A biphasic curve was obtained with a saturatable component at about 0.1 μM ouabain and a nonsaturatable or low affinity component measured in the presence of 20 mM K^+, i.e., under conditions for nonspecific binding. At saturation the number of sites per cell was 1.6 $\times$ 10^6 and the K_D value, 3.1 nM. The calculated receptor density or pump density was 720 molecules per μm^2.

1.   The binding was rapid ($t_{1/2}$ = 8 to 10 min) and was strongly affected by K^+, whereas the dissociation was slow ($t_{1/2}$ = 11 hr) and unaffected by K^+ or ouabain.
2.   Although only the high affinity sites are characterized to some degree, the K^+ influx is inhibited with an EC_{50} of 5 μM, which might correspond to a low affinity site.

C. Cells from Adult Rabbit, Dog, and Rat Hearts

Myers and co-workers have shown that cardiomyocytes from adult New Zealand white rabbits in a Ca^{2+}-free medium, containing either 6 mM or 1.2 mM K^+, bind 2.45 $\pm$ 0.4 and 4.72 $\pm$ 0.59 pmol/mg protein, respectively.[32]

Cardiomyocytes from adult dog bind ouabain specifically in the presence of Mg^{2+} and this binding is inhibited by K^+. Maximal binding capacity (B_{max}) was 7.4 $\times$ 10^5 molecules per cell. Two components of binding with high and low affinities, i.e., with K_Ds of 56 nM and 0.67 μM, respectively, were reported. Apparent negative cooperativity was suggested by a Hill coefficient of n = 0.72 and a $S_{0.5}$ value of 0.71 μM obtained from an average affinity profile constructed according to DeMeyts and Roth.[33] Onji and Liu[33] attributed the decrease in the apparent receptor affinity to increasing occupancy which was not due to a site to site interaction between receptors. They therefore proposed that there is only one class of ouabain receptors with two or more interconvertible conformational states.

In studies carried out by Adams and co-workers,[34] myocytes from adult rat myocardium only showed single class of sites with a $K_D = 0.124 \pm 0.09$ μM. A second class of sites was not detectable, although ouabain produced a biphasic effect in right ventricular strips with ED_{50} values of 0.5 and 19 μM, respectively.[34]

Although this chapter has been devoted to the characterization of the receptor(s), some correlation with physiologic parameters has been included in order to point out the relevant conclusions. Studies in intact cells cannot provide information on the molecular nature of the receptors. The only recognizable and characterized receptor for cardiac glycosides so far is Na-K-ATPase — the molecular expression of the sodium pump in the sarcolemma. Any binding parameters of cardiac glycosides which may be correlated with functions carried out by this pump support the contention that it is the digitalis receptor. The molecular nature and function of any other sites involved in the Na-K pump remain to be discovered.

V. ACTIVE TRANSPORT, FLUXES, CELLULAR CATION CONCENTRATIONS, AND CONTRACTILITY: MODE I AND MODE II

The immediate consequences of the interaction of cardiac glycosides with sarcolemmal receptor(s) are changes in certain properties of these receptors (i.e., activity) leading to direct or indirect responses. In general, the effects of any drugs which enhance the force of contraction of the heart are ultimately due to an increase in the availability of intracellular, free Ca^{2+}, which then interacts with the contractile proteins. Consequently, several major systems must be integrated in order to support a causal relationship between the interaction of cardiac glycosides with their receptor(s) and the increase in the force of contraction, i.e., positive inotropy.

Three systems are currently believed to be involved in this delicately balanced mechanism:

1. The classical Na^+-K^+ pump which electrogenically counter-transports three Na^+ ions out of the cells, two K^+ ions flowing into the cell in exchange — the energy for this process being derived from the hydrolysis of ATP catalyzed by the sarcolemmal Na-K-ATPase.
2. A Na^+-Ca^{2+} carrier or "exchanger system" was postulated and later shown to operate in intact cardiac tissue and can function in a reversible fashion. It couples the transport of Na^+ in one direction with that of Ca^{2+} in the opposite direction, by an electrogenic mechanism.
3. The source of the intracellular free Ca^{2+} has not been sufficiently well defined and controversy exists as to the site and mechanism of action for the immediate supply of free $[Ca^{2+}]_i$. Is it the same as for (2) or does it require additional stores? The control of transport processes for the three cations, Na^+, K^+, and Ca^{2+}, and maintenance of their intracellular concentrations, are considered to be intimately related to the mode of digitalis action.[14,15,35-44]

It is currently believed that cardiac glycoside interaction with cardiac cells may be mediated by the above mentioned systems in one of the following ways:

Mode I — Cardiac glycosides inhibit the sodium pump in the cardiac sarcolemma, which correspond to Na-K-ATPase, by what is known as the "sodium lag hypothesis". There are now some doubts as to whether inhibition occurs at inotropic, subtoxic drug concentrations or only at levels which are toxic. It may well be that only limited inhibition of the pump is beneficial, with a greater degree of inhibition possibly leading to a depolarization of the cells and irregularities eventually resulting in arrhythmia. Many studies show that in cardiac tissue exposed to concentrations of ouabain which increase the force of contraction, the sodium pump (measured as $^{86}Rb^+$ or $^{42}K^+$ influx) is also inhibited with a significant correlation. If the pump is inhibited, the transient increase in $[Na^+]_i$ due to the inward flow of

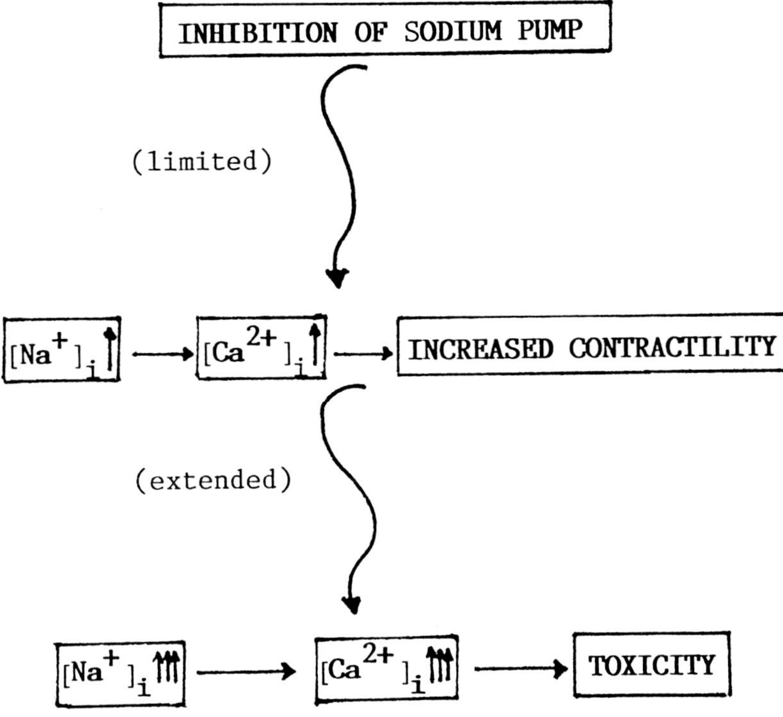

FIGURE 1. Mode I.

Na^+ during the course of each action potential, will persist and may either be confined to a compartment close to the pumping unit (experimentally difficult to assess), or may even lead to an increase in total $[Na^+]_i$. Increase in the intracellular Na^+ provides a stimulus for an augmented Ca^{2+} influx and a reduced or delayed Ca^{2+} efflux via the voltage-dependent Na^+-Ca^{2+} exchanger. The result is a net increase of intracellular Ca^{2+}.

This may be summarized as shown in Figure 1.

Mode II — A different approach to explaining the mode of digitalis action is based on observations in which concentrations of cardiac glycosides in the nanomolar range are sufficient to elicit small inotropic effects and are also associated with *stimulation* of the sodium pump rather than its inhibition.

Alternatively, the interaction of cardiac glycoside with its receptor (Na-K-ATPase) does not cause any measurable alterations in the normal Na^+ or K^+ gradients, but favors a conformational state of the receptor or perturbs the membrane components in a different fashion, enhancing the lability of the Ca^{2+} store associated with the sarcolemmal acidic phospholipids (Figure 2).

In both of these modes of action, the inhibition of the sodium pump at toxic concentrations of cardiac glycosides causes the toxic symptoms.

A number of experiments have been done with cultured cells in attempts to explain the mechanism and to obtain more clear-cut data on the unidirectional fluxes of cations.

A. Mode I

A detailed analysis was conducted by Smith's group employing chick embryo cells. Their results may be summarized as follows.[14,15,23,30] Cultured cells are suitable for studying the effects of cardiac glycosides using the following criteria:

1.   Positive inotropism, reflected in an increase in the amplitude and velocity of the motion

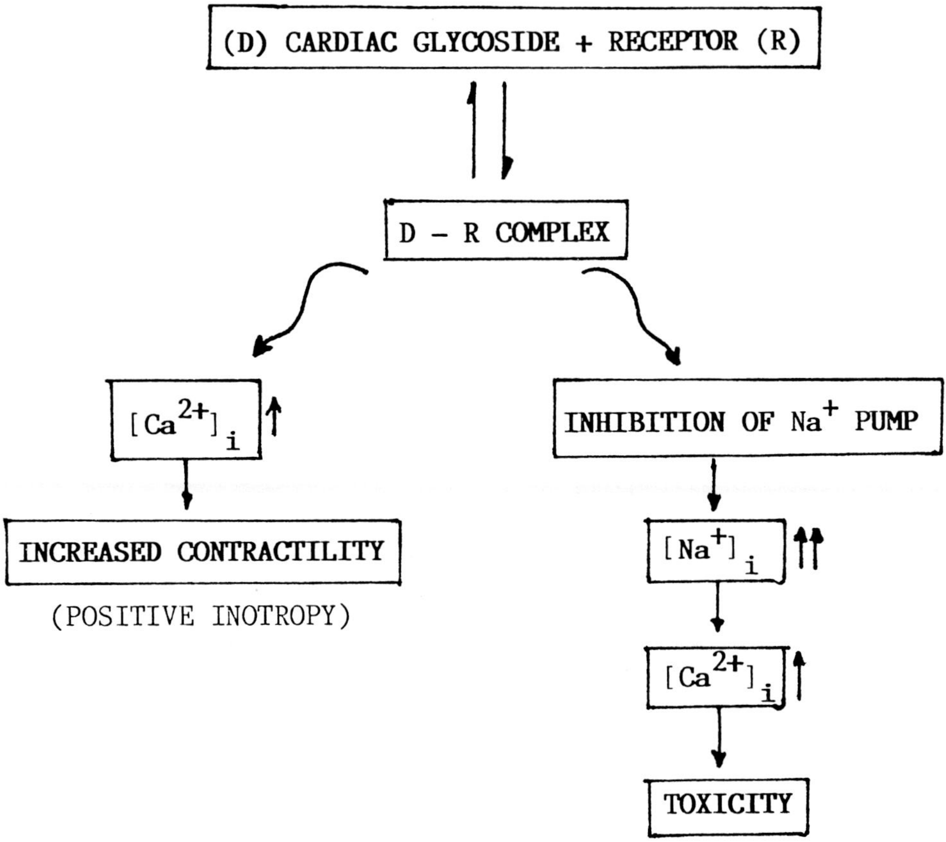

FIGURE 2.   Mode II.

of the cell wall, was apparent within 0.5 to 2 min of exposing the cells to 1 μM ouabain, and reached a plateau within 5 to 6 min, the effect being reversed within 5 min of ouabain washout.

2.   The degree of increase in cell wall amplitude correlated well with the inhibition of active $^{86}Rb^+$ or $^{42}K^+$ uptake in a concentration range from 0.1 to 1.5 μM ouabain.

A measurable reduction in the rate of monovalent cation transport occurred only when a certain, critical fraction of the sodium pump sites became occupied by ouabain. It was maintained that at 10% occupancy, the change in the pump flux per cell is probably compensated by an increase in turnover rate of the unblocked pumping sites, since no increase in $[Na^+]_i$ was detected. Only at 1 μM ouabain do sufficient sites become occupied in order to cause inhibition of the sodium pump with measurable elevation of $[Na^+]_i$.

The physiological $[Na^+]_i$ in these cells is close to the sodium pump (or sarcolemmal ATPase) K_M value. Therefore, conditions leading to an increase in $[Na^+]_i$ (e.g., $[K^+]_o \rightarrow$ 0 or subtoxic concentrations of cardiac glycosides which also cause positive inotropism) are expected to result in a transient elevation of $[Na^+]_i$ and consequently increase the sarcolemmal sodium pump activity. Elevation of $[Na^+]_i$ may not be detected under all conditions due to the lack of adequately sensitive equipment. Nevertheless, speculations suggesting localized or compartmentalized increase in $[Na^+]_i$ may explain the increase in transport activity by more than fourfold when chick myocytes are exposed to K^+-free medium. Such phenomena are frequently described by statements such as "these cells possess a substantial reserve capacity of sodium pumps".

It is well known that the sodium pump activity in intact cells or the activity of Na-K-ATPase in isolated sarcolemma, are regulated by the concentrations of monovalent cations, either in a hyperbolic (i.e., Michaelis-Menten) or in a sigmoidal (allosteric) type of activity vs. concentration relationship. Agents increasing the intracellular concentrations (i.e., of $[Na^+]_i$) at the relevant site of action may alter the activity of the pump by a simple substrate or ligand effect to a certain degree. Other explanations based on different considerations are also possible.[45]

$^{45}Ca^{2+}$ uptake by monolayers of cultured chick embryo cells may be resolved into two components: A rapidly exchangeable component which becomes evident between 1 sec and 1 min and accounts for about 37% of total calcium content, has a rate constant of 3.91 min^{-1} and half life $(t_{1/2})$ of 11 sec. The second, a slow, exchangeable component is completed within 120 min, accounting for 63% of Ca^{2+} content and has a rate constant of 0.069 min^{-1}. The rapidly exchangeable Ca^{2+} pool seems to be related to the contractile state, being sensitive to trans-sarcolemmal Na^+ gradients. The intracellular content of Na^+, K^+, and rapidly exchangeable Ca^{2+} pools change upon exposure of the cells to ouabain. The threshold concentration for inotropic response, i.e., 0.1 μM, caused no measurable changes in either $[Na^+]_i$ or in $[Ca^{2+}]_i$. Only in the range between 0.5 to 1 μM ouabain did the increases in intracellular Na^+ and Ca^{2+} of up to 130% of the control values correlate well with the measured inotropy.

Washout of inhibitory concentrations of ouabain (e.g., 1 μM) caused a rapid return of: (1) Monovalent cation transport ($^{42}K^+$ uptake: from 55% to 107% of the control value within 1 min); (2) Contractile state (within an additional 4 to 5 min); and (3) Rapidly exchangeable cellular Ca^{2+} (7 min after removal of ouabain) to baseline levels. On the other hand, intracellular concentrations of Na^+ and K^+ only returned to normal 60 min after the ouabain washout.

In studies on postnatal rat cells, McCall[17] showed that $^{24}Na^+$ efflux is also composed of a fast and a slow component. The fast component accounts for 80% of total exchangeable Na^+ (i.e., 27.5 pmol/cm^2 $\times$ sec), has a rate constant of 1.92 min^{-1} and $t_{1/2} = 22$ sec. The slow component has a rate constant of 0.234 min^{-1} and $t_{1/2} = 183$ sec. Omission from the medium of extracellular 5.4 mM K^+ reduced the total flux by 56%, but only affected the fast component, decreasing its rate constant to 0.84 min^{-1}, increasing its $t_{1/2}$ to 49.5 sec. $^{42}K^+$ flux is described by a single exponential curve with $t_{1/2} = 12$ min with a total rate of 13.9 pmol/cm^2 $\times$ sec. Both fluxes were inhibited by ouabain with an $EC_{50} = 5$ μM.

Taking all the evidence together, these findings suggest the existence of rapidly exchangeable pools of Ca^{2+} and of Na^+, which are somehow related to the bulk content of the cell and may be compartmentalized at subsarcolemmal sites. It is possible that the activity of Na^+ affects cellular Ca^{2+} content via a Ca^{2+}-Na^+ exchanger employing these two pools.

Evidence in both types of cells supporting the concept of causal correlation between the state of contractility (inotropic effects) and the effects of ouabain on the Na-K pump ($^{86}Rb^+$ or $^{42}K^+$ influx), and alterations in the intracellular content of Na^+, K^+, and $^{45}Ca^{2+}$ influx have been presented earlier. All of them occurred at EC_{50} values of ouabain in the micromolar range: chick embryonic cells — 0.6 to 1.0 μM[29] or 1 to 7 μM[12] and postnatal rat cells at 10 to 20 μM.[31]

B. Mode II

Concentrations of ouabain or other cardiac glycosides administered at therapeutic doses may be measured in the nanomolar range in the patients' plasma in the free form (i.e., not bound to the plasma proteins). Certain tissue preparations also respond to these low concentrations of ouabain with small but significant inotropic effects via mechanism(s) probably unrelated to the Na-K pump inhibition.[36] Evidently, in in vitro tissue preparations, nanomolar concentrations of cardiac glycosides "stimulate" the monovalent cation active transport, while inhibition of the pump only occurs in the presence of higher concentrations of these drugs. Other studies have shown small but reversible *negative* inotropic effects associated with the ''stimulatory'' effects caused by therapeutic doses (1 to 10 nM) of ouabain.[34,36,46]

Similar effects have also recently been observed in cultured cardiac muscle cells from postnatal rats in which the total [86]Rb$^+$ influx was measured in medium containing 5 mM RbCl.[47] A significant increase in influx of 10 to 50% over the values in control cells was obtained at nanomolar concentrations of ouabain, whereas the influx was inhibited by ouabain at EC$_{50}$ = 1 to 5 $\mu$$M$. The active component of the [86]Rb transport (i.e., the component which is inhibited by ouabain in the millimolar concentration range) was 50% inhibited by 0.1 $\mu$$M$ ouabain. At present, no other data have been published describing the effects of nanomolar concentrations of ouabain on contractility, other electrophysiologic parameters or on Na$^+$, K$^+$, or Ca^{2+} in cultured cardiac muscle cells. However, results of experiments on isolated atria (guinea pig) show apparent *stimulation* of the pump (measured as [86]Rb$^+$ uptake), and contractility and *decreased* Na$^+$ and K$^+$ gradients.[36] As has been emphasized, the dose-response curves of isolated heart preparations exposed to cardiac glycosides are dependent on several factors, including the nature of the cardiac glycoside used, composition of perfusion fluid, rate of electrical stimulation, anatomical origin of the tissue and the species from which the heart originated.[46,48]

Another important factor is the catecholamine content.[14,15] It is known that β-adrenergic agonists can stimulate active transport of monovalent cations. Mammalian myocardial tissue contains norepinephrine stores in the adrenergic nerve terminals.[14,15,49,50] These terminals seem to contribute a substantial portion of the total Na-K-ATPase activity of myocardial homogenates.[14,51]

The stimulation in pumping rate could be correlated with the effects of beta (but not alpha) adrenergic agonists. Direct application of β-adrenergic agonists (norepinephrine, isoproterenol), antagonists (propranolol), or endogenously released norepinephrine by tyramine, or previously depleted, catecholamine-depleted tissue by 6-hydroxydopamine or reserpine, all consistently showed a 25 to 35% stimulation of the [86]Rb$^+$ uptake.[49]

Two distinct, positive inotropic sites for ouabain are demonstrable in adult rat ventricular strips.[34] A higher affinity response (EC$_{50}$ = 0.5 $\mu$$M$) was well correlated with an apparent high affinity site for ouabain in intact myocytes prepared from these ventricles with K$_{D}$ = 0.12 $\mu$$M$, measured under equilibrium conditions. A low-affinity site for contractile force with an EC$_{50}$ = 19 $\mu$$M$ could not be detected by ouabain binding, but correlated quite well with an IC$_{50}$ = 50 $\mu$$M$ for the inhibition of the sarcolemmal Na-K-ATPase prepared from the ventricles.

The biphasic effect of ouabain on the isometric contractile force in adult rat ventricular strips was unaffected by β-adrenoreceptor blockers such as Sotalol or by pretreatment with reserpine.[34]

If the stimulatory effect by the nanomolar concentrations of ouabain is indeed mediated via catecholamines, it could be represented by the additional high affinity site for ouabain with K$_{D}$ values also in the nanomolar concentration range. Even so, one still is faced with several problems:

1.   How does the process in the nerve endings occur? If the release of catecholamines occurs at nanomolar concentrations of ouabain, would this be sufficient to inhibit the nerve ending Na-K-ATPase activity — a process thought to be responsible for norepinephrine re-uptake?[51-54] Indeed, recent studies have shown little support for an association between Na-K-ATPase activity and norepinephrine release or re-uptake.[54]
2.   Could this be related to the presence of two molecular forms of Na-K-ATPase (isoenzymes), with different affinities for cardiac glycosides, one of which may contain the second site for digitalis binding?[55]

3.   Would this effect be sufficient to explain the stimulatory action by nanomolar concentrations of ouabain of the sodium pump in cultures of myocytes from postnatal rat hearts? It should be borne in mind that such cultures are most probably devoid of cells of neuronal origin, and do not contain endogeneous catecholamines.[12]

The concept of calcium store(s) is a difficult one. Calcium is distributed in heart tissue over multiple compartments which are very difficult to assess quantitatively, so that their role in regulating calcium-dependent contractility is not easy to define. Simplification of some of these difficulties, by using monolayers of cultured cells thus providing less complicated models for unidirectional fluxes, was thought to be possible. It has been suggested that the Ca^{2+} available for the control of contractility comes from sarcolemmal sources. Alternatively, it may be bound to those sarcolemmal sites previously thought to be the acidic portions of the glycocalyx ("superficially located on the outer surface of the sarcolemma and La^{2+} displaceable" — identical with sialic acid). More recently, this issue has been reexamined and the acidic phospholipids (phosphatidyl-serine, -inositol, and -glycerol) located on the inner surface of the sarcolemma, facing the cytoplasm, are now thought to carry out this particular function.[43]

The effects of cardiac glycosides on the rapidly exchangeable calcium pool in chick embryo cells has already been discussed above.

VI. REGULATION OF RECEPTOR CONCENTRATION

Modulation of the pumping activity of the existing pumping units by ligands or by other substrates may be responsible for "short term" regulation of the monovalent cation transport in these cells. However, pumping units, or their enzymatic expression (Na-K-ATPase) are subject to metabolic turnover, which keeps their density at the cell surface at a steady level. The steady state may be modified in a process which alters the turnover rate by affecting the rates of either biosynthesis or degradation of the enzyme. This type of regulation is considered as being the "long-term regulation" mechanism. A detailed discussion on turnover studies of Na-K-ATPase in cultured cells, their synthesis, insertion into cell membrane, and degradation may be found in a recent overview.[56]

The "short" and "long" term regulation of the pump activity and its surface density, respectively, may be identified by:

1.   Changes in pumping rate per site following Na^+ loading which elevate the intracellular concentration ($[Na^+]_i$) above the K_m for Na^+ and consequently accelerate pumping, i.e., "short-term" regulation
2.   An increase or decrease of the actual total number of functional enzyme (or receptor) molecules following the growth of the cells in media either low in K^+ or containing nontoxic concentrations of cardiac glycosides, i.e., "long-term" regulation

According to a hypothesis based on the above data, partial blockade of sodium pumping units leads to the production of more sites by the cell. Furthermore, experiments were done to show that the cells were responding to an elevation of $[Na^+]_i$ rather than to $[K^+]_i$ losses.[57-59]

These observations are of great interest and correlate quite well with the observations made in patients suffering chronic hypokalemia or those subjected to chronic treatment with cardiac glycosides. Their erythrocytes had a greater number of binding sides for ouabain as well as elevated Na-K-ATPase activity.[60] The relevance of these findings lies in that they may have clinical implications related to the processes involved in the functional regulation of the heart and other tissues, especially for individuals under cardiac glycoside treatment or in those suffering from chronic hypokalemia over extended periods. Such a mechanism of adaptation may be of importance in weakening both toxic effects as well as the positive inotropic action of these drugs.

A. Girardi Cells

In Girardi heart cells, an established line derived from human atria,[61] grown in medium low in K^+ or without K^+ and in the presence of up to 50 nM ouabain for 24 hr, $[Na^+]_i$ content increases from 19 to 107 mM, and $[K^+]_i$ decreases from 172 to 7 mM, concomitantly with an increase in the number of ouabain molecules bound per cell. This number only represents the free sodium pumping units which increased from 2.2×10^6 to 3.82×10^6 molecules per cell.[57-59]

B. Chick Embryo Ventricular Cells

Conflicting results on the long-term regulation of pumping units have been obtained from ^{3}H-ouabain equilibrium binding studies:

1.   Lamb's group[62] maintains that cells grown for 24 hr in 10 nM ouabain have less free sites than normal cells, i.e., 0.88 vs. 2.17×10^5 molecules per cell, respectively. During this period $[Na^+]_i$ increased from 29 to 155 mM and $[K^+]_i$ decreased from 230 to 70 mM. Such down-regulation could be explained by the fact that measurements of the total ouabain binding sites had been too low since only the *free* sites, unblocked by molecules of unlabeled ouabain during growth, had been determined.
2.   Erdmann's group[63,64] demonstrated upregulation of sites in these cells following chronic exposure (20 to 72 hr) either to low K^+ (0.7 to 1 mM) or to ouabain (3 mM). The number of pumping units/cell was increased by 150 to 200% without altering their affinity for ouabain. In control cells there were 10^6 sites per cell.
3.   A more detailed study by Smith's group[30,65] confirmed and extended these results showing that cells grown for 48 hr in either low K^+ (1 mM), ouabain (1 μM), veratridine (1 μM) or triiodothyronine (T_3, 10 nM) increased the surface density of digitalis receptors by 60, 25, 20, and 61%, respectively. The first three agents also increased the steady state levels of $[Na^+]_i$ by 37, 18, and 35%, respectively, whereas T_3 *lowered* it to 80% of the control value.

They proposed that the signal controlling the surface density of the sodium pump sites was elevated $[Na^+]_i$. Agents causing an increase in $[Na^+]_i$ (i.e., low $[K^+]_o$ and ouabain), or agents such as veratridine (which do not act on pump activity) have a well-known direct effect on the sodium pump, in that they apparently regulate the density of the pumping sites *without* inhibiting the pump or altering intracellular levels of Ca^{2+} (over the range of concentrations tested). T_3 was an exception to this pattern. Although it *increased* the total number of functional sodium pump units, this was associated with a *decrease* of cellular sodium content, and *reduction* of both positive inotropic and toxic responses of ouabain.

We are still left with the problem of how to explain the mechanism(s) for the long-term regulation ATPase turnover in the sarcolemma of isolated myocytes.

VII. SUMMARY

Cultured cardiac cells have only been used to a limited extent in studies on the effects of cardiac glycosides on various physiological, pharmacological, biochemical, and regulatory processes. Prior to 1979, about 15 papers had been published on this subject,[66] and a few more have appeared since then.

The key factor governing the application of both cardiac glycosides and cardiac muscle cells in culture was, and still is, the crucial question: how is the level of intracellular Ca^{2+} controlled?

The detection of more than one binding site for cardiac glycosides with high affinity but low capacity has complicated matters still further, since no physiological relevance or function has been attributed to these so far. Short-term regulatory mechanisms of pump activity and the regulation of the pumping site density, both mediated by intracellular Na^+, yet with different time scales, are an attractive proposition, but these two systems may not be the only ones present.

ACKNOWLEDGMENTS

The author wishes to thank Mrs. S. Beck, Miss H. Hallaq, and Dr. H. Schwalb, without whose able assistance the studies done in his laboratory would not have been possible.

The results of studies performed in Professor Heller's laboratory reported in this review were supported in part by grants from the Chief Scientist, Ministry of Health, Israel and the Schoenbrun Foundation, The Hebrew University.

REFERENCES

1. **Withering, W.,** An Account of the Foxglove and some of its medical uses, with practical remarks on dropsy and other diseases, in *Classics of Cardiology,* Willus, F. A. and Keys, T. E., Eds., Henry Schuman Inc., New York, 1941, 231.
2. **Schatzmann, H.,** Herzglykoside als Hemmstoffe fur den aktivan kalium und Natrium transport durch die Erythrozyten membran, *Helv. Physiol. Pharmacol. Acta,* 11, 346, 1953.
3. **Skou, J. C.,** The influence of some cations on the ATPase from peripheral nerves, *Biochim. Biophys. Acta,* 23, 394, 1957.
4. **Repke, K. and Portius, H. J.,** Uber die Indentität der Ionenpumpen ATPase in der Zellmembran des Herzmuskels mit einem Digitalis Rezeptoren, *Experientia,* 19, 452, 1963.
5. **Cavanaugh, M. W.,** Pulsation, migration and division in dissociated chick embryo heart cells *in vitro, J. Exp. Zool.,* 128, 537, 1955.
6. **Harary, I., and Farley, B.,** In vitro studies of single isolated beating heart cells, *Science,* 131, 1674, 1960.
7. **Wollenberger, A. and Halle, W.,** Specificity of the effects of cardiac glycosides on the rhythmic contraction of single cultured cardiac muscle cells, *Nature,* 188, 144, 1960.
8. **Wollenberger, A. and Halle, W.,** Action of cardiac glycosides on automaticity and contractile activity in single cardiac muscle cells grown in vitro, *Proc. 1st Int. Pharmacol. Meeting,* 3, 87, 1963.
9. **Halle, W. and Wollenberger, A.,** Beiheft zum Forschungsfilm: Rhythmische Kontraktion Kultivierter isolierter Herzzellen und ihr verhalten gegenüber Digitoxin, Berlin Deutsches Zentralinstitut für Lehrmittel, 1962, 10.
10. **Biedert, S., Barry, W. H., and Smith, T. W.,** Inotropic effects and changes in sodium and calcium contents associated with inhibition of monovalent cation active transport by ouabain in cultured myocardial cells, *J. Gen. Physiol.,* 74, 479, 1979.
11. **Harary, I., Wallace, G., and Bristol, G.,** A video-computer for the chronotropic and inotropic measurements of the beating of cultured heart cells, *Cytometry,* 3, 367, 1983.
12. **Kazazoglou, T., Renaud, J. F., Rossi, B., and Lazdunski, M.,** Two classes of ouabain receptors in chick ventricular cardiac cells and their relation to Na-K-ATPase inhibition intracellular Na^+ accumulation, Ca^{2+} influx and cardiotonic effect, *J. Biol. Chem.,* 258, 12163, 1983.
13. **Werdan, K., Bauriedel, G., Bozsik, M., Krawietz, W., and Erdmann, E.,** Effects of vanadate in cultured rat heart muscle cells, *Biochim. Biophys. Acta,* 597, 364, 1980.
14. **Smith, T. W., Antman, E. M., Friedman, P. L., Blatt, C. M., and Marsh, J. D.,** Digitalis glycosides, mechanisms and manifestations of toxicity. Part II, *Prog. Cardiovas. Dis.,* 26, 495, 1984.
15. **Smith, T. W. and Barry, W. H.,** The role of Na-K-ATPase as a cardiac glycoside receptor, in *Receptor Science in Cardiology,* Haft, J. and Karliner, J., Eds., Mount Kisco, N.Y., 1979, 259.

16. **Powell, T.,** Methods for the preparation and characterization of cardiac myocytes, in *Methods in Studying Cardiac Membranes,* Vol. I, Dhalla, N. S., Ed., CRC Press, Boca Raton, Fla., 1984, 41.

17. **McCall, D.,** Cation exchange and glycoside binding in cultured rat heart cells, *Am. J. Physiol.,* 236, C87, 1979.

18. **Barry, W. H. and Smith, T. W.,** Mechanisms of transmembrane calcium movement in cultured chick embryo ventricular cells, *J. Physiol.,* 325, 243, 1982.

19. **Powell, T.,** Electrophysiological properties of isolated ventricular myocytes, *Basic Res. Cardiol.,* 80 (Suppl.), 87, 1985.

20. **Heller, M. and Beck, S.,** Interactions of cardiac glycosides with cells and membranes. I, *Biochim. Biophys. Acta,* 514, 332, 1978.

21. **Bailey, I. A., von Tscharner, V., and Harris, D. R.,** Non-invasive measurements of cell surface receptors, *Basic Res. Cardiol.,* 80 (Suppl.), 43, 1985.

22. **Friedman, I., Schwalb, H., Hallaq, H., Pinson, A., and Heller, M.,** Interactions of cardiac glycosides with cultured cardiac cells. II, *Biochim. Biophys. Acta,* 598, 272, 1980.

23. **Kim, D., Barry, W. H., and Smith, T. W.,** Kinetics of ouabain binding and changes in cellular sodium content,$^{42}K^+$ transport and contractile state during ouabain exposure in cultured chick heart cells, *J. Pharmacol. Exp. Therap.,* 231, 326, 1984.

24. **Scatchard, G.,** The attraction of proteins for small molecules and ions, *Ann. N.Y. Acad. Sci.,* 51, 660, 1949.

25. **Feldman, H. A.,** Mathematical theory of complex-ligand binding systems at equilibrium. Some methods for parameter fitting, *Anal. Biochem.,* 48, 317, 1972.

26. **Weidmann, M. Y., Erdelt, H., and Klingenberg, M.,** Adenine nucleotides translocation of mitochondria, *Eur. J. Biochem.,* 16, 313, 1970.

27. **Fehlmann, M., Morin, O., Kitahgi, P., and Freychet, P.,** Insulin and glucagon receptors of isolated rat hepatocytes: comparison between hormone binding and amino acid transport stimulation, *Endocrinology,* 109, 253, 1981.

28. **Erdmann, E., Ed.,** Cardiac glycosides receptors and positive inotropy evidence for more than one receptor: symposium, *Basic Res. Cardiol.,* 79 (Suppl.), 1984.

29. **Werdan, K., Wagenknecht, B., Zwissler, B., Brown, L., Krawietz, W., and Erdmann, E.,** Cardiac glycoside receptors in cultured heart cells. I. Characterization of one single class of high affinity receptors in heart muscle cells from chick embryos., *Biochem. Pharmacol.,* 33, 55, 1984.

30. **Kim, D., Marsh, J. D., Barry, W. H., and Smith, T. W.,** Effect of growth in low potassium medium or ouabain on membrane Na-K-ATPase, cation transport and contractility in cultured chick heart cells, *Circ. Res.,* 55, 39, 1984.

31. **Werdan, K., Wagenknecht, B., Zwissler, B., Brown, L., Krawietz, W., and Erdmann, E.,** Cardiac glycoside receptors in cultured heart cells. II. Characterization of high and low affinity binding sites in heart muscle cells from neonatal rats, *Biochem. Pharmacol.,* 33, 1873, 1984.

32. **Myers, T. D., Fraser, J. P., and Boerth, R. C.,** Binding of ^{3}H-ouabain by isolated cardiac myocytes, *Fed. Proc.,* 376, 913, 1978.

33. **Onji, T. and Liu, M. S.,** Ouabain receptor in isolated adult dog heart myocytes, *Arch. Biochem. Biophys.,* 207, 148, 1981.

34. **Adams, R. J., Schwartz, A., Grupp, G., Grupp, I., Lee, S. W., Wallick, E. T., Powell, T., Twist, V. W., and Gathiram, P.,** High affinity ouabain binding site and low dose positive inotropic effect in rat myocardium, *Nature,* 296, 167, 1982.

35. **Lüllmann, H. and Peters, T.,** Action of cardiac glycosides on the excitation-contraction coupling in heart muscle, *Prog. Pharmacol.,* 2, 1, 1979.

36. **Noble, D.,** Mechanism of action of therapeutic levels of cardiac glycosides, *Cardiovas. Res.,* 14, 495, 1980.

37. **Lüllmann, H., Peters, T., and Preuner, J.,** Mechanism of action of digitalis glycosides in the light of new experimental observations, *Eur. Heart J.,* 3 (Suppl.), 45, 1982.

38. **Lüllmann, H., Peters, T., and Preuner, J.,** Role of plasmalemma for Ca^{2+} homeostasis and E.C.C. in cardiac muscle, in *Cardiac Metabolism,* Drake-Holland, A. J. and Noble, M. I. M., Eds., John Wiley & Sons, New York, 1983, chap. 1.

39. **Lüllmann, H., Peters, T., and Ravens, U.,** Pharmacological approaches to influence cardiac inotropism., *Pharmacol. Therap.,* 21, 229, 1983.

40. **Lüllmann, H., Peters, T., Prillwitz, H. H., and Ziegler, A.,** Cardiac glycosides with different effects on the heart, *Basic Res. Cardiol.,* 79 (Suppl.), 93, 1984.

41. **Langer, G. A.,** Control of Ca^{2+} movement in the myocardium, *Europ. Heart. J.,* 4 (Suppl.), 5, 1983.

42. **Langer, G. A.,** The "sodium pump lag" revisited, *J. Mol. Cell. Cardiol.,* 15, 647, 1983.

43. **Langer, G. A.,** Calcium at the sarcolemma, *J. Mol. Cell. Cardiol.,* 16, 147, 1984.

44. **Repke, K. R. H. and Schonfeld, W.,** Na-K-ATPase as the digitalis receptor, *Trends Pharmacol. Sci.,* 393, 1984.

45. **Eisner, D. A., Vaughan-Jones, R. D., and Lederer, W. J.,** Comments on "Active transport and inotropic state in guinea pig left atrium, Lechat, P., Malloy, C. R., and Smith, T. W., Eds., *Circ. Res.*, 52, 411, 1983; *Circ. Res.*, 53, 834, 1983.
46. **Grupp, G., Grupp, I., Ghysel-Burton, J., Godfraind, T., and Schwartz, A.,** Effects of very low concentrations of ouabain on contractile force of isolated guinea pig, rabbit, and cat atria and right ventricular papillary muscle: An interinstitutional study, *J. Pharmacol. Exp. Ther.*, 220, 145, 1982.
47. **Heller, M., Beck, S., and Hallaq, H.,** Interaction of cardiac glycosides with cells and membranes, *J. Mol. Cell. Cardiol.*, in press.
48. **Godfraind, T.,** Subclassification of cardiac glycoside receptors, *Basic Res. Cardiol.*, 79 (Suppl.), 27, 1984.
49. **Hougen, T. J., Spicer, N., and Smith, T. W.,** Stimulation of monovalent cation active transport by low concentrations of cardiac glycosides, *J. Clin. Invest.*, 68, 1207, 1981.
50. **Rogus, E. M., Cheng, L. C., and Zierler, K.,** Beta-adrenergic effect on Na^+ and K^+ transport in rat skeletal muscles, *Biochim. Biophys. Acta*, 467, 347, 1977.
51. **Sharma, V. K. and Banerjee, S. P.,** Regeneration of ^3H-ouabain binding to Na-K-ATPase in chemically sympathectomized Ca^+ peripheral organs, *Mol. Pharmacol.*, 15, 35, 1979.
52. **Seifen, E.,** Evidence for participation of catecholamines in cardiac action of ouabain, *Eur. J. Pharmacol.*, 26, 115, 1974.
53. **Sharma, V. K., Pottick, L. A., and Banerjee, S. P.,** Ouabain stimulation of noradrenaline transport in guinea pig heart, *Nature*, 286, 817, 1980.
54. **Wyllie, M. G. and Wood, M. D.,** Modulation of synaptic noradrenaline levels by Na-K-ATPase, in Proc. 4th Int. Conf. on the Properties of Na-K-ATPase, London, 1984, Abst. 122.
55. **Sweadner, K. J.,** Possible functional differences between the two Na-K-ATPases of the brain, *Curr. Topics in Membranes and Transp.*, 19, 765, 1983.
56. **Karin, N. J. and Cook, J. S.,** Regulation of Na-K-ATPase by its biosynthesis and turnover, *Curr. Topics in Membranes and Transp.*, 19, 713, 1983.
57. **Lamb, J. F. and McCall, D.,** Effect of prolonged ouabain treatment on Na, K, Cl, and Ca concentration and fluxes in cultured human cells, *J. Physiol.*, 225, 599, 1972.
58. **Boardman, L. J., Lamb, J. F., and McCall, D.,** Uptake of ^3H-ouabain and Na pump turnover rates in cells cultured in ouabain, *J. Physiol.*, 225, 619, 1972.
59. **Boardman, L. J., Heutt, M., Lamb, J. F., Newton, J. P., and Polson, J. M.,** Evidence for the genetic control of the sodium pump density in Hela cells, *J. Physiol.*, 241, 771, 1974.
60. **Erdmann, E., Werdan, K., and Krawietz, W.,** Influence of digitalis and diuretics on ouabain binding sites on human erythrocytes, *Klin. Wochensch.*, 62, 87, 1984.
61. **Girardi, A. J., Warren, J., Goldman, C., and Jeffries, B.,** Growth and C. F. antigenicity of measles virus in cells derived from human heart, *Proc. Soc. Exp. Biol. Med.*, 98, 18, 1958.
62. **Aiton, J. F., Lamb, J. F., and Ogden, P.,** Down regulation of the sodium pump following chronic exposure of Hela cells and chick embryo heart cells to ouabain, *Br. J. Pharmacol.*, 73, 333, 1981.
63. **Werdan, K., Schneider, G., Krawietz, W., and Erdmann, E.,** Chronic exposure to low K^+ increases cardiac glycoside receptors in cultured cardiac cells: different responses of cardiac muscle and nonmuscle cells from chicken embryos, *Biochem. Pharmacol.*, 33, 1161, 1984.
64. **Werdan, K., Reithmann, Ch., Schneider, G., and Erdmann, E.,** Modulation of cardiac glycoside sensitivity of beating heart cells in culture, *Proc. 4th Int. Conf. on Properties of Na-K-ATPase*, London, 1984, Abst., 120.
65. **Kim, D. and Smith, T. W.,** Effects of thyroid hormone on sodium pump sites, sodium content, and contractile response to cardiac glycosides in cultured chick ventricular cells, *J. Clin. Invest.*, 74, 1481, 1984.
66. **Schanne, O. F. and Bkaily, G.,** Explanted cardiac cells: a model to study drug action?, *Can. J. Physiol. Pharmacol.*, 59, 443, 1981.

Chapter 21

ADRIAMYCIN TOXICITY IN HEART CELLS IN CULTURE

Maria W. Seraydarian and Chandrasekharam N. Nagineni

TABLE OF CONTENTS

I. INTRODUCTION

The anthracycline antibiotics, daunomycin[1®] (daunorubicin, DM) and adriamycin[2®] (doxorubicin, ADM), have been widely used as very effective chemotherapeutic agents in the treatment of a variety of tumors.[3,4] The major drawback in the use of these potent antitumor agents in cancer chemotherapy is the development of dose-dependent cardiotoxicity.[5-8] The antitumor action of the anthracyclines has been attributed to the intercalation of the drug between base pairs of DNA helices[9,10] and subsequent inhibition of DNA, RNA, and protein synthesis.[11] Therefore, in tumors and other rapidly dividing cells where the mitotic rate is high, anthracyclines inhibit cell division and growth. While these mechanisms are known to be common to most of the anticancer drugs, the toxic effects on the heart seem to be specific to the anthracyclines. ADM, which is more effective in cancer therapy than DM, has more severe cardiotoxic effects. Since after birth myocardial cells essentially cease to divide, the cardiotoxicity of anthracyclines is generally thought to operate via mechanisms other than the antimitotic effect.

Several investigators have focused their attention on the mechanisms of anthracycline-induced damage to the heart in an attempt to prevent or reduce their cardiotoxicity. Rabbit,[12] rat,[13] and mouse[14] have been used as animal models to study anthracycline cardiotoxicity in vivo and various cardiotoxic effects have been described. Freshly isolated myocytes[15] and subcellular organelles[16-18] have also been employed for in vitro studies in attempts to elucidate the mechanism of anthracycline effects at specific loci. However, there is as yet no conclusive evidence regarding the mechanism of the toxicity. Moreover, due to the multiplicity of animal species, drug concentrations, and animal protocols for drug treatment, it is often difficult to compare results obtained from different laboratories.

Possible mechanisms for the cardiotoxic effects of the anthracyclines extensively studied in both in vivo and in vitro experiments are as follows:

1. Elevation of cyclic AMP with a concomitant decrease in cyclic GMP levels;[19,20]
2. Stimulation of free radical formation that would damage biological membranes by lipid peroxidation;[17,21-24]
3. Effect on calcium fluxes across plasma,[18] mitochondrial,[25] and sarcoplasmic reticulum[26] membranes resulting in calcium overload.[27,28] Lack of involvement of calcium in cardiotoxicity has also been suggested;[29,30]
4. Inhibition of mitochondrial coenzyme Q_{10} and other enzymes leading to decrease in energy metabolism;[16,31,32]
5. Inhibition of DNA, mRNA and rRNA and protein synthesis;[33-36]
6. Ultrastructural changes which include vacuolization, myofibrillar disintegration, dilation of sarcoplasmic reticulum, mitochondrial swelling and degeneration, and nuclear damage.[27,37-39]
7. Inhibition of mitochondrial creatine kinase reassociation with mitochondrial membrane (see Chapter 14, Volume 2).

In general, in vivo experiments are complicated due to involvement of several organs at the same time, even if they each respond to a different degree. In some instances, it is difficult to identify whether toxic effects are produced by direct action of the drugs on the target tissue or are due to metabolites produced elsewhere. The experiments are rather expensive and time consuming. There is, therefore, an increasing interest in the use of cell cultures for studying the mechanisms of cytoxicity, carcinogenesis, and mutagenesis, and for screening toxic compounds.[40,41] Organ-specific primary cultures as well as established cell lines are frequently used to investigate the toxic mechanisms of drugs, heavy metals, and environmental pollutants.[42] The metabolic inhibition test (MIT) combined with microscopic observation of injury to cells are considered as satisfactory criteria for toxicity in large scale screening of various drugs.[43] Application of cell cultures for toxicological studies appears to be a promising field worthy of serious consideration, even though in vitro experimental results may not always correlate well with in vivo studies.

Cardiotoxic effects of several drugs and chemicals have been well documented;[44] however, the use of heart cells in culture for toxicity studies has not been explored in depth and has mostly been confined to anthracycline antibiotics. Heart cells in culture are of particular interest, since they are free from neurohumoral influences while exhibiting rhythmic contractions and other differentiated functions. Several direct effects of ADM on myocardial cells have been demonstrated by using heart cells in culture as a model system.

II. GENERAL CONSIDERATIONS REGARDING TOXICITY IN CELL CULTURES

The aim of this chapter is to discuss the state of the art regarding the mechanisms of the toxic effect of anthracycline antibiotics on cultured heart cells, and therefore, only pertinent information from other model systems will be discussed. Toxicity studies have been performed in cultured heart cells derived from chick embryos, neonatal mice, and neonatal and adult rats. The age of the cultures, concentration of the drugs, and duration of treatment and other experimental conditions varied in different studies. A summary of the anthracycline toxicity studies in heart cells in culture is outlined in Table 1.

In early studies, the general cytotoxic effects of ADM in cultured heart cells were described in terms of morphological changes, number of cells, and protein content per dish and rates of beating.[45,46] When ADM (1.7 μM) was present in the medium during seeding, heart cells failed to attach to the dish. On the other hand, ADM treatment for 3 hr resulted in cessation of growth.[45] The fast growing nonmuscle cells present in the cultured heart cells were more susceptible to ADM treatment than myocardial cells, as would be expected for rapidly dividing cells. Consequently, the cultures became enriched in myocardial cells, but growth of the cells was inhibited. Both the number of cells and protein content was decreased by ADM treatment (1 μg/mℓ) by more than twofold relative to the control (Figure 1A). ADM treatment for 1 hr at a higher concentration (10 μg/mℓ) resulted in structural changes and lysis of cells at 48 to 72 hr following exposure to the drug.[47] Exposure of cultures to a lower concentration of the drug (2 μg/mℓ) for 1 hr did not produce any toxic effects.

These general observations do not provide any information about the mechanisms of cardiotoxicity. It is not known whether ADM and DM affect the cells directly or whether they undergo metabolic conversion in rat heart cells; however, in rat embryo fibroblasts no such metabolic conversion could be demonstrated.[48]

III. UPTAKE AND SUBCELLULAR DISTRIBUTION OF ADRIAMYCIN IN HEART CELLS

Uptake and compartmentalization into subcellular fractions of ^{14}C labeled ADM and DM in cultured embryonic chick heart cells has recently been described.[49] ADM ($2 \times 10^{-6} M$) was taken up rapidly by the cells at a rate of 280 fmol/cell within minutes of exposure, reaching up to 480 fmol/cell at the end of a 30 min period. Thereafter, the amount of ADM incorporated into the cells showed no change for up to 4 hr. Although the rate of DM uptake was lower, it exhibited a similar pattern to that of ADM during the initial period. A calcium channel blocker verapamil ($5 \times 10^{-6} M$), and ATP ($10^{-5} M$), when present in the medium at the same time as ADM, decreased the amount of radioactivity incorporated into the cells by a factor of three. Uptake of DM was not affected by ATP. Therefore, it seems likely that ADM is bound to the Ca^{++} channels and to the purinergic receptors. It is not clear

Table 1
ANTHRACYCLINE TOXICITY STUDIES IN HEART CELLS IN CULTURE

Animal species	Age of cultures	Drug concentration	Duration of treatment	Time of observation	Nature of studies	Ref.
Chick embryos (7 day old)	5 days	2 μM (ADM and DM)	1—240 min	During treatment	Uptake and incorporation into subcellular fractions	49
			60 min		Ultrastructural studies	49
			6—24 hr		Incorporation of methionine to proteins	49
Neonatal mice	3—4 days	0.5 μg/mℓ (ADM)	Until the beats min reached 50% of the control		Electron microscopic ultrastructural studies	64
Neonatal mice	3—8 days	0.9 μM (ADM)	3—4 hr	During treatment	Frequency and rhythm of beating	53
Neonatal rats	1 day	1.7 μM (ADM)	3 hr	Up to 7 days	Cell growth and adenine phosphate metabolism	45
Neonatal rats	18 hr	1—10 μg/mℓ (ADM)	3 hr	19, 67, or 157 hr following treatment	DNA replicative and repair synthesis	61
Neonatal rats	2 days	1.7 μM (ADM)	3 hr	1 and 2 days following treatment	Growth, beating and energy metabolism in the presence or absence of adenosine	46
Neonatal rats	3—4 days	0.1—1 μg/mℓ (ADM)	30 hr or days	During treatment and up to 17 days following treatment	Frequency of beating with or without γ-irradiation	54
Neonatal rats	4—7 days	1.7 μg/mℓ (ADM)	24 hr	24 and 48 hr following treatment	Nuclear and mitochondrial structure	55
Neonatal rats	7 days	2—200 μg/mℓ (ADM)	1 hr	During and following treatment	Ultrastructural changes, beating rate and pattern recorded with polygraph	47
		0.05—10 μg/mℓ (ADM)	Continuous exposure for up to 2—10 days			
Neonatal rats	—	0.35 μM (DM)	2 days	Immediately after treatment	Induction of enzymes for metabolism of the drug	62
Neonatal rats	1 day	1 μg/mℓ (ADM)	3 hr	12—24 hr following treatment	Light and electron microscopic stereological analysis of ultrastructure	65

| Neonatal rats | 4 hr | 0.5—10 μM (ADM) | 1 hr | 1—4 hr following treatment | DNA, RNA, and protein syntehsis | 63 |
| Adult rats | 4—7 days | 1.8—18 μM (ADM and DM) | 6—24 hr | During treatment | Enzyme leakage from cells and the effects of cytoprotectants | 60 |

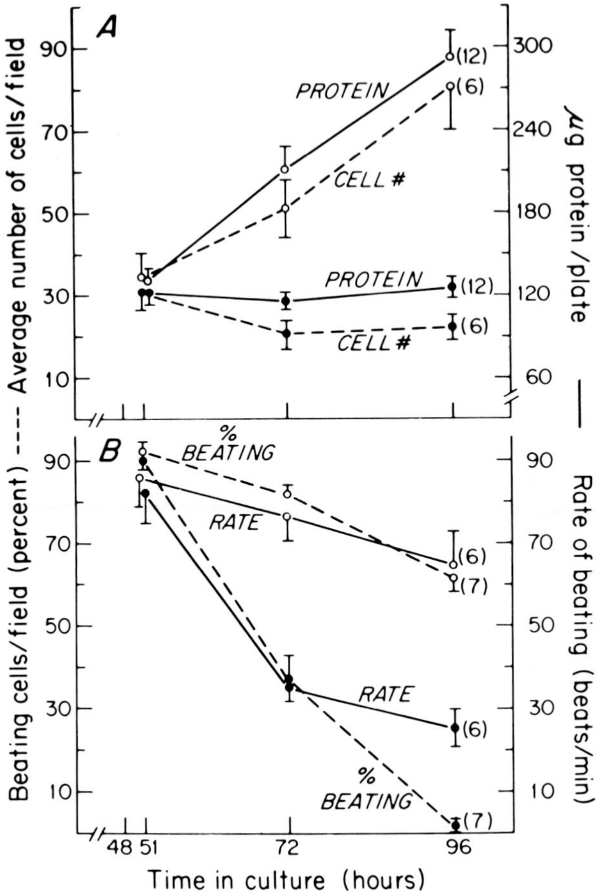

FIGURE 1.   Effect of ADM on number of cells, protein content, and beating of rat heart cells in culture; ADM (1 μg/mℓ medium) treatment at 48 to 51 hr. (A) average number of cells per field and protein content per plate in control and ADM-treated cells. Each point derived from an average number of cells from a random sampling of 10 fields/plate (×40 objective, inverted-phase microscope) and from μg protein/plate, respectively. Bars, S.E.; numbers in parentheses, number of experiments; ○, control; ●, ADM. (B) percentage of beating cells per field and rate of beating in control and ADM-treated cells. Each point derived from an avrerage number of beating cells per field from the same random sampling of 10 fields, as in (A) and from the rate of beating, respectively. Bars, S.E.; numbers in parentheses, number of experiments; ○, control; ●, ADM. (From Seraydarian, M. W. and Artaza, L., *Cancer Res.*, 39, 2940, 1979. With permission.)

whether ADM is transported into the cells through these channels in addition to the other routes for internalization. Twenty to 25% of the radioactively labeled ADM taken into the cells was recovered in a pellet consisting mostly of nuclear material. In the cytosolic component, ADM was detected mainly in the fraction free of plasma membranes, mitochondria, lysosomes, and endoplasmic reticulum. However, cardiotoxic effects of ADM at the plasma membrane cannot be ruled out, since ADM has been shown to be cytotoxic to murine cancer cells (line L1210) although it does not enter the cells.[50]

In rat embryo fibroblasts, uptake of both ADM and DM exhibited time-dependent saturation characteristics of carrier-mediated transport.[48] The highest concentration of the drug was localized in the lysosomes. Since ADM did not interfere with the simultaneous accumulation of DM in these cells, it is likely that these two anthracyclines are transported via different routes. Flow cytometry studies[51] have also indicated time and concentration dependent uptake of ADM in fibroblast cell lines. Half of the total fluorescence was associated with the nucleus. Similar carrier-mediated transport of ADM and DM was reported in Ehrlich tumor cells.[52]

In summary, ADM and DM are taken up from the medium into the cell, even at low concentrations of the drug, and incorporated into the nucleus, subcellular organelles, and cytosol. This may occur via carrier-mediated transport. The involvement of calcium channels and purinergic receptor sites in ADM transport is likely, in addition to the other routes for internalization of the drug.

IV. THE EFFECT OF ADRIAMYCIN ON SPONTANEOUS CONTRACTIONS IN HEART CELLS

Since heart cells in culture exhibit characteristic rhythmic contractions, force and rate of contraction are good indicators of functional integrity of the cells. Several investigators have studied the effects of ADM on the beating of heart cells in culture under various experimental conditions. The beating rate of rat myocardial cells was unaffected immediately following 3 hr of ADM (1.7 μM) treatment after 24 to 27 hr or 48 to 51 hr in culture.[45,46] In these studies, however, 24 hr after ADM treatment, the beating rate was considerably reduced, and after 48 hr almost complete cessation of beating was observed (Figure 1B). A decrease in ATP levels, adenylate energy charge, and total adenine nucleotides was correlated to the beating of cells.

In neonatal mouse heart cells in culture, ADM ($0.9 \times 10^{-6}M$) reduced the frequency of beating by more than 50% within 3 to 4 hr.[53] Glucagon ($10^{-7}M$) or ouabain ($10^{-7}M$) restored the beating rate of cells following ADM treatment to the control rate while oligomycin (10 $\mu g/m\ell$) had no effect. The positive chronotropic action of glucagon might be due to an increase of intracellular cyclic AMP levels. Since ouabain is known to bind to membrane Na^+/K^+-ATPase sites and alter intracellular Na^+ and Ca^{++} pools, the reversal of the negative inotropic effects of ADM might be mediated via an increase in the intracellular calcium. These authors suggest that the effects of ADM on the beating of cells are mediated via cellular membrane sites.

The effect on the beating of rat heart cells in culture of ADM (0.1 to 1.0 $\mu g/m\ell$) either alone or in combination with γ-irradiation has been studied.[54] Beating was recorded as the contraction rhythm of entire cultures grown in microplates, γ-irradiation alone of up to 10 K-rad had no effect either on beating or on the ADM-induced reduction in beating. Significant decrease in the number of wells (microplates) with beating cells was found after 4 days of continuous ADM (0.1 $\mu g/m\ell$) treatment. Two days after the withdrawal of drug from the medium, a significant number of wells with beating cells had resumed beating at the control level. In the cultures continuously exposed to ADM (1 $\mu g/m\ell$), a decrease in the number of wells with beating cells was noted on the third day. Restoration of beating was not found in this case following drug withdrawal. The actions of low concentrations of drugs were reversible, while irreversible effects were induced by high levels of the drug.

In a series of elegant experiments, Lampidis et al.[47] conducted a systematic study of the beating of heart cells exposed to a wide range of ADM concentrations. Pulsations of the cultures were monitored with a polygraph recorder. At high concentrations of the drug, i.e., 100 to 200 $\mu g/m\ell$, pulsations disappeared at 250 and 30 min of exposure, respectively. These effects could be temporarily abolished by removing the drug and washing the cultures with fresh medium. After 48 hr, the pulsating action ceased and cells began to lyse. The development of dose-dependent arrhythmias at 0.01 to 0.1 μg/mℓ in cultures was observed upon continuous exposure to the drug for two days. Under these conditions no ultrastructural changes were observed and the recovery was complete after withdrawal of the drug.

Most of the heart cell cultures, which served as a model system for studying the effects of ADM on beating rate and rhythm, contained nonmuscle cells, e.g., fibroblasts, yet the role of fibroblasts in mediating the ADM-induced effects has not been given serious consideration. The fibroblasts, which have differential susceptibility to ADM,[45,55] may be involved in synchronizing the beating rhythm of heart cells in culture.[56,57] Reversible effects induced by low concentrations of ADM on beating of cells are probably mediated via effects at membrane sites, i.e., binding to receptors and/or ion transport channels, as discussed in the previous section. Higher concentrations of ADM (1 μg/mℓ and above) might affect beating as a consequence of the reduced adenine phosphate metabolism and structural disorganization.

V. THE EFFECT OF ADRIAMYCIN ON METABOLISM IN HEART CELLS

A. Adenine Nucleotide Metabolism

The possible involvement of ADM in adenine nucleotide metabolism of myocardial cells in culture has been investigated.[45,46] In ADM-treated cultures both ATP and phosphocreatine (PCr) levels decreased significantly.[45] The presence of creatine (Cr) (5mm) in ADM treated cultures only restored PCr but not ATP to control levels. Since PCr can only be synthesized by the phosphorylation of Cr with ATP, it is unlikely that ATP synthesis via glycolysis or oxidative phosphorylation has been directly affected. A decrease in total adenine nucleotides levels (TAN) may explain why the addition of Cr to ADM-treated cells did not restore ATP to control levels. In further studies, Seraydarian and Artaza[46] reported significant decreases in ATP, TAN (Table 2), and adenylate energy charge, whereas the PCr mole fraction (PCr/total Cr) was not affected to the same extent.

Decreased ATP levels could be due to an imbalance in energy production and utilization. In rat myocardial cells, ADM (1 μg/mℓ) treatment for 3 hr (after 48 to 51 hr in culture) did not alter oxygen consumption even after 24 or 48 hr,[58] by which time ATP and TAN were significantly lower than in the corresponding controls.[46] In organ-cultured 3-day-old chick embryo hearts, ADM ($3.8 \times 10^{-4}M$) caused no change in oxygen consumption.[59] The effects of uncoupling oxidation and phosphorylation under these conditions have not been examined to date. Decrease in ATP levels due to elevated myosin ATPase activity as a result of calcium overload has been suggested by other workers.[28] Our own observations suggest that myosin ATPase activity and its isoenzyme pattern (Figure 2) remained unaltered even after 48 hr after ADM (1 μg/mℓ) treatment. Creatine kinase (CK) and adenylate kinase (AK), involved in ATP metabolism, were significantly decreased 24 and 48 hr following ADM (1 μg/mℓ) treatment for 3 hr (Figure 3). These enzymes were not detected in the culture medium, suggesting that the decrease in enzyme activity was not due to leakage from the cells. Since ADM was reported to interact with the ATP binding sites of purinergic receptors, the possibility of direct interaction of ADM with the catalytic sites of CK and AK was examined. However, addition of ADM (up to 10 μg/mℓ) to 3,000 × g supernatants from cell extracts had no effect on the activities of these enzymes.[58]

Newman et al.[60] used adult rat heart cells in culture to investigate the ADM and DM mediated release of enzymes into the medium. After 6 hr of continuous exposure to DM (1.8 μg/mℓ) there was no significant change in either the intracellular LDH or in the LDH released into the medium, but after 24 hr of continuous exposure, the decrease in intracellular LDH was significant. The effect of DM on cellular CK levels and on LDH released into medium appeared to be both dose and time dependent. When the cells were exposed to ADM or DM (18 μM) continuously for 12 hr, intracellular CK fell to about 25% of the

Table 2
EFFECT OF ADM AND ADENOSINE ON THE CONCENTRATION OF ADENINE NUCLEOTIDES IN RAT HEART CELLS IN CULTURE

Concentration (nmol/mg protein)

Conditions	ATP			ADP			AMP			Total adenine nucleotides		
	51 hr	72 hr	96 hr	51 hr	72 hr	96 hr	51 hr	72 hr	96 hr	51 hr	72 hr	96 hr
Control	15.5 ± 0.9[a]	15.7 ± 1.0	20.3 ± 0.8	3.8 ± 0.3	3.2 ± 0.4	3.0 ± 0.3	2.0 ± 0.3	1.7 ± 0.2	1.7 ± 0.1	21.3 ± 1.1	20.5 ± 0.9	24.9 ± 0.8
ADM[b] 1 mM	15.0 ± 1.3	10.7 ± 0.9	9.7 ± 1.8	4.3 ± 0.3	2.7 ± 0.1	2.6 ± 0.5	2.1 ± 0.3	2.8 ± 0.4	2.1 ± 0.2	21.4 ± 1.4	16.3 ± 1.1	14.4 ± 2.2
adenosine[c] ADM, 1 mM	21.7 ± 2.1	19.6 ± 1.5	24.1 ± 1.4	4.5 ± 0.5	3.6 ± 0.3	3.8 ± 0.1	1.8 ± 2.1	2.1 ± 0.3	1.3 ± 0.1	27.9 ± 1.9	25.4 ± 1.5	29.4 ± 1.4
adenosine	17.5 ± 1.8	14.6 ± 1.9	9.5 ± 1.9	4.8 ± 0.3	3.2 ± 0.4	2.6 ± 0.1	2.6 ± 0.2	2.0 ± 0.3	1.7 ± 0.1	24.9 ± 1.7	20.0 ± 2.1	13.8 ± 1.3

[a] Mean ± S.E.
[b] ADM (1 μg/mℓ medium) treatment of cells which have been in culture for 48 to 51 hr.
[c] Adenosine treatment throughout the duration of the experiment.

From Seraydarian, M. W. and Artaza, L., *Cancer Res.*, 39, 2940, 1979. With permission.

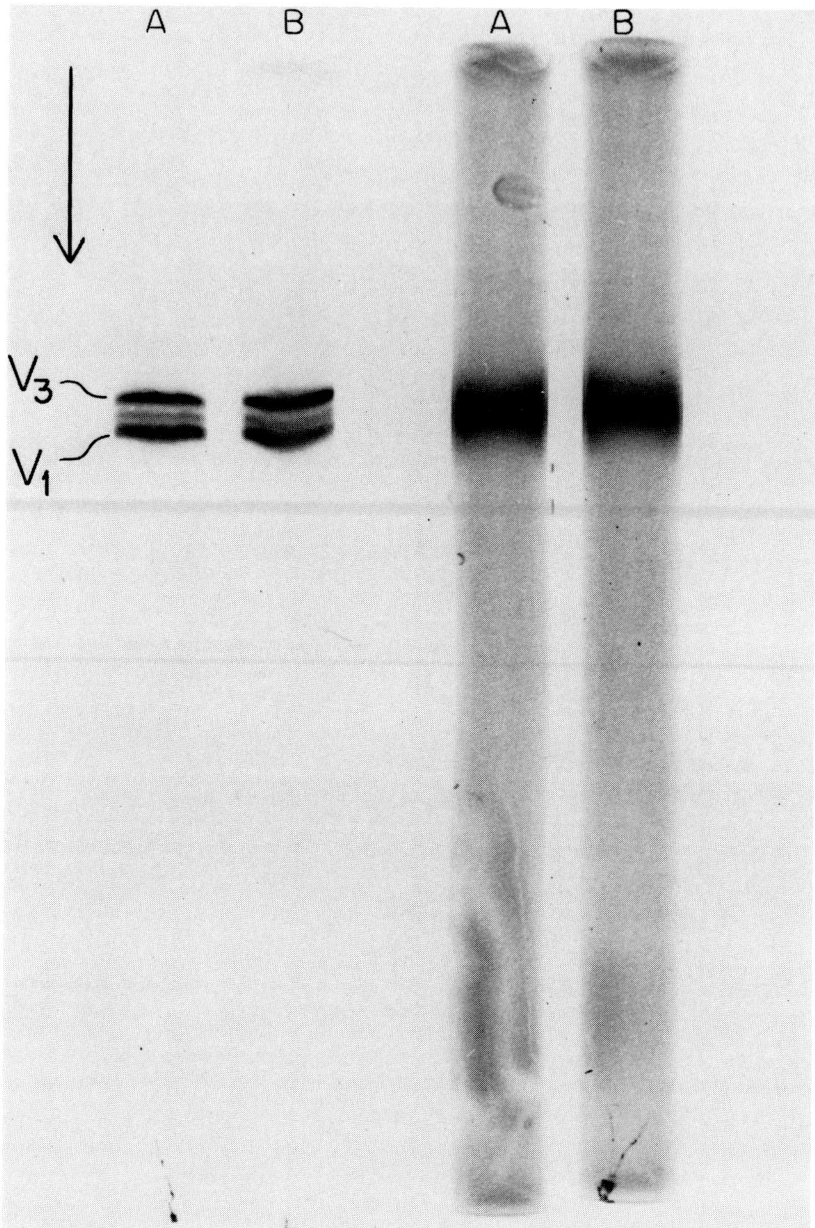

FIGURE 2.   Myosin isoenzyme pattern of control (A) and ADM treated (B) rat heart cells in culture. ADM (1 μg/mℓ) treatment for 3 hr, from 48 to 51 hr of culture age. Myosin was extracted from control and ADM treated cells of 96 hr culture age and subjected to polyacrylamide gel electrophoresis under nondissociating conditions following the procedure of d'Albis, A., Pantaloni, D., and Bechet, J. J., *Eur. J. Biochem.*, 99, 261, 1979. Direction of electrophoretic mobility is indicated by arrow. V$_1$ isoenzyme — fast migrative component; V$_3$ isoenzyme — slow migrative component. Left panel, protein staining; Right panel, ATPase staining.

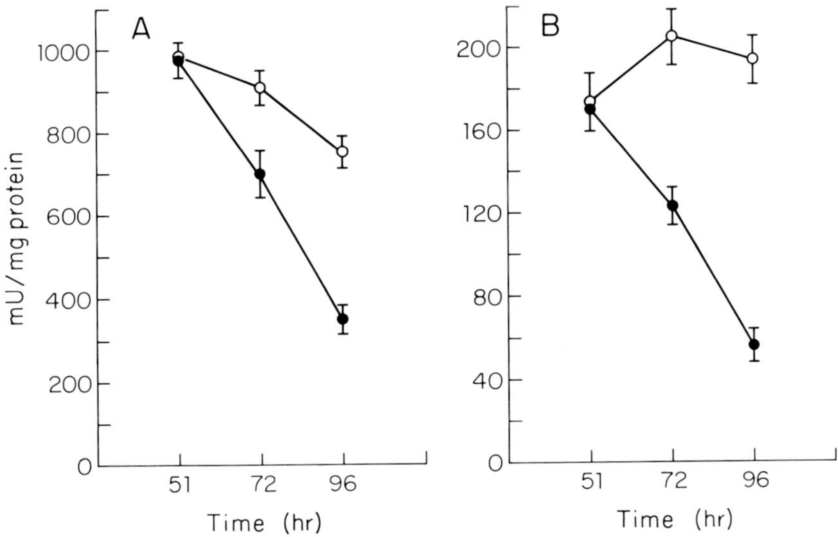

FIGURE 3.   Effect of ADM treatment on creatine kinase (A) and adenylate kinase (B) activities of rat heart cells in culture. ADM (1 μg/mℓ) treatment for 3 hr, from 48 to 51 hr of culture age. Cells were harvested, homogenized, and centrifuged at 3000 × Gr for 10 min and the supernate was used for enzyme assays according to the method of Oliver, I. T., *Biochem. J.*, 61, 116, 1955, with some modifications. Creatine kinase assay was performed in the presence of P^1-P^5-di-adenosine-5′-pentaphosphate (10 μM), a potent adenylate kinase inhibitor. ○—○, control; ●—●, ADM.

control value and LDH released into the medium was 400% of the corresponding control values. In spite of marked decreases in cytoplasmic enzyme levels, no significant reduction in cell viability was observed.

B. Nucleic Acid and Protein Synthesis

DNA replication and repair synthesis were studied in rat heart cells using thymidine (^{3}H) incorporation.[61] ADM (1 μg/mℓ) treatment reduced DNA replication by approximately 75% without any effect on the UV irradiation stimulated DNA repair synthesis. At higher concentrations of ADM (2 to 10 μg/mℓ), UV-irradiation stimulated DNA repair synthesis was also inhibited. It is suggested that at a low concentration of ADM, drug-induced effects were limited to a site(s) proximal to the intercalation sites in the DNA double helix. At higher concentrations, the number of drug binding sites increased and affected both DNA replication and repair. Since 75% inhibition of DNA replication was observed after 3 hr of drug treatment (Figure 4), it is possible that synthesis of specific proteins, coded by DNA close to drug binding sites, had been inhibited. If this were so, decreases in levels of some crucial enzymes with high turnover rates may affect adenine phosphate metabolism. As mentioned earlier, in our studies, neither the myosin ATPase activity nor the isozyme pattern were affected by ADM treatment,[58] but AK and CK activities decreased progressively with respect to time following ADM treatment; however, adenosine deaminase and adenylate deaminase remained at the pretreatment levels.[58]

Another line of evidence suggesting selective effects of anthracyclines on protein synthesis rather than overall inhibition comes from the studies of Galaris and Rydstrom.[62] In rat heart cells treated with DM (0.35 μM) for two days, DT-diaphorase and glutathione-s-transferase activities were increased by about threefold, while NADPH-cytochrome-*c*-reductase, glucose-6-phosphate dehydrogenase, and transhydrogenase activities remained unchanged. Induction of the former enzymes by DM was attributed to their effects on the metabolism of

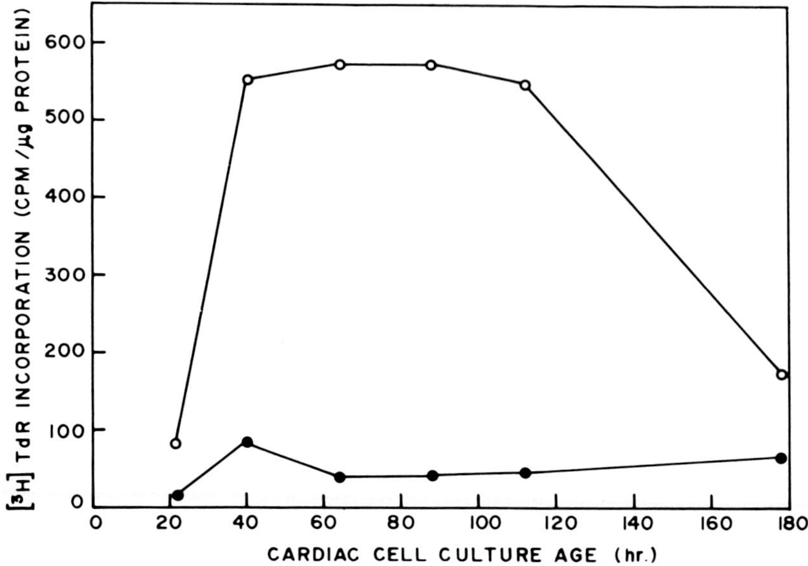

FIGURE 4.   Incorporation of [³H]dThd into the DNA fraction of heart cells in culture as a function of age in culture. DNA was measured immediately following a 3-hr pulse of [³H]dThd. One μg/mℓ ADM was administered at 18 to 21 hr. ○—○, Control; ●—●, ADM. (From Fialkoff, H., Goodman, M. F., Seraydarian, M. W., and Yang, J. J., *Adv. Myocardiol.*, 1, 565, 1980. With permission.)

the drug. In chick cardiac cells treated with ADM (2 μM) for 6 hr, incorporation of methionine-^{35}S into cytosolic and contractile proteins was reduced by about 70% relative to the control.[49] Actin synthesis was the least affected among the contractile proteins. The effect on methionine incorporation into proteins was not related to the methionine uptake mechanism. In rat heart cells treated with ADM (2 μM) for 1 hr, complete inhibition of RNA and DNA synthesis was observed with no significant effect on protein synthesis.[63] However, protein synthesis was totally blocked by higher concentrations of ADM (100 μM).

VI. ULTRASTRUCTURAL CHANGES IN HEART CELLS

Electron microscopic observation of mouse myocardial cells treated with ADM (0.5 μg/mℓ) revealed hypertrophy and swelling of the sarcoplasmic reticulum, with no alteration in the ultrastructure of mitochondria or contractile elements.[64] The number and total extension of gap junctions between myocardial cells increased. The increase in dimensions of the sarcotubular system of the cells was suggested to be due to the decrease in cytoplasmic calcium. In rat myocardial cells, 24 hr after ADM (1.7 μg/mℓ) treatment, severe nuclear damage, including nucleolar fragmentation, segregation, and chromatin clumping, was observed.[55] At this stage no ultrastructural changes in the mitochondria were detected. Using ultraviolet microbeam irradiation, the functional capacity of the mitochondria of ADM-treated cells was found to be lower than in controls. Light and electron microscopic examination of cells continuously exposed to a low dose of ADM (0.1 μg/mℓ) for 17 days showed a striking loss of muscle fibers.[47] Vacuolization and nuclear fragmentation as seen with the high dose of ADM were not observed in these studies. Chick myocytes treated with ADM (2 μM) for 60 min showed dispersed heterochromatin in the nuclei, swelling of mitochondria, and an increase in osmophilic vacuoles; sarcomeres were not affected.[49] Tobin and Abbott[65] examined the effect of ADM on the ultrastructure of rat heart cells in culture by light and electron microscopic stereological analysis. They suggested that ADM induces

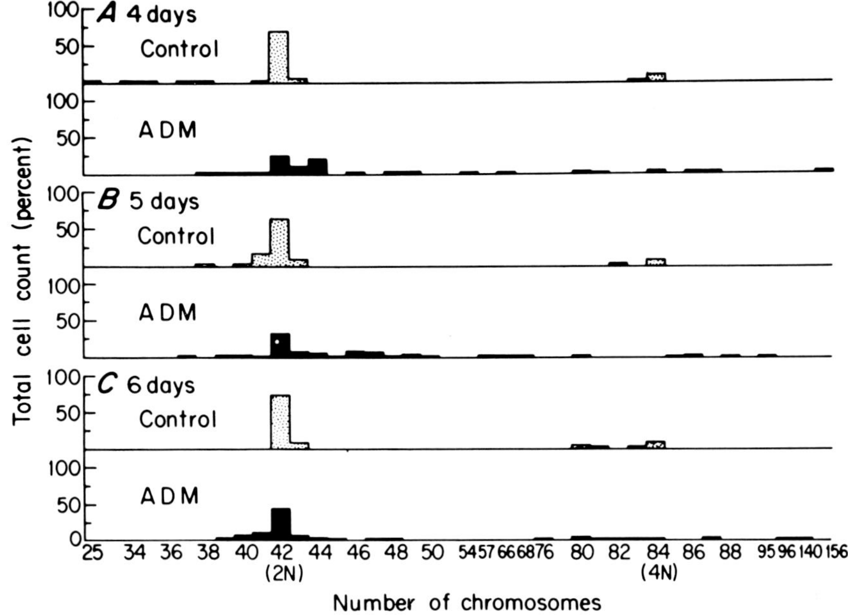

FIGURE 5. Chromosome distribution of control and ADM-treated rat heart cells in culture. Cultures were treated with ADM for 3 hr (from 48 to 51 hr of culture age) at a concentration of 1 μg/mℓ of culture medium. Cells were exposed to colcemid for 24 hr at indicated times and processed for chromosome analysis using 0.5% sodium citrate as lysing solution. (From Fialkoff, H., Goodman, M. F., Seraydarian, M. W. and Yang, J. J., *Adv. Myocardiol.*, 1, 565, 1980. With permission.)

nuclear pleomorphism and an increase in mitochondrial size, while the volumes of the cell and sarcoplasm were not altered. The ultrastructural changes described above indicate general injury to the cells. Initial effects on specific organelles depend upon the concentration of the drug and the species of animal used.

Chromosomal aberrations were detected in rat heart cells in culture exposed to ADM (1 μg/mℓ) for 3 hr,[66] aneuploidy and euploidy were the most frequent aberrations (Figure 5). Random breaks on the chromosomal arms and in the centromere region were also noted. The chromosomal aberrations were more numerous following 48 and 72 hr of ADM treatment, but had returned to almost normal ploidy by 96 hr. It has been suggested that the return of normal ploidy is due to either the restoration of repair synthesis and/or to the loss of extensively damaged cells from the culture, only leaving the less affected cells.

VII. PREVENTION OF ADRIAMYCIN TOXICITY IN HEART CELLS

Several agents such as vitamin E,[67] coenzyme Q_{10},[68] and N-acetylcysteine[37] have been shown to reduce cardiotoxicity of ADM in vivo. Attempts have also been made, using cell cultures as a model, to examine the possible role of putative myocardial protectants in reversing or reducing ADM injury to cells. Seraydarian and Artaza[46] studied the effect of adenosine. Presence of adenosine (1 mM) in the medium delayed the ADM-induced fall in ATP and TAN levels, in adenylate energy charge, and in the rate and the number of beating cells. However, the effects could not be completely reversed, adenosine only delaying cell death and lysis.

Newman et al.[60] examined the effects of adenosine, DL-α-tocopherol, carnitine, and N-acetylcysteine in preventing the ADM and DM mediated release of LDH from cells and the decrease in intracellular CK levels. These agents were added 24 hr before drug treatment. Of all the agents tested, only adenosine (1 mM) completely inhibited the LDH release from the cells, with intracellular CK activity remaining at the control level. Reversal of the effects of ADM and DM was shown to be dependent on the adenosine (10^{-6} M to 10^{-3} M) concentration. It is likely that the effects of adenosine in preventing enzyme leakage from cells are due to restoration of ATP levels by the increased TAN levels. Decrease in ATP levels in the heart is known to induce release of enzymes, while adenosine can restore ATP levels in myocardial cells treated with ADM.[46]

VIII. CONCLUSIONS

Several of the cardiotoxic effects of ADM, e.g., cardiac arrhythmias, membrane effects, as well as changes in adenine phosphate metabolism, macromolecular synthesis, and ultrastructure, observed in animal models and in isolated subcellular organelles could be demonstrated in heart cells in culture. In in vivo experiments it is often difficult to distinguish between pathological changes brought about by direct action of ADM on the myocardium and the effects of ADM metabolites formed in other organs such as the liver and the kidney. The use of heart cell culture, as a model system, offers the potential for the determination of the specific sites of interaction of ADM at the myocyte level. The plasma membrane transport mechanism, synthesis of DNA, and proteins are thus implicated as the major sites for direct interaction of ADM. However, investigations of ADM toxicity carried out so far, have not provided sufficient data to determine whether multiple mechanisms of action operate even at the cellular level. Further studies, using cell cultures might contribute to further elucidation of the mechanism(s) of ADM cardiotoxicity.

ACKNOWLEDGMENTS

The authors' work reported in this paper was supported by grants from the Public Health Service NHLBI #21080, the American Heart Association #81940, the American Heart Association Greater Los Angeles Affiliate #680-1G, and by a gift from the Adria Laboratories, Columbus, Ohio.

REFERENCES

1. **Di Marco, A., Gaetani, M., Dorigotte, L., Sodati, M., and Belline, O.,** Daunomycin: a new antibiotic with antitumor activity, *Cancer Chemother. Rep.,* 38, 31, 1964.
2. **Di Marco, A., Gaetani, M., and Scarpinato, B.,** Adriamycin (NSC 123127): a new antibiotic with antitumor activity, *Cancer Chemother. Rep.,* 53, 33, 1969.
3. **Tan, C., Tasaka, H., Yu, K-P., Murphy, M. L., and Karnofsky, D. A.,** Daunomycin: an antitumor antibiotic in the treatment of neoplastic disease, *Cancer,* 20, 333, 1967.
4. **Blum, R. H. and Carter, S. K.,** Adriamycin: a new anticancer drug with significant clinical activity, *Ann. Intern. Med.,* 80, 249, 1974.
5. **Bonadonna, G. and Monfardini, S.,** Cardiac toxicity of daunomycin, *Lancet,* 1, 837, 1969.
6. **Lefrak, E. A., Pitha, J., Rosenheim, S., and Gottlieb, J. A.,** A clinicopathologic analysis of adriamycin cardiotoxicity, *Cancer,* 32, 302, 1973.
7. **Minow, R. A., Benjamin, R. S., and Gottlieb, J. A.,** Adriamycin cardiotoxicity: an overview with determination of risk factors, *Cancer Chemother. Rep.,* 6, 195, 1975.
8. **Ferrans, V. J.,** Overview of cardiac pathology in relation to anthracycline cardiotoxicity, *Cancer Treat. Rep.,* 62, 955, 1978.
9. **Pigram, W. J., Fuller, W., and Hamilton, L. D.,** Stereochemistry of intercalation: interaction of daunomycin with DNA, *Nature New Biol.,* 235, 17, 1972.
10. **Zunino, F., Gambetta, R., DiMarco, A., and Zaccaro, A.,** Interaction of daunomycin and its derivatives with DNA, *Biochem. Biophys. Acta,* 277, 489, 1972.

11. **Momparler, R. L., Karon, M., Siegel, S. E., and Avila, F.,** Effect of adriamycin on DNA, RNA, and protein synthesis in cell-free systems and intact cells, *Cancer Res.,* 36, 2891, 1976.
12. **Jaenke, R. S.,** An anthracycline antibiotic-induced cardiomyopathy in rabbits, *Lab. Invest.,* 30, 292, 1974.
13. **Mettler, F. P., Young, D. M., and Ward, J. M.,** Adriamycin-induced cardiotoxicity (cardiomyopathy and congestive heart failure) in rats, *Cancer Res.,* 37, 2705, 1977.
14. **Rosenoff, S. H., Olsen, H. M., Young, D. M., Bostick, F., and Young, R. C.,** Adriamycin-induced cardiac damage in the mouse: a small animal model of cardiotoxicity, *J. Natl. Cancer Inst.,* 55, 191, 1975.
15. **Lowe, M. C. and Smallwood, J. I.,** Adriamycin and daunomycin dose-dependent effects upon contractions of isolated rat monocytes, *Cancer Chemother. Pharmacol.,* 5, 61, 1980.
16. **Ferrero, M. E., Ferrero, E., Gaja, G., and Bernelli-Zazzera, A.,** Adriamycin: energy metabolism and mitochondrial oxidations in the heart of treated rabbits, *Biochem. Pharmacol.,* 25, 125, 1976.
17. **Bachur, N. R., Gordon, S. L., and Gee, M. V.,** Anthracycline antibiotic augmentation of microsomal electron transport and free radical formation, *Mol. Pharmacol.,* 13, 901, 1977.
18. **Caroni, P., Villani, F., and Carabole, E.,** The cardiotoxic antibiotic doxorubicin inhibits the Na^+-Ca^{2+} exchange of dog heart sarcolemmal vesicles. *FEBS Lett.,* 130, 184, 1981.
19. **Levey, G. S., Levey, B. A., Ruiz, E., and Lehotay, D. C.,** Selective inhibition of rat and human cardiac guanylate cyclase in vitro by doxorubicin (adriamycin): possible link to anthracycline cardiotoxicity, *J. Mol. Cell. Cardiol.,* 11, 591, 1979.
20. **Unverferth, D. V., Fertel, R. H., Talley, R. L., Magorien, R. D., and Balcerzak, S. P.,** The effect of first dose doxorubicin on the cyclic nucleotide levels of the human myocardium, *Toxicol. Appl. Pharmacol.,* 60, 151, 1981.
21. **Myers, C. E., McGuire, W. P., Liss, R. H., Ifrim, I., Grotzinger, K., and Young, R. C.,** Adriamycin: the role of lipid peroxidation in cardiac toxicity and tumor response, *Science,* 197, 165, 1977.
22. **Goodman, J. and Hochstein, P.,** Generation of free radicals and lipid peroxidation by redox cycling of adriamycin and daunomycin, *Biochem. Biophys. Res. Commun.,* 77, 797, 1977.
23. **Adachi, T., Nagae, T., Ito, Y., Hirano, K., and Sugiura, M.,** Relation between cardiotoxic effect of adriamycin and superoxide anion radical, *J. Pharm. Dyn.,* 6, 114, 1983.
24. **Nohl, H. and Jordan, W.,** OH-generation by adriamycin semiquinone and H_2O_2: An explanation for the cardiotoxicity of anthracycline antibiotics, *Biochem. Biophys. Res. Commun.,* 114, 197, 1983.
25. **Moore, L., Landon, E. J., and Cooney, D. C.,** Inhibition of the cardiac mitochondrial calcium pump by adriamycin in vitro, *Biochem. Med.,* 18, 131, 1977.
26. **Villani, F., Piccinini, F., Merelli, P., and Favalli, L.,** Influence of adriamycin on calcium exchangeability in cardiac muscle and its modification by ouabain, *Biochem. Pharmacol.,* 27, 985, 1978.
27. **Olson, H. M., Young, D. M., Prieur, D. J., Leroy, A. F., and Regan, R. L.,** Electrolyte and morphologic alterations of myocardium in adriamycin-treated rabbits, *Am. J. Pathol.,* 77, 439, 1974.
28. **Azuma, J., Sperelakis, N., Hasegawa, H., Tanimoto, T., Vogel, S., Ogura, K., Awata, N., Sawamura, d A., Harada, H., Ishiyama, T., Morita, Y., and Yamamura, Y.,** Adriamycin cardiotoxicity: possible pathogenic mechanisms, *J. Mol. Cell. Cardiol.,* 13, 381, 1981.
29. **Klugmann, S., Klugmann, B. F., Decorti, G., Gori, D., Silvestri, F., and Camerini, F.,** Adriamycin experimental cardiomyopathy in Swiss mice. Different effects of two calcium antagonistic drugs on ADM-induced cardiomyopathy, *Pharmacol. Res. Commun.,* 13, 769, 1981.
30. **Rabkin, S. W., Otten, M., and Polimeni, P. I.,** Increased mortality with cardiotoxic doses of adriamycin after verapamil pretreatment despite prevention of myocardial calcium accumulation, *Can. J. Physiol. Pharmacol.,* 61, 1050, 1983.
31. **Iwamoto, Y., Hansen, J. L., Porter, T. H., and Folkers, K.,** Inhibition of coenzyme Q_{10}-enzymes succinoxidase and NADH oxidase by adriamycin and other quinones having antitumor activity, *Biochem. Biophys. Res. Commun.,* 58, 633, 1974.
32. **Muhammed, H., Ramasarma, T., and Ramakrishna Kurup, C. K.,** Inhibition of mitochondrial oxidative phosphorylation by adriamycin, *Biochem. Biophys. Acta,* 722, 43, 1982.
33. **Zahringer, J., Hofling, B., Raum, W., and Randolph, R.,** Effect of adriamycin on the polyribosome and messenger-RNA content of rat heart muscle, *Biochem. Biophys. Acta,* 608, 315, 1980.
34. **Dalbow, D. G. and Jaenke, R. S.,** In vivo RNA synthesis in the hearts of adriamycin-treated rats, *Cancer Res.,* 42, 79, 1982.
35. **Zahringer, J.,** The regulation of protein synthesis in heart muscle under normal conditions and in adriamycin cardiomyopathy, *Klin. Wochenschr.,* 59, 1273, 1981.
36. **Van Helden, P. D. and Wild, I. J. F.,** Effects of adriamycin on heart and skeletal muscle chromatin, *Biochem. Pharmacol.,* 31, 973, 1982.
37. **Doroshow, J. H., Locker, G. Y., Ifraim, I., and Myers, C. F.,** Prevention of doxorubicin cardiac toxicity in the mouse by N-acetyl cysteine, *J. Clin. Invest.,* 68, 1053, 1981.
38. **Porta, E. A., Joun, N. S., Matsumura, L., Nakasone, B., and Sablan, H.,** Acute adriamycin cardiotoxicity in rats, *Res. Commun. Chem. Pathol. Pharmacol.,* 41, 125, 1983.

39. **Taylor, A. L. and Bulkley, B. H.,** Acute adriamycin cardiotoxicity: morphologic alterations in isolated perfused rabbit heart, *Lab. Invest.,* 47, 459, 1982.

40. **Williams, G. M., Dunkel, V. C., and Ray, V. A., Eds.,** Cellular systems for toxicity testing, *Ann. N.Y. Acad. Sci.,* 407, 1983.

41. **Ekwall, B.,** Screening of toxic compounds in tissue culture, *Toxicology,* 17, 127, 1980.

42. **Nardone, R. M.,** Toxicity testing in vitro, in *Growth, Nutrition, and Metabolism of Cells in Culture,* Vol. 3, Rothblat, R. H. and Christofalo, V. J., Eds., Academic Press, New York, 1977.

43. **Wenzel, D. G. and Cosma, G. N.,** A quantitative metabolic inhibition test for screening toxic compounds with cultured cells, *Toxicology,* 29, 173, 1983.

44. **Balazs, T., Ed.,** *Cardiac Toxicology,* CRC Press, Inc., Boca Raton, Fla., 1981.

45. **Seraydarian, M. W., Artaza, L., and Goodman, M. F.,** Adriamycin: effect on mammalian cardiac cells in culture 1 cell population and energy metabolism, *J. Mol. Cell. Cardiol.,* 9, 375, 1977.

46. **Seraydarian, M. W. and Artaza, L.,** Modification by adenosine of the effect of adriamycin on myocardial cells in culture, *Cancer Res.,* 39, 2940, 1979.

47. **Lampidis, T. J., Henderson, I. C., Israel, M., and Canellos, G. P.,** Structural and functional effects of adriamycin on cardiac cells in vitro. *Cancer Res.,* 40, 3901, 1980.

48. **Noel, G., Peterson, C., Trouet, A., and Tulkens, P.,** Uptake and subcellular localization of daunorubicin and adriamycin in cultured fibroblasts, *Eur. J. Cancer,* 14, 363, 1978.

49. **Lewis, W., Galizi, M., and Puszkin, S.,** Compartmentalization of adriamycin and daunomycin in cultured chick cardiac myocytes. Effects on synthesis of contractile and cytoplasmic proteins, *Circ. Res.,* 53, 352, 1983.

50. **Tritton, T. R. and Yee, G.,** The anticancer agent adriamycin can be actively cytotoxic without entering cells, *Science,* 217, 248, 1982.

51. **Durand, R. E. and Olive, P. L.,** Flow cytometry studies of intracellular adriamycin in single cells in vitro, *Cancer Res.,* 41, 3489, 1981.

52. **Skovsgaard, T.,** Carrier-mediated transport of daunorubicin, adriamycin, and rubidazone in Ehrlich ascites tumor cells, *Biochem. Pharmacol.,* 27, 1221, 1978.

53. **Necco, A., Dasdia, T., Francesco, D. D., and Ferroni, A.,** Action of ouabain, oligomycin, and glucagon on cultured heart cells treated with adriamycin, *Pharmacol. Res. Commun.,* 8, 105, 1976.

54. **Petrovic, D., Brown, S. M., and Yatvin, M. B.,** Effects of adriamycin and irradiation on beating of rat heart muscle cells in culture *Int. J. Radiation Oncol. Biol. Phys.,* 2, 505, 1977.

55. **Lampidis, T. J., Moreno, G., Salet, C., and Vinzens, F.,** Nuclear and mitochondrial effects of adriamycin in singly isolated pulsating myocardial cells, *J. Mol. Cell. Cardiol.,* 11, 415, 1979.

56. **DeHaan, R. L. and Hirakow, R.,** Synchronization of pulsation rates in isolated cardiac myocytes, *Exp. Cell Res.,* 70, 214, 1972.

57. **Gross, W. O.,** Role of fibroblasts in synchronizing the beat rhythm of isolated heart muscle cells in culture, in *Perspectives in Cardiovascular Research,* Vol. 5, Pexieder, T., Eds., Raven Press, New York, 1981, 331.

58. **Nagineni, C. N., Yamada, T., Yang, J. J., and Seraydarian, M. W.,** Effects of adriamycin on enzymes of adenine nucleotide metabolism in heart cell cultures, *Res. Commun. Chem. Pathol. Pharmacol.,* 50, 301, 1985.

60. **Newman, R. A., Hacker, M. P., and Krakoff, I. H.,** Amelioration of adriamycin and daunorubicin myocardial toxicity by adenosine, *Cancer Res.,* 41, 3483, 1981.

61. **Fialkoff, H., Goodman, M. F., and Seraydarian, M. W.,** Differential effect of adriamycin on DNA replication and repair synthesis in cultured neonatal rat cardiac cells, *Cancer Res.,* 39, 1321, 1979.

62. **Galaris, D. and Rydstrom, J.,** Enzyme induction by daunorubicin in neonatal heart cells in culture, *Biochem. Biophys. Res. Commun.,* 110, 364, 1983.

63. **Okano, C., Kwok, O., Hokama, Y., and Chou, S. C.,** Effects of adriamycin on the macromolecular synthesis in rat myocardiocytes and KB cells, *Res. Commun. Chem. Pathol. Pharmacol.,* 45, 279, 1984.

64. **Necco, A., Dasdia, T., Cozzi, S., and Ferraguti, M.,** Ultrastructural changes produced in cultured, adriamycin-treated myocardial cells, *Tumori,* 62, 537, 1976.

65. **Tobin, T. P. and Abbott, B. C.,** A stereological analysis of the effect of adriamycin on the ultrastructure of rat myocardial cells in culture, *J. Mol. Cell. Cardiol.,* 12, 1207, 1980.

66. **Fialkoff, H., Goodman, M. F., Seraydarian, M. W., and Yang, J. J.,** Replicative and unscheduled DNA synthesis in adriamycin-treated myocardial cells, in *Advances in Myocardiology,* Vol. I, Tajuddin, M., Das, P. K., Tariq, M., and Dhalla, N. S., Eds., University Park Press, Baltimore, 1980, 565.

67. **Wang, Y. M., Madanat, F. F., Kimball, J. C., Gleiser, C. A., Ali, M. K., Kaufman, M. W., and Van Eys, J.,** Effect of vitamin E against adriamycin-induced toxicity in rabbits, *Cancer Res.,* 40, 1022, 1980.

68. **Bertazzoli, C. and Ghione, M.,** Adriamycin associated cardiotoxicity: Research on prevention with co-enzyme Q, *Pharmacol. Res. Commun.,* 9, 235, 1977.

Chapter 22

THE HEART CELL IN CULTURE: A MODEL OF CARDIAC OXYGEN RESTRICTION AND DEPRIVATION IN THE HEART

Arié Pinson

TABLE OF CONTENTS

I. INTRODUCTION

During the last few decades enormous progress has been made in advancing basic understanding of the roles played in cardiac function by oxidative phosphorylation, the sarcoplasmic reticulum, the T-tubule system, and release and sequestration of calcium. Intensive research efforts have been directed towards application of this knowledge aimed at shedding light on the processes occurring during ischemic injury, employing both in vivo and in vitro models. Heart cells in culture have also been utilized, not always successfully, for studying the effects oxygen restriction and deprivation. Indeed, cultured heart cells should be the system of choice for such studies, as they allow the experimental conditions to be controlled with ease, and, thus, constitute a model system well suited to the investigation of these phenomena at the cellular level.

This chapter aims to provide a brief discussion on some general aspects of cardiac ischemia, to review the data obtained using heart cells in culture, and, finally, to present a recently developed novel approach for using the cultured heart cell in a more efficient manner for studying the phenomenon of oxygen deprivation in the cardiac muscle cell.

II. MYOCARDIAL ISCHEMIC INJURY

The heart is basically an oxidative organ which uses fatty acids (FAs) as the major energy source.[1] The mechanical function of cardiac contractility seems to depend primarily upon oxidative phosphorylation as an energy source.[2,3] Indeed, under normal conditions, the heart synthesizes high-energy phosphates (HEPs) continuously to replace those utilized.[4] Oxygen restriction or deprivation occurs in vivo when the lumen of a coronary artery is totally or partially occluded; and ischemia when the blood flow to the heart region is reduced to such an extent that the volume of perfusing blood falls below that required for the maintenance of normal function. Thus, in ischemia, anoxia, and hypoxia, the oxygen levels in the heart are too low to prevent the shift from aerobic respiration to anaerobic glycolysis. The shift to anaerobic respiration is very rapid, and manifested in the cessation of contractions due to changes in the availability of both ATP and calcium ions (Ca^{2+}).[5]

In conditions of reduced perfusion rates, oxygen deprivation is followed by a concomitant fall in the tissue oxygen tension coupled with a rapid reduction in cellular glycogen content.[6] In addition, metabolic products are not cleared from the system. Creatine phosphate levels fall rapidly, creatine content is increased, and ATP breakdown products are accumulated.[7]

By contrast, in both hypoxia and anoxia, adequate perfusion is maintained; however, the oxygen supply is reduced, or absent, respectively.

Current knowledge indicates that cellular damage during anoxic injury is the result of a complex sequence of events — the contribution of each depending on the balance between energy requirements and consumption involved in a particular step.

During the initial (reversible) phases of ischemia, the heart is capable of resuming normal mechanical and electrical activity upon restoration of the blood circulation.[8,9] However, following long periods of ischemia, the damage is enhanced upon restoration of arterial blood flow, leading to swelling of the cells and increased necrosis.[8,9,10] This somewhat unexpected increase in damage has been termed the ''oxygen paradox''.[10]

The sequence of events which causes the transition from the reversible to the irreversible phase of cell damage, and leads to cell death, remains to be elucidated, primarily with respect to the molecular events, i.e., the intracellular biochemical changes leading to cell death under ischemic conditions. Many different schemes have been proposed to explain this functional change. Although some of the differences between the proposed schemes may be put down to variations in the experimental systems, the major factor seems to be the events that particular research groups have chosen to emphasize. Within seconds of the onset of ischemia, a shift from aerobic metabolism to anaerobic glycolysis begins,[7,11] and this process continues for a relatively long period. Anaerobic glycolysis is accompanied by a concomitant depletion of the glycogen stores, and is coupled with lactate accumulation — upon the decrease of endogenous glycogen levels, exogenous glucose becomes an important substrate.[12] Thus, the duration of ischemia relative to the rate of glycogenolysis may control the route of glucose utilization.

Neely and Morgan[13] reported that glycolysis proceeds at a high rate during the early stages of ischemic injury. Glycolysis rates are even higher during the initial stages of anoxia in the well perfused heart, and if glucose is added to the perfusate, it may provide sufficient ATP to maintain some contractile activity, and to prevent the decrease in ATP levels. Subsequently, glycolysis rates fall due to the inhibition of glyceraldehyde-3-phosphate-dehydrogenase activity by the high levels of lactate, which are not removed from the cell, and which lower the pH and NADH levels.[7,14-16] Although glycolysis may proceed slowly over a very long period, the ATP produced is not sufficient to prevent depletion of the adenine nucleotide pool.[17,18] According to Kübler and Spieckerman,[14] glycolysis ceases when the ATP levels are so low that the conversion of fructose-6-phosphate to fructose-1,6-diphosphate does not take place.

ATP is degraded to ADP, which may then be further utilized leading to the accumulation of AMP[18] — $2ADP \rightarrow ATP + AMP$. AMP may either be deaminated to inosine monophosphate (IMP) or dephosphorylated to adenosine, which is then rapidly further degraded to inosine and hypoxanthine and released from the cell. Within 60 to 90 min, most of the ATP and other adenine nucleotides are converted to purine bases.[17,18]

Electron microscopic studies[17,19] showed a gradual development of cellular damage. In the early stages, the mitochondria are already swollen and glycogen stores are greatly reduced. With continuation of ischemia, chromatin becomes condensed at the periphery of the nucleus, and mitochondrial deformation becomes more pronounced and amorphous dense bodies are formed. At this stage, the sarcolemmal inner leaflet becomes disrupted; however, the outer leaflet appears to remain intact. These sarcolemmal transformations are only detectable at a relatively late phase of ischemic injury.

Although membrane alterations may only be detected at a late stage of ischemia by electron microscopic techniques, the onset of sarcolemmal damage, as manifested in enzyme release,[20] and by Ca^{2+} accumulation[21] is thought to occur during the early stages of ATP depletion. Changes in the cation distribution (Ca^{2+}, Na^+, and K^+) are due to impaired function of the sarcolemmal pumps involved in Ca^{2+} extrusion and subsequent sequestration by the sarcoplasmic reticulum, and in the maintenance of membrane potential via the differences in Na^+ and K^+ concentrations on both sides of the membrane.

Many authors believe that calcium reperfusion is a crucial step in the progress towards irreversibility. Indeed, accumulation of large amounts of calcium may be detected in ischemic reperfused cardiac myocytes, and leads to exacerbation of ischemic injury (the ''calcium paradox'').[22-24] Furthermore, the gradient for calcium entry into the cell is very favorable (10^{-3} M extracellular, and 10^{-6} to 10^{-7} M intracellular Ca^{2+}). Lack of removal of cytosolic calcium by the Ca^{2+} transport mechanism, or the increase in slow-channel electrogenic Ca^{2+} influx through the plasma membrane, as manifested in the immediate depression of heart contractility, is followed by the inhibition of the cation pumps, and may lead to intracellular calcium overload. The deleterious effects of calcium overload on mitochondrial function include: (1) the uncoupling of oxidative phosphorylation, (2) alteration of the permeability of the inner mitochondrial membrane via the activation of membraneous phospholipases, and (3) it may also precipitate as calcium phosphate in the mitochondrial matrix. Calcium overload does not occur during the ischemic period, but only after perfusion is re-established and oxygen becomes available, which allows the mitochondria to take up intracellular calcium giving rise to overload. However, if calcium overload does not take place, mitochondrial function remains unaffected, and injury is reversible.[25]

It is generally held that sarcolemmal damage indicates irreversibility of injury; however, the nature of this damage remains to be elucidated. Activation of sarcolemmal phospholipase, either by the increased Ca^{2+} concentration or by release from the inhibitory effects of certain phosphorylated proteins on membraneous phospholipases, may explain some of the aspects of sarcolemmal damage.[17,26] Increased phospholipase activity may, in turn, affect the distribution pattern of sarcolemmal phospholipids and also bring about changes in membrane fluidity. Another possible explanation for sarcolemmal damage is that it is brought about via the release of lysosomal enzymes, mainly phospholipases, which may be more active at the low pH prevailing in the ischemic cell.[27,28] Whatever the cause of sarcolemmal damage, membraneous phospholipases, lysosomal hydrolytic activity towards sarcolemmal phospholipids, and the degradation of sarcolemmal phospholipids coupled with fact that these may not be replaced by phospholipid neosynthesis are both certainly important factors contributing to sarcolemmal damage.

More recently, many researchers have proposed that increased concentrations of free radicals, and consequently sarcolemmal lipid peroxidation, may cause sarcolemmal damage and irreversible cellular injury (for a review, see Reference 29).

Free radicals (i.e., species with unpaired electrons, including molecular oxygen) are involved in the physiological process of membrane renewal and in the synthesis of prostaglandins and leucotrienes via lipid peroxidation. In addition, activation of free radical formation and lipid peroxidation also occur in various organs, including the heart, in response to stress.[30] Moreover, the deleterious effects of high levels of oxygen have been variously attributed to the presence of oxygen radicals (for a review, see Reference 31) and was assumed in the theory of superoxide radicals (O_2) and hydrogen (OH) radicals.[31,32] Oxygen is more soluble in the hydrophobic compartment of the cell membrane, and superoxides are much more active in this phase than in the hydrophilic environment. In addition, the two hydroxyl radicals formed as a result of the homolytic fission of the oxygen-oxygen bond in the hydrogen peroxide molecule (the Fenton reaction)[33] readily react with most organic compounds. Superoxides produced in the hydrophobic bilayer could destroy phospholipids by attack on the carbonyl group of the ester bond,[34,35] and hydrogen radicals may initiate the chain reaction of lipid peroxidation.

During oxygen deprivation when the respiratory chain is inhibited, NADH accumulates. The oxygen dissolved in the lipid compartments may be incompletely reduced and form oxygen and hydrogen radicals, and thus may initiate the lipid peroxidation reaction. The increase in levels of lipid peroxides leads to drastic changes within the membrane, including loss of membranal enzymes, changes in viscosity, permeability, and, finally, in sarcolemmal disruption.

A summary of this brief bird's-eye view of the events occurring in cardiac tissue during ischemia is presented in Scheme 1, showing two main routes that lead to irreversible damage: lysosomal enzyme release, and sarcolemmal transformation.

III. CULTURED HEART CELLS AS A MODEL SYSTEM

In the early 1970s, researchers first began to make occasional use of cultured heart cells as a model system in studies on the effects of reduced oxygen tension. However, it was only at the end of that decade that this system was used in a more consistent manner. The lack of initial interest can be put down to the fact that results obtained with cultured cells bore little similarity to those found in vivo or in the perfused heart. The methodology employed generally followed two routes: cells were either deprived of oxygen by placing them in an environment of nitrogen, or by incubation in the presence of metabolic inhibitors, such as inhibitors of glycolysis and/or uncouplers of oxidative phosphorylation, in order to simulate conditions of oxygen deprivation. Often, these two approaches — oxygen restric-

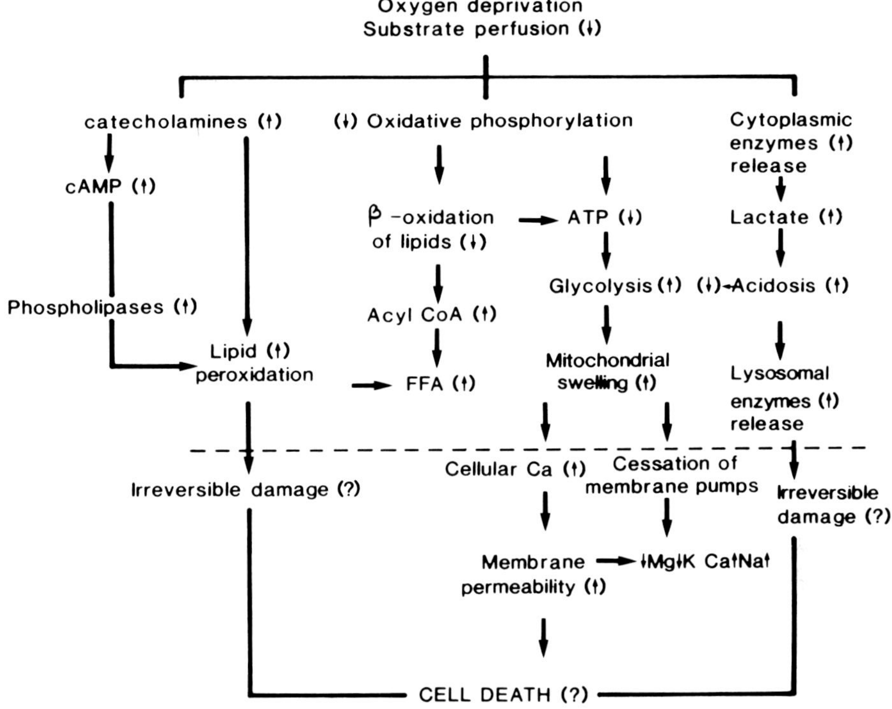

SCHEME 1.   Schematic representation of the possible sequence of events that takes place during oxygen and substrate deprivation.

tion, and metabolic inhibitors — were combined. Various aspects relating to these approaches will be reviewed.

A. Effect of Oxygen Tension

Early studies were essentially concerned with the various effects of pericellular oxygen tension on heart cells in culture. Hollenberg[36] found that in long-term hypoxia, both RNA and protein synthesis increased in inverse proportion to the fraction of oxygen, which was reduced to 2 to 5%. A high oxygen fraction (80%) led to impairment of both RNA and protein synthesis. These findings were confirmed by Karsten et al.,[37] who reported that a pericellular oxygen fraction of 83% was lethal to cultured cells, and that fractions ranging from 2% to 20% favored growth of these cells, as indicated by the increased content of muscle-specific proteins. Even at oxygen fractions as low as 0.08%, there were no differences in either ATP levels or beating rates from those observed in the presence of 5% oxygen, in spite of the lower levels of myosin. However, at the lower oxygen levels, there was a shift in the LDH isoenzyme pattern towards the M type. Morphological studies[38] showed that at less than 0.08% oxygen, there was an increase in the number of mitochondria without cristae, which were both larger than normal and irregularly shaped.

B. Simulation of Anoxia by Metabolic Inhibitors

The rationale behind the use of metabolic inhibitors for the simulation hypoxic or ischemic states is the attempt to reproduce either the toxicity of the cellular environment (low pH), or the impairment of cellular function at the same cellular sites as in anoxic injury.

Wenzel and Acosta[39] employed hypoxia (1% oxygen) over a period of 4 days, followed by incubation for various periods of time with either cyamide or 2-deoxyglucose (DOG), agents causing partial hypoxia, either alone or in combination with low pH (6.4). In spite of this drastic treatment, cellular injury, as manifested by increased mitochondrial or lysosomal permeability, was not very marked. However, metabolic inhibitors may be useful for studying the roles played by different cellular compartments in the maintenance of cardiac cell function. Thus, it has been shown[40] that in the presence of cyanide, a selective blocker of the electron transport chain, beating activity, membrane potential, and amplitude of action potentials were maintained for 24 hr, albeit at lower levels. However, during long-term hypoxia (24 to 48 hr), both the beating activity and the action potentials were completely inhibited. Thus, whereas long-term hypoxia led to impairment cellular function at various different levels, potassium cyanide was more selective.

Barry et al.[41] described a negative inotropic effect of hypoxia (1.5% O_2), which occurred within three to 4 min. By contrast, the beating rate was not significantly altered if normal glycolytic function was maintained. Inhibition of glycolysis by iodoacetate led to further depression of beating rates and contractile amplitudes; while, under normoxic conditions, iodoacetate had no effect on beating rates, which remained similar to those in controls. Thus, the fall in ATP levels primarily have a negative inotropic effect, and only more drastic decreases in cellular energy charge affect beating rates. Several research groups have proposed that glycolytically derived ATP plays an important role in the maintenance of membrane integrity and electrophysiological function.[42-49]

The relationship between the fall in high-energy phosphates content and the contractility of the cultured cells has been investigated further[48] by the utilization of various inhibitors of the mitochondrial electron transport chain (such as cyanide), or inhibitors of glycolysis (such as DOG and iodoacetate). Cyanide added to the medium at low concentrations (10^{-6} M to $1.5 \cdot 10^{-4}$ M), led to a progressive decline in the contractile amplitude (up to 18% of the control levels). The decrease in ATP and CP levels was accompanied by a concimitant decline in contractile function — further confirmation that cardiac function depends on mitochondrial oxidative capacity.[49] However, following an increase of cyanide concentration, HEP content did not decrease to below 50 to 60% of the control values, but a further decline in the amplitude of contractions was observed. In the presence of glycolytic inhibitors, HEP content decreased even on the addition of pyruvate or acetate which can maintain oxidative phosphorylation. However, contractile amplitude was unaffected in spite of the lower energy charge of the cardiac cell.[48] This observation lends further support to the concept that ATP is compartmentalized within the cell. It may also indicate that the balance between ATP production and utilization is more critical for the maintenance of mechanical activity than the overall energy charge of the cell.[48]

Alteration in cellular metabolic activity, primarily in the energy charge, affects the accessibility of membrane phospholipids to attack by phospholipases.[50,51] Indeed, in cultured heart cells, with the inhibition of glycolysis, and the concomitant fall in ATP levels, the sarcolemma becomes more susceptible to attack by exogenous lipases,[52,53] as manifested in LDH release, which is inversely correlated with both ATP content and glycolytic activity.

It has been shown[44,54,55] that the addition of various metabolic inhibitors to heart cells in culture led to the time-dependent release of cellular enzymes, such as lactic dehydrogenase (LDH), creatine phosphokinase (CPK), and α-hydroxybutyrate dehydrogenase. Thus, it was proposed that enzyme release is a marker for irreversible cell damage.[54,55] Other researchers, however, showed full recovery of hypoxic or metabolically inhibited cells in culture shortly after reoxygenation or washout of the inhibitor.[41,48]

In conclusion, the differences in the results obtained in various laboratories can probably be put down to the different protocols employed. Metabolic inhibitors are certainly of great value in studies on the roles that various cellular compartments play in maintaining cellular activities. However, although these data are important, their significance for elucidation of the mechanism of hypoxic and anoxic injury is questionable, since metabolic inhibitors have other effects on cardiac cells besides those which are directly related to oxygen restriction or deprivation.

C. Simulation of Anoxic Injury by Oxygen and Substrate Deprivation

Studies performed in the presence of metabolic inhibitors, although useful in some respects, introduce aspects that are not due to oxygen deprivation. During ischemic injury, the supply of both oxygen and substrates is either impaired or completely lacking. Some research groups attempted to simulate ischemic conditions in culture by deprivation of both oxygen and substrates in heart cell cultures.

Acosta and Puckett[56] subjected cultured heart cells to various levels of oxygen and substrate deprivation. However, even after 24 hr under conditions of severe oxygen and substrate deprivation, cellular damage, as reflected by morphological changes and beating function, were not very considerable and completely reversible. LDH release in these cultures reached 16%, and 20%, after 4 and 24 hr under ischemic conditions, respectively, as compared to 10% in control cultures. In addition, even after 24 hr of complete substrate and oxygen deprivation, the number of viable cells was similar in both ischemic and control cultures.[57-59]

Van der Laarse et al.[54] studied the effects of complete anoxia and metabolic substrate deprivation on the release of creatinine phosphokinase (CPK) and α-hydroxybutyrate dehydrogenase (αHBDH) from cardiac cells in culture over a period of 27 hr. During the first 3 hr, release of enzymes was hardly detectable; and, after 4 hr, it only reached 16% of the total activity. However, the release rates were subsequently accelerated, so that after 23 hr, there was almost complete loss (about 95%) of enzymic activity. In the presence of glucose in the medium, enzyme release was considerably reduced — to 20% after 23 hr. It has also been shown[60] that the release of aspartate amino transferase (AST) is delayed in the mitochondrial, as compared to the cytoplasmic compartment — 80% vs. 34%, following 13 hr of anoxia. This phenomenon may be explained by a higher proportion of bound, and hence ''immobile'' form of the enzyme, in the mitochondrial compartment.

In a more detailed study on the sequence of release of enzymes from various cellular compartments,[61] it was shown that following 10 hr under anoxic conditions approximately 80% of the cytoplasmic enzymes, 50% of the enzymes found in the sarcolemmal, lysosomal, and the outer mitochondrial membranes, and <10% of those in the inner mitochondrial membranes, were released. These workers also showed that raising the extracellular calcium concentrations during anoxia had a deleterious effect on cellular integrity, which is reflected in increased enzyme release, and probably brought about by partial inhibition of glycolytic flux. Lactate production and release, which also occurs in control cultures of heart cells (see Chapter 16), is increased during anoxia to reach almost on the same level as glucose utilization. Thus, lactate production serves as an indication of the glycolytic flux.[62-64] An increase in palmitate uptake, and its subsequent incorporation into TGs, under anoxic conditions, has also been reported.[65] In fetal heart organ cultures, protein synthesis was inhibited, probably due to reduced mRNA entry into polysomes.[66]

Important studies with respect to the sequence, and timing and magnitude of certain events in cardiac cell cultures, as compared to the in vivo and to the perfused heart, were carried out by De Luca et al.[67] In cultured heart cells under conditions of oxygen and substrate deprivation, they found that during the first 4 hr about 60 to 80% of the LDH and CPK was released, and was accompanied by almost total depression of ATP levels (see Figure 4 of Reference 67). The discrepancy between these data and those obtained by other researchers probably lies in greater attention to the control of the experimental conditions.

Although release of enzymes is a generally accepted marker for cellular damage in anoxic and ischemic injury, the fall in energy charge is the predominant feature, occurring at a much earlier phase. Thus, hypoxanthine, a product of ATP catabolism, is considered to be a marker for anoxic aggression.[68] It has been shown that hypoxanthine release reaches a maximum during the second hour of anoxia, and is followed by cellular enzyme release that reaches maximal levels during the fourth hour (cf. Figure 1 and 2, Reference 68). It has also been reported[69,70] that cholesterol release from the plasma membrane precedes the release of both sarcolemmal and cytoplasmic enzymes. Since cholesterol is a major constituent of the plasma membrane, affecting both viscosity and permeability,[74] it may be concluded that the fall in HEP content disrupts sarcolemmal structure and function, and that this occurs prior to the release of cellular enzymes.

With respect to heart cells in culture as a model of ischemia, the data using oxygen and substrate deprivation are no less significant than those obtained in the presence of metabolic inhibitors. However, the time scale for these events, 16 or 24 hr, giving rise to anoxic injury, differs significantly from that occurring in vivo or in the perfused heart. According to some researchers[63,70] enzyme release is linked with irreversible cell damage, as indicated by trypan blue exclusion staining. However, the large differences in release of enzymes from different compartments, as reported by several workers,[60,61] would not be expected to occur in dead cells, and, thus, the validity of the trypan blue exclusion test as a marker for irreversible anoxic damage is questionable (see below).

D. A New Approach for Studying Anoxia and Ischemia in Cultured Heart Cells

Various aspects relating to the use of cultured heart cells as a model system have been discussed above. Although some valuable findings have been reported utilizing this system, these did not simulate the in vivo events, mainly with respect to the timing, and, in particular the onset of irreversible damage to the cardiac cell (''the point of no return''). Simulation of hypoxia or anoxia in cultured cells may be achieved with relative ease; however, the validity of data obtained after 48, 24, or even 16 hr of anoxia,[54,56,57,62] is highly questionable. As mentioned previously,[36,37] cultured cells may grow and function for long periods at very low oxygen pressures, so that the equilibration of the extracellular medium in experiments performed under anoxic conditions must be rigorously controlled.

Is it possible to simulate true ischemic conditions in culture? As stated above, ischemia occurs when perfusion of the heart occurs at such a low rate that both the oxygen and substrate requirements, and the washout of metabolites, are inadequate. A technique has recently been developed in our laboratory in order to overcome this problem via reduction of the volume of the extracellular medium to the minimal amount, coupled with oxygen deprivation.[73,74] The minimal volume possible was found to be 0.2 mℓ for a Petri dish of 35 mm diameter, thus, the height of the liquid layer above the cells approximates to the size of a cardiac cell and constitutes a true micro-environment. The medium, Ham F10 without glucose, was rendered anoxic by pre-equilibration with 95% N_2/5% CO_2 which was bubbled through it for at least 30 min (see Figure 4, Reference 74). A specially adapted incubation unit with a minimal dead volume was also employed. This methodology should be equivalent to oxygen deprivation coupled with low perfusion rates — the prominent features of ischemia (substrates do not reach the cell, and there is inadequate washout of the metabolites from the vicinity of the cells). Figure 1 shows a plot of LDH release against volume of the extracellular medium. At volumes between 0.2 to 1 mℓ, these parameters are linearly correlated. In addition, at the volume of 0.2 mℓ (ischemia), release of LDH reached 40% within 2 hr. As also shown in Figure 1, under normoxic conditions, reduction of extracellular volume per se did not lead to increased LDH release, even when there was no glucose present in the medium. LDH release began shortly after the onset of ischemic conditions (see Figure 4, Reference 74). Reduction of the volume of extracellular fluid coupled with oxygen and substrate deprivation undoubtedly provide the conditions that best simulate ischemia in cultured heart cells.

The release of the lysosomal enzyme, β-galactosidase, lags behind that of LDH — lysosomal enzymes could only be detected in the medium after 2 hr of ischemia (5% of the total activity), rising dramatically to 18%, after 3 hr, and to 35% after 4 hr (see Figure 6, Reference 74). The onset of lysosomal enzyme release after 2 hr of ischemia implies that

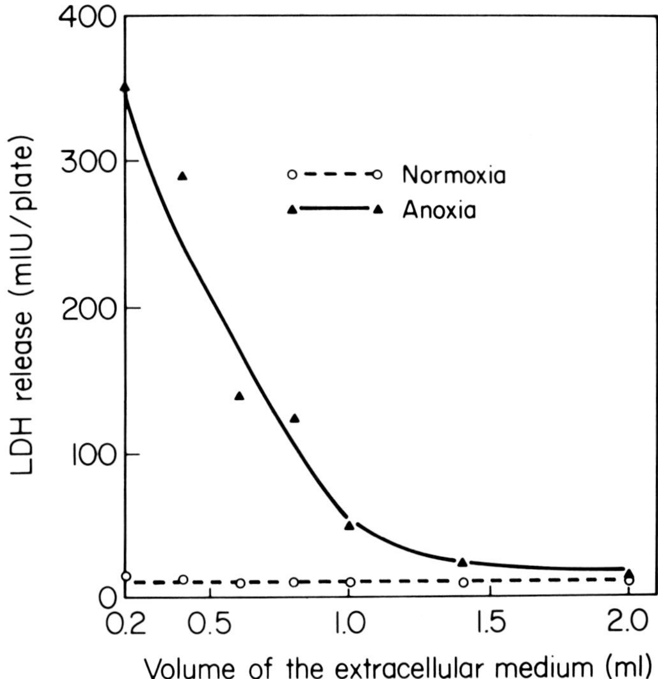

FIGURE 1.   The effect of the volume of the extracellular medium on LDH release from the cells during anoxia. Cells were incubated for 120 min in varying volumes of glucose-free medium under normoxic or anoxic conditions.

the hydrolytic enzyme activity is favoured in these cells in which lactate is produced via glycogenolysis.[75] Indeed, 2 hr after the onset of ischemic conditions, many functions of the cardiac cell are lost concomitantly with the release of lysosomal enzymes:[76] there is an acceleration of protein degradation (Figure 13, Reference 76); cholesterol release (Figure 23, Reference 76); Ca^{2+} accumulation (Figure 24, Reference 76); and Na^+ accumulation and K^+ release (Figures 25 and 26, Reference 76). Clearly, labilization of the lysosomes, accompanied by the release of hydrolytic enzymes, is a distinct stage in the progress of ischemic injury.

The presence of glucose in the medium, under both anoxia (in a volume of 2 mℓ of extracellular medium) and ischemia (0.2 mℓ of medium), had a marked beneficial effect, delaying enzyme release and other disruptive phenomena.[75] HEP levels were maintained via glycolysis for relatively long periods of time,[75,76] particularly during anoxia. Thus, anoxia in this system in the presence of glucose gives rise to data which are consistent with studies in the highly perfused heart, sufficient ATP is provided in order to maintain a certain measure of mechanical work.[13] Thus, with a volume of 2 mℓ of extracellular fluid, the presence of 6 mM glucose in the medium coupled with the cellular glycogen stores (240 μg per plate) leads to the maintenance of HEP levels for long periods of time.[76,77] However, at a volume of 0.4 mℓ of extracellular fluid, and even in the presence of glucose, HEP levels fell by more than 50% after 4 hr.[77] By contrast, in the absence of glucose, HEP levels were almost completely depleted after 2 hr,[76,77] with a concomitant acceleration of lysosomal enzyme release.

Sarcolemmal damage is indicated by the early release of cytoplasmic enzymes. Lipid peroxidation is believed to play a major role in the transition from the reversible to the irreversible phase of ischemic damage in the myocardial cell.[29] Malonyl dialdehyde (MDA) formation has been shown to be associated with free radical formation and lipid peroxidation.

In cultured heart cells, MDA formation was increased by 4 to 6% after 2 hr under conditions of oxygen deprivation, with sudden increases to 12%, and to 18%, in anoxia, and ischemia, respectively (Figure 27, Reference 76). Furthermore, the rise in MDA levels was shown to correlate with cholesterol depletion (see above), and changes in the lipid conformation at the sarcolemmal surface, as demonstrated by the method of glucose oxidase/lactoperoxidase radioiodination of the sarcolemmal lipids,[76,79,80] which allows the detection of subtle changes in the conformation of surface lipids, i.e., the degree of exposure of the lipid moieties in the outer sarcolemmal leaflet to the extracellular environment, which depends on the physical state of the sarcolemma.[80-82] No significant differences were observed between normoxic and anoxic cardiac cells in culture on analysis of the amount of label associated with the total lipid fraction following enzymatic radioiodination of the cell surface. However, ischemic cells showed a marked increase in the amount of label associated with the total lipid fraction — enhanced by 1.71-, 1.92-, and 1.84-fold, following 1, 2, and 3 hr of incubation, respectively. In particular, the radioactivity in the phospholipid (PL) classes was 1.3-fold higher in ischemia than in anoxia.

Further analysis of the distribution of radioactivity among the different classes of PL showed that major changes occurred in this respect after 1 hr of ischemia — labeling of phosphatidylethanolamine (PE) was increased by 2.1-fold, while the labeling of phosphatidylcholine (PC) was reduced by 40%. Consequently, the PE/PC labeling ratio, which was approximately 1.0 in normoxia and anoxia, rose to 2.2 in ischemic cells.

Asymmetric distribution of PE and PC between the two plasma membrane leaflets has been demonstrated — PC being the more abundant in the outer leaflet, and PE in the inner one. Taking into consideration the relatively high degree of freedom of motion of the fatty acyl chains in PE, due to their high degree of unsaturation,[83] these labeling experiments should reflect the dynamic state of the cell surface. Thus, variations in the labeling patterns of PE and PC should indicate the physical state of the membrane. These data, therefore, imply that reorganization of the sarcolemmal phospholipids occurs during ischemia. Changes of this nature, prominent in ischemia, but much less so during anoxia, may bring about the variations in permeability towards cations and larger molecules.

Do these membrane transformations and the release of cytoplasmic enzymes imply irreversible cell damage? Since reperfusion and reoxygenation of ischemic heart tissue have been shown to increase damage,[2,10] and anoxic damage may be measured by enzyme release, it may be assumed that once the irreversible phase of cell damage has been reached, enzyme release will persist even on reoxygenation ("oxygen paradox"). Indeed, on the reoxygenation of anoxic or ischemic cells (Figures 28 adn 29, Reference 76), enzyme release only continued if the point at which the HEPs had been exhausted coupled with increased lysosomal enzyme release had been reached. At earlier stages, enzyme release ceased upon reoxygenation, and the beating function was restored within 30 min.

IV. CONCLUSIONS AND PERSPECTIVES

Oxygen deprivation coupled with the restriction of the volume of the extracellular medium in heart cell cultures provides an invaluable system for studying anoxia and ischemia at the cellular level. It is particularly suitable for studies of the kinetics of the events associated with these phenomena.

Release of cytoplasmic enzymes can be detected shortly after the onset of ischemia and is probably related to sarcolemmal reorganization. During this period, HEP production slows down, but they do not become completely depleted. A definite sequence of events may be described; and these changes appear to be reversible.

At the stage at which CP and ATP are almost completely exhausted, lysosomal enzymes are released, and related catabolic and disruptive events occur. The cell enters the irreversible phase of ischemic damage. It is impossible to identify a definite sequence of events — the cell seems to lose all the cellular and functional properties simultaneously. Further definition of the parameters characterizing the reversible and irreversible phases of cell damage, in view of its obvious clinical significance, clearly deserves more detailed study.

The cultured heart cell model should provide important data concerning reoxygenational and reperfusional repair; and will certainly prove to be an invaluable tool for studying the effects of various cardioplegic solutions and drugs on functional repair in cardiac cells.

ACKNOWLEDGMENTS

Some of the research carried out in the author's laboratory, which is reported in this chapter, was partially supported by grants from the Chief Scientist of the Israel Ministry of Health; the Ministry of Education and Science of the state of Niedersachsen, FRG; the Mr. and Mrs. D. Vidal-Madjar Foundation for Heart Research, Paris, France; Mrs. F. Berk, Brussels, Belgium; Mr. and Mrs. S. Benador, Geneva, Switzerland; and Visiting Fellowships from INSERM, France, and the Dutch Heart Foundation.

REFERENCES

1. **Bing, R. J., Siegel, A., Ungar, I., and Gilbert, M.,** Metabolism of the human heart. II. Studies on fat, ketone, and fatty acid metabolism, *Am. J. Med.,* 16, 504, 1954.
2. **Hearse, D. J.,** Reperfusion of the ischemic myocardium, *J. Mol. Cell. Cardiol.,* 9, 605, 1977.
3. **Doorey, A. J. and Barry, W. H.,** The effects of the inhibition of oxidative phosphorylation and glycolysis on contractility and high energy phosphate content in cultured chick heart cells, *Circ. Res.,* 53, 192, 1983.
4. **Seraydarian, M. W., Artaza, L., and Abbott, B. C.,** The effect of adenosine on cardiac cells in culture, *J. Mol. Cell. Cardiol.,* 4, 477, 1972.
5. **Katz, A. M. and Hecht, H. H.,** The early ''pump'' failure of the ischemic heart, *Am. J. Med.,* 47, 497, 1969.
6. **Jennings, R. B.,** Early phase of myocardial injury and infarction, *Am. J. Cardiol.,* 247, 753, 1969.
7. **Braasch, W., Gubjarnason, S., Puri, P., Ravens, K. J., and Bing, R. J.,** Early changes in energy metabolism in the myocardium following acute coronary occlusion in anaesthetized dogs, *Circ. Res.,* 23, 429, 1968.
8. **Jennings, R. B. and Ganote, C. E.,** Structural changes in myocardium during acute ischemia, *Circ. Res.,* 35 (Suppl. III), 156, 1974.
9. **Wahlen, D. A., Jr., Hamilton, D. G., Ganote, G. E., and Jennings, R. B.,** Effects of a transient period of ischemia on myocardial cells: I Effects on cell volume regulation, *Am. J. Pathol.,* 74, 381, 1974.
10. **Hearse, D. J., Humphrey, S. M., Bullock, G. R.,** The oxygen paradox: Two facets of the same problem?, *J. Mol. Cell. Cardiol.,* 10, 641, 1978.
11. **Neely, J. R., Rovetto, M. J., and Whitmer, J. I.,** Rate-limiting steps of carbohydrate and fatty acid metabolism in ischemic hearts, *Acta Med. Scand., Suppl.,* 587, 9, 1976.
12. **Opie, L. H.,** Effects of regional ischemia on metabolism of glucose and fatty acids. Relative rates of aerobic and anaerobic energy production during myocardial infarction and comparison with the effects of anoxia, *Circ. Res.,* 38 (Suppl. I), 52, 1976.
13. **Neely, J. R. and Morgan, H. E.,** Relationship between carbohydrate and lipid metabolism and the energy balance of heart muscle, *Ann. Rev. Physiol.,* 36, 413, 1974.
14. **Kübler, W. and Spieckermann, P. G.,** Regulation of glycolysis in the ischemic and anoxic myocardium, *J. Mol. Cell. Cardiol.,* 1, 351, 1970.
15. **Rovetto, M. J., Lamberton, W. F., and Neely, J. R.,** Mechanisms of glycolytic inhibition in ischemic rat hearts, *Circ. Res.,* 37, 742, 1975.
16. **Neely, J. R. and Feuvray, D.,** Metabolic products and myocardial ischemia, *Am. J. Pathol.,* 102, 282, 1981.
17. **Jennings, R. B., Hawkins, H. K., Lowe, J. E., Hill, M. L., Klotman, S., and Reimer, K. A.,** Relation between high energy phosphate and lethal injury in myocardial ischemia in the dog, *Am. J. Pathol.,* 92, 187, 1978.

18. **Jennings, R. B., Reimer, K. A., Hill, M. L., and Mayer, S. A.,** Total myocardial ischemia *in vitro:* I Comparison of high energy phosphate production, utilization and depletion and adenine nucleotide catabolism in total ischemia in vitro vs. severe ischemia in vivo, *Circ. Res.,* 49, 892, 1981.

19. **Jennings, R. B. and Hawkins, H. K.,** Ultrastructural changes of acute myocardial ischemia, in *Degradative Processes in Heart and Skeletal Muscle,* Wildenthal, K., Ed., Elsevier/North Holland Biomedical Press, Amsterdam, 1980, 295.

20. **Kaltenbach, J. P. and Jennings, R. B.,** Metabolism of ischemic cardiac muscle, *Circ. Res.,* 8, 207, 1960.

21. **Jennings, R. B., Ganote, G. E., and Reimer, K. A.,** Ischaemic tissue injury, *Am. J. Pathol.,* 81, 97, 1975.

22. **Shen, A. C. and Jennings, R. B.,** Myocardial calcium and magnesium in acute ischemic injury, *Am. J. Pathol.,* 67, 471, 1972.

23. **Shen, A. C. and Jennings, R. B.,** Kinetics of calcium accumulation in acute ischemic injury, *Am. J. Pathol.,* 67, 441, 1972.

24. **Ruigrock, T. J. C., Boink, A. B. T. J., Spies, F., Blok, F. J., Maas, A. H. J., and Zimmerman, A. N. E.,** Energy dependence of the calcium paradox, *J. Mol. Cell. Cardiol.,* 10, 991, 1978.

25. **Nayler, W. G.,** The role of calcium in ischemic myocardium, *Am. J. Pathol.,* 102, 262, 1981.

26. **Nayler, W. G., Poole-Wilson, P. A., and Williams, A.,** Hypoxia and calcium, *J. Mol. Cell. Cardiol.,* 11, 683, 1979.

27. **Weglicki, W. B., Owens, K., Urschel, C. W., Serur, J. R., and Sonnenblick, E. H.,** Hydrolysis of myocardial lipids during acidosis and ischemia, *Recent Adv. Stud. Card. Struct. Metab.,* 3, 781, 1972.

28. **Wildenthal, K.,** Lysosomal alterations in the ischemic myocardium: Result or cause of myocellular damage?, *J. Mol. Cell. Cardiol.,* 10, 595, 1978.

29. **Meerson, F. Z., Kagan, V. E., Kozlov, Y. P., Belkina, L. M., and Arkipenko, Y. V.,** The role of lipid peroxidation in the pathogenesis of ischemic damage and the antioxidant protection of the heart, *Bas. Res. Cardiol.,* 77, 465, 1982.

30. **Meerson, F. Z., Kagan, V. E., Arkipenko, Y. V., Belkina, L. M., and Rozhitskaya, I. I.,** Prevention of acitivation of lipid peroxidation and myocardial antioxidative system damage in stress and experimental myocardial infraction, *Kardiologia,* 12, 55, 1981.

31. **Halliwell, B. and Gutteridge, J. M. C.,** *Free Radicals in Biology and Medicine,* Clarendon Press, Oxford, 1985.

32. **Fridovich, I.,** Superoxide radicals: an endogenous toxicant, *Ann. Rev. Toxicol.,* 23, 239, 1983.

33. **Walling, C.,** Fenton's reagent revisited, *J. Am. Chem. Soc.,* 8, 125, 1975.

34. **Niehaus, W. G.,** A proposed role of superoxide anion as a biological nucleophile in the deesterification of phospholipids, *Bioorg. Chem.,* 7, 77, 1978.

35. **Halliwell, B.,** Free radicals, oxygen toxicity and ageing, in *Age Pigments,* Sohal, R. S., Ed., Elsevier/North Holland, Amsterdam, 1981, 1.

36. **Hollenberg, M.,** Effect of oxygen on growth of cultured myocardial cells, *Circ. Res.,* 28, 148, 1971.

37. **Karsten, U., Kössler, Janiszewski, E., and Wollenberger, A.,** Influence of variations in pericellular oxygen tension on individual cell growth, muscle-characteristic proteins, and lactate dehydrogenase isoenzyme pattern in cultures of beating rat heart cells, *In Vitro,* 9, 139, 1973.

38. **Hermersdörfer, H., Karsten, U. and Schultze, W.,** Quantitative morphologische untersuchungen un hypoxisch kultivierten rattenhertz muskelzellen, *Acta Biol. Germ.,* 35, 254, 1976.

39. **Wenzel, D. G. and Acosta, D.,** Labilization of lysosomes and mitochondria *in situ* by hypoxia and hypoxia related factors, *Res. Commun. Chem. Pathol. Pharmacol.,* 12, 173, 1975.

40. **Auclair, M. C., Adolphe, M., Moreno, G., and Salet, C.,** Comparison of the effects of potassium cyanide and hypoxia on ultrastructure and electrical activity of cultured rat myoblasts, *Toxicol. Appl. Pharmacol.,* 37, 387, 1976.

41. **Barry, W. H., Pober, J., Marsh, J. D., Frankel, S. R., and Smith, T. W.,** Effects of graded hypoxia on contraction of cultured chick embryo ventricular cells, *Am. J. Physiol.,* 239, H651, 1980.

42. **Bricknell, O. L. and Opie, L. H.,** Effects of substrates on tissue metabolic changes in the isolated rat heart during underperfusion and on the release of lactic dehydrogenase and arrhythmias during reperfusion, *Circ. Res.,* 113, 102, 1978.

43. **Opie, L. H. and Bricknell, O. L.,** Role of glycolytic flux in the effect of glucose on decreasing fatty-acid-induced release of lactic dehydrogenase from isolated coronary ligated rat heart, *Cardiovasc. Res.,* 13, 693, 1979.

44. **Higgins, T. J. C., Allsopp, D., Bailey, P. J. and D'Souza, E. D. A.,** The relationship between fatty acid metabolism and membrane integrity in neonatal myocytes, *J. Mol. Cell. Cardiol.,* 13, 599, 1981.

45. **Girardier, L.,** Dynamic energy partition in cultured heart cells, *Cardiology,* 56, 88, 1971.

46. **McDonald, T. F. E., Hunter, E. G., and MacLeod, D. P.,** Adenosinetriphosphate partition in cardiac muscle with respect to transmembrane electrical activity, *Pflüger's Arch.,* 322, 95, 1971.

47. **McLeod, D. P. and Daniel, E. E.,** Influence of glucose on the membrane action potential of anoxic papillary muscle, *J. Gen. Physiol.,* 48, 887, 1965.
48. **Doorey, A. J. and Barry, W. H.,** The effects of inhibition of oxidative phosphorylation and glycolysis on contractility and high-energy phosphate content in cultured chick heart cells, *Circ. Res.,* 53, 192, 1983.
49. **Neely, J. R., Liebermeister, H., Battersby, E. J., and Morgan, E. J.,** Effect of pressure development on oxygen consumption of the isolated rat heart, *Am. J. Physiol.,* 212, 804, 1967.
50. **Gazitt, Y., Ohad, I., and Loyter, A.,** Changes in the phospholipid susceptibility towards phospholipases induced by ATP depletion in avian and amphibian erythrocytes, *Biochim. Biophys. Acta,* 382, 65, 1975.
51. **Gazitt, Y., Ohad, I., and Loyter, A.,** Phosphorylation and dephosphorylation of membrane proteins as a possible mechanism for structural rearrangement of membrane components, *Biochim. Biophys. Acta,* 436, 1, 1976.
52. **Higgins, T. J. C., Bailey, P. J., and Allsopp, D.,** The influence of ATP depletion on the action of phospholipase C on cardiac myocyte membrane phospholipids, *J. Mol. Cell. Cardiol.,* 13, 1027, 1981.
53. **Higgins, T. J. C., Bailey, P. J., and Allsopp, D.,** Interrelationship between cellular metabolic status and susceptibility of heart cells to attack by phospholipase, *J. Mol. Cell. Cardiol.,* 14, 645, 1982.
54. **Van der Laarse, A., Hollaar, L., Van der Valk, J. M., and Witteveen, S. A. G. J.,** Enzyme release from and enzyme depletion in rat heart cell cultures during anoxia, *J. Mol. Med.,* 3, 123, 1978.
55. **Altona, J. C. and Van der Laarse, A.,** Anoxia-induced changes in composition and permeability of sarcolemmal membranes in rat heart cell cultures, *Cardiovasc. Res.,* 16, 138, 1982.
56. **Acosta, D. and Puckett, M.,** Ischemic myocardial injury in cultured heart cells: Preliminary observations on morphology and beating activity, *In Vitro,* 13, 818, 1977.
57. **Acosta, D. and Cheng-Pei, L.,** Injury to primary cultures of rat heart endothelial cells by hypoxia and glucose deprivation, *In Vitro,* 15, 929, 1972.
58. **Acosta, D., Puckett, M., and McMillin, R.,** Ischemic myocardial injury in cultured heart cells: Leakage of cytoplasmic enzymes from injured cell, *In Vitro,* 14, 728, 1978.
59. **Acosta, D., Puckett, M., and Cheng-Pei, L.,** Reduction of cell injury in hypoxic culture of rat myocardial cells by methylprednisolone, *In Vitro,* 16, 93, 1980.
60. **Van der Laarse, A., Davids, H. A., Hollaar, L. and Hermens, W. Th.,** The enhanced release of mitochondrial aspartate aminotransferase (mAST) from anoxic rat heart cell cultures during reoxygenation. Comparison to plasma mast levels in patients after acute myocardial infarction and after cardiac surgery, *Cardiovasc. Res.,* 15, 11, 1981.
61. **Altona, J. C., Van der Laarse, A., and Bloys van Treslong, C. H. F.,** Release of compartment specific enzymes from neonatal rat heart cell cultures during anoxia and reoxygenation, *Cardiovasc. Res.,* 18, 99, 1984.
62. **Allsop, D., Bailey, P. J., and Higgins, T. J. C.,** The effects of incubation conditions on enzyme release from anoxic rat heart cell cultures, *Biochem. Soc. Trans.,* 8, 582, 1980.
63. **Higgins, T. J. C., Allsopp, D. N., and Bailey, P. J.,** The effect of extracellular calcium concentration and Ca-antagonist drugs on enzyme release and lactate production by anoxic heart cell cultures, *J. Mol. Cell. Cardiol.,* 12, 909, 1980.
64. **Higgins, T. J., Bailey, P. J., Allsopp, D., and Imhof, D. A.,** Cultured neomnate rate myocytes on a model for the study of myocardial ischaemic necrosis, *J. Pharm. Pharmacol.,* 33, 644, 1981.
65. **Bailey, P. J. and Higgins, T. J. C.,** Metabolism of palmitate by anoxic and reoxygenated heart cell cultures, *Cell Sci.,* 60, 209, 1983.
66. **Ouellette, A. J., Watson, R. K., Billmire, K., Dygert, M. K., and Ingwall, J. S.,** Protein synthesis in cultured fetal mouse heart: Effects of deprivation of oxygen and oxidizable substrate, *Biochemistry,* 22, 1201, 1983.
67. **De Luca, M. A., Ingwall, J. S. and Bittl, J. A.,** Biochemical responses of myocardial cells in culture to oxygen and glucose deprivation, *Biochem. Biophys. Res. Commun.,* 59, 749, 1974.
68. **Van der Laarse, A., Graf-Minar, M. L., and Witteveen, S. A. G. J.,** Release of hypoxanthine from and enzyme depletion in rat heart cell cultures deprived of oxygen and metabolic substrates, *Clin. Chim. Acta,* 91, 47, 1979.
69. **Altona, J. C. and Van der Laarse, A.,** Anoxia induced changes in composition and permeability of sarcolemmal membrane in rat heart cell cultures, *Cardiovasc. Res.,* 16, 138, 1982.
70. **Altona, J. C., Zoet, A. C. M., and Van der Laarse, A.,** Anoxia induced changes in rat heart cell cultures, in *Advances in Studies on Heart Metabolism,* Calderera, C. M. and Harris, P., Eds., CLUEB, Bologna, Italy, 1982, 69.
71. **Demel, R. A. and de Kruyff, B.,** The function of sterols in membranes, *Biochim. Biophys. Acta,* 457, 109, 1976.
72. **Van der Laarse, A., Hollaar, L., and Van der Valk, L. J. M.,** Release of alpha hydroxybutyrate from neonatal rat heart cell cultures exposed to anoxia and reoxygenation: comparison with impairment of structure and function of damaged cardiac cell, *Cardiovasc. Res.,* 13, 345, 1979.

73. **Yagev, S., Heller, M., and Pinson, A.,** The effect of oxygen and volume restriction of extracellular medium on enzyme release from heart cells in culture, *J. Mol. Cell. Cardiol.,* 15 (Suppl. 1), 135, 1983.
74. **Vemuri, R., Yagev, S., Heller, M., and Pinson, A.,** Studies on oxygen restriction in cultured cardiac cells, I.A model for ischemia and anoxia with a new approach, *In Vitro,* 21, 521, 1985.
75. **Vemuri, R., Heller, M., and Pinson, A.,** Studies of oxygen and volume restriction in cultured heart cells: II. The glucose effect, *Basic Res. Cardiol.,* 80 (Suppl. 2), 165, 1985.
76. **Vemuri, R.,** Biochemical alterations in cultured heart cells and their sarcolemma during ischemia, hypoxia and anoxia, Ph.D. thesis in Biochemistry, the Hebrew University of Jerusalem, January, 1986.
77. **Vemuri, R., de Jong, J. W., Hegge, J. A. J., Heller, M., and Pinson, A.,** High energy phosphate metabolism in cultured rat heart cells during anoxia and ischemia, submitted for publication, 1986.
78. **Guarnieri, C., Flamigi, F., and Calderera, C. M.,** Role of oxygen in cellular damage induced by re-oxygenation of hypoxic heart, *J. Mol. Cell. Cardiol.,* 12, 797, 1980.
79. **Vemuri, R., Mersel, M., Heller, M., and Pinson, A.,** Studies on oxygen and volume restriction in cultured cardiac cells: possible rearrangement of sarcolemmal lipid moieties during ischemia, in press.
80. **Benenson, A., Mersel, M., Heller, M., and Pinson, A.,** Radioiodination techniques for identification and localization of lipids and proteolipids in membranes of heart cells in culture, in *Methods in Studying Cardiac Membranes,* Vol. II, Dhalla, N. S., Ed., CRC Press, Boca Raton, 1984, 59.
81. **Benenson, A., Mersel, M., Pinson, A., and Heller, M.,** Radioiodination of pure and membrane-bound phospholipids catalyzed by lactoperoxidase, *Anal. Biochem.,* 101, 507, 1979.
82. **Mersel, M., Beneson, A., Pinson, A., and Heller, M.,** Phospholipid Asymmetry in mixed liposomes detected by enzymatic radioiodination, *FEBS Lett.,* 110, 69, 1980.
83. **Dessort, D., Mersel, M., Lepage, P., and Van Dorssellaer, A.,** Fast-heating mass spectrometry of phosphatidylcholine, lysophosphatidylcholine, phosphatidylethanolamine, and sphingomyelin, *Anal. Biochem.,* 142, 43, 1984.

Chapter 23

THE ADULT HEART CELL MAINTAINED IN CULTURE

Hans M. Piper, Irmelin Probst, Peter Schwartz, Rolf Spahr, and Paul G. Spieckermann

TABLE OF CONTENTS

I. THE ADULT MYOCYTE MODEL

Since the late 1970s, the number of publications dealing with isolated myocytes from the adult myocardium has grown rapidly (for a review see References 1 and 2). However, before that, half a century had elapsed between Burrows'[3] explant culture of embryonic chick heart muscle cells in 1910 and Powell and Twist's[4] success in isolating adult cardiac myocytes retaining viability for several hours at physiological Ca^{2+}-concentrations. Yet, it still remains unclear why it is so difficult to prepare "Ca^{2+}-tolerant" adult myocytes.

It has been suggested,[5] that the failure of many attempts to prepare Ca^{2+}-tolerant myocytes might be attributed to the low Ca^{2+}-concentrations necessary for tissue disintegration and because Ca^{2+}-withdrawal might elicit the so-called "Ca^{2+}-paradox".[6] Thus, Powell hypothesized[7] that his success was due to avoiding Ca^{2+}-concentrations[8] that were too low, during the disintegration process. But it is doubtful whether the Ca^{2+}-paradox phenomenon is generally responsible for difficulties in isolating Ca^{2+}-tolerant adult myocytes. First, during the isolation process, small sarcolemmal disruptions at the intercalated discs seem to occur and their repair probably requires a minimal Ca^{2+}-concentration.[9] Second, Ca^{2+}-tolerant adult myocytes, once they are isolated no longer remain sensitive to Ca^{2+}-removal. In contrast to perfused myocardium, they can even be incubated for some time (e.g., 30 min, at 37°C) in the presence of the Ca^{2+}-chelator EGTA (1 mM) without undergoing irreversible hypercontracture upon reintroduction of normal extracellular Ca^{2+}-levels.[10]

The use of proteolytic enzymes allows effective and rapid tissue disaggregation. However, as a prerequisite, the intercalated discs must be separated at low ($<$ 100 μM) Ca^{2+}-concentrations.[11] The suitability of commercially available enzyme preparations has yet to be tested empirically. So far, crude collagenase, hyaluronidase, and trypsin have been used in most cases.[11,12] Unpurified collagenase also contains considerable amounts of other proteolytic enzymes, crucial for the success of cell isolation. Purified collagenase alone does not bring about complete tissue disintegration.[13]

Adult heart cells have been isolated not only from ventricular myocardium but also from atrial tissue,[14] the Purkinje fibers,[15] and sinus[16,17] and the AV node.[16] In most studies, rat hearts have been used, but Ca^{2+}-tolerant myocytes have also been obtained from guinea pig,[16,18] rabbit,[19,20] cat,[21] dog,[22,23] cow,[24] and from resected human myocardium.[25]

A primary culture of isolated adult heart muscle cells was first described in a detailed study in 1977 by Jacobson.[26] In that system adult myocytes undergo rapid changes: within a week-span, they become transformed into flat, spontaneously beating cells, resembling immature muscle cells in many respects. But even after 2 months in culture, no cell divisions were observed.[26] Subsequently, similar results have also been reported by others.[27-31] In all these adult myocyte cultures, there was no early attachment of the cells. It was first shown by Piper et al.[32] in 1982 that cellular attachment can be achieved during the first 3 hr in culture. Under these conditions, the rapid morphological and functional alterations of the isolated cells that were originally described by Jacobson do not occur. Since the attachment process is selective for intact myocytes, this plating procedure also provided a way of overcoming the problems caused by the high proportions of damaged cells in freshly isolated material. After a few hours under culture conditions, these cells are in a metabolic steady state, as far as is known, maintaining all typical features of the adult myocardium. Thus, they represent an in vitro model of the adult myocardium with the main advantages of homogeneity of the environmental conditions and direct accessibility of all cells. Therefore, this model is very useful for studying the metabolic behavior of the myocardial cell under normal and pathological conditions.[33-41] In the absence of external stimulation, these adult ventricular myocytes are in a functional and metabolic state of rest,[32,37,41] but may be stimulated to contractions by depolarization of the cell membrane.[42,43]

II. ISOLATION AND PRIMARY CULTURE OF ADULT CARDIAC MYOCYTES

The method used for cell isolation was similar to the one proposed by Powell et al.[44] — 12-week-old female Sprague-Dawley rats were decapitated and the hearts rapidly excised, then transferred to ice-cold saline and mounted on the cannula of a Langerdorff perfusion system. The basic medium (KRB) was a calcium-free Krebs-Ringer bicarbonate buffer, consisting of 110.0 mM NaCl, 2.6 mM KCl, 1.2 mM KH$_2$ PO$_4$, 1.2 mM MgSO$_4$, 25.0 mM NaHCO$_3$, and 11.0 mM glucose, gassed with 5% CO$_2$ and 95% O$_2$.

Hearts were first perfused with a nonrecirculating KRB solution (5 min), then by recirculation of 50 mℓ KRB with 25 μM CaCl$_2$ plus 0.1% fatty acid-free bovine albumin and 0.06% collagenase (30 min). Thereafter, the ventricular tissue was chopped and incubated in the recirculation medium with 1% albumin. Release of myocytes was enhanced by gently pipetting this tissue suspension. Perfusion and incubation was performed at 37°C and all following steps at ambient temperature. The incubated material was filtered through a 200 μM mesh nylon gauze, then centrifuged twice (90 sec, 25 g) and resuspended in KRB with 25 μM CaCl$_2$ for the first wash and with 50 μM CaCl$_2$ and 1% albumin for the second wash. The calcium concentration was then raised to 0.5 mM. Subsequently, cells were centrifuged in KRB with 4% albumin and 1 mM CaCl$_2$ (75 sec, 25 g). The cell pellet was resuspended in M-199 medium. All steps were performed under sterile conditions.

The material thus obtained usually contains 70% rod-shaped cells and 30% round cellular remnants. Small cells and small-sized cellular detritus are almost absent, due to the "filtering" effect of the centrifugation steps. Therefore, the impurities in this cell preparation were damaged rather than nonmuscle cells.

The isolated rod-shaped cells have a normal ultrastructure.[32] The high-energy phosphate levels are high (5.6 μmol ATP/g$_{ww}$, 8.4 μmol CP/g$_{ww}$) and the oxygen demands are low (22 $\mu\ell$ O$_2$/g$_{ww}$ · min in Tyrode's solution with 5 mM glucose). These conclusions were reached by relating measured values for these parameters to the number of rod-shaped cells in a given population.[10,32]

The procedure for plating isolated cell material was carried out as follows: Tissue culture dishes (e.g., Falcon, 60 mm) were preincubated for 12 hr with M199 medium containing 4% fetal calf serum (FCS) in an atmosphere of 95% air and 5% CO$_2$, at 37°C. Before use, the medium was decanted. Then the cells, suspended in culture medium with 4% FCS, were plated. The culture medium consisted of M199 medium with Earle's salts, supplemented with 5 mM creatine, 0.2% bovine serum albumin, 10^{-9} M insulin, 100 IU/mℓ penicillin G and 100 μg/mℓ streptomycin. Cell culture dishes were incubated for 4 hr in an atmosphere of 95% air and 5% CO$_2$ at 37°C. After the attachment period (4 hr), the medium was changed.

Subsequently serum was omitted from the medium. After 4 hr incubation and an additional washing of the dishes, about 10^5 cells were attached per 60 mm dish. Irrespective of the purity of the cell material plated, more than 95% of the cells remaining on the dishes are intact rod-shaped cells. The maximal plating density is apparently limited by geometric factors: only cells that have a broad contact area with the dish surfaces and do not overlap other cells attach firmly. Therefore, in general, the attached cells do not have direct physical contact with their immediate neighbors.

The described procedure is easy to perform and gave the best results, compared with a variety of other methods tested (45). On some batches of plastic Petri-dishes from particular manufacturers, selective cell attachment without any pretreatment of the dishes was also observed — the cell density, however, was low. Various other surface coatings gave inferior yields, were toxic, or totally inefficient (e.g., cellophane, dinitrocellulose, agar, collagen I, fibronectin, lectins, and polyaminoacids). In polymerizing collagen or agar gels, cell material was simply entrapped. Precoating of the dishes with other sera (guinea pig, pig, horse, cow, or man) was less effective than with fetal, newborn calf, or rat serum. Maintaining FCS in the culture medium for longer time periods resulted in cellular rounding and detachment. Thus, 25 different FCS batches used, either without or following heat-inactivation, were tested. Other types of sera promoted cell rounding even earlier. Supplementing FCS with fatty acids, phospholipids, or cholesterin did not alter this effect.

In summary, the method described above enables early attachment of adult myocytes. Both early attachment and absence of certain serum ingredients seems to be necessary for maintaining the typical shape of the isolated adult myocyte.

III. MORPHOLOGICAL DEVELOPMENT OF ADULT CARDIAC MYOCYTES IN CULTURE

During low-Ca^{2+} tissue perfusion, fasciae and maculae adhaerentes are rapidly split open but nexus are not.[46,47] On cell separation, the intercellular connection of the nexus is apparently mechanically removed, as indicated by the adherence of double-faced nexus complexes to either of the two adjacent cells when separated.[48-52] However, the small number of such remnants found in freshly isolated cells implies that these structures undergo rapid disassembly. Some of them are also internalized in vesicular enclosures, but almost all of these residual structures disappear during the first 12 hr in culture.[53]

After 1 day in culture, the overall contours of the cells are still preserved, but the rippling of the intercalated disc areas is greatly reduced. The ultrastructure of these elongated cultured cells is still the same as in fresh myocardium. The mitochondria contain many small electron-dense matrix granules which are typical of energetically well preserved myocardium,[36,54,55] and abundant glycogen granules occupy the intermyofibrillar spaces. These ultrastructural observations agree with the biochemical findings that ATP content remains high while ADP and AMP are still at low levels after about 1 week in culture (after 6 days: 5.8 ± 0.7 µmol ATP/g_{ww}, 0.7 ± 0.1 µmol ADP/g_{ww}, and 0.3 ± 0.1 µmol AMP/g_{ww}.* Creatine phosphate is maintained at a high level (after 6 days: 9.0 ± 0.8 µmol/g_{ww}) only if creatine is added to the culture medium.[32]

A uniform sarcomere length of 1.85 ± 0.08 µm is found in all relaxed myocytes by phase contrast microscopy.[1,56,57] This value appears rather low compared with the sarcomere lengths of 2.2 µm or more, usually found in diastolically relaxed myocardium.[58,59] But unlike the cell in tissue, the isolated myocyte is free of external constraints and therefore the observed value seems to represent the true slack length of the myofibrils.

During the first 3 days in culture, the cell endings round off and the cell surface becomes indistinguishable from the lateral cell surface by conventional TEM. Many cells develop single pseudopodia that extend from the former intercalated disc areas.[32] A general cytoplasmic spreading, however, does not occur during the first week in culture. The morphological stability in this system differs greatly from the phenotypical transformations of adult myocytes under culture conditions, which do not allow early attachment.[26,28,30] As reported, those cells round off during the first few days and become transformed into a spread cell type after about 1 week. At an intermediate stage, myofibrils become disintegrated and mitochondria show severe signs of dissolution. Finally, cells exhibiting properties of an earlier stage of development are found attached to the culture dish surface. Like heart cells of the immature type these cells are spontaneously beating, show extensive cellular spreading, and contain myofibrillar bundles in random orientation.

In the culture system described above, intercellular distances and absence of cell spreading prevent the development of new cell-cell contacts for most cells. But occasionally, such formations occur when cells are gathered in a cluster. Indeed, even after 4 hr, adjoining

---

* All data are given as x̄ ± SD of at least five independent experiments. Statistical data analysis was performed using the Kruskal-Wallis test.

cells form contacts via thin cytoplasmic bridges with membrane thickening at opposing contact areas and with intermembraneous material deposits. In most cases, these contacts are end-to-side,[53] which is the most probable mutual orientation of two randomly oriented rods.

The fact that such end-to-side connections do not occur in heart tissue proves the *de novo* character of these contacts. Figure 1 shows a detail from a contact zone after 2 days in culture with several points of contact between cells in orthogonal mutual orientation. By day 4 some of these connections exhibit the characteristics of nexus.[53] This morphological evidence suggests that isolated cells may once again establish low conductivity contacts. However, in contrast to cell cultures of postnatal origin in which most cells start spontaneous beating on contact,[60-62] beating clusters are not observed during the first week in cultures of this type.

It is known from other cell types that the molecular components of the nexus complex may disassemble and reassemble under appropriate conditions even in the absence of active protein synthesis.[63] Other cells also develop cell contacts via long cytoplasmic projections, e.g., neonatal heart muscle cells.[69] The knowledge of the factors stimulating the cells to form mutual connections is as yet, very limited. The structural changes in the sarcolemma induced by close apposition to another cell seem to depend on the nature of surface glycoproteins and on the ionic milieu, especially on the presence of Ca^{2+} and Mg^{2+}.[63] Apparently, all regions of the surface of the myocyte are apt to form specific contact structures when in close cellular vicinity. But in tissue, the longitudinal surface areas of myocytes, which are parallel to the nearest myofibrils, do not develop contact structures with adjacent cells, in spite of a close spatial vicinity. It might be suggested that the formation of lateral contacts in tissue is prevented by the presence of extracellular matrix.

In a coculture of myocytes with hepatocytes,[65] a similar morphological and functional development takes place as in the "dedifferentiating" myocyte culture system, without the round intermediate stage. To establish the coculture, hepatocytes were first isolated from 12-weeks-old Spragues-Dawley rats and then plated as described elsewhere.[66] Ventricular muscle cells from the same kind of rat, isolated as described above, were plated on top of the 4-hr-old hepatocyte monolayer. After an attachment phase of 4 hr in the presence of 4% fetal calf serum, the serum was omitted and the culture maintained in M199 medium with 10^{-9} *M* dexamethasone and 10^{-9} *M* insulin.

During the first day of culture, myocytes are attached in random orientation on the hepatocyte substratum, still exhibiting the typical elongated shape of freshly isolated cells. However, cells in the former intercalated disc areas become smooth and round at edges. In fenestrations of the hepatocyte substratum, myocytes often gather in clusters. After 3 days on the hepatocyte monolayer, most myocytes are flattened and spread out in a large polygonal shape (Figure 2). Using light microscopy, their presence is only recognized by their cross-striation. Myocyte clusters frequently exhibit a network appearance. At this time, many myocytes spontaneously start beating, with frequencies up to 300 beats per min. Beating, however, mostly occurs in bursts with intervening silent periods. In cell cultures beating may be synchronized in a group of cells.[60,67] In contrast, in myocyte monocultures in which these cells undergo only minimal morphological change during the first week, beating can only be elicited by electrical stimulation.[32]

During the development of structural changes, first, the parallel structure of the myofibrils is lost. On day 2, myofibrils often appear distorted. Although different from similar orientations in hypercontracture, these myofibrils are still relaxed. After day 3, many cells exhibit disintegrated myofibrillar material with dissolution of Z-bands. However, even in these cells, morphological indications of a well preserved energetic state are present: glycogen is abundant and even markedly increases during the culture period, and mitochondria contain small matrix granules.

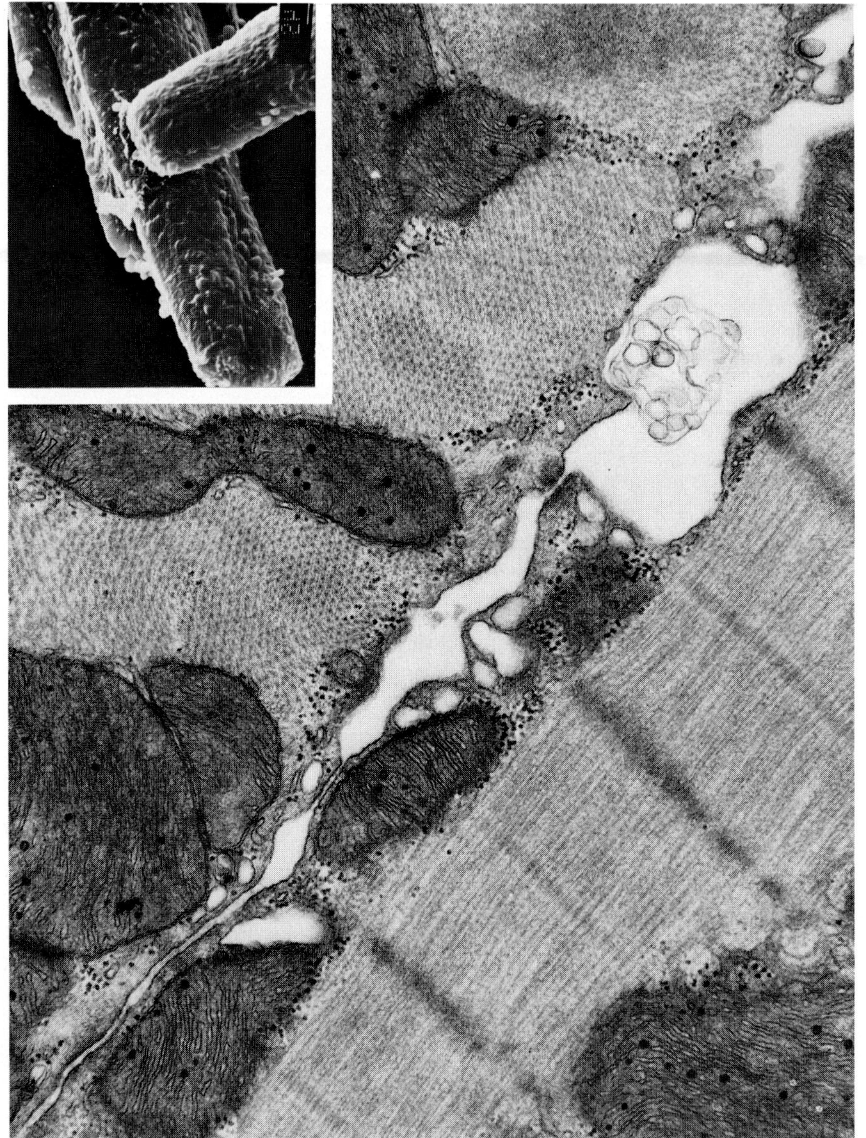

FIGURE 1.   *De novo* formation of intercellular contacts between adjacent myocytes in orthogonal mutual orientation after 48 hr in culture (transmission electron micrograph, ×56,000). A series of small contact spots connects the opposing sarcolemmal faces. The inset shows two contiguous cells in a scanning electron micrograph.

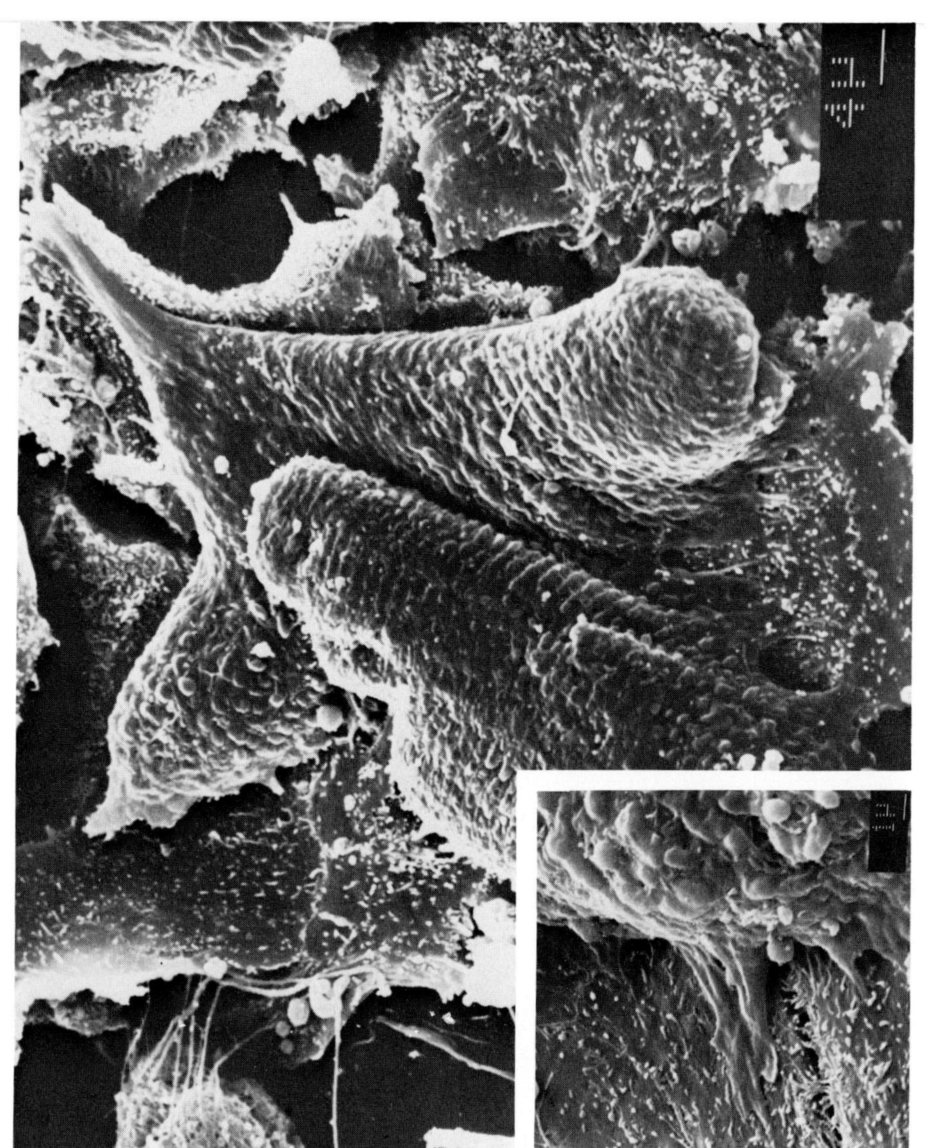

FIGURE 2. Two myocytes on top of underlying hepatocytes, after 48 hr in the myocyte-hepatocyte coculture (scanning electron micrograph). Cytoplasmic spreading starts around the whole cellular circumference (for a detail see inset), the cell on the right is already flattening. The hepatocyte substratum can be recognized by microvilli on its surface.

After 3 days in coculture, broad contact areas are formed not only between myocytes, but also between myocytes and hepatocytes. While it is difficult to categorize the nature of the immature contact structures formed between hepatocytes and myocytes at this stage, some of those formed between myocytes are apparently nexus. This coincidence suggests a causal relation between nexus connections and beating synchronicity,[62] even if beating synchronicity does not always depend on the existence of nexus.[68]

These observations demonstrate that, under appropriate environmental conditions, adult heart muscle cells may pass through a continuum of morphological alterations and thereby regain features of earlier developmental stages. Spontaneous beating is characteristic for cultured embryonic or postnatal cells. In agreement with the reports about the "transforming cultured cell" model,[27,29,31] we did not observe cell divisions of myocytes in culture. It is not yet known which factors are responsible for the phenotypical changes of heart muscle cells in this coculture. However, they may well be caused by hormones or growth factors from the hepatocytes. The phenotypical plasticity of adult heart cells demonstrates that under appropriate environmental conditions, the cellular elements of the myocardium can undergo a variety of structural alterations. There is some ultrastructural resemblance of the "dedifferentiated" cell to atypical muscle cell types found in cardiomyopathic hearts.[69] Therefore, identification of the environmental factors inducing a change in the mature cellular structure in vitro might also increase the understanding of cardiac pathology in vivo.

IV. THE ENERGETIC STATE OF ISOLATED ADULT CARDIAC MYOCYTES

In the absence of external stimulation, adult ventricular myocytes remain quiescent. They have a stable resting potential, which is of the order of 80 mV at 5.4 mM K$^+$, and can be stimulated to generate action potentials and active contractions.[20,42] In suspensions of freshly isolated cells the oxygen consumption is appropriately low: 22 $\mu\ell$ O$_2$/min $\cdot$ g$_{ww}$.[32] The same value can be inferred from substrate oxidation rates determined in 4-hr-old cultures of 100% viable myocytes (Table 1). In Ca^{2+}-free medium (0.5 mM EGTA) the oxygen consumption of suspended myocytes is even lower, i.e., 11 $\mu\ell$O$_2$/g$_{ww}$ $\cdot$ min. These values are near to the basic demand of an optimally arrested rat heart.[70] However, such values are only measured when cells are not mechanically agitated.[32,71] In other studies oxygen consumption of isolated myocytes was recorded in stirred or shaken cell suspensions. The reported data are usually much higher[5,72-76] and in most cases correspond to the demand of the unloaded beating heart.[77] (For a comparison of literature data see Reference 71.)

Thus the isolated ventricular myocyte is in a state of basic energy demand. The range of metabolic stimulation, however, is similar to that of myocardial tissue. In dog myocardium, the normothermic oxygen demands are about 8 $\mu\ell$ O$_2$/g$_{ww}$ $\cdot$ min under optimal cardioplegia, 40 for unloaded beating, 80 during physical rest, and 400 to 800 under maximal catecholamine stimulation.[77] For freshly isolated myocytes, we found a minimum of about 10 $\mu\ell$ O$_2$/g$_{ww}$ $\cdot$ min, 90 in a stirred cuvette, and 600 in the presence of the respiration-uncoupler 2,4-dinitrophenol (50 μM).[32,71] Haworth et al.[78] reported a close correlation of oxygen consumption and stimulated beating rate of isolated myocytes, with a similar demand per beat to that of the empty beating organ. When the temperature is lowered from 37°C to 27°C, the oxygen consumption decreases by about half.[32] The Q$_{10}$ value of 1.8 agrees well with similar values for the arrested rat heart (1.4),[70] and myocardial tissue slices in vitro (1.6).[79] The oxygen uptake of isolated heart cells in suspension is independent of the extracellular oxygen pressure above 3 mmHg O$_2$. For lowering oxygen consumption by half, oxygen tension has to drop below 0.2 mmHg.[75]

Freshly isolated rod-shaped cells contain high levels of energy-rich phosphates that even increase somewhat during a recovery period of 4 hr under culture conditions. The relations among the adenine nucleotides are typical for well-preserved myocardium. Thus, after 4 hr

Table 1
TURNOVER OF EXOGENOUS AND ENDOGENOUS SUBSTRATES IN CULTURED ADULT CARDIOMYOCYTES

	PDH$_a$ (%)	Glucose		Other substrates	Glycogen		Triglycerides		O$_2$-demand	
		C6	O$_2$	O$_2$	C6	O$_2$	FA	O$_2$	ΣO$_2$	ΣO$_2$
Glucose 5 mM	10.3	14.5	8.3	—	−0.3	0.2	−2.0	49.8	58.3	21.7
Glucose 5 mM + insulin 10^{-7} M	12.8	57.7	13.0	—	(+2.2)	—	−1.9	46.5	59.5	22.2
Glucose 5 mM + lactate 5 mM	22.3	4.1	1.6	20.1	(+0.6)	—	−1.6	39.0	60.7	22.7
Glucose 5 mM + DCA 5 mM	96.5	18.8	33.8	—	(+0.7)	—	−1.7	42.3	76.1	28.4
Glucose 5 mM + D$_{31}$-palmitate 100 μM	6.7	14.4	5.9	41.8	(+0.4)	—	−1.0 (+4.7)	24.1	71.8	26.8
Standard error (% of mean)	<10	<15	<10	<10	<15	—	<15	—	—	—
Initial contents (μmol/g$_{ww}$)	—	—	—	—	39.8 ± 5.3	—	12.4 ± 1.4	—	—	—

Notes: Substrate turnover was measured by incubating cells for 3 hr at 37°C under air in a modified Tyrode solution (125 mM NaCl, 2.6 mM KCl, 1.2 mM KH$_2$PO$_4$, 1.2 mM MgCl$_2$, 1.0 mM CaCl$_2$, 10 mM HEPES, pH 7.4) to which substrates (D-glucose, L-lactate, palmitate complexed 5:1 to albumin) were added.

Oxidation of exogenous substrates was determined by measuring CO$_2$ production from U-^{14}C-labeled substrates. Amount of fatty acids (FA) in triglycerides was determined by mass spectrometry (Dr. D. Hunneman, Dr. C. Schweickhardt, Göttingen), permitting to separate the deuterized D$_{31}$-analogue of palmitate (+4.7) from other long-chain fatty acids (−1.0). Values in brackets indicate an increase in endogenous substrates.

The percentage of the active a-form of pyruvate dehydrogenase (PDH$_a$) was determined in cell extracts according to Reference 89. The total activity of activatable PDH remained constant in the various experiments. Glycolytic flux was estimated by the production of ^{3}H$_2$O from 3-^{3}H-glucose. O$_2$-equivalents were calculated from (a) CO$_2$ production from exogenous substrates by using the theoretical RQ-values, (b) glycogen degradation by assuming a C6/O$_2$ ratio as for glucose, and (c) triglyceride degradation by assuming complete oxidation of the respective fatty acids.

All values are in μmol/hr · g$_{ww}$ except last column, which is expressed in μℓ O$_2$/min · g$_{ww}$.

in culture, CP is 8.7 $\pm$ 0.8, ATP 6.0 $\pm$ 0.6, ADP 0.8 $\pm$ 0.1, and AMP 0.3 $\pm$ 0.0 μmol/g_{ww}. It has been shown by Geisbuhler et al.[80] that in hypoxic isolated myocytes, the cascade of adenine nucleotide degradation proceeds no further than to the stage of inosine, whereas the hypoxic heart in addition produces hypoxanthine and uric acid. From this, xanthin oxidase appears not to be present in the cardiomyocytes, but exclusively in the endothelial cell.[81]

V. SUBSTRATE UTILIZATION OF ADULT CARDIAC MYOCYTES

Because Ca^{2+}-tolerant isolated ventricular muscle cells are mechanically at rest, the basal metabolic activity of the myocardium in the absence of contractile performance could be studied using the isolated cell system. This may allow more detailed investigation of metabolic processes with very small energy needs compared to the energy required for contraction. In contrast to the arrested heart, in the isolated cell system, this can be done in physiological media.

Using adult myocytes 4 hr after plating, some features of the interaction of carbohydrate and of fatty acid utilization in the myocardial cell have been investigated. Earlier studies of the metabolic behavior of isolated cardiocytes were partly performed with preparations that were not Ca^{2+}-tolerant, and in most cases shaken or stirred suspension were used.[2,72-74,76,82-86] Such mechanical agitation, however, may increase the cellular energy turnover (*v.s.*).[32] Therefore, it is difficult to relate the results of those studies to a defined physiological state. Additionally, the cell material used in those studies contained 10 to 40% severely damaged cells. Their contribution to the observed effects is difficult to estimate.[87]

When glucose (5 mM) is the only exogenous oxidizable substrate, the cells under air produce lactate and CO_2 linearly for hours at a constant molar ratio of 2.7 $\pm$ 0.5. This means that only one out of nine glucose molecules degraded is completely oxidized to CO_2. This ratio is not changed under 100% oxygen, thus the high lactate output is not due to hypoxic conditions. Isolated hearts can also produce lactate when glucose is the sole external carbon source and the reason for this is believed not to be cellular hypoxia.[88-91] However, in the beating heart, CO_2 production largely exceeds lactate formation. This discrepancy might partially be explained by a less active pyruvate dehydrogenase (PDH) in the resting cell as compared to the beating organ.[89,91] With 5 mM glucose present, 10% of this interconvertible enzyme were found in the active form (PDH$_a$, see Table 1). In the presence of dichloroacetate (DCA) which inhibits PDH phosphorylation[91] and thereby increases the proportion of the active form (Table 1), flux through pyruvate dehydrogenase can be significantly increased, up to fourfold with 10 mM DCA, whereas lactate production is unaffected (Figure 3). The lactate/CO_2 ratio therefore decreases below 1.

Whereas variation of glucose from 2 to 10 mM does not change the rate of glucose degradation, insulin at saturating concentration (10^{-7} M) leads to a 4.5-fold stimulation of lactate production and stimulates CO_2 formation by 60% (Figure 3). This again increases the lactate/CO_2 ratio up to sevenfold. The half maximal effective dose of insulin for the stimulation of CO_2 and lactate production is about 10^{-9} M,[40] which agrees with data reported for the stimulation of the glucose transporter.[92-94] Thus we can assume that the increased glycolytic flux is caused by an augmented glucose entry into the cell and presumably by subsequent activation of phosphofructokinase by fructose-2,6-biphosphate[95] which we found 5-fold increased with 10^{-7} M insulin vs. no insulin.[144] The increase in glycolytic flux preferentially leads to enhanced lactate formation, as this seems to be the main route by which cytosolic NADH can be oxidized. In the resting myocardium, the import of reduction equivalents into the mitochondrion is impaired by an increased mitochondrial NADH/NAD ratio.[89,97] This ratio being high may be causally related to the low activity of the PDH,[91] which is not significantly influenced by addition of 10^{-7} M insulin (Table 1).

There is no in vivo situation in which glucose is the only carbon and energy source; in

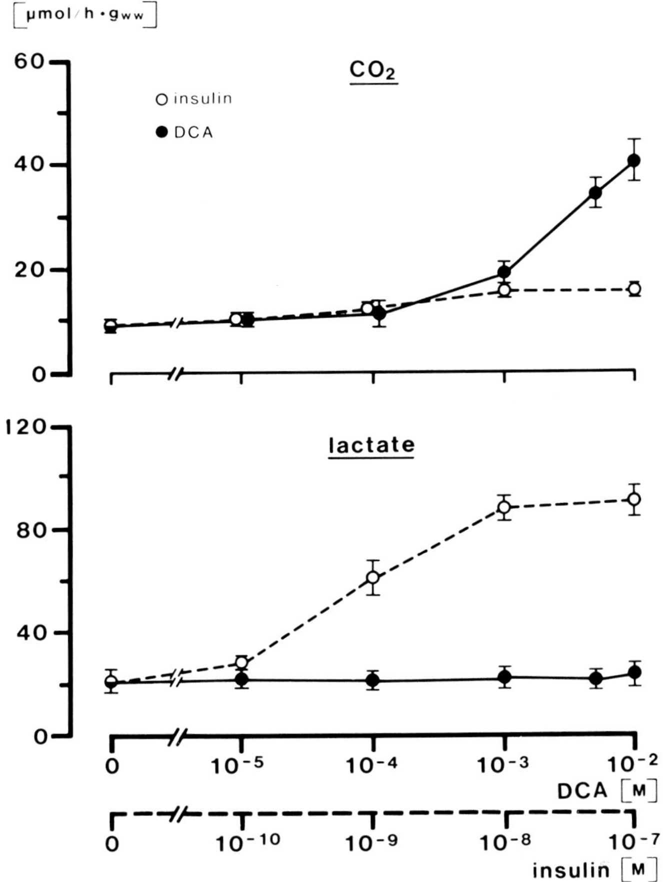

FIGURE 3.   Effect of DCA and insulin on the oxidation of U-^{14}C-glucose to lactate and CO_2. ($\bar{x}$ ± SD, n = 5). As described elsewhere in detail,[40] after 4 hr culturing, the cells were incubated in a modified Tyrode's solution (125 mM NaCl, 2.6 mM KCl, 1.2 mM KH_2PO_4, 1.2 mM $MgSO_4$, 1 mM $CaCl_2$, 10 mM HEPES at pH 7.4 and 37°C, air-saturated; 60 mm dishes containing 4.5 ± 0.6 mg cellular wet weight were filled with 3 mℓ).

fact, lactate and fatty acids are preferentially oxidized (Figure 4).[91,98-100] In cultured cells, glucose oxidation to CO_2 is more sensitive to low concentrations of exogenous lactate than glucose conversion to lactate, which again leads to a change in the ratio of lactate/CO_2 as the two main endproducts of glucose degradation. This ratio, measured in the presence of 10^{-7} M insulin, could be as high as 15 at 5 mM lactate.[40] The variability of the lactate/CO_2 ratio illustrates that in this cell system there is no close coupling between glycolytic flux and the oxidative degradation of pyruvate.

Exogenous lactate is oxidized by these cells at much higher rates than glucose (Figure 5). This seems partly due to an activation of the PDH (Table 1) and partly due to a reduction of endogenous lipolysis which contributes to energy production *(vide infra)*. Exogenous fatty acid supply, however, can reduce lactate oxidation (Figure 5, inset) as it reduces the oxidation of glucose (Table 1), apparently by reducing PDH activity at increased β-oxidation rates.

The oxidation of palmitate to CO_2 also exceeds that of glucose alone. Palmitate oxidation is saturated at 100 μM (Figure 6). No change in the palmitate substrate saturation curve is detected when the palmitate/albumin ratio is varied in the range between 1:1 and 5:1 (Figure 6, inset). This finding contrasts with the demonstration that in the working heart palmitate

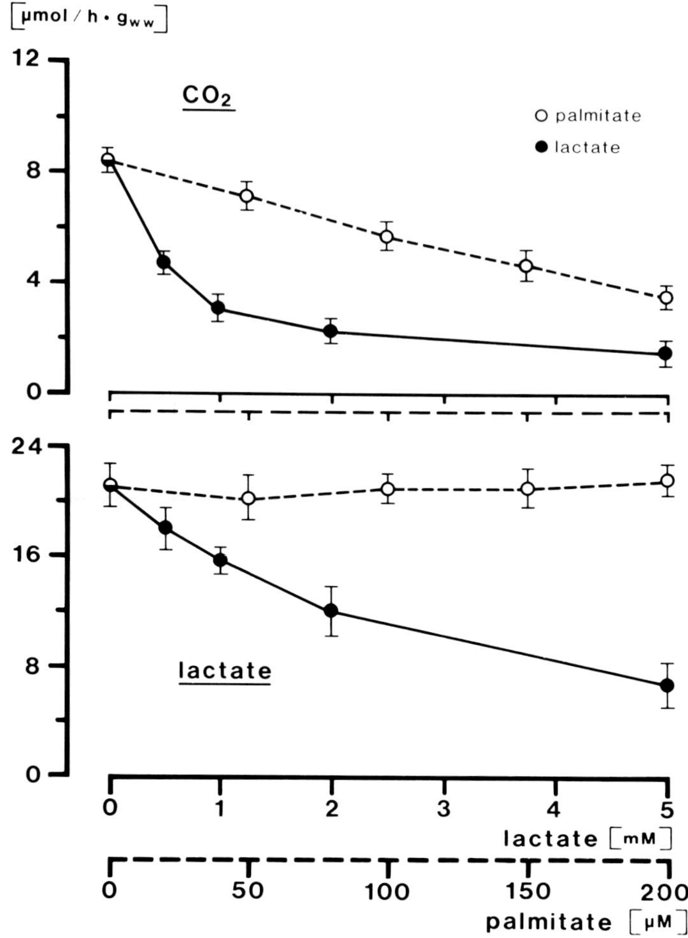

FIGURE 4.   Effect of palmitate and L-lactate on the oxidation of U-^{14}C-glucose to lactate and CO_2. ($\bar{x} \pm$ SD, n = 5). The Tyrode's solution contained 5 mM glucose plus the additional substrates, palmitate was complexed 5:1 to albumin.

oxidation does depend on the albumin-bound palmitate fraction. This latter finding led to the hypothesis that an albumin-mediated uptake is the rate limiting step in palmitate oxidation.[101] However, in the quiescent cell to a later step, for instance acetyl-CoA oxidation, might be limiting[102] and the specific transport mechanism in the plasma membrane might work far below its maximal capacity because of the much lower rate of palmitate oxidation in this system. Thus there may be enough time for complete equilibration of palmitate between the medium and the sarcoplasm via both a receptor-mediated uptake from the albumin-bound fraction and an uptake by diffusion from the unbound fraction.

With the pyruvate dehydrogenase being only poorly active, endogenous lipids are the main substrate for the resting myocardial cell (Table 1). In the presence either of 5 mM glucose solely, or of glucose plus 5 mM lactate or plus 100 μM palmitate, endogenous lipolysis shares in energy production by 80, 60, and 30%, respectively. When exogenous fatty acids are present, the quiescent myocytes exhibit a net synthesis of triglycerides from which, however, the continuing hydrolysis can be differentiated when the deuterized D[31]-analogue of palmitate is used as substrate (Table 1). The fatty acid content of the serum-supplemented culture medium, to which the cells are exposed during the preincubation period,

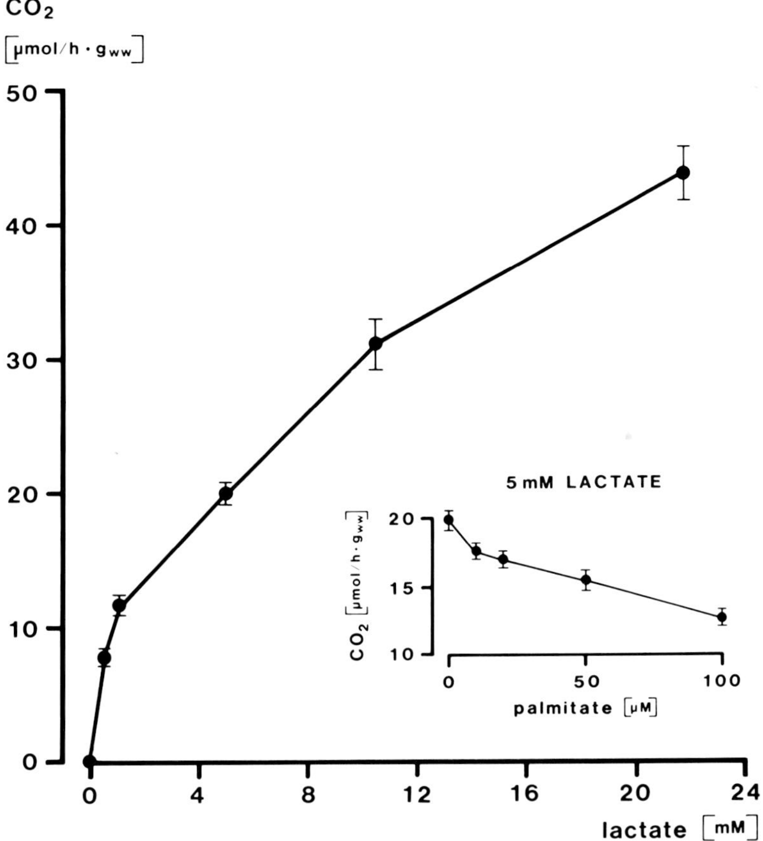

FIGURE 5. Oxidation of U-^{14}C-L-lactate in Tyrode's solution containing 5 mM glucose. ($\bar{x} \pm$ SD, n = 5). The inset shows the influence of palmitate on lactate oxidation at 5 mM L-lactate.

is responsible for the fourfold elevation of triglyceride content in cells after 4 hr in culture (12.4 µmol/g$_{ww}$ triglycerides) compared to that in the rat heart in vivo (3.5 µmol/g$_{ww}$).

It has been reported about freshly isolated adult heart cells that glucose or lactate may considerably lower the oxidation of exogenous fatty acids.[2,76,82] As suggested by Claycomb et al.,[103] this behavior appears to be specific for freshly isolated myocytes and indicative of some cellular damage. Indeed, from the studies using freshly isolated myocytes in suspension,[104,105] it seems likely that the cell material used was depleted of carnitine and, therefore, fatty acid activation was impaired. In the cultured cells a metabolic pattern of preferences among physiological substrates similar to the intact heart is observed, although the cells had but 4 hr to recover in a serum-, thereby also carnitine-containing medium after isolation. Supplementing isolation, culture, and experimental media with additional 5 mM carnitine had no effect on fatty acid oxidation in the described experiments with 4 hr cultured cells.

In conclusion, it might be stated that the cultured cell system shows the metabolic behavior of the myocardium in a resting state. This has not yet been sufficiently recognized in the current literature.

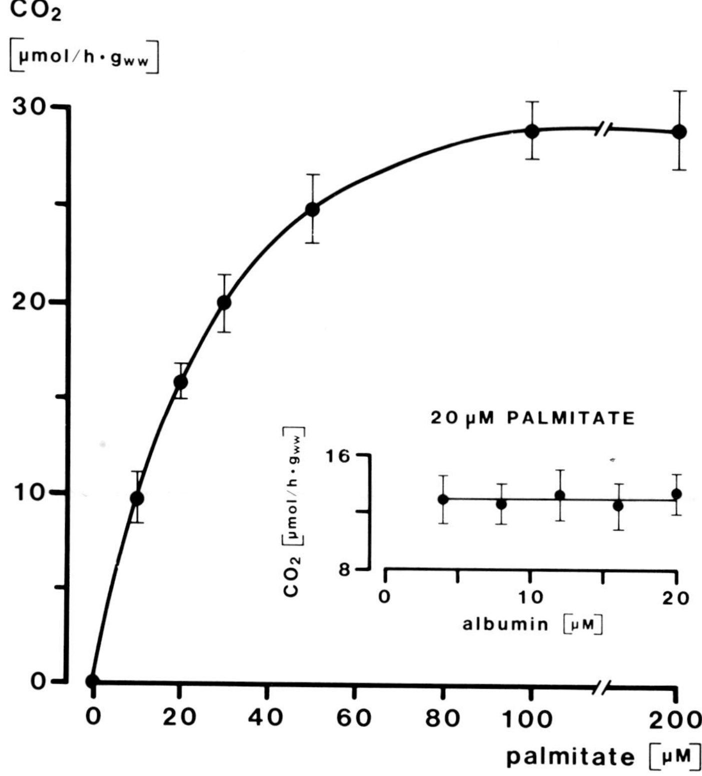

FIGURE 6. Oxidation of U-^{14}C-palmitate (complexed 5:1 to albumin) in Tyrode's solution containing 5 mM glucose. ($\bar{x}$ ± SD, n = 5). The inset shows that palmitate oxidation at 20 μM palmitate does not change when the fatty acid/albumin ratio is varied from 1:1 to 5:1.

VI. ENERGY METABOLISM AND ENZYME RELEASE OF ADULT CARDIAC MYOCYTES UNDER ANOXIA AND GLYCOLYTIC BLOCKADE

During hypoxia, release of enzymes is often regarded as a definite sign of the irreversible damage of individual cells in myocardial tissue.[106-108] A clear proof of this hypothesis, however, is still lacking, since in global or regional hypoxia cellular protein loss is only detectable with a temporal delay and diluted in the venous or lymphatic outflow. For a precise temporal analysis and for the detection of small initial protein losses, a culture system of isolated myocytes, in which all the cells are kept under homogenous environmental conditions in a single extracellular compartment, would be preferable.

The early phase of myocardial hypoxia is characterized by a drop in CP content and a rapid increase in glycolytic flux.[109] The consecutive decay of ATP only leads to a moderate increase in ADP and AMP, and thus the total content of adenine nucleotides also decline.[110,111] Adenosine is rapidly degraded and also lost from the cell by diffusion.[112,113] Since the *de novo* synthesis of adenosine is low,[114] in a recovering heart, loss of adenine nucleotides is compensated for only after a few days.[111,115] In contrast, the creatine reserves of the cardiac cells are high and the hypoxic losses are moderate. Therefore, on early reoxygenation, control levels of CP are rapidly restored.

There have been several attempts to study the hypoxic process in cellular models of the myocardium. In studies performed with immature muscle cells, it has to be recognized that these cells differ in their sensitivity from the adult myocardium.[116,117] On the other hand,

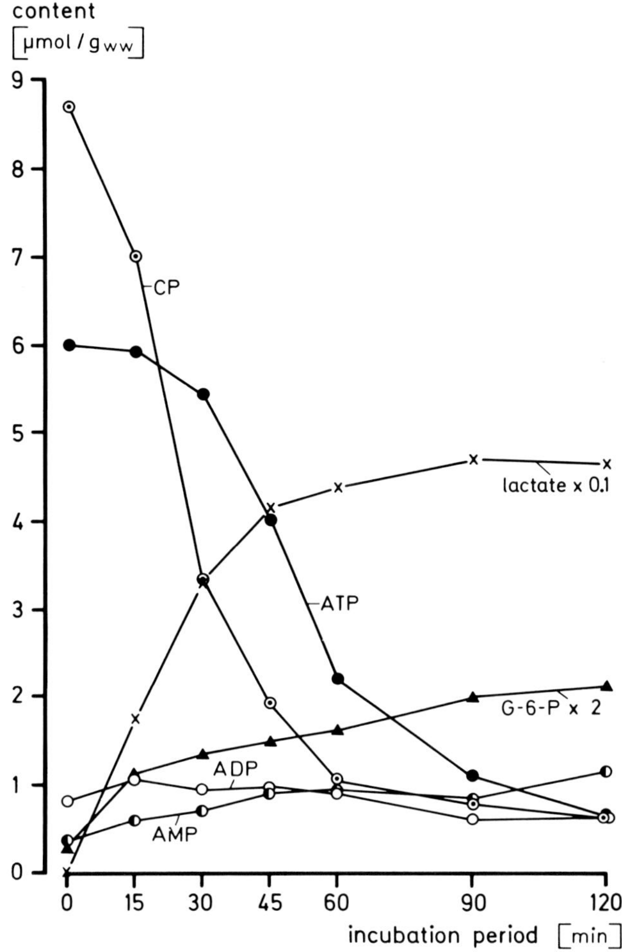

FIGURE 7. Synopsis of energetic changes during substrate-free anoxia. Creatine phosphate (CP), adenine nucleotides (ATP, ADP, AMP), glucose-6-phosphate (G-6-P) and lactate content on the culture dishes are related to the cellular wet weight. After 4 hr culturing the cells were incubated in a modified anoxic substrate-free Tyrode's solution (125 mM NaCl, 2.6 mM KCl, 1.2 mM KH$_2$PO$_4$, 1.2 mM MgSO$_4$, 1 mM CaCl$_2$, 25 mM HEPES at pH 7.4 and 37°C; 60 mm dishes filled with 1 mℓ). Mean values of 5 different experiments performed under each of the specified conditions. Methodological details are described elsewhere.[39]

isolated adult myocytes have been used in mechanically agitated suspensions containing 10 to 40% of already damaged cells.[22,57,118-123]

Compared to the oxygen deficient organ, the anoxic process in the cell culture system is delayed as in the arrested heart. Under aerobic control conditions, the cultured cells contain physiological levels of high-energy phosphates. Under anoxia (Figure 7), high-energy phosphate content falls and, concomitantly, activities of cytosolic enzymes (lactate and malate dehydrogenase, LDH, MDH) increase in the extracellular space (15 min: P < 0.01, 30 min: P < 0.001), while lysosomal (acid phosphatase) and mitochondria-specific (glutamate dehydrogenase) enzyme activities do not. Only half as much MDH, about half of which is cytosolic, is released as compared to the cytosolic marker, LDH. The anoxic release of cytosolic enzymes correlates well with the actual ATP content (Figure 8) and therefore slows

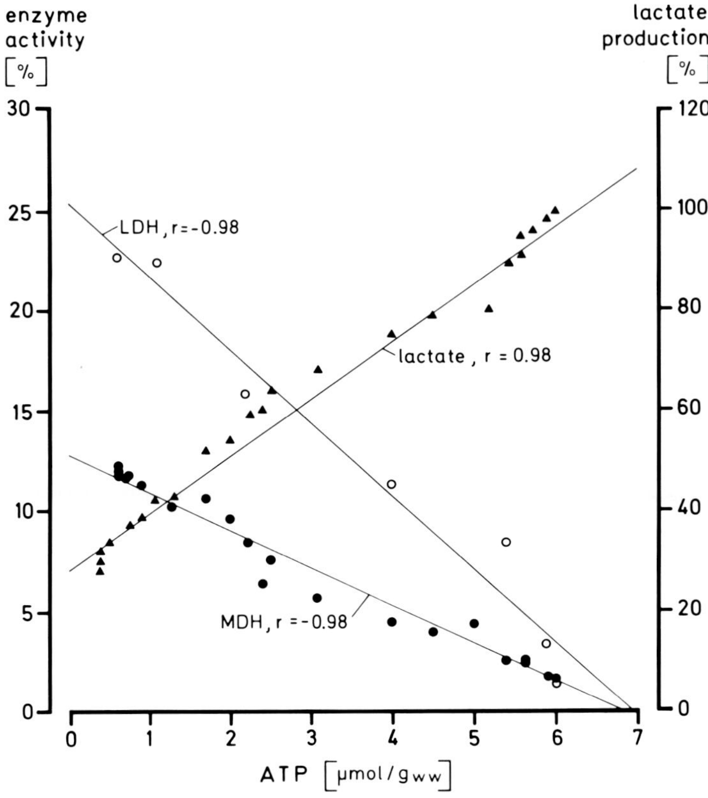

FIGURE 8.   Substrate-free anoxia. Correlation of the released activities of lactate dehydrogenase and malate dehydrogenase (LDH, MDH, expressed as percent of the initial total cellular activity, $p < 0.001$) and of the lactate production rate (expressed as percent of the maximal value of 1.2 μmol/g$_{ww}$ · min, $p < 0.001$) with the actual ATP content. Mean values of 5 different experiments.

down when ATP degradation decelerates. This deceleration is most pronounced at concentrations close to 2 μmol ATP/g$_{ww}$ (60 min anoxia). At this point, glycogenolysis also distinctly decelerates although only half of the initial amount of glycogen has been consumed (control: 43 ± 3, at 60 min: 22 ± 2, at 120 min: 18 ± 2 μmol glucose/g$_{ww}$). After 60 min anoxia, 30 min reoxygeneation restores normal CP contents (97 ± 7% of control), but ATP is only partially restored (39 ± 4% of control), because the amount of total adenine nucleotides is already markedly reduced.

At average ATP levels below 2 μmol/g$_{ww}$, more and more cells round off in hypercontracture. Initially, the cell population conatins 3.0 ± 1.6% round cells. Their number only increases after 90 min of anoxia (60 min: 3.6 ± 1.5%, 90 min: 14.1 ± 2.5%, 120 min: 29.4 ± 4.7%, 120 min under aerobic control conditions: 3.7 ± 2.9%). With up to 120 min anoxia, reoxygenation neither changes the number of round cells nor the amount of enzymes released during anoxia, indicating that a reoxygenation damage does not occur in this system.[35]

Thus, cytosolic enzyme release precedes irreversible cell death. Indeed, it even decelerates when single cells start rounding off. As indicated by its correlation to the ATP level, early enzyme release seems to depend on the ATP availability as for an active process of exocytosis. This suggestion is further emphasized by the finding that, in early anoxia, the number of subsarcolemmal vesicles (Figure 9) increases, and then, decreases again in reoxygenated

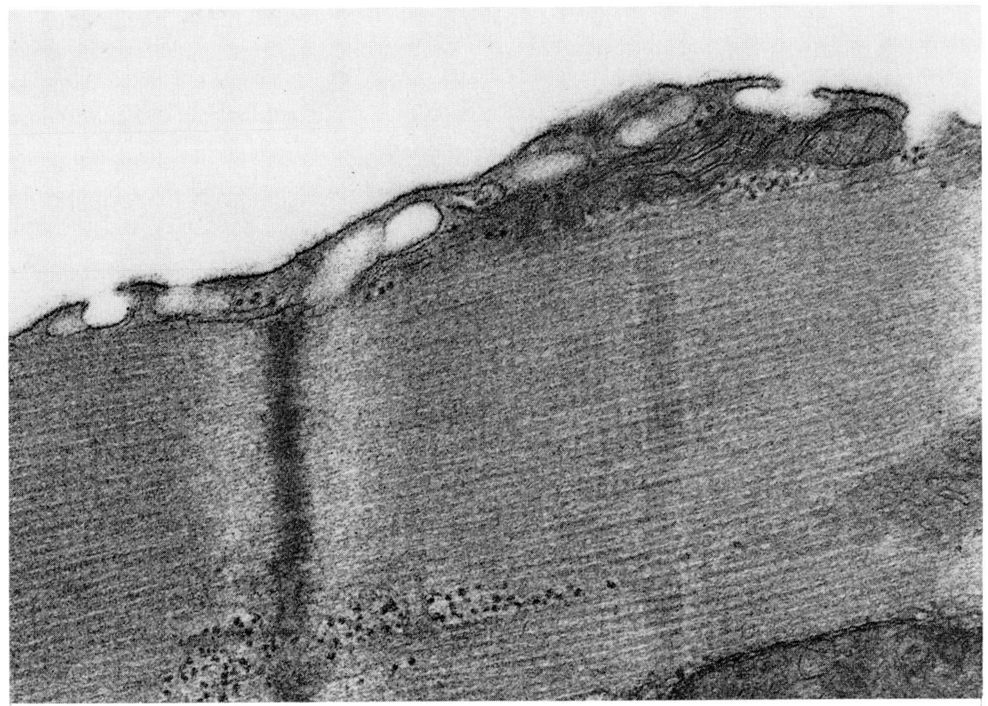

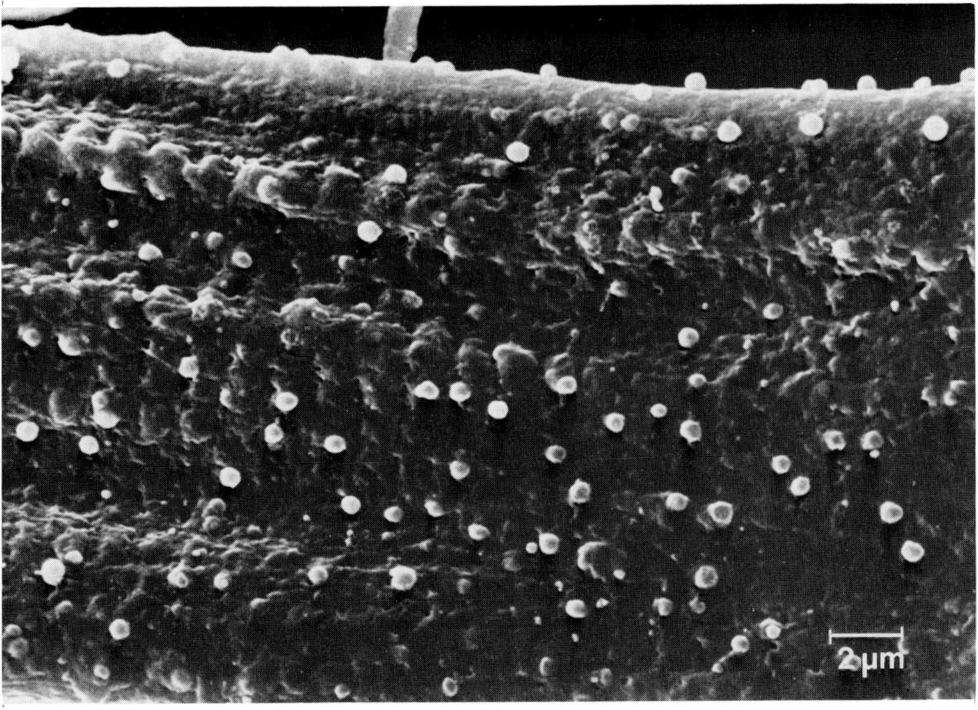

FIGURE 9.   Early anoxic sarcolemmal alterations after 30 min anoxia. Top: Subsarcolemmal space with many vesicular structures that are occasionally open to the extracellular space (transmission electron micrograph, ×80,300). Below: Numerous ''microbleb'' protrusions spread over the lateral cell surface (scanning electron micrograph).

cells. Since lysosomal enzymes in the external space do not increase at that time, a lysosomal origin for these vesicles seems unlikely.

During the early phase of ATP decay, the cells gradually shorten while preserving their overall polygone shape. Cell shortening is accompanied by the appearance of "microblebs" (1 μm in diameter, Figure 9), preferentially near the former intercalated disc areas. During the reversible stage of anoxia, these "microblebs" increase in number and spread over the entire surface. At 60 min of anoxia, they are spread over the surface of all polygonal cells. In reoxygenated polygonal cells, microblebs are only occasionally seen. During the reversible stage of anoxia (up to 60 min) they show no tendency to become confluent. In most cases, these microblebs contain one mitochondrion filling the cavity of the bleb.[36]

In size, microblebs differ considerably from large protrusions (10 to 30 μm in diameter) which are typical for round cells with overcontracted, often condensed myofibrillar masses, appearing on prolonged anoxia. These sarcolemmal pouches contain mixtures of cell organelles or their remnants. It can be demonstrated by ruthenium-red staining that the sarcolemma with its adherent surface coat is still continuous around the hypercontracted cell.[35]

This agrees with the observation that enzyme release is not accelerated when round cells appear. The ultrastructure of round cells does not differ before and after reoxygenation. Typical features of the "oxygen paradox" damage are absent, i.e., localized contraction bands, intramitochondrial amorphous densities, and sarcolemmal discontinuities.[124,125] It might be hypothesized that in tissue these ultrastructural changes and the massive enzyme loss, characteristic to the "oxygen paradox", are secondary to mechanical cell-cell interactions,[125,126] which are absent in an isolated cell system. The ultrastructural characteristics of the oxygen paradox have also been described in anoxic perfused hearts[127-129] and tissue slices[130] incubated in a large buffer volume. In both models, acidification plays only a minor role. Therefore, the buffered pH (7.4) of our system cannot be regarded as responsible for this deviant behavior.

In conclusion, the anoxic isolated cardiocyte may already lose a moderate amount of cytosolic proteins before cell injury becomes irreversible.[131] Irreversible cell injury, characterized by hypercontracture, is not necessarily accompanied by immediate cytolysis, and irreversible cell contracture does not lead to the rapid disintegration process of the oxygen paradox upon reoxygenation. Early microbleb information might be caused by local disruptions of the cytoskeleton when the intrinsic tension of the sarcolemma increases in a shortening cell.[36] This development might be favored by a rearrangement of the subsarcolemmal "cytoskeletal scaffold" that seems to take place rapidly after oxygen withdrawal.[39] In round cells covered with large protrusions, the cytoskeletal anchoring of the sarcolemma is probably almost completely destroyed, which might be a reason for irreversibility in this case. It has been hypothesized by Bricknell and Opie[132] that glycolytically generated ATP plays a major role in the maintenance of cell membrane integrity under conditions of energy depletion. Indeed, they found, that in low-flow ischemia with several substrates, those allowing higher glycolytic flux rates led to a lower enzyme release, even if compared at identical homogenate ATP levels. It was also observed, by the same group, that contracture development at a given total ATP content is less when glycolytic flux is high.[133] Therefore, the question might be asked whether high enzyme release at low rates of glycolysis is due to compartmentalized availability of ATP or secondary to contracture-induced cell damage. Since contracture in tissues may cause cell damage by mechanical cell-cell interaction, and such forces do not occur in an isolated cell system, use of isolated myocytes may help to differentiate between the aforementioned alternatives.

During the early phase of anoxia, characterized by a progressive decay of high-energy phosphate, the lactate production rate decelerates in correlation with the fall in ATP content. As the activities of cytosolic enzymes in the medium again correlate with the actual ATP levels (Figure 8), there is also a correlation between released cytosolic enzyme activities and lactate production rate (for LDH and MDH: r = − 0.97, P < 0.001). But this means that the rate of enzyme release decreases in a similar fashion with time as does the rate of lactate production, and it, therefore, does not support the hypothesis that low glycolytic activity accompanies high rates of enzyme release.

With exogenous glucose present, glycolysis is largely stimulated. When cells are incubated anaerobically in the rich M199 medium, containing 5.5 mM glucose, lactate production is maintained constant at a high level for 120 min: 3.2 ± 0.2 μmol/g$_{ww}$ · min. Under these conditions, enzyme release is largely postponed. At 120 min of anoxia, only 1.7 ± 0.2% of total control activity of MDH is released; while ATP levels still remain unaffected. The experiments described so far do not allow differentiation between the importance of glycolytic energy production and of the presence of a certain total ATP level, since states with low enzyme release exhibit both high rates of glycolysis and high ATP levels. Therefore, we investigated enzyme release under conditions of energy depletion by blocking glycolytic flow. Iodoacetate (IAA) is an inhibitor of glycolysis. Although it is actually a nonspecific alkylating agent, at lower concentrations its inhibitory effect on glyceraldehyde-3-phosphate dehydrogenase is rather specific.[134,135] Therefore, under aerobic conditions, ATP production can only be maintained when carbohydrates are available and enter the Embden-Meyerhof pathway below this site of blockade, or when acetate or fatty acids are available. In the absence of exogenous substrate, IAA leads to a rapid decline of high-energy phosphates (Figure 10), most probably because endogenous fatty acids are not sufficiently activated in the absence of cytosolic ATP production. Thus, after 30 min exposure, the CP content is 14 ± 2% and the ATP content is 25 ± 2% of the control values.

When pyruvate is used as an exogenous substrate, the energetic depletion is greatly delayed: after 30 min exposure CP is 45 ± 4% control, while ATP is still 91 ± 6%. Thus, pyruvate oxidation cannot satisfy the energetic needs even in resting muscle cells, which may be due to the fact that in the presence of IAA, huge amounts of phosphate are trapped in the increasing stores of glycolytic intermediates (Figure 10). After 30 min exposure to IAA without pyruvate, fructose-1, 6-bisphosphate has increased to 115 ± 14%, and with IAA plus pyruvate to 36 ± 3%. Thus, there may be too little inorganic phosphate for ADP rephosphorylation. While acid phosphatase and glutamate dehydrogenase activities in the medium, remained unchanged during 60 min exposure to IAA, the time course of cytosolic enzyme release followed the ATP decay both in the presence and in the absence of pyruvate. The relationship between enzyme release and total ATP content is identical in both cases (Figure 11). Lactate production was negligibly low under IAA ± pyruvate. When glycolysis is blocked, differences in cytosolic energy production cannot account for the differences in cytosolic enzyme release. In contrast, the actual total ATP level seems to determine the rate of early protein loss.

The absence of enzyme release from cell compartments other than the cytosol when taken together with the fact that the amount of enzyme activity released at a given ATP level is even lower than in the presence of IAA than during anoxia, indicates that the early enzyme release in this case is not due to cellular lysis. Lower enzyme release for a given total ATP level in the presence of IAA, as compared to that in anoxia, might indicate that the thermodynamic factors, rather than the amount of ATP present, determine the extent of this gradual protein loss. Due to the trapping of inorganic phosphate in glycolytic intermediates, the change in the free energy of ATP hydrolysis should be higher in the presence of IAA at given ATP and ADP levels.[136] Although these findings demonstrate that there is no direct connection between cytosolic protein release and glycolytic energy production in the isolated cell system, it may be possible to explain why such a relationship seems to hold in tissue. As reported for the myocardium under glycolytic blockade,[133] IAA causes pronounced contracture. In tissue, the development of contracture may lead to single cell disruptions caused by the mechanical cell coupling, consequently bringing about enhanced release of enzymes.

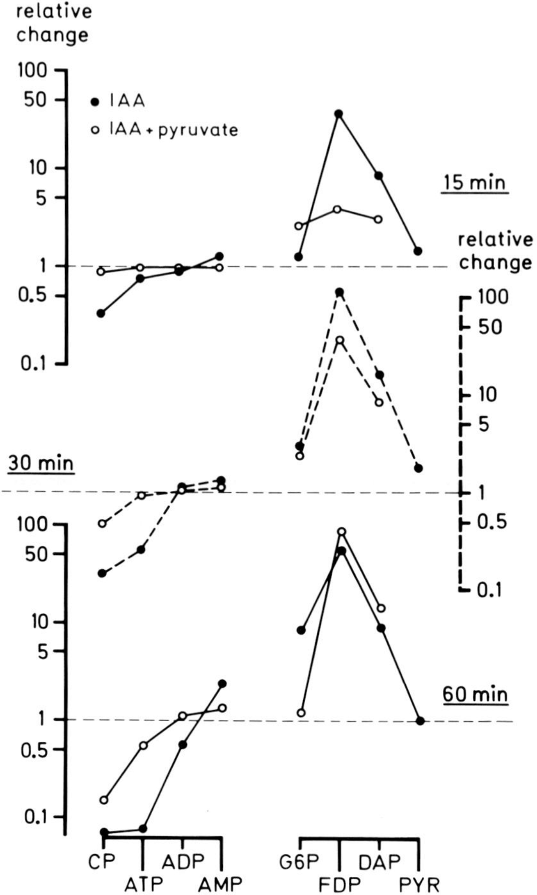

FIGURE 10.   Changes of metabolite content (expressed relative to control values) in aerobic modified Tyrode's solution (125 mM NaCl, 2.6 mM KCl, 1.2 mM KH₂PO₄, 1.2 mM MgSO₄, 1 mM CaCl₂, 5 mM glucose, 25 mM HEPES at pH 7.4 and 37°C) with 1 mM iodoacetate (IAA) + 5 mM pyruvate. Representation for 15, 30, and 60 min incubation. Contents of creatine phosphate (CP), adenine nucleotides (ATP, ADP, AMP) and glycolytic intermediary products: glucose-6-phosphate (G-6-P), fructose-1,6-bisphosphate (FBP), dihydroxyacetone phosphate (DAP) and pyruvate (PYR). Mean values of 5 different experiments. Methodological details described elsewhere.[38]

VII. SUMMARY AND PERSPECTIVES

With the recent progress in the techniques of isolating and culturing Ca^{2+}-tolerant adult cardiac myocytes, new directions in cardiological research have been opened. So far, one major reason for the use of culture models derived from the immature myocardium was the analytical advantages of a culture system per se. For a long time, isolated cell material from the adult myocardium was not available of sufficient quality for reliable metabolic studies. However, with the plating and culturing technique, described above, preparations of viable adult myocytes can be reproducibly obtained, allowing detailed investigations of the specific properties of the adult myocardium under in vitro conditions.

In physiological media, isolated myocytes from the adult ventricular tissue reside in a

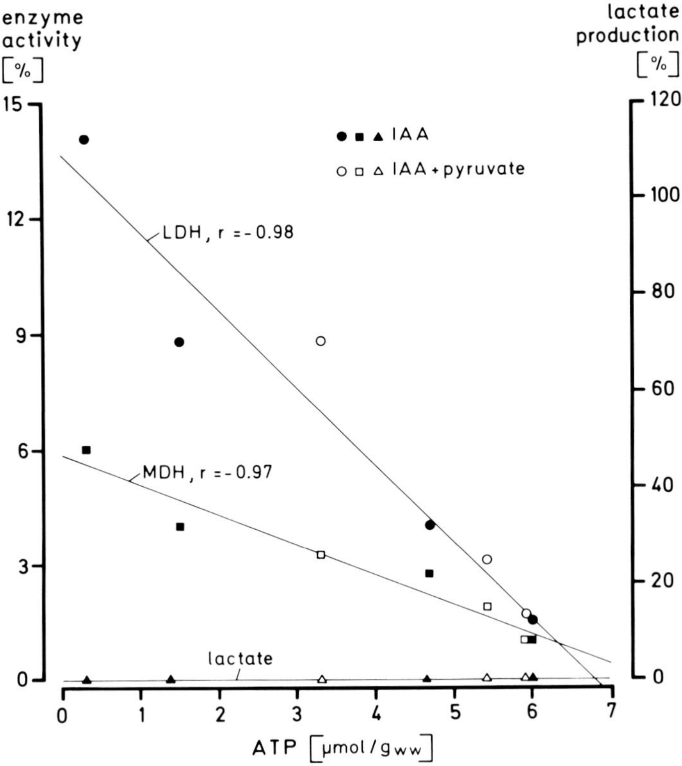

FIGURE 11. Blockade of glycolysis. Correlation of the released activities of lactate dehydrogenase and malate dehydrogenase (LDH, MDH, expressed as percent of the initial total cellular activity, $p < 0.001$) and of the lactate production rate (expressed as percent of 1.2 μmol/g_{ww} · min, no correlation) with the actual ATP content. Mean values of 5 different experiments.

metabolic and functional resting state. This has not always been sufficiently recognized. It is also important to know that in the in vitro models of the adult myocyte currently used direct intercellular communications are usually absent. As suggested by the studies of cells under anoxia, in some cases this difference may even allow particular insights into the role that cell-to-cell contact plays in tissue. For other problems in which the embedding of the individual cells in a cellular network is of importance, the immature cell type cultures may present be a better model.

As detailed in this chapter, the adult myocytes in culture has been investigated so far mainly with respect to structure and metabolism. The development of electrophysiological properties in culture has only been investigated in a single study, which tries to characterize the spontaneous activity of myocytes in long-term culture.[137] There are, however, a number of recent review articles summarizing the electrophysiological experiments with freshly isolated adult myocytes.[138-140] One might expect that most of these insights also apply to myocytes in a short-term culture, as long as they maintain structural similarity to the myocyte in tissue or to the state immediately after isolation.

As long as the adult myocytes cannot be stimulated to divide and still preserve their typical elongated shape, cell-to-cell contacts are likely to occur in vitro only if the seeding density can be greatly increased. As discussed above, this may not be possible when a single layer of cells covers the culture surface in random orientation. It might be speculated that the density could be improved by oriented surface coatings with laminin or collagen IV, since both these extracellular matrix components have recently been shown to mediate attachment of adult cardiac myocytes.[137] The observation that close spatial contact is sufficient for the formation of new intercellular contact structures in vitro gives rise to the hope that a cellular network can indeed be obtained by dense packing of isolated myocytes even when these maintain their typical shape.

Long-term primary cultures in which the isolated muscle cells undergo a structural transformation[26,28,30] may help to elucidate the developmental potency of the adult myocardial cell. This should increase the understanding of the causal factors inducing morphological and functional changes of the myocyte in so-called degenerative heart diseases and of the factors responsible for terminal differentiation in normal adult myocytes.[138]

For experimental research of human myocardium the use of isolated cardiocytes opens new possibilities. Human myocytes can be isolated from pieces of myocardium resected during cardiac surgery.[25,143] Thus, in principle, animal experiments are no longer the only possibility for testing the action of a drug before it is applied to the human body. However, apart from a limitation in the mass of human myocardium available as surgical waste material, also another caveat has to be considered, namely that such waste material usually consists of pathological tissue.

ACKNOWLEDGMENTS

This study was supported by the Deutsche Forschungsgemeinschaft, SFB 89 — Kardiologie Göttingen. Section VI is part of a joint project with A. Pinson and M. Heller, supported by the government of Niedersachsen. The help of B. Eickhoff, H. Haacke, E. Neumeyer, and R. Zöllner is gratefully acknowledged.

NOTE

This manuscript was completed early in 1985. Since then a number of new findings have been reported about cultured adult cardiomyocytes. Many of these are discussed in a recent review.[145]

REFERENCES

1. **Dow, J. W., Harding, N. G. L., and Powell, T.,** Isolated cardiac myocytes. I. Preparation of adult myocytes and their homology with the intact tissue, *Cardiovasc. Res.,* 15, 483, 1981.
2. **Dow, J. W., Harding, N. G. L., and Powell, T.,** Isolated cardiac myocytes. II. Functional aspects of mature cells, *Cardiovasc. Res.,* 15, 549, 1981.
3. **Burrows, M. T.,** The cultivation of tissues of the chick embryo outside the body, *JAMA,* 55, 2057, 1910.
4. **Powell, T. and Twist, V. W.,** A rapid technique for the isolation and purification of adult cardiac muscle cells having respiratory control and a tolerance to calcium, *Biochem. Biophys. Res. Commun.,* 72, 327, 1976.
5. **Clark, M. G., Gannon, B. J., Bodkin, N., Patten, G. S., and Berry, M. N.,** An improved procedure for the high-yield preparation of intact beating heart cells from the adult rat. Biochemical and Morphological study, *J. Mol. Cell. Cardiol.,* 10, 1101, 1978.
6. **Zimmerman, A. N. E., and Hülsmann, W. C.,** Paradoxical influence of calcium ions on the permeability of the cell membranes of the isolated rat heart, *Nature,* 211, 646, 1966.
7. **Powell, T.,** Isolation of cells from adult mammalian myocardium, *J. Mol. Cell. Cardiol.,* 11, 511, 1979.
8. **Crevey, B. J., Langer, G. A., and Frank, J. S.,** Role of Ca^{2+} in maintenance of rabbit myocardial cell membrane structure and functional integrity, *J. Mol. Cell. Cardiol.,* 10, 1081, 1978.
9. **De Mello, W. C.,** Membrane sealing in frog skeletal muscle fibres, *Proc. Natl. Acad. Sci. USA,* 70, 982, 1973.

10. **Piper, H. M.**, Isolierte Adulte Herzmuskelzellen als Myokardmodell, Eigenschaften und Anwendungen, Thieme, Stuttgart, 1985.
11. **Muir, A. R.**, Further observations of the cellular structure of cardiac muscle, *J. Anat.*, 99, 27, 1965.
12. **Haworth, R. A., Hunter, D. R., and Berkoff, H. A.**, Mechanism of Ca^{2+} resistance in adult heart cells isolated with trypsin plus Ca^{2+}, *J. Mol. Cell. Cardiol.*, 14, 523, 1982.
13. **Glick, M. R., Burus, A. H., and Reddy, W. J.**, Dispersion and isolation of beating cells from adult rat heart, *Anal. Biochem.*, 61, 32, 1974.
14. **Bechem, M. and Pott, L.**, Atrial muscle cells from hearts of adult guinea-pigs in culture: a new preparation for cardiac cellular electrophysiology, *Eur. J. Cell. Biol.*, 31, 366, 1983.
15. **Callewaert, G., Carmeliet, E., van der Heyden, G., and Vereecke, J.**, The pacemaker current in a single cell preparation of bovine cardiac Purkinje fibres, *J. Physiol. (London)*, 326, 66, 1982.
16. **Taniguchi, J., Kokobun, S., Noma, A., and Irisawa, H.**, Spontaneously active cells isolated from the sino-atrial and atrio-ventricular nodes of the rabbit heart, *Jpn. J. Physiol.*, 31, 547, 1981.
17. **Masson-Pevet, M., Jongsma, H. J., Bleeker, W. K., Tsjernina, L., van Ginneken, A. C. G., Treitel, B. W., and Bouman, L. N.**, Intact sinus node cells from the adult rabbit heart, *J. Mol. Cell. Cardiol.*, 14, 295, 1982.
18. **Trautwein, W., Taniguchi, J., and Noma, A.**, The effect of intracellular cyclic nucleotides and calcium on the action potential and acetyl choline response of isolated cardiac cells, *Pflügers Arch.*, 392, 307, 1982.
19. **Bustamante, O. J., Watanabe, T., and McDonald, T. F.**, Single cells from adult mammalian heart: isolation procedure and preliminary electro physiological studies, *Can. J. Physiol. Pharmacol.*, 59, 907, 1981.
20. **Isenberg, G. and Klöckner, U.**, Calcium tolerant ventricular myocytes prepared by preincubation in a "KB medium". *Pflügers Arch.*, 395, 6, 1982.
21. **Silver, L. H., Hemwall, E. L., Marino, T. A., and Houser, S. R.**, Isolation and morphology of calcium-tolerant feline ventricular myocytes, *Am. J. Physiol.*, 245, H891, 1983.
22. **Spanier, A. M., and Weglicki, W. B.**, Ca^{2+}-tolerant adult canine myocytes, preparation and response to anoxia/acidosis, *Am. J. Physiol.*, 243, H448, 1982.
23. **Hewett, K., Legato, M. J., Danilo, P., and Robinson, R. B.**, Isolated myocytes from adult canine left ventricle: Ca^{2+} tolerance, electrophysiology, and ultrastructure, *Am. J. Physiol.*, 245, H830, 1983.
24. **Isenberg, G. and Klöckner, U.**, Isolated bovine ventricular myocytes, Characterisation of the action potential, *Pflügers Arch.*, 395, 19, 1982.
25. **Powell, T., Sturridge, M. F., Suvarna, S. K., Terrar, D. A., and Twist, V. W.**, Intact individual heart cells isolated from human ventricular tissue, *Br. Med. J.*, 283, 1013, 1981.
26. **Jacobson, S. L.**, Culture of spontaneously contracting myocardial cells from adult rats, *Cell Struct. Function*, 2, 1, 1977.
27. **Schwarzfeld, T. A. and Jacobson, S. L.**, Isolation and development in cell culture of myocardial cells of the adult rat, *J. Mol. Cell. Cardiol.*, 13, 563, 1981.
28. **Claycomb, W. C. and Palazzo, M. C.**, Culture of the terminally differentiated adult cardiac muscle cell. A light and scanning electron microscope study, *Dev. Biol.*, 80, 466, 1980.
29. **Claycomb, W. C. and Bradshaw, H. D.**, Acquisition of multiple nuclei and the activity of DNA polymerase α and reinitiation of DNA replication in terminally differentiated adult cardiac muscle cells in culture, *Dev. Biol.*, 99, 331, 1983.
30. **Nag, A. and Cheng, M.**, Adult mammalian cardiac muscle cells in culture, *Tiss. Cell*, 13, 515, 1981.
31. **Nag, A. C., Cheng, M., Fischman, D. A., and Zak, R.**, Long-term cell culture of adult mammalian cardiac myocyte. Electron microscopic and immunofluorescent analysis of myofibrillar structure, *J. Mol. Cell. Cardiol.*, 15, 301, 1983.
32. **Piper, H. M., Probst, I., Schwartz, P., Hütter, J. F., and Spieckermann, P. G.**, Culturing of calcium stable adult cardiac myocytes, *J. Mol. Cell. Cardiol.*, 14, 397, 1982.
33. **Piper, H. M., Hütter, J. F., and Spieckermann, P. G.**, Relation between enzyme release and metabolic changes in irreversible anoxic injury of myocardial cells, *Life Sci.*, 35, 127, 1984.
34. **Piper, H. M., Schwartz, P., Spahr, R., Hütter, J. F., and Spieckermann, P. G.**, Early enzyme release from myocardial cells is not due to irreversible cell damage, *J. Mol. Cell. Cardiol.*, 16, 385, 1984.
35. **Piper, H. M., Schwartz, P., Spahr, R., Hütter, J. F., and Spieckermann, P. G.**, Absence of reoxygenation damage in isolated heart cells after anoxic injury, *Pflügers Arch.*, 401, 71, 1984.
36. **Schwartz, P., Piper, H. M., Spahr, R., and Spieckermann, P. G.**, Ultrastructure of adult myocardial cells during anoxia and reoxygenation, *Amer. J. Pathol.*, 115, 349, 1984.
37. **Piper, H. M., Schwartz, P., Hütter, J. F., and Spieckermann, P. G.**, Energy metabolism and enzyme release of cultured adult rat muscle cells during anoxia, *J. Mol. Cell. Cardiol.*, 16, 995, 1984.
38. **Piper, H. M., Schwartz, P., Hütter, J. F., and Spieckermann, P. G.**, Enzyme release and glycolytic energy production, *Basic Res. Cardiol.*, 80 (Suppl. 1), 143, 1985.

39. **Finch, S. A. E., Piper, H. M., Spieckermann, P. G., and Stier, A.,** Anoxia influences the lateral diffusion of a lipid probe in the plasma membrane of isolated cardicac myocytes, *Basic Res. Cardiol.,* 80 (Suppl. 1), 149, 1985.

40. **Spahr, R., Probst, I., and Piper, H. M.,** Substrate utilization of adult cardiac myocytes, *Basic Res. Cardiol.,* 80 (Suppl. 1), 1985.

41. **Piper, H. M., Spahr, R., and Probst, I.,** Glucose, lactate, and palmitate as substrates for the resting cardiac myocyte, *Basic Res. Cardiol.,* 80 (Suppl. 2), 78, 1985.

42. **Pelzer, D., Trube, G., and Piper, H. M.,** Low resting potentials in single isolated heart cells due to membrane damage by the recording microelectrode, *Pflügers Arch.,* 400, 197, 1984.

43. **Piper, H. M., Hütter, J. F., and Spieckermann, P. G.,** Temperature dependence of nifedipine action, *J. Mol. Cell. Cardiol.,* 16, 277, 1984.

44. **Powell, T., Terrar, D. A., and Twist, V. W.,** Electrical properties of individual cells isolated from adult rat ventricular myocardium, *J. Physiol. (London),* 302, 131, 1980.

45. **Piper, H. M., Spahr, R., Probst, I., and Spieckermann, P. G.,** Substrates for the attachment of adult cardiac myocytes in culture, *Basic Res. Cardiol.,* 80 (Suppl. 2), 149, 1985.

46. **Muir, A. R.,** The effects of divalent cations on the ultrastructure of the perfused rat heart, *J. Anat.,* 101, 239, 1967.

47. **McNutt, S. T.,** Ultrastructure of intercellular junctions in adult and developing cardiac muscle, *Am. J. Cardiol.,* 25, 169, 1970.

48. **Nag, A. C., Fischman, D. A., Aumont, M. C., and Zak, R.,** Studies of isolated adult rat heart cells. The surface morphology and the influence of extracellular calcium ion concentration on cellular viability, *Tiss. Cell,* 9, 419, 1977.

49. **Bishop, S. P., and Drummond, J. L.,** Surface morphology and cell size measurements of isolated rat cardiac myocytes, *J. Mol. Cell. Cardiol.,* 11, 423, 1979.

50. **Fry, D. M., Scales, D., and Inesi, G.,** The ultrastructure of membrane alterations of enzymatically dissociated cardiac myocytes, *J. Mol. Cell. Cardiol.,* 11, 1151, 1979.

51. **Flegler-Baron, C., and Behrendt, H.,** Effects of Ca^{2+} deficiency, collagenase, and mechanical dispersion on the ultrastructure of cardiac myocytes of adult rats, *Eur. J. Cell. Biol.,* 27, 262, 1982.

52. **Severs, N. J., Slade, A. M., Powell, T., Twist, V. W., and Warren, R. L.,** Correlation of ultrastructure and function in calcium-tolerant myocytes isolated from the adult rat heart, *J. Ultrastruct. Res.,* 81, 222, 1982.

53. **Schwartz, P., Piper, H. M., Spahr, R., Hütter, J. F., and Spieckermann, P. G.,** Development of new intercellular contacts between adult cardiac myocytes in culture, *Basic Res. Cardiol.,* 80 (Suppl. 1), 75, 1985.

54. **Paulussen, F., Hübner, G., Grebe, D., and Bretschneider, H. J.,** Die Feinstruktur des Herzmuskels während einer Ischämie mit Senkung des Energiebedarfs durch spezielle Kardioplegie, *Klin. Wochenschr.,* 46, 165, 1968.

55. **Hatt, P. Y. and Moravec, J.,** Acute hypoxia of the myocardium. Ultrastructural changes, *Cardiology,* 56, 73, 1971.

56. **Krueger, J. W., Forletti, D., and Wittenberg, B. A.,** Uniform sarcomere shortening behaviour in isolated cardiac muscle cells, *J. Gen. Physiol.,* 76, 587, 1980.

57. **Haworth, R. A., Hunter, D. R., and Berkoff, H. A.,** Contracture in isolated adult rat heart cells. Role of Ca^{2+}, ATP, and compartmentation, *Circ. Res.,* 49, 1119, 1981.

58. **Page, S. G.,** Measurements of structural parameters in cardiac muscle, in the Physiological Basis of Starling's Law of the Heart, *Ciba Found. Symp.,* 24, 13, 1974.

59. **Jewell, B. R.,** A re-examination of the influence of muscle length on myocardial performance, *Circ. Res.,* 40, 221, 1977.

60. **Harary, I. and Farley, B.,** *In vitro* organisation of single beating rat heart cells into beating fibres, *Science,* 132, 1839, 1960.

61. **De Haan, R. L. and Sachs, H. G.,** Cell coupling in developing systems: the heart-cell paradigm, in *Current Topics in Developmental Biology,* Vol. 7, Moscana, A. A., and Monroy, A., Eds., Academic Press, New York, 1972, 193.

62. **Griepp, E. B., Peackock, J. H., Bernfield, M. R., and Revel, J. P.,** Morphological and functional correlates of synchronous beating between embryonic heart cell aggregates and layers, *Exp. Cell Res.,* 113, 273, 1978.

63. **Lowenstein, W. R.,** Junctional intercellular communication: the cell-to-cell membrane channel, *Physiol. Rev.,* 61, 829, 1981.

64. **Shimata, Y., and Fischman, D. A.,** Cardiac cell aggregation by scanning electron microscopy, in *Development and Physiological Correlates of Cardiac Muscle,* Liebermann, M. L. and Sano, T., Eds., Raven Press, New York, 1975, 81.

65. **Spahr, R., Piper, H. M., Probst, I., Schwartz, P., and Spieckermann, P. G.,** Morphological dedifferentiation of adult cardiac myocytes in coculture with hepatocytes, *Basic Res. Cardiol.,* 80 (Suppl. 1), 83, 1985.
66. **Probst, I., Schwartz, P., and Jungermann, K.,** Induction in primary culture of "glyconeogenic" and "glycolytic" hepatocytes resembling periportal and perivenous cells, *Eur. J. Biochem.,* 126, 271, 1982.
67. **Mark, G. E. and Strasser, F. F.,** Pacemaker activity and mitosis in cultures of newborn rat heart ventricle cells, *Exp. Cell Res.,* 44, 217, 1966.
68. **Sperelakis, N.,** Propagation mechanisms in heart, *Ann. Rev. Physiol.,* 41, 441, 1979.
69. **Davis, M. J.,** The cardiomyopathies: a review of terminology, pathology and pathogenesis, *Histopathology,* 8, 363, 1984.
70. **Arnold, G. and Lochner, W.,** Die Temperaturabhängigkeit des Sauerstoffverbrauchs stillgestellter künstlich perfundierter Warmblüterherzen zwischen 34°C und 4°C, *Pflügers Arch.,* 284, 169, 1967.
71. **Spieckermann, P. G. and Piper, H. M.,** Oxygen demand of calcium-tolerant adult cardiac myocytes, *Basic Res. Cardiol.,* 80 (Suppl. 2), 59, 1985.
72. **Burns, A. H. and Reddy, W. J.,** Amino acid stimulation of oxygen and substrate utilization by cardiac myocytes, *Am. J. Physiol.,* 234, E461, 1978.
73. **Burns, A. H. and Montini, J.,** Myocardium in hypertrophy: oxygen consumption by isoplated cardiac myocytes and working hearts from spontaneously hypertensive rats, *Life Sci.,* 30, 29, 1982.
74. **Frangakis, C. J., Bahl, J. J., McDaniel, H., and Bressler, R.,** Tolerance to physiological calcium by isolated myocytes from the adult rat heart: an improved cellular preparation, *Life Sci.,* 27, 815, 1980.
75. **Wittemberg, B. A. and Robinson, T. F.,** Oxygen requirements, morphology, cell coat and membrane permeability of calcium tolerant myocytes from hearts of adult rats, *Cell Tiss. Res.,* 216, 231, 1981.
76. **Montini, J., Bagby, G. J., and Spitzer, J. J.,** Importance of exogenous substrates for the energy production of adult rat heart myocytes, *J. Mol. Cell. Cardiol.,* 13, 903, 1981.
77. **Bretschneider, H. J. and Hellige, G.,** Pathophysiologie der Ventrikel kontraktion — Kontraktilität, Inotropie, Suffiziengrad und Arbeitsökonomie des Herzens, *Verh. dtsch. Ges. Kreislcufforsch.,* 42, 14, 1976.
78. **Haworth, R. A., Hunter, D. R., Berkoff, H. A., and Moses, R. L.,** Metabolic cost of the stimulated beating of isolated rat heart cells in suspension, *Circ. Res.,* 52, 342, 1983.
79. **Fuhrman, G. J., Fuhrman, F. A., and Field, J.,** Metabolism of rat heart slices with special reference to effects of temperature and anoxia, *Am. J. Physiol.,* 163, 642, 1950.
80. **Geisbuhler, T., Altschuld, R. A., Trewyn, R. W., Ansel, A. Z., Lamka, K., and Brierley, G. P.,** Adenine nucleotide metabolism and compartmentalization in isolated adult rat heart cells, *Circ. Res.,* 54, 536, 1984.
81. **Jarasch, E. D., Grund, C., Bruder, G., Heid, H. W., Keenan, T. W., and Franke, W. W.,** Localization of xanthine oxidase in mammary-gland epithelium and vascular endothelium, *Cell,* 25, 67, 1981.
82. **Chen, V., Bagby, G. J., and Spitzer, J. J.,** Exogenous, substrate utilization by isolated myocytes from chronically diabetic rats, *Am. J. Physiol.,* 245, C46, 1983.
83. **Kao, R. L., Christman, E. W., Luh, S. L., Krauhs, J. M., Tyers, G. F. O., and Williams, E. H.,** The effects of insulin and anoxia on the metabolism of isolated mature rat cardiac myocytes, *Arch. Biochem. Biophys.,* 203, 587, 1980.
84. **Montini, J., Bagby, G. J., Burus, A. H., and Spitzer, J. J.,** Exogenous substrate utilization in Ca^{2+}-tolerant myocytes from adult rat hearts, *Am. J. Physiol.,* 240, H659, 1981.
85. **Tamboli, A., O'Looney, P., van der Maten, M., and Vahouny, G. V.,** Comparative metabolism of free and esterified fatty acids by the perfused rat heart and rat cardiac myocytes, *Biochim. Biophys. Acta,* 750, 404, 1983.
86. **Walker, E. J., Burns, J. H., and Dow, J. W.,** Amino acid transport and protein synthesis in energetically stable calcium-tolerant isolated cardiac myocytes, *Biochim. Biophys. Acta.,* 721, 280, 1982.
87. **Long, W. M., Bagby, G. J., and Spitzer, J. J.,** Contribution of viable and nonviable heart myocytes to substrate oxidation, *Am. J. Physiol.,* 238, H740, 1980.
88. **Opie, L. H.,** Adequacy of oxygenation of isolated perfused rat heart, *Basic Res. Cardiol.,* 79, 300, 1984.
89. **Hiltunen, J. K. and Hassinen, I. E.,** Energy linked regulation of glucose and pyruvate oxidation in isolated perfused rat heart. Role of pyruvate dehydrogenase, *Biochim. Biophys. Acta.,* 440, 377, 1976.
90. **Kobayashi, K. and Neely, J. R.,** Control of maximum rate of glycolysis in rat cardiac muscle, *Circ. Res.,* 44, 166, 1979.
91. **Randle, P. J. and Tubbs, P. K.,** Carbohydrate and fatty acid metabolism, in *Handbook of Physiology,* Sect. 2, Vol. 1, Berne, R. M., Sperelakis, N., and Geiger, S. R., Eds., American Physiological Society, Bethesda, 1979, 805.
92. **Haworth, R. A., Hunter, D. R., and Berkoff, H. A.,** Insulin stimulates deoxyglucose transport in adult rat heart cells in the absence of Ca^{2+}, *FEBS Lett.,* 141, 37, 1982.

93. **Lindgren, C. A., Paulson, D. J., and Shanahan, M. F.,** Isolated cardiac myocytes. A new cellular model for studying insulin modulation of monosaccharide transport, *Biochim. Biophys. Acta.,* 721, 385, 1982.

94. **Eckel, J., Pandalis, G., and Reinauer, H.,** Insulin action on the glucose transport system in isolated cardiocytes from adult rat, *Biochem. J.,* 212, 385, 1983.

95. **Claus, T. H., El-Maghrabi, M. R., Regen, D. M., Stewart, H. B., McGrane, M., Kountz, P. D., Nyfeler, F., Pilkis, J., and Pilkis, S. J.,** The role of fructose 2,6-bisphosphate in the regulation of carbohydrate metabolism, in *Current Topics in Cellular Regulation,* Vol. 23, Horecker, B. S. and Stadtman, E. R., Eds., Academic Press, London, 1984, 57.

96. **Williamson, J. R., Ford, C., Illingworth, J., and Safer, B.,** Coordination of citric acid cycle activity with electron transport flux, *Circ. Res.,* 38 (Suppl. 1), 39, 1976.

97. **Fath, P. A. and Kako, K. J.,** Glycolytic and tricarboxylic acid cycle intermediates during cardiac arrest and recovery in eu-, hyper-, and hypothyroid rats, *J. Mol. Cell. Cardiol.,* 5, 359, 1973.

98. **Neely, J. R. and Morgan, H. E.,** Relationship between carbohydrate and lipid metabolism and the energy balance of heart muscle, *Ann. Rev. Physiol.,* 36, 413, 1974.

99. **Randle, P. J., Newsholme, E. A., and Garland, P. B.,** Regulation of glucose uptake by muscle. 8. Effects of fatty acids, ketone bodies and pyruvate, and of alloxan-diabetes and starvation, on the uptake and metabolic fate of glucose in rat heart and diaphragm muscles, *Biochem. J.,* 93, 652, 1964.

100. **Williamson, J. R.,** Effects of insulin and diet on the metabolism of L(+)-lactate and glucose by the perfused rat heart, *Biochem. J.,* 83, 377, 1962.

101. **Hütter, J. F., Piper, H. M., and Spieckermann, P. G.,** Kinetic analysis of myocardial fatty acid oxidation suggesting an albumin receptor mediated uptake process, *J. Mol. Cell. Cardiol.,* 16, 219, 1984.

102. **Idell-Wenger, J. A. and Neely, J. R.,** Regulation of uptake and metabolism of fatty acids by muscle, in *Disturbances in Lipid and Lipoprotein Metabolism,* Dietschy, J. M., Gotto, A. M., and Ontko, J. A., Eds., American Physiological Society, Bethesda, 1978, 269.

103. **Claycomb, W. C., Burns, A. H., and Shepherd, R. E.,** Culture of terminally differentiated ventricular cardiac muscle cells. Characterisation of exogenous substrate oxidation and the adenylate cyclase system, *FEBS Lett.,* 169, 261, 1984.

104. **Liu, M. S. and Spitzer, J. J.,** Oxidation of palmitate and lactate by beating myocytes isolated from adult dog heart, *J. Mol. Cell. Cardiol.,* 10, 415, 1978.

105. **Glatz, J. F. C., Jacobs, A. E. M., and Veerkamp, J. H.,** Fatty acid oxidation in human and rat heart. Comparison of cell-free and cellular systems, *Biochim. Biophys. Acta.,* 794, 454, 1984.

106. **Shell, W. E., Kjekshus, J. K., and Sobel, B. E.,** Quantitative assessment of the extent of myocardial infarction in the conscious dog by means of analysis and serial changes in serum creatine phosphokinase activity, *J. Clin. Invest.,* 50, 2614, 1971.

107. **Hearse, D. J. and Chain, E. B.,** Effect of glucose on enzyme release from, and recovery of, the anoxic myocardium, in *Myocardial Metabolism: Recent Advances in Studies on Cardiac Structure and Metabolism,* Vol. 3, Dhalla, N. S., Ed., University Park Press, Baltimore, 1973, 763.

108. **Ahmed, S. A., Williamson, J. R., Roberts, R., Clark, R. E., and Sobel, B. E.,** The association of increased plasma MB CPK activity and irreversible ischemic myocardial injury in the dog, *Circulation,* 54, 187, 1976.

109. **Kübler, W., and Spieckermann, P. G.,** Regulation of glycolysis in the ischemic and the anoxic myocardium, *J. Mol. Cell. Cardiol.,* 1, 351, 1970.

110. **Danforth, W. H., Nägele, S., and Bing, R. J.,** Effect of ischemia and reoxygenation on glycolytic reactions and ATP in the heart muscle, *Circ. Res.,* 8, 965, 1960.

111. **Isselhard, W., Mäurer, W., Stremmel, W., Krebs, J., Schmitz, H., Neuhof, H., and Esser, A.,** Stoffwechsel des Kaninchenherzens *in situ* während Asphyxie und in der postasphyktischen Erholung, *Pflügers Arch.,* 316, 164, 1970.

112. **Gerlach, W. and Deuticke, B.,** Bildung und Bedeutiung von Adenosin in dem durch Sauerstoffmangel geschädigten Herzmuskel unter dem Einfluss von 2,6-Bis(diaethanolamino)-4,8-dipiperidino-pyrimido (5,4-d) pyrimidin, *Arzneim. Forsch.,* 13, 48, 1963.

113. **Katori, M. and Berne, R.,** Release of adenosine from anoxic hearts. Relationship to coronary flow, *Circ. Res.,* 19, 420, 1966.

114. **Zimmer, H. G., Trendelenburg, C., Kammermeier, H., and Gerlach, E.,** De novo synthesis of myocardial adenine nucleotides in the rat. Acceleration during recovery from oxygen deficiency, *Circ. Res.,* 32, 635, 1973.

115. **Kammermeier, H.,** Verhalten von Adenin-Nucleotiden und Kreatinphosphat im Herzmuskel bei funktioneller Erholung nach länger dauernder Asphyxie, *Verh. dtsch. Ges. Kreislaufforsch.,* 30, 206, 1964.

116. **Hoerter, J. A., and Opie, L. H.,** Perinatal changes in glycolytic function in response to hypoxia in the incubated or perfused rat heart, *Biol. Neonate,* 33, 144, 1978.

117. **Harary, I.,** Biochemistry of cardiac development: in vivo and in vitro studies, in *Handbook of Physiology,* Sect. 2, Vol. 1, Berne, R. M., Sperelakis, N., and Geiger, S. R., Eds., American Physiological Society, Bethesda, 1979, 43.

118. **Rajs, J., Sundberg, M., Harm, T., Grandinsson, M., and Soderlund, V.,** Interrelationship between ATP levels, enzyme leakage, gluconate uptake, trypan blue exclusion and length/width ratio of isolated rat cardiac myocytes subjected to anoxia and reoxygenation, *J. Mol. Cell. Cardiol.,* 12, 1227, 1980.

119. **Altschuld, R. A., Hostetler, J. R., and Brierley, G. P.,** Response of isolated rat heart cells to hypoxia, reoxygenation and acidosis, *Circ. Res.,* 49, 307, 1981.

120. **Hohl, C., Ansel, A., Altschuld, R., and Brierley, G. P.,** Contracture of isolated rat heart cells on anaerobic to aerobic transition, *Am. J. Physiol.,* 242, H1022, 1982.

121. **Murphy, M. P., Hohl, C., Brierley, G. P., and Altschuld, R. A.,** Release of enzymes from adult rat heart myocytes, *Circ. Res.,* 51, 560, 1982.

122. **Cheung, J. Y., Thomson, I. G., and Bonventre, J. V.,** Effects of extracellular calcium removal and anoxia on isolated rat myocytes, *Am. J. Physiol.,* 243, C184, 1982.

123. **Cheung, J. Y., Leaf, A., and Bonventre, J. V.,** Mechanism of protection by verapamil and nifedipine from anoxic injury in isolated cardiac myocytes, *Am. J. Physiol.,* 246, C323, 1984.

124. **Jennings, R. B., Hawkins, H. K., Lowe, J. E., Hill, M. L., Klotman, S., and Reimer, K. A.,** Relation between high energy phosphate and lethal injury in myocardial injury in the dog, *Am. J. Physiol.,* 92, 187, 1978.

125. **Ganote, C. E.,** Contraction band necrosis and irreversible myocardial injury, *J. Mol. Cell. Cardiol.,* 15, 67, 1983.

126. **Ganote, C. E. and Kaltenbach, J. P.,** Oxygen-induced enzyme release: early events and a proposed mechanism, *J. Mol. Cell. Cardiol.,* 11, 389, 1979.

127. **Ganote, C. E., Seabra-Gomes, R., Nayler, W. G., and Jennings, R. B.,** Irreversible myocardial injury in anoxic perfused rat hearts, *Am. J. Pathol.,* 80, 419, 1975.

128. **Hearse, D. J., Humphrey, S. M., Nayler, W. G., Slade, A., and Border, D.,** Ultrastructural damage associated with reoxygenation of the anoxic myocardium, *J. Mol. Cell. Cardiol.,* 7, 315, 1975.

129. **Feuvray, D. and DeLeiris, J.,** Ultrastructural modifications induced by reoxygenation in the anoxic isolated rat heart without exogenous substrate, *J. Mol. Cell. Cardiol.,* 7, 307, 1975.

130. **Ganote, C. E., Jennings, R. B., Hill, M. L., and Grochowski, E. C.,** Experimental myocardial ischemic injury II. Effects of in vivo ischemia on dog heart slice function in vitro, *J. Mol. Cell. Cardiol.,* 8, 189, 1976.

131. **Gebhard, M. M., Denkhaus, H., Sakai, K., and Spieckermann, P. G.,** Energy metabolism and enzyme release, *J. Mol. Med.,* 2, 271, 1977.

132. **Bricknell, O. L. and Opie, L. H.,** Effects of substrates on tissue metabolic changes in the isolated rat heart during underperfusion and on release of lactate dehydrogenase and arrhythmias during reperfusion, *Circ. Res.,* 43, 102, 1978.

133. **Bricknell, O. L., Daries, P. S., and Opie, L. H.,** A relationship between adenosine triphosphate, glycolysis and ischemic contracture in the isolated rat heart, *J. Mol. Cell. Cardiol.,* 13, 941, 1981.

134. **Williamson, J. R.,** Metabolic control in the perfused rat heart, in *Control of Energy Metabolism,* Chance, B., Estabrook, R. W., and Williamson, J. R., Eds., Academic Press, London, 1965, 333.

135. **Gercken, G., and Hürter, P.,** Stationäre Metabolitkonzentrationen im insuffizienten Säugetierherzen nach Monojodazetat- und Natrium-fluoridvergiftung, *Pflügers. Arch.,* 292, 100, 1966.

136. **Kammermeier, H., Schmidt, P., and Jüngling, E.,** Free energy change of ATP-hydrolysis: a causal factor of early hypoxic failure of the myocardium, *J. Mol. Cell. Cardiol.,* 14, 267, 1982.

137. **Jacobson, S. L., Kennedy, C. B., and Mealing, G. A. R.,** Evidence for functional sodium and calcium ion channels in the membrane of cultured cardiomyocytes of the adult rat, *Can. J. Physiol. Pharmacol.,* 61, 1312, 1983.

138. **Powell, T.,** Electrophysiological properties of isolated ventricular myocytes, *Basic Res. Cardiol.,* 80 (Suppl. 1), 87, 1985.

139. **Isenberg, G. and Klöckner, U.,** The electrophysiological properties of the isolated adult heart cell: an overview, *Basic Res. Cardiol.,* 80 (Suppl. 2), 51, 1985.

140. **Bkaily, G., Sperelakis, N., and Doane, J.,** A new method for preparation of isolated single adult myocytes, *Am. J. Physiol.,* 247, H1018, 1984.

141. **Lundgren, E., Borg, T., and Mardh, S.,** Isolation, characterization and adhesion of calcium-tolerant myocytes from the adult rat heart, *J. Mol. Cell. Cardiol.,* 16, 355, 1984.

142. **Zak, R.,** Development and proliferative capacity of cardiac muscle cells, *Circ. Res.,* 34/35 (Suppl. II), 17, 1974.

143. **Jacobson, S. L., Banfalvi, M., and Schwarzfeld, T. A.,** Long-term primary cultures of adult human and rat cardiocytes, *Basic Res. Cardiol.,* 80, Suppl. 1, 79, 1985.

144. **Probst, I., Spahr, R., Schweichkhardt, C., Hummeman, D. H., and Piper, H. M.,** Carbohydrate and fatty acid metabolism of cultured adult cardiac myocytes, *Am. J. Physiol.,* 250, H853, 1986.

145. **Jacobson, S. L. and Piper, H. M.,** Cell cultures of adult cardiomyocytes as models of the myocardium, *J. Mol. Cell. Cardiol.,* 18, 661, 1986.

Chapter 24

MICROINJECTION INTO CULTURED CHICK HEART CELLS

Hans M. Eppenberger, Beat Schäfer, and Jean-Claude Perriard

TABLE OF CONTENTS

I. INTRODUCTION

In the course of differentiation, myogenic cells undergo multiple changes which may be characterized at the morphological and the molecular level. A prominent event in the developmental transition of skeletal muscle cells is the fusion of mononucleated myoblasts to form multinucleated myotubes. These morphological changes are coordinated with the expression of a set of muscle-specific genes and are accompanied by isoprotein transitions. In our laboratory we study a key enzyme of energy metabolism, creatine kinase (EC 2.7.3.2), which catalyzes the regeneration of ATP from ADP and creatine phosphate. This enzyme exists in multiple molecular forms and in higher vertebrates two types of subunits are found: M-CK and B-CK, which combine to form the dimeric active isoenzymes MM-CK, MB-CK, and BB-CK. In mitochondria a third type of subunit is present. The more ubiquitous form of creatine kinase, BB-CK, is found in adult brain, smooth muscle, and heart as well as in many other tissues and is also the predominant form in embryonic skeletal muscle cells.[1,2] During muscle differentiation synthesis of the M-CK subunits is induced and M-CK-containing enzymes become the prominent forms of CK in cell cultures and completely replace B-CK in adult muscle.[3] Part of the MM-CK present in muscle cells is found in the electron-dense M-band of skeletal muscle myofibrils,[4,5] which has also been shown to contain two other M-band proteins, myomesin and M-protein.[6] This transverse structure might be associated with registering thick filaments in skeletal muscle A-bands and MM-CK and may also be involved in the formation of primary m-bridges. In mammalian heart a similar structure is observed and M-CK was shown to be present as well. In chick heart, however, the typical switch to M-CK does not take place during development[1] nor is there a typical dense M-band, while the other two M-band components known, myomesin and M-protein, have been shown to be present.[16] Thus, as far as is known, favorable conditions exist in the chick heart M-band region for MM-CK association, if M-CK would be expressed in these cells.

Preliminary experiments aimed at answering these questions have been done on isolated myofibrils. Glycerinated chick heart myofibrils were incubated with exogenous purified MM-CK. Weak, positive, immunofluorescent staining for MM-CK could be observed at the M-band region, along with a little fluorescence in the Z-line regions. Similar experiments have recently been used to study how contractile proteins become organized into sarcomeric units.[7] Isolated rabbit myofibrils were exposed to fluorescently labeled actin, α-actinin, and tropomyosin. These data were not conclusive since the decoration pattern with labeled α-actinin, for instance, did not show Z-disc specificity. In order to test for the active processes of incorporation during myofibril formation it appears to be necessary to introduce directly into cells by the technique of microinjection either proteins or mRNA coding for particular proteins. The introduction of mRNA should make proteins available in the native state and, thus, the partial denaturation and degradation observed in purified preparations should not occur. Microinjection experiments of fluorescently labeled proteins or of mRNA into cultures of living cells have been carried out by several investigators.[8-10] We used the microinjection technique either with MM-CK purified from chick skeletal muscle enriched or with translatable mRNA from skeletal muscle with M-CK mRNA to probe for possible association sites in the chick heart myofibrils.

II. MICROINJECTION OF PURIFIED MM-CREATINE KINASE

Heart cell cultures were prepared from 8-day-old chick embryos according to standard techniques.[12] Glutamine was omitted from the culture medium in order to reduce overgrowth of the myocardial cultures with fibroblasts.[13] After 2 to 4 days in culture myocytes were rather easily distinguishable from other cell types present in the culture dish by the appearance of myofibrils with the ability of spontaneous contraction. Microinjection into both contracting and noncontracting myocytes was performed. On the one hand, microinjection into beating cells resulted in the temporary cessation of beating activity, while, on the other hand, nonbeating cells were sometimes induced to contract. Glycogen deposits, typically found in cultured myocytes, were displaced by the injected liquid, but this apparently had no effect on the organization of the existing myofibrils. In the first of a series of experiments, purified chick MM-CK was injected into the cytoplasm of the heart cells close to the nucleus, using known methods.[14] About 10 pℓ containing 10 ng of protein was injected. Cells were then incubated for periods from between 5 min to several hours in a humid atmosphere of 5% CO_2/95% air at 37°C and then fixed with 3% para-formaldehyde. The cells were then incubated with polyclonal affinity-purified rabbit anti-chick antibodies against MM-CK together with the monoclonal B_4 mouse anti-chick antibody against the M-band protein, myomesin.[6] This was followed by incubation with secondary antibodies against rabbit IgG (labeled with rhodamine) and mouse IgG (labeled with FITC). Double immunofluorescence staining allowed identification of the M-band in the myofibrils in both injected and noninjected cells (Figure 1, a to c). Incorporation of MM-CK into the myofibrillar structure is more difficult to discern due to the fluorescence of the cytoplasmic antigen and careful extraction of the soluble M-band M-CK may be necessary. Alternatively, injected cells were immunostained after very short postinjection periods (5 min) and a cross-striated pattern along the myofibrils could be detected, indicating possible associations of the MM-CK to the M-band. As a control, either fluorescently labeled bovine serum albumin or IgG was injected into the cells and resulted in nonspecifically distributed uniform fluorescence (not shown). Similar results were obtained by other workers after injection of rhodamine-labeled α-actinin into chick myocytes and rapid incorporation of fluorescence into the I-bands of the myofibrils was observed.[15] In a control, no specifically distributed fluorescence was observed when fluorescently labeled ovalbumin was injected into the heart cells (not shown).

III. MICROINJECTION OF POLY (A$^+$) RNA

In order to administer native, undegraded MM-CK to the cells, the mRNA for M-CK was injected into these cells to test the behavior of newly synthesized MM-CK. RNA was isolated from 18-day-old chick breast muscle and chromatographed twice on an oligo (dT) cellulose column. Subsequently, 2.5 OD_{20} units of poly (A$^+$) RNA was resolved on an isokinetic sucrose gradient. Fractions were analyzed by the "northern blot" technique, and those containing mRNA for M-CK were pooled. The enriched RNA for M-CK mRNA was injected into cardiac cells at a concentration of 1 mg/mℓ of RNA. After 1 to 3 hr of incubation the cells were fixed and analyzed by the double immunofluorescence technique described above for the protein injection experiments (Figure 1, d to f). The abundant fluorescence in Figure 1f demonstrated that M-CK had, indeed, been synthesized by the injected heart cells. This is an indication that translation of mRNA for skeletal muscle proteins in heart cells is relatively unrestricted. Again, in spite of the strong background fluorescence caused by nonassociated MM-CK, a labeling pattern indicating its association with the M-band region may be clearly seen in the injected cells (see arrow, Figure 1f). Uninjected cells, on the other hand, contain myomesin in the M-band, and no fluorescence staining for MM-CK can be demonstrated (as indicated by N in Figure 1, c and f). The experiments described show that myogenic cardiac cells may serve as "test tubes" for transferred proteins or their mRNAs in processes such as regulation, assembly, function, and maintenance of myofibrils and their components.

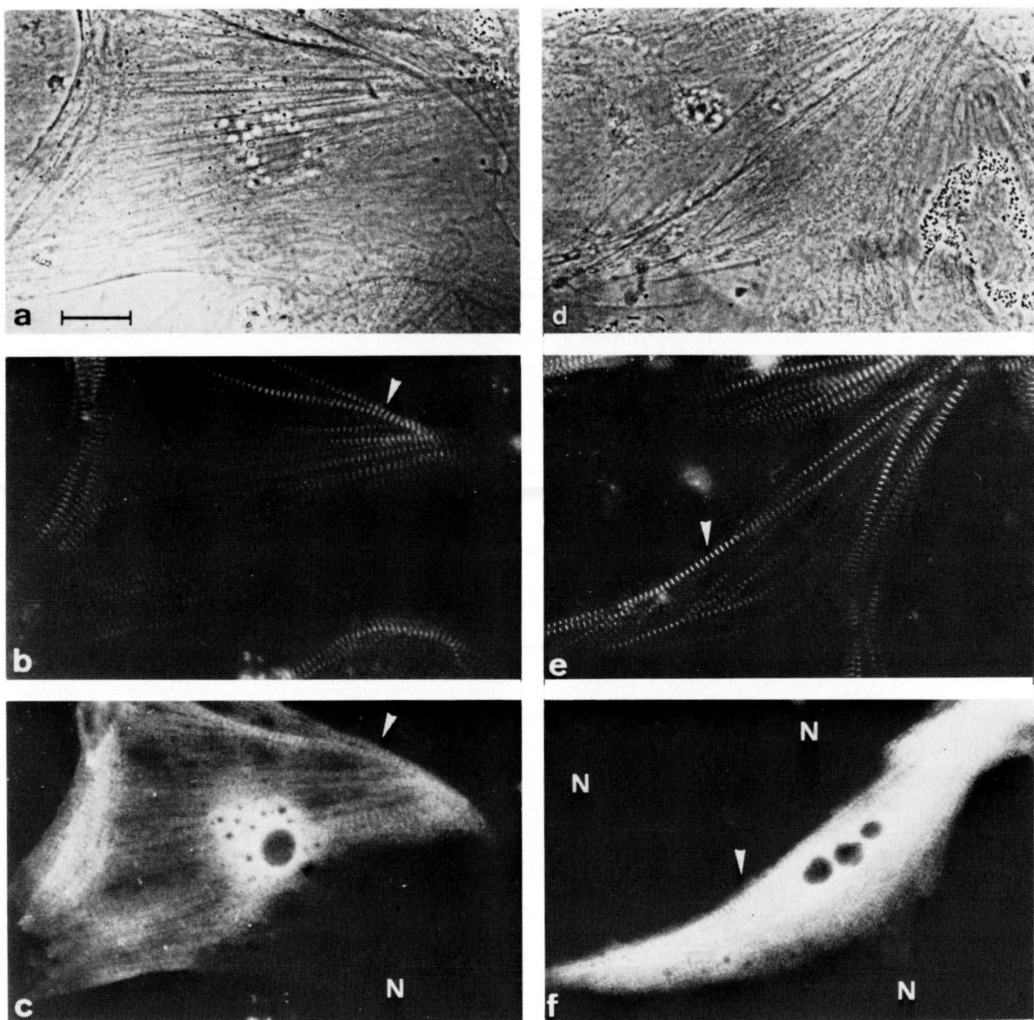

FIGURE 1.   (a to c) About 10 pℓ of purified MM-CK (4 mg/mℓ) was microinjected, according to the method of Graessmann,[14] into 5-day-old cells from primary heart muscle cultures. After 2 to 3 hr the cells were prepared for double immunofluorescence staining. All cells stain for myomesin (b); only injected cells are positive for MM-CK (c) and contain some MM-CK associated with the myofibrils (arrowhead). (d to f) About 10 pℓ of poly (A$^+$) RNA (1 mg/mℓ) from breast muscle, enriched for M-CK mRNA on a sucrose density gradient, was microinjected into the cells which were subsequently processed by the double immunofluorescence technique. All cells stain for myomesin (e), while only injected cells are positive for translated MM-CK (f). Arrows point to MM-CK associated with myofibrils. N — noninjected cells; a, d — phase contrast (bar 20 μm); b, e — incubated with monoclonal antibody B4 (anti-myomesin); c, f — incubated with rabbit antichicken MM-CK.

REFERENCES

1. **Eppenberger, H. M., Dawson, D. M., and Kaplan, N. O.,** The comparative enzymology of creatine kinases. I. Isolation and characterization from chicken and rabbit tissues, *J. Biol. Chem.,* 242, 204, 1967.
2. **Perriard, J. C., Caravatti, M., Perriard, E. R., and Eppenberger, H. M.,** Quantitation of creatine kinase isoenzyme transitions in differentiating chick embryonic breast muscle and myogenic cell cultures by immunoadsorption, *Arch. Biochem. Biophys.,* 191, 90, 1978.

3. **Perriard, J. C.,** Developmental regulation of creatine kinase isoenzymes in myogenic cell cultures from chicken, *J. Biol. Chem.,* 254, 7036, 1979.

4. **Turner, D. C., Wallimann, T., and Eppenberger, H. M.,** A protein that binds specifically to the M-line of skeletal muscle is identified as the muscle form of creatine kinase, *Proc. Natl. Acad. Sci.,* 70, 702, 1973.

5. **Wallimann, T., Kuhn, H. J., Pelloni, G., Turner, D. C., and Eppenberger, H. M.,** Localization of creatine kinase isoenzymes in myofibrils. II. Chicken heart muscle, *J. Cell Biol.,* 75, 318, 1977.

6. **Grove, B. K., Kurer, V., Lehner, C., Doetschmann, T. C., Perriard, J. C., and Eppenberger, H. M.,** A new 185,000 dalton skeletal muscle protein detected by monoclonal antibodies, *J. Cell Biol.,* 98, 518, 1984.

7. **Sanger, J. W., Mittal, B., and Sanger, J. M.,** Analysis of myofibrillar structure and assembly using fluorescently labeled contractile proteins, *J. Cell Biol.,* 98, 825, 1984.

8. **Kreis, T. E. and Birchmeier, W.,** Microinjection of fluorescently labeled proteins into living cells with emphasis on cytoskeletal proteins, *Int. Rev. Cytol.,* 75, 209, 1982.

9. **Franke, W. E., Schmid, E., Mitnacht, S., Grund, Ch., and Jorcumo, J. L.,** Integration of different keratins into the same filament system after microinjection of mRNA for epidermal keratins into kidney epithelial cells, *Cell,* 36, 813, 1984.

10. **Stacey, D. W. and Allfrey, V. G.,** Microinjection studies of chick globin mRNA translation in human and avian cells, *Cell,* 9, 725, 1976.

11. **Schafer, B., Perriard, J. C., and Eppenberger, H. M.,** Appearance of M-band attached MM-creatine kinase in differentiating chicken heart cells after injection of M-type isoprotein or poly (A^+)-RNA enriched for M-type creatine kinase message, *Basic Res. Cardiol.,* Suppl. 2, 107, 1985.

12. **Bogenmann, E. and Eppenberger, H. M.,** DNA-synthesis and polyploidization of chicken heart muscle cells in mass cultures, *J. Mol. Cell Cardiol.,* 12, 17, 1980.

13. **Clark, W. A. and Fischman, D. A.,** Analysis of population cytokinetics of chick myocardial cells in tissue culture, *Dev. Biol.,* 97, 1, 1983.

14. **Graessmann, H. and Graessmann, A.,** Microinjection of tissue culture cell, in *Methods in Enzymology,* Vol. 101, Colowick, S. P. and Kaplan, H. O., Eds., Academic Press, New York, 1983, 482.

15. **Glacy, S. D.,** Pattern and time course of rhodamine-action incorporation in cardiac myocytes, *J. Cell. Biol.,* 95, 1164, 1983.

16. **Grove, B. K., Cerny, L., Perriard, J. C., and Eppenberger, H. M.,** Myomesin and *M*-protein: expression of two M-band proteins in pectoral muscle and heart during development, *J. Cell Biol.,* 101, 1413, 1985.

Chapter 25

INTRACELLULAR OXYGEN DELIVERY IN HEART CELLS

Beatrice A. Wittenberg

TABLE OF CONTENTS

I. PREPARATION AND PROPERTIES OF FRESHLY ISOLATED ADULT HEART CELLS

The preparation of functionally intact, freshly isolated adult heart cells is a very recent technical achievement. Here we will limit the discussion to cardiac myocytes freshly isolated from the mature rat. The early literature has been reviewed,[1-3] and some representative preparations[4-11] illustrate both the diversity and conserved aspects of successful methods. Although the disruption of the heart into its constituent myocytes is prerequisite for both cultured and freshly isolated heart cell preparations, preparation of heart cells in culture permits a 12- to 24-hr "healing" or "recovery" period after the disruption of the heart tissue into isolated cells, while freshly isolated cells must be fully recovered within several hours of tissue dissociation.

To free individual cells from the tissue matrix it is necessary to remove not only extracellular collagen networks, blood capillaries, and neural connections, but also cell-to-cell connections including both gap junctions and desmosome structures. Histological evidence[12,13] demonstrates that the desmosomes are cleaved between separated cells, but both halves of a gap junction are retained by one member of a pair of previously connected cells under the conditions currently used to prepare functionally intact cells. The removal of the gap junctional structure, with some adhering nonjunctional membrane from one of a pair of previously connected cells, suggests that small "holes" or "gaps" are torn in the membrane of freshly dissociated cells. If each cell is connected to at least four other cells in the tissue, then the chance of any one cell being victorious in tearing all the gap junctions from all its neighbors is slight. A major emphasis of techniques for the preparation of representative populations of freshly isolated adult heart cells has been to minimize sarcolemmal damage and optimize sarcolemmal healing. Since the yield of intact myocytes from rat heart in our experiments is 30 to 50% of the myocytes of the ventricle, based on recovery of both protein and intracellular myoglobin, these cell membranes must be healed before there is significant loss of intracellular proteins or excessive ingress of extracellular constituents, particularly calcium.

The extraordinary sensitivity of freshly isolated adult cardiac myocytes to calcium overload, characterized by loss of myofilament alignment, structural integrity, and effective synchronous contractile activity, was a serious obstacle in the development of preparations of adult cardiac myocytes. Current methods[1-11] generally use "calcium tolerance", the ability to maintain intracellular calcium homeostasis[14] in the presence of physiological extracellular calcium, as the first criterion of the viability of freshly isolated cardiac cells. Although it is generally assumed that calcium overload in this preparation is a result of changed sarcolemmal properties, the relative contribution of increased sarcolemmal permeability and/or decreased activity of sarcolemmal Ca ATPase[15] has not been established.

The freshly isolated adult heart cell model exploits the advantages of minimal functional structural and biochemical changes after the transition from intact tissue to intact cell suspension. Sarcolemmal integrity is a basic prerequisite for functionally intact freshly isolated myocytes. Other crucial functional criteria which must be met include structural, functional, and biochemical integrity. Functionally intact, individual, freshly isolated adult heart cells retain myofilament alignment and sarcomere organization, show no spontaneous contractile activity in the presence of extracellular calcium, show specific synchronous contraction in response to electrical stimulation (3 msec duration, 3 to 12 V strength),[16] and maintain normal resting potential and action potential (see, for example, References 4, 7, 10, 11, 17, and 18). Pairs of isolated cells show gap junctional conductance which may be measured unambiguously.[18,19] Biochemical parameters which have been established in a population of isolated heart cells include the maintenance of: (1) high energy phosphate levels both at rest[6,8,20-22] and in the face of continuous ATP breakdown,[22,23] (2) physiologically low intracellular sodium levels,[8,9,24] (3) substrate utilization,[25] and (4) many others. Freshly isolated adult heart cells can now be maintained for about 24 to 48 hr without detectable change in calcium tolerance, structural integrity, sarcomere alignment or electrophysiological properties,[18] or high energy phosphate levels.[21]

II. OXYGEN SUPPLY

Oxygen delivery to the intracellular mitochondria of cardiac myocytes is of vital functional importance to the continuously beating heart which is largely dependent on aerobic energy supplied by oxidative phosphorylation. The oxygen requirement for maximal oxidative phosphorylation by isolated mitochondria is satisfied at extremely low oxygen pressure — approximately 0.1 to 0.2 torr. Nonetheless, the venous oxygen pressure is maintained at about 20 torr, suggesting a very large oxygen pressure gradient from capillary blood to intracellular mitochondria of about 20 torr. This has led some workers to suggest *intracellular* oxygen gradients from the inner mitochondrial membrane to the sarcoplasm,[26,27] while others have implicated *extracellular* diffusion barriers.[28-30] The preparation of freshly prepared isolated adult heart myocytes offers unique advantages for the study of intracellular oxygen delivery:

1. Modulations of oxygen supply by vascular changes are absent. The oxygen pressure of the medium surrounding the myocytes may be accurately defined and determined.
2. The intracellular architecture and compartmentation is not disrupted. Intracellular myoglobin and cytochrome oxidase content are comparable to that of the intact heart.[22]
3. The heart cells can be studied both at rest and during continuous stimulation.
4. Several endogenous oxygen indicators are present in specifically defined intracellular compartments and may be used to establish the volume average oxygen pressure sensed in these compartments during defined steady states of extracellular oxygen pressure, rate of intracellular respiration, and high energy phosphate production.

There have been numerous studies on the response of freshly isolated adult rat cardiac myocytes to oxygen deprivation. The response resembles that of cells in the intact heart to high-flow hypoxia in many respects: lactate accumulation is increased, respiratory rate is decreased, oxidative phosphorylation is decreased, and contractility ceases.[22,31-35] The creatine phosphokinase system provides a "spatial and temporal buffer" of high energy phosphate levels,[36] but when the rate of utilization of ATP exceeds the rate of ATP production in the absence of oxygen, adenine nucleotide breakdown leads to loss of the intracellular adenine as adenosine and inosine, which may be recovered in the extracellular medium.[37] In contrast to cells in the beating heart, however, resting heart cells can maintain viability and ionic gradients, albeit with decreased high energy phosphate levels, for several hours in the absence of oxygen.

III. INTRACELLULAR OXYGEN GRADIENTS

As a result of the foregoing it has become possible to address the question: are there significant oxygen pressure differences in different intracellular compartments of respiringcardiac myocytes?

Monoamine oxidase,[38] a flavoprotein enzyme located at the outer mitochondrial membrane, was used to monitor the intracellular oxygen pressure at this site. The sensitivity of this indicator is greatest at about 7 torr ($\sim$9 μM O_2), the oxygen pressure at which the rate of substrate oxidation by the enzyme is half-maximal. The oxygenation of intracellular myoglobin was used to assay the local chemical activity of dissolved oxygen in the sarcoplasmic compartment in the oxygen pressure range from 0.2 to 3 torr.[22,39] Oxygen-sensitive indicators at the inner mitochondrial membrane, including fractional oxidation of cytochrome

Table 1

A COMPARISON OF OXYGEN PRESSURE IN TWO DIFFERENT INTRACELLULAR COMPARTMENTS OF ISOLATED MYOCYTES AS FUNCTIONS OF EXTRACELLULAR OXYGEN PRESSURE AND RESPIRATORY RATE[a]

| | Extracellular pO_2 (torr) | | | | |
| | Basal | | Stimulated | | |
Treatment	0.1 Control	0.83 Control	0.83 2 μM CCCP	1.3 2 μM CCCP	3.3 2 μM CCCP
Respiratory oxygen uptake, μmol O_2/min/10^6 cells	0.01	0.02	0.04	0.05	0.07
Sarcoplasmic pO_2 reported by myoglobin (torr)	—	0.8	0.2	0.3	1.3
pO_2 at inner mitochondrial membrane reported by half-maximal respiration and cytochrome oxidase oxidation (torr)[a]	0.05	—	0.05	—	—
Oxygen gradient between extracellular medium and inner mitochondrial membrane (torr)	0.05	—	0.8	—	—
Oxygen gradient between extracellular medium and sarcoplasm (torr)	—	0.03	0.6	1.0	2.0

[a]   Oxygen pressure at the inner mitochondrial membrane is estimated only at that pressure where cytochrome oxidase is half oxidized. Under this condition respiratory oxygen uptake is also half maximal.

Data from Wittenberg, B. A. and Wittenberg, J. B., Oxygen pressure gradients in isolated cardiac myocytes, *J. Biol. Chem.*, 260, 6548, 1985.

oxidase, fractional oxidation of cytochrome c, the rate of respiratory oxygen uptake, and the rate of lactate accumulation, reflect balances between the availability of electrons flowing in the mitochondrial electron transport chain and the availability of oxygen reacting with cytochrome oxidase. These balances are sensitive to the availability of oxygen at the inner mitochondrial membrane when electron flow is unchanging, with the half-maximal response to oxygen near to 0.05 torr in isolated mitochondria.[40] The steady-state extracellular oxygen pressure and steady-state respiratory rate were monitored with an oxygen-sensitive polarographic electrode simultaneously with the spectrophotometric measurements of intracellular oxygen-sensitive changes. The results are summarized in Table 1.

Well-stirred suspensions of cardiac myocytes respiring under basal conditions showed only barely detectable oxygen pressure differences between the extracellular medium and volume-average intracellular medium and volume-average intracellular compartments. At extracellular oxygen pressures near 7 torr the local chemical activity of dissolved oxygen at the outer mitochondrial membrane was 5 to 7 torr, not more than 2 torr ($\sim$3 μM O_2) lower than the extracellular oxygen pressure at 95% confidence limits.[38] When the intracellular volume-average myoglobin-reported oxygen pressure was 0.2 to 3 torr, the extracellular pO_2 was nearly the same. Table 1 shows that at 0.1 and 0.83 torr, oxygen gradients between the extracellular medium and the intracellular oxygen sensors are barely detectable, being about 0.03 to 0.05 torr.

Respiration of cell suspensions was stimulated 3.5-fold by the addition of the uncoupler of mitochondrial phosphorylation, CCCP. At the concentration of CCCP chosen the oxygen consumption of the cells is comparable to that of the beating rat heart performing work under moderate load, and a new stable and readily reversible steady state is established in which the level of high energy phosphates is about half that of aerobic resting cells. Cell suspensions with respiration stimulated by CCCP required significantly higher extracellular oxygen pressure to maintain intracellular pO_2 values between 0.2 and 1.3 torr (see Table 1).The difference between extracellular oxygen pressure and myoglobin-reported intracellular sarcoplasmic oxygen pressure in the stimulated cells was about 2 torr at maximal respiratory rates and at 30°C. The magnitude of the oxygen pressure difference between the extracellular medium and the intracellular myoglobin-containing compartment was proportional to the respiratory rate.[22] In the presence of 2 μM CCCP, half-maximal respiration and maximal sensitivity of the oxygen sensors at the inner mitochondrial membrane both occur at a volume-average myoglobin-sensed pO_2 of 0.2 torr and at an extracellular pO_2 of 0.83 torr. It is apparent (Table 1) that most of the oxygen pressure gradient from the medium to the inner mitochondrial membrane (0.8 torr) occurred between the extracellular medium and the sarcoplasm (0.6 torr), while the difference of oxygen pressure between the sarcoplasm and the inner mitochondrial membrane was only 0.2 torr. These data suggest the presence of an oxygen diffusion barrier such as an unstirred layer around the cell boundary.

It was concluded that gradients of oxygen pressure within the cardiac myocyte are very shallow, with an upper limit of 2 torr. This suggests that most of the large differences in oxygen pressure, about 20 torr, between the capillary lumen and mitochondria of working heart must be extracellular.

IV. MYOGLOBIN AND OXYGEN DELIVERY

It is relevant to conclude with a discussion of the role of myoglobin in oxygen delivery, having treated myoglobin as a useful intracellular oxygen sensor. Myoglobin occurs, also, in aerobic skeletal muscle and in the root nodules of leguminous plants where its presence is required for nitrogen fixation.[41,42] It is tempting to speculate that myoglobin functions to maintain oxygen supply for mitochondrial oxidative phosphorylation when the oxygen supply becomes limiting.[40] In skeletal muscle several experiments have suggested such a role.[43,44] The mechanism of such an effect has not been demonstrated but has been the subject of much theoretical modeling[45] and speculation. Mechanisms such as long-term oxygen store, short-term oxygen store, temporal and spatial buffering,[35] and facilitated diffusion[46] of oxygen have all been proposed. The experiments of Wittenberg and Wittenberg[22] show that oxygen availability at the inner mitochondrial membrane does not become limiting until the intracellular myoglobin becomes largely deoxygenated. Under the conditions of these experiments at steady states of oxygen pressure, any oxygen storage function of intracellular myoglobin disappears. The suggestion that myoglobin functions by delivering oxygen at the optimal oxygen pressure and optimal rate for efficient oxidative phosphorylation is consistent with experimental findings in heart, skeletal muscle, and root nodules and is a useful working hypothesis.

ACKNOWLEDGMENTS

This work was supported in part by USPHS grant NIH HL 19299 and a grant in aid from the New York Heart Association.

REFERENCES

1. **Wittenberg, B. A. and Robinson, T. F.,** Oxygen requirements, morphyology, cell coat and membrane permeability of calcium-tolerant myocytes from hearts of adult rats, *Cell Tissue Res.,* 216, 231, 1981.
2. **Dow, J. W., Harding, N. G., and Powell, T.,** Isolated cardiac myocytes. I. Preparation of adult myocytes and their homology with intact tissue, *Cardiovasc. Res.,* 15, 483, 1981.
3. **Farmer, B. B., Mancina, M., Williams, E. S., and Watanabe, A. M.,** Isolation of calcium tolerant myocytes from adult rat hearts: review of the literature and description of a method, *Life Sci.,* 33, 1, 1983.
4. **Powell, T., Terrar, D. A., and Twist, V. W.,** Electrical properties of individual cells isolated from adult rat ventricular myocardium, *J. Physiol.,* 302, 131, 1980.
5. **Haworth, R. A., Hunter, D. R., and Berkoff, H. A.,** The isolation of Ca^{++} tolerant myocytes from the adult rat, *J. Mol. Cell. Cardiol.,* 12, 715, 1980.
6. **Kao, R. L., Christman, E. W., Luh, S. L., Krauhs, J. M., Tyers, G. F. O., and Williams, E. H.,** The effects of insulin and anoxia on the metabolism of isolated mature rat cardiac myocytes, *Arch. Biochem. Biophys.,* 203, 587, 1980.
7. **Isenberg, G. and Klockner, U.,** Calcium tolerant ventricular myocytes prepared by preincubation in a "KB medium", *Pflüger's Arch.,* 395, 6, 1982.
8. **Cheung, J. Y., Thompson, I. G., and Bonventre, J. V.,** Effects of extracellular calcium removal and anoxia on isolated rat myocytes, *Am. J. Physiol.,* 243, C184, 1982.
9. **Hohl, C. M., Altschuld, R. A., and Brierley, G. P.,** Effects of calcium on the permeability of isolated adult rat heart cells to sodium, *Arch. Biochem. Biophys.,* 221, 197, 1983.
10. **Bkaily, G., Sperelakis, N., and Doane, J.,** A new method for preparation of isolated single adult myocytes, *Am. J. Physiol.,* 247, H1018, 1984.
11. **Bustamente, J. O., Watanabe, T., and McDonald, T. F.,** Nonspecific proteases: a new approach to the isolation of adult cardiocytes, *An. J. Physiol. Pharmacol.,* 60, 997, 1981.
12. **Slade, A. M., Severs, N. J., Powell, T., and Twist, V. W.,** Isolated calcium-tolerant myocytes and the calcium paradox: an ultrastructural comparison, *Eur. Heart J.,* 4 (Suppl. H), 113, 1983.
13. **Mazet, F., Wittenberg, B. A., and Spray, D. C.,** Fate of intercellular junctions in isolated adult rat cardiac cells, *Circ. Res.,* 56, 195, 1980.
14. **Caroni, P. and Carafoli, E.,** The Ca^{2+} pumping ATPase of heart sarcolemma, *J. Biol. Chem.,* 256, 3263, 1981.
15. **Williamson, J. R., Williams, R. J., Coll, K. E., and Thomas, A. P.,** Cytosolic free Ca^{2+} concentration and intracellular calcium distribution of Ca^{2+}-tolerant isolated heart cells, *J. Biol. Chem.,* 258, 13411, 1983.
16. **Krueger, J. W., Forletti, D., and Wittenberg, B. A.,** Uniform sarcomere shortening behavior in isolated cardiac muscle cells, *J. Gen. Physiol.,* 76, 587, 1980.
17. **Trautwein, W., Taniguchi, J., and Noma, A.,** The effect of intracellular cyclic nucleotides and calcium on the action potential and acetylcholine response of isolated cardiac cells, *Pflüger's Arch.,* 392, 307, 1982.
18. **White, R. L., Spray, D. C., Campos de Carvalho, A. C., Wittenberg, B. A., and Bennett, M. V. L.,** Some electrical and pharmacological properties of gap junctions between adult ventricular myocytes, *Am. J. Physiol.,* 249, C447, 1985.
19. **Kameyama, M.,** Electrical coupling between ventricular paired cells isolated from guinea-pig heart, *J. Physiol.,* 336, 345, 1983.
20. **Walker, E. J., Burns, J. H., and Dow, J. W.,** Amino acid transport and protein synthesis in energetically stable calcium-tolerant isolated cardiac myocytes, *Biochem. Biophys. Acta,* 721, 280, 1982.
21. **Piper, H. M., Probost, I., Schwartz, P., Hutter, F. J., and Spieckermann, P. G.,** Culturing of calcium stable adult cardiac myocytes, *J. Mol. Cell. Cardiol.,* 14, 397, 1982.
22. **Wittenberg, B. A. and Wittenberg, J. B.,** Oxygen pressure gradients in isolated cardiac myocytes, *J. Biol. Chem.,* 260, 6548, 1985.
23. **Haworth, R. A., Hunter, D. R., Berkoff, H. A., and Moss, R. L.,** Metabolic cost of stimulated beating of isolated adult rat heart cells in suspension, *Circ. Res.,* 52, 342, 1983.
24. **Wittenberg, B. A. and Gupta, R. K.,** NMR studies of intracellular sodium ions in mammalian cardiac myocytes, *J. Biol. Chem.,* 260, 2031, 1985.
25. **Burns, A. H. and Reddy, W. J.,** Amino acid stimulation of oxygen and substrate utilisation by cardiac myocytes, *Am. J. Physiol.,* 235, E461, 1978.
26. **Tamura, M., Oshino, N., Chance, B., and Silver, I. A.,** Optical measurements of intracellular oxygen concentration of rat heart in vitro, *Arch. Biochem. Biophys.,* 191, 8, 1978.
27. **Jones, D. P. and Kennedy, F. G.,** Intracellular O_2 gradients in cardiac myocytes. Lack of a role for myoglobin in facilitation of intracellular O_2 diffusion, *Biochem. Biophys. Res. Commun.,* 105, 419, 1982.
28. **Sies, H.,** Oxygen gradients during hypoxic steady states in liver. Urate oxidase and cytochrome oxidase as intracellular O_2 indicators, *Hoppe-Seyler's Z. Physiol. Chem.,* 358, 1021, 1977.

29. **Steenbergen, C., DeLeeuw, G., Barlow, C., Chance, B., and Williamson, J. R.,** Heterogeneity of the hypoxic state in perfused rat heart, *Circ. Res.,* 41, 606, 1977.
30. **Honig, C. R., Gayeski, T. E. J., Federspiel, W., Clark, A., and Clark, P.,** Muscle O_2 gradients from hemoglobin to cytochrome: new concepts new complexities, *Adv. Exp. Med. Biol.,* 169, 23, 1984.
31. **Altschuld, R. A., Hostetler, J. R., and Brierly, G. P.,** Response of isolated rat heart cells to hypoxia, re-oxygenation, and acidosis, *Circ. Res.,* 49, 307, 1981.
32. **Haworth, R. A., Hunter, D. R., and Berkoff, H. A.,** Contracture in isolated adult rat heart cells. Role of Ca^{2+}, ATP and compartmentation, *Circ. Res.,* 49, 1119, 1981.
33. **Koke, J. R., Wills, M. A., and Bittar, N.,** Contractile activity and oxidative metabolism in single adult myocytes during normoxia and hypoxia, *Cytobios,* 35, 149, 1982.
34. **McDonough, K. H. and Spitzer, J. J.,** Effects of hypoxia and reoxygenation on adult rat heart cell metabolism, *Proc. Soc. Exp. Biol. Med.,* 173, 519, 1983.
35. **Piper, H. M., Schwartz, P., Spahr, R., Hutter, J. F., and Spieckermann, P. G.,** Absence of reoxygenation damage in isolated heart cells after anoxic injury, *Pflüger's Arch.,* 401, 71, 1984.
36. **Meyer, R. A., Sweeney, H. L., and Kushmerick, M. J.,** A simple analysis of the "phosphocreatine shuttle", *Am. J. Physiol.,* 246, C365, 1984.
37. **Geisbuhler, T., Altschuld, R. A., Trewyn, R. W., Ansel, A. Z., Lamka, K., and Brierley, G. P.,** Adenine nucleotide metabolism and compartmentalization in isolated adult rat heart cells, *Circ. Res.,* 54, 536, 1984.
38. **Katz, I. R., Wittenberg, J. B., and Wittenberg, B. A.,** Monoamine oxidase, an intracellular probe of oxygen pressure in isolated cardiac myocytes, *J. Biol. Chem.,* 259, 7504, 1984.
39. **Wittenberg, B. A., Wittenberg, J. B., and Katz, I. R.,** Oxygen transport in isolated cardiac myocytes, *Basic Res. Cardiol.,* Suppl. 2, 75, 1985.
40. **Cole, R. P., Sukanek, P. C., Wittenberg, J. B., and Wittenberg, B. A.,** Mitochondrial function in the presence of myoglobin, *J. Appl. Physiol.,* 53, 1116, 1982.
41. **Wittenberg, J. B.,** Utilization of leghemoglobin-bound oxygen by *Rhizobium* bacteroids, in *Nitrogen Fixation II,* Orme-Johnson, W. H. and Newton, W. E., Eds., University Park Press, Baltimore, 1980, 53.
42. **Appleby, C. A.,** Leghemoglobin and *Rhizobium* respiration, *Annu. Rev. Physiol.,* 35, 443, 1984.
43. **Wittenberg, B. A., Wittenberg, J. B., and Caldwell, P. R. B.,** Role of myoglobin in the oxygen supply to red skeletal muscle, *J. Biol. Chem.,* 250, 9038, 1975.
44. **Cole, R. P.,** Skeletal muscle function in hypoxia: effect of alteration of intracellular myoglobin, *Respir. Physiol.,* 53, 1, 1983.
45. **Jacquez, J. A.,** The physiological role of myoglobin: more than a problem in reaction-diffusion kinetics, *Math. Biosci.,* 68, 57, 1984.
46. **Wittenberg, J. B. and Wittenberg, B. A.,** Facilitated oxygen diffusion by oxygen carriers, in *Oxygen and Living Processes: An Interdisciplinary Approach,* Gilbert, D. L., Ed., Springer-Verlag, New York, 1981, 177.

Chapter 26

COENZYME Q_{10} IN CULTURED HEART CELLS

Noboru Suzuki, Tetsuya Nakamura, Hideyuki Ishida, and Kiyoshi Hosono

TABLE OF CONTENTS

I. COENZYME Q_{10}

Coenzyme Q_{10} (CoQ_{10}), one of the coenzymes occurring in living cells, was isolated by Crane et al. from heart muscle in 1957.[1] CoQ_{10}, a lipid-soluble material, is located in the inner mitochondrial membrane and plays an important role in the electron transport chain and, thereby, in ATP generation in the cell. CoQ_{10} accepts electrons from NADH or FADH and transfers them to cytochrome. In the final step, four electrons are conveyed per oxygen molecule, producing two molecules of water and generating ATP.

The clinical administration of CoQ_{10} in diseases thought to be brought about by mitochondrial dysfunction has been particularly well developed in Japan. Folkers et al.[2] have suggested that in some heart diseases CoQ_{10} levels are reduced in cardiac muscle mitochondria. It has already been shown that exogenous CoQ_{10} is incorporated into the mitochondrial inner membrane.[3] In this minireview of the physiological significance of CoQ_{10}, as a supplier of CoQ_{10} to the endogenous CoQ pool in mitochondria, a membrane stabilizer, and an antioxidant, will be considered.[4] The four proceedings[5-8] of international symposia on CoQ_{10} held in 1976, 1979, 1981, and 1983, respectively, serve as useful aids to understanding the recent progress in the biomedical and clinical aspects of CoQ. The experiments, using a novel technique to be described below, have confirmed previous findings on the role of exogenous CoQ_{10}.

II. HEART CELL CULTURE AND EFFECT OF COENZYME Q_{10}

Recently, great progress has been made in techniques of heart cell culture, stimulating studies on the structure and function of myocardium at a cellular level,[9-12] in particular, electrophysiological studies on the mechanism of myocardial beating, since the action potentials in these cells can be monitored in the absence of nervous or hormonal stimulation.[13,14] Recent studies by Goshima[9] and Yoneda and Toyama[15] were designed to clarify the mechanisms involved in the development of arrhythmia in cultured myocardial cells.

Figure 1 shows the isolation and culture methods for mouse embryonic heart muscle. Mouse embryos were used for making cultures because embryonic heart cells maintained more regular beating than adult ones. After trypsinization of heart muscle, cells were collected and then cultured in an atmosphere of 5% CO_2 and 95% air at 37°C in Ham F-12 medium supplemented with 10% fetal calf serum. Cultures consisted either of single cells with no contact with one another, or of sheets of cells (monolayers), depending on the cell concentration.

After 48 hr of culture in an atmosphere of 5% CO_2 and 95% air, the myocardial cells began to beat and this was monitored by means of a television set connected to an inverted phase microscope. Changes in light transmission caused by contractions and relaxations of myocardial cells were recorded using a phototransistor. The interval between contractions was compared to the mean contraction interval for a period of 2 or 3 min. Changes of greater than $\pm 20\%$ were defined as irregular beating and expressed either as beats per minute or as a percentage of the control level. The ratios of regularly to irregularly beating cells observed are shown in Table 1. Single cell cultures contained less beating cells than cultures in monolayers of myocardial cells. About 50% of the cells were beating irregularly. The percentage of cells with irregular beating was much higher than in monolayer cultures.

CoQ_{10} solution was added to the medium in these different types of culture and the ensuing changes in beating followed.[16] On addition of vehicle alone the mean beating rate changed to $83 \pm 3\%$ after 5 min and to $103 \pm 7\%$ after 60 min (mean $\pm$ SE) of the original level, while 5 min after the administration of 50, 100, 200, and 400 $\mu g/m\ell$ of CoQ_{10} it increased to 131 ± 18, 129 ± 13, 144 ± 22, and $161 \pm 7\%$, respectively. Sixty minutes after the addition of CoQ_{10}, the beating rate reached about 150% of the control levels throughout the entire range of CoQ_{10} concentrations. The difference in the beating rate between the vehicle-

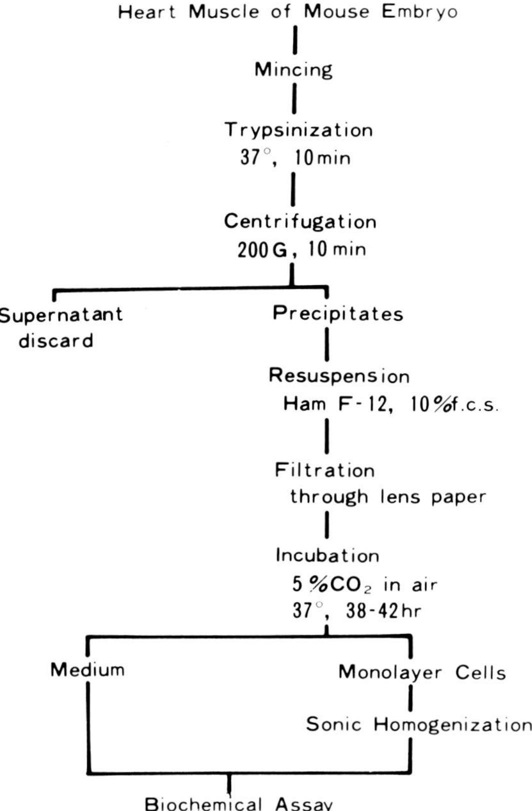

FIGURE 1.   Flow sheet of the preparation of mouse embryonic myocyte cultures. (From Hosono, K., Ishida, H., Okuda, M., Suzuki, N., and Nakamura, T., *Biomedical and Clinical Aspects of Coenzyme Q*, Vol. 3, Folkers, K. and Yamamura, Y., Eds., Elsevier/North-Holland, Amsterdam, 1981, 269. With permission.)

and CoQ_{10}-treated myocardial cells was significant ($p < 0.01$). Furthermore, the frequency of irregular beating was not changed by the vehicle alone after 60 min. However, 5 min after the addition of CoQ_{10}, the irregular beating frequency decreased to 79 ± 7, 59 ± 9, and 67 ± 10% of the original frequency at concentrations of 100, 200, and 400 μg/mℓ, respectively. The frequency of irregular beating after 60 min reached 60% of the control levels. At these concentrations, the decreased frequency of irregular beating was statistically significant ($p < 0.02$).

III. THE EFFECTS OF COENZYME Q_{10} ON A CELL CULTURE MODEL OF ISCHEMIC INJURY

The ischemic injury model was produced by preliminary incubation of the cells under the normal conditions prior to placing them in an atmosphere of 5% CO_2 and 95% N_2 gas for 60 min. Before hypoxic treatment, the beating rate of myocardial cells was 62.8 ± 22.6 beats/minute and irregular beating was 2.3 ± 1.3 beats/minute. Sixty minutes after the induction of hypoxic conditions, the beating rate in the myocardial cells was reduced to 46.6 ± 22.1 beats/minute and the irregular beating frequency increased to 13.7 ± 4.1 beats/minute (Figure 2). After the addition of CoQ_{10} to the medium, there was no change in the

Table 1
COMPARISON OF BEATING CHARACTERISTICS OF CULTURED SINGLE MYOCARDIAL CELLS AND MONOLAYER CELLS

	Myocardial cells (% of the total population)	Beating cells (% of total cardiac cells)	Irregular beating cells (% of the total beating cells)
Single cell (n = 16)	82 ± 3.1	67 ± 1.8	49 ± 2.9
Monolayer cells (n = 7)	81 ± 1.4	87 ± 2.3	1 ± 0.4

From Hosono, K., Ishida, H., Okuda, M., Suzuki, N., and Nakamura, T., *Biomedical and Clinical Aspects of Coenzyme Q*, Vol. 3, Folkers, K. and Yamamura, Y., Eds., Elsevier/North-Holland, Amsterdam, 1981, 269. With permission.

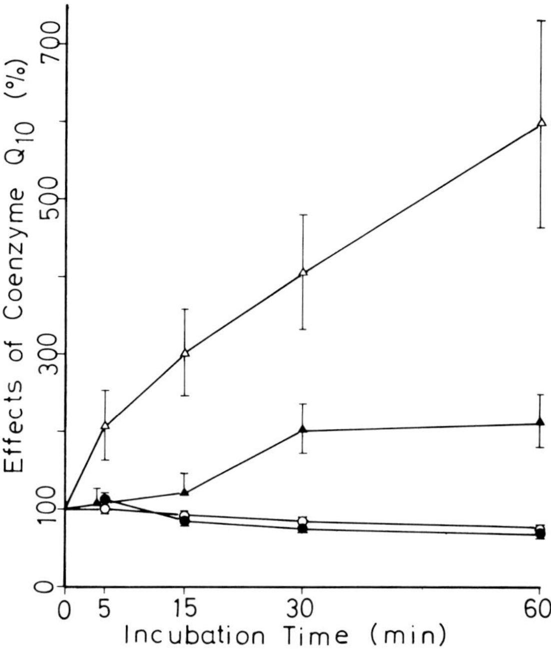

FIGURE 2.   Effect of coenzyme Q_{10} on the beating rate and irregular beating frequency of the single myocardial cells under the hypoxic conditions. —○—, beating rate during hypoxia; —●—beating rate after the addition of CoQ_{10} (200 μg/mℓ) during hypoxia; —△—, irregular beating frequency during hypoxia; —▲—, irregular beating frequency after the addition of the CoQ_{10} during the hypoxia. The values represent the mean ± SE of 12 experiments. (From Suzuki, N. et al., *Chem. Pharm. Bull.*,33, 2896, 1985. With permission.)

beating rate as compared with control values, but the rate of irregular beating dramatically decreased (Figure 2). Sixty minutes after the addition of CoQ_{10} the irregular beating frequency was about 200% of control.

cAMP concentrations in the myocardial cells were measured both before and after hypoxic conditions were imposed. Cardiac cAMP levels decreased in cells under hypoxic conditions during a 60-min observation period ($p < 0.01$). It is interesting to note that in the presence of CoQ_{10}, cAMP levels were increased under normal conditions but tended to decrease in cells maintained in hypoxic ones.

IV. DISTRIBUTION OF ENDOGENOUS AND EXOGENOUS COENZYME Q

The distribution of endogenous and exogenous CoQ in the cultured myocardial cells was determined. In order to measure the endogenous and exogenous CoQ in the same sample, CoQ_{10} labeled with deuterium (CoQ_{10}-d_5) was prepared (see Figure 3).[3] After incubation with 200 μg/mℓ of CoQ_{10}-d_5, the myocardial cells were washed three times with saline solution to remove unincorporated labeled material. Washed cells were scraped from the plate, centrifuged at 800 to 1000 rpm, and then resuspended by gentle homogenization using three strokes of Teflon® homogenizer. The homogenate was fractionated by continuous density gradient centrifugation at 50,000 rpm for 180 min using Hitachi 55p-2 Automatic Preparative Ultracentrifuge (RP-50 rotor). Fractions of 0.5 mℓ in volume were collected for analysis and the lipids were extracted from cach fraction. Endogenous CoQ_9, CoQ_{10}, and

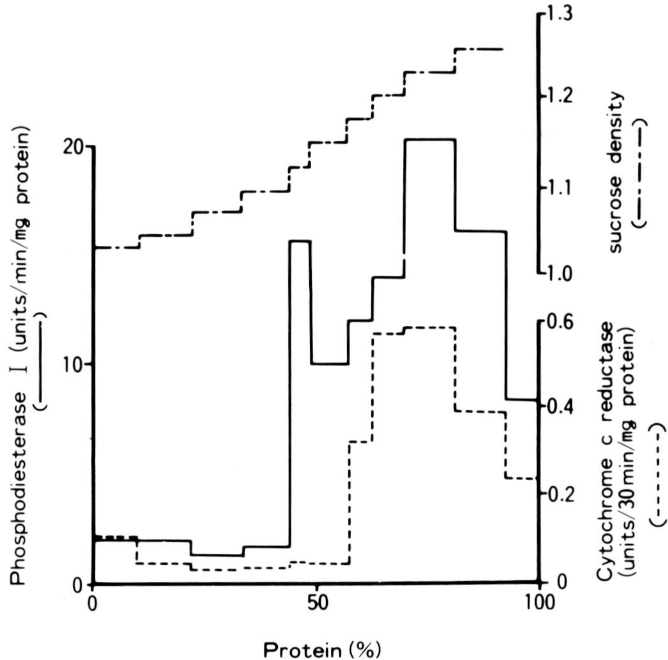

FIGURE 3.   Structure and labeled sites of CoQ_{10}-d_5.

FIGURE 4.   Concentration of phosphodiesterase I and cytochrome *c* reductase
in subcellular fractions obtained from cultured myocardial cells. The abscissa
indicates percentage of total protein. (From Suzuki, N. et al., *Chem. Pharm.
Bull.*, 33, 2896, 1985. With permission.)

exogenous CoQ_{10}-d_5 were determined simultaneously by the method described in a previous
report with minor modifications.[3] The extracts were hydrogenated in an atmosphere of
hydrogen in the presence of a platinum catalyst and then applied to a direct inlet-integrated
mass fragmentography (DI-IMF).

After fractionation of myocardial cells, protein concentration and the specific activities
of phosphodiesterase I[17] and cytochrome *c* reductase[18] were determined (Figure 4).

Phosphodiesterase I activity, a marker for plasma membrane, was found in two fractions
with specific gravities of 1.13 and 1.23. However, this enzyme has also been reported to
occur in lysosomes.[19]

Cytochrome *c* reductase, a marker enzyme for mitochondria, was found in fraction with
specific gravities ranging from 1.21 to 1.23. The specific gravity of the lysosomol fraction
was very close to that of the mitochondrial fraction, and, therefore, the fractions with specific
gravity ranging from 1.21 to 1.23 contained two marker enzymes. So, we concluded that

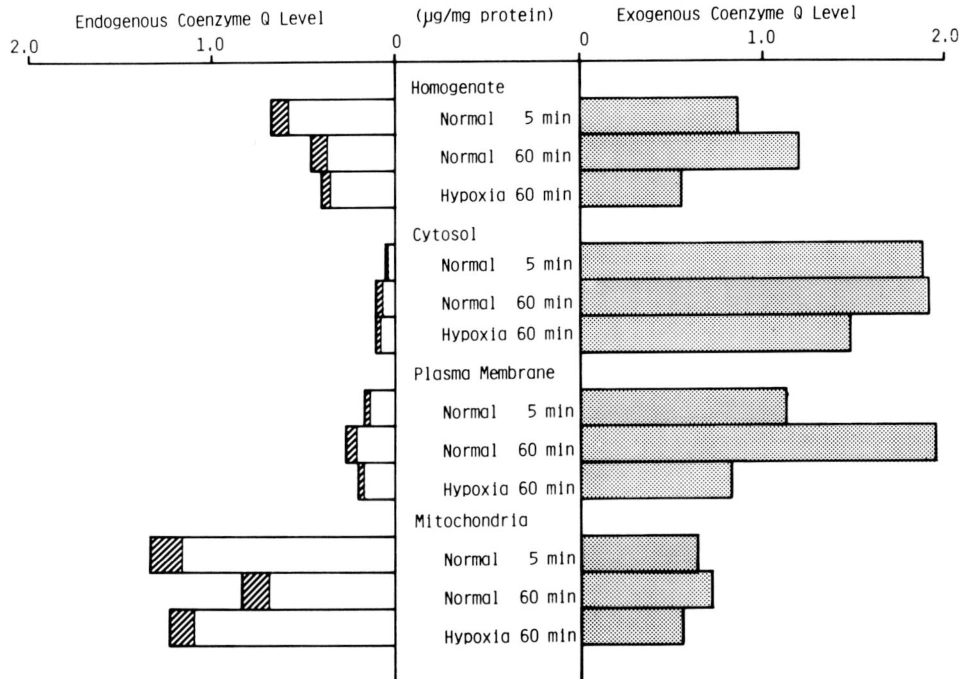

FIGURE 5.   Profile of endogenous CoQ_9, CoQ_{10}, and exogenous CoQ_{10}-d_5 concentrations in subfractions of cultured myocardial cells. □, endogenous CoQ_9; ▨, endogenous CoQ_{10}; ▦, exogenous CoQ_{10}-d_5. The value represents the mean of three experiments. (From Suzuki, N. et al., *Chem. Pharm. Bull.*, 33, 2896, 1985. With permission.)

this fraction contained both mitochondria and lysosomes. Further separation of these organelles was difficult under the experimental conditions used. The fractions with specific gravities from 1.21 to 1.23 were considered as being mitochondrial, while the fraction of specific gravity of 1.13 consisted chiefly of cell membranes.

The presence of endogenous CoQ_9 and CoQ_{10}, and exogenous CoQ_{10}-d_5 in various fractions was determined by DI-IMF (Figure 5). Endogenous CoQ_9 and CoQ_{10} were mainly found in mitochondrial fraction in both types of culture. CoQ_{10}-d_5 was incorporated in amounts ranging from 0.86 to 1.2 μg/mg protein under normal conditions, but only to the extent of 0.56 μg/ mg protein in hypoxic ones, within a 60-min time period. The apparent decrease of incorporated exogenous CoQ_{10}-d_5 indicated functional damage of the cells during hypoxia. However, these results also showed that CoQ_{10} was incorporated into the myocardial cells during hypoxia in amounts which should have been sufficient to reduce hypoxic damage. The intracellular distribution profiles of endogenous CoQ homologues were the same under both conditions — CoQ_{10}-d_5 was found in the mitochondria, plasma membrane, and cytosol. Large amounts of CoQ_{10}-d_5 were incorporated into the plasma membrane and cytosol fractions.

V. CONCLUSIONS

Cultured cells serve as a very useful model for determining the effects of drugs on cells because of the simplicity of the system.[20,21] The effects of CoQ_{10} on myocardial cells were determined in cultured heart cells. The results reported above indicated that CoQ_{10} had acted rapidly on myocardial cells, converting irregular to normal beating. Deuterium-labeled CoQ_{10} (CoQ_{10}-d_5) was synthesized to determine the concentration of incorporated exogenous CoQ_{10} and allowed simultaneous determinations of exogenous CoQ_{10} and endogenous CoQ_9 and CoQ_{10} to be made. Following incubation with CoQ_{10}-d_5, labeled material was found in the mitochondria, cytosol, and plasma membranes. Exogenous CoQ_{10} incorporation into the plasma membrane and mitochondrial fractions may explain CoQ_{10}-mediated modulation of cAMP levels and normalization of irregular beating.

REFERENCES

1. **Crane, F. L., Hatefi, Y., Lester, R. I., and Widmer, C.,** Isolation of a quinone from beef heart mitochondria, *Biochim. Biophys. Acta,* 25, 220, 1957.
2. **Folkers, K., Littarru, G. P., Ho, L., Runge, T. M., Havanonda, S., and Cooly, D.,** Evidence for deficiency of coenzyme Q_{10} in human heart disease, *Int. J. Vitam. Res.,* 40, 380, 1970.
3. **Nakamura, T., Sanma, H., Himeno, M., and Kato, K.,** Transfer of exogenous coenzyme Q_{10} to the inner membrane of heart mitochondria in rats, in *Biomedical and Clinical Aspects of Coenzyme Q,* Vol. 2, Yamamura, Y., Folkers, K., and Ito, Y., Eds., Elsevier/North-Holland, Amsterdam, 1980, 3.
4. **Nakamura, T.,** Trends in coenzyme Q research, *J. Synth. Org. Chem. (Jpn.),* 43, 1, 10, 1985.
5. **Folkers, K. and Yamamura, Y., Eds.,** *Biomedical and Clinical Aspects of Coenzyme Q,* Vol. 1, Elsevier/North-Holland, Amsterdam, 1977.
6. **Yamamura, Y., Folkers, K., and Ito, Y., Eds.,** *Biomedical and Clinical Aspects of Coenzyme Q,* Vol. 2, Elsevier/north-Holland, Amsterdam, 1980.
7. **Folkers, K. and Yamamura, Y., Eds.,** *Biomedical and Clinical Aspects of Coenzyme Q,* Vol. 3, Elsevier/North-Holland, Amsterdam, 1981.
8. **Folkers, K. and Yamamura, Y., Eds.,** *Biomedical and Clinical Aspects of Coenzyme Q,* Vol. 4, Elsevier/North-Holland, Amsterdam, 1984.
9. **Goshima, K.,** Arrhythmic movements of myocardial cells in culture and their improvement with antiarrhythmic drugs, *J. Mol. Cell. Cardiol.,* 8, 217, 1976.
10. **Norwood, C. R., Castaneda, A. R., and Norwood, W. I.,** Heterogeneity of rat cardiac cells of defined origin in single cell culture, *J. Mol. Cell. Cardiol.,* 12, 201, 1980.
11. **Ahumada, G. G., Sobel, B. E., and Needleman, P.,** Synthesis of prostaglandins by cultured rat heart myocytes and cardiac mesenchymal cells, *J. Mol. Cell. Cardiol.,* 12, 685, 1980.
12. **Hosono, K., Ishida, H., Okuda, M., Suzuki, N., and Nakamura, T.,** Protective effects of coenzyme Q_{10} against arrhythmia and its intracellular distribution: a study on the cultured single myocardial cell, in *Biomedical and Clinical Aspects of Coenzyme Q,* Vol. 3, Folkers, K. and Yamamura, Y., Eds., Elsevier/North-Holland, Amsterdam, 1981, 269.
13. **Schanne, O. F., Ruiz-Ceretti, E., Payet, M. D., and Deslauriers, Y.,** Influence of varied $(Ca^{2+})_0$ and $(Na^+)_0$ on electrical activity of clustered cardiac cells from neonatal rats, *J. Mol. Cell. Cardiol.,* 11, 477, 1979.
14. **Koidl, B. and Tritthart, H. A.,** The effects of ouabain on the electrical and mechanical activities of embryonic chick heart cells in culture, *J. Mol. Cell. Cardiol.,* 12, 663, 1980.
15. **Yoneda, S. and Toyama, S.,** Scanning electron microscopic studies on cultured heart cells after the induction of arrhythmia and fibrillation, *Cardiovasc. Res.,* 12, 201, 1978.
16. **Ohhara, H., Kanaide, H., Yoshimura, R., Okada, M., and Nakamura, M.,** A protective effect of coenzyme Q_{10} on ischemia and reperfusion of the isolated perfused rat heart, *J. Mol. Cell. Cardiol.,* 13, 65, 1981.
17. **Brightwell, R. and Tappel, A. L.,** Subcellular distributions and properties of rat liver phosphodiesterases, *Arch. Biochem. Biophys.,* 124, 325, 1968.
18. **Azzi, A., Montecucco, C., and Richter, C.,** The use of acetylated ferricytochrome c for the detection of superoxide radicals produced in biological membranes, *Biochem. Biophys. Res. Commun.,* 65, 597, 1975.
19. **Yamamoto, K., Ikehara, Y., Kawamoto, S., and Kato, K.,** Characterization of enzymes and glycoproteins in rat liver lysosomal membranes, *J. Biochem.,* 87, 237, 1980.
20. **Langer, G. A. and Nudo, L. M.,** Addition and kinetic characterization of mitochondrial calcium in myocardial tissue culture, *Am. J. Physiol.,* 239, H769, 1980.
21. **Wakabayashi, S. and Goshima, K.,** Kinetic studies of sodium-dependent calcium uptake by myocardial cells and neuroblastoma cells in culture, *Biochim. Biophys. Acta,* 642, 158, 1981.

Chapter 27

METABOLISM OF GLYCOSAMINOGLYCANS IN CULTURED HEART CELLS

T. Scott-Burden and Wieland Gevers

TABLE OF CONTENTS

I. INTRODUCTION

In considering the metabolism of glycosaminoglycans by cultured heart cells two aspects have to be defined and delineated. The first is what are glycosaminoglycans, and, secondly, what types of cells are to be considered when we think of cultured heart cells? After all, only some 30% of the cells in an adult heart are cardiomyocytes, the remainder being fibroblasts, smooth muscle cells, endothelial cells, and other elements.[1] Since the heart is a complex organ made up of many cellular entities, the analysis of the types of glycosaminoglycans found in cardiac tissue can give us only a superficial understanding of their metabolism, especially in relation to the relevance of such processes to a specific cell type. It is clear, therefore, that any discussion of the metabolism of glycosaminoglycans by heart cells in culture must clearly address the questions raised above and may not give much insight into the in vivo physiological events that take place in cardiac tissue as a whole.

II. GLYCOSAMINOGLYCANS: CHEMISTRY AND STRUCTURE

It is generally accepted today that glycosaminoglycans are found, not as free entities, but with the exception of hyaluronic acid are synthesized as proteoglycans.[2] These molecules are a special class of glycoprotein and consist of a polypeptide core to which linear carbohydrate chains (glycosaminoglycans) are attached via, in most cases, a specific tetrasaccharide of xylose-galactose-galactose-glucuronic acid, the xylose moiety being itself linked to the polypeptide via the hydroxyl group of a serine (most commonly) or a threonine residue.[2,3] To date, several separate classes of glycosaminoglycans have been identified and their differences are due to the variation in the types of hexosamine and/or hexuronic acid that make up the repeat disaccharide unit of the individual chains (Table 1).[2] With the exception of hyaluronic acid, glycosaminoglycan chains contain the sulfate residues normally associated with the hexosamine residues with the consequence that these molecules act as large polyanions with the ability to interact with other macromolecules associated with the extracellular milieu.[4,5]

Proteoglycans/glucosaminoglycans are found associated both with the truly extracellular compartment of the extracellular matrix as well as with the cell surface/pericellular compartment. They have been shown to play important roles in extracellular matrix physiology as well as acting as cell surface receptors for molecules such as lipoprotein lipase.[6]

In recent years, the concept that the extracellular matrix is an inert, three-dimensional scaffold for the attachment of cells and their organization in multicellular organisms has been modified to accommodate the fundamental role that the matrix plays in the modulation of differentiative processes leading to phenotypic expression.[7] Since glycosaminoglycans/proteoglycans are major components of such matrix material, this leads to the inevitable conclusion that their individual metabolism must *a priori* be of great importance to the overall development and maintenance of the cells associated with them.[8,9]

Before any consideration of the metabolism of glycosaminoglycans by cultured cells can be undertaken it is important to have some knowledge of the content of such macromolecules in the parent tissue itself.

In early development the heart consists of a double tubular arrangement of two layers of epithelia separated by an extensive extracellular matrix referred to as cardiac jelly, the inner layer being endocardium and the outer one, the myocardium.[10,11] The latter tissue has been shown to be largely responsible for the biosynthesis of cardiac jelly and a large component of this material has been shown to be proteoglycan/glycosaminoglycan in nature.[12,13] Markwald et al. have shown that in very early chick embryos, cardiac jelly is made up of large quantities of hyaluronic acid, the remainder being sulfated glycosaminoglycans.[14] With the onset of endocardial cushion tissue development, which is accompanied by the migration

Table 1
DISACCHARIDE COMPOSITION OF
GLYCOSAMINOGLYCANS SHOWN TO BE PRESENT IN
CARDIAC TISSUE

Galactosaminoglycans	Disaccharide repeat unit	Sulfate moiety
Chondroitin-4-sulfate	D-Glucuronic acid	*O*-Sulfate
	D-*N*-Acetyl galactosamine	
Chondroitin-6-sulfate	D-Glucuronic acid	*O*-Sulfate
	D-*N*-Acetyl galactosamine	
Dermatan sulfate	L-Iduronic/D-glucuronic	*O*-Sulfate
	L-Iduronic/D-glucuronic	
Glucosaminoglycans		
Hyaluronic acid	D-Glucuronic acid	
	D-*N*-Acetyl glucosamine	
	L-Iduronic/D-glucuronic acid	
Heparan sulfate/heparin	D-*N*-Acetyl	*O*-Sulfate
	D-*N*-Acetyl glucosamine[a]	*N*-Sulfate

[a]   The amino function of the disaccharide may be substituted with a sulfate group in place of an acetyl group.

of hypertrophying endocardial cells into the matrix (cardiac jelly), there is a decrease in the hyaluronic acid content of the matrix (± 80 to $\pm 45\%$) with a concomitant increase in the content of heparan sulfate and chondroitin-6-sulfate, 7 to 14% and 6 to 34%, respectively.[14]

Adult human cardiac tissue retains high levels of hyaluronic acid ($\pm 60\%$) with moderate levels of heparan sulfate and chondroitin sulfate detectable on whole tissue extraction (± 20 or 15%, respectively).[13] Only small amounts of the iduronic acid-containing glycosaminoglycan, dermatan sulfate, have been found to be associated with cardiac tissue.[15,16]

III. CULTURED HEART CELLS — DEFINITION OF CELL TYPES

Although cardiomyocytes have been successfully cultured in several laboratories,[17-19] studies on their glycosaminoglycan metabolism have not been undertaken. This is largely due to problems that arise during long-term culture of heart muscle cells, not the least of these being overgrowth by connective tissue cells; for a review of such problems the reader is referred to Coetzee.[20]

The cell types from hearts that have been studied with respect to metabolism of glycosaminoglycans are fibroblasts[21] and smooth muscle cells,[22] the former from 14-day chick embryos and the latter isolated from neonatal rat hearts. Some work has also been carried out on glycosaminoglycan synthesis by cushion tissue cells[23] obtained from developing chick embryos in shell-less culture.[24]

IV. BIOSYNTHESIS OF GLYCOSAMINOGLYCANS BY CULTURED CHICK EMBRYO FIBROBLASTS

Conrad et al. studied the biosynthesis of glycosaminoglycans by fibroblast-like cells from 14-day chick embryos hearts as part of a comparative study with fibroblasts isolated from other tissues of the embryo.[21] As these workers pointed out, the identification of a cell as a fibroblast is somewhat subjective and we shall refer to this identification problem again in relation to our own work with neonatal rat heart smooth muscle cells in a later discussion.[22]

The fibroblasts referred to above were established in monolayer cultures using the differential plating technique of Blondel et al.[25] The cultures were shown to be free of cardiomyocytes and other cell contaminants by this technique.

After culture for 8 days most (about 85%) of the radiolabeled sulfated macromolecules were found to be associated with the culture medium and the remainder associated in some manner with the cell layer. Similar results were obtained when [³H]-glucosamine was used as the radioactive precursor for glycosaminoglycan synthesis.

Analysis of the glycosaminoglycans associated with the two culture compartments yielded the following findings. Chondroitin-6-sulfate was the predominant type of glycosaminoglycan associated with both medium and cell layer, taking up about 40% and about 30% of the total, respectively. The culture medium contained approximately 19% glycosaminoglycans in the form of chondroitin-4-sulfate with only trace amounts (about 4%) of dermatan sulfate and about 10% of material susceptible to Streptomyces hyaluronidase digestion.[21] Cell layers contained (about 20%) chondroitin-4-sulfate in amounts similar to those found in the culture medium, low levels of dermatan sulfate (about 10%), and some hyaluronic acid (about 6%).

About 15% of the sulfated macromolecules found in both culture compartments were sensitive to nitrous acid degradation and were, thus, classified as heparan sulfate or heparin.[26] There was also material that was not susceptible to any enzymatic or chemical degradation utilized by the authors and as such could not be assigned to a specific class of glycosaminoglycan.

The fact that such low levels of hyaluronic acid were secreted by heart fibroblasts in culture could be interpreted in two possible ways. They may not represent the principal hyaluronate-synthesizing cell type in the intact tissue or they may have lost their capacity to carry out such synthesis once isolated and cultured. This latter possibility may also be a reflection of the fact that cells at high density in culture appear to synthesize less hyaluronic acid than when they are in sparse culture and proliferating rapidly.[27] Further discussion of this point will be made in the concluding paragraphs.

V. METABOLISM OF GLYCOSAMINOGLYCANS BY SMOOTH MUSCLE CELLS DERIVED FROM NEONATAL RAT HEARTS

A number of aspects of glycosaminoglycan metabolism were studied in cultured rat heart cells isolated from neonatal (1- to 3-day-old) rats.[28-30]

The cells were isolated by enzymatic disruption of cardiac tissue and plated using the differential cell plating technique referred to earlier.[25] Cells were either cloned using a single cell plating technique or micro-well cluster dishes or mass cultures were used. Both systems gave rise to cultures that contained only smooth muscle cells after one or two passages. Identification of these cells as modulated smooth muscle cells, as defined by Chamley-Campbell et al.,[31] was carried out using electron microscopy and by characterization of the connective tissue macromolecules synthesized by the cultures.[22,32]

Initial studies involved looking at the effect of ascorbic acid supplementation of cultures in relation to their glycosaminoglycan synthetic patterns.[27] These studies were prompted by our previous findings related to changes in the content of cross-linked elastin and collagen in cultures that were supplemented or deprived of vitamin C during culture. In the absence of the vitamin, the cross-linked connective tissue content of matrices was mainly elastin and glycoprotein, whereas in the presence of ascorbic acid, elastin levels fell to 50% of the scorbutic levels and the collagen content rose.[32]

On studying the effects of the vitamin on glycosaminoglycan synthesis, it was found that 37% of the macromolecules labeled with radioactive sulfate were associated with the culture medium in the presence of ascorbic acid, whereas in its absence as much as 63% was found in this culture compartment (Table 2). In addition, ascorbic acid affects the sulfation levels

Table 2
**THE EFFECTS OF ASCORBATE ON THE
INCORPORATION OF RADIO-LABELED SULFATE
INTO MARCOMOLECULAR MATERIAL ASSOCIATED
WITH DIFFERENT CULTURE COMPARTMENTS**

	Incorporation of $Na^{35}SO_4$ dpm/µg cell protein[a]		
	Medium	**Pericell/matrix**	**Intracell**
Ascorbic supplementation	85 ± 7.2	136.1 ± 12.4	8.0 ± 3.2
Ascorbate deprivation	193 ± 11	104.4 ± 9.4	8.1 ± 3.3

Note: The material associated with the pericellular/matrix compartment is defined as that material which is releasable by proteolytic digestion.

[a] Figures represent mean ± standard deviation.

Table 3
**COMPOSITIONAL ANALYSIS OF SULFATED GLYCOSAMINOGLYCANS
ASSOCIATED WITH THE PERICELLULAR/MATRIX COMPARTMENT OF
CULTURED RAT CELLS IN THE ABSENCE AND IN THE PRESENCE OF
ASCORBIC ACID**

Glycosaminoglycan type	Ascorbate addition	Radioactivity (dpm × 10^{-3})		Ratio ^{3}H/^{35}S
		^{3}H-glucosamine	$^{35}SO_4$	
Heparan sulfate	+	59.95 ± 4.37	30.96 ± 6.08	1.93
	−	154.00 ± 38.15	27.33 ± 2.11	5.60
Dermatan sulfate	+	26.01 ± 2.96	18.46 ± 0.88	1.41
	−	7.65 ± 1.28	4.67 ± 0.20	1.63
Chondroitin-4-sulfate	+	116.55 ± 11.72	18.36 ± 1.89	2.53
	−	48.78 ± 6.36	5.67 ± 6.35	5.27
Chondroitin-6-sulfate	+	46.53 ± 5.99	59.31 ± 6.79	1.97
	−	31.86 ± 2.16	12.42 ± 0.97	3.98

of some of the glycosaminoglycan synthesized by supplemented cultures (Table 3), and the observed increased incorporation of radioactive sulfate into matrix-associated glycosaminoglycans was not related to increased collagen synthesis.[27]

These results have recently been confirmed by other workers,[33] but we have as yet no explanation for the detailed mode of action of vitamin C in relation to its effect on sulfation levels of glycosaminoglycans. A mechanism suggested previously, involving ascorbic acid-2-sulfate as an active intermediate in the metabolic pathway of glycosaminoglycan sulfation, seems to have fallen from favor.[34,35]

Our data (Table 3) with respect to the types of glycosaminoglycans synthesized by our culture system are similar to that of Conrad et al.[21] in that we also find very little synthesis of hyaluronic acid by heart-derived smooth muscle cells. Again, the same arguments as used above may be put forward to explain why so little hyaluronic acid is synthesized by these cells in culture. Further comments in relation to hyaluronic acid synthesis will be made below.

Supplementation of culture medium with levels of hyaluronic acid of up to 400 µg/mℓ did not have any effects on the synthesis of sulfated glycosaminoglycan species in the rat smooth muscle cultures.[36]

In other studies which are pertinent to this review, we found that the rat cells have an enzyme(s) capable of desulfating the sulfated glycosaminoglycans without any further degradation of the carbohydrate backbone on their surface.[29] These ectosulfatases may be important in the first step of a process involving sulfated proteoglycan turnover by a receptor-mediated endocytotic pathway. We have shown that the protein core is a requirement for this desulfation process and that internalization of labeled carbohydrate chain occurs, albeit slowly. The latter findings are consistent with the elegant work of Kresse et al.[37,38] on the metabolism of glycosaminoglycans by cultured fibroblasts.

VI. GLYCOSAMINOGLYCAN SYNTHESIS BY CUSHION TISSUE CELLS

Bernanke and Markwald[23] have studied the metabolism of glycosaminoglycans during cardiogenesis by cushion tissue cells, using chick embryo shell-less culture.[23] They studied the proliferation of cushion cells in the cardiac jelly when they arose from endocardial tissue by hypertrophy and migration into the predominantly glycosaminoglycan-rich material. There were changes in the species of glycosaminoglycans associated with atrioventricular canals (AV) isolated by microdissection from various developmental stages of chick embryos. These findings suggested that once there was clear microscopic evidence of cushion cell seeding into the cardiac jelly, there was a decline in the hyaluronic acid content of that matrix, with a concomitant increase in the level of sulfated glycosaminoglycans, particularly chondroitin-6-sulfate. The period studied was from 48 to 96 hr of development (stage 23) during which time many cushion cells had migrated into the cardiac jelly matrix material.

Using isolated cushion cells seeded onto synthetic collagen matrices, Bernanke and Markwald[23] showed that addition of hyaluronic acid to these cultures resulted in increased migration of cells into the three-dimensional collagen matrix. Their conclusion was that the high hyaluronate content of cardiac jelly was a requirement for the promotion of cell migration from the endocardium and the establishment of cushion tissue.

VII. CONCLUDING REMARKS

The universal recognition of the fact that the extracellular matrix, composed as it is of a particular blend of macromolecules for each specific tissue, plays a fundamental role in the maintenance of the specific tissue phenotypic expression, has led to new and exciting discoveries.

When the study of a composite tissue like that of the heart is undertaken, it is of even greater interest to dissect out and elucidate the secretion patterns of the individual component cells in that tissue. This is of necessity a very complex problem, but it seems that the study of cultured heart muscle cells for the purposes of assessing their glycosaminoglycan metabolism is long overdue.

The culture of chick presumptive myoblasts on substrates either rich in hyaluronic acid or in culture media containing elevated levels of the nonsulfated glycosaminoglycans inhibited myogenesis.[9,39] This was assessed both by morphological criteria and by measurement of creatine phosphokinase (CPK) levels in the cultures. Both removal of hyaluronic acid and/or the culture of cells in the presence of chondroitin sulfate promoted myogenesis as evidenced by the formation of lenticular myotubes and elevated CPK levels.[9,40] Such observations with skeletal muscle cells make pursuit of related studies with cardiomyocytes all the more important and fascinating.

The identification of specific cell types in culture still presents problems as does the assessment of whether, in culture, their metabolic functions truly reflect the in vivo situation. However, a start has to be made, since it is through studies of this type that we already know that hyaluronic acid is the glycosaminoglycan generally associated with the surfaces of undifferentiated cells and that as they go through their path of terminal differentiation, the synthesis of specific sulfated glycosaminoglycans takes over from that of hyaluronic acid. Hyaluronic acid acts as the "oil" for cell migration, but once cells reach their site of choice the sulfated macromolecules act as the "glue" to hold them in their microenvironment.[5]

REFERENCES

1. **Gevers, W.,** Protein metabolism of the heart, *J. Mol. Cell. Cardiol.,* 16, 3, 1984.
2. **Hardingham, T.,** Proteoglycans: their structure, interactions and molecular organization in cartilage, *Biochem. Soc. Trans.,* 9, 487, 1981.
3. **Roden, L. and Smith, R.,** Structure of the neutral trisaccharide of the chondroitin-4-sulphate-protein linkage region, *J. Biol. Chem.,* 241, 5949, 1966.
4. **Muir, H.,** Proteoglycans as organisms of the intercellular matrix, *Biochem. Soc. Trans.,* 11, 613, 1983.
5. **Hook, M., Kjellen, L., Johansson, S., and Robinson, J.,** Cell-surface glycosaminoglycans, *Annu. Rev. Biochem.,* 53, 847, 1984.
6. **Cheng, C.-F., Oosta, G. M., Bensadoun, A., and Rosenberg, R. D.,** Binding of lipoprotein lipase to endothelial cells in culture, *J. Biol. Chem.,* 256, 12893, 1981.
7. **Kleinman, H. K., Klebe, R. J., and Martin, G. R.,** Role of collagenous matrices in the adhesion and growth of cells, *J. Cell Biol.,* 88, 473, 1981.
8. **Poole, B. P.,** *Cell Biology of Extracellular Matrix,* Hay, E., Ed., Plenum Press, New York, 1981, chap. 9.
9. **Kujawa, M. J. and Tepperman, K.,** Culturing chick muscle cells on glycosaminoglycan substrates: attachment and differentiation, *Dev. Biol.,* 99, 277, 1983.
10. **Manasek, F. J.,** Macromolecules of the extracellular compartment of embryonic and matrix hearts, *Circ. Res.,* 38, 331, 1976.
11. **Manasek, F. J.,** Embryonic development of the heart. I. A light and electron microscopic study of myocardial development in the early chick embryo, *J. Morphol.,* 125, 329, 1968.
12. **Manasek, F. J.,** Histogenesis of the embryonic myocardium, *Am. J. Cardiol.,* 25, 149, 1970.
13. **Gessner, I. H., Lorincz, A. E., and Bostrom, H.,** Acid mucopolysaccharide content of the cardiac jelly of the chick embryo, *J. Exp. Zool.,* 160, 291, 1965.
14. **Markwald, R. R., Funderberg, F. M., and Bernanke, D. H.,** Glycosaminoglycans: potential determinants of cardiac morphogenesis, *Tex. Rep. Biol. Med.,* 39, 253, 1979.
15. **Ohkawa, S.-I., Sugira, M., Hata, R., and Nagai, Y.,** Acidic glycosaminoglycans in urine, serum and myocardium of aged patients with myocardial infarction, *J. Mol. Cell. Cardiol.,* 9, 541, 1977.
16. **Dalferes, E. R., Radhakrishnamurthy, B., and Berenson, G. S.,** Glycosaminoglycans of cardiac tumours, *Proc. Soc. Exp. Biol. Med.,* 157, 461, 1978.
17. **Polinger, I. S.,** Separation of cell types in embryonic heart cell culture, *Exp. Cell. Res.,* 63, 78, 1970.
18. **Polinger, I. S.,** Growth and DNA synthesis in embryonic chick heart cells *in vivo* and *in vitro, Exp. Cell Res.,* 76, 253, 1973.
19. **Coetzee, G. A. and Gevers, W.,** Myosin in primary cultures of hamster heart cells, *Dev. Biol.,* 63, 128, 1977.
20. **Coetzee, G. A.,** *Hamster Heart Cells in Culture,* Ph.D. thesis, University of Stellenbosch, South Africa, 1977.
21. **Conrad, G. W., Hamilton, C., and Haynes, E.,** Differences in glycosaminoglycans synthesized by fibroblast-like cells from chick cornea, heart and skin, *J. Biol. Chem.,* 252, 6861, 1977.
22. **Jones, P. A., Scott-Burden, T., and Gevers, W.,** Glycoprotein, elastin and collagen secreted by rat smooth muscle cells, *Proc. Natl. Acad. Sci. USA,* 76, 353, 1979.
23. **Bernanke, D. H. and Markwald, R. R.,** Effects of hyaluronic acid on cardiac cushion tissue cells in collagen matrix cultures, *Tex. Rep. Biol. Med.,* 39, 271, 1979.
24. **Dunn, B. E. and Fitzharris, T. P.,** Differentiation of the chorionic epithelium of chick embryos maintained in shell-less culture, *Dev. Biol.,* 71, 216, 1979.
25. **Blondel, B., Roigen, I., and Cheneval, J. P.,** Heart cells in culture: a simple method for increasing the proportion of myoblasts, *Experimentia,* 27, 356, 1971.
26. **Levy, P., Picard, J., and Bruel, A.,** Glycosaminoglycans biosynthesis in arterial wall: sulphation of heparan sulphate in cell membranes of aortic media-intima, *Eur. J. Biochem.,* 115, 397, 1981.

27. **Cohn, R. H., Cassiman, J.-J., and Bernfield, M. R.,** Relationship of transformation, cell density, and growth control of the cellular distribution of newly synthesized glycosaminoglycans, *J. Cell Biol.,* 71, 280, 1976.

28. **Scott-Burden, T., Murry, E., Diehl, T. S., and Gevers, W.,** Glycosaminoglycan synthesis by smooth muscle cells cultured in the absence and presence of ascorbic acid, *Hoppe-Seyler's Z. Physiol. Chem.,* 364, 61, 1983.

29. **Diehl, T. S., Scott-Burden, T., and Gevers, W.,** Sulphate turnover of surface proteoglycans in cultured rat smooth muscle cells, *Biochem. Int.,* 6, 29, 1983.

30. **Murray, E., Scott-Burden, T., Ferguson, P., and Gevers, W.,** Increased sulphation level and altered composition of glycosaminoglycans synthesized by cultured smooth muscle cells in the presence of β-D-xylosides, *Biochim. Biophys. Acta,* 763, 299, 1983.

31. **Chamley-Campbell, J., Campbell, G. R., and Ross, R.,** The smooth muscle cell in culture, *Physiol. Rev.,* 59, 1, 1979.

32. **Scott-Burden, T., Davies, P. J., and Gevers, W.,** Elastin biosynthesis by smooth muscle cells cultured under scorbutic conditions, *Biochem. Biophys. Res. Commun.,* 91, 739, 1979.

33. **Edwards, M. and Oliver, R. F.,** Ascorbate increases the sulphation of glycosaminoglycans synthesized by human skin fibroblasts, *Biochem. Soc. Trans.,* 12, 304, 1984.

34. **Hatanaka, H. and Egami, F.,** Sulphate incorporation from ascorbic-2-sulphate into chondroitin sulphate by embryonic chick cartilage epiphyses, *J. Biochem.,* 80, 1215, 1976.

35. **Shapiro, S. S. and Poon, J. S.,** Apparent sulphation of glycosaminoglycans by ascorbic acid-2-[^{35}]-sulphate: an explanation, *Biochem. Biophys. Acta,* 385, 221, 1975.

36. **Diehl, T. S.,** *Biosynthesis and Degradation of Proteoglycans in Cultured Smooth Muscle Cells,* M.Sc. thesis, University of Cape Town, South Africa, 1982.

37. **Truppe, W. and Kresse, H.,** Uptake of proteoglycans and sulphated glycosaminoglycans by cultured skin fibroblasts, *Eur. J. Biochem.,* 85, 351, 1978.

38. **Prinz, R., Schwermann, J., Buddecke, E., and von Figura, K.,** Endocytosis of sulphate proteoglycans by cultured skin fibroblasts, *Biochem. J.,* 176, 671, 1978.

39. **Elson, H. F. and Ingwall, J. S.,** The cell substratum modulates skeletal muscle differentiation, *J. Supramol. Struct.,* 14, 313, 1980.

40. **Singley, C. T. and Salursh, M.,** The spatial distribution of hyaluronic acid and mesenchymal condensation in embryonic chick wing, *Dev. Biol.,* 84, 102, 1981.

Chapter 28

THE EFFECTS OF THYROID HORMONES ON CULTURED CARDIOMYOCYTES

Amirav Gordon

TABLE OF CONTENTS

I. INTRODUCTION

The heart is a major target organ for thyroid hormone action and both excess and deficiency of thyroid hormones induce profound cardiovascular changes. These changes may result from a number of mechanisms, those related to the direct effects on the cardiac muscle and those secondary to the cardiovascular demands associated with the effects of thyroid hormones on metabolic activity in the whole body. An example of the latter is the comparison of the effect of excess thyroid hormone on skeletal and cardiac muscle. Excess thyroid hormone results in skeletal muscle atrophy probably secondary to an increase in protein degradation, whereas in the heart muscle mass increases and protein degradation is either unaffected (myofibrillar fraction) or even reduced (sarcoplasmic fraction).[1] Since, in skeletal muscle, work induces hypertrophy with a concomitant decrease in protein degradation,[2] the effects of thyroid hormone excess on the heart may be secondary to work overload rather than a direct effect on the heart. Cultured cardiomyocytes can be a useful system for the solution of such problems. In the pages that follow experiments conducted with cultured cardiomyocytes will be described and the data compared with the known in vivo effects.

II. EFFECTS OF THYROID HORMONES VIA NUCLEAR INTERACTION

Thyroid hormone appears to be a pleiotypic hormone affecting the target cell by several independent mechanisms. There is evidence for independent action at the level of the nucleus, the mitochondria and the plasma membrane (see review, Reference 3).

The evidence that thyroid hormones affect cells through nuclear interaction stems from the pioneering work of Tata and co-workers (reviewed in Reference 4), who demonstrated that the stimulation of the basal metabolic rate (BMR) by thyroxine is preceded by an increase in protein synthesis, and that the effect on the BMR could be blocked by pretreatment with puromycin. The first biochemical event shown by these authors was the increased labeling of a rapidly turning-over nuclear RNA fraction in the rat liver, occurring within 4 to 6 hr after T3 injection. A major advance in the understanding of the nuclear effects of T3 was the demonstration of specific nuclear receptors for T3 by Oppenheimer et al.[5] Since then voluminous literature has been published on this subject.

At the present time the following general scheme may be proposed. T3, the major cellular thyroid hormone, reaches the nucleus and there it binds to its receptor, a nonhistone acid protein. As a result of this binding mRNA, with a rapid turnover, is produced which codes for a protein that stimulates RNA polymerase. This is followed by mRNA synthesis and the subsequent production of proteins and enzymes responsible for the final expression of the effects of T3.[3]

The experiments to be described in the following subsections deal with data that are clearly related to protein synthesis or with observations made after a prolonged lag period and, therefore, are presumed to occur through T3-nuclear interaction.

A. Myosin Isoenzyme Profiles

Myosin isoenzyme V_1 predominates in the euthyroid rat ventricle, whereas myosin V_3 is the major form in the hypothyroid heart.[6,7] Thyroid hormone replacement shifts the predominant myosin form from V_3 to V_1. Myosin V_1 contains two α-myosin heavy chains (MHC), while myosin V_3 contains two β-MHC. Indeed, peptide mapping of mRNA translational products showed that mRNA from hypothyroid hearts coded primarily for β-MHC, while products of mRNA from euthyroid hearts predominated in α-MHC.[6,7]

As a counterpart to these elegant in vivo studies, Nag and Cheng[8] studied the effect of thyroid hormones on the embryonic maturation of these isoenzymes in isolated ventricular cells from both rats and chicks. The 20-day-old rat embryonic heart shows a predominance of slow migrating V_3 myosin isoenzyme, while the adult ventricles contain mostly the V_1 isoenzyme. This transition during maturation resembles the isoenzyme changes seen when the hypothyroid adult rat is treated with thyroid hormones.[6,7]

Nag and Cheng, therefore, sought to determine whether these changes are directly thyroid hormone dependent. Cells were cultivated in a serum-free medium either in the presence or in the absence of 10 nM T4. In the rat the embryonic pattern was transformed to the adult profile after 4 days of exposure to T4, while in the chick, the embryonic and adult patterns were similar and, therefore, T4 had no effect on the chick enzyme profile. These results indicate that thyroid hormone in the rat may play a role in the transition from the neonatal ventricular myosin isoenzyme profile to the adult pattern.

B. The Sodium Pump and Its Sensitivity to Digitalis Glycosides

Another problem associated with hyperthyroidism is the decreased sensitivity of heart muscle to digitalis glycosides.

It is also well known that thyroid hormones, in vivo, increase the Na, K-ATPase content and activity in various tissues, including the heart.[9] It is not clear at the present whether the decreased sensitivity to digitalis is a result of the increase in the Na-pump activity or whether it is due to the decreased availability of the drug secondary to an increased clearance of digitalis.

Kim and Smith[10] investigated this problem in cultured chick embryo heart cells exposed to T3 in a serum-free medium. After 48 hr in the presence of T3 there was a dose-dependent increase in sodium pump units, as indicated by ^{3}H-ouabain binding. This effect was maximal at 10^{-8} M T3 and was associated with a parallel decrease in the intracellular Na^+ content. The response of the T3-treated cells to 1 μM ouabain was decreased both with respect to the effects of this drug in elevating the cellular Na^+ content and to its inotropic action. The dose-response curve to ouabain was shifted to the right by T3 treatment. This in vitro study confirms that the effect of T3 is mediated via its stimulation of the Na-K ATPase rather than via a change in the bioavailability of ouabain.

C. The Catecholamine Thyroid Interaction

This subject has been reviewed extensively,[11] and a brief summary of this review is presented here. Clinically, excess thyroid hormones result in cardiac symptoms which resemble the effects of sympathetic stimulation. The amelioration of these symptoms with β-adrenergic blockers suggested a close interrelationship between the peripheral effects of thyroid hormones and the adrenergic system.

Physiologic data, however, on the increased activity of the sympathetic system in hyperthyroidism is rather inconclusive. It is apparent that hyperthyroidism is not associated with an increase in the level of circulating catecholamines, that thyroid hormones do not interact directly with the β-receptors, and, finally, that the ameliorative effect of propranolol is not associated with a drastic decrease in the circulating levels of T3. These observations, therefore, suggest that thyroid hormones increase the cardiac response to catecholamines.

At the biochemical level, there is no agreement as to whether the cAMP response to catecholamines is augmented in membranes isolated from hearts of hyperthyroid animals. There is, however, a uniformity in the reports in that they show augmented phosphorylase activity in such membranes. Since phosphorylation is a step which is subsequent to cAMP elevation, it is probable that the lack of consistent results with respect to the cAMP system is due to the assay methods.

Further support for the possible increase in the sensitivity of hyperthyroid hearts to catecholamines comes from receptor studies. Williams et al.[12] have shown that there is a twofold increase in the β-adrenergic receptor density in hyperthyroid cardiac plasma membranes, as measured by dihydroalprenolol binding, and a corresponding decrease of these receptors in hypothyroidism.[11] The situation with respect to the α-receptors is not very clear. Several, but not all, studies show a decrease in dihydro-ergocryptine binding in hyperthyroid cardiac membranes. The reciprocal changes in hypothyroidism are also not well established.[11] It is difficult at present to evaluate the possible contribution of the change in α-adrenergic receptor density and the cardiac response to catecholamines. The problems dealing with catechol-amine-thyroid interaction have been the subject of many in vitro studies in isolated muscle preparation taken from animals with altered thyroid activity, but in only one study were tissue culture techniques used. Tsai and Chen[13] cultivated heart cells of newborn rats in a medium containing 10% hypothyroid calf serum. Under these conditions, the addition of T3 resulted in a dose-dependent increase in dihydro-alprenolol binding sites within 24 hr. The cAMP response to 1 μM ($-$) adrenaline was also increased in parallel to the increase in the β-adrenergic receptor density. This in vitro study is in keeping with the clinical observations of increased sympathetic effects in thyrotoxicosis. The improved tissue culture techniques now available should prove useful in assessing the effect of thyroid hormone on α-receptor number and turnover.

D. Effects of Thyroid Hormones on Carbohydrate Metabolism

Thyroid hormones are known to raise total body oxygen consumption. This is accompanied by an increase in carbohydrate utilization. The effect of T3 on carbohydrate metabolism in cardiac cells in tissue culture was first described by Schwartz and Gordon.[14] In their system, heart cells were prepared from 8-day-old chick embryos. They were grown for the first 24 hr in 1% serum-supplemented medium and subsequently transferred to a serum-free medium for the next 48 hr. Experiments with T3 were started in confluent, synchronously contracting monolayers 72 hr after plating. Under these conditions, 24 hr after the addition of T3 there was a dose-dependent increase in lactic acid production by these cells. The increase in lactic acid production was associated with an increase in the uptake rate of 2-deoxy-glucose, indicating that sugar transport was stimulated as well. These data led them to conclude that T3 raises glucose utilization in heart cells primarily via the increase in glucose availability. Similar data were obtained in cultured newborn rat cardiac cells.[15,16]

Another facet of the effect of thyroid hormone on carbohydrate metabolism is the stimulation of the mitochondrial FAD-linked α-glycerol-3-phosphate dehydrogenase (FAD-G3PDH; EC 1.1.99.5). This enzyme, together with the cytosolic NAD-linked α-glycerol-3-phosphate dehydrogenase (NAD-G3PDH; EC 1.1.1.8), is responsible for the transfer of NADH reducing equivalents across the mitochondrial membrane via the glycerol-3-phosphate shuttle. The levels of FAD-G3PDH in rat liver and kidney mitochondria were elevated on thyroid hormone treatment, while the NAD-G3PDH activity remained unchanged.[17] The same effects were also shown for the adult rat heart.[18]

Recently, Freerksen et al.[16] studied the effects of T3 on carbohydrate metabolism in cultured neonatal rat heart cells and the enzymatic profiles associated with this effect. Glucose utilization, measured after 4 days exposure to 3×10^{-8} M T3, was significantly enhanced. Analysis of 11 enzymes involved in various phases of glucose utilization showed that the increase in glycolytic activity was not associated with changes in the hexokinase, phos-phofructokinase, or the pyruvate kinase activity, or with the altered rate of the pentose shunt activity, tricarboxylic acid cycle, or electron transport activity as determined by measuring the activity of six other enzymes. Only the enzymes associated with the glycerol-3-phosphate shunt were affected. T3 raised the activity of the mitochondrial FAD-G3PDH, confirming previous in vivo studies.[17,18] Surprisingly, T3 depressed the activity of the cytosolic NAD-G3PDH in a dose-dependent manner. The latter finding is a novel observation and is hard to interpret in terms of the increase in glucose utilization. This study shows that, in vitro, T3 is also associated with a selective reduction in the activity of a specific enzyme and is consistent with the in vivo data related to mRNA studies, where T3 was also shown to augment or depress the production of specific mRNA products.[7]

Table 1
**EFFECT OF ACTINOMYCIN D AND T3 ON THE 2-DEOXY-GLUCOSE
UPTAKE IN CULTURED CHICK EMBRYO HEART CELLS**

Actinomycin D (μg/mℓ)	Exposure[a] (hr)	2-Deoxy-D-[1-³H]glucose uptake (³H dpm/45 min/μg protein[b])		
		Control	T3 (1 $\times$ 10⁻⁸ M)	p[c]
0	6	20.9 $\pm$ 0.9	27.1 $\pm$ 0.8	
10	6	21.1 $\pm$ 0.9	27.4 $\pm$ 1.1	>0.8
0	24	19.6 $\pm$ 2.0	36.3 $\pm$ 3.3	
0.02	24	20.9 $\pm$ 1.8	25.6 $\pm$ 1.4	<0.02

[a]   Five-day-old culture plates were incubated with or without 10⁻⁸ M T3. Actinomycin D was added at the beginning of the incubation to both the control and the T3 plates, where indicated. At the end of the experiment the uptake rate of 2-deoxy-[³H]-glucose was measured.

[b]   Mean $\pm$ SEM: N> = six plates per group.

[c]   p Values for the comparison between the T3-treated group to the T3 + actinomycin D-treated cultures were determined by using the student's *t* test. The sugar uptake in the T3-treated cells was always significantly increased over the appropriate control cultures.

From Segal, J. and Gordon, A., The effects of actinomycin D, puromycin, cycloheximide and hydroxyurea on 3′,5,3,-triiodothyronine stimulated 2-deoxy-D-glucose uptake in chick embryo heart cells in vitro, *Endocrinology*, 101, 150, 1977. With permission.

III. NONNUCLEAR EFFECTS OF THYROID HORMONES

Thyroid hormones were shown to have several effects that are not dependent on nuclear interaction.[5] One of these, the effect of thyroid hormone on sugar transport, has been studied in great detail in our laboratory, using cultured chick embryo heart cells as the model system.[19-22] In these experiments, cells derived from 10-day-old chick embryo hearts were grown for the first 48 hr in an Eagle's basal glucose-supplemented medium (EBGM)[23] containing 4% inactivated chick serum. The medium was subsequently changed every 24 hr with and EBGM medium which did not contain chick serum. Cells were regularly used at 96 hr after plating when they had reached confluency and were contracting synchronously and rhythmically. Under these conditions, T3 (1 $\times$ 10⁻⁸ M) stimulated the uptake of 2-deoxy-glucose in a linear fashion for the first 6 hr. Sugar uptake rate was further increased, in a nonlinear fashion, reaching a plateau around 24 hr. Leucine incorporation into TCA precipitable products was also stimulated by T3, but with a lag period of at least 6 hr.[20] These facts suggested that the stimulation of 2-DOG uptake for the first 6 hr did not require neosynthesis of proteins. Indeed, in experiments in which protein synthesis was blocked by either puromycin, cycloheximide, or actinomycin-D, the effect of T3 in the first 6 hr was unchanged by these drugs.[20] The additional increase in 2-DOG uptake, observed between 6 and 24 hr, was effectively blocked by these inhibitors of protein synthesis (Table 1). The data, therefore, indicate that the stimulation of 2-DOG uptake by T3 is a biphasic phenomenon, initially not requiring protein synthesis, but having a later component which is dependent on both translational and transcriptional events. A likely locus for this initial interaction of T3 was the plasma membrane and experiments were, therefore, conducted to test whether T3, when given in a noninternalizable form, will also affect glucose uptake.[21] Cultured cells were treated with T3 which was covalently bound to human erythrocytes (T3-RBC). As may be seen in Table 2, T3-RBC stimulated sugar uptake for the first 6 hr to the same extent as an equivalent concentration of free T3. At 24 hr, however, the free T3 caused a further increase in the 2-DOG uptake, while the T3-RBC could not elicit an effect beyond the level achieved during the first 6 hr. These data indicated that for the initial effect it is sufficient that T3 be present on the outside of the cell, while in the latter, the protein

Table 2
THE EFFECT OF ACTINOMYCIN D ON 2-DEOXY-GLUCOSE UPTAKE RATES IN RESPONSE TO 20 PMOL FREE OR RBC-BOUND T3

	T3/control (mean $\pm$ SEM)	
Treatment	Free T3	RBC-T3
6 hr None	1.13 $\pm$ 0.04	1.18 $\pm$ 0.04
24 hr None	1.32 $\pm$ 0.06[a]	1.12 $\pm$ 0.05
Act-D	1.16 $\pm$ 0.03	1.13 $\pm$ 0.05

Note: Actinomycin (20 ng/mℓ) was added to the culture plates 30 min before the exposure to T3, either RBC bound or free. The appropriate control groups were also treated with Act-D. Six or 24 hr later 2-deoxy-glucose uptake was determined. Six wells were used for control or T3-treated group. The ratio T3/control was calculated for each pair out of the six. Pairing between the treated and the control wells was sequential according to their position in the plate. 2-[^{3}H]-DOG uptake ranged between 9.0 $\pm$ 0.3 to 10.6 $\pm$ 0.6 dpm/45 min/μg protein (mean $\pm$ SEM) for the control group without Act-D and between 15.7 $\pm$ 0.6 to 18.5 $\pm$ 0.4 in the controls in the presence of the inhibitor.

[a] $p < 0.05$ as compared to the 6-hr free T3 group. In all groups the T3/control was significantly greater than 1.00.

From Dickstein, Y., Schwartz, H., Gross, S., and Gordon, A., Stimulation of sugar transport in cultured heart cells by triiodothyronine (T3) covalently bound to red blood cells and by T3 in the presence of serum, *Endocrinology*, 113, 381, 1983. With permission.

synthesis-dependent phase, T3 must be in a free, diffusible form. Further analysis of these phases was conducted by measuring the kinetic parameters of the uptake of two glucose analogs, 3-*O*-methyl-glucose (3-OMET) and 2-deoxy-glucose (2-DOG).[22] Both analogs use the same transport system as glucose and competitively inhibit each other's transport. The 3-OMET analog is transported into the cell, but does not undergo any further modification and was used to measure transport parameters only. The 2-DOG analog, on the other hand, is phosphorylated by hexokinase but is not metabolized. The kinetic parameters of 2-DOG uptake, therefore, reflect both the transport and the phosphorylating capacity of the cell. The results of our study are shown in Table 3. T3 did not have any effect on the Km for either analog. T3 in the first 6 hr increased the V_{max} for both analogs but resulted in a further increase, after a 24-hr exposure, only in the V_{max} of 2-DOG. Since 3-OMET measures transport alone, it is clear that the nonprotein-dependent phase involves an increase in the transport capacity alone. Indeed, in recent experiments, cytochalasin B, a drug known to block glucose uptake via the glucose transporter system, was also shown to block the early effect of T3.[24] The further increase in the V_{max} of 2-DOG, observed between 6 and 24 hr, should also reflect the phosphorylating capacity of the cells. Indeed, the data obtained showed an increase in this capacity after 24 hr of exposure to T3. These data were interpreted by us as indicating an increased hexokinase activity, possibly due to neosynthesis of enzyme molecules.

In a recent set of experiments,[24] we grew heart cells in a serum-free defined medium. This medium was essentially the same as described by Bauer et al.,[25] except that thyroxine was omitted. It contains insulin (200 nM) and hydrocortisone (10 nM), but not fetuin. Cells grown in this medium show an increased sensitivity of 2-DOG uptake in response to picomolar concentrations of T3. This increased sensitivity and response to T3 is dependent on the presence of both insulin and hydrocortisone in the medium.[24]

Thus, the in vitro tissue culture technique helped to unravel a hidden effect of T3, i.e.,

Table 3
THE EFFECT OF T3 (10^{-8} M) ON THE KINETIC PARAMETERS OF 2-DEOXY-GLUCOSE AND 3-O-METHYL-GLUCOSE UPTAKE IN CULTURED CHICK EMBRYO HEART CELLS

| | | 2-Deoxy-D-glucose | | | 3-O-Methyl-D-glucose | |
| | | V_{max}[b] (μmol/min/g protein) | Km (mM) | | V_{max} (μmol/min/g protein) | Km (mM) |
Treatment	n[a]			n		
Control	119	17.5 ± 0.8	3.00 ± 0.30	144	325.5 ± 13.9	31.8 ± 2.5
T3, 6 hr	72	20.4 ± 0.8[c]	3.04 ± 0.27	88	422.6 ± 20.6[d]	36.2 ± 3.3
T3, 24 hr	104	27.1 ± 0.8[d,e]	3.08 ± 0.24	69	426.0 ± 15.5[d]	32.5 ± 2.4

[a] n = Number of plates tested.
[b] All the results are presented as the mean $\pm$ SEM.
[c] $p < 0.05$ as compared to the control group.
[d] $p < 0.001$ as compared to the control group.
[e] $p < 0.001$ as compared to the T3, 6-hr group.

From Segal, S. and Gordon, A., The effect of 3,5,3′-triiodo-L-thyronine on the kinetic parameters of sugar transport in cultured chick embryo heart cells, *Endocrinology*, 101, 1468, 1977. With permission.

its interaction with the plasma membrane. This effect was achieved at physiological concentrations of free T3 and exhibits the same thyroid hormone analog specificity that is seen in vivo.[19] For these reasons we believe that this novel site of action is physiologically significant.

IV. SUMMARY

In this review, experiments were summarized in which tissue culture techniques have been applied to the study of thyroid hormone action on cardiac tissue. Since the available material encompasses no more than a dozen studies, it was relatively easy to cover the published material to date. These studies addressed the major problems of thyroid hormone action, problems that had previously been defined by in vivo experimentation. In addition, we have described the experiments that established the plasma membrane as a target for thyroid hormone action, a system that has as yet not been studied in vivo.

The reasons behind this limited utilization of tissue cultures stem, from two considerations:

1. The culture medium. Until recently cardiac cells could be grown only in serum-supplemented media, containing endogenous T4 and T3. In order to obtain baseline conditions in which thyroid hormone is absent, the serum had to be stripped from the hormones by adsorption.[26] This procedure introduces variables with respect to other serum fractions that are lost along with the adsorbed hormones.
2. The source of cells. In the absence of cardiomyocyte cell lines with defined properties, primary cell cultures were used. These, until recently, were of embryonic chick or neonatal rat origin. These cells, derived from nonmature tissue, would limit the interpretation of the results and its applicability to the adult situation. Indeed, in one in vitro study, use was made of neonatal rat heart cells to show the differential action of thyroid hormones on enzyme profile during maturation.[8]

With the advent of the new tissue culture techniques that allow the use of adult rat cardiac cells grown in defined serum-free media, the stage is set for wider applicability of these systems to the study of hormone action. These techniques will allow us to study problems which are too complex to be clearly analyzed in the in vivo situation, where effects secondary to the action of the hormones in other parts of the body are superimposed on direct cardiac effects.

ACKNOWLEDGMENTS

I wish to thank Mrs. H. Schwartz and Prof. J. Gross for the help extended in the preparation of this manuscript.

This work was supported by funds from the J. Hamburger Center for Thyroid Disease.

REFERENCES

1. **Carter, W. J., Van der Weijden Benjamin, W. S., and Fass, F. H.,** Effect of experimental hyperthyroidism on protein turnover in skeletal and cardiac muscle, *Metabolism,* 29, 910, 1980.
2. **Goldberg, A. L.,** Protein turnover in skeletal muscle. I. Protein catabolism during work-induced hypertrophy and growth induced with growth hormone, *J. Biol. Chem.,* 244, 3217, 1969.
3. **Muller, M. J. and Seitz, H. J.,** Pleiotypic action of thyroid hormones at the target cell level, *Biochem. Pharmacol.,* 33, 1579, 1984.
4. **Tata, J. R.,** Growth and developmental action of thyroid hormones at the cellular level, in *Handbook of Physiology,* Vol. 3, Greep, R. O., Astwood, E. B., Greer, M. A., and Solomon, D. H., Eds., American Physiological Society, Washington, D.C., 1974, chap. 26.
5. **Oppenheimer, J. H., Koerner, D., Schwartz, H. L., and Surks, M. I.,** Specific nuclear triiodothyronine binding sites in rat liver and kidney, *J. Clin. Endocrinol. Metab.,* 35, 330, 1972.
6. **Dillmann, W. H., Berry, S., and Alexander, N. B.,** A physiological dose of triiodothyronine normalizes cardiac myosin adenosine triphosphate activity and changes myosin isoenzyme distribution in semistarved rats, *Endocrinology,* 112, 2081, 1983.
7. **Dillmann, W. H.,** Influence of thyroid hormone on messenger RNA levels in the rat heart, in *Endocrinology,* Labrie, F. and Proulx, L., Eds., Elsevier, Amsterdam, 1984, 973.
8. **Nag, A. C. and Cheng, M.,** Expression of myosin isoenzymes in cardiac-muscle cells in culture, *Biochem. J.,* 221, 21, 1984.
9. **Guernsey, D. L. and Edelman, I. S.,** Regulation of thermogenesis by thyroid hormones, in *Molecular Basis of Thyroid Hormone Action,* Oppenheimer, J. H. and Samuels, H. H., Eds., Academic Press, New York, 1983, chap. 10.
10. **Kim, D. and Smith, T. W.,** Effects of thyroid hormone on sodium pump sites, sodium content and contractile responses to cardiac glycosides in cultured chick ventricular cells, *J. Clin. Invest.,* 74, 1481, 1984.
11. **Williams, R. S. and Lefkowitz, R. J.,** The effects of thyroid hormone on adrenergic receptors, in *Molecular Basis of Thyroid Hormone Action,* Oppenheimer, J. H. and Samuels, H. H., Eds., Academic Press, New York, 1983, chap. 11.
12. **Williams, A. T., Lefkowitz, R. J., Watanabe, A. M., Hathaway, D. R., and Besch, H. R.,** *J. Biol. Chem.,* 252, 2787, 1977.
13. **Tsai, J. S. and Chen, A.,** Effect of l-triiodothyronine on $(-)$ ^{3}H-dihydroalprenolol binding and cyclic AMP response to $(-)$ adrenaline in cultured heart cells, *Nature (London),* 275, 138, 1978.
14. **Schwartz, H. and Gordon, A.,** Effect of triiodothyronine on chick embryo heart cells maintained in tissue culture, *Isr. J. Med. Sci.,* 9, 877, 1975.
15. **Tsai, J. S. and Chen, A.,** Thyroid hormones: effect of physiological concentration on cultured cardiac cells, *Science,* 194, 202, 1976.
16. **Freerksen, D. L., Schroedl, N. A., and Hartzell, C. R.,** Triiodothyronine depresses the NAD-linked glycerol-3-phosphate dehydrogenase activity of cultured neonatal rat heart cells, *Arch. Biochem. Biophys.,* 228, 474, 1984.
17. **Lee, Y. P., Takemori, A. E., and Lardy, H.,** Enhanced oxidation of alpha-glycerophosphate by mitochondria of thyroid-fed rats, *J. Biol. Chem.,* 234, 3051, 1959.

18. **Isaacs, G. H., Sacktor, B., and Murphy, T. A.,** The role of the alpha-glycerophosphate cycle in the control of carbohydrate oxidation in the heart in the mechanism of action of thyroid hormones, *Biochem. Biophys. Acta,* 177, 196, 1969.
19. **Segal, J., Schwartz, H., and Gordon, A.,** The effect of triiodothyronine on 2-deoxy-D-[1-^{3}H]glucose uptake in cultured chick embryo heart cells, *Endocrinology,* 101, 143, 1977.
20. **Segal, J. and Gordon, A.,** The effects of actinomycin D, puromycin, cycloheximide and hydroxyurea on 3',5,3,-triiodothyronine stimulated 2-deoxy-D-glucose uptake in chick embryo heart cells *in vitro, Endocrinology,* 101, 150, 1977.
21. **Dickstein, Y., Schwartz, H., Gross, J., and Gordon, A.,** Stimulation of sugar transport in cultured heart cells by triiodothyronine (T3) covalently bound to red blood cells and by T3 in the presence of serum, *Endocrinology,* 113, 391, 1983.
22. **Segal, J. and Gordon, A.,** The effect of 3,5,3'-triiodo-L-thyronine on the kinetic parameters of sugar transport in cultured chick embryo heart cells, *Endocrinology,* 101, 1468, 1977.
23. **Macpherson, I. and Stoker, M.,** Polyoma transformation of hamster cell clones — an investigation of genetic factors affecting cell competence, *Virology,* 16, 147, 1962.
24. **Gordon, A., Schwartz, H., and Gross, J.,** unpublished data.
25. **Bauer, R. F., Arthur, L. O., and Fine, D. L.,** Propagation of mouse mammary tumor cell lines and production of mouse mammary tumor virus in a serum free medium, *In Vitro,* 12, 558, 1976.
26. **Mulaisho, C. and Utiger, R. D.,** Serum thyroxine-binding globulin: determination by competitive ligand-binding assay in thyroid disease, *Acta Endocrinol. (KBH),* 85, 314, 1977.

Chapter 29

IRON OVERLOAD IN CULTURED CARDIOMYOCYTES: A MODEL OF IRON TOXICITY IN THE HEART

Chaim Hershko, Gabriella Link, and Arié Pinson

TABLE OF CONTENTS

I. INTRODUCTION

Although iron toxicity resulting in extensive myocardial fibrosis is a major cause of death in thalassemia and other chronic anemias requiring long-term transfusional therapy,[1,2] its mechanism at a cellular and biochemical level is poorly understood. In the intact animal, one group of workers failed to detect any of the expected clinical manifestations of hemo-chromatosis — cirrhosis, diabetes, or cardiomyopathy — after more than 7 years of parenteral iron loading in dogs.[3] Rats subjected to the parenteral administration of ferric nitroloacetate developed reversible diabetes mellitus, but none of the more critical cardiac or hepatic manifestations of hematochromatosis.[4] The use of cultured cell systems, on the other hand, has allowed some insights into the site and mechanism of action of iron toxicity.

Several studies employing hepatocytes indicate that the primary site of damage during iron loading is located to the mitochondria. This is manifested in decreased ATP production, inhibition of enzymes of the Krebs cycle resulting in lactic acidosis, and enhanced malonyl dialdehyde (MDA) formation, suggesting that damage to the mitochondrial membrane occurs via iron-induced lipid peroxidation.[5-8] Other workers have proposed that lysosomes obtained by hepatic biopsies from hemochromatotic patients are damaged by iron overload, resulting in the intracellular release of lysosomal enzymes.[9,11] However, these findings have not been supported by subsequent studies.[11] In vitro studies of Chang cell cultures showed that transfer of iron to lysosomes (siderosomes) may increase cell tolerance to high concentrations of iron following induction of ferritin synthesis, providing a mechanism for protecting other cellular organelles against iron toxicity.[12] The nuclear[13] and microsomal[14,15] membranes may also be sites of iron-induced toxicity.

These effects on membranes are all probably mediated via small amounts of low molecular weight chelate iron, present in the plasma when the capacity for carrying iron in a nontoxic transferrin-bound ferric state has been exceeded.[16,17] Studies by the Cardiff group employing Chang cell cultures showed that iron-nitriloacetate, in contrast to iron bound to transferrin, is located following cellular uptake to the membrane-rich subcellular fraction and its uptake into cells is much more rapid than of transferrin-iron. Such low molecular weight iron complexes may be responsible for the generation of hydroxyl-radical-induced lipid peroxidation.[12,16-20] Indeed, there is direct evidence supporting increased in vivo lipid peroxidation, as reflected by elevated MDA or conjugated diene levels in iron-overloaded spleens from thalassemic patients[21] and in the liver and pancreas of iron-loaded rats and mice.[7,9,14,15,22] Furthermore, iron-induced tissue damage may be prevented by vitamin E, a potent antioxidant, or by iron chelators such as desferral.[15,23] Myocardial iron uptake, on the other hand, is enhanced during hypoxia, an interaction which may explain the increased incidence of myocardial failure in anemic as compared to hypertransfused thalassemic patients.[24]

II. THE CULTURED HEART CELL SYSTEM

Rat myocardial cell cultures, in contrast to most other models of iron toxicity, provide a system in which impairment of function, namely, myocyte contractility, may be studied in conjunction with structural and biochemical changes, with complete control of the extra-cellular environment. The contractility of such cultured cells, their metabolism, electron microscopic structure, and their electric function resemble those of the intact heart. In some recent preliminary studies of iron uptake by myocardial cell cultures, it has been shown that they are able to assimilate iron into ferritin and membrane-bound structures. About one half of ferric ammonium citrate assimilated by these cells is incorporated into ferritin.[25,26] Electron microscopic examination of these iron-loaded cells showed endocytotic vesicles and lyso-somes containing iron bound to ferritin, the concentration of which increased with time of incubation and in proportion to extracellular iron concentrations.[25] Iron-chelating drugs, such as deferioxamine and agrobactin, inhibited iron uptake from transferrin and other sources.[27]

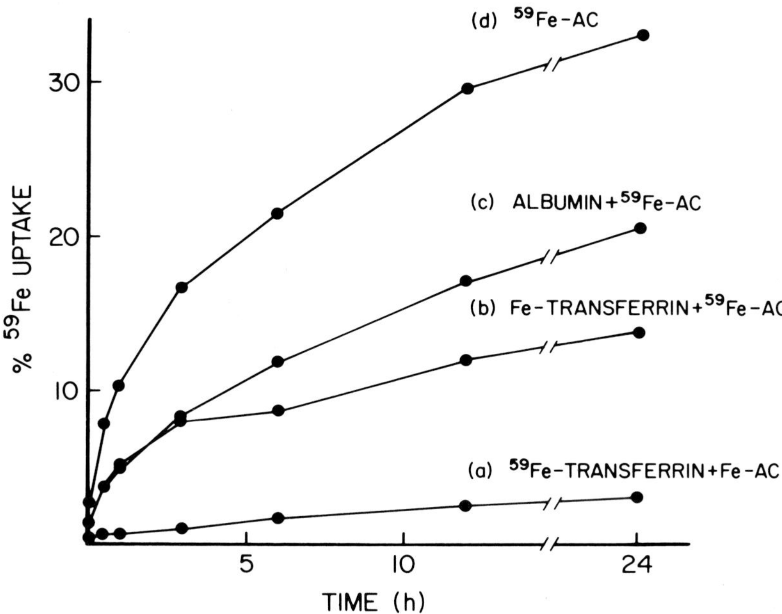

FIGURE 1. Effect of culture medium composition on the iron uptake. Iron (FeAC) concentration in all the cultures was 20 μg iron per milliliter. (a) Tracer ^{59}Fe was allowed to bind to transferrin prior to saturation with FeAC; (b) saturation of transferrin with FeAC prior to the addition of ^{59}FE; (c) serum-free medium supplemented with albumin; (d) serum-free medium. In (c) and (d) transferrin was not present in the incubation mixture.

Our studies have been designed to address three questions related to the use of beating myocardial cell cultures as an in vitro model of iron toxicity: (1) to define the conditions permitting iron loading in short-term studies of iron toxicity; (2) to examine whether or not these conditions may be associated with increased lipid peroxidation as evidenced by the generation of malonyldialdehyde; and (3) to study the ability of deferioxamine (DF) to mobilize iron in vitro from iron-loaded myocardial cells and to examine whether iron chelation is associated with a decrease in lipid peroxidation. The methods employed in these studies have been described in detail in a recent publication.[28]

These studies have shown that the single most important factor determining the rate of myocardial cell iron uptake is the composition of the culture medium. As shown by Fava et al.,[29] cardiac myocytes similar to many other cells normally acquire iron from diferric transferrin through a process involving high affinity, specific binding to membrane receptors. The rate of iron uptake from transferrin is low. Our data indicate that transferrin ^{59}Fe uptake is only one-tenth the uptake of ^{59}Fe-ammonium citrate from serum-free medium (Figure 1, a and d). Considering the fact that the maximal transferrin iron binding capacity of 20% serum-supplemented medium was 0.6 μg/mℓ as against the 20-μg/mℓ total iron content of the incubation mixture, the difference in rates of iron uptake from the transferrin and nontransferrin compartments was more than 1:300.

When transferrin in serum-supplemented medium was presaturated with FeAC and then followed by the radio-iron label which was now diluted in a large nontransferrin iron pool (Figure 1b), radio-iron uptake was higher than ^{59}Fe-transferrin uptake, but lower than ^{59}Fe uptake from serum-free medium. As the reduction in ^{59}Fe uptake can clearly not be attributed to iron binding to transferrin, we assumed that it may reflect a nonspecific effect of albumin and other serum proteins which may have formed a loose complex with nontransferrin iron, thereby reducing the rate of its uptake by cultured cells. This hypothesis is strongly supported by results of the next study (Figure 1c) in which albumin at a concentration representing

Table 1
THE EFFECT OF TEMPERATURE ON ^{59}FeAC UPTAKE

Time (hr)	Medium plus 20% serum			Serum-free medium		
	37°C	28°C	10°C	37°C	28°C	10°C
	(n)			(n)		
0	(3) 0.48 ± 0.06[a]	—	—	(3) 0.91 ± 0.10	—	—
0.5	(3) 1.82 ± 0.14	1.35 ± 0.24	1.07 ± 0.11	(3) 1.83 ± 0.25	1.44 ± 0.17	1.12 ± 0.11
1.0	(3) 2.49 ± 0.15	2.02 ± 0.33	1.19 ± 0.03	(3) 3.66 ± 0.37	2.19 ± 0.15	1.33 ± 0.17
3.0	(3) 3.30 ± 0.25	2.50 ± 0.30	1.47 ± 0.15	(3) 6.44 ± 0.64	4.95 ± 1.08	1.33 ± 0.02

[a] Mean ± I.S.D.

about 20% serum has been added to serum-free medium. The reduction in ^{59}Fe uptake following the addition of albumin to serum-free medium indicates that proteins other than transferrin are, indeed, able to slow down iron uptake from serum-free medium. It is possible that some of the variations in the rate of ^{59}Fe uptake from serum-supplemented medium may have reflected differences in the composition of nontransferrin proteins in the sera employed. Since we have shown that iron uptake under these conditions represents true uptake and not adherence to the sarcolemma,[28] it would seem that myocardial cells preferentially take up iron which is not bound to transferrin. This assumption is strongly supported by other clinical and laboratory observations. In congenital atransferrinemia small amounts of nontransferrin iron exist in the plasma.[31] These patients, if untreated, die of myocardial hemosiderosis in the first decade of their life. Similarly, in β-thalassemia major with severe transfusional siderosis, serum transferrin is saturated and a small fraction of the total serum iron appears as a nontransferrin fraction.[17] Similar to atransferrinemia, the most important cause of mortality in thalassemic patients is myocardial siderosis.[1] It should be emphasized, however, that the concentrations of nontransferrin iron employed in the present studies were about a hundred times higher than those observed in congenital atransferrinemia and thalassemia major. Our studies may, therefore, be more representative of acute than chronic iron poisoning.

The temperature dependence of iron uptake described in Table 1 is probably explained by reduced membrane fluidity at low temperatures, although inhibition of active transport cannot be ruled out. A similar reduction of iron uptake at low temperatures has been observed by Iacopetta and Morgan in rabbit reticulocytes.[32] The inability of cycloheximide to inhibit iron uptake from serum-supplemented and serum-free medium may indicate that iron uptake does not require the *de novo* production and (or) consumption of a membrane receptor.

Morphologic evidence of cell damage by iron overload was furnished by electron micrographs of cardiomyocytes previously exposed to medium containing high iron concentrations for 3 or 24 hr. These studies showed the accumulation of mitochondrial-dense bodies and iron-rich ferritin particles in membrane-bound structures (probably lysosomes) preceded by the formation of iron-containing particles in the cytosol. Nonmuscle cells such as fibroblasts present in the same cultures, although capable of iron uptake and ferritin synthesis, were less affected by iron overload.[25,33] These findings were further extended to Mössbauer spectroscopy of iron-loaded cells showing a gradual shift of iron particles in the cell from small particles (<10 Å) to larger ones (>25 Å) typical of ferritin aggregates, over a period of 24 hr.[34] It was further shown by Mössbauer spectroscopy that in contrast to transferrin iron which is transported in the reduced ferrous form (Fe^{2+}), iron uptake from FeAC at high extracellular concentrations occurs in the ferric (Fe^{3+}) form and is followed by the incorporation of trivalent iron into ferritin molecules.[34]

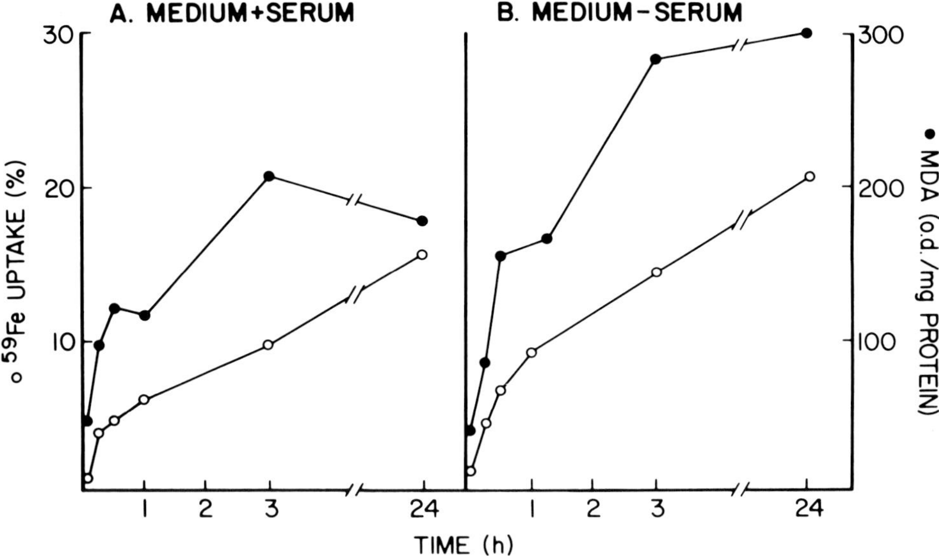

FIGURE 2.   Correlation between (○) radio-iron uptake and (●) cellular MDA formation. The iron-containing incubation medium was either (A) serum-supplemented or (B) serum-free.

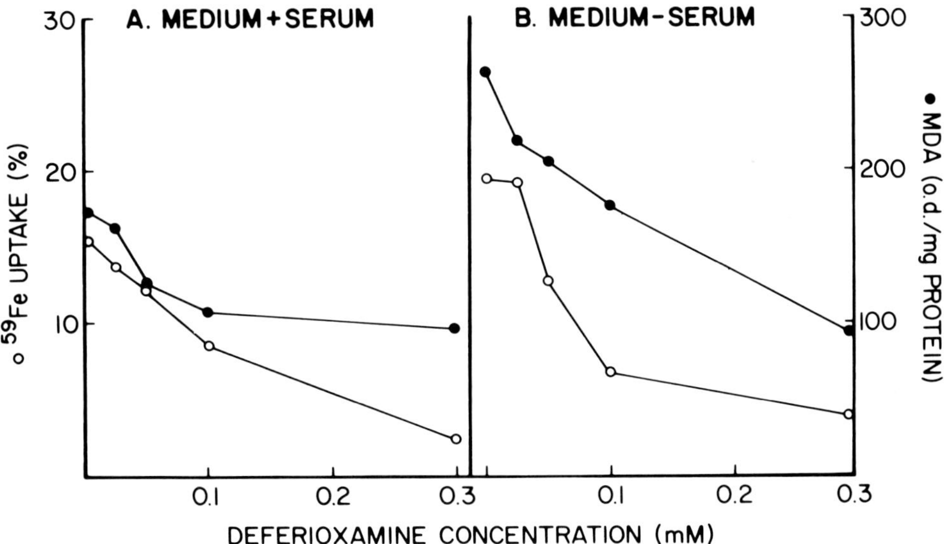

FIGURE 3.   Effect of deferioxamine on iron chelation and MDA levels. Correlation between (○) residual radioactivity and (●) cellular MDA concentrations following incubation for 24 hr with increasing concentrations of deferioxamine. The iron-containing incubation medium was either (A) serum-supplemented or (B) serum-free.

An important aspect of our studies was the correlation between myocardial cell iron loading and cellular malonyldaldehyde concentrations: iron loading resulted in a rapid increase in cellular MDA concentrations with a doubling of baseline values at 15 min and with maximal levels reached within 3 hr (Figure 2). Conversely, iron mobilization by in vitro desferioxamine treatment resulted in normalization of cellular MDA concentrations (Figure 3). These findings may be relevant to our previous observation that in vitro iron loading of myocardial cells results in a gradual reduction of contractility and increased rates and irregularity of heart cell beats and that these effects are reversed by iron chelation.[35,36] Although it is tempting to interpret these correlations as evidence of a cause-and-effect relation between iron-induced membrane lipid peroxidation and impaired cellular function, such a relation cannot be taken for granted until more direct evidence concerning the mechanism of impaired cellular contractility is provided.

III. CONCLUSIONS AND NEW PERSPECTIVES

These studies provide the background for future investigations in which the relation between iron loading and altered myocardial cell function may be explored. Such studies should provide a link between free radical formation resulting from iron loading, peroxidation of sarcolemmal and other membrane lipids, and a reversible functional derangement of cultured, beating myocardial cells. Although a primary sarcolemmal effect is the most likely mechanism explaining the functional impairment of iron-loaded myocardial cells in culture, alternative mechanisms will have to be explored. Thus, reduced mitochondrial phosphorylation or intracellular release of lysosomal enzymes will have to be carefully documented, and the time relation between such possible effects of in vitro iron loading and the anticipated changes in lipid membrane structure should be studied.

The model presented above, the only in vitro model of iron overload affecting cellular function, describes the acute effects of iron loading. It would clearly be instructive if the chronic effects of iron overload could be simulated in such a system, particularly if iron chelators proved able to attenuate these toxic effects. Further studies on the mechanism ofimpaired automatic rhythmic contractions in the cardiomyocyte and the related changes in sarcolemmal composition and fluidity should increase our understanding of cardiac dysfunction in hemochromatosis.

ACKNOWLEDGMENT

Studies from the authors' laboratories reported in this review were supported by National Institutes of Health grant no. 34062-01 awarded to C.H. and grants for Basic Research of the Israel Academy of Sciences and Humanities and the Rivka and Salomon Benador Foundation for Heart Research awarded to A.P. and C.H.

REFERENCES

1. **Buja, L. M. and Roberts, W. C.,** Iron in the heart: etiology and clinical significance, *Am. J. Med.,* 51, 209, 1971.
2. **Engle, M. A., Erlandson, M., and Smith, C. H.,** Late cardiac complications of chronic, severe, refractory anemia with hemochromatosis, *Circulation,* 30, 698, 1964.
3. **Brown, E. B., Dubach, R., and Smith, D. E.,** Studies in iron transportation and metabolism. X. Long term iron overload in dogs, *J. Lab. Clin. Med.,* 50, 862, 1957.
4. **Awai, M., Narasaki, M., Yamanoi, Y., and Seno, S.,** Induction of diabetes in animals by parenteral administration of ferric nitriloacetate, *Am. J. Pathol.,* 95, 663, 1979.
5. **Ganote, C. E. and Nahara, G.,** Acute ferrous sulphate hepatotoxicity in rats. An electronmicroscopic and biochemical study, *Lab. Invest.,* 28, 426, 1973.
6. **Reissmann, K. R. and Coleman, T. J.,** Acute intestinal iron intoxication. II. Metabolic, respiratory and circulatory effects of absorbed iron salts, *Blood,* 10, 46, 1955.
7. **Hanstein, W. G., Heitman, T. D., and Sandy, H. I.,** Effects of hexachlorobenzene and iron loading on rat liver mitochondria, *Biochem. Biophys. Acta,* 678, 293, 1981.
8. **Hanstein, W. G., Sacks, P. V., and Muller-Eberhard, U.,** Properties of liver mitochondria from iron loaded rats, *Biochem. Biophys. Res. Commun.,* 67, 1175, 1975.

9. **Peters, T. J. and Seymour, C. A.,** Acid hydrolase activities and lysosomal integrity in liver biopsies from patients with iron overload, *Clin. Sci. Mol. Med.,* 50, 75, 1976.
10. **Seymour, C. A. and Peters, T. J.,** Organelle pathology in primary and secondary hemochromatosis with special reference to lysosomal changes, *Br. J. Haematol.,* 40, 239, 1978.
11. **Fitchett, D. H., Coltart, D. J., and Littler, W. A.,** Cardiac involvement in secondary hemochromatosis: a catheter biopsy study and analysis of myocardium, *Cardiovasc. Res.,* 14, 719, 1980.
12. **Jacobs, A., Hoy, T., and Humphreys, J.,** Iron overload in Chang cell cultures. Biochemical and morphological studies, *Br. J. Exp. Pathol.,* 59, 489, 1978.
13. **Sanyal, S. K., Johnson, W., and Jalalakshmamma, B.,** Fatal "iron heart" in an adolescent: biochemical and ultrastructural aspects of the heart, *Pediatrics,* 55, 336, 1975.
14. **Bacon, B. R., Tavill, A. S., and Brittenham, G. M.,** Hepatic lipid peroxidation *in vivo* in rats with chronic iron overload, *J. Clin. Invest.,* 71, 429, 1983.
15. **Wills, E. D.,** Lipid peroxide formation in microsomes. The role of non-haem iron, *Biochem. J.,* 113, 315, 1969.
16. **Halliwell, B.,** Superoxide-dependent formation of hydroxyl radicals in the presence of iron salts is a feasible source of hydroxyl radicals *in vivo, Biochem. J.,* 205, 461, 1982.
17. **Hershko, C., Graham, G., Bates, G. W., and Rachmilewitz, E. A.,** Non-specific iron in thalassemia: an abnormal serum iron fraction of potential toxicity, *Br. J. Haematol.,* 40, 255, 1978.
18. **White, G. P. and Jacobs, A.,** Iron uptake by Chang cells from transferrin, nitriloacetate and citrate complexes. The effects of iron loading and chelating with desferrioxamine, *Biochem. Biophys. Acta,* 543, 217, 1978.
19. **Beamish, M. R., Walker, R., and Miller, F.,** Transferrin iron, chelatable iron and ferritin in idiopathic hemochromatosis, *Br. J. Haematol.,* 27, 219, 1974.
20. **Goto, Y. and Listowsky, I.,** Ferritin synthesis and iron incorporation in cells grown in trasnferrin-free media, in *Structure and Function of Iron Storage and Transport Proteins,* Urushizaki, I., Ed., Elsevier/North-Holland, Amsterdam, 1983, 117.
21. **Nienhuis, A. W., Benz, E. J., and Propper, R.,** Thalassemia major: molecular and clinical aspects, *Ann. Intern. Med.,* 91, 883, 1979.
22. **Yamanoi, Y., Matsuura, R., and Awai, M.,** Mechanism of iron toxicity in the liver and pancreas after a single injection of ferric nitroloacetate, *Acta Haematol. Jpn.,* 45, 1229, 1982.
23. **Goldberg, L. and Smith, J. P.,** Changes associated with the accumulation of excessive amounts of iron in certain organs of the rat, *Br. J. Exp. Pathol.,* 39, 59, 1958.
24. **Necheles, T. F., Beard, M. E. J., and Allen, D. M.,** Myocardial haemosiderosis in hypoxic mice, *Ann. N.Y. Acad. Sci.,* 165, 167, 1969.
25. **Cox, P. G., Harvey, N. E., Sciortino, C., and Byers, B. R.,** Electron-microscopic and radioiron studies of iron uptake in newborn rat myocardial cells *in vitro, Am. J. Pathol.,* 102, 151, 1981.
26. **Byers, B. Y., Sciortino, C. V., Cox, P., Robinson, P., and Hillibert, S.,** Iron transport and distribution in cultures of beating heart cells, in *The Biochemistry and Physiology of Iron,* Saltmann, P. and Hegenauer, J., Eds., Elsevier/North-Holland, Amsterdam, 1982, 179.
27. **Sciortino, C. V., Byers, B. R., and Cox, P.,** Identification of a potential drug for drug therapy, *J. Lab. Clin. Med.,* 96, 6, 1980.
28. **Link, G., Pinson, A., and Hershko, C.,** Heart cells in culture: a model of myocardial iron overload and chelation, *J. Lab. Clin. Med.,* 106, 147, 1985.
29. **Fava, R. A., Comeau, R. D., and Woodworth, R. C.,** Specific membrane receptors for differic transferrin in cultured rat skeletal myocytes and chick-embryo cardiac myocytes, *Biosci. Rep.,* 1, 377, 1981.
30. **Cook, J. D., Hershko, C., and Finch, C. A.,** Storage iron kinetics. V. Iron exchange in the rat, *Br. J. Haematol.,* 25, 695, 1973.
31. **Goya, N., Miyazaki, S., and Kodate, S.,** A family of congenital atransferrinemia, *Blood,* 40, 239, 1972.
32. **Iacopetta, B. J. and Morgan, E. H.,** The kinetics of transferrin endocytosis and iron uptake from transferrin in rabbit reticulocytes, *J. Biol. Chem.,* 258, 9108, 1983.
33. **Iancu, T. C., Shilo, H., Link, G., Bauminger, E. R., Pinson, A., and Hershko C.,** Ultrastructural pathology of iron-loaded rat myocardial cells in culture, *Br. J. Exp. Pathol.,* 68, 53, 1987.
34. **Bauminger, E. R., Iancu, T. C., Link, G., Pinson, A., and Hershko, C.,** Iron overload in cultured rat myocardial cells, *Hyperfine Interactions,* 33, 249, 1987.
35. **Link, G., Urbach, J., Hasin, Y., Pinson, A., and Hrshko, C.,** Beating rat heart cell cultures: an *in vitro model* of iron toxicity and chelating therapy, *Blood,* 62(5) (Suppl. 1) (Abstr. No. 38a), 9, 1985.
36. **Moreb, J., Hershko, C., and Hasin, Y.,** Effect of acute iron loading on contractility and spontaneous beating rate of cultured rat myocardial cells, submitted.

Chapter 30

DIFFERENTIAL SENSITIVITY OF HEART FIBROBLASTS AND MYOCYTES TO 7β-HYDROXYCHOLESTEROL

M. Mersel, H. Hietter, and B. Luu

TABLE OF CONTENTS

I. SUMMARY

Primary cultures derived from neonatal rat hearts were treated with 7β-hydroxycholesterol (7βOH-CH, 18 μ*M*). The fibroblasts no longer adhered to the solid substratum and were all removed by simply washing the cell cultures. Myocardial cells still remained attached to the solid substratum and their beating rhythm, which had changed, was restored by replacing the 7βOH-CH-containing medium with fresh culture medium.

Treatment with 7βOH-CH of pure cardiac fibroblast subcultures and enriched primary myocardial cultures had highly toxic effects to the former and reversibly affected the beating rhythm of the latter.

[^{14}C-4]7βOH-CH was synthesized so that it could be used to follow its metabolic pathway in myocardial and fibroblast cultures. A major ^{14}C-labeled metabolite could be extracted from fibroblast cultures after 12 hr of treatment, and its concentration remained constant for 48 hr. In contrast, in enriched myocardial cultures, only a small amount of ^{14}C-labeled metabolite could be detected. This metabolite was almost absent in the culture medium and dead fibroblasts did not continue to metabolize 7βOH-CH.

Extraction and purification of this metabolite and subsequent treatment of the cultures hardly affected the fibroblasts and not at all the myocardial cells. Preliminary results using GLC-mass spectrometry revealed that this compound appeared to be a steroidal derivative.

These data demonstrate: (1) 7βOH-CH is differentially metabolized by cardiac fibroblasts and myocytes and (2) the 7βOH-CH metabolite is toxic to his own genitor cell.

II. INTRODUCTION

Newborn rat heart cultured myocytes possess the main morphological and biochemical characteristics of the cardiac tissue in vivo capable of beating spontaneously and synchronously.[1,2] Thus, such heart cell cultures provide a suitable model for studying morphological and biochemical differentiation, regulation of the synthesis and spatial coordination of contractile proteins, ion transport and channeling. responses to external chemical or physical stimuli, and the initiation of the action potential and its propagation along and across the sarcolemma. However, these highly differentiated cells are not suited for investigations of the relationship(s) between differentiation, aging, and spontaneous transformation. Any attempt to obtain subcultures of both newborn and embryonic rat heart resulted either in the loss of the cardiac tissue properties[1] or in the subcultured cells becoming similar to skeletal muscle cells.[3]

On one hand, from a pharmacological point of view, heart cell primary cultures are not used for research in cancer chemotherapy, since heart cancer has rarely been reported; on the other hand, these cultures may serve to test the possible side effects of various anticancer compounds.[4,5] 7βOH-CH has been shown to be toxic to HTC and Zajdela ascitic hepatoma cells,[6] whereas it had little effect on normal mouse fibroblasts.[7] Studies on its mechanism of action, at both cellular and molecular levels,[7-10] as well as on its in vivo antitumor properties,[11] have been reported. Primary cultures of neonatal rat hearts contain both myocytes and fibroblasts (mixed cultures) and a selective method may be employed to obtain enriched heart myocyte cultures.[12] We showed, using these two culture systems, that heart fibroblasts and myocytes behaved differentially towards 7βOH-CH.[13] Further investigations indicated that the active compound was not 7βOH-CH itself, but a metabolite specifically synthesized by fibroblasts and only toxic to its genitor cells.

This report discusses the possibility that newborn rat heart cell cultures may serve as a useful model for studying: (1) the regulation of enzyme(s) involved in steroidal metabolism at a cellular, biochemical, and molecular level; and (2) the relationship of this regulation to cellular aging and differentiation.

III. MATERIALS AND METHODS

A. Synthesis of [^{14}C-4]-7β-Hydroxycholesterol 7βOH-CH)

Microsynthesis of [^{14}C-4]-7βOH-CH was carried out according to Rong[14] with the following modifications: the starting material used was [^{14}C-4]cholesterol purchased from CEA France (specific activity, 50 mCi/mmol) and diluted 1:1 (v/v) with unlabeled cholesterol (Roth, Germany). All the reagents were of analytical grade (Prolabo, France). The yield of the β-isomer was increased by reducing 7-oxo-cholesterol acetate with Dibal (Aldrich, U.S.), at −78°C for 30 min. Purification of 7βOH-CH was achieved by thin-layer chromatography (TLC) on silica gel-coated plates (60F-254, Merck, Darmstadt, Germany), developed either in diethylether/methylene chloride (1:1) or ethyl acetate/acetic acid (99:1). Labeled 7βOH-CH was detected by a β-scanning analyzer (Berthold, Germany), the radioactive band was scraped off, and 7βOH-CH was extracted with methylene chloride/methanol (2:1). Before every experiment, 7βOH-CH was checked for purify by TLC.

B. Cell Cultures

1. Primary cardiac mixed cell cultures containing both fibroblasts and myocytes were obtained essentially according to Pinson et al.[2]
2. Enriched myocardial cell primary cultures were obtained in one of two ways: either (1) only ventricular apices were dissociated enzymatically before cell seeding or (2) advantage was taken of the fact that 7βOH-CH was almost exclusively cytotoxic to fibroblasts under the experimental conditions used, and mixed cultures were treated in the following manner. Three-day-old mixed cultures were treated with 7βOH-CH at a final concentration of 18 μM. When most of the fibroblasts had been injured, they were removed from the cultures by rinsing several times with fresh culture medium. The myocardial cells, remaining in culture, recovered their synchronous beating ability after 30 hr.
3. Pure cardiac fibroblast secondary cultures were obtained as follows. After removing the culture medium, primary cardiac mixed cell cultures were dissociated by adding 0.05% Trypsin-EDTA (Gibco, 2 mℓ/35 mm Ø Falcon culture dish) for 10 min at room temperature, the cells were collected and suspended using a 20-mℓ syringe, the cell suspension filtered through a 82-μm nylon sieve, and it was finally centrifuged at 800 × g for 15 min. The pellet was then suspended in DMEM (Fibco) supplemented with 3% glucose and 20% FCS, the cells seeded on 35 mm Ø culture dishes (Falcon), and the cells kept in an atmosphere of 5% CO_2 at 37°C.

The medium in all types of cell cultures was changed every 2 days, unless otherwise indicated.

C. Treatment of Cells with [^{14}C-4]7β-Hydroxycholesterol

[^{14}C-4]7βOH-CH was generally added as an ethanolic solution to 3- to 8-day-old cultures to a final concentration of 18 μM. It was checked that ethanol addition (5 to 7 $\mu\ell$/2 mℓ culture medium) had no effects either on cell morphology or on the synchronous beating of myocardial cells. In order to prove that the 7βOH-CH metabolites were synthesized exclusively by living cells, we added [^{14}C-4]7βOH-CH, either to dead cells which remained attached to the culture dish (the cell medium was removed and the cultures heated at 100°C for 9 hr) or to culture dishes containing medium only. The incubation times were 12, 24, 36, and 48 hr.

D. Cell Viability

The cell viability was tested by exclusion of Trypan blue dye.

E. Cell Counting

Cell cultures were rinsed with a phosphate buffered saline solution, incubated with 0.5 mℓ of 0.1% Trypsin in a Tyrode solution for 10 min at 37°C, and the cells collected and then counted in a Neubauer chamber.

F. Morphological Studies

Cells were observed with a Nikon phase contrast microscope and photographed with a Nikon camera.

G. Extraction, Separation, and Identification of the 7βOH-CH Derivatives

At the end of each incubation, cells or culture medium were extracted according to Folch et al.[15] After partitioning with 0.74% KCl aqueous solution, the organic phase was evaporated under reduced pressure, and the lipidic residue spotted onto TLC plates (silica gel 60F-254, Merck, Darmstadt, Germany) and developed in ethyl acetate/acetic acid (99:1). The radio-labeled bands were located using a β-scanner TLC analyzer, the radioactive bands scraped off separately, and the material extracted overnight with methylene chloride/methanol (2:1) at room temperature with magnetic stirring (about 90% of the labeled material was recovered by extraction). The radioactivity was counted in Rotiszint 22 (Roth, Karlsruhe, Germany) fluid using a Beckman 2000 scintillation counter.

H. Proteins

Protein concentrations were determined according to the method of Lowry et al.[16]

I. Mass Spectrometry

The methylene chloride-methanol extract was analyzed on an LKB 9000 mass spectrometer coupled to a gas chromatograph (separator temperature, 270°C; ion source temperature, 290°C, electron energy, 70 eV; and acceleration voltage, 3500 V) equipped with a computer system that allowed the recording of one mass spectrum every 5 sec.

IV. RESULTS AND DISCUSSION

Methods of obtaining myocardial cultures derived from neonatal rat heart are well established. However, if the cells to be seeded originate from the whole organ, myocardial cultures are contaminated by fibroblast-like cells (mixed cultures) and, in some cases, fibroblasts represent up to 50% of the cells in culture. Enriched myocardial cell cultures may be produced by a refinement of the seeding techniques: either by trypsinization of the heart apices exclusively, prior to cell seeding, or by the "preplating" technique.[12,13] The cultures yielded are highly enriched in myocardial cells, with thick synchronously beating syncytia. Sub-culturing primary cultures of heart cardiac cells provided pure "fibroblast" cultures (Figure 1a), since the highly differentiated myocardial cells were unable to undergo further cell division to generate "newborn" myocardial cells. In our culture conditions, heart fibroblast secondary cultures reached confluency on the 5th day in culture, were up to 95% viable, and ceased to undergo cell division (approximately 800 μg proteins and 3 × 10⁵ cells per culture dish).

Treatment with 7βOH-CH of either mixed primary cultures (4 days in culture) or fibroblast secondary cultures (3 days in culture), to a final concentration of 18 μ*M*, had drastic effects on the fibroblasts. Within 24 hr of incubation, cells underwent progressive morphological changes (Figure 1b), no longer adhering to the solid substratum, and were removed by simply washing the culture dish either with a Tyrode solution or with simple DMEM fresh culture medium. When enriched myocardial cell cultures were treated similarly, detachment of fibroblasts was not observed, probably due to the thick syncytia entraping the fibroblastic areas.

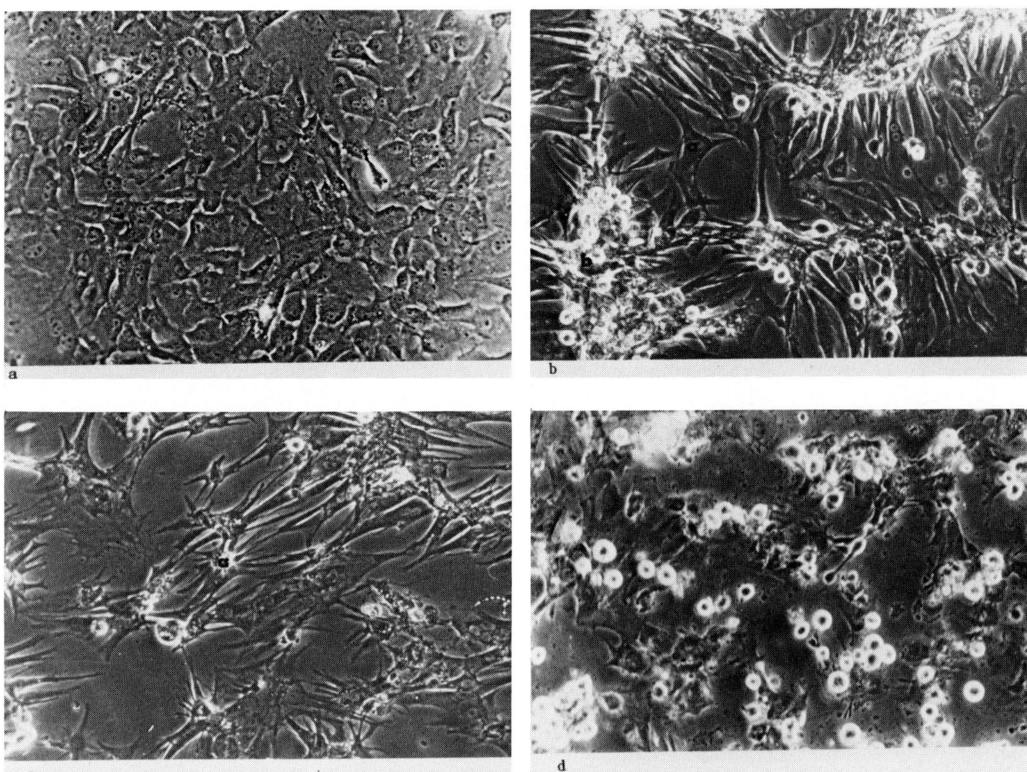

FIGURE 1.   (a) Phase contrast micrograph of rat heart subcultures (second passage, 3 days in culture); note the absence of myocardial cells and the presence of fibroblasts in division. (b) Phase contrast micrograph of cultured heart cells treated with 7βOH-CH (18 μM); note the retraction of the fibroblasts (a) and the presence of cellular debris adhering to the syncyntia (b). (c) Phase contrast micrograph of the same culture 24 hr after removing the 7βOH-CH containing medium and adding simple culture medium; only a few fibroblasts were detected; (a) and the remaining syncytia recovered their synchronous beating rhythm. (d) Phase contrast micrograph of fibroblast secondary cultures (3 days old) treated with the 7βOH-CH metabolite. Note the cytotoxic effect of this compound after 3 hr of treatment.

Myocardial cells behave differently on treatment with 7βOH-CH. They continue to adhere to the support, but their beatings progressively become arrhythmic and asynchronous, and some of the syncytia even stopped beating completely. However, washing the treated myocardial cells with Tyrode solution or DMEM and subsequent incubation of the cultures with culture medium in the absence of 7βOH-CH restored the syncytia with respect to morphological criteria (Figure 1c) and the synchronous beating within 24 hr.

The above observations raised key questions at a molecular and functional level: is 7βOH-CH itself or a related metabolite, specifically, cytotoxic to fibroblasts? What is the molecular link between 7βOH-CH and the perturbation of the beating rhythm in the treated myocardial cells?

We focused our studies on the metabolic pathways of 7βOH-CH when added to enriched myocardial cells, mixed cells, and fibroblast cultures. Therefore, we initially used labeled 7βOH-CH diluted with nonlabeled (cold) 7βOH-CH, added to the cell cultures at a concentration of 18 μM.

We took also advantage of the fact that 7βOH-CH was specifically cytotoxic to fibroblasts; thus, by first treating mixed cultures with unlabeled 7βOH-CH, removing the dead fibroblasts, and replacing fresh culture medium, we were able to obtain almost pure cultures of myocardial cells. These cultures (''cold-treated mixted cultures'') (Figure 1c) were then treated with labeled 7βOH-CH.

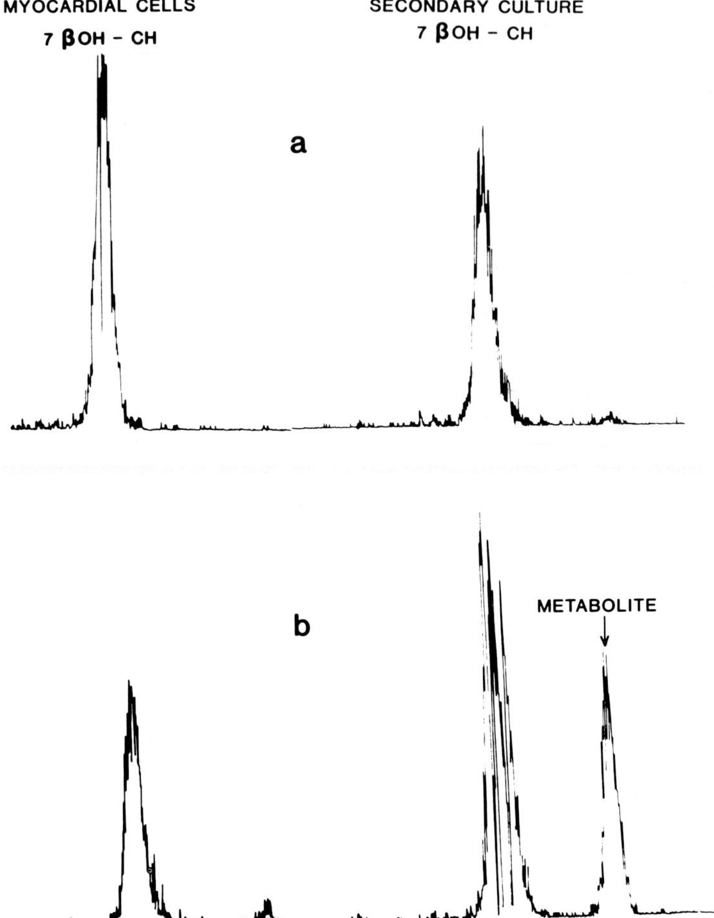

FIGURE 2.   β-Thin-layer chromatogram. Mixed cell cultures, fibroblast secondary cultures, and enriched myocardial cell cultures were treated with radioactive 7βOH-CH, the hydrophobic material was then extracted, purified, and chromatographed on silica gel plates (for more details, see Section III). 7βOH-CH: 7β-hydroxycholesterol. (a) After 10 min of incubation; (b) after 12 hr of incubation; front of the chromatographic run: right side of the β-chromatograms.

There was no evidence of labeled metabolites other than 7βOH-CH, up to 48 hr after treatment of either enriched myocardial cell cultures or of "cold-treated mixed cultures" with labeled 7βOH-CH, followed by extraction, purification, and detection of the labeled substances (Figure 2). However, on treatment of mixed cell cultures or fibroblast secondary cultures, a labeled compound was detected after 12 hr and persisted until no fibroblasts remained attached to the culture dish (Figure 2). This substance, which represents 10% of the amount of incorporated 7βOH-CH, was less polar than 7βOH-CH according to its chromatographic properties, and preliminary mass spectrometric data suggest a steroidal structure.

Addition of 7βOH-CH either to culture medium alone or to dead cells (see Section III) did not lead to formation of this labeled metabolite.

These data clearly demonstrate that in our culture conditions, 7βOH-CH was essentially metabolized by fibroblasts. Since myocardial cells were unable to metabolize 7βOH-CH, it may be assumed that 7βOH-CH itself affected the cardiac beating rhythm. However, determination of the cellular target which binds 7βOH-CH and the relationship between 7βOH-CH and cell contraction require further investigation.

The cytotoxic action of the metabolite upon fibroblasts was demonstrated as follows. The metabolite, after extraction and purification, was added to fibroblast secondary cultures. The cells underwent morphological changes after 3 hr and became detached from the support after 6 hr of treatment (see Figure 1d). Similar treatment of enriched myocardial cell cultures did not affect the cell morphology, but led to slight alteration of the beating rhythm.

In summary, our results demonstrate that (1) 7βOH-CH is differentially metabolized by heart fibroblast-like and myocardial cell cultures and (2) the cytotoxic effect of 7βOH-CH on the fibroblasts is not due to 7βOH-CH itself, but to a metabolite.

These data, which describe the synthesis of a metabolite which is cytotoxic to its genitor cell, raise many questions. Is there a link between the selective killing effect of Ca ionophore A 23 187 and the cytotoxic effect of 7βOH-CH on heart fibroblasts?[17] Are both 7βOH-CH and Ca ionophore interfering with membrane-bound protein kinase C which seems to be involved in calcium-dependent phenomena of proliferation?[18] Why was 7βOH-CH not metabolized by myocardial cells? Is it because it does not cross the sarcolemma or because the enzymatic machinery metabolizing 7βOH-CH is blocked or not expressed under the conditions of our experiments? Are the myocardial cells unable to metabolize 7βOH-CH because they are highly differentiated? Would treatment of such cells with oncogeneses or cell division-stimulating factors initiate metabolism of 7βOH-CH? Would the ability to metabolize 7βOH-CH be inhibited in fibroblasts forced to differentiate or to age?[19] In which subcellular fraction is the enzymatic machinery localized?

Such topics, on which the spade work remains to be done, indicate that studies of the effects of 7βOH-CH on heart cell cultures may serve as a promising model of differentiation, aging, and transformation at both cellular and molecular levels.

ACKNOWLEDGMENTS

We wish to thank Dr. A. Pinson for his valuable advice regarding this study.

REFERENCES

1. **Harary, I. and Farley, B.,** In vitro studies on single beating rat heart cells, *Exp. Cell Res.,* 29, 451, 1963.
2. **Pinson, A. and Padieu, P.,** Les cultures de cellules isolées: technique croissance et application, *Cah. Nut. Diet.,* 9, 287, 1975.
3. **Kimes, B. W. and Brandt, B. L.,** Properties of a clonal muscle cell line from rat heart, *Exp. Cell Res.,* 98, 367, 1976.
4. **Dagani, R.,** In vitro methods may offer alternatives to animal testing, *Chem. Eng. News,* 12, 25, 1984.
5. **Seraydarian, M. W. and Artaze, L.,** Modification by adenosine of the effect of adriamycin of myocardial cells in culture, *Can. Res.,* 39, 2940, 1979.
6. **Cheng, K. P., Nagano, H., Luu, B., Ourisson, G., and Beck, J. P.,** Chemistry and biochemistry of chinese drugs. I. Sterol derivatives cytotoxic to hepatoma cells isolated from the drug *Bombyx cum Botryte, J. Chem. Res.,* (S), 217, 1977; (M), 2501, 1977.
7. **Bergmann, C.,** Activité Cytotoxique et Antitumorale de Stérols Hydroxylés Isolés à Partir de Drogues Extrême-Orientales, Thesis, Université Louis Pasteur, Strasbourg, France, 1982.
8. **Richert, L., Bergmann, C., Beck, J. P., Rong, S., Luu, B., and Ourisson, G.,** The importance of serum lipoproteins in the cytolytic action of 7β-hydroxycholesterol on cultured hepatoma cells, *Biochem. Biophys. Res. Commun.,* 117, 851, 1983.

9. **Richert, L., Castagna, M., Beck, J. P., Rong, S., Luu, B., and Ourisson, G.,** Grown-rate-related and hydroxysterol induced changes in membrane fluidity of cultured hepatoma cells: correlation with 3-hydroxy-3-methyl glutaryl CoA reductase activity, *Biochem. Biophys. Res. Commun.,* 120, 192, 1984.

10. **Hietter, H., Trifilieff, E., Richert, L., Beck, J. P., Luu, B., and Ourisson, G.,** Antagonistic action of cholesterol towards the toxicity of hydroxysterols on cultured hepatoma cells, *Biochem. Biophys. Res. Commun.,* 120, 657, 1984.

11. **Rong, S., Bergmann, C., Luu, B., Beck, J. P., and Ourisson, G.,** Activité antitumorale in vivo de dérivés hydrosolubles de 7-hydroxy-cholesterols, *C.R. Acad. Sci. Paris,* 300, 89, 1985.

12. **Blonlel, B., Roijen, I., and Cheneval, J. P.,** Heart cells in culture: a simple method for increasing the proportion of myoblasts, *Experientia,* 27, 356, 1971.

13. **Mersel, M., Luu, B., and Ourisson, G.,** Action differentielle du 7β-hydroxycholesterol sur des cellules myocardiales et des fibroblastes. Obtention de cultures primaires de cellules myocardiales pures, *C.R. Acad. Sci. Paris,* 299, 221, 1984.

14. **Rong, S.,** Activité Antitumorale de Dérivés Hydrosolubles de Stéroides Polyoxygénés, Thesis, Université Louis Pasteur, Strasbourg, France, 1983.

15. **Folch, J., Lees, M., and Sloane-Stanley, G. H.,** A simple method for the isolation and purification of total lipids from animal tissues, *J. Biol. Chem.,* 226, 497, 1957.

16. **Lowry, O. H., Rosebrough, N. J., Farr, A. L., and Randall, R. J.,** Protein measurement with Folin phenol reagent, *J. Biol. Chem.,* 119, 265, 1951.

17. **Ki Kaneko, K. and Goshima, K.,** Selective killing of fibroblast like cells in cultures of mouse heart cells by treatment with A Ca ionophore, A 23187, *Exp. Cell Res.,* 146, 142, 1982.

18. **Donnelly, T. E., Sittler, R., Jr., and Scholar, E. M.,** Relationship between membrane bound protein kinase C activity and calcium dependent proliferation of BALB/c 3T3 cells, *Biochem. Biophys. Res. Commun.,* 126, 741, 1985.

19. **Bell, E., Marek, L. F., Levinstone, D. S., Merrill, C., Sher, S., Young, I. T., and Eden, M.,** Loss of division potential in vitro: aging or differentiation, *Science,* 202, 1158, 1978.

Chapter 31

LASER MICROIRRADIATION OF HEART CELLS IN CULTURE

Margarita C. Kitzes and Michael W. Berns

TABLE OF CONTENTS

I. INTRODUCTION

Bessis et al.[1] first used a ruby laser to alter subcellular organelles in 1962. Since then, the technique of laser-microbeam irradiation has been principally applied to studies of nuclear and cytoplasmic functions. Several reviews have been published on this subject.[2-6] During the last 12 years, the development of biotechnology in combination with the use of the highest quality optical components has perfected this technique.[6,7] Selective alteration of part of a subcellular organelle can be confined to an area of less than 0.25 μm. This capability has made laser-microbeam irradiation, in combination with other techniques, an important tool in the study of an array of specific and basic problems in cell and developmental biology.[6,7]

II. REVIEW

The structural and functional effects of mitochondrial microirradiation have been extensively studied in neonatal rat cultured myocardial cells.[8-20] In these cells, single mitochondria can be selectively irradiated and are affected by laser wavelengths without prior treatment with photosensitizing agents. Cytochromes act as natural chromophores with specific absorption for the laser light.

Laser microirradiation of a single mitochondrion in cultured myocardial cells can result visually in specific alterations in the normal pattern of contractility in the target cell. These changes include an increase in contraction frequency, a cessation of contraction, and fibrillation. In a large number of cases, the cells return to pre-irradiation contraction rates and rhythmicity. This recovery occurs within a few minutes. The evoked alterations in contractility have been shown, both with phase-contrast and transmission-electron microscopy, to correlate with specific types of damage to the target mitochondrion.[9,11,13] Ultrastructural studies by Rattner et al.[13] demonstrated that a lack of contractile response was associated with ''weak lesions'' in the target organelle, while ''moderate lesions'' were associated with alterations in the normal spontaneous pattern of contractility. The morphology of the moderate lesions suggested a possible thermal effect that could be responsible for precipitating a change in contractility. Indeed, heat transfer to the overlying cell membrane may result in transient alteration of its physical characteristics and normal permeability. Nathan et al.[21] observed a direct effect of UV light on embryonic chick heart cell aggregates. Their results suggested a possible direct alteration of membrane-channel conductance by the incident light. Morphological studies provided information on the subcellular sites and characteristics of laser damage associated with changes in contractility, but did not clarify the physiological basis underlying them.

The intracellular study of the ongoing electrical activity prior, during, and after laser microirradiation was carried out in order to elucidate any alterations in transmembrane activity associated with or underlying the observed changes in contractility.[15] These studies demonstrated that microirradiation of a mitochondrion, indeed, results in characteristic alterations of the intracellular electrical activity in the myocardial cells. These alterations are accompanied by changes in their rhythmic contractility. The activity of the impaled and irradiated cell typically showed a depolarization at the time of irradiation and subsequent gradual recovery to its normal resting membrane potential and activity. These electrophysiological changes were accompanied, visually, by altered contractility with a return to rhythmic contraction. Furthermore, although all myocardial ventricular cells (pacemaker and non-pacemaker) were depolarized by laser microirradiation, the ultimate response elicited seemed to depend upon the type of cell impaled. Typical fibrillation electrical activity was induced mainly in pacemaker cells. These findings demonstrated a definite effect upon the cell's normal transmembrane electrical characteristics with irradiation and onset of contractile alterations. The nature of the observed effects clearly indicated a definite membrane effect. Localized heating affecting the overlying cell membrane and/or interruption of cell membrane integrity would account for the characteristics of the observed depolarization. When cardiac muscle is injured by physical means, it depolarizes due to interruption of cell membrane integrity. Recovery or "resealing" is accomplished by restoration of the cell membrane and its selective permeability.[22,23]

These findings indicated that the observed effects could not be mitochondrion specific. Indeed, Strahs et al.,[24] using a Q-switched, neodymium-YAG laser, demonstrated that myofibrillar lesions resulted in similar visual contractility changes that were triggered as frequently as by mitochondrial lesions. Furthermore, the irradiation of carbon particles adhering to the cell membrane surface resulted in contractile changes at 1/600 of the energy required to produce such changes by irradiating either mitochondria or myofibrils.[24]

Further investigation of the contractile behavior and cell-surface morphology of cultured neonatal rat heart cells following irradiation of a single mitochondrion was carried out with phase contrast and scanning electron microscopy.[16] It was concluded that when the contractile activity of the cell remained unchanged, the morphology of the cell membrane overlying the target organelle appeared normal even though the organelle was visibly damaged as observed under phase contrast microscopy. By increasing the intensity of the laser energy, more severe mitochondrial lesions resulted that were paralleled by both contractile changes and alterations in the surface morphology of the cell membrane overlying the target organelle. These studies further corroborated earlier phase microscopy[8,9,11,13] and transmission electron microscopy studies[13] in which specific alterations in the structure and ultrastructure of the irradiated organelle were correlated with laser energy density. Both electrophysiological findings[15] and scanning electron microscopy studies[16] demonstrate that the contractile alterations triggered in cultured myocardial cells following laser microirradiation of mitochondria are due mainly to an effect on the cell membrane.

Other effects cannot be disregarded. Salet et al.[18,19] demonstrated that photochemical effects occur at the level of mitochondria when myocardial cells are irradiated in the visible range without appreciable temperature change. According to these investigators, a laser-induced change in ATP levels may be involved in triggering a stimulation in contractility.

The laser microbeam has also been used to study motility related cytoplasmic cell structures[24,25] and mechanisms of cell repair and regeneration.[26] The cytoplasmic stress fibers and bands of intermediate filaments (100 Å) of subcultured fibroblasts or "nonmuscle" cells derived from neonatal rat myocardial primary cultures are amenable to selective microirradiation.[25,26] Intracellular motility patterns have been studied by placing multiple 0.25-μm lesions in preselected regions of bands of 100-Å filaments and then examining the relative movement of the lesion sites.[25] Mechanisms of cellular repair and regeneration have been investigated in severed single-stress fibers.[26] Selective alteration of such cellular structures permits elucidation of the role of cytoskeletal elements in cell migration and cellular shape changes.

Another approach in the use of the laser for the study of enzymatically dispersed myocardial cells was used by Leung.[27,28] Laser diffraction studies of single intact cardiac muscle cells at rest were consistent with a model in which the myofibrils are randomly packed in the cell and each myofibril acts as a cylindrical diffractor of one-dimensional order.[27] Similarly, a specially designed diffractometer with high spatial and temporal resolution was used for the study of sarcomere dynamics in single myocardial cells.[28]

The laser microbeam has also been used to stimulate localized fluorescence in submicrometer regions of single mitochondria in cultured myocardial cells.[29] Because of the coherence, monochromaticity, and single mode structure of the laser beam, it is possible to focus it to an intense spot approaching diffraction limits and, therefore, monitor fluorescence in localized regions of 0.25 to 1 μm in a single cell.[30] Electron microscopic examination of serial sections of mitochondria in which laser-stimulated fluorescence was analyzed showed no alterations in the normal ultrastructure.[29] Thus, dynamic variations in fluorescence can be analyzed quantitatively and qualitatively *in situ* under normal or experimental conditions in specific regions of the cytoplasm or subcellular organelles. This approach with the use of the appropriate fluorescent probes could give valuable information, for example, in the study of localized fluctuations in cytosolic and membrane-bound Ca^{2+} during normal contractility as well as during induced arrhythmias.

REFERENCES

1. **Bessis, M., Gires, F., Mayer, G., and Nomarski, G.,** Irradiation des organites cellulaires à l'aide d'un laser à rubis, *C.R. Acad. Sci.,* 225, 1010, 1962.
2. **Berns, M. W. and Salet, C.,** Laser microbeams for partial cell irradiation, *Int. Rev. Cytol.,* 33, 131, 1972.
3. **Berns, M. W.,** Recent progress with laser microbeams, *Int. Rev. Cytol.,* 39, 383, 1974.
4. **Berns, M. W.,** *Biological Microirradiation,* Prentice-Hall, Englewood Cliffs, N.J., 1974.
5. **Berns, M. W., Kitzes, M., Rattner, J. B., Burt, J., and Meredith, S.,** Laser microirradiation of cells, *Acta Biol. Med. Ger.,* 38, 1285, 1979.
6. **Berns, M. W., Aist, J., Edwards, J., Strahs, K., Girton, J., McNeill, P., Rattner, J. B., Kitzes, M., Hammer-Wilson, M., Liaw, L.-H., Siemens, A., Koonce, M., Peterson, S., Brenner, S., Burt, J., Walter, R., Bryant, P. J., van Dyk, D., Coulombe, J., Cahill, T., and Berns, G. S.,** Laser microsurgery in cell and developmental biology, *Science,* 213, 505, 1981.
7. **Berns, M. W. and Walter, R. J.,** Laser and computer video optical microscopy in cell analysis, in *Cell Analysis,* Vol. 1, Catsimpoolas, N., Ed., Plenum Press, New York, 1982, 33.
8. **Berns, M. W., Gamaleja, N., Olson, R., Duffy, C., and Rounds, D. E.,** Argon laser microirradiation of mitochondria in rat myocardial cells in tissue culture, *J. Cell. Physiol.,* 76, 207, 1970.
9. **Berns, M. W., Gross, D. C. L., Cheng, W. K., and Woodring, D.,** Argon laser microirradiation of mitochondria in rat myocardial cells in tissue culture. II. Correlation of morphology and function in irradiated single cells, *J. Mol. Cell. Cardiol.,* 4, 71, 1972.
10. **Berns, M. W., Gross, D. C. L., and Cheng, W.,** Argon laser microirradiation of mitochondria in rat myocardial cells in tissue culture. III. Irradiation of multicellular groups, *J. Mol. Cell. Cardiol.,* 4, 427, 1972.
11. **Adkisson, K. P., Baic, D., Burgott, S. L., Cheng, W., and Berns, M. W.,** Argon laser microirradiation of mitochondria in rat myocardial cells in tissue culture. IV. Ultrastructural and cytochemical analysis of minimal lesions, *J. Mol. Cell. Cardiol.,* 5, 559, 1973.
12. **Berns, M. W. and Cheng, W. K.,** Argon laser microirradiation of mitochondria in rat myocardial cells in tissue culture. V. Pacemaker versus non-pacemaker cells, *Life Sci.,* 12, 469, 1973.
13. **Rattner, J. B., Lifsics, M., Meredith, S., and Berns, M. W.,** Argon laser microirradiation of mitochondria in rat myocardial cells. VI. Correlation of contractility and ultrastructure, *J. Mol. Cell. Cardiol.,* 8, 239, 1976.
14. **Waymire, K., Kitzes, M., Meredith, S., Twiggs, G., and Berns, M. W.,** Argon laser microirradiation of mitochondria in rat myocardial cells in tissue culture. VII. Fibrillation in ventricle and auricle cells, *J. Cell. Physiol.,* 89, 345, 1976.
15. **Kitzes, M., Twiggs, G., and Berns, M. W.,** Alteration of membrane electrical activity in rat myocardial cells following selective laser microbeam irradiation, *J. Cell. Physiol.,* 93, 99, 1977.
16. **Burt, J. M., Strahs, K. R., and Berns, M. W.,** Correlation of cell surface alterations with contractile response in laser microbeam irradiated myocardial cells. A scanning electron microscope study, *Exp. Cell Res.,* 118, 341, 1979.
17. **Salet, C.,** Accélération par micro-irradiation laser du rythme de contraction de cellules cardiaques en culture de tissu, *C.R. Acad. Sci. Paris,* 272, 2584, 1971.
18. **Salet, C.,** A study of beating frequency of a single myocardial cell. I. Q-switched laser microirradiation of mitochondria, *Exp. Cell Res.,* 73, 360, 1972.
19. **Salet, C., Moreno, G., and Vinzens, F.,** A study of beating frequency of a single myocardial cells. III. Laser microirradiation of mitochondria in the presence of KCN or ATP, *Exp. Cell Res.,* 120, 25, 1979.
20. **Salet, C., Moreno, G., and Lampidis, T. J.,** Effects of microirradiation of heart cells in culture at the mitochondrial level, *Biol. Cell,* 37, 195, 1980.

21. **Nathan, R. D., Pooler, J. P., and De Haan, R. L.,** Ultra-violet-induced alterations of beat rate and electrical properties of embryonic chick heart cell aggregates, *J. Gen. Physiol.,* 67, 27, 1976.

22. **Délèze, J.,** The recovery of resting potential and input resistance in sheep heart injured by knife or laser, *J. Physiol.,* 208, 547, 1970.

23. **De Mello, W. C.,** The healing-over process in cardiac and other muscle fibers, in *Electrical Phenomena in the Heart,* De Mello, W. C., Ed., Academic Press, New York, 1972, 323.

24. **Strahs, K. R., Burt, J. M., and Berns, M. W.,** Contractility changes in cultured cardiac cells following laser microirradiation of myofibrils and the cell surface, *Exp. Cell Res.,* 113, 75, 1978.

25. **Strahs, K. R. and Berns, M. W.,** Laser microirradiation of stress fibers and intermediate filaments in non-muscle cells from cultured rat heart, *Exp. Cell Res.,* 119, 31, 1979.

26. **Koonce, M. P., Strahs, K. R., and Berns, M. W.,** Repair of laser-severed stress fibers in myocardial non-muscle cells, *Exp. Cell Res.,* 141, 375, 1982.

27. **Leung, A. F.,** Laser diffraction of single intact cardiac muscle cells at rest, *J. Muscle Res. Cell. Motil.,* 3, 399, 1982.

28. **Leung, A. F.,** Sarcomere dynamics in single myocardial cells as revealed by high-resolution light diffractometry, *J. Muscle Res. Cell. Motil.,* 4, 485, 1983.

29. **Siemens, A., Walter, R., Liaw, L.-H., and Berns, M. W.,** Laser-stimulated fluorescence of submicrometer regions within single mitochondria of rhodamine-treated myocardial cells in culture, *Proc. Natl. Acad. Sci. U.S.A.,* 79, 466, 1982.

30. **Berns, M. W.,** Fluorescence analysis of cells using a laser light source, *Cell Biophys.,* 1, 1, 1979.

INDEX

A

B

C

Oxygen supply, 85
Oxygen tension, 38—39, 56

P

Pacemaker cells, 134
Palmitate, 59—62
PDH, see Pyruvate dehydrogenase
Pentose shunt, 110
Petri dish, coating of, 51—52
Phosphatidylcholine, 44
Phosphatidylethanolamine, 44
Phosphatidylglycerol, 13
Phosphatidylinositol, 13
Phosphatidylserine, 13
Phosphocreatine, 26
Phosphodiesterase I, 96
Phosphofructokinase, 58, 110
Phospholipase, 37—40
Phospholipid, 44
Phosphorylase, 109
Plasma membrane, 44, 108, 111, 113, 135
Plating, 51
Plating density, 51
Ploidy, 31
Poly(A⁺)RNA, injection into cultured cells, 79
Postnatal heart cells, cardiac glycoside binding to,
    6—7
Potassium fluxes, 4, 8—13, 37, 43
Potassium gradient, 12
Propranolol, 12, 109
Prostaglandins, 38
Protein kinase C, 131
Proteins
  degradation of, 43, 108
  synthesis of
      anthracyclines and, 20, 24, 29—30
      oxygen tension and, 39
      thyroid hormones and, 111—112
Proteoglycans, 100
Pseudopodia, 52
Purine bases, 37
Purinergic receptor, 21, 25—26
Puromycin, 111
Pyruvate dehydrogenase (PDH), 58—60
Pyruvate kinase, 110

R

Receptor
  α-adrenergic, 109—110
  β-adrenergic, 109—110
  cardiac glycoside, 2, 4—8, 13—14
  purinergic, 21, 25—26
Reoxygenation, 40, 44, 62, 64
Reperfusional repair, 45
Reserpine, 12
Respiratory rate, 86—87
RNA synthesis
  anthracyclines and, 20, 29—30

oxygen tension and, 39

S

Sarcolemma
  in anoxia, 65
  enzyme release from, 41
  ischemic injury and, 37—38, 43—44
Sarcomere length, 52
Sarcoplasmic reticulum
  hypertrophy of, 30
  oxygen in, 85—87
  swelling of, 30
Scatchard plot, 5
Seeding density, 69
Serum, 51—52
Sialic acid, 13
Siderosome, 118
Sodium-calcium exchanger, 8, 11
Sodium fluxes, 4, 8—13, 37, 43
Sodium gradient, 12
Sodium lag hypothesis, 8
Sodium-potassium pump
  cardiac glycosides and, 8—14
  effect of thyroid hormones on, 109
  reserve capacity of, 10
Sotalol, 12
Stress fiber, 135
Substrate deprivation, 41—42
Sugar transport, thyroid hormones and, 111—113
Superoxide radical, 38

T

T3, see Thyroid hormones
T4, see Thyroid hormones
Thalassemia, 118, 120
Thyroid hormones
  cardiac glycosides and, 14, 109
  effect of
      on carbohydrate metabolism, 110—111
      on heart cell culture, 107—114
      on myosin isozyme patterns, 108—109
      on sodium pump, 109
  interactions with catecholamines, 109—110
  nonnuclear effects of, 111—113
  nuclear effects of, 108—111
Thyrotoxicosis, 110
Thyroxine, see Thyroid hormones
Tissue disaggregation, 50
α-Tocopherol, 31—32
Toxic compounds
  in heart cell culture, 21—23
  screening for, 20—21
Toxicity, iron, 117—122
Transferrin, 118—120
Transforming cultured cell model, 56
Transfusional therapy, 118
Transhydrogenase, 29
Triiodothyronine, see Thyroid hormones